Molecular
Physics

Molecular Physics

Theodore Buyana

Mediterranean College, Athens

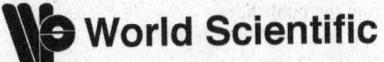

World Scientific

NEW JERSEY · LONDON · SINGAPORE · BEIJING · SHANGHAI · HONG KONG · TAIPEI · CHENNAI

Published by

World Scientific Publishing Co. Pte. Ltd.
5 Toh Tuck Link, Singapore 596224
USA office: 27 Warren Street, Suite 401-402, Hackensack, NJ 07601
UK office: 57 Shelton Street, Covent Garden, London WC2H 9HE

Assignment Problems 112, 113, 115, 116, 117, and 118 in this book are adaptations from Problems 4.1, 4.6, 4.7, 4.2, 4.5, and 4.4(b) respectively, in J. D. Jackson's *Classical Electrodynamics*, 2nd ed., John Wiley & Sons, Inc., New York, © 1962, 1975, by John Wiley & Sons, Inc. Modified and reprinted by permission of John Wiley & Sons, Inc.

Assignment Problems 87, 88, 90, and 91 in this book are adaptations from Problems 16.1, 16.2, 16.3, and 16.6 respectively, in C. Kittel's *Introduction to Solid State Physics*, 5th ed., John Wiley & Sons, Inc., New York, © 1953, 1956, 1966, 1971, 1976, by John Wiley & Sons, Inc. (The problems appear also in the current 7th edition.) Modified and reprinted by permission of John Wiley & Sons, Inc.

The author thanks John Wiley & Sons, Inc. for its kind permission.

Library of Congress Cataloging-in-Publication Data
Buyana, Theodore.
    Molecular physics / Theodore Buyana.
       p. cm.
    Includes bibliographical references and index.
    ISBN-13 9789810208301 -- ISBN-10 9810208308
    ISBN-13 9789810208318 (pbk.) -- ISBN-10 9810208316 (pbk.)
    1. Molecules. 2. Nuclear magnetic resonance. I. Title.
    QC173.3.B89   1997
    539'.6--dc21   97-17367
                                                        CIP

British Library Cataloguing-in-Publication Data
A catalogue record for this book is available from the British Library.

Printed in Singapore

ΑΦΙΕΡΩΣΙΣ

Τό παρόν πόνημα ἀφιεροῦται τῷ σεβαστῷ μοι πνευματικῷ πατρί, Δεσπότῃ, συμπατριώτῃ καί εὐεργέτῃ ἐκ Μεγάλου Ῥεύματος Βοσπόρου Κωνσταντινουπόλεως

ΘΕΟΦΙΛΕΣΤΑΤΩ ΕΠΙΣΚΟΠΩ
ΜΕΛΟΗΣ κυρίῳ ΦΙΛΟΘΕΩ ΚΑΡΑΜΗΤΣΩ

Βοηθῷ Ἐπισκόπῳ τῆς Ἱερᾶς Ὀρθοδόξου Ἀρχιεπισκοπῆς Ἀμερικῆς, οὗ τῇ ἠθικῇ τε καί ὑλικῇ προστασίᾳ ἐπραγματοποιήθησαν αἱ ἐν Ἀμερικῇ σπουδαί μου.

Ἡ παροῦσα ἀφιέρωσις ἔστω δεῖγμα ἐλάχιστον καί σπονδή εὐγνωμοσύνης ἅμα τε καί τιμῆς, οὐ μήν καί υἵκῆς ἀγάπης ἀντίδωρον.

Ὁ πονήσας

Μηνί Ὀκτωβρίῳ αϠϟα´

"Are these things then necessities?
Then let us meet them like necessities."

2 Henry IV, III, i.

PREFACE

This book is designed to serve the needs of a college course of the same title. It addresses a mixed class of graduate and upper undergraduate physics students where chemistry and biochemistry students can join, too. The material is mainly theoretical, but important experimental aspects are discussed as well, and details are given when necessary. The presentation of the material follows traditional lines. The ratio of words to mathematical expressions is kept reasonable. Excess in mathematical abstraction has always been outside my intention; I think it would be repellent to a physicist to have to deal with a mass of mathematics produced by a purist and rigorous *école*. (The succession of theorems, proofs, corollaries, axioms, postulates etc. in a physics book is not an attractive feature.) Perhaps rigor is valuable *per se* (and in scientific and other investigations), but what has priority is the mastering of the physical concepts and ideas first. There was a time when a fool and his money were soon parted. Now with the IRS it happens to everybody! Similarly, there was a time when only mathematicians dealt with highly abstract mathematical formalisms. Now with quantum mechanics, physicists themselves struggle with the difficulty of understanding the quantum mechanical formalism of their own discipline, a formalism which soars to ethereal heights of scholarly mathematical punctuality. For this reason, tedious mathematical calculations and harsh derivations are avoided here, in an effort to stress the physical background of the phenomena (instead of suppressing it for the sake of sheer mathematical rigor and exactness). I tried to depart from the bad habit (or should I say vanity) of hiding the beautiful essence of physics behind formidable mathematical formulas and dry expressions. Indeed, in theoretical physics the physical content is usually hidden behind discouraging mathematics (which is necessarily used as a tool and an elegant language of brevity), just like a beautiful music which is hidden behind the insipid looking music notes in a partiture.

Each chapter in this book is a self-contained unit. The chapters are arranged almost independently: Omission of a chapter should not affect much the reader's progress or the continuity of the treatment.

I believe that this text will, to some extent, relieve the student from the worry of taking notes in class, and thus render him/her more attentive and receptive to the discussion. Having the textbook handy for future reference, the student will feel comfortable as he/she will be able to concentrate on the material taught, and focus attention on the *spoken* lecture rather than on note taking which makes him/her miss the substance while he/she is trying to catch what is being said.

Given the scientific merit of molecular physics today (as related to biochemistry and molecular biology), this book is written to primarily motivate and help students who are interested in these fields and wish to specialize in them. Non-student readers (instructors, researchers, interested scientists, and armchair readers) can also use it as a reference.

This book intends to teach, not to comment on scientific and academic knowledge. The material is kept as diverse as possible. The major stress is on resonance methods (NMR, NQR, EPR, and ENDOR) whose theory, experimental apparatus, techniques, utility, and applications are discussed, together with an adequate number of examples.

The treatment has the character of a general review. It includes the necessary elements of instruction concerning the subject, but does not exhaust them, and has no claims for any originality besides the fact that it has the advantage of combining the diverse spirit of several similar works in a single concise volume. Thus the student will not have to resort to several separate books and get loaded with unimportant details outside the scope of the course. To the convenience of the student (and the general interested reader), I have skimmed all necessary and useful information and instruction material and packed them as an extract into one volume at the proper level, wishing to enable the student to find what he/she needs in just one handy book, adequate for the course purposes.

One of the defects of this book is the insufficient number of assignment problems per topic. But this is intentional, because I want to urge the instructor to make up and assign homework problems of his/her own choice and taste. So, if he/she considers to adopt this book as the formal text of the course, I suggest that he/she hands out additional problems regularly. A word of caution here: It is true that problems and exercises consolidate acquired knowledge (actually, no physics can be learnt without solving problems and by just reading physics books like novels* or staying with the theory only), but if there is an ill training in mathematics, problems and exercises may sometimes waste the physicist instead of teaching and edifying him/her — unless, of course, he/she is determined to pick up mathematics and ambitiously attack "athletic" exercises.

Another flaw of the book is its uneven bibliography. The relevant bibliography is so vast, that when one includes certain authors and leaves out others, one could commit an unfair discrimination against the latter. Anyway, a thorough bibliography would not be of appreciable help, nor could it possibly contribute much to the student's output. Moreover, the judgement among the works to be included would be subjective and too personal, even if I were competent to judge. Hence I opted for keeping the list of references shorter than what I had in mind, hoping that the instructor will recommend professional reading and reference material, according to the needs and level of the class, and pertinent to the topics that he/she wants to emphasize.

On the students's part, knowledge of quantum mechanics and atomic physics is assumed. I also assume that the reader knows some elementary nuclear physics.[†]

I should affirm that I am not an expert in many topics included in the book. For instance, my personal quantum mechanics arsenal was, until recently, no richer than what could be extracted from Merzbacher's and Park's book. However, I undertook physics teaching mostly for having a chance to learn the subject and benefit personally, and thus be fair to myself, honest to my students, and consistent with my major. From this experience, I confess, I learnt much more than what I had shallowly encountered at college when I was a student. The challenge and enthusiasm grew larger into writing this book,

---

*Physics books are to be *studied*, not just read.

[†]It is assumed that the reader has been exposed to a typical, at least one-semester quantum mechanics course and to some (undergraduate-level) atomic physics. However, we suggest and hope that during the progress of the course, the instructor will briefly and remedially revise some basic facts of quantum mechanics, as necessary, in order to bridge gaps and fill missing knowledge on the students' part.

on a subject that I had been practically ignorant of before. It was a process and experience of self-edification that took me more than five years. As I was writing, it was necessary for me to first master the involved hardships, page for page. Now I am aware of them more than ever. And the final rewarding moral satisfaction of having tamed many facts of the substance counts more than anything else.

By no means do I claim that this is a perfect textbook on molecular physics. In fact I feel apologetic, as I impertinently venture to present such a mediocre book to the academia. Indeed, my book does not purport to be a work that discusses the subject of molecular physics in dept and esoterically. It is only the materialization of a five-year dream and longing, in a most modest way. If this motivation is not taken into account, it becomes evident that the defects and shortcomings of this work are many. For instance, some topics had to be discussed very briefly, in principle only and without details, while others had to be overcondensed, and still others totally omitted for space reasons.[‡] Certain parts may lack clarity. In any case, I regret the omission of any topic, the inclusion of which the reader might deem indispensable. I also invoke the tolerance of the reader, given that my talent as a reader is much stronger than that (almost nonexistent one) as an author, and further, given the fact that I had to write this book while working in an overseas country where the access to sources and references and the availability of physics facilities are still limited.

Some molecular physics facts I repeat here and there intentionally over and over, in order to educate the unexperienced reader, and lest he/she misses something important if he/she omits some chapters. I hope this will not bore the reader.

I had to hand draw all the figures of this book myself, with a painstaking and time-consuming effort, in an exhausting championship of meeting deadlines; that is why the figures lack artistic professionality — I seek the reader's understanding here, too.

My heartiest wish is the following: May this book become an efficient source of inspiration for other authors who intend to write its flawless counterparts, and a motivation for students who desire to do research in this field. On my part, I am not a physics author; I am a *worker of physics*.

November 1991

Theodore Buyana

[‡]For example, I could have included whole chapters dedicated to topics like molecular magnetism, molecular beams, macroscopic molecular phenomena, and LASER.

ACKNOWLEDGEMENTS

I record my deep gratitude to World Scientific Publishing Co Pte Ltd for carrying out the whole process of the production of this book and making this publication a reality. I acknowledge and express my gratitude to Dr. K.K. Phua, Editor-in-Chief, and Mr. Steven Patt, Editor, of said company, to whom I present my special thanks in public, for their personal brotherly care that transcended cold businesslike formality and materialized as a kind interest, first-class professional help, and valuable advice. I am indebted to the administrative and technical staff of the company for their polite response, patience, and understanding. I wish to thank the typesetters and designers for the wonderful job they did in performing their difficult and demanding task. A hearty praise goes to Ms. Elsie Tan, typesetter, for her commendable diligence, patience and meticulousness in this production.

I am also grateful to my dear students whose attentiveness and pleasant responses I should praise. Their enthusiastic interest proved a valuable driving force and a source of inspiration for me.

I extend my thanks to my fellow colleagues for all their help. A specific debt of gratitude is due to Professors Benjamin Bederson, Bernardo Jaduszliwer, Henry Stroke, Howard H. Brown, Lawrence A. Bornstein, Robert W. Richardson, Engelbert Schücking, Alberto Sirlin, Leonard Yarmus, Edward J. Robinson, Paul R. Berman, Burton Budick, James H. Christenson, John Sculli, and V.T. Rajan of New York University; Professors Ahmed Yüksel Özemre, Hayati

Budak, Nezihe Taşköprülü, Gediz Akdeniz, and Harutyun Agopyan of Istanbul University; Professor John Scarlatos of Bosphorus University (formerly American Roberts' College, Istanbul); Professor Altan M. Ferendeci of Case Western Reserve University, Ohio; Professors William R. Schultz and Spyridon Taraviras of Mediterranean College, Athens; and Professor Cleanthes Nicolaides of the Greek National Research Institute, Athens, together with the librarians of said institute.

I gratefully submit my thanks to my spiritual father His Eminence Archbishop Iakovos of New York, N.Y., for his paternal care and blessing. I acknowledge moral help and technical support from the Honorable Daniel Oliver Newberry, formerly U.S. Consul General in Istanbul; Messrs David Marshall and Raymond M. Nowakowski of the U.S. Department of State, Thomas Hastings Edelblute of Charleston, West Virginia, Frank Bates (OIC/ESC, U.S. Embassy, Athens), Alexander Papadopoulos of Athens; Mrs. Vicky Grivas of the Greek-British Center, Athens, and Miss Dorothy Fitzsimmons, Education Officer and formerly Field Administrator of the once U.S. Air Force Base, Athens. I should cite friendly help from Miss Irene Port and Miss Sylvia Falcon of New York, N.Y.; Mrs. Ute Mügge-Lauterbach of Hamburg; Prof. Larissa Bonfante of New York, N.Y.; Prof. Helen Ioannides of the University of Patras; Dr. Alkiviades Michalis of Athens; Messrs René R. Espina, Jr. of New Haven, Conn., Saim Bengi Gören of Ankara, George S. Voutsinas, Dimitri Bakas, Photios Hadjis, Mrs. Theodora Matheopoulos and Miss Leda Dimitriou of Athens. I also want to thank the American Institute of Physics and the New York Academy of Sciences.

Last but not least, I all-heartedly thank my mother Maria Buyana and my good sister Antigone Buyana who all patiently endured with me the agony of the preparation of the manuscript.

T. Buyana

TO THE INSTRUCTOR

Is it worth teaching a course titled Molecular Physics? Would there be students interested in taking it? I think yes, given the indisputable scientific importance of this field nowadays. The ambitious student, especially the one who intends to specialize or follow an academic career, takes courses by which he/she receives a good training and strengthens the basis of his/her professional future. On the other hand, the specialist instructor has a chance to present the latest developments in the field, via in-class lectures.

I feel that the usefulness of this book will be enhanced if it is used as a text-book by students who take a course in molecular physics and *simultaneously* use the book. Sitting in class and attending the course has a strong bearing on the optimization of the efficiency of a textbook. Starting off from these reflections, I would like to make the following suggestions:

(1) This book can be equally used as a reference or a formal textbook for a college course in molecular physics taken by both graduate and upper undergraduate students in a mixed class. The level is carefully kept in between. I avoided a rigid line that separates the graduate from the undergraduate level. Nevertheless the material can also be taught at a purely undergraduate level. If the instructor feels that some chapters or sections are too hard for a class of undergraduates, he/she can omit or simplify them. On the other hand, if he/she finds some parts too low for a graduate class, he/she can elevate

the level and enrich it with supplementary material. The instructor should feel free to reduce or augment the material, all by his/her own initiative.

(2) The presentation allows flexibility. For this purpose, certain topics are broken down to subtopics by separate chapters, to allow for expansions or contractions in the discussion, as the level and interests of the class may dictate. Also, an altogether omission of some parts like the removal of chain links is possible, without disturbing the pedagogical order or losing the sequential continuity. Admittedly, some chapters are necessarily sequential, but the instructor can still be selective among the contents and fit the material to the requirements and objectives of the course. Nevertheless, for pedagogical reasons, it is suggested that Chapters 1 and 3 be taught first. Chapter 2 can be totally omitted in favor of the expansion of other chapters and elaborations on other topics.

(3) The material can be fitted to a one-semester course consisting of two bi-hourly lecture sessions plus an hourly recitation session per week. The latter should be devoted to problem solving and discussing extra topics. Since one semester means about thirty-two lecture sessions, the instructor should cover about a chapter a session, and leave some sessions for midterm exams and extra work. It is also possible to expand the material to fill a two-semester course, by dwelling on the topics more elaborately. In that case, there must be a proportionate stretch-out of the discussion with more details, and the pace can be slower. At the end of the Bibliography there is a medley of special topics, to help the students make choices in preparing term papers or finding reading material. Also, interested students may even choose topics for specific research from the given list.

(4) Student participation should be encouraged, and topic titles for term papers may be suggested. Reading material outside the formal text (articles, book sections, reprints and other literature) should be handed out by the instructor.

(5) In my effort and wish to improve this book by possible future editions, I will gladly await opinions, comments, suggestions, advisory recommendations, and corrective criticism from the readers.

CONTENTS

Chapter 1

GETTING TO KNOW MOLECULAR PHYSICS

1.1. Introduction

Molecular physics is a vast subject with a vast volume of research. It is a science in itself, with a huge library of books, essays, research articles, reports, and publications. When I set off to write this textbook, I realized that it was like having to push and drag a boulder up a steep hill, because it was extremely discouraging to face the fact that one had to tame an overwhelmingly vast material and enclose it in a book of a limited number of pages. Of course, one could relax the constraint by narrowing down the subject and specializing in particular topics, but this does not work with textbooks. To my dismay, I had to subjugate a formidable size of raw material based on my personal research and studies plus a library work that was the crop of getting deep into analyzing and classifying research material amassed over the years in the vast library of molecular physics. Extensive notes taken when at college, constituted the third component of the substance.

The dilemma was awful and the decision painful: What (and how much) was to be included, and what was to be left out? Would I do justice to the material by favoring certain topics and excluding others? It was difficult to judge what to pick up and what to sacrifice. A fair selectivity does not go by the importance always, because in a science like this, what is important to me may not be so to others, and vice versa. Nevertheless a comforting thought rushed to my relief: If the story of the creation of the world is told in six hundred words in the Bible, this is enough to eventually arm one with courage to

1

sit down and write the book. Further, poet K. Cavafy's words convinced me that one can certainly dare:

> "If a story that should be told in fifty pages is written in thirty, it will be better ... That is, the artist will leave something out, but that is not a fault ... But if he gives it in a hundred pages, it is a fearful fault."

1.2. Historical Remark

Molecular theory and the concept of the molecule appeared on the stage before the advent of atomic theory in its modern scientific sense. Although the atomic idea started in the fifth century B.C. in ancient Greece, it faced doubts and was not established and fully accepted until the molecular theory was first developed.

In ancient Greece, philosophers Democritus, Leucippus, and Epicurus believed in the discontinuity of matter and its corpuscular structure. These philosophers were the first persons to ever mention the word *atom* (in the sense of the ultimately small and indivisible particle of matter).[1] They held out that matter was discontinuous. However, their speculations were purely theoretical and philosophical, lacking experimental support, because no experiment was of question during those times. On the other hand, Aristotle rejected the atomic idea, and so did the philosophers and scientists of the Middle Ages.

Centuries later, the subject was revisited and brought up by Boyle who retrieved the idea from the dust of the ancient Greek stochasm. So, in the seventeenth century, the atom was discussed in its modern context and meaning: Boyle believed that matter consisted of atoms, and mentioned them in his book. The first experimental attempt was made by the British chemist and naturalist Dalton who actually brought the subject to the current interest of his times. He worked with gases and dealt with the subject of gas atoms, clearing the way for both theoretical and experimental scientific research.

However, the question of existence of atoms caused doubts first. The assertions were initially welcomed with suspicion and unpopularity, because with the novel acceptance of the atomic existence, some disaccordances arose concerning chemical reactions. The issue was resolved by Avogadro whose reasoning was as follows: If we take two gases whose volumes are V_1 and V_2 respectively, and

---

[1]Literally, *atom* means indivisible, in Greek. Although today we know that the atom is not indivisible, the name remained as a scientific term.

whose numbers of atoms are N_1 and N_2 respectively, the composite gas produced by the chemical reaction for their union will have volume V and number of atoms N, where there must be a certain relation (a simple arithmetical one) between N_1, N_2, and N. So the atoms of V must be called *compound atoms*,[2] namely *molecules* or *radicals*. These were atomic groups that behaved as single particles in chemical reactions, and that could be divided and reshuffled during the reactions.

The word *molecule* — as a scientific term — was first used by Avogadro in 1811. He published articles and proposed his famous hypothesis: "For a given volume, and under the same pressure and temperature conditions, all gases have the same number of molecules".

Avogadro's theory restored the defects of Dalton's atomic theory, but at first it faced indifference. The Italian chemist Cannizzaro, a proponent of the atomic theory, suggested a full return to it, in 1858. But the ensued development of the kinetic theory of gases — thanks to the contributions of Maxwell, Boltzmann, and Gibbs — strengthened the molecular theory for good.

The first experimental observation of molecules was made by the French physicist Jean Perrin in 1908. He observed the *Brownian motions*, that is, the motions of any tiny particles of matter suspended in a medium, *e.g.*, smoke in the air, or dye pigments in water. The same thing holds for the thermal disorderly molecular motions in gases, as the molecules collide with each other (like couples dancing on a dancing floor where fatter dancers cause more prominent collisions owing to their size, just as the molecules do).

The Brownian motions were first observed by Brown in 1827, but of course, not with gas molecules. Brown observed them under the microscope, with tiny particles in water. As the particles collide with each other, every one of them executes a random motion, following a zigzag path (Fig. 1.1).

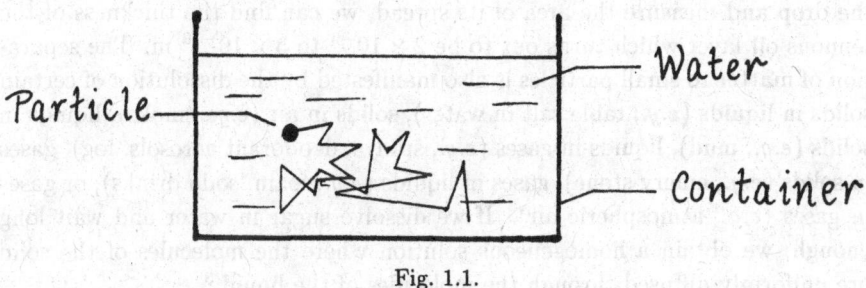

Fig. 1.1.

---

[2]The term *compound atom* was the initial name of the molecule. It was taken to be the smallest part of a chemical compound, consisting of atoms.

After Perrin's work, the corpuscular (molecular) structure of matter was eventually considered seriously and without doubts any more. The advent of quantum mechanics which was developed in 1927 and thereafter, furthered the advance of both the atomic and molecular theories. Quantum chemistry proved very useful in providing explanations about chemical reactions and liquid molecules.

As far as the atomic theory is concerned, its development was briefly as follows: The Greeks thought that the atom was the fundamental particle of matter and indeed indivisible, unaware of the possibility that they might be proven wrong in the future. Galileo vaguely believed in the atomic idea, and Boyle criticized it in the 1660s. Newton supported the view about the atomic structure of matter, whereas Leibnitz opposed to it. Dalton formally proposed the atomic theory in 1808, based on the ideas of Boyle and Lavoisier (including the definition of the chemical element in its modern hypostasis). The progress of chemistry and the construction of the periodic system by Mendeleev in the nineteenth century consolidated the atomic theory. The indivisibility of the atom was first challenged by J. J. Thomson and Ernest Rutherfold. Next came Niels Bohr's work, followed by that of a pantheon of atomic scientists of our century.

1.3. Definitions

How far can we mechanically divide a solid body down to its constituent particles without changing its chemical characteristics? This limit of mechanical division lies at about 10^{-6} to at most 10^{-7} m roughly. But the division of matter is not infinitely continuous forever. For example, if we leave off an oil drop on the surface of water, it spreads over the surface (since its density is less than that of water), forming a very thin film. If we know the volume of the drop and measure the area of its spread, we can find the thickness of the tenuous oil layer which turns out to be 2×10^{-9} to 3×10^{-10} m. The separation of matter to small particles is also manifested by the dissolution of certain solids in liquids (*e.g.*, table salt in water), solids in air (*e.g.*, smoke), liquids in solids (*e.g.*, mud), liquids in gases (*e.g.*, sprays, deodorant aerosols, fog), gases in solids (*e.g.*, emery stone), gases in liquids (*e.g.*, foam, soda drinks), or gases in gases (*e.g.*, atmospheric air). If we dissolve sugar in water and wait long enough, we obtain a homogeneous solution where the molecules of the solid are uniformly diffused through the molecules of the liquid.[3]

---

[3]The same thing is true for macromolecules (*e.g.*, starch, gum etc.), too, which form colloidal solutions exhibiting the Tyndall effect, *i.e.*, scattering of light of short wavelengths.

The smallest particle of matter that can be found in a free state and that still retains the chemical properties of the bulk is called *molecule*.

Another definition of the molecule is the following: A molecule is the union (or group) of atoms that constitutes the least material unit of a chemical compound; it is the smallest quantity of a chemically pure body, that can exist in a free state.

As a system of atoms orderly bound together, the molecule is the smallest and infinitesimally tiny part (or division) of matter that maintains the chemical properties of the material. It is the smallest unit of a chemical compound.

The bulk consists of (a collection of) many molecules bound together. If solid, the bulk is formed by the repetition of orderly patterns of molecules (or, sometimes, atoms) called *crystals*. In solids the bonds are stronger than those in liquids. (Solids have definite shapes, whereas liquids take the shape of their container.) In gases there is no cohesion between the molecules which fly loose and free.

Chemical compounds consist of molecules: They are formed by the composition (unification) of more than one chemical element at definite ratios. Each molecule is a group of (at least two) atoms bound together. In other words, several atomic nuclei bound together along with their electron clouds around, form a molecule (Fig. 1.2).

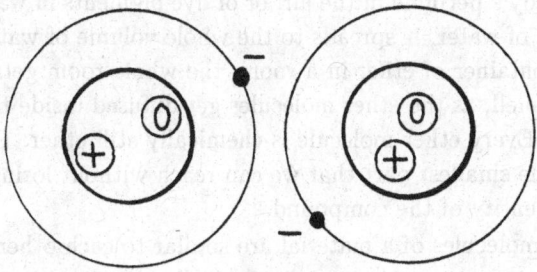

Fig. 1.2.

Chemical elements are monatomic substances consisting of atoms. In monatomic substances (*e.g.*, iron) the atom and the molecule are functionally the same entity. That is, the molecules of a chemical element are the atoms themselves.

The smallest particle of matter that can be reached by chemical ways only is called *atom*. It is not the fundamental unit of matter (because it consists of subatomic particles). It is just the smallest amount of a chemical element which enters chemical changes and participates in chemical compositions that are formed by the element together with other elements. An atom cannot be resolved to simpler parts by chemical methods, but it can be carried from a compound to another by chemical reactions.

The atom is the building block that characterizes and identifies the element. The bulk of the element consists of atoms in large quantities.

1.4. Molecular Properties

(1) Molecules consist of atoms (*i.e.*, chemical compounds consist of chemical elements). A molecule is a group of atoms bound together where the interatomic forces form linkages that are called *bonds*.

(2) A molecule consists of at least two atoms.

(3) A molecule cannot be further divided (broken down) by mechanical means (cutting, crushing, filing, cleaving, grinding, pulverizing, ripping, straining, sieving etc.)

(4) Every molecule maintains the characteristics and the chemical properties of the material (even after the dissolution of the material inside another material[4]). For example, when we dissolve sugar in water, the solution maintains the sweet taste of the sugar, an indication that there is a uniform diffusion of sugar (molecules) in water (among water molecules). Similarly, we have the diffusion of a lady's perfume in the air, or of dye pigments in water. If we drop ink into a glass of water, it spreads to the whole volume of water in the glass. If we open a container of ether in a room, the whole room gets filled with its characteristic smell, as the ether molecules get diffused inside the room, to fill all its volume. Every ether molecule is chemically still ether: Out of the bulk material it is the smallest part that we can reach without losing or disturbing the chemical identity of the compound.

(5) All the molecules of a material are similar to each other.

(6) There are as many kinds of molecules as the existing chemically pure materials.

(7) The molecules of solids are orderly arranged in crystalline structures, and they vibrate about equilibrium positions without moving away (or else we would have observed self-deformation of solid bodies). In liquids some molecules may fly off the surface (and some of those escaped may return again to the liquid). The molecules of liquids and gases are in an endless motion which is random and disorderly, towards all directions. This is called *molecular*

---

[4]Beware of the difference between a solution and a compound.

thermal motion. The speed of the moving (and elastically colliding with each other and with the walls of the container) molecules is great and becomes still greater as the temperature rises.

(8) The size of a molecule is typically of the order of a millimicron (10^{-7} cm) or less (Ångström, 10^{-8} cm). The size of an atom equals a few Ångströms.

1.5. Molecular Sizes

Although molecules are very tiny, one can measure their size by finding how many molecules there are in a certain amount of material. The number of molecules per mol is given by Avogadro's number

$$N_A = 0.6022 \times 10^{24} \text{ molecules/mol}.$$

If M stands for one mol of matter, then the mass of a molecule is

$$m = \frac{M}{0.6022 \times 10^{24}} \text{ gr} = 1.65M \text{ Ångström-mass}.$$

One *mol* (or *mole*) is the amount of a molecular material (chemical compound) in grams, numerically equal to its molecular weight. The term *mol* comes from the Latin word *moles* (structure), not from the word *molecule*. Thus one mol of O_2 is just $16 \times 2 = 32$ gr of O_2.

Because of the small size of the molecules, the unit of Å is used to express sizes. So, since 1 Å= 10^{-8} cm, the volume unit is $(10^{-8})^3 = 10^{-24}$ cm^3, and the mass unit becomes 10^{-24} gr.

Thus the mass of an O_2 molecule is $1.65 \times 32 = 52.8$ Ångström-mass $= 52.8 \times 10^{-24}$ gr.

The size (dimension) of a molecule can be found from the density of its solid material. For example, the density of solid oxygen at $-225°$C is 1.43 gr/cm^3. So, the volume of 32 gr of solid oxygen is $(32/1.43) = 22.4$ cm^3.[5] And hence the volume of a single molecule is

$$\frac{22.4}{0.6022 \times 10^{24}} = 37 \times 10^{-24} \text{ cm}^3 = 37 \text{ Ångström-volume}.$$

Since a molecule is too small, we can approximately take it as a cube. Thus the cube root of 37 Ångström-volume is 3.3×10^{-8} cm which is the size of the O_2 molecule.

The table below gives the sizes of some molecules.

---

[5]This figure has no relation with the 22.4 lt. It is an accidental numerical result.

Molecular sizes

Molecule	Dimension $\times 10^{-8}$ cm
Argon	2.9
Krypton	2.69
Nitrogen	3.3
Hydrogen bromide	2.76
Oxygen	3.0

These sizes are not constant. As the atoms of a molecule get excited, the orbits of their electrons start bulging, thereby changing the molecular diameters.

To get an idea about the molecular size, we give the analogy used by Lord Kelvin: If we enlarged a water drop to the size of the Earth, the molecules it contained would become larger than a pea but smaller than a tennis ball. To get another idea about molecular sizes and crowds, consider the following: Suppose that we somehow mark all the water molecules in a glass of water (so that they become discernible among other water molecules). We pour the marked water into the sea and wait long enough, until it spreads uniformly over all the seas of the world. If we now retrieve a glass of sea water from *any* ocean (from another part of the world), we are going to find *at least five thousand* marked molecules in the glass, belonging to the initial glassful of water!

1.6. What is Molecular Physics?

Molecular physics, atomic physics, and nuclear physics are subjects next to each other, according to the hierarchy of sizes. They study the microscopic structure of matter. Nuclear physics deals with the nucleus only (Fig. 1.3). Atomic physics deals with *free* atoms, that is, a single nucleus *and* the electrons around it (Fig. 1.4).[6] *Molecular physics* is the branch of physics dealing with molecules (systems consisting of more than one bound nuclei and electrons around them).

Molecular physics is (a) *theoretical* and (b) *experimental*. The former is based on quantum mechanics and deals with molecular phenomena and results

---

[6]A neutral atom has as many electrons as the protons in the nucleus, and hence it is electrically neutral because the total positive charge is equal to the total negative charge, and the net charge is zero. An *ion* is a charged atom, either with a surplus of negative charge, or with a net positive charge by the loss of one or more electrons.

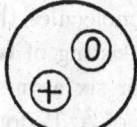

Fig. 1.3.

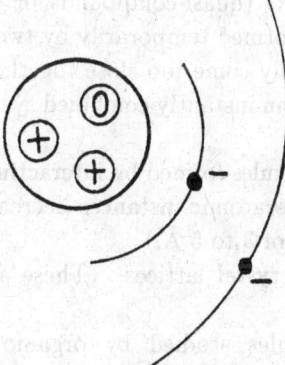

Fig. 1.4.

that can be calculated, explained, and interpreted by quantum physics. It involves much theory with pomp and circumstance, and requires complicated calculations and computer programming. The latter employs sophisticated laboratory apparatus, experimental techniques and formalism, and computerized processes of data analysis. It involves measurements, instrumental diagnostics and maintenance, laboratory strategy, improvement design, error analysis, identification of experimental parameters, service and troubleshooting, technical optimization methods etc.[7]

Molecular physics concerns energies of the order of eV because the molecular energies (binding forces) are of the order of 1 to 5 eV. That is why molecular physics can be actually called "very low-energy physics".

Molecules are formed by the binding of two or more atoms. In general, the size of a molecule is of the order of Å. Molecules dealt with by molecular physics have dimensions of the order of a few Ångströms.

---

[7]In the experimental physics the physicist is simultaneously a physicist and a janitor, a repairman, a technician, and a machine shop specialist. As he enters the lab to do physics, perform an experiment, or collect data, he also has to fix an instrument, design and set up the apparatus, know some engineering and electronics for this purpose, and have a solid experience about lab matters.

Molecular physics deals with molecules that (a) are *stable* under normal conditions, and (b) are *small*, consisting of two to (at most) ten atoms (and sometimes of no more than five or six atoms), because such molecules have interatomic distances of the order of Å. It does not deal with *macromolecules* (like DNA) and organic molecules which include many atoms.

Specifically, molecular physics does *not* deal with the following:

(1) Atomic combinations (quasi-compounds or superatoms of very high atomic number Z) formed temporarily by two *free* atoms which collide and thus momentarily come too close together during the collision interval, constituting an instantly combined system (which is an unstable pair).

(2) Van der Waals molecules formed by interacting atoms too close to each other (when the interatomic distances decrease). (The Van der Waals forces have a range of 3 to 5 Å.)

(3) Solid crystals and crystal lattices. (These are studied by solid state physics.)

(4) Very heavy molecules studied by organic chemistry, and macromolecules (*i.e.*, large molecules consisting of many atoms) which are subjects of chemistry, biology, biochemistry, molecular biology, and genetics (*e.g.*, DNA and other such molecules).

Those who expect discussion on these topics should snip off their wings of hope, for these topics do not pertain to molecular physics.

1.7. Objective of Molecular Physics

The main objective of molecular physics is the study of the following:

(1) Sizes, shapes, and structure of molecules: Molecules are not dull and compact entities. They have a structure to be examined, described, and mapped. For example, one can measure the interatomic distances inside the molecule, the crystal lattice constants, and the angles between the nuclei in the molecular configuration.

(2) Molecular symmetry: It is the first thing to be considered when we study the structure of a molecule. Molecules exhibit remarkable spatial symmetries.

(3) Binding energies, binding forces, and molecular ionization.

(4) Internal energy states of molecules (they give an idea about the molecules themselves and the structure of matter in general).

(5) Optical, electrical, and magnetic properties of molecules.

(6) Construction of molecular models and formulation of the theory of real molecules.

(7) Development of methods of probing and surveying molecules.

(8) Applications of the above and extension of the importance of the subject to technology, biology, biochemistry, and medicine.[8]

1.8. Methods of Molecular Physics

Molecular structure is studied mainly by three experimental methods: Molecular spectroscopy, diffraction methods, and resonance methods. We will discuss these methods in their own chapters ahead.[9] In this section we will outline them as follows:

(1) *Spectroscopy*: It can be done with visible and invisible light. So we have

   (a) *Optical spectropscopy* (with visible wavelengths).
   (b) *Infrared spectroscopy*.
   (c) *Raman method*.
   (d) *Microwave spectra*.

   The spectroscopic methods helps us determine molecular energies. For example, the microwave, the far infrared, and the high-resolution Raman spectra are good for the determination of rotational states. The linear vibrational energy levels are given by the near infrared spectroscopy. The electronic energy levels (and the dissociation energy) are obtained by optical, ultraviolet, and mass spectrometry. We can also determine molecular moments of inertia and interatomic separations by spectroscopy, in general. Spectroscopy (and diffraction) is a good means of determining the distribution of atoms inside a molecule.

(2) *Diffraction methods*:

   (a) *X-ray diffraction*: This method is used to study crystals and molecular structure of solids, especially materials made of atoms of large Z. X-rays have a good penetration ability. So, we can determine interatomic distances in molecules consisting of atoms of large Z.

---

[8]This last item is in fact a subject for a separate book.

[9]The diffraction methods will be discussed briefly, qualitatively, and descriptively, lest the treatment bulge at the expense of other chapters.

(b) *Electron diffraction*: This method is used to investigate gaseous molecules, solid body surfaces, and thin layers (films) of solids. It is used wherever X-rays are useless.

(c) *Neutron diffraction*: This method is used to figure out the structure of molecules consisting of atoms of low Z (*e.g.*, hydrogen). This method is good for molecules of light atoms and hydrogen whose structural parameters (interatomic distances and angles) can thus be found. As they pass through a medium, neutrons leave energy around. Neutron beams are used usually for low-Z materials and wherever the use of X-rays is not possible.

It is interesting to note that by the method of *Laue diffraction* of X-rays by crystals we can examine the structure of the crystal if we know about the X-rays used, and conversely, we can *study X-rays* if we already know the structural properties of the crystal.[10]

(3) *Resonance methods*:

(a) *NMR*: It pertains to the magnetic resonance of the nuclear spin (at frequencies 50 kHz to 50 MHz). It gives information about the magnetic properties of the molecule. Combined with spectroscopic methods and the study of the Zeeman effect, this method provides detailed information about the fine structure of atomic nuclei in molecules.

(b) *ESR*: It pertains to the measurement of the energy changes of the electron spin in a strong magnetic field. This resonance occurs at frequencies $\sim 10^4$ MHz which lie in the microwave region. (Thus the microwave methods include microwave spectroscopy and ESR.)

(c) *ENDOR*: It is the combination of NMR and ESR.

(d) *NQR*: It pertains to the interaction of the nuclear quadrupole moment with the electric field gradient, *i.e.*, $\overset{\leftrightarrow}{Q} \cdot \vec{\nabla}\vec{\mathcal{E}}$, which is equal to a quantized energy of the atomic nucleus.

Other method of molecular physics are the following:

(1) *Molecular beams*.
(2) *Modern radio-frequency spectroscopy*.
(3) *Mass spectroscopy*.
(4) *Macroscopic measurements*.

---

[10]In the X-ray diffraction we also get interference between the ray scattered by a crystal molecule of the surface layer and the ray scattered by a molecule of the second layer right below.

(5) *Classical stereochemistry*.

(6) *LASER* and *MASER* in applied molecular physics.

The resonance methods and the radio-frequency spectroscopy are modern methods to study the fine structure of molecular nuclei, molecular structure, bonds, electric dipole and quadrupole moments, intramolecular fields, electron clouds and charge distributions etc. The macroscopic measurements constitute another source of information: One can measure macroscopic properties like dielectric constants, optical polarization, and molecular refraction. Molecular beams and mass spectrometry can be used to study isolated molecules and controlled collisions. They both are sensitive methods.

Furthermore, there are thermodynamic and acoustical methods that can usefully supply us with information about molecules (but these methods will not be discussed in this book).

Modern techniques and advanced technology (*e.g.*, LASER, radio-frequency technology, sensitive spectrometers and auxiliary equipment — like sophisticated oscilloscopes, beam machines, and other modern scientific devices) helped a lot towards the progress and development of experimental molecular physics.

The following table summarizes the utility of the methods of molecular physics. If the method works in the determination of the pertinent property

Name of method	Study of molecular		
	Symmetry	Bonds	Parameters
X-ray diffraction	Yes	Yes	Yes
Electron diffraction	Yes	Yes	Yes
Neutron diffraction	Yes	No	Yes
Vibration spectrum	Yes	No	No
Vibration Raman spectrum	Yes	No	No
Rotation Raman spectrum	Yes	No	Yes
Pure rotation spectrum	Yes	No	Yes
Radio-frequency spectrum	Yes	Yes	Yes
NMR	Yes	Yes	Yes
Stereochemistry	Yes	No	No
Magnetic measurements	Yes	No	No
Dipole-moment measurements	Yes	Yes	No

of interest, we mark it by "Yes"; otherwise, by "No". Notice that all the experimental methods give information about symmetries, but not all of them are fit for informing us about molecular parameters. Spectral methods are the best for studying molecular structure.

1.9. Molecular Parameters

Molecules consist of several atoms located at certain distances (from each other) that make certain angles between them. These distances and angles are called *molecular parameters*.

If the molecule consists of N atoms, $3N-6$ parameters are enough to determine its structure. However, if a molecule has symmetries, the number of the

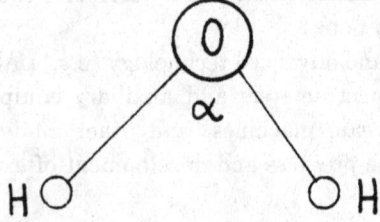

Fig. 1.5.

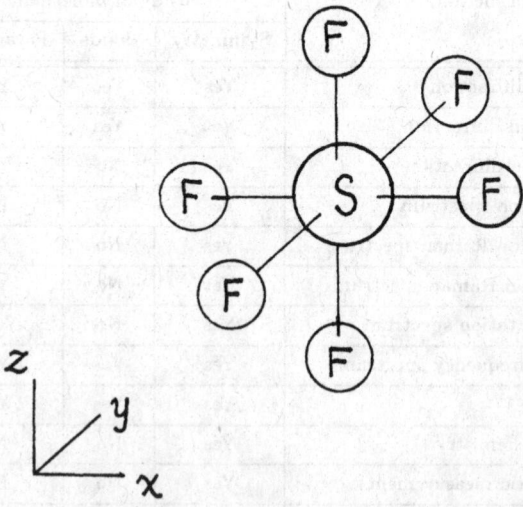

Fig. 1.6.

needed parameters decreases (in proportion with the number of the symmetries). For example, consider the water molecule (Fig. 1.5). Only two parameters are enough (that is, the O–H distance and the angle α), because the molecule is symmetric. In the SF_6 molecule (Fig. 1.6) only one parameter (the S–F distance) is enough because the rest is symmetry: There is a symmetry by structure (the angles are either 90° or 180°), and any additional piece of information is redundant.

1.10. Divisibility of Matter

By using physical mechanical means (cutting, crushing, dissolving, pulverizing etc.), we can divide all bodies into tiny parts without depriving them of their characteristic properties. For example, sugar dissolved in water gives a solution that still retains the sweet taste of sugar, which means that sugar has been diffused throughout the volume of water uniformly, in tiny parts. Amounts of other materials (perfumes, ether) can be felt by their characteristic smell in the air which testifies to the fact that the substance has suffered a fine partitioning and a uniform diffusion in the air.

But the division of matter is not indefinite. Matter is not continuous. Physical and chemical phenomena show that every body consists of small separate particles, the *molecules*. Every molecule retains the characteristic properties of the body. All molecules of the same material are alike (identical), and there are as many kinds of molecules as there are chemically pure bodies (compounds).

Chemical research proves that molecules consist of still smaller particles, called *atoms*. If a molecule and an atom coincide in identity, we have a chemical *element*. If a molecule consists of many kinds of atoms (or more than one atoms of the same kind), we have a chemical *compound*. Therefore, the following statements hold:

(1) The *atom* is the least amount of a chemical element that participates in chemical compounds that the element forms with other elements.
(2) Atoms in a molecule are held together by (molecular) forces that arise from the electronic charge of the atoms, *i.e.*, the forces are of electrostatic origin ultimately.
(3) The molecule is the smallest amount of a chemically pure body that can exist in a free state.

There are 105 chemical elements (92 in natural state and 13 artificially produced in the laboratory). There are as many kinds of atoms as there are elements.

The fact that matter — although it appears continuous macroscopically[11] — consists in fact of many separate particles was introduced as a hypothesis 2500 years ago by Democritus who gave the name *atoms* to these particles. Theoretical and experimental research laid the foundations of the theory of the discontinuity of matter.

Many physical phenomena are due to the molecular structure of bodies. How large is the size, and how great is the number of molecules? It is known that there are about 6×10^{23} water molecules in 1 mol (18 grams) of water. So, in 1 gr of water there are 33×10^{21} molecules. This huge number of molecules tells us how many particles are stuffed in a mass of water that fills a volume of just 1 cm^3. Thus one can imagine how small a molecule is.

The order of 10^{22} molecules/cm^3 in a sample (say, molecular beam) is a typical number density encountered in the laboratory. It is a very large number, making the system statistical, *i.e.*, the number of particles is too large to be treated by deterministic dynamics (which keeps track of the equation of motion and the trajectory of each individual particle). There are too many particles around, and when their number is so large, statistics takes over, that is, the *overall behavior* of the collection of particles is studied, not the behavior of a single particle. The same thing happens when we are to watch the behavior of a large crowd of people in a demonstration, or to predict the response of the population of a city; we do not care about — and cannot follow — the motion of every single person, because the number is so large that it is beyond our control. We rather deal with the *gross behavior* of the populace, taken collectively (disregarding extreme cases of statistical exceptions and freaks). And if we apply statistics, the results about the rounded off macroscopic behavior of the system are satisfactory and serve our needs when compared to the actual case observed experimentally. In molecular physics all molecules are identical and indistinguishable (they have no labels and carry no individual identity), like an impersonal populace and unlike a phone book where there is a large number of subscribers but every subscriber is identifiable and a significant personality, because he has a separate label (name and phone number).

The molecules of a gas are in a ceaseless motion (quite disorderly) towards all directions, and collide with each other and with the walls of the container. The randomness (disorder) here means that upon each collision, memory of the previous velocity state (speed and direction of motion) is lost. This explains the diffusion of dye molecules in water or perfume in air. There is a distribution of speeds and velocities in the gas.

---

[11] Aristotle supported the idea of continuity.

So, the molecules of a body constitute a huge population and have too small sizes. They are in a continuous motion whose speed increases with temperature.

During *physical phenomena* (water → ice) the energy changes, not the constitution of the body that takes part in the phenomenon; its molecules remain the same. During *chemical phenomena* ($H_2 + 1/2\ O_2 \rightarrow H_2O$) the substance (constitution of matter) is altered, being affected by the phenomenon. Chemical phenomena involve *essential changes*, and are accompanied by the release or absorption of heat or another form of energy. The molecules change, and new bodies with different properties are produced.

At the atomic level, there is no sharp and definite separation line between physics and chemistry.

What do we call *properties* of a body? All the characteristic features that make that body unique and different from another one, *e.g.*, color, hardness, taste, smell, density, solvability, melting and boiling points, chemical behavior, electric properties (conductivity, dielectric constant), thermal properties (specific heat), optical properties (index of refraction, scattering dispersion, transparency), and magnetic characteristics (magnetization, susceptibility). Notice that the properties are physical and chemical.

In a physical phenomenon these characteristics do not change along with the change. The essence of the body does not change (no new body is produced). Physical properties are characteristics that are often used to distinguish a material from another, because they directly affect our senses.

In a chemical phenomenon, the change creates a new body (material). It is through the chemical properties that a body participates in chemical changes.

1.11. Binding Energy

Consider two (nonrelativistic) atoms of energies E_1 and E_2. Classically, the total energy of the system (of these two atoms taken together) is $E = E_1 + E_2$, if the atoms are free and unbound (but just considered together).

If the two atoms are *bound* and considered again together as a single system, then $E < E_1 + E_2$. The "missing" energy is stored as order (bond). The binding together is mediated by the attractive forces of the two atoms, and a single, composite system is formed.

To break the system into its separate components requires work, namely addition of energy to the system from outside. The break-up of the bound system (molecule) is shown schematically in Fig. 1.7 where $E_{binding}$ is the energy to be added to the compound system (molecule) in order to completely

separate the parts (atoms). The energy needed to break the system is equal
to the energy that binds the system, *i.e.*, the *binding energy*. We have

$$E + E_{\text{binding}} = E_1 + E_2 .$$

If the system is bound and $E_{\text{binding}} > 0$, then $E < E_1 + E_2$, *i.e.*, the energy
of the bound system is less than the sum of the energies of the individual atoms
when separated at infinity. Less energy means more order and stability.

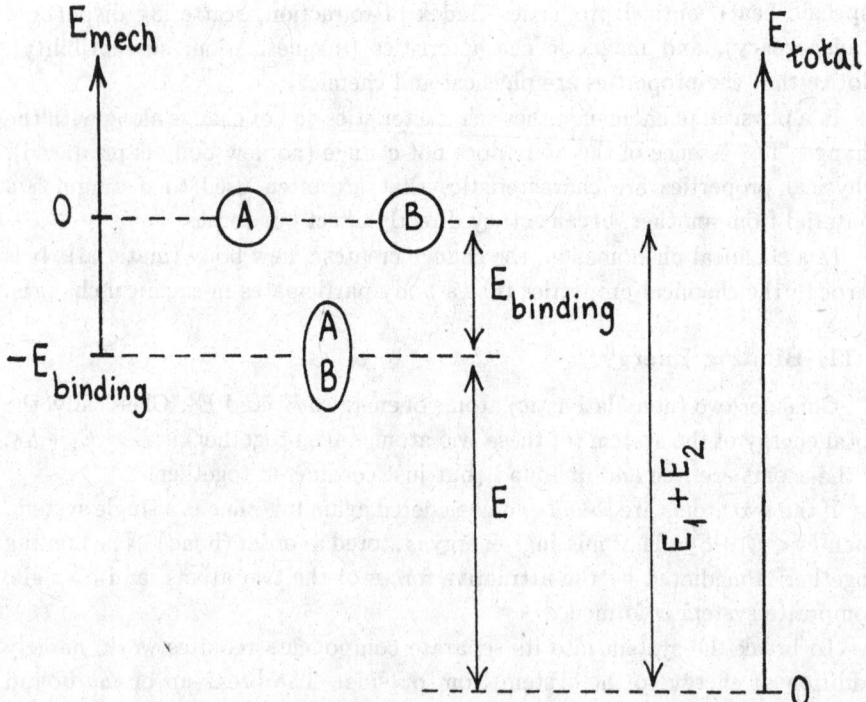

Fig. 1.7.

Fig. 1.8.

We are free to choose the zero level of the total (kinetic and potential) *mechanical* energy E_{mech} of a system of particles at any value. The most practical choice for particles attracting one another is 0 when the particles are infinitely separated and at rest (purely potential energy). When they are bound together, then E of the bound system is *negative*, because energy must be added to the compound to separate the atoms and bring the system's energy up to 0. Figure 1.8 shows the energy, level diagram of the energy of two atoms, both bound together and separated. On the left is the mechanical energy; on the right, the total energy.

These considerations hold classically. In the relativistic case, relativistic corrections (including the rest masses of the atoms) must be made. For a well-written relativistic treatment, see R. T. Weidner and R. L. Sells, *Elementary Physics: Classical and Modern*, Allyn and Bacon, Boston, Mass. (1975), ISBN: 0-205-04647-9, pp. 604–607.

1.12. Molecular Potential

Given a potential $V(x) = V_0 \delta(x - a)$, and assuming that $\psi(x)$ is continuous at $x = a$, we will show that

$$\lim_{\varepsilon \to 0} \left\{ \left(\frac{d\psi}{dx} \right)_{x=a+\varepsilon} - \left(\frac{d\psi}{dx} \right)_{x=a-\varepsilon} \right\} = \frac{2m}{\hbar^2} V_0 \psi(a) .$$

This is a one-dimensional problem (where ψ and V are functions of the single coordinate x).

The Schrödinger equation is

$$-\frac{\hbar^2}{2m} \frac{d^2\psi}{dx^2} + V(x)\psi(x) = E\psi(x)$$

or

$$-\frac{\hbar^2}{2m} \nabla^2 \psi + V_0 \delta(x - a)\psi = i\hbar \frac{\partial \psi}{\partial t} .$$

To get the answer, we should integrate $\int$ (The equation) dx, i.e.,

$$\int_{x_1-\varepsilon}^{x_1+\varepsilon} \frac{d^2\psi}{dx^2} dx = \frac{2m}{\hbar^2} \int_{x_1-\varepsilon}^{x_1+\varepsilon} [V(x) - E]\psi(x)dx .$$

Integrate the left-hand side and have

$$\frac{d\psi}{dx}\bigg|_{x_1+\varepsilon} - \frac{d\psi}{dx}\bigg|_{x_1-\varepsilon} = \frac{2m}{\hbar^2} \int_{x_1-\varepsilon}^{x_1+\varepsilon} [V_0 \delta(x - a) - E]\psi(x)dx ,$$

$$\lim_{\varepsilon \to 0} \left\{ \frac{d\psi}{dx}\bigg|_{x_1+\varepsilon} - \frac{d\psi}{dx}\bigg|_{x_1-\varepsilon} \right\} = \lim_{\varepsilon \to 0} \left\{ \frac{2mV_0}{\hbar^2} \int_{x_1-\varepsilon}^{x_1+\varepsilon} \delta(x-a)\psi(x)dx \right.$$

$$\left. - \underbrace{\frac{2m}{\hbar^2} \int_{x_1-\varepsilon}^{x_1+\varepsilon} E\psi(x)dx}_{\text{zero}} \right\}$$

$$= \frac{2mV_0}{\hbar^2} \underbrace{\int_{x_1-\varepsilon}^{x_1+\varepsilon} \delta(x-a)\psi(x)dx}_{\psi(a)}$$

$$= \frac{2mV_0}{\hbar^2}\psi(a) = \text{const.}$$

Given now the potential $V(x) = V_0\delta(x)$ with $V_0 < 0$, we will show that there exists one and only one *bound state* for this potential, and we will find its energy and eigenfunction.

We start again from the Schrödinger equation

$$-\frac{\hbar^2}{2m}\psi''(x) + V_0\delta(x)\psi(x) = E\psi(x)$$

where $\delta(x) \equiv \delta(x-0)$. We have

$$-\psi''(x) + \underbrace{\frac{2mV_0}{\hbar^2}}_{\equiv -\alpha^2}\delta(x)\psi(x) = \underbrace{\frac{2mE}{\hbar^2}}_{\equiv -k^2}\psi(x),$$

$$\psi''(x) + \alpha^2\delta(x)\psi(x) = k^2\psi(x).$$

For $x \neq 0$, $\psi'' = k^2\psi$ yields $\psi = \mathcal{A}e^{kx} + \mathcal{B}e^{-kx}$.

Refer to Fig. 1.9. In region I ($x < 0$) we have $\psi_\text{I} = \mathcal{A}e^{kx}$. In region II ($x > 0$) we have $\psi_\text{II} = \mathcal{B}e^{-kx}$. The conditions are the following:

(a) ψ must be continuous at $x = 0 \Rightarrow \psi_\text{I}(x)\big|_{x=0} = \psi_\text{II}(x)\big|_{x=0} \Rightarrow \mathcal{A} = \mathcal{B}$.

(b) At $x = 0$ we have

$$\int_{-\varepsilon}^{\varepsilon} \psi''dx + \alpha^2\underbrace{\int_{-\varepsilon}^{\varepsilon} \delta(x)\psi(x)dx}_{\psi(0)=\mathcal{A}} = E\underbrace{\int_{-\varepsilon}^{\varepsilon} \psi dx}_{0 \text{ if lim as } \varepsilon \to 0 \text{ is taken}},$$

$$\psi'(\varepsilon) - \psi'(-\varepsilon) + \alpha^2\psi(0) = 0,$$

$$\psi_I'(0) - \psi_I'(0) + \alpha^2\psi(0) = 0\,,$$

$$-kBe^{-kx} - kAe^{kx}\big]_{x=0} + \alpha^2 A = 0\,.$$

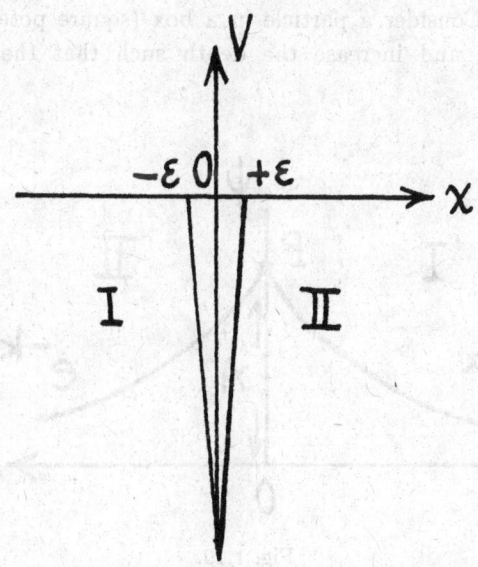

Fig. 1.9.

We can drop A and $B = A$. We will have $2k = \alpha^2$ or

$$2\sqrt{-\frac{2mE}{\hbar^2}} = \frac{2mV_0}{\hbar^2}\,,$$

a relation satisfied by one E only. We have

$$-\frac{2mE}{\hbar^2} = \left(\frac{mV_0}{\hbar^2}\right)^2 \quad \text{and} \quad E = -\frac{mV_0^2}{2\hbar^2}\,.$$

$$\psi(x) = Ae^{-kx} \quad \text{for} \quad x > 0\,.$$

$$\psi(x) = Ae^{kx} \quad \text{for} \quad x < 0\,.$$

To fix the constant A, we require that ψ be normalized to unity:

$$\int_{-\infty}^{\infty} |\psi|^2 dx = 1\,.$$

Here, in this problem, ψ is continuous, but its derivative is not, because of the δ function. At point P in Fig. 1.10 there is a smooth transition between regions I and II (ψ is continuous), but $d\psi/dx$ is discontinuous.

The case is well-treated in Merzbacher's and Gasiorowicz's books on quantum mechanics. Consider a particle in a box (square potential, Fig. 1.11). Shrink the width and increase the depth such that the "area" remains

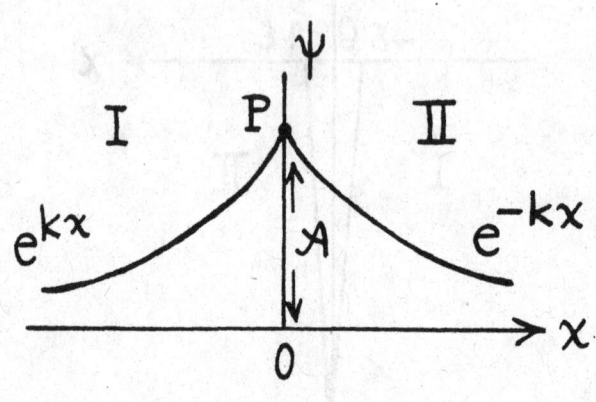

Fig. 1.10.

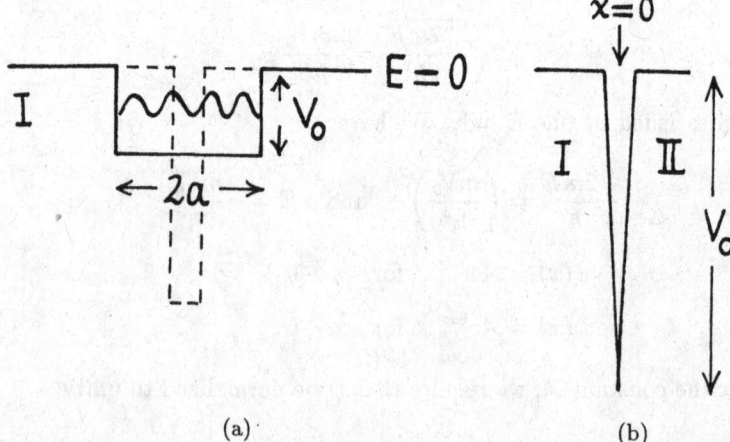

(a)   (b)

Fig. 1.11.

constant: $2aV_0 = $ const. By the limiting procedure, as $V_0 \to \infty$ *and* $2a \to 0$, we get a delta function potential. Since $V_0 < 0$, it is a potential well (trap).

In region I, $\psi_{\mathrm{I}} = \mathcal{B} \exp(\sqrt{2mE}x/\hbar) + \mathcal{C} \exp(-\sqrt{2mE}x/\hbar)$ where the second term must be rejected because it blows up as $x \to \infty$ (which is physically meaningless and far from reality). In region II, $\psi_{\mathrm{II}} = \mathcal{A} \exp(-\sqrt{2mE}x/\hbar) + \mathcal{D} \exp(\sqrt{2mE}x/\hbar)$ where the second term must be rejected because of the discontinuity at $x = 0$. If E is the energy of the particle in regions I and II, and $E < 0$, the states are discrete. (And if $E > 0$, the spectrum is continuous.)

The equation

$$\frac{d^2\psi}{dx^2} + \frac{2mE}{\hbar^2}\psi = 0$$

has solutions $\psi_{\mathrm{I}} = \mathcal{B}e^{(\cdots)}$ and $\psi_{\mathrm{II}} = \mathcal{A}e^{(\cdots)}$. At the boundary $x = a$, $\psi_{\mathrm{I}} = \psi_{\mathrm{II}}$, *i.e.*, $\psi_{\mathrm{I}}(a) = \psi_{\mathrm{II}}(a) \equiv \psi(0)$ because of the δ function. Hence $\mathcal{B} = \mathcal{A}$ (from matching boundary conditions at $x = 0$). The second condition is

$$\frac{d}{dx}\psi_{\mathrm{I}}(0) - \frac{d}{dx}\psi_{\mathrm{II}}(0) = \frac{-2mV_0}{\hbar^2}\psi(0) = \frac{-2mV_0}{\hbar^2}\mathcal{A}.$$

$$\frac{\mathcal{A}}{\hbar}\sqrt{2mE} - \left(-\frac{\mathcal{A}\sqrt{2mE}}{\hbar}\right) = -\frac{2mV_0}{\hbar^2}\mathcal{A},$$

$$2\sqrt{2mE} = -\frac{2mV_0}{\hbar} \quad \text{whence} \quad E = \frac{-mV_0^2}{2\hbar^2}.$$

Only *one* value of energy for the state is found and exists.

$$\psi = \mathcal{A}\exp\left(-\frac{m|V_0|}{\hbar^2}|x|\right).$$

Normalize ψ and get $\mathcal{A} = \sqrt{m|V_0|/\hbar^2}$.

This δ-function potential is a localized, sharp (narrow) and deep attractive potential well ($V_0 < 0$) located at a certain coordinate ($x = 0$), and zero everywhere else. If doubled as in Fig. 1.12, it represents the potential of a molecule where the two dips (negative peaks) represent the potentials of the two ion nuclei at their locations. V_0 is the (negative) potential strength, and the δ function shows its location in space. The region in between (where $V = 0$) represents the potential wall between the ions. Of course, this is a

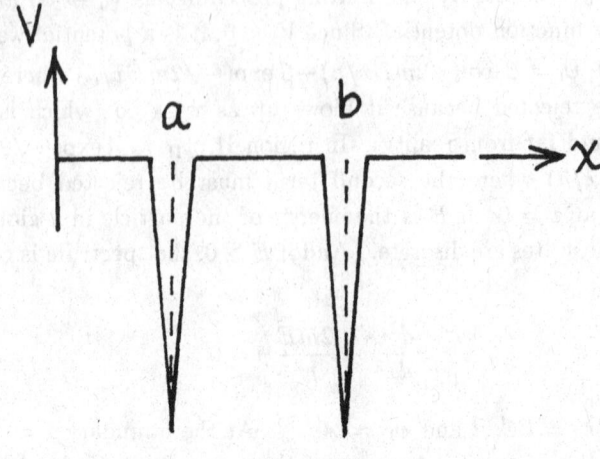

Fig. 1.12.

convenient, rough, and neatly idealized modelization of the actual situation, in a mathematical framework.

1.13. Interatomic Interaction Forces

The interaction forces between atoms are mainly (with respect to strength and order of magnitude) of electrical character. The problem is complicated even for simple atoms as they approach each other, because the picture concerns a many-body problem. The wave character of the electrons (which mediate the bonds) must be considered and the Schrödinger equation must be solved.

If the interaction energy is less than the total energy when the atoms are isolated apart, the interaction produces attractive forces, and thus a stable molecule is formed out of these atoms, by the chemical reaction $A + B \rightarrow AB$. But if the energy of the system increases upon approach, repulsive forces set up (because higher energy means instability), and thus no molecule can be formed.

The forces governing the interaction can be of gravitational, magnetic, and electric nature, but this does not mean that they all contribute to the binding (molecule formulation), just like the fact that not all partners can contribute to the financial establishment of a company, because the capital contribution of some of the partners may not be quantitatively appreciable.

Let us first estimate how big the gravitational potential can be. We have

$$V_{\text{grav}} = -G\frac{(m_H A)^2}{r}$$

where m_H is the hydrogen mass, and G is the gravitational constant. For a heavy atom, we can take $m_H A = 250 \times 1.66 \times 10^{-24}$ gr and $r \sim 3$ Å to have (with $G = 6.7 \times 10^{-8}$ units) $V_{\text{grav}} = -3.9 \times 10^{-44}$ erg $= -2.4 \times 10^{-32}$ eV. So, in forming molecules, F_{grav} is too weak to play a role.

Let us turn now to the order of magnitude of the magnetic interaction potential (and forces). We consider two magnetic moments (each equal to the Bohr magneton), mutually interacting. The interaction energy of two elementary magnetic dipoles is

$$V_{\text{magn}} = -\frac{2\mu_B^2}{r^3} = -\frac{2(9.27 \times 10^{-21})^2}{(3 \times 10^{-8})^3} = -7 \times 10^6 \text{ eV} = -0.2 \text{ cal/mol}.$$

Although $V_{\text{magn}} > V_{\text{grav}}$, it is still less than the energy that binds and holds the atoms together. (However, since $V_{\text{magn}} > V_{\text{grav}}$, V_{magn} can be measured. For example, the experimental value of V_{magn} for the magnetite (Fe_3O_4) crystal is 0.1 cal/mol.)

Finally, the electrostatic interaction energy (of two elementary charges at 3 Å) is

$$V_{\text{el}} = -\frac{e^2}{r} = -\frac{(4.8 \times 10^{-10})^2}{3 \times 10^{-8}}$$

$$= -8 \times 10^{-12} \text{ erg} = -5 \text{ eV} = -220 \text{ kcal/mol}.$$

Thus the interatomic electrostatic interaction energy is comparable to the chemical energy that binds atoms and contributes to the binding. This interaction is mediated by the outermost orbit electrons of the atoms. The valence electrons contribute to bonds, *i.e.*, chemical bonds are due to them.

1.14. Atomic and Molecular Structure

The state of motion (orbital) of an atomic electron is determined by the quantum numbers n, l, m_l, and m_s, as $|n\, l\, m_l\, m_s\rangle$. (The quantum number s is not a label; since an electron is a Fermion, its spin is 1/2 and we always have $s = 1/2$ for an electron.) According to the Pauli principle, no two electrons with exactly the same (values of these) quantum numbers can exist in the same orbital in an atom. The electronic structure in atoms is as follows: In order of increasing energy, the possible orbitals are

$\psi_{100,1/2}, \quad \psi_{100,-1/2}$

$\psi_{200,1/2}, \quad \psi_{200,-1/2}$

$\psi_{210,1/2}, \quad \psi_{210,-1/2}, \quad \psi_{211,1/2}, \quad \psi_{211,-1/2}, \quad \psi_{21,-1,1/2}, \quad \psi_{21,-1,-1/2}$

$\psi_{300,1/2}, \quad \psi_{300,-1/2}, \quad \psi_{310,1/2}, \ldots$ etc.

But states with large l (for a given n) shift to higher energies if the system is perturbed (as when many electrons are present). Hence we have

$\psi_{100,1/2}, \quad \psi_{100,-1/2}$

$\psi_{200,1/2}, \quad \psi_{200,-1/2}$

$\psi_{210,1/2}, \quad \psi_{210,-1/2}, \quad \psi_{211,1/2}, \quad \psi_{211,-1/2}, \quad \psi_{21,-1,1/2}, \quad \psi_{21,-1,-1/2},$

$\psi_{300,1/2}, \quad \psi_{300,-1/2}$

$\psi_{310,1/2}, \ldots$ etc.

Each electron occupies the lowest possible energy; therefore in H, the single electron goes to the orbital $\psi_{100,1/2}$ or $\psi_{100,-1/2}$. In He, one electron occupies $\psi_{100,1/2}$ and the other one $\psi_{100,-1/2}$ (except if we ignore spin, in which case both electrons occupy ψ_{100}, and we have two electrons in the same orbital — two at most — neglecting spin). In Li, two electrons occupy these two orbitals, and the third will necessarily go to ψ_{200}, the next higher energy. If we continue this way, we see how atoms are built.

This rule holds for any closed quantum system. Each particle of the system should have a unique wavefunction (and hence its state can be uniquely labelled and characterized by it). It is all due to the Pauli principle (in other words, to the existence of spin), or else the whole universe would be made of atoms of just one kind, *i.e.*, one kind of matter, and it would be too dull to live in such a universe! Of course, the deep and esoteric physical reasons of the principle are not known; they remain as mysterious as Pauli himself.

Noble gases do not react or interact because they are stable, as they have all their orbitals and shells filled (by the electrons needed to neutralize the nuclear charge). They are called noble because they do not deign to chemically interact with other elements and form compounds. They are snobbish at that. (The rest of the elements try to liken themselves to them, by taking or giving electrons, but they cannot thus get noble, just like the ordinary people who cannot become noble by trying to become so via an elegant attire.)

Atoms with one excessive electron (than the number needed to fill a level) tend to lose it and get a positive ion. Atoms which are one electron short of filling the level tend to receive a foreign electron and get negatively ionized.

The groups of orbitals (energy levels) are the "shells". The K-shell contains the two orbitals where $n = 1$. The L-shell is the group of the eight wavefunctions with $n = 2$. The M-shell has the orbitals $n = 3$, and so on, *i.e.*, the electronic configuration in shells is 2, 8, 18, ... etc., that is, $2n^2$ electrons per shell (where the factor 2 is the spin multiplicity, since the spin function can be $\chi_+ = |\uparrow\rangle \equiv |m_s = 1/2\rangle$ or $\chi_- = |\downarrow\rangle \equiv |m_s = -1/2\rangle$ — "up" and "down", two possibilities — for each group of n, l, m_l.[12]) Why do we start from K-shell and not from A-shell? Because scientists initially thought it wise to leave some room for possible lower energies — which of course do not exist.

Therefore, we have

$$K\text{-shell:} \quad \psi_{1s}\chi_+, \ \psi_{1s}\chi_- \quad \text{ or } \quad 1s^2$$

$$L\text{-shell:} \quad \psi_{2s}\chi_+, \ \psi_{2s}\chi_-, \ \psi_{2p_y}\chi_+, \ \psi_{2p_y}\chi_-, \ \psi_{2p_x}\chi_+, \ \psi_{2p_x}\chi_-,$$
$$\psi_{2p_z}\chi_+, \ \psi_{2p_z}\chi_- \quad \text{ or } \quad 2s^2 2p^6$$

$$M\text{-shell:} \quad \psi_{3s}\chi_+, \ \psi_{3s}\chi_-, \ \psi_{3p_y}\chi_+, \ \psi_{3p_y}\chi_-, \ \psi_{3p_x}\chi_+, \ \psi_{3p_x}\chi_-,$$
$$\psi_{3p_z}\chi_+, \ \psi_{3p_z}\chi_- \quad \text{ or } \quad 3s^2 3p^6$$

$$\cdots\cdots\cdots$$

where the superscript shows the number of electrons in that state; ψ is the spatial (depending on coordinates of the Euclidean spatial configuration) and χ is the spin part (depending on the spin coordinates) of the total wave function which is written as a product $\psi\chi$.

The electronic structure is thus

$$\text{He: } 1s^2$$

$$\text{Li: } 1s^2 2s^1$$

$$\text{F: } 1s^2 2s^2 2p^5 .$$

---

[12]The quantum number n determines the shell (or the radius = distance from the nucleus) of the orbit, l determines its shape, and m_l its orientation (with respect to an external field), while m_s has no classical analogue — it refers to the spin, an intrinsic property associated with the particle, not necessarily its literal spinning about its own axis. The energy of each shell goes as $E_n \sim 1/n^2$, and the radius as $r_n \sim n^2$, *i.e.*, the radii get spaced out more and more as n increases.

Li excited may be $1s^2 2s^0 2p^1$. Li^+ (ion) is $1s^2 2s^0$. F^- ion is $1s^2 2s^2 2p^6$ etc. The radial density graph (with the peaks) tells the story more eloquently (Fig. 1.13).

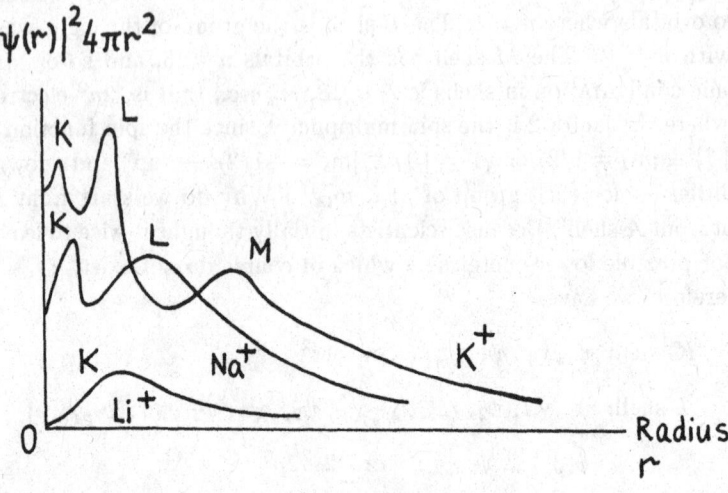

Fig. 1.13.

Now let us get to the molecules. The electronic motion here is not as simple as in the atom. Electrons jump from atom to atom, as they are shared in common by them. They simultaneously belong to two or more atoms (just like a mother bird in a nest, flying over her nestlings in turn, and thus preventing them from flying off from the nest). But it is possible to determine this motion by forming linear combinations of occupied atomic orbitals, so that the system's potential energy be a minimum, subject, of course, to the Pauli principle as well. To approximate, consider two hydrogen atoms A and B that are close together and in their ground state (of lowest energy). How would they behave?

Each atom has

$$\psi_{100,1/2} = \frac{1}{\sqrt{\pi}} \left(\frac{Z}{a_0} \right)^{3/2} e^{-\frac{Z}{a_0}r} \chi_+$$

$$\psi_{100,-1/2} = \frac{1}{\sqrt{\pi}} \left(\frac{Z}{a_0} \right)^{3/2} e^{-\frac{Z}{a_0}r} \chi_-$$

where in our case $Z = 1$ (hydrogen). One electron can occupy either orbital. Hence four linear combinations are possible:

$$\psi_{100,1/2}(A) + \psi_{100,1/2}(B)$$

$$\psi_{100,-1/2}(A) + \psi_{100,-1/2}(B)$$

$$\psi_{100,1/2}(A) + \psi_{100,-1/2}(B)$$

$$\psi_{100,-1/2}(A) + \psi_{100,1/2}(B).$$

The Pauli principle dictates that the first two are repulsive states (as the electrons are in the same type of orbital, the atoms repeal one another). The last two are attractive states (as the electrons are in different types of orbitals): The atoms are mutually tied, and each one shares the electron of the other. Figure 1.14 shows the partial (component) and total potential for the first two cases (left) and the last two cases (right). The latter has a potential minimum and thus the formation of a molecule (stable equilibrium) is possible.

We can plot the polar angular graphs of the two situations (Fig. 1.15) where the contours are lines of the same (constant) density (because the density of the compound probability clouds $\propto$ |Sums of the four cases|^2). Notice that the closed shell repulsion distorts the atoms (Fig. 1.15(a)), while in the attraction (elementary molecule in Fig. 1.15(b)) there is a negative charge concentration between the two nuclei (covalent bond).

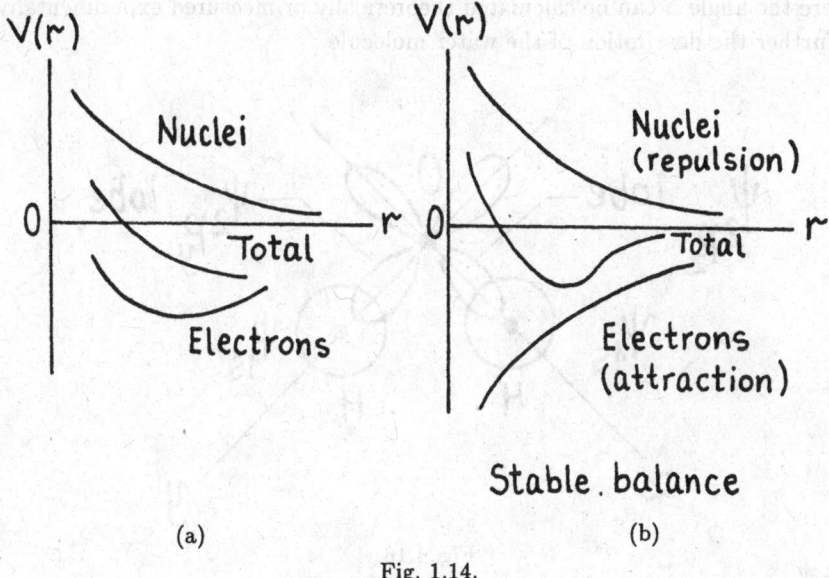

Fig. 1.14.

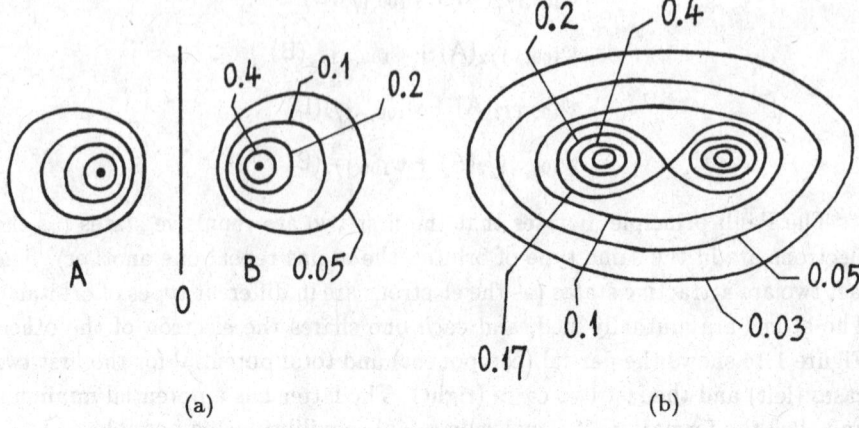

Fig. 1.15.

If we superimpose the atomic orbitals likewise, we can make descriptions of complex molecules, too. In H_2O, the O atom has symmetric $2p_y$ and $2p_z$ orbitals, *i.e.*, $1s^2 2s^2 2p^4$, which offer themselves to bonds (Fig. 1.16) with the $1s$-orbital of two H atoms. Therefore, the two H atoms take positions on the y and z-axis, as shown, with their orbitals overlapping with the y and z-lobes of the O atom, so that two bonds are formed. As a result, we have Fig. 1.17 where the angle α can be calculated theoretically or measured experimentally, to further the description of the water molecule.

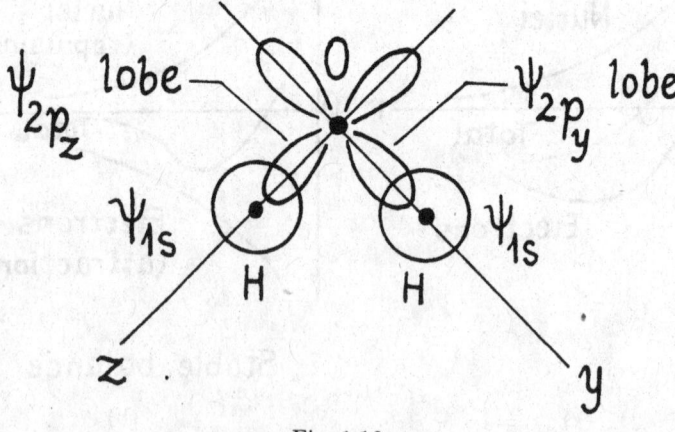

Fig. 1.16.

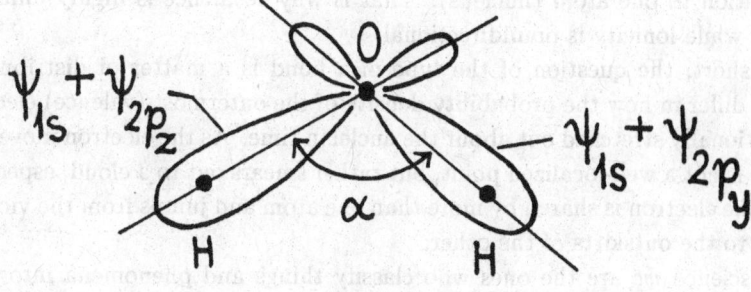

Fig. 1.17.

In a covalent bond the electron is partially transferred from an atom to another, like volleyball players who throw the ball between them and thus are kept in touch by the game. How much of the electron is transferred? (By this we mean stretching of the density of the charge cloud $|\psi|^2$, as if the electron spent more time close to one atom and less time close to the other). This depends on the charge composition of the atoms bound. For example, if the atoms are far from the column of the noble gases, the answer is half the electron. If the atoms are close to the noble column (*e.g.*, alkali atoms), the answer is the whole electron. That is why the former make strong covalent bonds, whereas the latter get bound more cautiously and superficially, making ionic bonds (where either atom is less participant of the joining and more selfish).

In a definite ionic bond, the two atoms exchange an electron (which is transferred from one to the other), become ions, and thus attract each other electrostatically (as ions of opposite charge) to form the bond. (The atom which loses its electron becomes a positive ion, and the atom which receives it, a negative ion.) Of course this attraction is not alone (otherwise the atoms would merge together); nuclear and (inner) closed shell repulsion are in opposition and provide equilibrium. This exchange allows electronegative and electropositive atoms to complete and close their shells; at the same time compounds are formed this way.

Notice that the interaction (binding) forces are ultimately electrostatic; so, whether the bond is covalent or ionic depends only on how much $|\psi|^2$ is stretched among the atoms. That is, ionic and covalent bonds differ in the location of the shared binding electrons (time-averaged) relative to the nuclear locations. In a covalent bond the participant atoms are generous enough to send their electrons to the region between them; the electrons spend a great deal of time there. In an ionic bond the electron pledges allegiance and association to one atom (nucleus). That is why covalence is highly unidirectional, while ionicity is omnidirectional.

In short, the question of the type of a bond is a matter of distribution: Bonds differ in how the probability density of the outermost (valence) electron is positionally stretched out about the nuclei in time. As the electron moves, its charge is not a well-localized point, but rather smears out to a cloud, especially when the electron is shared by more than one atom and jumps from the vicinity of one to the outskirts of the other.

In science *we* are the ones who classify things and phenomena into rigid categorizations, just to keep track well of them — it is a matter of human logistics. But in nature things work not within rigid walls; there are smooth and gradual transitions from one form to the other. That is why intermediate features, forms, and schemes are possible. Thus instantaneously, intermediate types of bonds are possible. What we categorize is usually the overall, average, gross behavior of the systems in a coarse picture.

There are three more kinds of bonds. The Van der Waals bond is due to induced dipoles in mutual interaction. Temporary electronic configurations create spatial separation of positive and negative charge concentrations, which in turn creates dipoles in nearby atoms or molecules. These induced dipoles interact with the original (inducing) dipole, and the interaction gives rise to forces. Noble gas atoms interact (and can be joined together) through this interaction only — it is the only one in them. These magnetic forces exist in all molecules (and crystals), because there are instantaneous dipoles *i.e.*, even averaged symmetrical electronic configurations may lose their symmetry for a moment). Further, some molecules have permanent dipoles (charge separations) giving rise to magnetostatic forces, though their binding strength is weak.

The hydrogen bond — a little bit stronger — can be seen in asymmetrical molecules where the charge cloud moves such that at some instant it exposes the H nucleus bare (because the screening offered by the electron cloud is swiftly off). Then excess negative ions can feel an electrostatic attraction towards the molecule. This happens in H_2O (and this molecular property of H_2O accounts for its different ways of crystallization when it becomes ice).

Finally, the metallic bond is observed in metallic crystals where the outer electrons of an atom move over all the other atoms (not any particular one), and hence the obtained picture is a crowd of electrons moving freely through the crystal (and giving rise to thermal and electrical conductivity). The crowd of the interstitial negative charges and the positive metal ions attract each other electrostatically. If the outer shells of the ions are not filled completely, additional covalent bonds may be formed (*e.g.*, in Fe, Co, Ni etc.). Thus the energy of the metallic bond (which has an intermediate strength) is increased.

Ionic and metallic bonds are omnidirectional. This explains crystal structures, a topic that concerns solid state physics, not molecular physics. The outer electrons in metals are also responsible for light scattering that gives rise to the metallic shiny appearance.

The motion of the electrons (negative charge clouds) determines the atomic, molecular, and crystalline structure of matter, as well as the optical, electric, and magnetic properties thereof. Since atoms and molecules are more or less stable configurations of smaller units (charges), in classical dynamics they can be viewed as particles or point objects.

1.15. Mutual Interatomic Interaction and Molecular Formation

Consider two atoms A and B, with their nuclei at their center. The atomic states are $|n\ l\ m\rangle_1$ and $|n\ l\ m\rangle_2$, or in short ψ_A and ψ_B, two wavefunctions that describe the free atomic behavior.

The probability of finding an atomic electron at distance r from the nucleus, inside a spherical shell (between two spheres of radii) r and $r + dr$ is $\rho(r) = f^2(r)r^2 dr$ where $\rho(r)$ is the electronic charge density, f^2 is the probability density (per unit length of space), $r^2 dr$ is the radial (volume) element, and $f(r)$ is the radial part of ψ_{atomic} (either ψ_A or ψ_B).[13]

The principal maximum of f occurs at $r = r_0$ which is the mean radius of the electronic orbit; it gives the position r_0 where the electron spends most of its time (= the average distance away from the nucleus where it is most probable — f peaks there — to find the electron). This gives the energy, too.

Let the atomic radii (of free atoms) be r_A and r_B. The atomic radius r_0 stays almost the same for all atoms. For example, for H, it is 0.5 Å ($= a_0$), and for heavy atoms is ~ 1.5 Å. As r increases, $\rho(r)$ decreases fast exponentially and touches the r axis asymptotically at infinity (that is, in quantum mechanics there is a finite probability to find the bound electron even at $r \to \infty$, though fantastically small).

Let us call the interatomic distance $r = R_{AB} \equiv R$. No interaction between A and B is monitored if $R =$ several atomic r_0 or larger. So, there is a minimum R of approach (for the two atoms) before an interaction sets up. There is

---

[13] $f(r)$ can be nasty, *i.e.*, hard to manage, because its explicit form as a function can be of any crazy algebraic change in r, unlike the multiplicative nice and elegant angular part $Y(\theta, \varphi)$. So, $f(r)$ is the wild part and $Y(\theta, \varphi)$ is the tame part.

no interaction for $r \geq R_{\min}$. After that $(r < R_{\min})$, as the atoms approach each other, the interaction starts, because their atomic wavefunctions intermix (overlap) and disturb each other (the outermost electron clouds interpenetrate and distort each other, thereby changing the electron distributions and probabilities). The principal maxima of the curves start overlapping if $R \sim r_A + r_B$. Upon this close approach, the two neighbor atoms affect each other by their Coulombic potentials. So, the motion of their valence electrons is perturbed, and that in turn changes the wavefunctions ψ_A and ψ_B to new (perturbed) forms.

If the valence electrons of the two atoms find themselves in the volume between the two nuclei, they are shared simultaneously by the two nuclei (principle of covalent bond), because the electric fields of the two nuclei on the electrons are almost the same. Then the electronic motion and behavior are not any more the same as they were when the atoms were free. The valence electrons move under the common influence of the two atoms, within the molecular volume. A new, common $\psi_{\text{molecular}}$ replaces ψ_A and ψ_B, and describes now the motion of the electrons better.

This electron sharing makes the electronic distribution function (of the two charge densities) take on a new form.

The total energy of the system is now $E_{\text{molecular}} \equiv E \neq E_A + E_B$, *i.e.*, no longer equal to $E_A + E_B$ when the atoms were separated far apart and noninteracting. E is now still quantized; the system's energy has discrete values. If the system's energy reduces to $E < E_A + E_B$, the atoms attract each other. But upon too much approach, this time repulsion (between their nuclei) starts. So, there is an optimum R where E becomes minimum, and equilibrium comes about: $F_{\text{repulsive}} = -F_{\text{attractive}}$. Then a stable molecule is realized. The system is an AB type molecule.

A molecule is not just the sum of two atoms of which it is formed and consists. It is more than that: It is a new entity, a new system with a new identity, with its own physical and chemical properties and data, both qualitatively and quantitatively. It is a compound with new characteristics, quite different than those of either atoms A or B when free. It is a chemical compound that is not the same as the two separate elements. H_2O is neither H_2 nor O_2 with respect to its properties. Whereas Na is an alkali and Cl_2 is a poisonous gas, NaCl is a compound that makes up our innocent table salt to have our French fries sprinkled with; NaCl has completely new properties: It is a salt, a *new*, chemically inert entity.

When the atomic wavefunctions overlap, the arising forces are the *cohesion forces* which mediate the chemical affinity and binding. To a first approximation, their range is ~ 1 to 3 Å, *i.e.*, the sum of the two atomic radii.

$E_{\text{chemical binding}} = E_A + E_B - E_{\text{molecular}}$. If $E_{\text{binding}} > 0$, the reaction $A + B \rightarrow AB$ is possible, and energy is released to the environment upon the formation of the molecule.

It should be noted that interactions exist also between molecules, whether the latter form crystals or are semi-free as in liquids, or free as in gases. Cohesive bonds exist also between molecules.

Molecules (and crystals) consist of nuclei and electrons (Fig. 1.18) where letters identify nuclei, and numbers, electrons — only single valence electrons are shown per atom, for simplicity). Let R_{jk} = internuclear distances (between all possible pairs of nuclei, *i.e.*, between the j-th and k-th nucleus). These distances form the molecular structural frame. Let r_{ij} = vector modulus of the distance between a nucleus and an electron, not necessarily its own only, but also to any other electron (*all* distances between the i-th electron and *any* j-th nucleus). The time-independent Schrödinger equation of the system is $\mathcal{H}\psi = E\psi$ where ψ is the total (molecular) wavefunction of the system (see Section 4.1), that depends on the coordinates of all the particles in the system: $\psi = \psi(x_1, y_1, z_1, \ldots, x_n, y_n, z_n, X_1, Y_1, Z_1, \ldots, X_N, Y_N, Z_N)$, if there are n electrons and N nuclei.

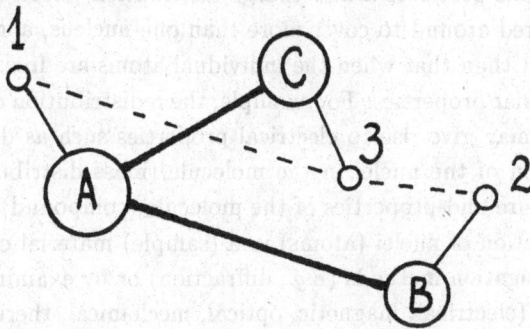

Fig. 1.18.

If we solve the Schrödinger equation, we see that its solutions have discrete eigenvalues which are the quantized E values. If these eigenergies are found, the molecular energy states (eigenstates) can be found. $|\psi|^2$ is the probability of finding the particles involved at different places (coordinates).

Since $m \neq M_j$, the electronic and nuclear motions are different. Electrons describe closed orbits (though the orbit paths may cross). While in free atoms they are at rest, nuclei in molecules vibrate (about the equilibrium, like oscillators). During the time an electron revolves many times around a nucleus, the nucleus moves only up to a small amplitude away from the equilibrium. The equilibrium position is that where $\overline{F} = 0$ (where $\overline{F}$ is the average force acting on the molecule). If the displacement is small, $\overline{F} \propto$ displacement when the nucleus is away from equilibrium. The displacement is $r - r_0$. Under the effect of $\overline{F}$, the nuclei vibrate harmonically (about their equilibrium) with small amplitude ($\ll R_{jk}$, the distance between them). The typical frequency of this vibration is $\sim 10^{13}$ Hz. So, in crystals the interatomic (actually internuclear) distance is taken between the centers of the atoms, *i.e.*, between the nuclear equilibrium positions. The vibrations are quantized, and even at $T = 0°K$ there is still a nuclear vibration; the nucleus is never at rest in equilibrium.[14]

If we assume that $\overline{R}_{jk} = $ constant in time in a molecule, it becomes possible to discuss about a definite atomic structure of the molecule. One can draw vectors between the nuclei. These straight lines give the direction of the bonds between atoms, and consequently, the rigid frame of the molecule. Electrons move without causing distortion to this frame. They move within the molecular volume, around their own nucleus and the neighbor nuclei (while inner-core electrons move only about their own nucleus). This motion of the outermost (valence) electrons produces a new charge distribution (electron cloud) in the molecule, smeared around to cover more than one nucleus, and therefore perturbed, different than that when the individual atoms are free. This situation gives the molecular properties. For example, the redistribution of the electronic charge density may give rise to electrical properties such as dipole moments; the configuration of the nuclei in the molecule (mass distribution) gives the chemical structure and properties of the molecule (compound), and so on.

The distribution of nuclei (atoms) in a (sample) material can be found by structure investigation methods (*e.g.*, diffraction) or by examining the macroscopic physical (electrical, magnetic, optical, mechanical, thermal) and chemical properties of it.

In dealing with the interatomic interactions, one relies on experimental results most of the time, after having assumed the structure as more or less theoretically known beforehand. (This way of action helps especially when

---

[14]This is nature's privacy; you can never remove the entire energy from the system. Some energy still remains around. So, the absolute zero cannot be reached exactly.

there is allotropy or polymorphism, *i.e.*, when chemically the same compound exhibits various structures. For example, ozone (O_3) is an allotropic form of O_2.) This means that one can simplify the Schrödinger equation, as it is done in Section 4.1. Notice that $R_{jk} = $ constant means that the nuclear term in the equation becomes zero, and the internuclear Coulombic repulsion term is constant. The rest is treated in Section 4.1. Approximate calculations are necessary, because the obtained Schrödinger equation cannot be solved exactly, even for simple molecules.

1.16. Bonds and Molecular Structure

When two (or more) atoms approach each other to within a relatively small (compared to their size) distance, attractive forces appear between them, which many times cause the holding together of the atoms. These forces provide the *bonds*. So, a bond has an ontological meaning as much as a force does. Such permanent (stable) complexes of two or more atoms (sometimes of ions, too) are called *molecules*.

During the creation of a molecule, despite the existence of attractive forces, the atoms do not interpenetrate each other; they are held apart at some distance determined by the appearing *repulsive* forces when the atoms attempt a too close approach. So, nature preserves its privacy, forbidding atomic *perichoresis*. That is, as soon as the electronic shells of the associating atoms start touching each other, the nuclei repel each other electrostatically. When the repulsion just balances the attraction, a stable bond is possible (for a separation $r = r_0$). This behavior is shown in Fig. 1.19 where the factors that determine the distance r_0 (for which the atoms are held together permanently) are explained. To simplify things, we assume a molecule consisting of a positive and a negative ion. The Coulombic forces between them are inversely proportional to the square of the separation distance r. When the atoms get close, the attractive forces do work, and hence their potential energy decreases (just like the decrease in the gravitational potential energy of a falling body under the influence of the Earth's attraction — the body's weight — as the body approaches the Earth; that the body does work can be demonstrated by placing a glass panel underneath it — it will break it on its way down). If we *define* the potential energy of the two atoms when they are infinitely separated apart (both at infinity) as zero potential, it will be obvious that their potential energy is always negative.

Repulsion increases drastically and sharply to strong values upon close approach beyond a certain value r' where the electronic ψ functions begin to overlap. The resulting repulsive potential energy (that derives a force)

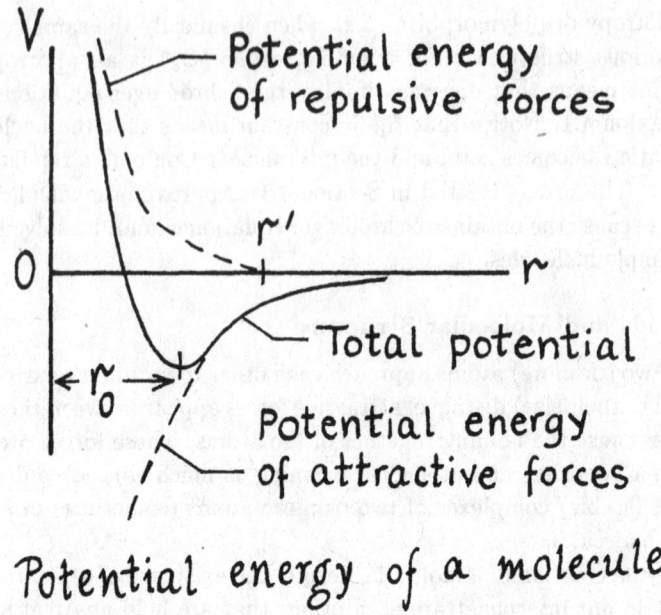

Potential energy of a molecule

Fig. 1.19.

is positive (in accordance with our definition) and shown in Fig. 1.19 by a dotted line. This positive-valued repulsion appears as soon as the separation becomes less than r'. Any further reduction of the distance r results in a considerable increase in the potential energy. No repulsion exists from $r = \infty$ to r'; it sets up from r' on.

Figure 1.19 shows the *total* potential between the two atoms (solid line) which is the algebraic sum (combination or superposition) of the attractive and repulsive parts. The curve exhibits a minimum at $r = r_0$; above r_0 it is attractive, and below r_0, strongly repulsive as $r \to 0$ (as the atoms tend to have their nuclei fall one on top of the other[15]). According to the relevant theorem of mechanics, the atoms reach stable equilibrium at r_0. If — for whatever reason — r slides to values a bit larger or less than r_0, the potential energy increases (notice the well), *i.e.*, a force shows up, calling the atoms back to their initial r_0 and thus restoring equilibrium, by driving the atoms back to the state of the least potential (stable equilibrium or lowest point of the well).

---

[15]Here r is the relative distance (or separation) between the atoms. Or, equivalently, we can take one atom standing still, with its nucleus at the origin, and the other one approaching from the right.

Depending on the nature of the attractive forces that hold the atoms together in the molecule, there are several kinds of bonds, but only few of them are important.

(1) To understand the *heteropolar bond*, we take two suitable atoms (*e.g.*, Na and Cl) that meet in space by chance. Na has two filled shells (with 2 and 8 electrons) plus one electron (valence electron) in the outermost shell $(Z = 11)$.[16] Cl $(Z = 17)$ has a shortage of just one electron by which the outermost shell would be completely full, because this shell can hold 8 electrons. Since all atoms tend to form complete shells (or subshells), the Na atom gives its outermost electron to the Cl atom. Thus two ions are formed: Na^+ and Cl^-, by the loss and gain of an electron respectively. These ions have now their shells all completely full, resembling those of Ne $(Z = 10)$ and Ar $(Z = 18)$. Now these two ions attract each other electrostatically (via Coulombic forces) in an heteropolar *ionic bond*. So, the atoms of Na and Cl show compatibility and willingness for a bond and molecule formation. When the outermost electronic shells begin entering one another's volume, the attractive forces get counteracted by the appearance of repulsive forces. A balancing equilibrium is achieved at a distance where these two opposite forces become equal and equilibrate each other.

Molecules whose atoms are kept together via heteropolar bonds are rare. NaCl is one of them, especially in vapor state. If the NaCl vapor condenses and liquefies, the molecules lose their selfhood and decompose into ions which move independently of each other. If molten NaCl solidifies, it crystallizes, *i.e.*, NaCl crystals are formed where the ions settle (they can vibrate only), occupying definite positions, and so we cannot talk about molecules in a strict sense. The HCl molecule, whose atoms were once considered as held together via an heteropolar bond, does not belong to this category, as we know today. Although the heteropolar bond is seldom encountered in molecules, it causes stable conglomerates in many solids (*e.g.*, ionic lattices).

Heteropolar valence is the number of electrons transferred by one atom of the molecule to the other. So, it is equal to the charge of the created ion (in units of e). The transfer occurs towards the creation of complete outer shells, because such states are energetically favorable, that is, states of the least energy. Thus the valence of Na is 1. That of Mg $(Z = 12)$ is 2, because if the M-shell loses two electrons, there remains a stable ion whose K and L-shells are complete.

---

[16]For this reason the Na atom resembles atomic hydrogen. All the inner core (nucleus + 10 electrons) can be taken as analogous to the nucleus of H.

In some cases an electron from a *complete* shell is extracted in order to have an ion formed! This happens during the formation of bivalent ions of copper (Cu^{++}), trivalent ions of gold (Au^{+++}) etc.

In these examples we meant *electropositive valence*, because the ions are positive (positively charged via electron loss), as they fire electrons. However, there are cases (*e.g.*, in Cl, S etc.) where the outermost shell gets filled by the *recruitment* of electrons, thereby producing negative ions (Cl^-, S^- etc.). These atoms (elements) have *electronegative valence*.

Noble gases have a fully completed outer shell; they neither take nor give electrons. So, their valence is zero. They do not form compounds. They do not enter into chemical marriage — they are hence called "noble". They are chemically inert.

This is in short the electronic theory of valence. From the foregoing discussion it is apparent that the valence depends on the position of a chemical element in the periodic table, and on the group the element belongs to.

(2) A large number of the known molecules is characterized by *homopolar bonds*. These molecules are formed by atoms which are not held by electrostatic forces, but rather by forces of different nature. If two atoms approach each other so that their shells mutually penetrate, the electronic motion in either atom changes: The electrons move as if they belonged commonly to both atoms. Figure 1.20 is a schematic representation of the form of the electronic shells of two free atoms (above) and the resulting homopolar molecule (below) out of them. Due to the commonly shared electrons, there appears an interatomic force called *exchange force* which provides the homopolar bond. It is substantial then, that common electrons exist to create an homopolar bond

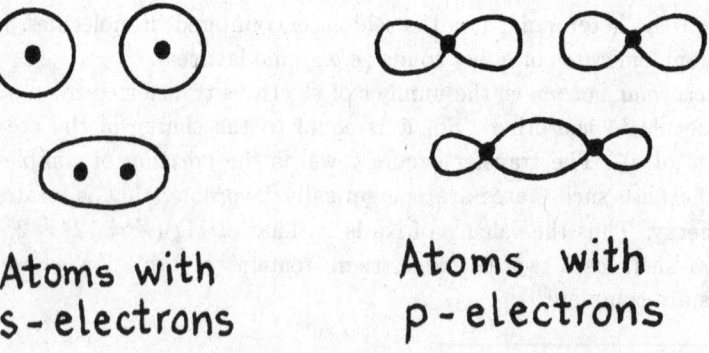

Atoms with
s - electrons

Atoms with
p - electrons

Fig. 1.20.

and account for it. This bond is strong to the extent of the interpenetration of the atomic shells. Molecules such as N_2, Cl_2, CCl_4 etc. are homopolar.

The mechanism of production of the exchange force cannot be explained by classical physics, nor by a simple quantum theory, but rather by *wave mechanics*. A rigorous analysis takes too long to explain. A simple and non-sumptuous explanation — which is not so satisfactory, to be honest — is as follows: During the formation of the homopolar bond, a concentration of electricity (electron charge density) is created on the average in the region between the two nuclei. An electron in this region feels the attraction of both nuclei. Conversely, the two nuclei are attracted by this commonly shared electron, and as a result, a stable bond arises between them.

As far as H_2 is concerned, it is found that when two H atoms get close to each other, attraction arises if the spins of the two electrons are antiparallel. If they are parallel, then repulsion is the result. So, formation of a H_2 molecule is possible only if the two spins are antiparallel.[17] The extension of this reasoning to the homopolar bond between atoms with *many* electrons leads us to the conclusion that those outer electrons partake of the bond, whose spin is not compensated for or paired off by the spin of another electron of the same atom. Such electrons are called *lonely electrons*.

In a *simple bond*, each atom contributes one lonely electron. The bond is characterized by this pair of lonely electrons. The pair becomes common for both atoms. An example to a simple bond is the Cl_2 molecule ($Z = 17$). It consists of two Cl atoms, each of which has 7 electrons in the (outermost) M-shell. Six out of these 7 electrons are paired off (their spins are compensated for in dyads). There remains one lonely electron. When the molecule sets up, the two lonely electrons belong to both atoms together, spending most of their time in the region between the two nuclei:

$$: \overset{\bm{\cdot\cdot}}{\underset{\bm{\cdot\cdot}}{Cl}} \cdot \; + \; : \overset{\bm{\cdot\cdot}}{\underset{\bm{\cdot\cdot}}{Cl}} \cdot \; \rightarrow \; : \overset{\bm{\cdot\cdot}}{\underset{\bm{\cdot\cdot}}{Cl}} : \overset{\bm{\cdot\cdot}}{\underset{\bm{\cdot\cdot}}{Cl}} :$$

In a *double* or *triple bond*, each atom contributes two or three lonely electrons in order to form the corresponding number of electron pairs. We have a triple bond in N_2 ($Z = 7$) where there are 5 electrons in the (outermost) L-shell. Each atom contributes three electrons:

---

[17]Actually, the attraction or repulsion is not due to the interaction of the spins (or magnetic moments) itself, but it is rather a consequence of the Pauli exclusion principle which forbids the presence of two electrons of the same velocity in a given region of space, except if they have antiparallel spins.

$$: \overset{\bullet}{\underset{\bullet}{N}} \bullet \; + \; : \overset{\bullet}{\underset{\bullet}{N}} \bullet \; \rightarrow \; : \overset{\bullet}{\underset{\bullet}{N}} \overset{\bullet}{\underset{\bullet}{N}} :$$

The electronic shells do not exhibit spherical symmetry always; they sometimes stretch out towards a principal direction (*e.g.*, in the case of p-electrons).[18] Figure 1.20 (above, right) shows atoms with such electronic shells. Upon the formation of the molecule, the common electrons form a peculiar (in shape) shell (same figure, below right). So, p-electrons make a bond which is *directional*.

We call *homopolar valence* (of an atom) the number of H atoms (or any other equivalent atoms, *e.g.*, F) with which the atom in question can form an homopolar bond. Since a simple homopolar bond requires a lonely electron from each atom, the homopolar valence is equal to the number of the lonely electrons of the atom.

To find the valence of an atom we, resort to the observation that in an homopolar bond an electronic configuration results about each atom, which — if we include the shared electrons as well — resembles a full shell which in turn constitutes a very stable configuration. So, in the H_2 molecule the two common electrons can be considered as constituting the completely filled shell of He. In the Cl_2 molecule the homopolar valence is 1 because only if the common electrons are two (one lonely electron per atom), is the formation of

---

[18]In wave mechanics, the notion of the orbit (or shell) is replaced by an electronic cloud around the nucleus. The charge and mass of the electron(s) is distributed throughout this cloud whose shape varies according to the atomic state. Figure 1.21 shows the charge distribution (of the electronic cloud) about the nucleus in the H atom. The shape is a measurement of the intensity of the waves, and hence — according to Schrödinger — of the electron charge distribution. The states are characterized by their quantum numbers. The letters s, p, d, f denote the values of the angular momentum ($l = 0, 1, 2, 3$), and m is the magnetic quantum number. (The intensity boundaries are not so sharp and definite as drawn here, but rather blurred, with a gradual fade-out.) In wave optics there is an equation relating the amplitude of an optical wave to the index of refraction of the medium, which changes spatially, depending on the coordinates (x, y, z). This equation helps us calculate complex phenomena of refraction and diffraction. The Schrödinger equation is analogous to this equation; it governs the propagation of probability waves or matter waves that escort a particle, not the particle's motion. Its solution gives the function $\psi(x, y, z)$ where $|\psi|^2$ is the probability (actually, the probability distribution function) of observing the particle at (x, y, z) in an experiment. The equation has a solution for every value of the constant parameter E in some cases, while in other cases, only for certain values of E. This mathematical property is nothing new, given that it appears in other similar cases, too (*e.g.*, standing waves produced in vibrating chords have definite frequencies — the fundamental and its harmonics). So, the acceptance of the quantum theory (where the electron's total energy can have only certain values) appears in wave mechanics as a consequence of the mathematical examination of the Schrödinger equation, without the necessity of any other assumption. In classical wave theory, ψ is the field of the wave in space.

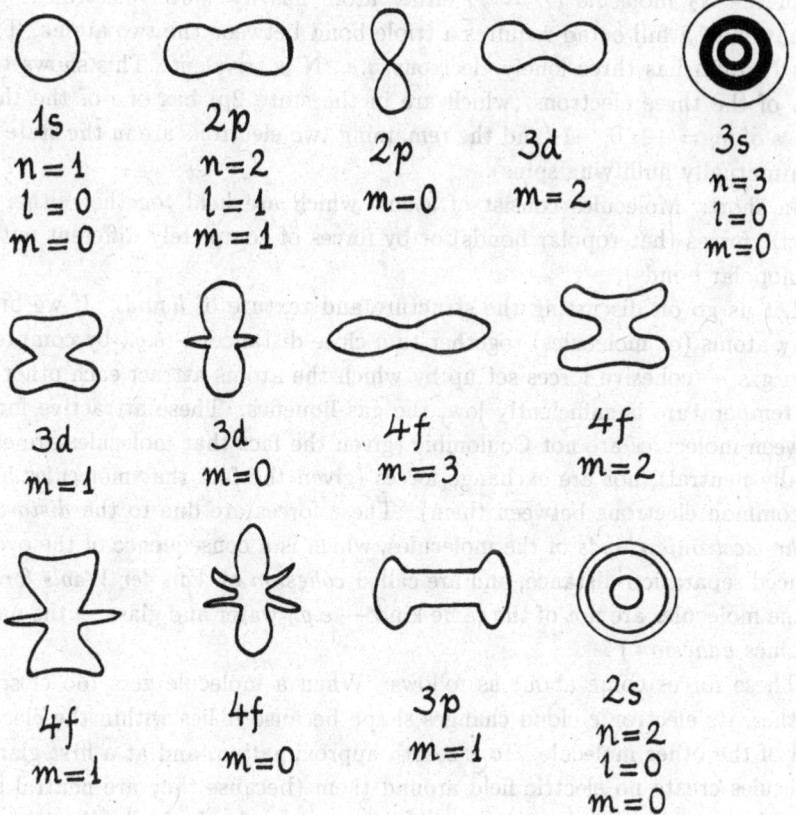

1s
$n=1$
$l=0$
$m=0$

2p
$n=2$
$l=1$
$m=1$

2p
$m=0$

3d
$m=2$

3s
$n=3$
$l=0$
$m=0$

3d
$m=1$

3d
$m=0$

4f
$m=3$

4f
$m=2$

4f
$m=1$

4f
$m=0$

3p
$m=1$

2s
$n=2$
$l=0$
$m=0$

Fig. 1.21.

a full shell possible about each of the two nuclei. This shell contains then an *octad* of electrons.

The way by which the seven outer electrons of the Cl atom arrange themselves so that one lonely electron remains, can be described as follows: Two out of these seven electrons are in the state 3s ($l = 0$), and hence their spins nullify each other. The remaining five are in the state 3p ($l = 1$), having $m_l = +1, 0, -1$. These threes values allow room for 6 electrons (because of the two values of $m_s = \pm 1/2$). Thus, if we distribute five electrons to six available positions, four of them will have two of the three values of m_l (causing saturation), while the third value will be gotten by one electron which will remain lonely.

In the N_2 molecule $(Z = 7)$ either atom has five outer electrons. The formation of a full octad requires a triple bond between the two atoms. Thus each N atom has three lonely electrons, *i.e.*, N is trivalent. This shows that each of the three electrons (which are in the state $2p$) has one of the three values of $m_l = +1, 0, -1$ (and the remaining two electrons are in the state $2s$, with mutually nullifying spins).

In short: Molecules consist of atoms which are held together either by electric forces (heteropolar bonds) or by forces of completely different nature (homopolar bonds).

Let us go on discussing the structure and texture of *liquids*. If we bring many atoms (or molecules) together to a close distance — *e.g.*, by compressing a gas — cohesive forces set up by which the atoms attract each other. If the temperature is sufficiently low, the gas liquefies. These attractive forces between molecules are not Coulombic (given the fact that molecules are electrically neutral), nor are exchange forces (given the fact that molecules have no common electrons between them). These forces are due to the *distortion of the electronic clouds* of the molecules, which is a consequence of the overly reduced separation distance, and are called *cohesion* or *Van der Waals forces*. (If the molecules are not of the same kind — *e.g.*, water and glass — the name becomes *adhesion*.)

These forces come about as follows: When a molecule gets too close to another, its electronic cloud changes shape because it lies within the electric field of the other molecule. To a zeroth approximation, and at a first glance, molecules create no electric field around them (because they are neutral and their electronic orbits about the nuclei are evenly and spherically distributed). But in fact electrons revolve about the nuclei, forming a rapidly pulsating electric moment. It is within the electric field of this "dipole" that the other molecule gets distorted (*polarization*), thereby feeling an attractive force as a result of electrification by induction. This force is called *force by polarization* or *Van der Waals force*,[19] and is as strong as the easiness by which the molecule undergoes distortion.

Such forces can act between atoms, too, but they are weak, because the electronic shells of atoms get distorted with difficulty, as they exhibit small and weak polarization. So, these forces are overshadowed by forces of heteropolar or homopolar nature which are much stronger. The only case in which the Van der Waals forces between atoms become noticeable and make themselves felt is the case of the noble gases where liquefaction is due to *interatomic* forces, given

---

[19] J. Van der Waals, Dutch physicist, Nobel Prize winner (1910).

the fact that these gases form no molecules, or equivalently, their molecules are monatomic, just the atoms themselves.

Owing to the fact that the Van der Waals forces between atoms are feeble, the boiling point of noble gases is low. Thermal motion in them is already adequate at very low temperatures to overcome the Van der Waals forces and start the boiling out. The boiling point of inert gases rises with the increasing atomic number, because the outer shells of the atoms get more easily distorted as their size increases.

In this discussion we are not going to deal with the structure of crystals (solids) and notions like primitive cells, axes, lattices, lattice planes, indices etc. which are beyond our scope and refer to solid state physics anyway. But we will discuss briefly the forces inside crystals.

The positions of the building blocks in crystals are positions of equilibrium occupied under the action of forces exerted by the rest of the building blocks. These forces can be described by curves like that in Fig. 1.19. If these blocks stay a little away from their position, attractive and repulsive forces develop, bringing them back to their proper equilibrium position (*elastic solids*). The

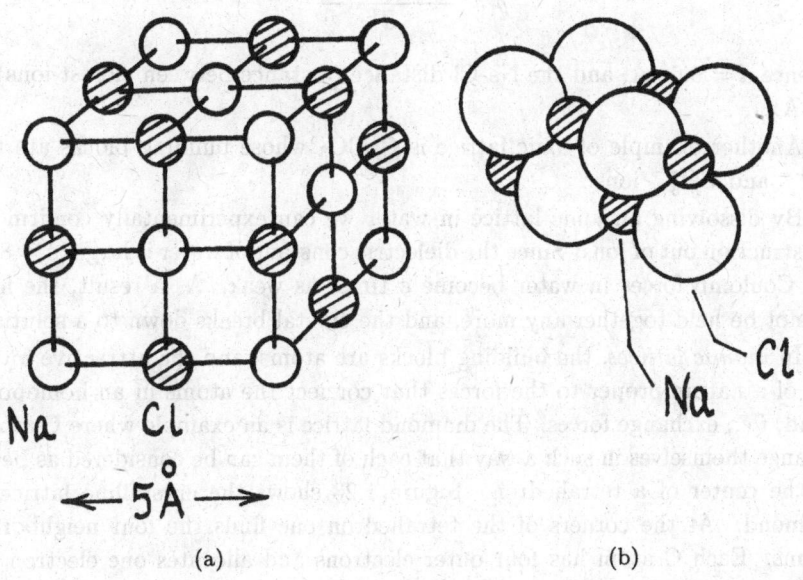

(a)   (b)

Fig. 1.22.

repulsive forces arise from the interpenetration of the building blocks, while the attractive ones are due to various reasons, depending on the kind of the blocks.

In *ionic lattices*, the building blocks are ions (*e.g.*, the NaCl grid). The restraining forces (that keep the ions grouped) are of electric (Coulombic) nature, because the ions have opposite charge. In Fig. 1.22(a) the ions are shown like small spheres positioned apart from each other (at the corners of a cubic frame). This picture is artificial and schematic for clarity, so that the positions of the ions be seen well. In fact, the ions are so large that they "touch" each other. The correct arrangement of the ions in the NaCl grid is shown in (b).

The shape and size of a unit cell can be determined by modern methods using X-rays. In the case of NaCl, the crystal structure is simple and can be identified by its geometric (and other) properties. The unit cell of NaCl is face-centered cubic. The lattice constant d can be found as follows: Each cell has 4 ions of Na and 4 ions of Cl. The mass of each ion is found from $m_{Na} = 23\ m_H$ and $m_{Cl} = 35.5\ m_H$ where the factors 23 and 35.5 are the atomic weights of Na and Cl respectively, and $m_H = 1.67 \times 10^{-24}$ gr is the hydrogen mass. The mass of the ions in a cell divided by its volume d^3 gives the density $\rho = 2.17$ gr/cm^3 which is known:

$$\rho = \frac{4m_{Na} + 4m_{Cl}}{d^3}$$

whence $d = 5.68$ Å, and the Na-Cl distance (distance between closest ions) is 2.8 Å.

Another example of ionic lattice is $CaCO_3$ whose building blocks are the Ca^{++} and CO_3^{--} ions.

By dissolving an ionic lattice in water we can experimentally confirm its construction out of ions: Since the dielectric constant of water is large ($\varepsilon = 81$), the Coulomb forces in water become ε times as weak. As a result, the ions cannot be held together any more, and the crystal breaks down to a solution.

In *atomic lattices*, the building blocks are atoms, and the attractive forces are of a nature proper to the forces that connect the atoms in an homopolar bond, *i.e.*, exchange forces. The diamond lattice is an example where C atoms arrange themselves in such a way that each of them can be considered as being at the center of a tetrahedron. Figure 1.23 shows the crystalline lattice of diamond. At the corners of the tetrahedron one finds the four neighboring atoms. Each C atom has four outer electrons and allocates one electron for each neighboring atom, forming an homopolar bond with it.

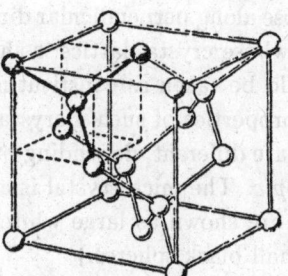

Fig. 1.23.

In *molecular lattices*, the elementary building blocks are molecules held together by Van der Waals forces. Organic compounds such as anthracene have such lattices. An anthracene cell containing two molecules is shown in Fig. 1.24.

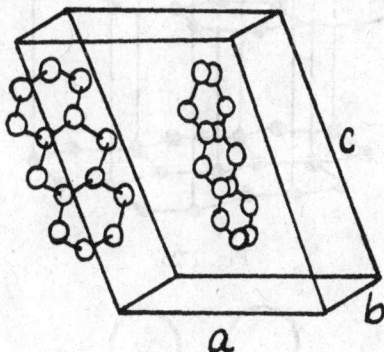

Fig. 1.24.

In *metallic lattices*, the building blocks are ions originating from atoms, from which one or more electrons have been removed. They differ from ionic lattices, because these electrons cannot be considered as belonging to a definite atom; they rather belong to the whole grid and move throughout it freely (travelling *free electrons*). In a metallic lattice the ions are kept at their equilibrium positions by repulsive forces (exerted between positively charged ions) *and* by attractive forces exerted by the "gas" of free electrons on positive ions (electronic theory of metals).

Many crystals do not belong to any of these categories in a strict sense. They are intermediate cases. So, in certain crystals the interatomic forces in a plane are stronger than those along perpendicular directions (*lattices in layers*), as it happens in graphite whose crystal lattice is shown in Fig. 1.25. Here all the atoms in a plane should be taken as constituting a huge two-dimensional molecule. Therefore, the properties of such a crystal (magnetic susceptibility, electric conductivity etc.) are different, depending on the direction considered, *i.e.*, the crystal is *anisotropic*. The mica crystal is another example (Fig. 1.26 where the potassium ions are shown as large white spheres weakly bound to oxygen atoms shown as small black spheres.)

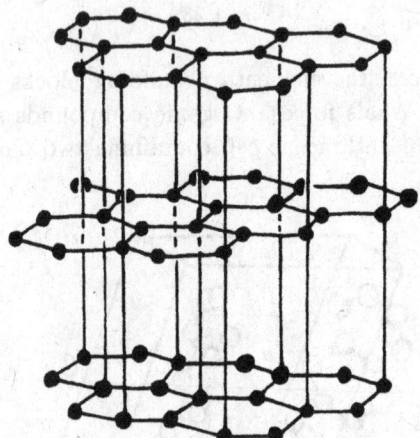

Fig. 1.25.

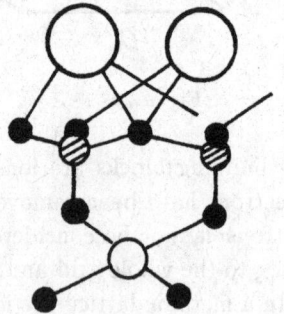

Fig. 1.26.

Crystals have the property of being rent or split in flat surfaces (*schismogenic planes*) by the action of external forces. This cleaving occurs in lattice planes which are rich in building blocks, as long as the bonds that connect the blocks on either side of a schismogenic plane are weak. The perfect schistose cleavage of mica is due to the fact that mica consists of layers in which building blocks abound, but the layers themselves are too weakly bound to one another. In graphite, too, cleavage is possible along certain planes only. Along these planes, lattice planes parallel to them slide very easily, and that is why graphite powder is used as grease whenever the use of mineral oil is not possible.

Another phenomenon is the *plastic deformation*: Crystallites of a solid slide (under the action of external forces) along certain lattice planes without any interruption in their continuity.

What are the factors that determine the crystalline structure? From a general thermodynamic point of view we can say that the building blocks form such a crystal that its *free energy* $F = U - TS$ (where U is the internal energy, T the absolute temperature, and S the entropy) be minimum. Such factors are the packing number,[20] the chemical valence, the radii of the atoms or ions etc.

(1) Chemical elements: In nonmetals the building blocks are neutral atoms or molecules; in metals the blocks are ions, because some electrons detach themselves from the atoms and become free.

(a) *Nonmetals*: The atoms of monovalent elements (*e.g.*, halogens) lack one electron so that a complete shell be realized. By the union of such atoms via a simple (homopolar) bond, stable molecules are created. Then the crystal consists of a grid of diatomic molecules. (Example: Iodine (Fig. 1.27(a)).) In bivalent elements each atom can form two simple (homopolar) bonds directed towards different directions, *i.e.*, long molecules can be formed in the form of closed chains. (Example: S_8 (Fig. 1.27(b)).) Each closed chain constitutes a building block of the lattice. In certain bivalent elements a double bond is formed (instead of two simple ones); in that case the lattice is molecular. (Example: Solid O_2.) In trivalent elements each atom can form three simple (homopolar) bonds directed towards different directions. Thus the formed lattice planes can be taken as huge planar molecules. The thus formed layers are held together by Van der Waals forces. (Example: Sb (Fig. 1.27(c)).) In tetravalent elements the crystal is thought as consisting of only one single and huge three-dimensional molecule. Each atom is surrounded by four neighbors to which it is connected by four simple (homopolar) bonds. (Example: Diamond.) However,

---

[20]That is, the number of the closest neighboring building blocks around a building block. In a simple cubic system, for example, each block has six nearest neighbors. In body-centered cubic systems the number is 8; in face-centered cubic systems, 12.

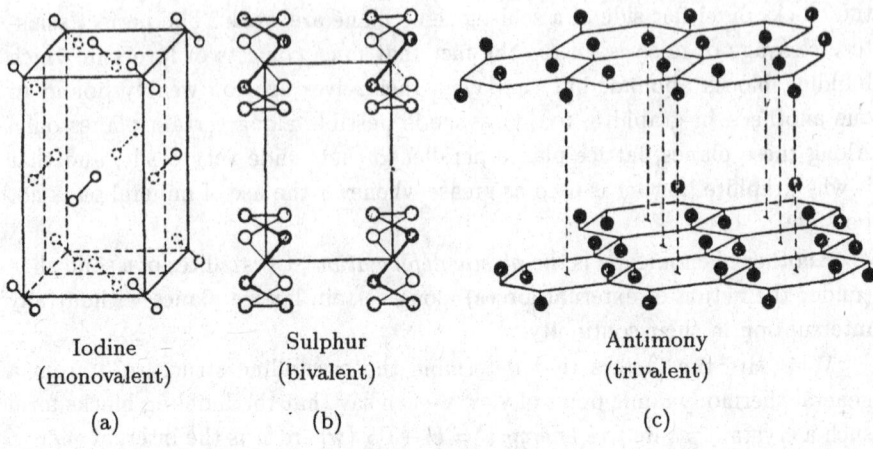

Iodine	Sulphur	Antimony
(monovalent)	(bivalent)	(trivalent)
(a)	(b)	(c)

Fig. 1.27.

it is possible for each atom to be surrounded by three neighbors of which two form a simple bond, while the third one forms a double bond. (Example: Graphite.) In that case, layers are formed which keep together thanks to quite weak forces.

(b) *Metals*: The crystal structure of metals does not depend on the valence, but rather on the tendency to create a grid packed as closely as possible. Thus most of the metals crystallize in the face-centered cubic system or in the hexagonal system of the highest packing density.

(2) Chemical compounds: The structure of compounds depends on the valence of each element and on the size of atoms or ions. When two elements have the same valence but different polarity (one electropositive and the other

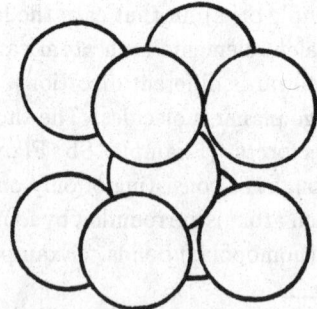

Fig. 1.28.

electronegative), an ionic lattice is formed. The dimensions of the ions affect the grid; for example, in CsCl the two ions have almost equal radii, and hence each ion of one kind can be surrounded by eight ions of the other kind. The resulting grid is shown in Fig. 1.28. In NaCl (where the radius of the Na^+ ion is much smaller than that of Cl^-), the Na^+ ion cannot touch eight Cl^- ions; here each ion of one kind touches only six ions of the other kind.

By the theory of the crystal structure we can explain some physical properties of solids, *e.g.*, thermal expansion, melting, electric conductivity (insulators, conductors-metals, semiconductors), elastic properties etc.

1.17. Molecular Spectra

On the Earth and around us, matter consists mostly of molecules, not individual atoms. Quantum theory explains molecular bonds and spectra which are important topics in physics, chemistry, medicine, biology, and science in general.

The number of (valence) electrons in the outermost atomic state and the strength of their bonds with the atom determine the way the atom reacts chemically. A diatomic molecule (two atoms bound together) is formed when one (or more) electron(s) of one atom get(s) to states in which it (they) can perambulate around in a manner orbiting around both nuclei, so that the compound consists of two coexisting atoms (Fig. 1.29). For example, in H_2 two H nuclei get bound together by the mutual orbit of two electrons. (So, in chemistry we speak of the *valence* of an atom, *i.e.*, the number of electrons — or electron spaces available — in the outermost level. A valence of +1 means that the atom can easily give an electron to be shared with another atom in forming a molecule or compound. A valence of −2 means that an atom can accommodate two more electrons in its outermost level and hence it can react readily with another atom which can give one or two electrons.)

Upon mutual approach, the electron cloud of one atom changes shape because of the effect of the Coulombic forces of the charges (protons and

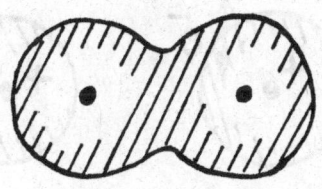

Fig. 1.29.

electrons) of the other atom (and vice versa). When the approach is close enough, the outer electrons can tour around both nuclei. This electron (cloud) sharing among the nuclei is called *molecular bonding* or *molecular orbital*. Obviously, an electron is exchanged by transfer from one atom to the other. All the types of bonds are intermediate cases — more or less — of two extreme (and basic) types of strong bonds: Covalent and ionic.

In a covalent bond, the electrons are directly shared by the two nuclei, without much ado. For example, in H_2 the two nuclei are just protons, and the two electrons form a covalent bond between the two H atoms. Here the Pauli principle concerning the spin of the two electrons enters. Figure 1.30 shows the two *different* electron probability distributions when the spins are parallel and antiparallel. In the former case (a) we have $m_s = +1/2$ for both electrons; according to the exclusion principle the electrons cannot get to the same state. They are kept (by nature) apart, and it is not possible to form a bond. In the latter case (b) we have $(m_s)_1 = +1/2$ and $(m_s)_2 = -1/2$ whence the electrons occupy different quantum states in the $1s$ sublevel, and a covalent bond is possible. The electron cloud (shaded area) gravitates to the space between the two nuclei, as the two electrons can come close together and spend most of their time in that space, giving rise to a bond, as if coming from a glue. The nature of the bond is merely the Coulombic attraction between the negatively charged electrons and the positively charged nuclei. So, quantum theory explains the binding (via the quantum effects and the exclusion principle). Other diatomic molecules (such as N_2 and O_2, and to some extent H_2O, CO_2, and CO) are good examples to the covalent bond, too. In Fig. 1.30 the shaded area is the electron cloud. In (a) there is no bond; in (b) the bond is covalent, with the cloud concentrated between the two nuclei. An equal sharing produces a *pure* covalent bond.

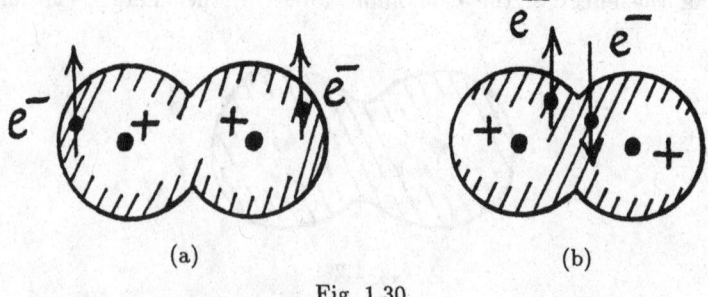

(a)   (b)

Fig. 1.30.

On the other hand, an uneven distribution of electrons produces an ionic bond. In NaCl the valence electron of Na is snatched by Cl to form the bond. The Na atom is an alkali whose electron configuration is $1s^2 2s^2 2p^6 3s^1$, *i.e.*, it has an odd (extra) electron ($3s$-electron) *outside* a perfectly closed shell. Cl is a halogen whose atom structure is $1s^2 2s^2 2p^6 3s^2 3p^5$, *i.e.*, it misses one $3p$-electron to complete (fill) the outer shell. So, it can take what it lacks from Na which is willing to give, so that both atoms fulfill their objective. The two atoms interact, and the energies involved can be computed in the quantum framework. The energy of the NaCl molecule is lower than the sum of the energies of the free atoms; this means that if the two atoms find themselves together, the reaction is energetically feasible, and hence favored and possible.

When Na and Cl get together, an electron is transferred: Cl becomes Cl^- (it has now a net negative charge), and Na becomes Na^+, *i.e.*, they become *ions*. And the Coulombic force holds them bound: The molecule owes its existence to it. Notice that in quantum theory closed shells are favored.

In AB or A_2B type molecules, features of both covalent and ionic bonds are present (CO, H_2O etc.). The (shared) electrons prefer to stay longer close to one nucleus than the other. Then obviously, the molecule is *polar*, and the bond partially ionic. One part of the molecule has an excess positive charge, and another part an excess negative charge. A dipole moment develops between these two parts due to the charge separation.

Weak (intermediate) chemical bonds are possible due to leftover Coulombic attractions in polar molecules, or to the fact that electron clouds may get polarized upon mutual approach. The bond strength is determined by how much potential energy is stored in the static field, *alias* the *bond energy*. It is the energy input to be supplied to the molecule to dissolve it by breaking the bond(s). Strong covalent or ionic bonds require 1 to 5 eV, and weak ones 0.03 to 0.3 eV.[21] For example, 0.03 eV corresponds to $T = E/k = 350°K$. So, weak bonds can be broken by collisions between molecules moving at reasonable speeds at room temperature; this allows for new reactions, *e.g.*, biochemical reactions of large macromolecules (protein formation, DNA replication, enzyme reactions etc.). That is why weak bonds pertain to medicine and biology.

A molecule — although itself a collection of atoms — has energies that free atoms and ions do not have. Because apart from the electron energies (of the electrons in their allowed and accessible states), the molecule has E_{vibr} and E_{rot}. The former comes from the oscillation of the atoms about equilibrium

---

[21]Since the latent heat (of fusion or vaporization) is just the energy required to break bonds, we can find it from the binding energy.

position (just like point masses tied to massless springs and loading them). Figure 1.31 shows the molecular vibration where the internuclear bond that holds the atoms together is elastic, and the nuclei vibrate (about their equilibrium position) with an equivalent spring constant κ. As the spring (bond) stretches and contracts, E_{vibr} alternates between potential and kinetic energy. The potential energy stored in the spring is $(1/2)\kappa x^2$ as the spring stretches or gets compressed a length x from its relaxed (equilibrium) position which for a molecule is the bond length r_0 (normal interatomic separation). E_{rot} is kinetic and internal, as the molecule moves (rotates) about an axis (or many axes). Collisions impart E_{vibr} and E_{rot} to a molecule.

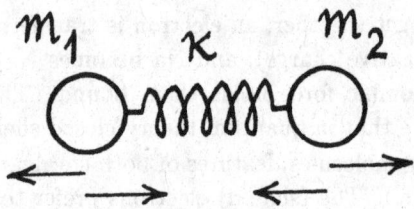

Fig. 1.31.

E_{vibr} and E_{rot} are quantized (like E_{electron}), because a molecule can take only discrete states and nothing in between. It can be in certain energies only, of specified amounts. A molecular energy state is specified as $|\alpha \; \text{v} \; J\rangle$ where index α absorbs the electronic quantum numbers, and the other two quantum numbers fully specify the rest, *i.e.*, the vibrational and rotational states or degrees of freedom.

Consider a diatomic molecule, *e.g.*, O_2 or CO, and its quantized vibrations, using the idealized model of a simple linear harmonic oscillator. The natural oscillation frequency is $\omega_o = \sqrt{\kappa/\bar{\mu}}$ where $\bar{\mu}$ is the effective (equivalent or reduced) mass. The oscillation period is $\tau = 2\pi/\omega_0 = 1/\nu_0$.

The total energy of a molecule is *negative*, because *we* must supply energy to the molecule to rend it apart. Since atomic vibrations are little positive amounts of energy, E_{total} is reduced by them; and if the vibrations become too energetic, they can exceed E_{total} and rip the molecule asunder. As long as the molecule exists, the atoms are in a parabolic potential well $V = (1/2)\kappa x^2$. The quantized vibrational energies (levels) in this well (due to the frequencies) are $E(\text{v}) = \hbar\omega_0 \left(\text{v} + \frac{1}{2}\right)$ with v $= 0, 1, 2, 3, \dots$ where v $= 0$ gives the ground state, and the rest are the higher (excited) states which are half-integer multiples (harmonics) of the fundamental amount $\hbar\omega_0$. Figure 1.32 shows the lowest few

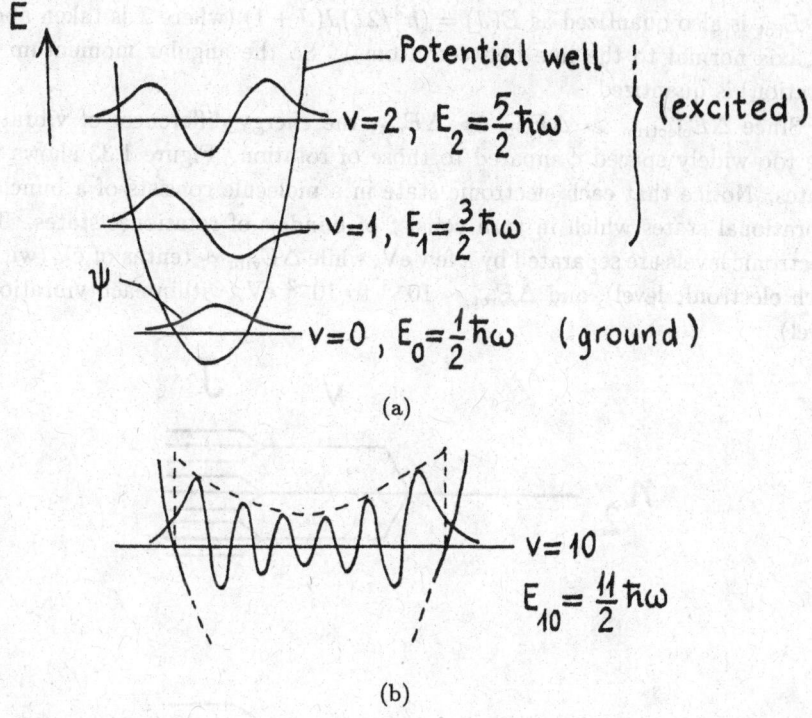

Fig. 1.32.

levels, their associated wavefunctions, and the case v = 10. The vibrations are in principle confined inside the well, but the wavefunctions — as quantum mechanics dictates — possibly get somehow beyond the well walls, allowing for a slim chance (probability) of finding the atoms outside the boundary, as nothing is strictly impossible in nature with quantum theory. In other words, there is some uncertainty in the atom's position, giving rise to *tunnelling*; the atom tunnels out through the walls, if we wait long enough.

This is the oscillator model of the molecule. If we know $\tilde{\mu}$ and κ, we can reach E_{vibr}, because $\hbar\omega_0 = \hbar\sqrt{\kappa/\tilde{\mu}}$. And even in the lowest state (v = 0) the molecule still has a residual zero-point energy $(1/2)\hbar\omega_0$ that cannot be extracted from it (nature's privacy!), as predicted by quantum theory and the uncertainty principle. That is, we can never fully immobilize the oscillating molecule: If we cut the motion we fully know the position (Δx = uncertainty = 0) and lose knowledge of momentum ($\Delta p \to \infty$, infinitely large uncertainty in p).

E_{rot} is also quantized as $E(J) = (\hbar^2/2\mathcal{I})J(J+1)$ (where $\mathcal{I}$ is taken about an axis normal to the line between atoms). So the angular momentum (of rotation) is quantized.

Since $\Delta E_{\text{electron}} \gg \Delta E_{\text{vibr}} \gg \Delta E_{\text{rot}}$, the energy differences of vibration are too widely spaced compared to those of rotation. Figure 1.33 shows the states. Notice that each electronic state in a molecule consists of a bunch of vibrational states which in turn consist of bundles of rotational states. The electronic levels are separated by a few eV, while $\Delta E_{\text{vibr}} \sim$ tenths of eV (within each electronic level), and $\Delta E_{\text{rot}} \sim 10^{-3}$ to 10^{-2} eV (within each vibrational level).

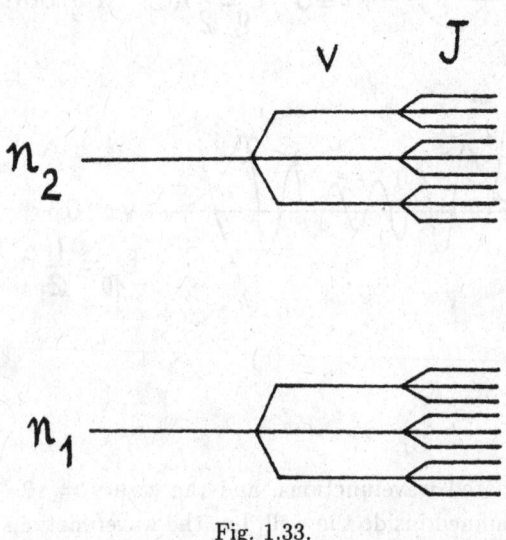

Fig. 1.33.

Similarly to the atomic case, electronic transitions between the levels cause absorption or emission of visible or ultraviolet radiation, while transitions between vibrational levels correspond to frequencies of infrared photons. Rotational transitions fall in the microwave and radio regions of the spectrum. The resulting molecular spectrum consists of bands (*band spectrum*), because each electronic state has a fine structure, and fine levels are regularly spaced — in general — so that each band consists of many closely-spaced lines that appear in the spectrogram. Figure 1.34 shows a wavelength spectrogram of a molecule where transitions between rotational states (within each vibrational state) give a regular line appearance of the spectrogram. Since by looking at the spectrum we can identify the compound that emits it, a molecular spectrum is, so to say, the fingerprint of its source material.

Fig. 1.34.

According to the selection rules, v and J change if a molecule changes state by making a transition from n to n'. For a given Δn (change in n), there are bands of various possible changes in v which are regularly-spaced (in the spectrum), because the vibrational levels are equally-spaced (within each n level). And each band has many lines because there are various changes in J. Figure 1.35 shows a vibration band with possible (allowed) transitions between energy levels: Each vibrational state consists of three rotational ones (here, in this case; actually there are more). The series of closely-spaced lines forms a band in the spectrogram. So, the molecular spectra are quite complex.

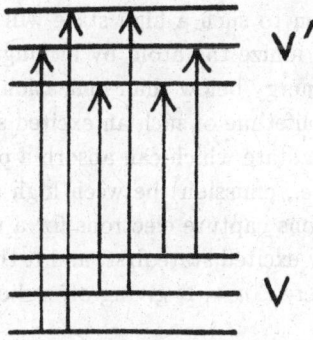

Fig. 1.35.

What is the use of studying spectra? We can learn about molecular structure, because energy levels and potential wells are mutually related. One can argue that by using quantum mechanics only, one can theoretically calculate the energy levels without resorting to any experimental spectroscopy. True in principle, but very hard in practice. When it comes to the actual calculation of all possible energies, the procedure stiffens up. It is possible with only a few simple cases, like CO and H_2.

Radio and infrared spectral lines are emitted — among others — by molecules, because the latter have levels very closely spaced together (like a fine structure). Physical processes can create spectral lines in any region of the spectrum, but infrared lines come from transitions between highly excited levels (in atoms), and when electrons make transitions between fine-structure states (states with the same n, l, and s, but different total angular momentum $l + s$, in atoms again). In fact these transitions are not allowed (because l must change), but if the density of the atomic gas is low, other radiative processes make these transitions possible. For example, a collision may cause an electron to get to an excited fine-structure state; if there is enough time before the next collision (since the density is low), a spontaneous de-excitation can take place with an infrared emission (because ΔE between fine-structure levels is less than 1 eV). On the other hand, radio spectral lines may be due to either Δn or changes in fine-structure levels (in atoms). For example, if n is high, a transition between $n = 101$ and $n' = 100$ in H will give

$$\frac{1}{\lambda} \propto \left(\frac{1}{100^2} - \frac{1}{101^2} \right)$$

which gives $\lambda = 4.6$ cm. We do not expect such an absorption, however, because to get the electron to such a high state will require energy to excite it, but not as much as to ionize the atom by kicking the electron out, *i.e.*, it will require just enough energy below than ionization energy by an amount of 0.0013 eV. Moreover, the lifetime of such an excited state is too short to have enough atoms around in a state which can absorb a photon of 4.6 cm. On the contrary, the opposite (*i.e.*, emission) between high states is not difficult. If the gas is ionized, some ions capture electrons for a recombination where the electron lodges in a highly excited state first, and it then cascades down to the ground state *in steps* where $\Delta n = 1$, giving off radio emission lines in series, *alias recombination lines*.

This is not the only mechanism for radio lines. We also have the famous 21-cm line due to the spin-flip transition in H, when the electron spin changes by itself (its quantum number m_s or state) from $+1/2$ to $-1/2$.[22]

We will continue discussing the ways of molecular excitation. While a free atom can be excited by elevating one of its electrons to a higher level (from a lower one), *i.e.*, by increasing its energy, a molecule has two more possibilities of excitation (getting from a state to another, of higher energy). The case of a diatomic molecule (say, N_2) is simple, because it can be likened to a dumbbell

---

[22]Such transitions occur in elements other than H, too, but the conditions do not favor their observation. Since the transition is forbidden — in fact strongly — we must have a collection of too many atoms in order to get an intense line (because only few of the excited atoms emit at a time). The universe is rich in H, and so we get the line even from space. Spin-flip transitions occur also in C, the fourth most abundant element in the universe (after H, He, and O).

made of two spheres connected together by a spring. (The mechanical analogue of such a molecule is shown in Fig. 1.31.) If we pull the spheres from the equilibrium position and release them, they will execute harmonic oscillations or vibrations. But the system can also (simultaneously) execute rotational motion about axes. So, the overall molecular energy will be the sum of the electronic plus the other kinetic energies. The radiation emitted by an excited molecule gives a spectrum which is more complex than an atomic spectrum, consisting of a large number of spectral lines which are arranged in such a way as to appear as bands. Figure 1.36 shows the spectrum of N_2.

Fig. 1.36.

Classically, a system of two masses (atoms) connected by elastic forces can vibrate, if excited, with a certain natural frequency ν_0, but with any amplitude. So, the total vibrational energy can take any value. Quantum theory, however, requires that E_{vibr} takes an integer multiple by $h\nu_0$, not any value: $E_{\text{vibr}} = Nh\nu_0$ with $N = 0, 1, 2, \ldots$ (integer). But wave mechanics gives the exact formula $E_{\text{vibr}} = \left(v + \frac{1}{2}\right) h\nu_0$. The least energy that a vibrating system can have is nonzero, equal to $(1/2)h\nu_0$. Even at absolute zero ($T = 0°K$) there will be a nonzero remnant of vibrational energy in the system.

The *free rotation* of the system occurs about an axis through the center of mass of the system, in fact only about a principal axis of inertia, which in the case of a diatomic molecule is perpendicular to the line joining the atoms. When rotationally excited, the molecule rotates about this axis. In classical mechanics the angular speed of the motion can take any value, but not so in quantum considerations. The amplitude J of the angular momentum must be an integer multiple of $\hbar$ in quantum theory: $J = N_J h/2\pi$ where $N_J = 0, 1, 2, \ldots$ (integer).

We conclude that an excited molecule will have various values of *total* energy, depending on whether the excitation occurs via an electronic jump only, or by a simultaneous appearance of vibration and rotation as well. Figure 1.37 gives the energy levels of a molecule. In (a) we have excitation by an electronic jump only. Since for a given energy E_n the molecule may be executing vibrations as well, each level in column (a) must be analyzed into further levels of greater energy (b) spaced from E_n by the amount of E_{vibr}. Finally, molecules with the same electronic energy and E_{vibr} may have different E_{rot} values. So,

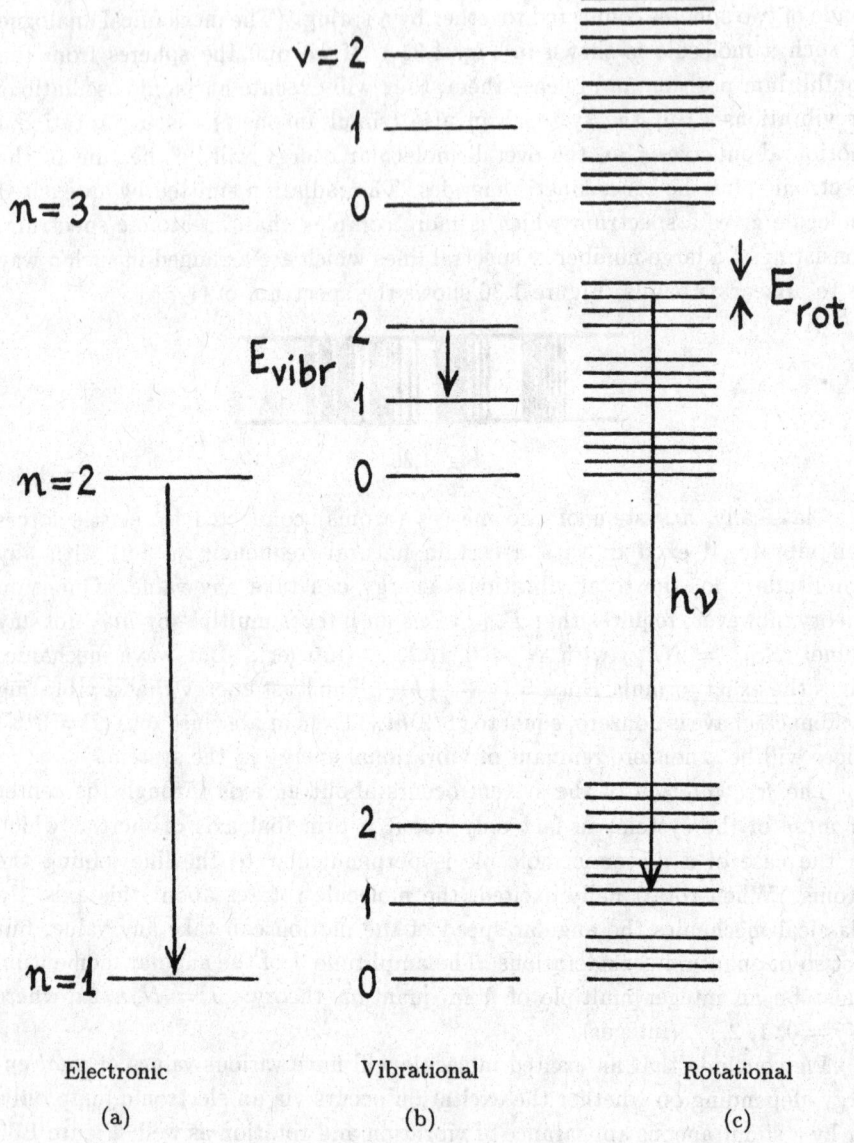

Fig. 1.37.

each vibrational level is in turn analyzed into other levels of greater energy (c), lying above E_{vibr} by the amount of E_{rot}.

The radiation emitted by an excited molecule can be due to changes in $E_{electronic}$ of E_{vibr} or E_{rot}, or all of them. In this last case, the energy of the emitted photon (per molecule) will be

$$h\nu = (E_n - E_{n'}) + (E_v - E_{v'}) + (E_J - E_{J'}) \tag{1.1}$$

where primes refer to final states in each category. So, excited molecules produce *molecular spectra* made of a large number of regularly arranged lines. All the lines that correspond to the first two parentheses in Eq. (1.1) but to different values of the third parenthesis, constitute one band. The lines of each band get accumulated towards one end of the band, forming the *band head*. The various bands correspond to the same value of ΔE_n, but to different values of ΔE_v.

Regularity appears not only in the array of the lines of a band, but also in the array of the bands themselves. Thus, if we inspect the bands that correspond to the same E_v value but different $E_{v'}$ values, we will get a *group of bands* where the separation between two successive bands gets smaller and smaller towards decreasing λ. Figure 1.36 shows six such groups of bands which belong to the same *system of groups*, because they all correspond to the same difference (change) $\Delta E_n = E_n - E_{n'}$. Depending on the value of ΔE_n, there are different group systems which are all similar to each other, but they are simply shifted. The system of band groups which arises when the parenthesis $E_n - E_{n'}$ in Eq. (1.1) is equal to zero (*i.e.*, when there is no change in $E_{electronic}$, but there is change in E_{vibr} and E_{rot}), constitutes the *vibration-rotation spectrum*. As seen from the diagram of energy levels in Fig. 1.37, the differences in E_{vibr} and E_{rot} are small, and hence the lines of vibration-rotation spectrum (more precisely, *vibration and rotation spectrum*) will fall in the infrared region.

We will close this section with a few words about the *Raman phenomenon*.[23] If we illuminate a transparent material (*e.g.*, benzene) with monochromatic radiation and analyze the light scattered[24] by the material, via a spectrograph (Fig. 1.38), we will observe that apart from the excitation line (*Rayleigh line*),[25] there are other lines, too, on its either side. This is the Raman phenomenon, explained by quantum theory, not classical physics. The incident photon is

---

[23]C. Raman, Hindu physicist, Nobel laureate (1930).
[24]The intensity of the scattered radiation is too weak to be perceptible by naked eye. Its detection is possible by extended photographic exposure only.
[25]Lord Rayleigh, British scientist, Nobel laureate (1904).

scattered by a molecule, but it can sometimes transfer an amount of energy ΔE to the molecule. This energy is consumed into the increase of E_{vibr} or E_{rot}. So, the scattered photon will have *less* energy (than the initial) by ΔE, and hence a new line will appear in the spectrum (formerly absent), having a frequency ν' which is less than that of the incident radiation. The bookkeeping equation is $h\nu' = h\nu - \Delta E$. Since there are many possible values for E_{vibr} and E_{rot}, it is possible to observe new lines in the spectrum, more than one. Besides the lines that appear with frequencies less than the initial frequency (*Stokes lines*), there appear lines of greater frequencies, too (*anti-Stokes lines*), which are due to the scattering of photons off molecules which had been already vibrating or rotating and which transferred part of *their* E_{vibr} or E_{rot} to the photon. Therefore, the scattered photons in the latter case will have *greater* energy, according to the equation $h\nu' = h\nu + \Delta E$, and hence lines of frequencies greater than the excitation frequency will show up.

By studying Raman spectra, we can determine the levels of E_{vibr} and E_{rot} of various substances and investigate the structure of their molecules.

In Fig. 1.38 we see the arrangement for the observation of the Raman phenomenon. The bent side of container C prevents the reflection of light by the rear wall.

Figure 1.39 shows a Raman spectrum where we have the spectrum of the exciting radiation (a) and the spectrum after scattering (b).

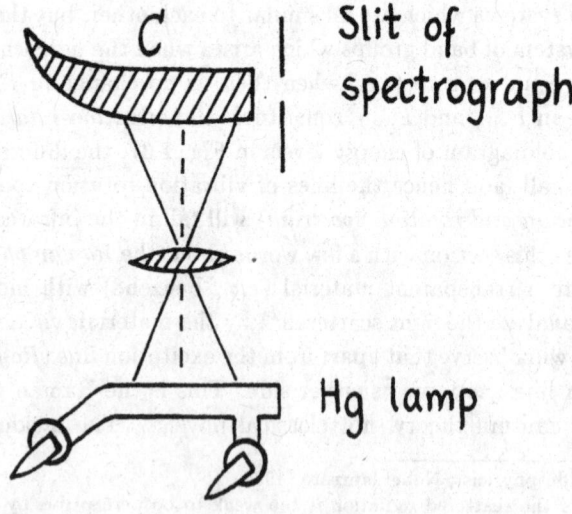

Fig. 1.38.

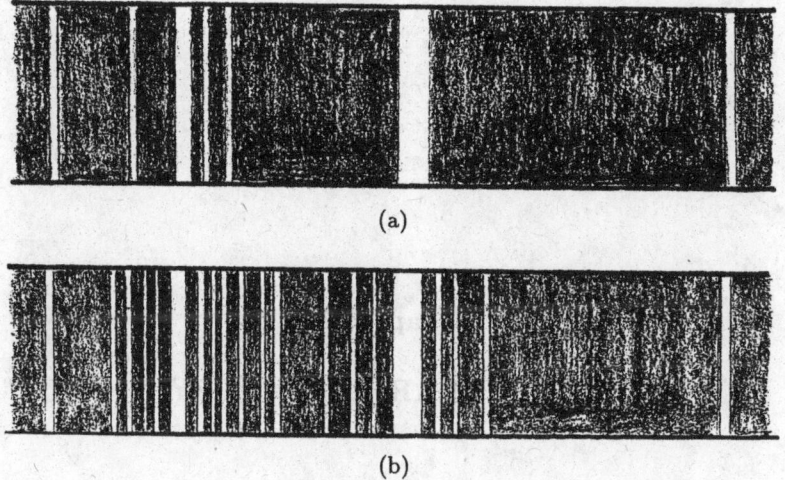

(a)

(b)

Fig. 1.39.

1.18. Remarks

In this book (since it is a textbook) we avoid the detailed presentation of interesting pertinent research results, purely for space reason. Only basic facts and important extracts are included, for education purposes.

Chapter 2

MOLECULAR SYMMETRY

2.1. Symmetry in Molecules

The concept of *symmetry* has always amazed scientists and artists since the times of ancient Greece. It has been a notion associated with the sense of beauty and measure, and hence its extensive use and significant position in geometry and fine arts. The symmetrical ornamentations of a baroque ceiling attract our attention and admiration; we observe symmetry in architecture where we can talk about the symmetry of a building (Fig. 2.1(a)), a wrought-ironscroll railing (b), or other decorations. But symmetry exists also in nature: If we neglect minor physiological asymmetries, the human body is symmetric with respect to a vertical axis through the middle (c). There is a beautiful natural

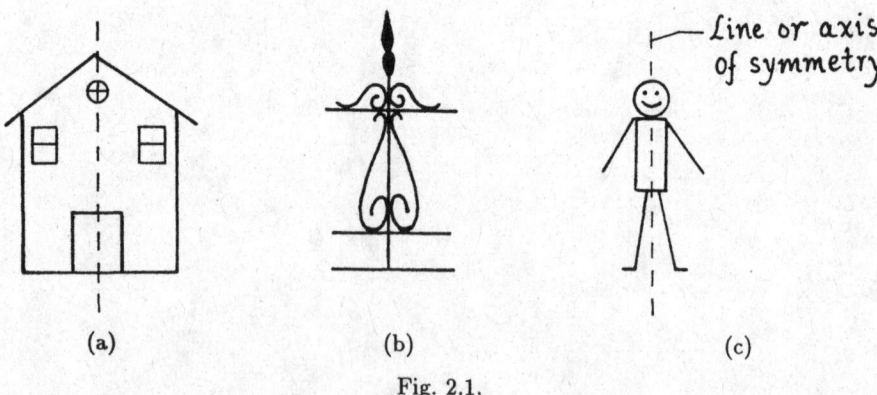

Fig. 2.1.

symmetry in the snow crystals. In nature we also have the symmetry of the opposites: Particles and antiparticles, integration and differentiation, the positive and negative charge, the positive and negative numbers in algebra, the four seasons, addition and subtraction etc.

Apart from the esthetic value and appearance of a symmetrical structure — whether in nature or in man-made fine arts — the property of symmetry has become an object of study in science, particularly in molecular physics and solid state physics. The latter uses symmetry to determine crystal structure; the former examines possible symmetries in molecules in order to determine molecular dimensions.

Some molecules have structural symmetry and some do not.[1] The study of the symmetry properties of molecules is an important branch of molecular physics because it leads to the extraction of information about molecular structure. *The more the symmetries in a molecule the easier the determination of its parameters.* If there is no symmetry, it is hard to find parameters like interatomic distances and angles. So, whenever a scientist comes across a new molecule, the first thing to ask (and look for) is whether it has a symmetry.

Symmetry is a property that depends on the molecular structure. It is a virtual notion. It does not depend on the space; it depends on the material itself.

To study symmetry in molecules, we must know the *symmetry elements*. Fortunately, their number is limited. There are two ways to study symmetry elements and analyze symmetries in molecules:

(1) Use of group theory: This method is elegant and rigorous, but too mathematical and scholastic.
(2) Use of simple schematics: This method is descriptive and less tedious. It uses simple diagrams and is a way out for those who fear the rigor of group theory. It is practical, but still detailed.

In this chapter we will use the second method.

2.2. Definition of Symmetry

In molecular physics, *symmetry is the orderly array (or repetition) of identical objects in space.* In other words, it is a certain configurational arrangement

---

[1]Asymmetric molecules cause optical activity, *i.e.*, they turn the polarization plane of the polarized light. These molecules do not have any symmetry plane or a symmetry center. They have only pure rotational symmetry.

(or line-up) of atomic nuclei.

Consider the H_2 molecule (Fig. 2.2). Two H atoms are separated by a distance d. If we rotate one of the H atoms (or the whole molecule) about the midpoint (center of mass) A through 180°, we obtain the same configuration: The molecule coincides with itself. We say that the H_2 molecule is symmetric (upon rotation by 180° about its center of mass).

Now consider the HCl molecule (Fig. 2.3). If we do the same thing about midpoint A (which in this case is not the center of mass), the molecule does not coincide with itself: We obtain a different situation. So, we say that the HCl is not symmetric (with respect to the same operation).

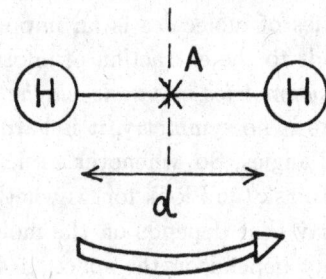

Fig. 2.2.

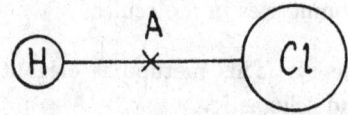

Fig. 2.3.

The operation performed (or imagined) in order to find possible symmetries is a *symmetry operation*. In the examples above the symmetry operation was a rotation through 180°.

Symmetry is a property of the atomic groups. In terms of symmetry, we have two kinds of atomic groups:

(1) *Point groups* (in molecules).
(2) *Space groups* (in crystals).

We will deal with points groups because they regard molecular physics. (Space groups regard solid state physics.) In point groups, every nucleus is considered as, and can be represented by a geometrical point.

For example, consider the BF_3 molecule (Fig. 2.4). The angle between the F atoms is α. If we rotate any F atom through α about an axis perpendicular to the plane of the paper and passing through B, it coincides with another F atom. This is a point symmetry. Now consider an array of BF_3 molecules shown in Fig. 2.5. If we rotate a molecule about the center point A through an angle 2α, the triple molecular group coincides with itself. This is a space group symmetry.

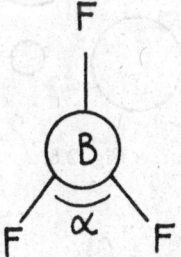

Fig. 2.4.

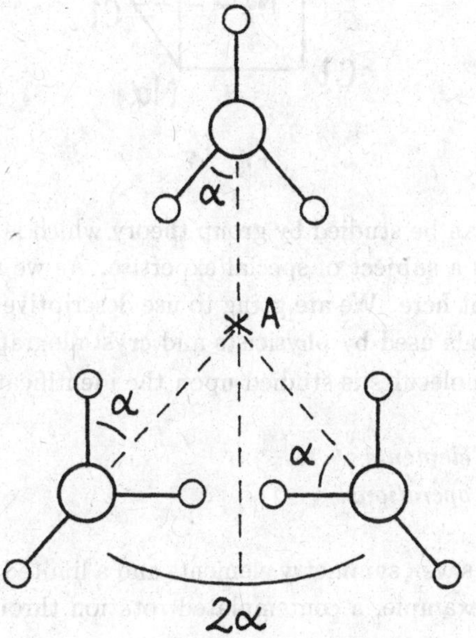

Fig. 2.5.

Consider the NaCl crystal (Fig. 2.6). It has a symmetry because the arrangement of atoms follows a definite order. The structure is cubic (Fig. 2.7). Orderly arranged Na and Cl atoms form a cubic crystal (the atoms are at the corners of the cube). This is also a space symmetry because we have space groups here.

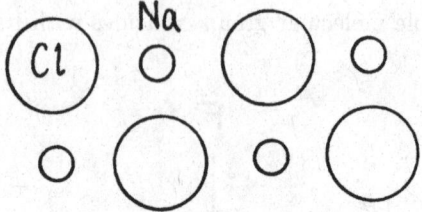

Fig. 2.6.

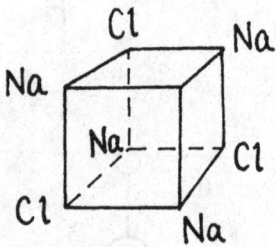

Fig. 2.7.

Space groups can be studied by group theory which is based on mathematical methods. It is a subject of special expertise. As we mentioned above, we will not deal with it here. We are going to use descriptive and crystallographic schematical methods used by physicists and crystallographers.

Symmetry in molecules is studied upon the identification of two things:

(1) *Symmetry elements.*
(2) *Symmetry operations.*

There are only seven symmetry elements and a limited number of symmetry operations. (For example, a contemplated rotation through an angle of 2α is an operation.)

Symmetries are denoted by two different notations:

(1) The *Schönflies method*: It is easier and chronologically precedes the second one. It applies to point groups only.
(2) The *Hermann-Mauguin method* (or *international system*): It was developed later and applies both to point and space groups.

2.3. Symmetry Elements

There are seven symmetry elements:

(1) *Symmetry center* (or *center of symmetry*): In the Schönflies[2] method it is denoted by C_i, and in the international method by $\bar{1}$. To illustrate this symmetry element, consider a coin. We could mark its two faces by the letters H and T (which stand for "heads" and "tails" respectively), but since these two letters are self-symmetric in appearance, the reader might think that this was intentional and get confused (because half of each face would be symmetric with the other half). Of course, we can mark the faces by the letters P and Q which are asymmetric in themselves, but let us be more realistic and mark the faces as in Fig. 2.8, just like in a real coin.

Fig. 2.8.

Now this coin is an example for an entity which has no symmetry. It is an asymmetric system or unit.

Consider Fig. 2.9. We place the symmetry center $\bar{1}$ at the origin of the Cartesian coordinate system. Then a coin at position $(+)$ (at a distance d along the $+x$-axis and in front of the yz-plane) can be connected to an identical coin at position $(-)$ (at a distance d along the $-x$-axis and behind the yz-plane) by an inversion with respect to $\bar{1}$. Each point of one coin is inverted with respect to $\bar{1}$, to reach a corresponding point of the other coin, that is, every point of one coin

---

[2]It is curious that this name appears with this spelling in literature. The correct German spelling is *Schönfliess* (= nice flow).

has a distance to $\bar{1}$ which is repeated equally on the other side, so that the counterpart point on the other coin is reached. All distances on either side of $\bar{1}$ are equal, from point to point between the two coins. The two coins together can represent a molecule which is symmetric with respect to $\bar{1}$ (by inversion with respect to $\bar{1}$). We say

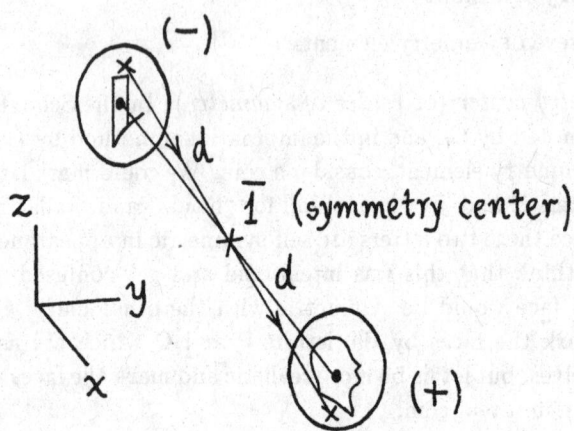

Fig. 2.9.

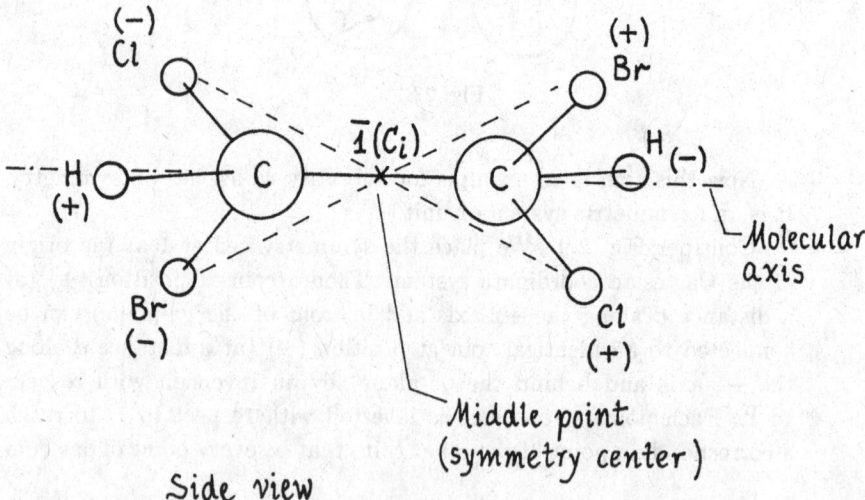

Fig. 2.10.

that such a molecule has a symmetry center: One half of the molecule can be obtained by inverting the other half with respect to $\bar{1}$. A good example for this case is the $C_2H_2Br_2Cl_2$ molecule (Fig. 2.10) whose side view is shown. All pluses denote positions *above* the plane of the paper and towards the reader; all minuses denote positions *below* the plane of the paper and away from the reader. For example, the straight line that connects Br($-$) with Br($+$) passes through $\bar{1}$. Notice that there is as much matter above the plane of the paper as there is below. Figure 2.11 shows the isometric view of the molecule.

(2) *Mirror symmetry* (or *mirror plane*): In the Schönflies system it is denoted by C_s, and in the international system by m. If we place our two hands side by side (Fig. 2.12 or Fig. 2.13) we observe that they are mirror-symmetric (with respect to a plane mirror placed between them and perpendicular to the plane of the paper). The corresponding distances d on either side of the mirror are equal. To obtain the image of the object, from every point of the object we draw a line perpendicular to the mirror plane and we repeat it equally (we take an equal distance) on the other side of the mirror. So, all drawn lines are parallel (whereas in the symmetry center the distance lines converge to $\bar{1}$ and diverge beyond it, because we have inversion, not mirror reflection).

The mirror symmetry is reflective, not rotational or invertive. Indeed, if we rotate our hand in Fig. 2.13 by 180°, we do not obtain its image on the other side: It will not coincide with its image. The exact likeness between the object and its image in a mirror involves *folding*, not rotation or inversion.

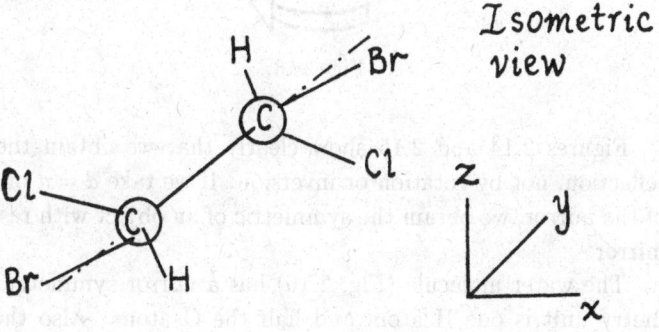

Fig. 2.11.

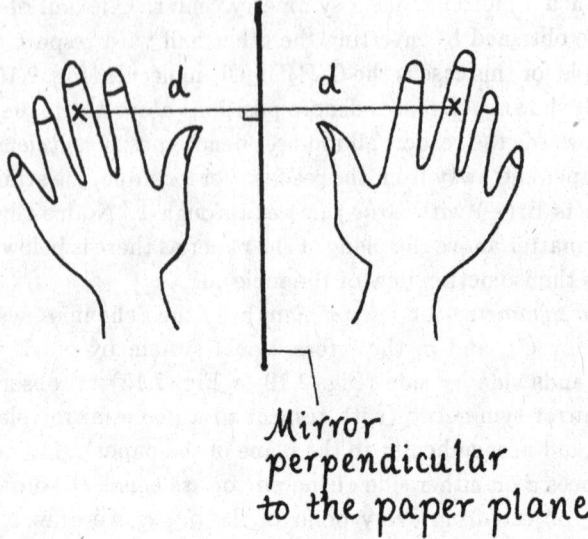

Mirror
perpendicular
to the paper plane

Fig. 2.12.

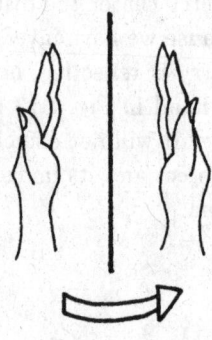

Fig. 2.13.

Figures 2.14 and 2.15 show clearly that we obtain the image by reflection, not by rotation or inversion. If we take $d = d$ on both sides of the mirror, we obtain the symmetric of an object with respect to the mirror.

The water molecule (Fig. 2.16) has a mirror symmetry. The symmetry unit is one H atom and half the O atom. Also the CH_2ClBr (monochlorobromomethane) molecule is mirror-symmetric: A mirror,

pendicular to the plane. The upper passes through the centres of (1,2) Cl_2, and H_2 (Fig. 2.14.)

Rotation axis. The denotes an axis of a sphere a about the turning of ce the rotation. The axis is a straight line (Fig. 2.15) about which we rotate the molecule by some amount and it coincides with itself with an initial set or distribution of a symmetry axis a an atom.

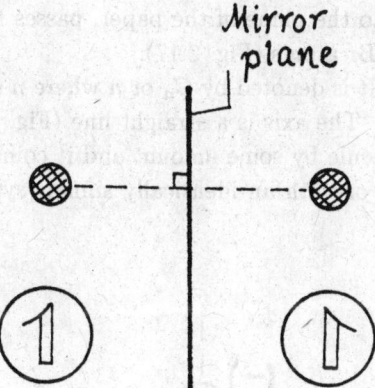

Fig. 2.14.

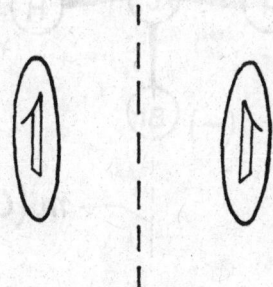

Fig. 2.15.

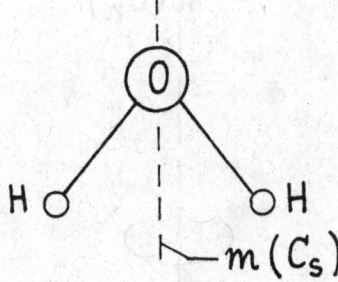

Fig. 2.16.

perpendicular to the plane of the paper, passes through the centers of the C, Cl, and Br atoms (Fig. 2.17).

(3) *Rotation axis*: It is denoted by C_n or n where n shows the multiplicity of the rotation. The axis is a straight line (Fig. 2.18) about which we rotate the molecule by some amount and it coincides with itself (with its initial state or with an identically similar symmetry unit, *e.g.*, an atom).

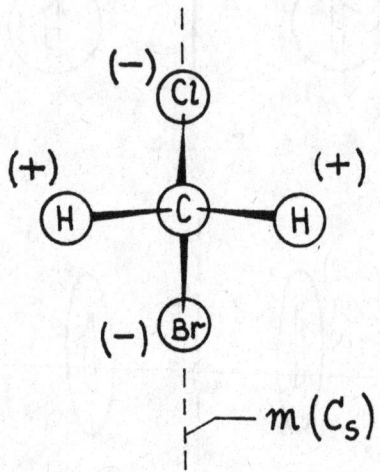

Fig. 2.17.

Fig. 2.18.

Here n is an integer (because when we say an n-fold rotation it does not make sense to have a noninteger n). We rotate the molecule (or an atom of it) about the axis (of the rotational symmetry) through an angle α given by

$$\alpha = \frac{2\pi}{n}.$$

Then the rotated atom reaches another atom of the same kind (or the molecule coincides with itself). (Of course, instead of an atom, the rotated symmetry unit could be a group of atoms.) So, if $\alpha = 120°$, then $n = 3$ and vice versa, *i.e.*, a threefold rotation means rotation through 120°.

The case $n = 1$ is the identity rotational operation: If we rotate the molecule by 360°, we obtain the same molecule, *i.e.*, its initial state itself. This is true for *all* molecules, whether symmetric or not. Every object in space has this property, and so it is not actually considered a symmetry property.

For $n = 2$ we have a twofold rotational symmetry or a *dyad*. For $n = 3$ we have a threefold symmetry axis or a *triad*. For $n = 4$ we have a fourfold axis or a *tetrad*. The case $n = 5$ is not encountered in nature: There is no fivefold rotational symmetry in molecules, that is, if we rotate a molecular atom by an angle of $\alpha = 2\pi/5$, we do not obtain any atom at the position reached after rotation. The same is true for $n = 7$. For $n = 6$ we have a sixfold rotation or an *hexad*.

Theoretically, n can run from 1 to ∞ (infinite number of axes of possible rotations), but we need not go beyond $n = 6$ to rotational symmetries of higher multiplicity, because they are very rarely encountered in molecules.

A dyad is shown as C_2 in the Schönflies system, and as just 2 in the international system. Schematically, it is shown in Fig. 2.19 (by a black oval shape). Figure 2.20 shows C_2 for a coin. Notice that the coin stays in the plane of the paper (upon rotation).

As a molecular example, consider the oxalate ion (Fig. 2.21 where the second carbon atom is not seen because it is behind the plane of the paper, hidden by the first carbon atom). This molecule has a twofold axis. The situation is seen better in Fig. 2.22.

The BF_3 molecule (Fig. 2.4) has a $3(C_3)$ axis. If we rotate any F atom, about an axis passing through the B atom (the molecule is planar) by 120°, it comes to the position of another F atom. In the

Fig. 2.19.

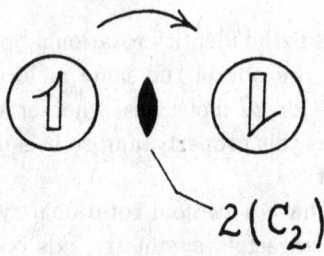

$2(C_2)$

Fig. 2.20.

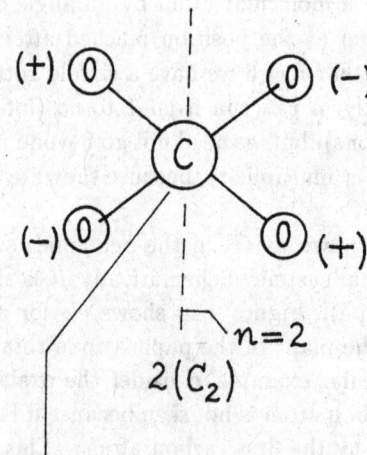

$n = 2$

$2(C_2)$

Two carbon atoms
(the other one is behind)

Fig. 2.21.

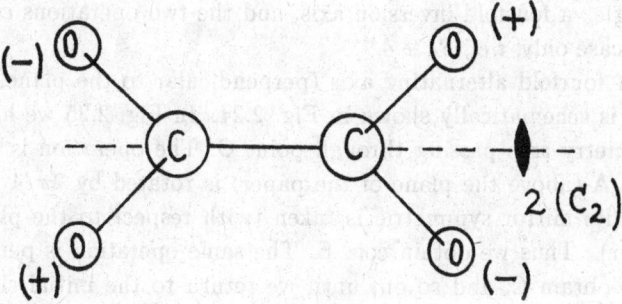

Fig. 2.22.

arrangement shown in Fig. 2.5 there is also a rotational symmetry for the whole molecule about point A.

In linear molecules (Fig. 2.23), the axis that passes through their atoms is automatically a symmetry axis for a rotation of ∞ multiplicity: If we rotate the molecule about this axis by *any* angle, it stays the same, if we assume that its atoms are perfect, structureless spheres without gears and wheels.

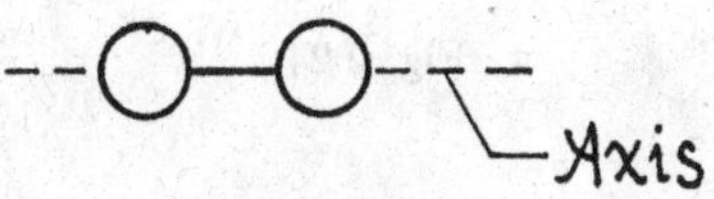

Fig. 2.23.

(4) The fourth symmetry element takes different names, depending on whether we use the Schönflies or the international system. In the former system it is called *alternating symmetry axis* or *rotation-reflection axis* and denoted by S_n. In the latter system it is called *inversion (symmetry) axis* and denoted by $\bar{n}$ (where we have a rotation plus an inversion). These two symmetry operations are *different*, except for the case of $n = 4$ where the two coincide and become identical.

So, we have two different operations in the two systems. An alternating axis means an n-fold rotation plus a mirror symmetry. An inversion axis means an n-fold rotation plus an inversion with respect to a symmetry center. Clearly, a fourfold alternating axis is the same thing as a fourfold inversion axis, and the two operations coincide for this case only, *i.e.*, $S_4 \equiv \bar{4}$.

A fourfold alternating axis (perpendicular to the plane of the paper) is schematically shown in Fig. 2.24. In Fig. 2.25 we have such a symmetry axis passing through point O. The operation is as follows: Coin A (above the plane of the paper) is rotated by $2\pi/4 = 90°$ and then its mirror symmetric is taken (with respect to the plane of the paper). Thus we obtain coin B. The same operation is performed on B to obtain C, and so on, until we return to the initial situation A. Notice that B is below the plane of the paper. If we rotate B by 180°, we obtain its diagonally symmetry D, as if we did an inversion with respect to the symmetry center O. That is why $\bar{4} \equiv S_4$. This operation follows the route A $\to$ B $\to$ C $\to$ D $\to$ A, *i.e.*, A returns back to A. The mirror symmetry gives A $\rightleftarrows$ B and D $\rightleftarrows$ C, that is, each position goes over to the next one. The inversion follows the diagonals: A $\rightleftarrows$ C and B $\rightleftarrows$ D.

Fig. 2.24.

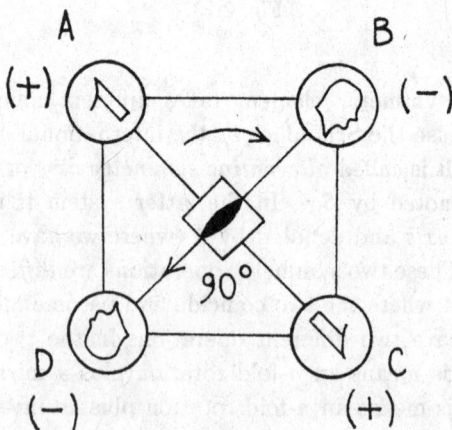

Fig. 2.25.

Notice that $\bar{2} \neq S_2$ and that $\bar{3}$ corresponds to S_6. On the other hand, S_3 corresponds to $\bar{6}$. We have $\bar{4} = S_4$. The operation $\bar{2}$ corresponds to S_1. Finally, $S_2 = \bar{1}$ (it is the inversion center itself).

These four symmetry elements hold for both point and space groups. There are three more symmetry elements (or operations) that hold for space groups only:

(5) *Translation operation*: In a single, isolated molecule we do not observe translational symmetry. But it gains importance in crystals. For example, consider the crystal lattice of NaCl (Fig. 2.26). If we translate a Na atom along an axis (or in a plane) in a specific direction and by the same amount in each step, we find another atom of the same kind (another Na atom).

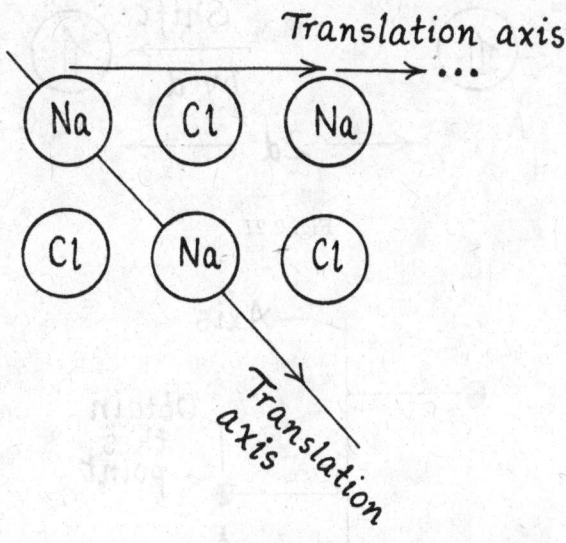

Fig. 2.26.

(6) *Sliding plane*: It is a mirror reflection plus a shift parallel to the mirror plane. The operation is shown in Fig. 2.27. Coin A is mirror reflected and then shifted by d along the sliding plane to reach coin B. If the same operation is applied to B, we reach C which is located a distance $2d$ from A. So, the sliding plane symmetry involves a translation (reflection, recession etc.) plus a sliding along a plane.

(7) *Screw axis*: It is an operation regarding space symmetries. It involves a rotation plus a parallel sliding. The screw (drilling) axes can be twofold, threefold, fourfold etc. One rotates an atom about an axis through an angle φ and then one translates it in a direction parallel to the same axis. Figure 2.28 shows a twofold screw axis. Of course, the shift

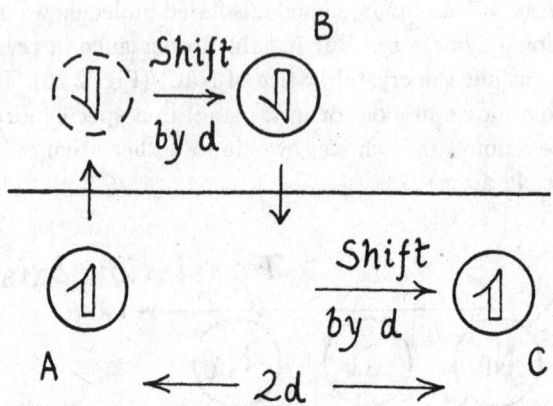

Fig. 2.27.

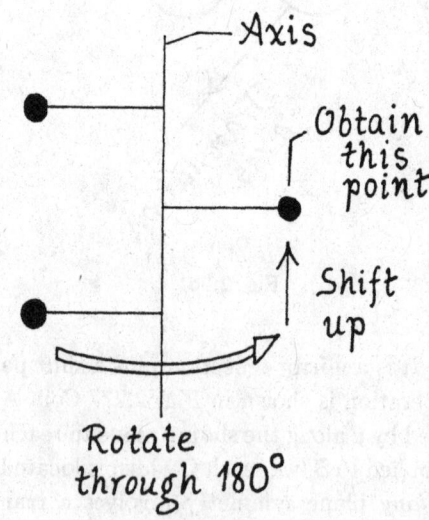

Fig. 2.28.

cannot be more than the size (length) of the unit cell. The rotation can be n-fold, but the rotation alone does not get us to the position of an atom of the same kind; a shift is necessary after rotation. That is, the symmetry unit is first rotated and then shifted, so that it coincides with another one of the same kind (identical).

The screw axis is denoted by n_p where n is the multiplicity of the rotation axis ($n = 2, 3, 4, 6$) and p can be $p = 1, 2, \ldots, n-1$. So, if $n = 2$, we can have only a 2_1 screw axis. This means that we rotate by 180° and we then shift by $p/n = 1/2$, half the length of the unit cell. Figure 2.29 shows 2_1. We rotate coin A by 180° and we then shift it by

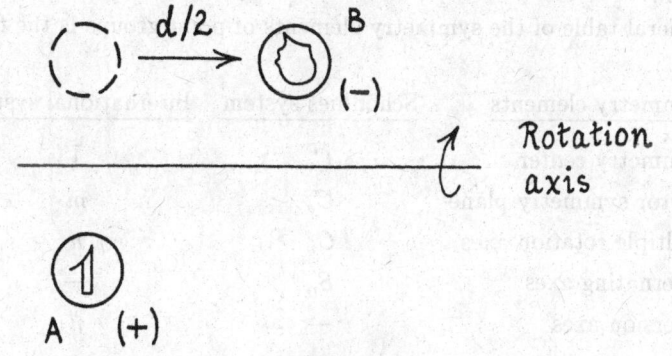

Fig. 2.29.

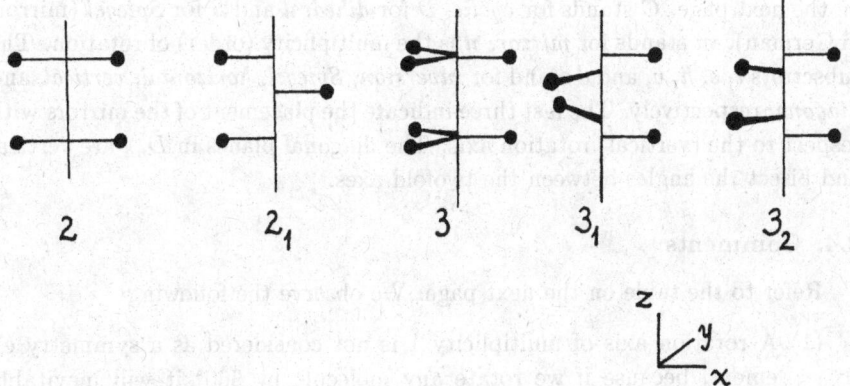

Fig. 2.30.

$d/2$ (where d is the length of the unit cell) to obtain coin B. Obviously, if A is in the plane of the paper, so is B, and if A is above the plane of the paper (by some height), B is below the plane (by the same height). Figure 2.30 shows the screw axes 2, 2_1, 3, 3_1, and 3_2. Clearly, 3_1 means rotation by 120° and translation by 1/3 of the unit cell. Since n can be at most $n = 6$, we can have the following screw axes: 2, 2_1, 3, 3_1, 3_2, 4, 4_1, 4_2, 4_3, 6, 6_1, 6_2, 6_3, 6_4, 6_5.

It goes without saying that knowledge of this symmetry means knowledge about the distribution of mass per unit cell.

Of course, the shape of a crystal may not reveal its symmetry elements immediately, but the sliding planes and the screw axes can be found via X-ray diffraction.

A general table of the symmetry elements of point groups is the following:

Symmetry elements	Schönflies system	International system
Symmetry center	C_i	$\bar{1}$
Mirror symmetry plane	C_s	m
Multiple rotation axes	C_n	n
Alternating axes	S_n	—
Inversion axes	—	$\bar{n}$

Notice that S_n is not the same as $\bar{n}$. These are the minimum symmetry elements for point groups. There are 32 point groups in total, listed in the table on the next page. C stands for *cyclic*, D for *dihedral* and S for *Spiegel* (mirror, in German). m stands for *mirror*. n is the multiplicity (order) of rotation. The subscripts i, s, h, v, and d stand for *inversion, Spiegel, horizontal, vertical*, and *diagonal* respectively. The last three indicate the placement of the mirrors with respect to the (vertical) rotation axis. The diagonal planes in D_{nd} are vertical and bisect the angles between the twofold axes.

2.4. Comments

Refer to the table on the next page. We observe the following:

(1) A rotation axis of multiplicity 1 is not considered as a symmetry element, because if we rotate any molecule by 360° it will inevitably coincide with itself. So, C_1 is not a symmetry.

THE THIRTY-TWO POINT GROUPS

Symbol	Symmetry center $\bar{1}(C_s)$	Mirror plane $m(C_s)$	Rotation symmetry axes			
			$2(C_2)$	$3(C_3)$	$4(C_4)$	$6(C_6)$
$1\ (C_1)$	—	—	—	—	—	—
$2\ (C_2)$	—	—	One	—	—	—
$3\ (C_3)$	—	—	—	One	—	—
$4\ (C_4)$	—	—	—	—	One	—
$6\ (C_6)$	—	—	—	—	—	One
$m\ (C_h)$	—	One	—	—	—	—
$2/m\ (C_{2h})$	One	One	One	—	—	—
$3/m\ (C_{3h})$	—	One	—	One	—	—
$4/m\ (C_{4h})$	—	One	—	—	One	—
$6/m\ (C_{6h})$	—	One	—	—	—	One
$222\ (D_2)$	—	—	Three	—	—	—
$32\ (D_3)$	—	—	Three	One	—	—
$42\ (D_4)$	—	—	Four	—	One	—
$62\ (D_6)$	—	—	Six	—	—	One
$mmm\ (D_{2h})$	One	Three	Three	—	—	—
$\bar{6}2m\ (D_{3h})$	—	Four	Three	—	—	—
$4/mmm\ (D_{4h})$	One	Five	Four	—	—	—
$6/mmm\ (D_{6h})$	One	Seven	Six	—	—	One
$\bar{1}\ (S_2)$	One	—	—	—	—	—
$\bar{4}\ (S_4)$	—	—	One	—	—	—
$\bar{3}\ (S_6)$	One	—	—	One	—	—
$\bar{4}2m\ (D_{2d})$	—	Two	Three	—	—	—
$\bar{3}m\ (D_{3d})$	One	Three	Three	One	—	—
$mm\ (C_{2v})$	—	Two	One	—	—	—
$3m\ (C_{3v})$	—	Three	—	One	—	—
$4mm\ (C_{4v})$	—	Four	—	—	One	—
$6mm\ (C_{6v})$	—	Six	—	—	—	One
$23\ (T)$	—	—	Three	Four	—	—
$43\ (O)$	—	—	Six	Four	Three	—
$m3\ (T_h)$	One	Three	Three	Four	—	—
$m3m\ (O_h)$	One	Nine	Six	Four	Three	—
$\bar{4}3m\ (T_d)$	—	Six	Three	Four	—	—

(2) An n-fold rotation means obtaining the same molecular configuration after a rotation by an angle of $2\pi/n$. So, a twofold rotation involves a rotation by $180°$.

(3) An alternating axis means a rotation (of the molecule) by $2\pi/n$ and then a mirror (m) symmetry. If the molecule coincides with itself, we say that it has an alternating axis (or symmetry). For example, consider the CH_4 molecule (Fig. 2.31). The carbon atom is at the center of a fictitious cube (Fig. 2.32). The four hydrogen atoms are at the corners of the cube. In this molecule, the axis that passes through the C atom is both an alternating and an inversion axis.

(4) The multiplicity of the inversion axis is shown by the corresponding number and a bar over it, *e.g.*, $\bar{4}$. Schematically, a fourfold inversion

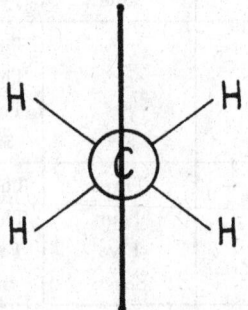

Fig. 2.31.

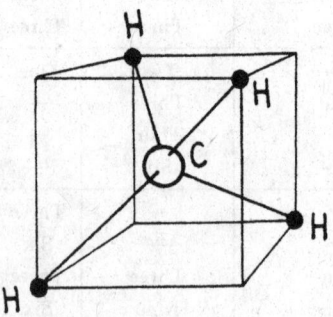

Fig. 2.32.

axis is symbolized by the shape in Fig. 2.33(a). The symbol in (b) is for a twofold rotation axis; that in (c) is for a threefold rotation axis; and that in (d) is for a fourfold rotation axis.

(5) The symmetry center is a onefold inversion operation. In Fig. 2.34, $\otimes$ is a point belonging to the molecule, of coordinates (x, y, z). If we place the inversion center (or symmetry center) C_i at the origin, we take the distance d between $\otimes$ and the origin and we repeat it beyond the origin, in the same direction. Thus we reach the symmetric point by inversion (with respect to C_i). The coordinates of the new point are $(\bar{x}, \bar{y}, \bar{z})$. The operation of finding such a symmetric point

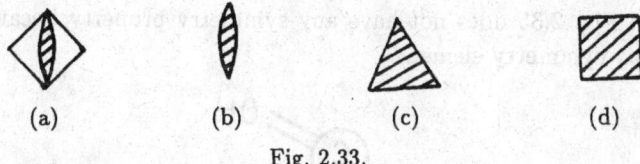

(a)   (b)   (c)   (d)

Fig. 2.33.

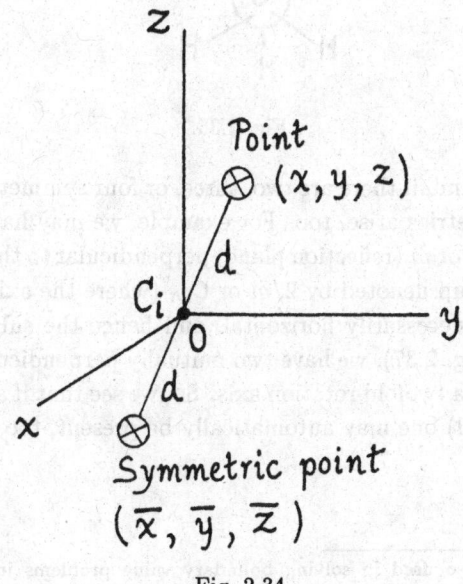

Fig. 2.34.

(with respect to some center) is called *inversion*.[3]

(6) A fivefold rotation is encountered in some molecules, but very rarely. (It is an exception.) However, there is *no* fivefold symmetry at all in space groups (crystals). We always consider a certain part of the crystal, called *unit cell*. We never encounter fivefold symmetry in any unit cell.

(7) An alternating axis and an inversion axis are two different operations (symmetry elements) in two separate systems of notation.

2.5. Combinations of Symmetries

In the preceding sections we presented the minimum symmetry elements. There are molecules which do not possess any of them. For example, the molecule in Fig. 2.35 does not have any symmetry property because it does not have any symmetry element.

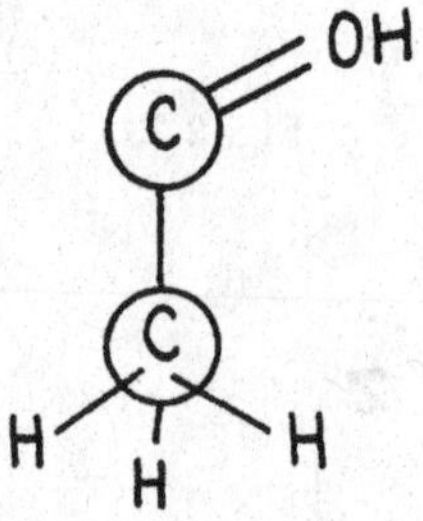

Fig. 2.35.

On the other hand, if there are two, three, or four symmetries in a molecule, other (new) symmetries arise, too. For example, we may have a twofold rotation axis and a mirror m (reflection plane) perpendicular to this axis (Fig. 2.36). This is a point group denoted by $2/m$ or C_{2h} (where the axis is taken vertical and the mirror is necessarily horizontal, and hence the subscript h). In the water molecule (Fig. 2.37), we have two mutually perpendicular mirrors whose intersection is just a twofold rotation axis. So, we see that if a symmetry exists, another (additional) one may automatically be present, too.

---

[3]This operation is also used in solving boundary value problems in electrostatics. See J.D. Jackson, *Classical Electrodynamics*, John Wiley, New York (1962), p. 35.

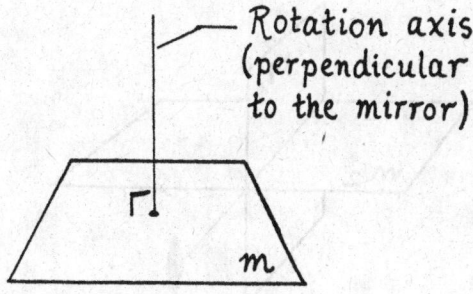

Fig. 2.36.

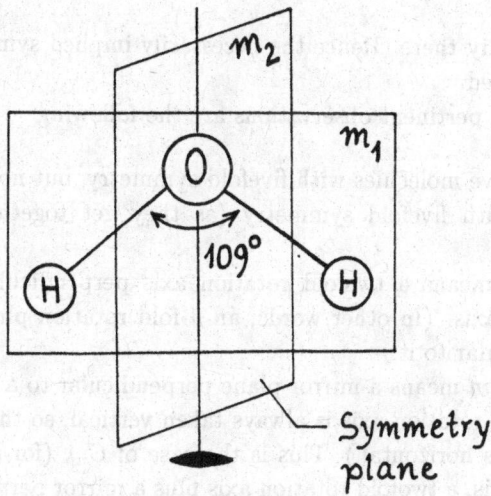

Fig. 2.37.

One might think that the number of combinations of the symmetry elements is ∞. But one can use group theory to prove that the number of such combinations is finite and limited, not infinite.

If one stops at $n = 6$ (because higher n values give outcomes identical with the first five), one realizes that there are only 32 point groups. (These have four basic symmetry elements. The space groups have three more.)

Notice that whenever two mirrors intersect (Fig. 2.38), their intersection (which is a straight line) is a rotation axis. So, it is redundant to mention the axis when we have mm (*i.e.*, two mirror planes mutually perpendicular),

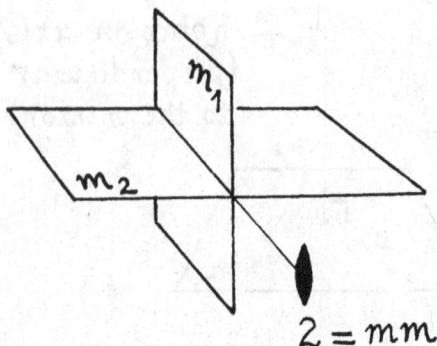

Fig. 2.38.

because it is already there. Hence the necessarily implied symmetry elements need not be denoted.

The important pertinent observations are the following:

(1) We can have molecules with fivefold symmetry, but no molecules form crystals with fivefold symmetry (as they get together and become neighbors).

(2) $n2$ or $\bar{n}2$ means a twofold rotation axis perpendicular to an n-fold (vertical) axis. (In other words, an n-fold rotation plus a twofold one perpendicular to it.)

(3) n/m or $\bar{n}/m$ means a mirror plane perpendicular to a (vertical) n-fold axis. (The rotation axis is always taken vertical, so that the mirror in this case is horizontal.) This is the case of C_{nh} (for example, $C_{2h} \equiv 2/m$, that is, a twofold rotation axis plus a mirror perpendicular to it). In the C_h group there is one m (always perpendicular to the rotation axis).

(4) n/mm means two perpendicular mirrors, of which one is the plane to which the n-fold axis belongs.

(5) nm or $\bar{n}m$ means a mirror plane and an n-fold axis belonging to it. (In other words, we have an n-fold axis and a mirror plane carrying this axis.) (Fig. 2.39.) For example, consider the water molecule (Fig. 2.40). It has a twofold axis (if we rotate the molecule by 180° about this axis, we obtain the same shape, *i.e.*, the molecule falls upon itself). But there is also a mirror plane m to which the axis belongs. This plane is normal to the plane of the paper. This is the case of C_{2v}. We have

a twofold rotation axis and a (vertical) mirror plane to which the axis belongs at the same time (that is why we put the subscript v — vertical — for the mirror) (Fig. 2.39). The axis is also a mirror. In general, we have C_{nv} which means an n-fold rotation axis and n mirror planes

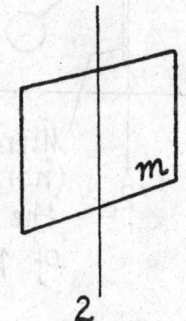

Fig. 2.39.

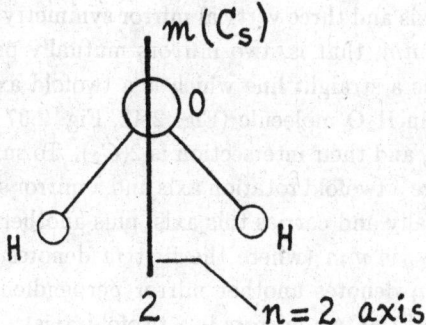

Fig. 2.40.

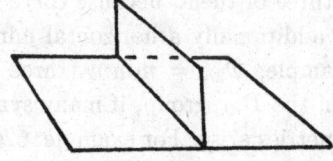

Fig. 2.41.

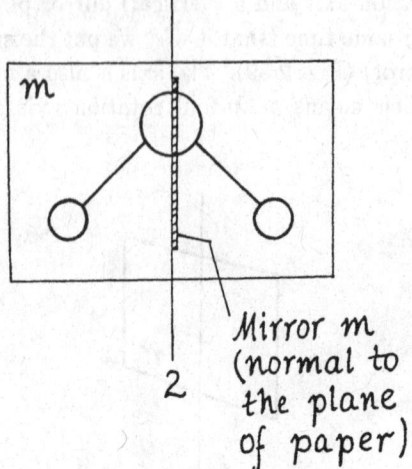

Fig. 2.42.

passing all through this axis vertically. For example, $C_{3v} \equiv 3m$ means a threefold axis and three vertical mirror symmetry planes. But we also have $C_{2v} \equiv mm$, that is, two mirrors mutually perpendicular. Their intersection is a straight line which is a twofold axis (Fig. 2.41). This can be seen in H_2O molecule (Fig. 2.42, Fig. 2.37 and Fig. 2.38). We have $m \perp m$, and their intersection is $2(C_2)$. To summarize, in the C_{2v} group we have a twofold rotation axis and a mirror symmetry plane that passes vertically and carries this axis, plus another m perpendicular to the first. C_{2v} is mm (where the first m denotes a mirror plane and the second m denotes another mirror perpendicular to the first; the intersection of the two mirrors is a twofold axis).

(6) The D_n group means an n-fold vertical axis plus n twofold symmetries perpendicular to it. For example, $D_3 \equiv 32$, that is, a threefold axis and a twofold one normal to it. (But there are two more twofold axes present, in total three of them, because there is a threefold symmetry axis.) If there is additionally a horizontal mirror, we denote the case by D_{nh}. For example, $D_{2h} = mmm$ (three mutually perpendicular mirror planes). In the D_{nh} group, if many symmetry elements get together, new symmetries arise. For example, $\bar{6}2m$ means a sixfold inversion plus a twofold rotation plus a perpendicular reflection symmetry. The diagonal horizontal planes are denoted by D_{nd}. The notation D_p

(dihedral) means that there is a p-fold symmetry axis and p twofold axes normal to it and at equal angles to each other. D_{ph} means a p-fold axis and p vertical mirrors carrying it, plus a mirror perpendicular to them. There is a material exhibiting a symmetry group of $\bar{8}2m$ or D_{4d}, but this symmetry is outside the 32 groups.

(7) T stands for *tetragonal* and O for *octagonal*.

(8) S_n means n-fold rotation-reflection axes (with n even). Clearly, $S_2 \equiv C_i$.

2.6. Linear Molecules

Linear molecules are the ones whose atoms are lined up along a straight line in one direction. This case is denoted by ∞/mm in the international system. The notation for symmetric linear molecules in the other system is $D_{\infty h}$.

The H_2 molecule is linearly symmetric (Fig. 2.43). For an asymmetric linear molecule (Fig. 2.44) the notation is ∞m or $C_{\infty v}$, that is, $n \to \infty$.

As we see in Fig. 2.45, if we rotate a linear molecule about its own longitudinal axis through any angle, it stays the same and coincides with itself.

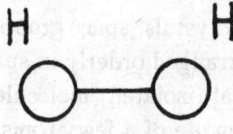

Fig. 2.43.

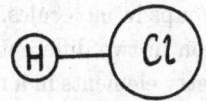

Fig. 2.44.

Fig. 2.45.

2.7. Summary

The idea here is that symmetry plays an important role in the determination of molecular structure. For example, consider the SF_6 molecule (Fig. 2.46). It has an hexagonal structure. It exhibits a sixfold symmetry. It is cubic. If we determine the length d and know its symmetry, these pieces of information are sufficient for us in our attempt to know its structure.

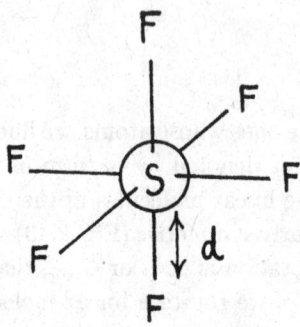

Fig. 2.46.

Solid state physics studies crystals (space groups), *i.e.*, systems consisting of many molecules (or atoms) arranged orderly in space. Molecular physics studies point groups and individual (isolated) molecules. The subject of molecular physics is a simple molecule made of a few atoms, and its physical properties (binding energies, internal energy, optical properties etc.). If a simple molecule has symmetry properties, it can be studied easily. That is why one looks for symmetry elements of point groups in molecules. These elements are discussed above, along with their notation in two different systems.

If the number of the symmetry elements in a molecule is more than one, further symmetries are produced. For example, in *mm* there is also a twofold axis which is the intersection of the two mutually perpendicular mirrors. (Example: H_2O molecule.)

The point groups concern symmetries of a single, isolated molecule. The space groups concern symmetries of crystals (assemblies of identical molecules). The space groups are important because all crystals are three-dimensional and consist of large numbers of atoms.

A table of noncubic point groups appears on the next page. On the page after the next, there appears a table of the space groups.

NONCUBIC POINT GROUPS

Schönflies	International	Hexagonal	Tetragonal	Trigonal	Orthorhombic	Monoclinic	Triclinic
C_n	n	C_6 6	C_4 4	C_3 3	—	C_2 2	C_1 1
C_{nv}	nmm (even n) nm (odd n)	C_{6v} 6mm	C_{4v} 4mm	C_{3v} 3m	C_{2v} 2mm	—	—
C_{nh}	n/m	C_{6h} 6/m	C_{4h} 4/m	—	—	C_{2h} 2/m	—
	$\bar{n}$	C_{3h} $\bar{6}$	—	—	—	C_{1h} m ($\bar{2}$)	—
S_n	$\bar{n}$	—	S_4 $\bar{4}$	S_6 $\bar{3}$ (C_{3i})	—	—	S_2 $\bar{1}$ (C_i)
D_n	$n22$ (even n) $n2$ (odd n)	D_6 622	D_4 422	D_3 32	D_2 222 (V)	—	—
D_{nh}	$\frac{n}{m}\frac{2}{m}\frac{2}{m}$ (n/mmm)	D_{6h} 6/mmm	D_{4h} 4/mmm	—	D_{2h} (mmm) 2/mmm (V_h)	—	—
	$\bar{n}2m$ (even n)	D_{3h} $\bar{6}2m$	—	—	—	—	—
D_{nd}	$\bar{n}\frac{2}{m}$ (odd n)	—	D_{2d} (V_d) $\bar{4}2m$	D_{3d} ($\bar{3}m$) $\bar{3}\frac{2}{m}$	—	—	—

THE 230 SPACE GROUPS

Crystallic system	Symbol	Minimum symmetry elements		Number of space groups per system
		In symbol	In words	
Triclinic	$1\ (C_1)$ $\bar{1}\ (S_2)$	— $\bar{1}$	None One inversion	2
Monoclinic	$2\ (C_2)$ $m\ (C_h)$ $2/m\ (C_{2h})$	$\frac{2}{m}$	One twofold rotation axis and/or a mirror plane	13
Orthorhombic	$222\ (D_2)$ $mmm\ (D_{2h})$ $mm\ (C_{2v})$	mmm	Three mutually perpendicular rotation axes or mirror symmetries	59
Rhombohedral or Trigonal	$3\ (C_3)$ $32\ (D_3)$ $\bar{3}\ (S_6)$ $\bar{3}m\ (D_{3d})$ $3m\ (C_{3v})$	$\bar{3}m$	One threefold rotation axis	25
Hexagonal	$6\ (C_6)$ $3/m\ (C_{3h})$ $6/m\ (C_{6h})$ $62\ (D_6)$ $\bar{6}2m\ (D_{3h})$ $6/mmm\ (D_{6h})$ $6mm\ (C_{6v})$	$6/mmm$	At least one sixfold rotation axis or a sixfold inversion axis (6 or $\bar{6}$)	27
Tetragonal	$4\ (C_4)$ $4/m\ (C_{4h})$ $42\ (D_4)$ $4/mmm\ (D_{4h})$ $\bar{4}\ (S_4)$ $\bar{4}2m\ (D_{2d})$ $4mm\ (C_{4v})$	$4/mmm$	At least one fourfold rotation or inversion axis	68
Cubic	$23\ (T)$ $43\ (O)$ $m3\ (T_h)$ $m3m\ (O_h)$ $\bar{4}3m\ (T_d)$	$m3m$	At least four threefold axes	36
			Total:	230

There are 32 point groups for molecules if we include the four symmetry elements and combine them together. (They combine in 32 different ways.) There are 230 space groups, *i.e.*, if we now include the three symmetry elements for space groups as well, we obtain 230 space groups (for solid bodies), or possible combinations, in total. These space groups are distributed to seven crystallic systems as shown in the table.

Since the molecules are three-dimensional, it is hard to mentally visualize their symmetries, even to draw them on the paper, especially when they are complicated. For this reason, one builds three-dimensional structures (laboratory models) for the purpose of instruction and illustration (Fig. 2.47). There are stereograms made of steel balls (to imitate and represent the atoms and their position) connected by rigid wires (to represent the bonds and the interatomic distances). By looking at them we comprehend the spatial construction and appearance of molecules (or crystals). We can make the model of any crystal and hold it in front of us.

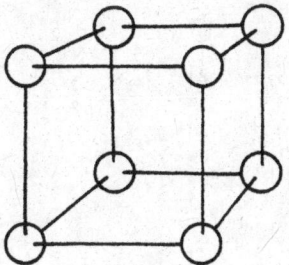

Fig. 2.47.

There are ready tables showing the symmetry elements of known molecules. But when we have to deal with an unknown molecule, we first have to find and study its symmetry. So, we need these steel ball-and-wire models to work on. When we have to study and work on a crystal or molecule, our job gets much easier if we know its symmetry in advance. (So, no wonder why we spy first on the enemy and collect intelligence information about him and his power before we draw up our strategy to fight him! The more we know about the subject the better we handle it. This is true universally, not only in molecular physics.)

The role of stereograms can be likened to that of written music. Music can be recorded in two ways: We either record it on tape (and then play the cassette in order to reproduce it), or we can *write it down* in musical symbols (notes) on the staff (whence we can also reproduce it by *reading* the written music). Similarly, ballet can be recorded in two ways: We can either record it in the form of a video movie (and replay it to reconstruct the whole thing), or we can *write* it (alongside with its music scores), using commonly accepted symbols, *i.e.*, a symbolography (notation system) that represents the three-dimensional movements of the dancers. By reading the written symbols and signs, we can reconstruct the choreography. Likewise in molecular physics we have two systems of notation whereby we can symbolically represent (in writing) the three-dimensional structure of a molecule, and conversely, by looking at the symbol we can visualize the appearance of the molecule in space (in our mind or by making a toy model or a computer simulation).

Chapter 3

QUANTUM MECHANICS IN MOLECULAR PHYSICS

3.1. General

Quantum mechanics is extensively used in molecular physics. In this chapter we do not intend to teach the reader quantum mechanics. Throughout this book it is assumed that the reader has already got a solid background in quantum theory. So, we are not going to repeat material already included in standard quantum mechanics texts, but we will rather stress points that are relevant to molecular physics, and important quantum facts employed thereby. Still, our treatment will not be all-inclusive.

Every physicist knows about the Schrödinger equation, the notion of operators in quantum mechanics, and the quantum mechanical discussion of the hydrogen atom; so, we will not cover these topics thoroughly here. Also, we will not dwell on topics like the quantization of angular momentum, the vector model of atoms, or the coupling schemes of angular momenta in atoms (and when each scheme is appropriate). For these topics the reader should consult a standard quantum mechanics text. In this book, knowledge on such topics on the part of the reader will be assumed as given. There is no doubt that quantum mechanics cannot be stuffed into the tiny capsule of a chapter, nor can one learn quantum theory by just reading a chapter — anyway, the best way of learning quantum mechanics is to already know it!

This chapter will refresh memories by underlining some basic facts only, concerning quantum theory, as related to molecular physics.

Quantum mechanics is a closed theory for tiny particles of microscopic sizes (molecular, atomic, and subatomic nuclear and elementary-particle level).

When we characterize a size as "microscopic", *i.e.*, as belonging to the microcosm (world of molecules and atoms), we compare with the wavelength λ of visible light. The more we approach sizes comparable to λ the more prominent the quantum mechanical character of the object in question. So, for large sizes $(d \gg \lambda)$, the classical character and behavior prevail. What we observe in the macrocosm (our world around us which involves sizes comparable to our size or greater) is the macroscopic manifestation of collective quantum phenomena, *i.e.*, bulk behavior of matter and other forms of energy producing phenomena which we grossly see as, and study by, classical physics. For a crude example, consider Fig. 3.1. When the size of the boat is much larger than the sea waves, the classical theory works better in describing the boat, its motion, and its behavior in general. But when we have a tiny boat, smaller than the sea waves, this time a proper "quantum theory for boats" will represent the boat better, because the boat has reduced from an object to a tiny particle compared to λ. This is a crude and bad example, but nevertheless illustrative.

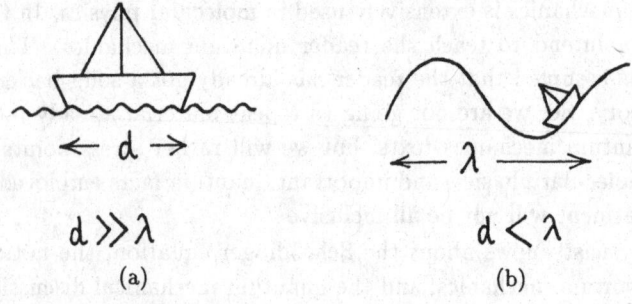

Fig. 3.1.

The *correspondence principle* in quantum physics states that as we go to large sizes, quantum mechanics yields smoothly to classical physics, so that the two disciplines in fact agree and complete each other as we move along the scale of sizes.

Quantum mechanics has a powerful and admirable mathematical formalism. Mathematics is the soul and life of quantum mechanics and is extensively used in it.[1] But sometimes the actual physical content is hidden behind

---

[1] Mathematics is a very useful tool for physicists in solving problems and describing phenomena. Mathematicians are needed, too, to help physicists solve difficult problems. There is truth in the words of a physics professor who said: "Mathematicians know math, but they don't know what they do it for; physicists know physics and the value of math, but they don't know math". Indeed there are good physicists who dislike mathematics, preferring to stress rather the physical underlayment at any given problem.

complicated mathematical expressions. A mathematical meditation and operation may lead to dry and dreadful expressions which obscure the physical meaning of the examined issue. True, mathematics is a poetically elegant way of expression and a language of symbolism and brevity (saving us pages of wordy and tedious extravagances of difficult prosaic texts), but it may become forbidding and discouraging if we lose sight of physics and enter the domain of mathematical typolatry at the expense of the physical essence which is hidden and suppressed. Quantum mechanics is very vulnerable in this aspect, because it is conceptual and employs hard and complex mathematical means and abstractions to express physics and reasoning symbolically, in its esthetic perfection. These means cannot be softened enough unless one has a strong and formal mathematical training and is able to read between the lines and see through the heaps of algebra, as one tries to interpret the results and reach physical conclusions.

So, one of the properties of quantum mechanics is that the connection between the physics problem (taken from real life and the real world) and the mathematical method is weak.

Quantum mechanics explains many physical problems, but in some cases, when we apply the quantum formalism to the problem to be solved, it is quite difficult to follow the way by which we get into the substance of the problem. However, the application of quantum theory to spectroscopy is easy; indeed, spectroscopic phenomena can be easily calculated by using quantum mechanics. This is so because the experimental spectroscopic results caused the birth of quantum mechanics. That is why a physicist who likes to become a specialist in molecular physics should study quantum mechanics *and* elementary spectroscopy together — these two go together.

A word of advice here: Quantum mechanics cannot be mastered unless one solves many problems. Reading the theory alone is not enough.[2]

Good books on quantum mechanics, by author, are the following: Park (1st and 2nd ed., pleasant to read), Schiff (rigid, rigorous, and demanding), Rojansky, Bohm, Saxon (partially good treatment), Anderson (spells out operator matrices), Davydov, Ruei (clear and very readable), Gasiorowicz (clear treatment, good examples), Powell and Crasemann (good in piecewise potentials), Merzbacher (of classical value), Fong (simple, lower-level, and clear), Messiah (generously detailed and too formal), Dirac, Baym, Landau & Lifshitz (full

---

[2] What guarantees that what we write as mathematical analysis on a piece of paper corresponds to the real world out there? This is a philosophical question with which we will not deal.

series), Fermi (extract), Pauling & Wilson (good applications and problems), Gottfield, Kramers, Kemble, Tomanaga, von Neumann, Dicke & Wittke, Kursunoglu, H. Smith and others. Problem books: Constantinescu & Magyari, Flügge, ter Haar, and Goldman & Krivchenkov.

3.2. Basic Postulates of Quantum Mechanics

(1) Every dynamical variable regarding the motion of a particle can be described by a quantum mechanical *operator*. For example, consider the following table:

Variable	Its operator	
$\mathbf{r}$	$\mathbf{r}$	[3]
$\mathbf{p}$	$-i\hbar\vec{\nabla}$	
p_x	$-i\hbar\dfrac{\partial}{\partial x}$	
p_x^2	$-\left(\dfrac{h^2}{4\pi^2}\right)\dfrac{\partial^2}{\partial x^2}$	
E	$i\hbar\dfrac{\partial}{\partial t}$ or $-\dfrac{\hbar^2}{2m}\vec{\nabla}^2 + V$	
V	V	
L_z	$-i\hbar\dfrac{\partial}{\partial \varphi}$	

(2) Every operator obeys an eigenvalue equation: If an operator Q acts on its eigenfunction u_n, it gives a scalar constant number a_n times *the same* function u_n:

$$Qu_n = a_n u_n$$

where a_n are the (labeled) eigenvalues of Q. For example,

$$L_z\psi_{nlm} = m_l\hbar\psi_{nlm}. \tag{3.1}$$

---

[3]Since $\mathbf{r}$ is not a differential operator, its operation means just multiplication by $\mathbf{r}$. The same thing holds for V.

(3) A sensitive experimental measurement of the dynamical variable of Q gives one of its eigenvalues.[4] The axiomatic rule in quantum mechanics is that for every physical quantity we (experimentally) measure the *eigenvalues* of the operator that represents this quantity, *not* the operator itself! (The operator *per se* is a mathematical entity, not a physical one.) So, in Eq. (3.1) an experiment gives only a value that m_l can take (is allowed quantum mechanically to take).

(4) The eigenfunctions are *orthonormal* to each other, forming an orthonormal set. That is,

$$\int_{-\infty}^{\infty} u_n^* u_n \, dV = 1 \quad \text{(normalization condition)}$$

and

$$\int_{-\infty}^{\infty} u_n^* u_m \, dV = 0 \quad \text{(orthogonalization condition)}$$

which can be summarized as

$$\int_{-\infty}^{\infty} u_n^* u_m \, dV = \delta_{nm} \quad \text{(Krönecker delta)}.$$

(5) The well-behaving eigenfunctions of the Hamiltonian operator $\mathcal{H}$ are the *wavefunctions* ψ_n of the system, and their corresponding eigenvalues are the stationary (equilibrium) energy states E_n:

$$\mathcal{H}\psi_n = E_n \psi_n.$$

If an energy eigenvalue (eigenenergy) corresponds to *several* independent wavefunctions (eigenfunctions) $\psi_1, \psi_2, \ldots$, this energy value is *degenerate*.

(6) Any function ψ can be expanded into (and expressed as) a series in terms of eigenfunctions u_n:

$$\psi = \sum_{n=1}^{\infty} c_n u_n \tag{3.2}$$

---

[4]It does *not* give its average (or mean) value $\langle Q \rangle$ which is just a prescription to make a scalar number out of a tensor (matrix, operator):

$$\langle Q \rangle = \int_{-\infty}^{\infty} u_n^* Q u_n \, dV.$$

We never measure $\langle Q \rangle$ in a single measurement; to obtain expectation values, we should prepare the system many times and repeat the measurements which should be averaged.

where c_n are the coefficients of the expansion. So, the eigenfunctions form a *complete orthonormal set (basis set)*.[5] If each of the u_n are particular solutions to the Schrödinger equation, the linear combination ψ in Eq. (3.2) is also a solution of the same equation.

(7) If a system is represented and described by a function ψ, a measurement done to measure a_n gives a numerical result which is proportional to the amplitude square of c_n (which is the coefficient of u_n in the expression for ψ):

$$a_{\text{expected}} \simeq |c_n|^2 .$$

Operators show mathematical operations performed on functions which are called *operands*. An operator applying to an operand gives a scalar quantity multiplied by the same operand. For example:

$$\widehat{\mathcal{H}}\psi = E \times \psi .$$

In quantum mechanics, operands are *well-behaving* eigenfunctions of the operators considered, *i.e.*, these eigenfunctions:

(1) Stay always finite (do not go to infinity).
(2) Do not vanish identically.
(3) Are single-valued.
(4) Have continuous derivatives.

3.3. The Schrödinger Equation of a Diatomic Molecule (Two-Body Problem)

In classical mechanics we can say that the moving body (or particle) is here or there, but in quantum mechanics this does not hold valid: The position of the particle is not well-defined; we can speak only of the *probability* for the particle to be somewhere in space.

When we deal with one particle only, the situation is still more difficult, because it is hard to estimate where a single particle might go, whereas if we observe a collection of many particles, it is easier to predict the motion, just as it is easier to tell the behavior of a mass of people in a demonstration rather than trying to follow what a single individual is doing in the crowd.

---

[5] Just like a vector which is expressible as a sum of components along axes where the unit vectors along the axes form a complete set of independent basis vectors (an orthonormal basis).

In molecular physics we usually study diatomic molecules, in which case we take a system consisting to two particles to apply quantum mechanics to. Such a system is easy and simple to handle and start our study with. Let the masses of the particles be m_1 and m_2. The (quantum mechanical) equation of motion of the system is the *Schrödinger equation.* If the total potential energy of the system is $V(r)$ and the kinetic energies of the particles are E_1 and E_2, the system's total energy is

$$E = (\text{Kinetic energy of 1st particle}) + (\text{Kinetic energy of 2nd particle})$$

$$+ (\text{Potential energy}),$$

or, in symbols,

$$E = E_1 + E_2 + V(r),$$

and the system's Hamiltonian (operator) is

$$\mathcal{H} = \frac{p_1^2}{2m_1} + \frac{p_2^2}{2m_2} + V(r) = -\frac{\hbar^2}{2m_1}\vec{\nabla}_1^2 - \frac{\hbar^2}{2m_2}\vec{\nabla}_2^2 + V(r)$$

where the quantities $\mathcal{H}$, p_i, and V are *operators* (in quantum mechanics). If $\mathcal{H}$ operates on the wavefunction ψ, it gives the eigenenergy E times ψ:

$$\mathcal{H}\psi = E\psi,$$

which is the Schrödinger equation. Here $\mathcal{H}$ *acts on* ψ (left-hand side) and gives the right-hand side which is a product (E *multiplied by* ψ). E is the eigenvalue of the operator $\mathcal{H}$ and includes the stationary values of the energy.[6]

---

[6]When we talk about a particular state (labeled) n of the system, then we take the wavefunction (eigenfunction) ψ_n (or $|n\rangle$, in Dirac notation) and have $\mathcal{H}\psi_n = E_n\psi_n$ where E_n is the energy of that state. Here ψ is normalized to unity, so that

$$\int_{-\infty}^{\infty} \psi^*\psi \, d\text{V} = 1$$

where the integration runs over all space.

So, we have

$$\mathcal{H}\psi = \underbrace{\left(-\frac{\hbar^2}{2m_1}\vec{\nabla}_1^2 - \frac{\hbar^2}{2m_2}\vec{\nabla}_2^2 + V(r)\right)}_{\text{Acts on the function}} \psi = E\psi,$$

$$-\frac{\hbar^2}{2m_1}\vec{\nabla}_1^2\psi - \frac{\hbar^2}{2m_2}\vec{\nabla}_2^2\psi + V(r)\psi = E\psi,$$

$$-\frac{\hbar^2}{2m_1}\vec{\nabla}_1^2\psi - \frac{\hbar^2}{2m_2}\vec{\nabla}_2^2\psi + [V(r) - E]\psi = 0,$$

$$\frac{\hbar^2}{2m_1}\vec{\nabla}_1^2\psi + \frac{\hbar^2}{2m_2}\vec{\nabla}_2^2\psi + [E - V(r)]\psi = 0. \tag{3.3}$$

This is the *Schrödinger wave equation* for *two particles* (*two-body problem*). Its solution pertains to two cases:

(1) Hydrogen atom (an atomic nucleus and an electron around it — where the nucleus is just a proton; in hydrogenlike atoms we have the nucleus with the core electrons taken as one body, and the valence electron taken as the second body).
(2) Diatomic molecule (two-body system).

Both cases can be jointly solved by a common (the same) mathematical method which works for both. The only difference is that in (1) we have the Coulombic force for hydrogen, whereas in (2) we have the valence binding force. Since the forces are different, the ways for the two cases part from the moment we put in the forces (potentials) for further operations. So, until further notice, both the atom and the diatomic molecule will be considered as a two-body system each, with an internal potential $V(r)$.

Equation (3.3) represents these two cases, but in molecular physics we deal with case (2) only. The *Laplacian operator* $\vec{\nabla}^2$ is

$$\Delta \equiv \vec{\nabla}^2 = \frac{\partial^2}{\partial x^2} + \frac{\partial^2}{\partial y^2} + \frac{\partial^2}{\partial z^2},$$

so that

$$\vec{\nabla}_1^2\psi = \frac{\partial^2\psi}{\partial x_1^2} + \frac{\partial^2\psi}{\partial y_1^2} + \frac{\partial^2\psi}{\partial z_1^2},$$

in Cartesian coordinates. Similarly, $\vec{\nabla}_2^2$ is with respect to the coordinates of the second mass (particle).

The coordinates 1 and 2 are independent of each other. In such a system we use the *center-of-mass coordinates* (Fig. 3.2):

$$X = \frac{1}{M}(m_1 x_1 + m_2 x_2)$$

$$Y = \frac{1}{M}(m_1 y_1 + m_2 y_2)$$

$$Y = \frac{1}{M}(m_1 z_1 + m_2 z_2)$$

where $M = m_1 + m_2$. Also, another coordinate system is needed, defined as

$$\left.\begin{array}{l} x = x_1 - x_2 \\ y = y_1 - y_2 \\ z = z_1 - z_2 \end{array}\right\} \mathbf{r} = \mathbf{r}_1 - \mathbf{r}_2$$

which are the *relative coordinates* and describe the motion of one particle with respect to (relative to) the other. These (two sets of) new coordinates are necessary because the old ones $m_1(x_1, y_1, z_1)$ and $m_2(x_2, y_2, z_2)$ are inappropriate to describe the motion: The system may move translationally (keeping the interparticle distance constant) and rotate about its center of mass.

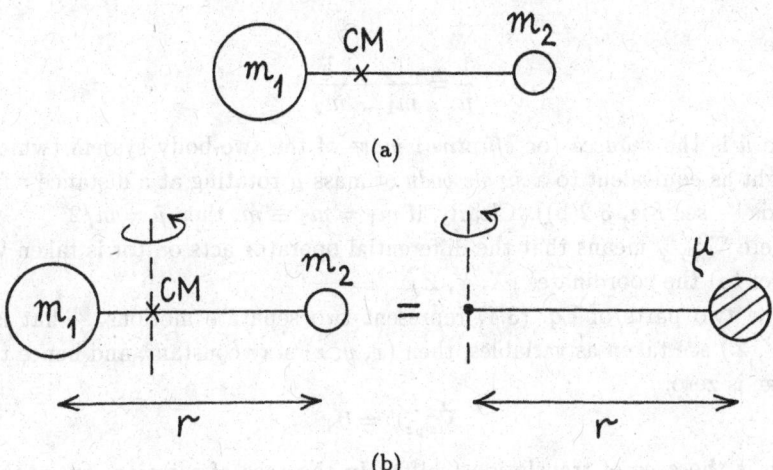

Fig. 3.2.

The new coordinates are conveniently taken differently for different motions of the system under study, in order to simplify and economize the geometry and algebra. If there is a translational motion (if the molecule drifts along as it stands), then (x, y, z) stay constant, while (X, Y, Z) are taken as variable. In this case the center of mass is translated in space (the molecule is shifted without any change in its shape and internal configuration). If there is a rotation or linear vibration (Fig. 3.3), then (X, Y, Z) stay constant, while the *difference coordinates* (x, y, z) vary.

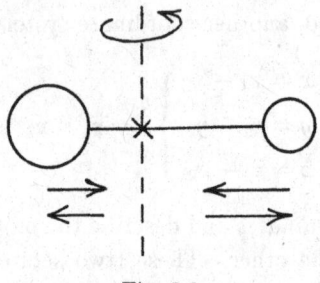

Fig. 3.3.

Now we plug the new variables into the Schrödinger equation to have

$$-\frac{\hbar^2}{2M}\vec{\nabla}^2_{XYZ}\psi - \frac{\hbar^2}{2\tilde{\mu}}\vec{\nabla}^2_{xyz}\psi + [V(r) - E]\psi = 0 \qquad (3.4)$$

where

$$\frac{1}{\tilde{\mu}} = \frac{1}{m_1} + \frac{1}{m_2}$$

where $\tilde{\mu}$ is the *reduced* (or *effective*) *mass* of the two-body system (which is thought as equivalent to a *single body* of mass μ rotating at a distance r from an axis — see Fig. 3.2(b)). Clearly, if $m_1 = m_2 \equiv m$, then $\tilde{\mu} = m/2$.

Here $\vec{\nabla}^2_{XYZ}$ means that the differential operator acts on (or is taken with respect to) the coordinates (X, Y, Z).

The two parts of Eq. (3.4) represent two separate motions. That is, if (X, Y, Z) are taken as variables, then (x, y, z) stay constant, and hence their change is zero:

$$\vec{\nabla}^2_{xyz}\psi = 0\,,$$

which is the case of translation (shift). In the case of rotation, (x, y, z) are variable and

$$\vec{\nabla}^2_{XYZ}\psi = 0$$

where (X, Y, Z) stay constant. Due to this independence between (X, Y, Z) and (x, y, z), the solution to Eq. (3.4) is the wavefunction

$$\psi(X, Y, Z, x, y, z) = \psi_I(X, Y, Z)\psi_{II}(x, y, z),\qquad(3.5)$$

i.e., a *product* (because we have *independent* degrees of freedom) of two functions ψ_I and ψ_{II}, one depending only on (X, Y, Z) and the other one only on (x, y, z).[7]

It is possible to find such a solution which is the product of two functions. This is just the *method of separation into variables* (or factors), well-known in theoretical physics and the theory of mathematical methods. If we substitute this ψ into Eq. (3.4) we obtain two relations: One for ψ_I and another one for ψ_{II}.

Notice that $V(r)$ is a function of r only, not $\mathbf{r}$. That is, V does *not* depend on the direction, but only on the relative separation of the two particles (their mutual *distance* r — which of course may vary due to vibrations). Figure 3.4 shows the distance r between the two particles, and the position vectors $\mathbf{r}_1$ and $\mathbf{r}_2$ of the two particles (bodies). Since $r = r(x, y, z)$, we also have $V = V(x, y, z)$. Therefore, the equation becomes

$$\underbrace{-\frac{\hbar^2}{2M}\frac{1}{\psi_I(X, Y, Z)}\vec{\nabla}^2_{XYZ}\psi_I}_{\text{Constant}} \underbrace{-\frac{\hbar^2}{2\tilde{\mu}}\frac{1}{\psi_{II}(x, y, z)}\vec{\nabla}^2_{xyz}\psi_{II}}_{\text{Constant}} + [V(x, y, z) - E] = 0$$

where E is constant and M is the total mass. In other words, we divided the equation by ψ. For the left-hand side to be equal to zero, each term must be equal to a (separate) constant:

$$-\frac{\hbar^2}{2M}[\cdots] \equiv c_I$$

$$-\frac{\hbar^2}{2\tilde{\mu}}[\cdots] \equiv c_{II}$$

where constants c_I and c_{II} are arbitrary, depending on the coordinates. We can write $c_I + c_{II} = E$ where c_I represents the translational energy and c_{II} represents the vibration + rotation energy: $c_I = E_{\text{transl}}$ and $c_{II} = E_{\text{vibr,rot}} = E_{\text{vibr}} + E_{\text{rot}}$.

---

[7]Independence means a product of the probabilities $|\psi|^2$ and a *sum* of the energies (where the contribution of each degree of freedom is added up): $E_{\text{total}} = E_{\text{translation}} + E_{\text{rotation}} + E_{\text{vibration}}$.

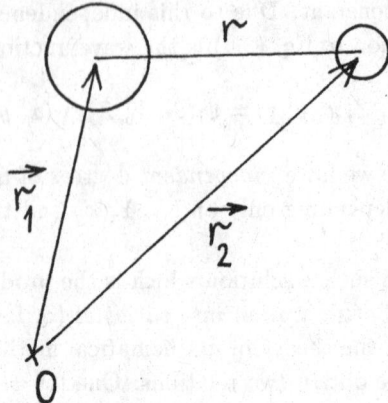

Fig. 3.4.

Notice that it is not important to know where the molecule is in space; as long as there are two atoms at a (separation) distance r between them, forming a molecule, we do not care about where the molecule will go.

The second part of the equation includes internal energies (E_{vibr} and E_{rot} and intramolecular energies), whereas the first part is the translational kinetic energy of the center of mass of the molecule (where the molecule moves as a whole).[8] To find a solution, $V(r)$ must be known explicitly. If $V(r)$ is *radial* (has spherical symmetry, *i.e.*, is independent of the polar angles θ and φ, and hence independent of the orientation and direction in space, depending on the distance r only), then one can use polar coordinates to solve the problem.[9] Molecular physics is interested in the second part of the equation only.

Here we have a system of two atoms (or particles, in general) of masses m_1 and m_2, at a distance r between them. But it could also be a hydrogen atom, as well as another atom with Z positive charges (protons) in the nucleus (Fig. 3.5) and a single outermost electron outside the electron core (Na atom). This system can also be thus studied. (For $Z = 1$ we have H; for $Z = 3$ we have Li etc.) As far as molecular physics is concerned, we can study H_2, HCl and other similar two-atom systems (molecules) without taking their electron clouds (internal atomic structure) into account.

---

[8]Vibration and rotation are internal molecular phenomena, and their corresponding degrees of freedom are internal ones.

[9]Radial means r-dependence only, and independence of the direction (angles). The dependence is on r, not **r**.

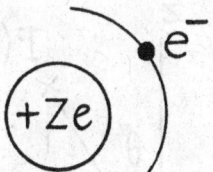

Fig. 3.5.

3.4. Solution of the Schrödinger Equation

In the previous section we saw that $\vec{\nabla}^2_{XYZ}$ operates on the coordinates (X, Y, Z). This part of the equation concerns the motion of the center of mass of the molecule, which we are not interested in. In other words, after writing ψ as in Eq. (3.5) and plugging it into Eq. (3.4), we obtain two terms of which the first is out of interest. What we are looking for is the *internal molecular energy*, and that is the second term. So, what interests us is the equation

$$-\frac{h^2}{8\pi^2\tilde{\mu}}\frac{1}{\psi_{\text{II}}(x, y, z)}\left[\frac{\partial^2\psi_{\text{II}}}{\partial x^2} + \frac{\partial^2\psi_{\text{II}}}{\partial y^2} + \frac{\partial^2\psi_{\text{II}}}{\partial z^2}\right] + V(r) = E$$

where $E = E_{\text{internal}}$. Since the translational kinetic energy of the two-atom system is not of interest, we will replace ψ_{II} by ψ from now on, to get rid of the subscript. We will know that ψ is in fact ψ_{II}.

The Schrödinger equation is

$$\vec{\nabla}^2_{xyz}\psi(x, y, z) + \frac{8\pi^2\tilde{\mu}}{h^2}[E - V(r)]\psi(x, y, z) = 0$$

where V is radial, depending on r only, not on **r**. We can solve this differential equation in spherical polar coordinates (Fig. 3.6), in which case we should write the Laplacian operator $\vec{\nabla}^2$ in spherical coordinates:

$$\vec{\nabla}^2 = \frac{1}{r^2}\frac{\partial}{\partial r}\left(r^2\frac{\partial}{\partial r}\right) + \frac{1}{r^2\sin\theta}\frac{\partial}{\partial\theta}\left(\sin\theta\frac{\partial}{\partial\theta}\right) + \frac{1}{r^2\sin^2\theta}\frac{\partial}{\partial\varphi^2}.$$

Again, to solve the equation, we will separate it into its variables r, θ, and φ. We will use the same method, *i.e.*, we will write ψ as a product of three functions f_i where i = 1 to 3 (we assume that it can be written so):

$$\psi = f_1(r)f_2(\theta)f_3(\varphi)$$

where f_1 depends only on r, f_2 only on θ, and f_3 only on the variable φ. Here f_i are three *different* functions, but each of one variable only. We accept ψ as such a product and proceed, plugging it into the equation:

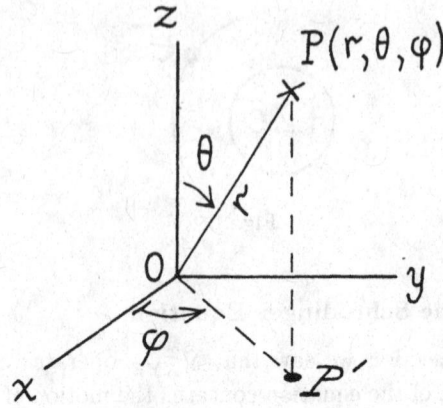

Fig. 3.6.

$$\underbrace{\frac{1}{f_3}\frac{\partial^2 f_3}{\partial \varphi^2}}_{\alpha} + \frac{\sin^2 \theta}{f_1}\frac{\partial}{\partial r}\left(r^2\frac{\partial f_1}{\partial r}\right) + \frac{\sin \theta}{f_2}\frac{\partial}{\partial \theta}\left(\sin \theta \frac{\partial f_2}{\partial \theta}\right)$$

$$+ \underbrace{\frac{8\pi^2\tilde{\mu}}{h^2}r^2 \sin \theta [E - V(r)]}_{-\alpha} = 0$$

where the equation is separated into its variables as a sum of independent terms. (Here subscripts 1 to 3 refer to the variables r, θ, and φ, *not* to the labels of the two masses 1 and 2.) Each term contains one variable only. For this equation to be possible (equal to zero), each term must be equal to a constant. So, we take the first term equal to α (= constant), and the last term equal to $-\alpha$. Thus we have a second-order differential equation with constant coefficients. For good and well-behaving results, we should have by definition

$$\alpha = -m^2$$

where m is a real number. Then the f_3 part (solution) is

$$f_3(\varphi) = \mathcal{A}e^{im\varphi} + \mathcal{B}e^{im\varphi}.$$

Since the wavefunction must be single-valued, it should take the same value upon a turn by $\varphi = 2\pi$:

$$f_3 = \mathcal{A}e^{im(\varphi+2\pi)} + \mathcal{B}e^{-im(\varphi+2\pi)}$$

where we must have m = ±1, ±2,..., (integer). Only then does the function coincide with itself upon a full turn through 2π. Here m is the *magnetic quantum number* or *azimuthal quantum number* (because it is related to the azimuthal angle φ). Its first name is due to the fact that it quantizes the orientation of the angular momentum vector in space, with respect to the direction of an external magnetic field. (The externally applied field can be along any of the three Cartesian axes, but it is customary to take it along z — though any other axis might do as well. So, m is the projection of the vector on the z-axis. This projection can be of definite quantized lengths only, determined by m.)

If we divide by $\sin^2 \theta$, the remaining expression is

$$\frac{1}{f_1}\frac{\partial}{\partial r}\left(r^2\frac{\partial f_1}{\partial r}\right) + \frac{8\pi^2\tilde{\mu}r^2}{h^2}[E - V(r)] + \frac{1}{f_2 \sin\theta}\frac{\partial}{\partial\theta}\left(\sin\theta\frac{\partial f_2}{\partial\theta}\right) - \frac{m^2}{\sin^2\theta} = 0$$

(3.6)

where we call

$$\frac{1}{f_1}\frac{\partial}{\partial r}\left(r^2\frac{\partial f_1}{\partial r}\right) + \frac{8\pi^2\tilde{\mu}r^2}{h^2}[E - V(r)] \equiv \beta = \text{constant}$$

(3.7)

$$\frac{1}{f_2 \sin\theta}\frac{\partial}{\partial\theta}\left(\sin\theta\frac{\partial f_2}{\partial\theta}\right) - \frac{m^2}{\sin^2\theta} \equiv -\beta = \text{constant},$$

(3.8)

so that $\beta - \beta = 0$, and the equation is verified. Note that Eq. (3.7) includes r-dependent terms, and Eq. (3.8) includes θ-dependent terms only. The differential equation can be solved by using series. We obtain a series

$$f_2 = P_l^m(\cos\theta)$$

where l is the index of the series, and $\beta = l(l+1)$. Here m is a superscript, not a power. There is a mathematical trick here to provide expediency. These so-defined P_l functions are called *Legendre polynomials* (or *Legendre functions*). If we call $\cos\theta \equiv u$ (for the sake of brevity), we have

$$f_2 = (1 - u^2)^{|m|/2}\frac{d^{|m|}P_l(u)}{du^{|m|}}$$

where we take the |m|-th derivative of P_l. If m = ±2, then we have $d^2 P_l(u)/du^2$.[10] If m is nonzero, it labels the Legendre polynomials as P_l^m,

---

[10]Note that P_l involve $\cos\theta$ and hence are quite usable especially when the solution to the problem contains powers of $\cos\theta$ and combinations thereof.

in which case we have the *associated Legendre polynomials* (when there is no azimuthal symmetry, *i.e.*, when the solution depends on φ, so that $m \neq 0$).

We have

$$P_l(u) = \frac{1}{2^l l!} \frac{d^l (u^2 - 1)^l}{du^l}.$$

These are devious results due to the difficulty of the problem, but they nevertheless give us what we want.

If $|m| > l$, we have $P_l^m = 0$. Therefore, the condition is $l \geq |m| \geq 0$ where l is the *orbital quantum number* or *angular momentum quantum number*.

We are interested in the polynomials $P_l(u)$ because we must know the atomic charge density (electron distribution or cloud) if we want to know which directions the atomic bonds are in. Here we have

$$\int_{\text{All space}} f_2 f_2^* \, dV = 1 \tag{3.9}$$

which means that the functions f_2 are normalized (to unity). The integration runs from $-\infty$ to $+\infty$. According to l and m values,[11] P_l take the following values (*e.g.*, $P_0 = \sqrt{2}/2$):

l	m (or m_l)	P_l (normalized) or $f_2(\theta)$
0	0	$\sqrt{2}/2$
1	0	$\dfrac{\sqrt{6}}{2} \cos\theta$
	± 1	$\dfrac{\sqrt{3}}{2} \sin\theta$
2	0	$\dfrac{\sqrt{10}}{4}(3\cos^2\theta - 1)$
	± 1	$\dfrac{\sqrt{15}}{2} \sin\theta\cos\theta$
	± 2	$\dfrac{\sqrt{15}}{4} \sin^2\theta$
3	0	$\dfrac{314}{4}\left(\dfrac{5}{3}\cos^3\theta - \cos\theta\right)$
	± 1	$\dfrac{42}{8}(5\cos^2\theta - 1)\sin\theta$

[11]This m is associated with l in atoms, and thus it is subindexed as m_l (to show that it is the projection of the electronic orbital angular momentum l along z, as distinguished from m_s, the projection of spin). In atoms we have l and m_l; but in molecules, since l (or L) does not exist, l (or L) goes over to J, and m_l goes over to M_J.

These are the first few P_l, in their already normalized form. (As they have to do with charge probability densities, they must be normalized to 1, by Eq. (3.9)).

Notice that P_l is a function that varies with θ as above. Now Eq. (3.7) becomes

$$\frac{\partial}{\partial r}\left(r^2 \frac{\partial f_1}{\partial r}\right) + \frac{8\pi^2 \tilde{\mu} r^2}{h^2}[E - V(r)]f_1 - l(l+1)f_1 = 0 \qquad (3.10)$$

where our only worry now is to know the potential $V(r)$ explicitly, in order to proceed further and solve the equation. Equation (3.10) is the *radial equation* (radial part of ψ); it can be obtained by letting $\beta = l(l+1)$ in Eq. (3.7), and can be sovled if we know the explicit function $V(r)$ which we plug in.

Now we put the proper $V(r)$, depending on whether we want a solution for the hydrogen atom or for the diatomic molecule. Equation (3.10) will look different in these two cases, because $V(r)$ will be different for each.

3.5. Simple Cases

The hydrogen atom and the diatomic molecule are the two simple cases to which Eq. (3.10) applies.

In the *hydrogen atom* $V(r)$ is easy: It is just the Coulombic (electrostatic) potential between the nucleus (p^+) and the revolving electron:

$$V(r) = -\frac{Ze^2}{r}$$

(see Fig. 3.5). The same thing holds for hydrogenlike atoms. For a *molecule*, $V(r)$ includes potentials that correspond to vibrations and rotations.

For hydrogen and hydrogenlike atoms, we put $V(r)$ into the equation and call

$$\frac{8\pi^2 \tilde{\mu} E}{h^2} \equiv -k^2 = \text{constant} \qquad (^{12})$$

$$\frac{4\pi^2 \tilde{\mu} Z e^2}{h^2 k} \equiv \eta = \text{another constant}.$$

We also change variables, adopting the new variable $x = kr$. Then we define a new function as

$$\mathcal{X}(r) \equiv \frac{f_1}{r}.$$

---

[12]Do not mistake this constant k as the Boltzmann constant. The choice of the square with the minus sign is deliberate, to facilitate the mathematics. Since the constant is arbitrary, we are free to choose it as we want.

After these expedient tricks, the equation becomes solvable. It is a second-order differential equation with constant coefficents:

$$\frac{d^2\mathcal{X}}{dx^2} + \left[\frac{2\eta}{x} - \frac{l(l+1)}{x^2} - 1\right]\mathcal{X} = 0$$

or

$$\frac{d^2\mathcal{X}}{dr^2} + \frac{8\pi^2\tilde{\mu}}{h^2}\left[E + \frac{Ze^2}{r} - \frac{\hbar^2 l(l+1)}{r^2}\right]\mathcal{X} = 0$$

where for $x \to \infty$ and $r \to \infty$ we have $\mathcal{X} \to 0$ (the function goes slowly to zero). So, we should look for a solution of the form

$$\mathcal{X} = \mathcal{A}x^n e^{-x}$$

which we put into the differential equation to have

$$\left\{\frac{1}{x^2}[n(n-1) - l(l+1)] + \frac{2}{x}(\eta - n)\right\}\mathcal{A}x^n e^{-x} = 0\,.$$

Since $\mathcal{A} = $ constant $\neq 0$, the quantity within the big brackets { } must be equal to zero, *i.e.*, the coefficients of $1/x^2$ and $2/x$ must vanish. This means that we must have $n = \eta$ and $n - 1 = l$ or $n = l + 1$, which is the condition. Here $n = 1, 2, \ldots$ must be a positive integer, because $l = 0, 1, 2, 3, \ldots$. Therefore, E is quantized as E_n, and n is the *principal quantum number* by which we may label the quantized energy states (levels) of the atom (actually of the atomic electron).

Since η turns out to be $\eta = n$, we replace k by its value to have

$$-\frac{8\pi^2\tilde{\mu}E}{h^2} = k^2 = \frac{16\pi^4\tilde{\mu}^2 Z^2 e^4}{h^4 n^2}\,,$$

$$E = -\frac{2\pi^2\tilde{\mu}Z^2 e^4}{h^2 n^2}$$

and thus $E_n \propto 1/n^2$ for the *bound* electron, where e is the electron charge and Z is the atomic number (number of protons in the nucleus). That is, we have the same result as in classical Bohr theory.[13] This is what the physics

---

[13]From the energies E_n we find that the radius of the energy levels (shells) in an atom goes as $r_n \sim n^2$, *i.e.*, orbits get spaced out as n increases. The atom resembles a planetary system, with the electrons revolving about the nucleus in correspondence with the planets revolving around the Sun in their orbits. Huge distances separate the orbits from the nucleus (compared to the nuclear size), so that the atom is not a compact object. A novice pupil is hence tempted to ask: Could we not hold a thin pin between the orbits? The answer is no, because even the thinnest pin end consists of so many atoms.

is all about here; the rest is mathematics, not physics. In this analysis that we follow to reach the results, the reader must be able to distinguish between where the physical meaning lies and what the mathematical technique is.

Figure 3.7 shows the energy levels for the hydrogen atom (actually, for its bound electron). They are *quantized*, whereas in the continuum (free electron) every kinetic energy is continuously possible (accessible to the electron and allowed for it to take). The electron which is bound to an atom takes discrete energies, not continuous. Nature imposes that the electron's energy values should depend on n. The electron can have stepwise energies only (with disallowed values in between). The energy changes by *quanta* ΔE (discrete definite dosages) (Fig. 3.8). The energy of an electron cannot be less than this quantum ΔE, except if it is at a higher n where $(\Delta E)_2 < (\Delta E)_1$ (Fig. 3.7).[14] Electrons of high n can be detached from the atom more easily, because ΔE becomes less and can be overcome by an incident energy supplied externally (*e.g.*, by radiation). Further, the electron cannot absorb *any* amount of energy; out of the incident energy it picks the proper ΔE only to get excited (jump on to an upper energy level). So, if we want to excite an atom, we should send the right energy, resonant to the needed (quantized) ΔE.

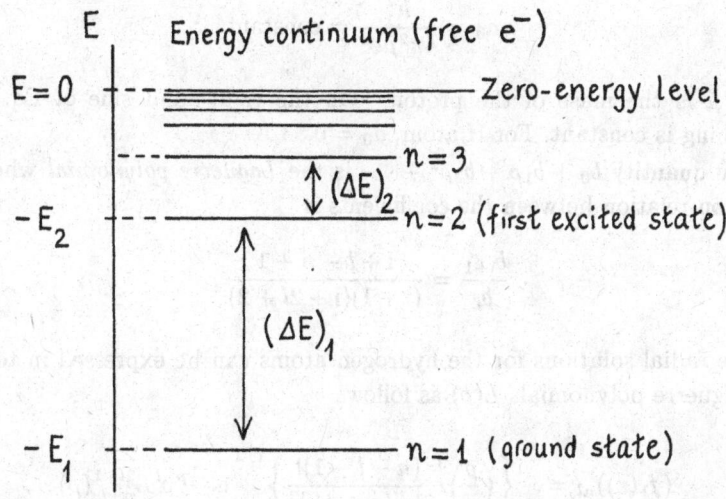

Fig. 3.7.

---

[14]A die is also quantized: We cannot obtain an outcome 4.5 when we throw it. Stairs are quantized: We cannot stand on the 2.5th step! Other quantized things in daily life are people and money.

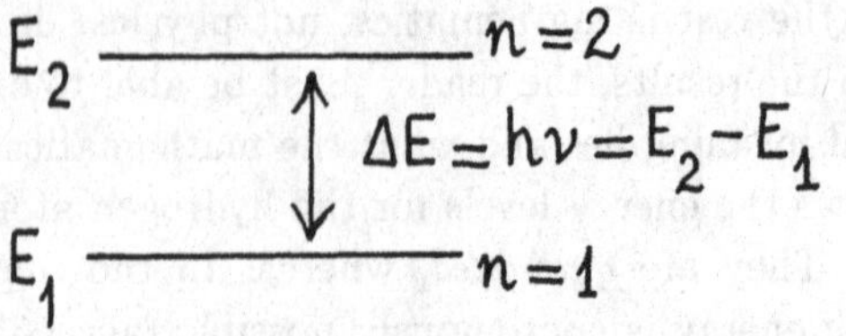

$$E_2 \underline{\hspace{3cm}} n = 2$$
$$\updownarrow \; \Delta E = h\nu = E_2 - E_1$$
$$E_1 \underline{\hspace{3cm}} n = 1$$

Fig. 3.8.

Here $f_1(r)$ is a series:

$$f_1(r) = \rho^l e^{-\rho/2}(b_0 + b_1\rho + b_2\rho^2 + \cdots)$$

where b_i are constant coefficients to be determined, and ρ is

$$\rho = \rho(r) = \frac{2Zr}{a_0 n}$$

where a_0 is the *Bohr radius* (circular orbit of the electron corresponding to $n = 1$):

$$a_0 = \frac{h^2}{4\pi^2 \tilde{\mu} e^2} = \text{constant} \tag{3.11}$$

where $\tilde{\mu}$ is the mass of the proton. On the right-hand side of Eq. (3.11) everything is constant. For H atom, $a_0 = 0.53$ Å.

The quantity $b_0 + b_1\rho + b_2\rho^2 + \cdots$ is the *Laguerre polynomial* where the recursion relation between the coefficients is

$$\frac{b_{i+1}}{b_i} = \frac{i + l - n + 1}{(i+1)(i+2l+2)}.$$

So, the radial solutions for the hydrogen atoms can be expressed in terms of the Laguerre polynomials $L(\rho)$ as follows:

$$(f_1(r))_{nl} = -\left\{ \left(\frac{\rho}{r}\right)^3 \frac{(n-l-1)!}{2n[(n+1)!]^3} \right\}^{1/2} e^{-\frac{1}{2}\rho} \rho^l L_{n+l}^{2l+1}(\rho).$$

Notice that for $n = 1$, l can be 0 only; for $n = 2$, l can be $l = 0$ or 1, and so on. (We will not discuss selection rules for interstate electronic transitions in atoms here. The selection rules in Section 3.9 are for molecules.)

The first few radial wavefunctions are the following:

n	l	$f_1(r)$ or $(f_1)_{nl}$
1	0	$2\left(\dfrac{Z}{a_0}\right)^{3/2} e^{-\rho/2} = (f_1)_{10}$
2	0	$\dfrac{1}{\sqrt{2}}(f_1)_{10}\left(1 - \dfrac{\rho}{2}\right) = (f_1)_{20}$
	1	$\dfrac{1}{\sqrt{6}}(f_1)_{10}\rho = (f_1)_{21}$
$\vdots$	$\vdots$	$\vdots$

where we notice that $b_0 = 1$ and $b_1 = -\rho/2$ in $(f_1)_{20}$. For $n = 2$, the wavefunctions are different for $l = 0$ and 1. Since $m_l = \pm 1$, there are four wavefunctions for $n = 2$: We have ψ_{200}, ψ_{210}, ψ_{211}, and $\psi_{21,-1}$. This is the *degeneracy* of the wavefunction: There are four different wavefunctions corresponding to the same energy (shell n); but the degeneracy can be lifted and the same energy can be split into levels (fine structure) under an externally applied electric or magnetic field. The energy will be split into four different energies.

In general, *if the principal quantum number is n, there are n^2 wavefunctions ψ (for every n), labeled as ψ_{nlm_l} (or $\psi_{n,l,m}$)*. In Dirac notation, the state wavefunction is the ket $|n\ l\ m_l\rangle$ where the quantum numbers serve as labels of the state of the electron. So, ψ_{100} is shown as $|1\ 0\ 0\rangle$.

The solutions for the hydrogen atom have the following properties:

(1) The electronic energy E is quantized as $E \propto 1/n^2$.

(2) The wavefunctions are degenerate, but the degeneracy can be removed by the application of an external field that splits the energy levels.

(3) The orbital angular momentum **L** is also quantized: It can take only certain definite values *and* orientations in space.

(4) The state of the bound electron(s) can be *specified* (labeled) by the quantum numbers n, l, and m_l, if we ignore electronic spin. If we include spin, there is an additional quantum number, the spin magnetic quantum number m_s, which can be only $m_s = \pm 1/2$ for the electron. The quantum numbers have definite (well-defined) values — without any measuring error involved.

(5) The function $f_2(\theta)$, *i.e.*, the θ-dependent part of ψ (which varies with θ) gives the values of the *interatomic bonds*.

(6) The bound atomic electrons are called s, p, d, f, g, h-electrons in spectroscopy, if their l is respectively $l = 0, 1, 2, 3, 4, 5$. For a given n, l can take integer values from 0 up to $n - 1$, one number less than n.

(7) If ψ_1 and ψ_2 are two degenerate wavefunctions (with the same energy E), then $\mathcal{H}\psi_1 = E\psi_1$ and $\mathcal{H}\psi_2 = E\psi_2$, but also

$$\mathcal{H}(\psi_1 + \psi_2) = E(\psi_1 + \psi_2),$$

and in general, if c_1 and c_2 are two constants,

$$\mathcal{H}(c_1\psi_1 + c_2\psi_2) = E(c_1\psi_1 + c_2\psi_2) = c_1 E\psi_1 + c_2 E\psi_2,$$

i.e., if ψ_1 and ψ_2 are two solutions to the Schrödinger equation, their linear combination $c_1\psi_1 + c_2\psi_2$ is also a solution.

(8) If ψ_1 and ψ_2 belong to different energies, they are orthogonal to each other:

$$\int_{\text{All space}} \psi_1\psi_2^* \, d\mathrm{V} = 0. \tag{3.12}$$

But if they belong to the same energy, they may not satisfy the mutual orthogonality condition (Eq. (3.12)). (Why do we choose orthogonal functions?)

The quantum number n is a measure of the atomic radius (the distance of the orbiting electrons from the nucleus). This average distance is

$$\bar{r} = \frac{n^2 a_0}{Z}\left[1 + \frac{1}{2}\left(1 - \frac{l(l+1)}{2}\right)\right], \tag{3.13}$$

a result whose derivation is left to the reader as an exercise. Notice that in Eq. (3.13) the importance of the quantum number l is not as strong as that of n. So, we can write

$$\bar{r}_n \simeq \frac{n^2 a_0}{Z} \propto n^2.$$

Since a theory must predict estimations for experimental results to some extent, one may ask if we can predict energy values. For which quantities can we find average values and for which ones definite values? If q is a (classical) physical quantity with a corresponding quantum mechanical operator Q, and if $Q\psi = a\psi$ (where a is a constant), then a is a definite value. If, on the other and, $Q\psi \neq a\psi$, then we must take average values for Q:

$$\langle Q \rangle = \frac{\displaystyle\int \psi^* Q\psi \, d\mathrm{V}}{\displaystyle\int \psi^*\psi \, d\mathrm{V}} \tag{3.14}$$

where the average of many experiments gives the value we find. In Eq. (3.14) the denominator gives the normalization. If the integration is over all space ($-\infty$ to $+\infty$), the denominator gives 1 (sum of probabilities); but if we integrate over a certain limited region, the result is < 1.

$Q\psi = a\psi$ is an operational equation where Q acts on a wavefunction (solution) ψ which itself has no physical meaning, but $|\psi|^2$ does: It is the probability of finding something somewhere (and if the particle is charged, it gives the charge density or distribution). So, ψ gives *information* about the *state* of the system and helps us find the system's energy.

So far in this section we have been discussing the case of the hydrogen atom as a two-particle system.[15] We applied quantum mechanics to it and saw that the solution to the Schrödinger equation is the wavefunction which is written as a product:

$$\psi = f_1(r)f_2(\theta)f_3(\varphi),$$

i.e.,

$$\psi = (\text{Radial part}) \times (\text{Angular part})$$

where further,

$$\text{Angular part}(\theta, \varphi) = (\theta\text{-dependent}) \times (\varphi\text{-dependent}). \tag{16}$$

For the hydrogen atom, $f_3(\varphi) = e^{im\varphi}$ is the azimuthal part (with m being the azimuthal or magnetic quantum number). For $f_2(\theta)$ we obtain a more complicated result, with l and n being the appropriate quantum numbers for the hydrogen atom (where l is the orbital angular momentum quantum number and n is the principal quantum number).[17] Finally, $f_1(r)$ gives $E \propto 1/n^2$ for the energy of the bound electron where E_n are discrete and definite values.

For the hydrogen atom, $V(r)$ is the radial Coulombic potential.

---

[15]The discussion of the hydrogen atom is both atomic physics and quantum mechanics.

[16]The angular part in quantum mechanics problems is usually neat and elegant (easy and symmetric), but the radial part $f_1(r)$ can be nasty (a peculiar and complicated algebraic function that shows the r-dependence which, depending on the problem, could be whimsical.

[17]In the H atom **L** is due to the revolution of the electron in an orbit, with its quantum number l where $|\mathbf{L}| = L = \sqrt{l(l+1)}\hbar$. The left-hand side is the *length* of the *vector* operator **L** which is due to the orbital motion, whereas l on the right-hand side is the *quantum number* associated with the angular momentum **L**. If there is only one electron, then we take the vector sum $\mathbf{l} + \mathbf{s} = \mathbf{j} = $ *total electronic angular momentum*, where **s** is the spin angular momentum of the electron. If there are many electrons in the atom, then the individual vectors **l** and **s** of each are added (they couple among themselves) to give $\mathbf{L} = \mathbf{l_1} + \mathbf{l_2} + \cdots$ and $\mathbf{S} = \mathbf{s_1} + \mathbf{s_2} + \cdots$, and then $\mathbf{J} = \mathbf{L} + \mathbf{S}$ (in atoms). In molecules the total molecular angular momentum (ignoring nuclei) is **J**. If we include nuclear spins, it is **N**.

Now consider the second simple case, that of a *diatomic molecule*. Again, the solution to the Schrödinger equation is written as

$$\psi = f_1(r)f_2(\theta)f_3(\varphi),$$

the product of three factors (functions), each depending on one variable (coordinate) only. The solution $f_1(r)$ is obtained when we know the internal molecular potential $V(r)$.

The equation is

$$\frac{1}{f_1}\frac{\partial}{\partial r}\left(r^2\frac{\partial f_1}{\partial r}\right) + \frac{8\pi^2\tilde{\mu}r^2}{h^2}[E - V(r)] = l(l+1) = \text{constant}, \qquad (3.15)$$

or

$$\frac{1}{r^2}\frac{\partial}{\partial r}\left(r^2\frac{\partial f_1}{\partial r}\right) + \frac{2\tilde{\mu}}{\hbar^2}[E - V(r)]f_1 = l(l+1)\frac{f_1}{r_2},$$

with the proper $V(r)$. Here $l = $ integer (constant), but in the case of a molecule (since **L** is for an atom, and is meaningless for a molecule) l goes over to the molecular quantum number K, or alternatively, J. It would be better to use K, but we can use J, provided that we do not confuse this J with the total electronic angular momentum quantum number that corresponds to $\mathbf{J} = \mathbf{L} + \mathbf{S}$ in atoms (when we include electronic spin). In molecules, J is the *rotational* quantum number of the *molecule*. And we reserve K for something else (see Section 3.8). So, Eq. (3.15) becomes

$$\frac{1}{f_1(r)}\frac{\partial}{\partial r}\left(r^2\frac{\partial f_1(r)}{\partial r}\right) + \frac{8\pi^2\tilde{\mu}r^2}{h^2}[E - V(r)] - J(J+1) = 0 \qquad (3.16)$$

where if we set $f_1(r) = R(r)/r$ and keep in mind that $\hbar = h/2\pi$ and that

$$\frac{\partial}{\partial r}\left(r^2\frac{\partial}{\partial r}\right) = \frac{\partial r^2}{\partial r}\frac{\partial}{\partial r} + r^2\frac{\partial^2}{\partial r^2} = 2r\frac{\partial}{\partial r} + r^2\frac{\partial^2}{\partial r^2},$$

we obtain the radial equation

$$\frac{1}{R(r)}\frac{\partial^2 R(r)}{\partial r^2} + \frac{2\tilde{\mu}r^2}{\hbar^2}[E - V(r)] - J(J+1) = 0,$$

$$\frac{\partial^2 R(r)}{\partial r^2} + \frac{2\tilde{\mu}R(r)}{\hbar^2}[E - V(r)] - J(J+1)\frac{R(r)}{r^2} = 0 \qquad (3.17)$$

where $R(r)$ is what we call $f(r)$ in Section 9.3 ahead. Getting from Eq. (3.16) to Eq. (3.17) is not that obvious, and the reader should do the intermediate algebra explicitly to satisfy himself about the resulting Eq. (3.17). We actually make a variable change $\rho = \alpha r$ where

$$\alpha \equiv \sqrt{\frac{2\tilde{\mu}(V - E)}{\hbar^2}}$$

to reach

$$\frac{d^2 f_1}{d\rho^2} + \frac{2}{\rho} \frac{df_1}{d\rho} + \left[1 - \frac{J(J + 1)}{\rho^2} \right] f_1 = 0$$

because

$$\vec{\nabla}^2 f_1(r) = \frac{d^2 f_1}{dr^2} + \frac{2}{r} \frac{df_1}{dr} = \frac{1}{r} \frac{d^2}{dr^2} (r f_1).$$

Eisberg (p. 301) and Richtmyer (p. 375) (both listed in the References) approach the mathematical operations slightly differently, but the substance is the same. See also Schiff, p. 84.

The analysis from here onwards about the molecule and how to find its energies will not be presented here, because it will be dealt with in detail in its proper place, that is, in Chapter 8 and Section 9.3 ahead.[18]

The total energy of the molecule is

$$E = E_{\text{el}} + E_{\text{vibr}} + E_{\text{rot}}$$

where E_{el} is the binding energy of the electrons to the atoms, E_{vibr} is the atomic vibrational energy in the molecule, and E_{rot} is the molecular rotation energy. We take the molecule as a system consisting of two rigid masses, ignoring internal atomic structure. Thus the results we obtain are approximate and rough; they roughly give an idea about the energy states of the molecule (Fig. 3.9). The molecular state may be found at any of the levels (quantized steps) shown in Fig. 3.9. The state includes vibrations and rotations of the molecule.

The energy that corresponds to the binding of the electron to the atom (or molecule) is $E_n \propto 1/n^2$, so that the energy difference is

$$\Delta E_n \propto \left(\frac{1}{n_1^2} - \frac{1}{n_2^2} \right)$$

in an electronic transition between states (labeled) n_1 and n_2.

---

[18]However, it is recommended to the reader to read Park's Chapter 16 first, for a better understanding of the material to follow. The book is listed in the References, and its Chapter 16 includes topics like interatomic forces, the hydrogen molecule-ion, mechanics of a molecule, and forces in molecules.

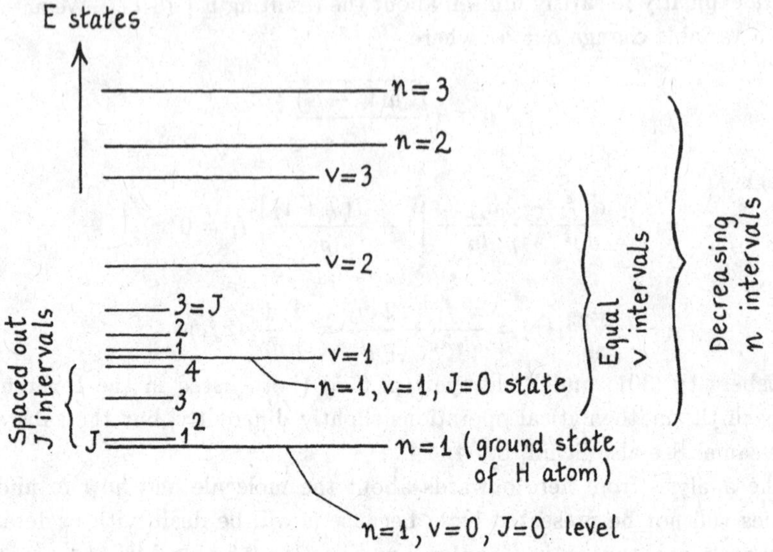

Fig. 3.9.

The energy difference between two closest vibrational levels is

$$\Delta E_{\text{vibr}} = \Delta E_{\text{v}} = h\nu_0 \left[\left(v_1 + \frac{1}{2} \right) - \left(v_2 + \frac{1}{2} \right) \right]$$

$$= h\nu_0 [v_1 - v_2]$$

where ν_0 is the natural vibration frequency. Here v = 0, 1, 2, ... (integer).

The rotational energy interval is

$$\Delta E_{\text{rot}} \propto 2BJ$$

where B is the *rotation constant*. As J increases, the interval ΔE_{rot} increases (for higher J values the rotational energy intervals get spaced out), as shown in Fig. 3.9.

Notice that $\Delta E_{\text{vibr}} \gg \Delta E_{\text{rot}}$. In Fig. 3.8, $\Delta E = \Delta E_{\text{el}}$: If the electron takes enough energy, it gets to the upper level (absorption); if it falls back to its original lower-energy level, it emits energy as a photon of frequency ν. The difference is the quantum $\Delta E_{\text{el}} = h\nu$.

Due to the Boltzmann statistics, if ΔE is too small, the transition probability from a state to the next is strong. So, the transition rule $\Delta J = \pm 1$ may break down and be violated (and a transition may be possible, say, from $J = 4$ of v = 0 to $J = 0$ of v = 1) if the energy interval ΔE is too small (Fig. 3.10).

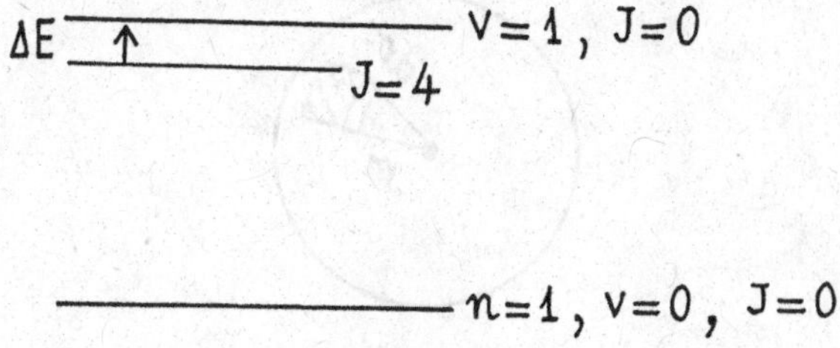

$$\Delta E \overline{\qquad \uparrow \qquad} \quad V = 1 \;,\; J = 0$$
$$\overline{\qquad\qquad} J = 4$$

$$\overline{\qquad\qquad\qquad} n = 1 \;,\; V = 0 \;,\; J = 0$$

Fig. 3.10.

The orders of magnitude are as follows:

$$E_{el} \sim 2 \text{ to } 10 \text{ eV}$$

$$E_{vibr} \sim 0.2 \text{ to } 2 \text{ eV}$$

$$E_{rot} \sim 10^{-5} \text{ to } 10^{-3} \text{ eV}$$

where we should rather write ΔE instead of E, because we never measure the energies themselves. We always measure the intervals (differences) ΔE. Also, we cannot measure the lowest $Ev = (1/2)h\nu_0$ directly; this is a limit value.

3.6. Dependence of the Wavefunctions on r and θ

The product $\psi^*\psi$ is a probability (or a probability wave) for a microscopic object (electron, subatomic particle, atom, molecule) to be somewhere. The wavefunctions in the molecular case exhibit a change with r, θ and φ. In molecules, the probability for an electron to be found in certain directions around the nucleus is related to $f_1(r)$ and $f_2(\theta)$. If we study these two functions we can find the probability for an electron to be located at a particular distance from the nucleus, in a particular direction, and at a particular angle about the nucleus.

Consider a sphere of radius r, with a surface area $4\pi r^2$. Consider a spherical layer (shell) of thickness dr at r (Fig. 3.11). What is the probability for an electron to be found within this layer? That is, we want the probability for an orbit depending only on r. For example, for $n = 2$ and $l = 0$ we obtain two maxima (two possible orbits where the electron probability peaks), that is, there are two possible radii for electronic orbits. These are $2s$-electrons. For electrons with $n = 2$ and $l = 0$ the answer is sketched in Fig. 3.12. The vertical axis is that of the function $r^2 f_1^2$; it is customary to plot $r^2 f_1^2$ (the probability of the electronic presence) instead of f_1. For $n = 2$, $l = 1$ we obtain the plot

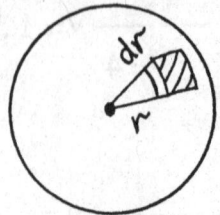

Fig. 3.11.

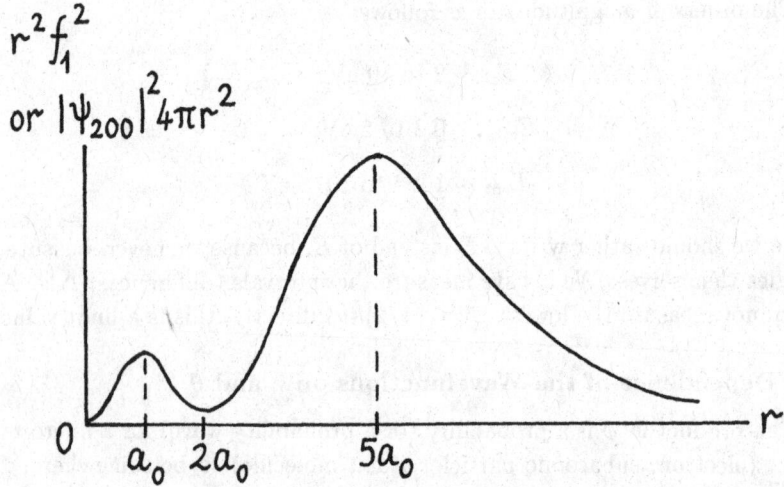

Fig. 3.12.

shown in Fig. 3.13 which is for 2p-electrons. For $n = 1$, $l = 0$ we have the curve in Fig. 3.14 (which gives the Bohr radius).

Notice that at certain distances r the probability to find an electron becomes maximum, depending on the quantum numbers. Also notice that $|\psi|^2$ is the radial probability *density* for the position of the electron. Since the volume of the layer is $4\pi r^2 dr$, the *total probability* to find the electron inside the shell is $|\psi|^2 4\pi r^2 dr$. The area under all curves is the same (if the function is integrated over all r values), since the curves are probability distributions. That is, the same amount of a charge cloud distributes and redistributes itself in each case, according to the plotted intensity versus r.

In all of the above cases there is no φ-dependence ($m = 0 \Rightarrow$ azimuthal symmetry); the distribution looks the same for all φ. In ψ_{100} and ψ_{200} there

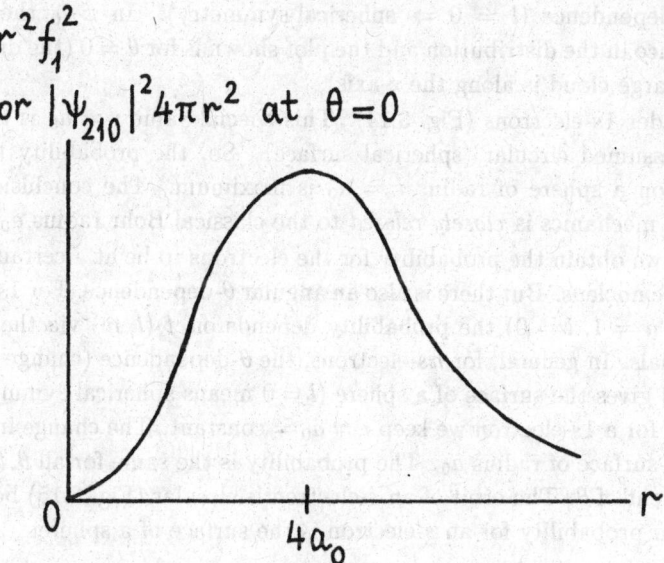

$$r^2 f_1^2$$

$$\text{or } |\psi_{210}|^2 4\pi r^2 \text{ at } \theta = 0$$

Fig. 3.13.

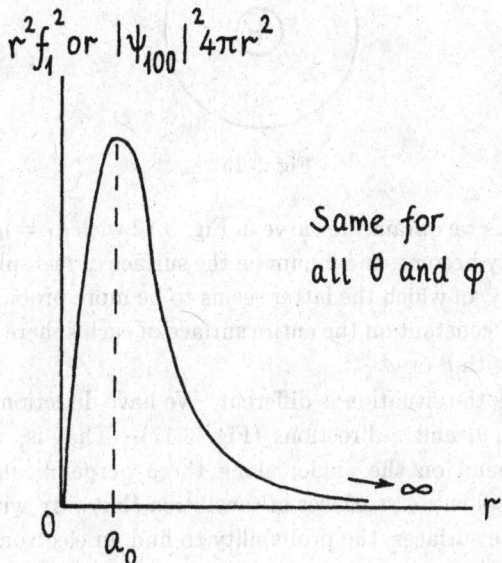

$$r^2 f_1^2 \text{ or } |\psi_{100}|^2 4\pi r^2$$

Same for all θ and φ

Fig. 3.14.

is no θ-dependence ($l = 0 \Rightarrow$ spherical symmetry). In ψ_{210} there is a θ-dependence in the distribution and the plot shown is for $\theta = 0$ (the distribution of the charge cloud is along the z-axis).

Consider $1s$-electrons (Fig. 3.14). This special value $r = a_0$ is the atomic radius, assumed circular (spherical surface). So, the probability to find an electron on a sphere of radius $r = a_0$ is maximum. The conclusion is that quantum mechanics is *closely related* to the classical Bohr radius a_0.

Thus we obtain the probability for the electrons to be at a certain distance r from the nucleus. But there is also an angular θ-dependence. For $1s$-electrons (state of $n = 1$, $l = 0$) the probability depends on $f_2(l, m)$ via the Legendre polynomials. In general, for ns-electrons, the θ-dependence (change as a function of θ) gives the surface of a sphere ($l = 0$ means spherical symmetry). For example, for a $1s$-electron we keep $r = a_0 = $ constant. The change in θ gives a spherical surface of radius a_0. The probability is the same for all θ, and hence independent of θ. The orbit of an s-electron is circular (Fig. 3.15) because the maximum probability for an s-electron is the surface of a sphere.

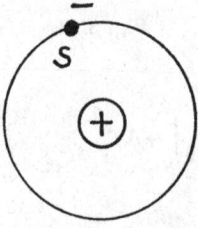

Fig. 3.15.

For $2s$-electrons we obtain the curve in Fig. 3.12 with $a_1 = a_0$ and $a_2 = 5a_0$, *i.e.*, the probability becomes maximum on the surface of *two* spheres, of radii a_1 and a_2 respectively, of which the latter seems to be more probable or preferred. The probability is constant on the entire surface of each sphere (Fig. 3.16), *i.e.*, it does not vary with θ or φ.

For p-electrons the situation is different. We have directional *lobes* of maxima along the x, y, and z-directions (Fig. 3.17). That is, we obtain three changes that depend on the angle, along three perpendicular directions in space (three ellipsoids, *i.e.*, surfaces of revolution that vary with θ in the form of $\cos\theta$). On these surfaces, the probability to find an electron is maximum.[19]

---

[19]In quantum mechanics we never say that the electron is definitely and exactly there. We speak of probabilities. We say that it is most likely to find the electron there.

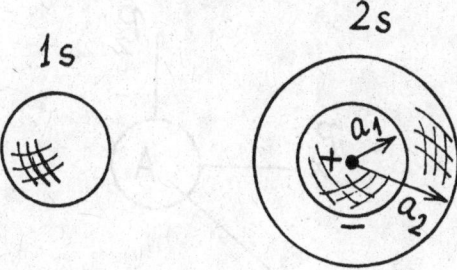

Fig. 3.16.

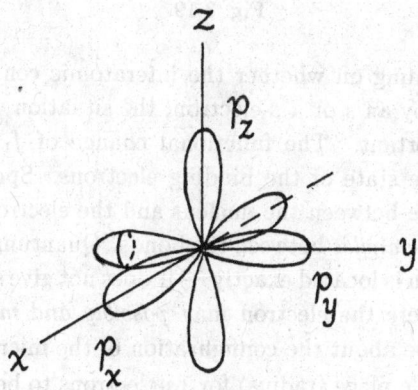

Fig. 3.17.

Consider two atoms (Fig. 3.18) bound by an *s*-electron. The angle of the bond can take any value. But if the electron is a *p*-electron, the two atoms form only a triple bond along three directions perpendicular to each other, and the angles between the bonds are approximately 90° each (Fig. 3.19), So, in this case the bond cannot be at any angle.

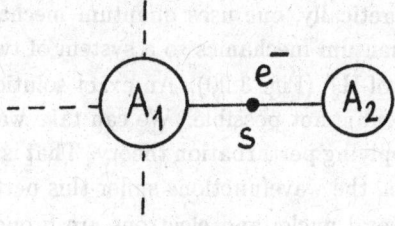

Fig. 3.18.

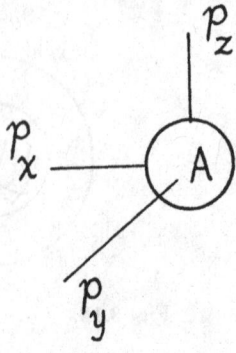

Fig. 3.19.

Therefore, depending on whether the interatomic connection (bond) in a molecule is effected by an s or a p-electron, the situation differs. In molecular bonds $f_2(\theta)$ is important. The functional change of $f_1(r)$ and $f_2(\theta)$ gives us an idea about the state of the binding electrons. Specifically, $f_1(r)$ tells us about the *distance* between the nucleus and the electron, while $f_2(\theta)$ gives information about the *angles* between the bonds. Quantum mechanics does not say where the electron is located exactly — it does not give such an information. But it estimates where the electron may *possibly and most likely* be found. Thus we get a picture about the configuration of the microcosmos.

The most probable place (radius) for $1s$-electrons to be found at, is a_0, the Bohr radius. For $2s$-electrons we have two peaks (radii), shown in Fig. 3.12. For s-electrons (*i.e.*, if the binding is mediated by $l = 0$ electrons), the bonds that are formed by these valence (outermost-orbit) electrons can be in any direction in space (spherical symmetry).

3.7. Angular Momentum of a Molecule

It is a complicated business to define the angular momentum of a molecule experimentally. Theoretically, one uses quantum mechanics, but to simplify things, one applies quantum mechanics to a system of two particles only.

Consider the case of H_2^+ (Fig. 3.20). An exact solution of the Schrödinger equation for this system is not possible. We can take wavefunctions only and superimpose them, applying perturbation theory. That is, we say the following: Under such conditions, the wavefunctions suffer this perturbation.

In a molecule, several nuclei and electrons are brought together. So, the procedure is complicated and quite difficult. However, a *vector sum* can be

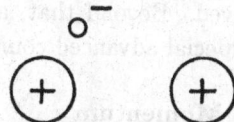

Fig. 3.20.

Fig. 3.20.

taken: One defines and expresses the total angular momentum **J** for a molecule (ignoring nuclear spin); its corresponding magnetic quantum number is M_J. Thus the radial equation (Eq. (3.10)) for a diatomic molecule can be solved.

In the case of an atom (a many-electron atom), the total electronic angular momentum is $\mathbf{J} = \mathbf{L} + \mathbf{S}$ (the combination of the orbital angular momentum **L** of the electrons and the electronic spin **S**), and the total atomic angular momentum is $\mathbf{F} = \mathbf{L} + \mathbf{S} + \mathbf{I}$ where we include the nuclear spin **I**, too, and we so define **F**. The effect of **I** (due to the — presence of — nucleus) can be managed by a perturbation approach.[20] The *total* angular momentum of a *molecule* is denoted by **N** and includes everything (nuclear spin and molecular rotations as well). So, **N** for a molecule is whatever **F** is for an atom — this is the correspondence.

In molecules, the total Hamiltonian of a molecule is

$$\mathcal{H} = (\text{Kinetic energy}) + V + V_{\text{cryst}} + \mathcal{H}_{L,S} + \mathcal{H}_{I,I}$$

where V is the potential energy of the molecule, V_{cryst} is the crystal potential (added because it is appreciably large),[21] $\mathcal{H}_{L,S}$ is the $\mathbf{L} \cdot \mathbf{S}$ (spin-orbit) interaction energy (small), and $\mathcal{H}_{I,I}$ is the part (energy term or contribution) that comes from the $\mathbf{I} \cdot \mathbf{I}$ (spin-spin) interaction (between the spins of the nuclei) (this is very small and can be treated as a perturbation).

Quantum mechanics can be applied to at most two particles. Even in the case of two hydrogen atoms bound by an electron, the Schrödinger equation cannot be solved exactly; one can consider only the amount of perturbation on ψ in such a case. By ignoring the nuclear spin, the radial equation for a

---

[20]The effect of **I** is usually very small and can be ignored, so that in atoms one is content with **J** only, instead of **F**. The effect of nucleus on the electronic energies is not considered, unless there is a reason for its inclusion.

[21]V_{cryst} is due to neighboring ions (nuclei or atoms) in the crystal (crystalline structure where we consider an ion site). These ions create electrostatic fields (due to their charge) which may be inhomogeneous. So, the electron of the considered ion is influenced — it feels these fields. Consequently, its ψ (and hence the charge distribution) changes. This effect must be included in $\mathcal{H}_{\text{total}}$; solid state physics deals with this in detail.

diatomic molecule can be solved. Beyond that, approximation methods are used, but these are topics of special advanced courses in quantum mechanics.

3.8. Notation for Angular Momentum

In this section we will juxtapose — for comparison and distinction — the notation for *atomic* and *molecular* angular momenta.

For a single atomic electron, $j = l + s$. For a many-electron atom, $J = L + S$ where

$$L = \sum_i l_i \quad \text{and} \quad S = \sum_i s_i \,.$$

The nuclear spin is I, and the *total atomic* angular momentum (including nuclear spin) is $F = J + I = L + S + I$.

If we ignore the nuclear spin, the total angular momentum for an atom *and* for a molecule is J, the total electronic angular momentum. In *molecules*, there are two cases for the projection of J:

(1) The projection of J in space, along the z-axis: It is denoted by M_J (like in atoms).

(2) The projection of J on the molecular axis: It is denoted by Ω.

The orbital angular momentum L (of the electrons) in atoms has a projection M_L on a fixed space axis (z-axis) both in atoms and molecules. Its projection on the *molecular* axis is Λ.

The total electronic spin is S both in atoms and molecules. Its projection on the z-axis is M_S in both cases. In molecules, its projection on the molecular axis is Σ.

Further, in molecules we have the vector O, representing the angular momentum of the nuclear motion (rotation of the molecule). So the grand total angular momentum that includes the rotation of the molecule is N. (This is the vector sum of all the angular momenta, including the rotation of the very molecule itself.) Its projection on the molecular axis is K.

In molecules, the projection of J, S, and I on the z-axis is M_J, M_S, and M_I respectively (the same as in atoms). N includes molecular rotation.

3.9. Selection Rules

The difference between any two energy levels corresponds to a quantum. But not all transitions between energy levels are possible, because there are some *selection rules* that forbid certain interlevel transitions, and hence we do

Summary Table

Atoms

Angular momentum	Symbol	Its projection on z-axis
Orbital (electronic)	**L**	M_L
Spin (electronic)	**S**	M_S
Total electronic (ignoring nucleus)	$\mathbf{J} = \mathbf{L} + \mathbf{S}$	M_J
Total atomic (with nucleus)	$\mathbf{F} = \mathbf{J} + \mathbf{I}$	M_F
Nuclear spin	**I**	M_I

Molecules

Angular momentum	Symbol	Its projection on molecular-axis
Orbital (electronic)	$\vec{\Lambda}$	Λ or M_Λ
Total electronic (ignoring nucleus)	**J**	Ω
Spin (electronic)	**S**	Σ
Nuclear rotation (orbital)	**O**	—
Total molecular	**N**	K
Nuclear spin	**I**	—

not obtain all the possible spectral lines in a spectrum. The selection rules determine which transitions are allowed by mother nature.

In classical electrodynamics, if an electric dipole oscillates, it emits electromagnetic waves. If the frequency of its periodic oscillation is f, that would be also the frequency of the emitted electromagnetic waves. A rotating dipole also emits electromagnetic energy in waves. So, if a molecule has no permanent dipole, it does not give a rotation spectrum. But if its dipole is $\mathfrak{p} \neq 0$, it emits a pure rotational spectrum.

Nevertheless molecules with a zero permanent dipole ($\mathfrak{p} = 0$) can exhibit microwave absorption. For example, the CO_2 molecule is linear. But if it vibrates as in Fig. 3.21 (lower part), it acquires a $\mathfrak{p} \neq 0$ for a moment (by virtue of the instantaneous shape it takes). So these vibrations cause emission (or absorption) of electromagnetic waves.

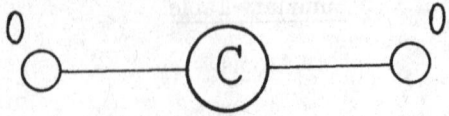

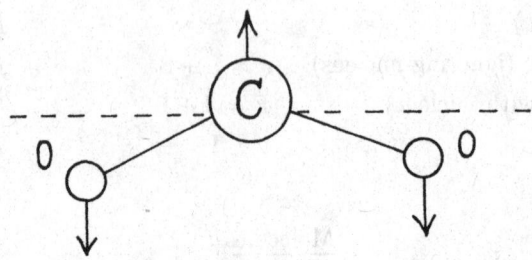

Fig. 3.21.

A molecule consisting of N atoms has $3N - 6$ modes (different shapes or ways) of normal molecular vibrations, out of which those with $\mathfrak{p} = 0$ do not cause emission of electromagnetic waves.

The selection rules for electronic transitions cannot be derived classically. But all the selection rules can be derived quantum mechanically, namely by perturbation theory, as follows:

The energy of a dipole (moment) $\vec{\mathfrak{p}}$ in a field $\vec{\mathcal{E}}$ is

$$E = \vec{\mathfrak{p}} \cdot \vec{\mathcal{E}},$$

represented by a Hamiltonian (operator) $\mathcal{H}'$ which depends on the field and the position of atoms. This is the perturbation term of the total Hamiltonian $\mathcal{H}$ of the system, that is,

$$\mathcal{H} = \mathcal{H}_0 + \mathcal{H}'$$

where $\mathcal{H}_0$ is the unperturbed part, and $\mathcal{H}'$ is the perturbation to $\mathcal{H}$.

Since $\vec{\mathfrak{p}}$ depends only on the intramolecular local coordinates, we have

$$\mathcal{H}' = \vec{\mathfrak{p}}(x, y, z) \cdot \vec{\mathcal{E}}(X, Y, Z) = \mathfrak{p}_x \mathcal{E}_X + \mathfrak{p}_y \mathcal{E}_Y + \mathfrak{p}_z \mathcal{E}_Z.$$

The interlevel transitions give

$$(\mathfrak{p}_x)_{n'n} = \int \psi_{n'}^*(x, y, z)\mathfrak{p}_x(x, y, z)\psi_n(x, y, z)dx\,dy\,dz$$

$$(\mathfrak{p}_y)_{n'n} = \int \psi_{n'}^*(x, y, z)\mathfrak{p}_y(x, y, z)\psi_n(x, y, z)dx\,dy\,dz$$

$$(\mathfrak{p}_z)_{n'n} = \int \psi_{n'}^*(x, y, z)\mathfrak{p}_z(x, y, z)\psi_n(x, y, z)\,dx\,dy\,dz \ .$$

Whichever of these expressions is nonzero, gives a possible transition. The selection rules can be thus obtained. For example, for hydrogen, $\mathfrak{p}_x = ex$. So, the selection rules for the hydrogen atom are the following:

$$\Delta n \text{ can take all integer values}$$
$$\Delta l \text{ can change as } \Delta l = \pm 1$$
$$\Delta m_l \text{ can change as } \Delta m_l = 0, \pm 1$$
$$\Delta s \text{ must be } \Delta s = 0 \ .$$

For a diatomic molecule, the selection rules are $\Delta n = \pm 1$, $\Delta J = \pm 1$, and $\Delta m_J = 0, \pm 1$.

Quantum mechanics poses other limitations as well, besides these selection rules. The selection rules affect also macroscopic properties (*e.g.*, molecular spectra).

3.10. Absorption and Emission of Electromagnetic Radiation

Let us consider a two-level system (*i.e.*, an atom or a molecule where only two energy levels are of interest) with energies E_n and $E_{n'}$ where $E_{n'}$ is the upper level: $E_{n'} > E_n$.

If we send a quantum $\Delta E = E_{n'} - E_n = h\nu$ to the system, the system gets excited from state n to state n' (if it is initially in state n), and then two processes can occur: Spontaneous emission or/and induced emission (Fig. 3.22). But let us take things from the beginning.

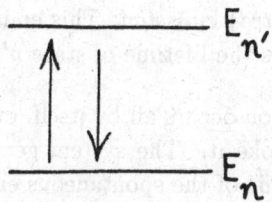

Fig. 3.22.

Let n be the ground state (although it does not have to be so; it can be any other excited state but, anyway, *lower than n'*). If we send a quantum of radiation (energy), three processes may occur in total:

(1) *Absorption*: The system is (actually, its electron is) in state n and gets to n' (Fig. 3.23).

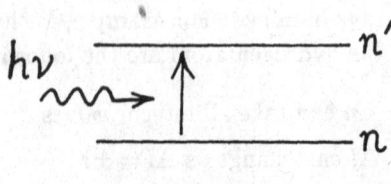

Fig. 3.23.

(2) *Spontaneous emission* (after the absorption): The (excited) state n' has a certain lifetime τ; the system does not stay in n' forever. After a while it decays back to n all by itself (spontaneously). The system naturally wants to return from n' to n, to get back to its initial level of lower energy (and hence of more stability). As it does so, it gives off the absorbed quantum as a photon (Fig. 3.24).

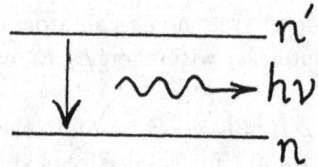

Fig. 3.24.

(3) *Induced* (or *stimulated*) *emission*: This emission is caused by external means and decreases the lifetime of state n'.

The spontaneous emission occurs all by itself, even if there are no external means to induce and provoke it. The system gets from n' to n (the initial state). The rate (per second) of the spontaneous emission is

$$w_{\mathrm{spont}}(n' \to n) = A_{\mathrm{spont}} N_{n'}$$

where A_{spont} is the *Einstein coefficient* for spontaneous emission, and $N_{n'}$ is the number of molecules in state n' as they leave it. Since $A_{\mathrm{spont}} = 1/\tau$, the quantity w_{spont} gives the probability (per unit time) of a spontaneous emission.

The induced emission is instigated by external means, *e.g.*, by black body radiation, and is proportional to the black body radiation intensity or density $\rho(\nu)$:

$$w_{\text{ind}}(n' \to n) = A_{\text{ind}} N_{n'} \rho(\nu).$$

But the system may also absorb black body radiation and get excited:

$$w_{\text{abs}}(n \to n') = B_{\text{abs}} N_n \rho(\nu)$$

where B_{abs} is the Einstein coefficient for absorption, and N_n is the number of molecules in the initial state n.

In equilibrium, $w_{\text{abs}}(n \to n') = w_{\text{spont}}(n' \to n) + w_{\text{ind}}(n' \to n)$.

With T being the black body temperature, the black body radiation density (at frequency ν) is

$$\rho(\nu) = \frac{8\pi h \nu^3 / c^3}{e^{h\nu/kT} - 1}.$$

The three coefficients are given by[22]

$$A_{\text{spont}} = \left(\frac{64\pi^4 \nu^3}{3hc^3} \right) \mathbf{p}_{nn'}^2$$

$$A_{\text{ind}} = \left(\frac{8\pi^3}{3h^2} \right) \mathbf{p}_{nn'}^2$$

$$B_{\text{abs}} = \left(\frac{8\pi^3}{3h^2} \frac{g_{n'}}{g_n} \right) \mathbf{p}_{nn'}^2$$

where g_n and $g_{n'}$ are the statistical weights of the corresponding states (*i.e.*, the degeneracy, say, of state n is g_n-fold).

We have

$$\frac{N_{n'}}{N_n} = \left(\frac{g_{n'}}{g_n} \right) e^{-h\nu/kT} = \left(\frac{g_{n'}}{g_n} \right) e^{(E_n - E_{n'})/kT}$$

and $\mathbf{p}_{nn'}^2 = \mathbf{p}_x^2 + \mathbf{p}_y^2 + \mathbf{p}_z^2$ where

$$\langle \mathbf{p}_x^2 \rangle = \int_{\text{All space}} \psi_{n'}^*(x, y, z) \mathbf{p}_x^2(x, y, z) \psi_n(x, y, z) dx \, dy \, dz$$

which is the expectation value $\langle n' | \mathbf{p}_x^2 | n \rangle$, or equivalently, the matrix element $(\mathbf{p}_x^2)_{nn'}$ of the operator $\mathbf{p}_x^2$. Do not confuse the matrix element (which is a number) with the operator itself.

---

[22]Various texts do not agree as to their exact expression. For example, out of the References listed in this book, compare Fermi (p. 104), Eisberg (p. 460), and Sargent (p. 22).

We further have

$$\frac{w_{\text{spont}}(n' \to n)}{w_{\text{ind}}(n' \to n)} = \frac{A_{\text{spont}} N_{n'}}{A_{\text{ind}} N_{n'} \rho(\nu)} = e^{-h\nu/kT} - 1.$$

In the microwave region, $h\nu/kT \sim 0$, and hence $w_{\text{ind}} \gg w_{\text{spont}}$. In the optical (visible) region, $h\nu/kT \sim 1$, and the two coefficients are of comparable importance.

The induced emission is coherent with the incident radiation, but the spontaneous emission is not; the spontaneous emission spreads towards all directions.

Since the rays of the induced emission are coherent with the incident rays, the two of them add up, an action that diminishes the absorption. The absorption efficacy of a spectrograph is given by

$$w_{\text{abs}} - w_{\text{ind}} = B_{\text{abs}} N_n \rho(\nu)(1 - e^{-h\nu/kT}),$$

since A_{ind} and B_{abs} are related (as it is seen from the relations written above for these coefficients). This equation is used in the measurements regarding the absorption spectral lines. It holds true for all spectral regions.

The absorption spectrographs are about 10^2 times more effective in the optical region than in the microwave region. In the latter region the induced emission diminishes the absorption too much (down to 1%).

3.11. Spectral Line Width

There is no ideally monochromatic (single and pure-wavelength) light or spectral line. Light of every color, even a narrow spectral line has a *line width* (spread), however small. Figure 3.25 shows the graph of the light intensity as a function of the wavelength λ, for a given $\lambda = \lambda_0$. Notice that λ_0 is the *nominal wavelength*, that is the wavelength at which the intensity is maximum and peaks. There is a spread (of rapidly decreasing intensity) on either side of λ_0. The interval $\Delta\lambda$ corresponding to the points where the maximum intensity is halved, is called *full width at half maximum*. It is measured across the points where I_{max} drops to $(1/2)I_{\text{max}}$.

For example, let us take the red line of Cd whose width is too narrow: $\Delta\lambda = 0.03$ Å. Since

$$\frac{\Delta\lambda}{\lambda} = \frac{\Delta\nu}{\nu},$$

for $\lambda = 6000$ Å we have $\Delta\nu/\nu = 10^{-6}$. And since $\nu \sim 10^{14}$ sec^{-1}, we obtain $\Delta\nu \sim 10^8$ sec^{-1}. This means that even one single spectral line includes a very

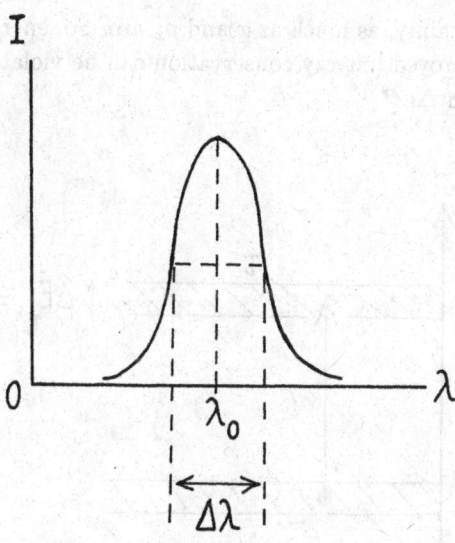

Fig. 3.25.

large amount of monochromatic and continuous waves lined up side by side, one next to the other.

3.12. Energy Width

The energy of any energy level is not exactly a sharp value; there is a spread (thickness) Γ called *energy width* (or *uncertainty*) (of the state) and is determined by the *uncertainty principle* which says that in quantum mechanics we cannot have too much knowledge of a state (especially when we narrow the time limits): We cannot know the energy and the lifetime τ of a state exactly (with the same exactitude) at the same time. Therefore, we have

$$\Gamma\tau \sim \hbar$$

where τ is the lifetime of the state in spontaneous de-excitation (Fig. 3.26).

The uncertainty (or indeterminancy) principle states that

$$\Delta E \Delta t \gtrsim \hbar$$

where the shorter the time interval of observation (as Δt shrinks to zero, $\Delta t \to 0$) the larger the uncertainty in E ($\Delta E \to \infty$). And conversely, the better we know the energy ($\Delta E \to 0$) the wider the time interval (the exact instant becomes largely indeterminate, so as Δt grows too wide). E and t are conjugate quantities in uncertainty, as much as x and p_x are. So, energy *can* be lent out or created and destroyed (energy conservation can be violated), provided that this happens within Δt.[23]

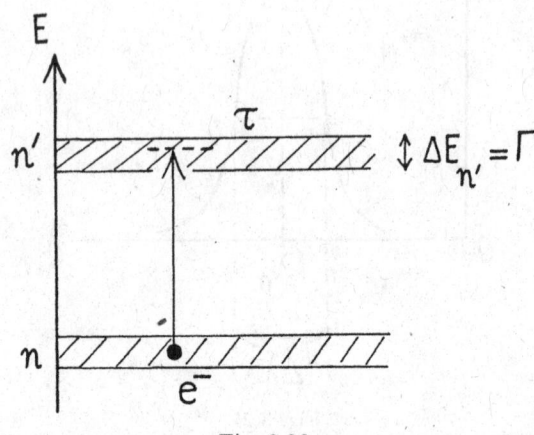

Fig. 3.26.

3.13. Electric Character of the Interatomic Forces

When two atoms approach each other, they interact. A mutual interaction starts (sets up) which is hard to analyze, even in the case of simple atoms.[24] The picture is complicated because we should take into account the individual behavior of many particles simultaneously and solve the Schrödinger equation for each — of course by an approximation method.

If the interaction energy of two atoms is less than the total energy obtained when they are free and separated, the mutual interaction produces attractive forces which in turn produce a stable AB type molecule. This interaction is represented by the simple chemical reaction

$$A + B \rightarrow AB.$$

But if the system's energy increases when the two atoms get close to each other, repulsive forces are produced, and it is impossible to form a molecule.

---

[23]An employee can embezzle an amount of money (ΔE) from the company's fund from Friday evening to Monday morning (Δt), *i.e.*, within a time interval between two audits, and return it before the shortage is noticed. Within this Δt no control is possible.

[24]One can imagine the difficulty even for *free* atoms. For example, for uranium one has to solve 92 Schrödinger equations simultaneously, one for each electron!

Interatomic forces could be *gravitational, electric,* or *magnetic.* A simple comparison of magnitudes convinces us that they are mainly of electric character. Indeed, if we compare these three, we see that the potential energy due to the gravitational forces is too weak. For example, consider two heavy atoms in a mutual gravitational attraction at $r \sim 3$ Å. Let, for example, $m_1 = m_2 \equiv m = 250 m_H$ (where m_H is the mass of the hydrogen atom — nearly the mass of a proton). We have

$$V_{grav} = -G \frac{(250 m_H)^2}{r} \sim 10^{-32} \text{ eV}$$

where G is the universal gravitational attraction constant. This amount (order of magnitude) tells us that the gravitational attraction is too weak to play an efficient role in the formation of molecules, and hence too insignificant and self-effacing to be reckoned in molecular physics.

The potential of the magnetic interaction is[25]

$$V_{magn} = -\frac{2\mu_B^2}{r^3} \sim 0.2 \text{ cal/mol} \tag{25}$$

where we took two interacting magnetic dipoles (each of a magnetic moment equal to the Bohr magneton) at 3 Å. We have $V_{magn} > V_{grav}$, but still the magnetic potential is less than the binding energy and stays dim.

Finally, let us estimate the electrostatic potential of two elementary charges at 3 Å:

$$V_{el} = -\frac{e^2}{r} \sim 5 \text{ eV}$$

where $e = 4.8 \times 10^{-10}$ statCb. This energy is of the order of the chemical binding energy. And chemical binding is effected by the outermost (valence) electrons in interacting atoms. Indeed, the valence of an atom is a measure of its willingness to interact or to contribute to a chemical interaction and a bond with another atom.

So, *the interatomic forces are mainly of Coulombic nature.* Forces, their electric origin, and bonds will be discussed in detail in the following chapters, but let us state them here synoptically:

(1) *Homopolar covalent (non-ionogenic) bonds.*
(2) *Homopolar metallic bonds.*

---

[25]The magnetic energy for crystals can also be measured experimentally.

(3) *Heteropolar ionic (ionogenic) bonds.*

(4) *Intermediate bonds.*[26]

As a molecule stores order, it does not violate the principle of the conservation of energy because the order (energy expenditure) is drawn from the
environment, not created out of nothing.[27] Energy can be taken from the
electric field around. A *bond* means *attachment to* another atom.

In all of the above cases the nature of the interaction is the same: Electrostatic. But the *manner* of the interaction is different, because the structure
of the electron cloud is different for different element groups. That is why we
have differences in the physical and chemical properties of compounds: The
chemical bondings are different.

How is the electronic structure in metallic atoms? Their outermost valence
band is partially filled. In alkali metals we have one s-electron in the valence
band, while the neighboring p-state is empty. In the *heavy* metals (at the end
of the periodic table) the energy states s, p, d, and f are filled, but some lower
states are empty. In metaloids most of the states are full and only a few ones
are empty, and the closest empty upper-energy state is too high compared to
the chemical binding energy; that is why when two atoms find themselves one
near the other it is almost impossible (energywise) to share electrons.

A few words about intermediate bonds: The ionic and covalent bonds are
extreme (limiting) cases, because in fact many intermediate type bonds are
possible and they do occur. In A_2 type homopolar molecules the atoms share
the valence electrons which revolve in an orbit that encompasses both nuclei. The result is easy to imagine: Between neighboring nuclei a symmetrical electron charge density without irregularities or asymmetries is formed
(even distribution of charge). In heteropolar bonds with atoms of different
electronegativity, the center of mass of the electron cloud shifts towards the
more electronegative atom, as if the electrons spent more time around it. And
if the difference in electronegativity is too large, there is a marked distortion (smear) of the electron cloud: The valence electron of the electropositive
atom gets completely into the electronegative atom, in which case we have the
extreme situation of the ionic bond.

---

[26]Nature is not rigid to exhibit strict categorizations. It allows for cases smoothly in between.
Categories are products of our rationalistic and categorizing mind, in an effort to study nature
orderly.

[27]Entropy (chaos) decreases locally by this increase in order, but this is possible because the
molecule is an *open* system, interacting with the environment.

3.14. Quantum Theory of Interatomic Interaction

Atomic wavefunctions mix (overlap), *i.e.*, interpenetrate each other in molecules. Consider a linear AB type diatomic molecule. Let ψ_A and ψ_B be the wavefunctions that describe the behavior of the valence electron of atoms A and B (when they are free apart). The probability for an atomic electron to be at a radial distance r from the nucleus, within a spherical shell of thickness dr (between r and $r + dr$) is

$$|\psi|^2 = \rho(r) = R^2(r)r^2 dr$$

where ρ is the (spherical) density of the electron cloud (electronic charge density), R is the radial (r-dependent) part of ψ, and $r^2 dr$ is the radial volume element. The maxima of $\rho(r)$ indicate the average distances from the nucleus where the electron might be found most of the time (average electron orbit radii). These distances correspond to the orbit radii in Bohr theory and give the energies. That is, the electron is at r_A in atom A and at r_B in atom B. So, r_A and r_B are the radii of the two atoms when free. This radius does not change much; for all atoms in the periodic table, it is up to a few Å (the typical size of an atom). For example, for hydrogen it is $r_0 = 0.5$ Å (Bohr radius), and in many-electron atoms it is ~ 1.5 Å. As r increases, $\rho(r)$ decreases fast, exponentially and asymptotically. The interatomic distance $r_{AB} \lesssim$ several atomic radii, when we have neutral non-interacting (and well-separated) atoms around.

When the atoms approach each other up to $r_{AB} \simeq r_A + r_B$ (sum of their radii), their ψ interpenetrate (gradually, upon approach), starting from the outer-orbit electrons first. Then the radii maxima overlap when $r_{AB} = r_A + r_B$ exactly. At this separation distance (with so much approach) a perturbation starts: The orbital motion of the valence electron(s) changes by the strong action of the electric fields of the neighboring atoms. *Cohesive forces* show up upon the close approach of the two atoms as their ψ mix; these are the *chemical affinity* (binding) forces.

When the atomic electrons are in the space between the two nuclei, the electric fields acting on them by the nuclei are almost equal to each other. So, the valence electrons are shared by the two nuclei. Therefore, the electronic motion has changed much compared to the motion in the free-atom case. The valence electrons move under the common influence of the fields of *both* atoms (nuclei), and this motion is now described by a new wavefunction, the *molecular* ψ_{mol}. This new function is *necessary*, because the old functions ψ_A and ψ_B are not valid any more, as they cannot describe the new situation.

By the sharing of the valence electrons, the electron densities take up new shapes.[28] Also, the system's new energy E_{mol} is *not* the sum of the two atomic energies of the two free atoms: $E_{mol} \neq E_A + E_B$.

However, the quantization of energy does not change, no matter how many atoms get together. The energy values are discrete. If $E_{mol} < E_A + E_B$ (if the energy decreases when ψ_A and ψ_B mix), an attractive force arises between the atoms. When the two atomic nuclei are at a certain separation, E_{mol} takes a minimum value E_{min}. There, at this separation, $F_{attr} = -F_{rep}$ (equal and opposite) and this implies stability. This is an equilibrium situation whose qualitative interpretation is that we have a *molecule*, a *different material* than the two separate atoms (elements). We have a *chemical compound*. Therefore, a molecule is not merely the sum of the two constituent atoms. Its physical and chemical properties differ from those of either atom (element). For example, K is an alkali and Cl_2 is a poisonous gas (taken separately), but KCl is a *new body and material* with completely different properties: A chemically inert salt.

The range of cohesive forces is ~ 1 to 3 Å (about the sum of the radii of the free atoms at most). The *chemical binding energy* is the difference

$$E_{binding} = (E_A + E_B) - E_{min} \equiv \Delta E$$

or

$$\Delta E = \text{(Energy sum of the energies of two free atoms)} - \text{(Molecular energy)}.$$

If $\Delta E > 0$, the formation of the molecule (AB) releases energy, and the reaction $A + B \rightarrow AB$ becomes possible.

In the chapters to follow we will discuss the Schrödinger equation of a molecule (system of atoms).

3.15. Remarks

Molecular physics studies molecules made of a few atoms. Its methods and activities involve the study of primarily the following:

(1) Interatomic distances, angles, and molecular parameters.
(2) Internal energies.
(3) Electric and magnetic properties.

---

[28]This smear of the electron clouds between the atoms provides the bond, that is, a merger through the fusion of the electronic clouds, widely speaking. In political science we have a *power diffusion* as we go from a monarchy to a parliamentary democracy where the power smears out from one person to a many-person legislative body.

(4) Interatomic mutual interaction forces (and bonds).
(5) Molecular spectrograms (where the spectra to be read and deciphered are as intricate as the choreographic steps of a ballet).

Quantum mechanics is necessary if we are to know the molecular spectra and the molecular energy states. It is a mathematical method whose results can be applied to experimental and practical ways, after proper and convenient transformations. Needless to say, theoretical molecular physics, too, employs the powerful mathematical formalism of quantum mechanics.

The substance of quantum mechanics in molecular physics can be condensed to the following:

(1) Assign an operator to every physical quantity, and set off to work. (This discipline operates with quantum mechanical operators — to each classical quantity there corresponds one.)
(2) Start with the consideration of a system that consists of two pointlike bodies — the oversimplified basis of the diatomic molecule.
(3) Use the quantum mechanics results for atoms as a valuable experience and arsenal in studying molecules (getting a step beyond).

When one studies quantum physics, one begins with textbook cases which are simplified and idealized. One thus gets a stepwise training little by little, moving from simple cases on to more complex ones, in stages. But by the time one acquires the necessary knowledge and formal conditioning to deal with real cases (*e.g.*, molecules, structures etc.) — the semester ends! In a typical quantum mechanics class one usually does not get to real cases, except if the class is a bit advanced and includes special topics on quantum theory. That is why education on real cases is left to be picked up from talks, research articles, scientific journals, dissertations, and conference proceedings. Today, it is better to learn physics from a recent article than from a book, because the scientific progress and the developments are so rapid, and the scientific paper production so massive, that by the time the textbook is on the market it may become obsolete in terms of latest updated data! (This is so, except of course, basic instruction material of permanent and diachronic scientific merit.)

In quantum mechanics, only a few simple cases are solved exactly: Only the hydrogen atom (actually, the two-body problem[29]) has an exact solution,

---

[29]It can be adopted for a diatomic molecule, too, if we ignore internal atomic structure and take the atoms of the molecule as two plain solid masses.

and a few other cases, too, are manageable. The rest is hard, because anything beyond two masses has to do with the many-body problem and requires approximations.[30]

Approximations can be made in quantum mechanics, and in real cases the approach is difficult. Sometimes other interactions that come into play make the overall real situation hopelessly difficult to describe exactly. Actually, what makes the Schrödinger equation hard to solve exactly is the explicit form of the potential function $V(r)$ that comes as an additive term in the differential equation, rendering it complicated.

Textbooks present contrived cases and staged problems whose solutions are usually known, for education, illustration, instruction, and training purposes. As we said above, these are ideal and sublimely simplified situations. A real-life

---

[30]Of late, however, there has been work going on with *exact* methods for multielectronic atoms or molecules. These methods (used in theoretical physics and molecular chemistry) concern gas molecules excited by LASER beams. That is, we have interaction between matter (at the molecular level) and electromagnetic energy (in the form of coherent visible light of high intensity and density). The gas is excited and de-excited as it interacts with LASER pulses (time-dependent phenomena), and by this perturbation that causes electron excitations one obtains an energy spectrum which, as printed, works also for time-independent systems. Apart from the approximation methods, the exact ones use the time-dependent Schrödinger equation to obtain and record the spectrum in the time domain (which by a Fourier transform can be translated into the energy domain, to get the energy levels). The theory finally agrees with what is observed experimentally, namely a picture (spectrum) in the time domain where at the instant $t = 0$ the molecule (or atom) appears somewhere, at a later time t' somewhere else beyond, and still at a later time t'' farther away.

problem is far harder — and sometimes it may not have a solution. Such cases constitute topics for doctoral dissertations, post-doctoral work, or scientific research that take years.[31]

---

[31] A thorough and rigorous experimental research project with all its scholastic details and severe scientific requirements is a grand program that involves a whole research team and takes about three decades. It involves a professional laboratory with sumptuous equipment, scientific methodology, data collection and analysis, and an in-depth interpretation and evaluation of the programmed experiment and its results, with the final conclusions published in journals. The work may be fragmented into partial assignments for better handling, even with theoretical ramifications and subroutines, if necessary, where everything is brought and sewn together at the end, under the supervision of the team leader, the professor or scientist who gives management to the project. (In this case his task is like that of a great encyclopaedia editor who brings together the entries of the contributing specialists.) In doctoral dissertations, however, the task is lighter, and of course, not meant to last this much! There we have a scientific research that could be completed in about three to four years' time. But still, the result must be a *new word*, an academic study with original, first-class material that opens new paths in science, and with challenging questions it puts forth, invites future scientists to further work.

A researcher scientist must be able to think beyond the data he has at hand, like a natural scientist who can describe an animal from a bone only. He/she must be a good thinker, able to guess the following and preceding links in a chain, if he is supplied only with one single link. For this purpose, he/she must (a) be a good observer, (b) have enormous knowledge, (c) be well-informed about the available sources, (d) have a scientific mind and a careful searching attitude, (e) work systematically, (f) evaluate all the data he/she has got, (g) pull in his/her creative imagination, like an investigator/detective, and (h) have faith and use both reasoning and objective intuition, neither eliminating the humanistic element, nor yielding to subjective judgement.

Chapter 4

INTERATOMIC INTERACTION

4.1. Quantum Theory of Interatomic Interaction

Ideally and for the sake of simplicity, when we say molecule it is customary to consider a system of two point masses, a diatomic molecule. But the situation in reality is by far more complex, because there are molecules with more than two atoms. Further, the nuclei have some extent and they are not point masses. However, before attacking the real complicated cases — which are hopelessly discouraging in terms of the involved difficulty — it pays off to consider ideal cases and build a sensible theory that would help us progress ahead.

The simplest molecule is the ionized hydrogen molecule: H_2^+. It is the simplest case (Fig. 4.1). We have three bodies (particles): Two protons and an electron binding them together. This is a three-body problem which has not been solved; it can be attacked only by mathematical ways and models.[1]

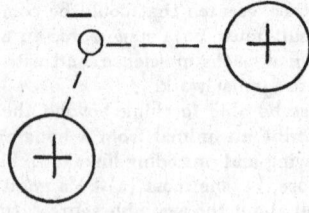

Fig. 4.1.

---

[1]The two-body problem has been fully solved.

When two atoms are far away from each other, they are two independent, individual atoms. (The mutual interaction still exists — it is nonzero because it dies out exponentially — but since the range is long, it is feeble and not appreciable.) But when they approach each other, the mutual interaction starts being marked. Upon further mutual approach (as the interatomic distance becomes smaller and smaller), repulsion shows up. In other words, complications arise upon mutual approach.

As we know from quantum mechanics, the atomic wavefunction $\psi(r)$ has a maximum at some r where the probability to find the orbiting electron is the highest, and that gives the radius of the orbit of the electron. This atomic radius of a free atom is of the order of a few Å and does not change much with Z, *i.e.*, it is almost the same for all atoms. $\psi(r)$ decreases asymptotically with r, which means that the atom has no definite boundary, but there is a probability — however slight — to find the electron even at infinity. This is what quantum mechanics says.

Figure 4.2 shows $\psi(r)$ for a 1s-electron of atom A and for a 2p-electron of another atom B next to it, where the maxima (radii) are r_A and r_B respectively. Notice that the distribution changes slower for p-electrons. The distance r is measured from the nucleus placed at the origin. The quantity $e|\psi|^2 = \rho(r)$ gives the distribution of charge density as a function of r.

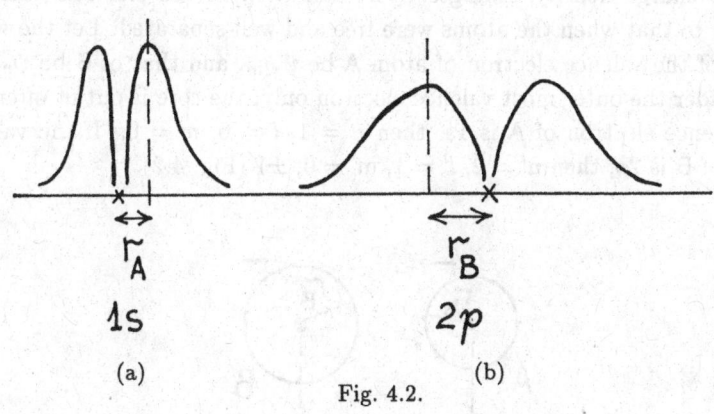

r_A

1s

(a)

r_B

2p

(b)

Fig. 4.2.

Molecules can be of A_2 type or of AB type, if diatomic. The former consists of two atoms of the same kind (identical), and its center of mass is at the

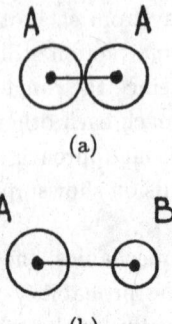

(a)

(b)

Fig. 4.3.

midpoint of the internuclear separation (Fig. 4.3(a)). The latter consists of two different atoms (Fig. 4.3(b)).

The distance where the probability of finding the electron is maximum is taken as the radius of the atom: r_A for A and r_B for B.

Consider now the atoms A and B in Fig. 4.4. When they approach each other to a distance less than their internuclear separation R, their atomic wavefunctions overlap, causing perturbations. Overlap means they share their valence electrons (and maybe others, too, from the core of the electron cloud). So ρ, the charge density, changes: The condition of the electrons changes compared to that when the atoms were free and well-separated. Let the wavefunction of the valence electron of atom A be ψ_{nlm}, and that of B be $\psi_{n'l'm'}$. (We consider the outer most valence electron only; the core is out of interest.) If the valence electron of A is $1s$, then $n = 1$, $l = 0$, $m = 0$. If the valence electron of B is $2p$, then $n' = 2$, $l' = 1$, $m' = 0, \pm1$ (Fig. 4.2).

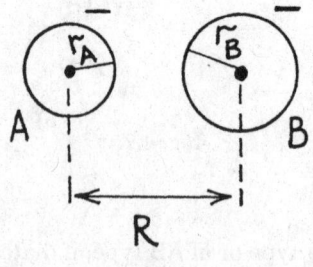

Fig. 4.4.

Here R (or r_0) is the equilibrium separation where the formation of a molecule is possible before repulsion sets up and after sufficient attraction has been achieved. If $R \gg r_A + r_B$, there is no molecule, because the atoms are far away from each other. If $R = r_A + r_B$, a molecule is formed. The case of $R < r_A + r_B$ is not possible because of the repulsion (to which we devote the next chapter).

For separated atoms ($R \gg r_A + r_B$), we have two distinct wavefunctions, one for each: ψ_{nlm} and $\psi_{n'l'm'}$. But if there is a stable molecular situation ($R = r_A + r_B$), these two functions are no longer valid, because the situation has changed: A new *molecular wavefunction* ψ_{mol} must be defined to express the molecular situation, again one for each atom constituting the molecule. In other words, the two atomic functions must change a bit. In the limit, we have the following approach:

$$\lim_{r \to \infty} \psi_A(\text{molecular}) \to \psi_{nlm}(\text{atomic}) \equiv \psi_A$$

$$\lim_{r \to \infty} \psi_B(\text{molecular}) \to \psi_{n'l'm'}(\text{atomic}) \equiv \psi_B .$$

For either atom, $\psi_{mol} = \psi_{electronic}\psi_{rotational}\psi_{vibrational}$, *i.e.*, it is a product of three parts (factor functions). No exact solutions can be found for ψ_{mol}; one can look for approximate solutions only.

Energywise the situation is as follows: The energy of the system (molecule) is $E_{mol} \neq E_A + E_B$ (where E_A and E_B are the energies of the separate individual atoms). If $E_{mol} < E_A + E_B$, the situation is unstable. If $E_{mol} > E_A + E_B$, as the two atoms come close together, the energy decreases, and at the minimum point of the potential well there is a stable situation where the molecule is formed. As the energy reaches a minimum, the corresponding distance is R. We have $R \simeq 1$ to 3 Å in molecules.

For $E < E_A + E_B$ the force is attractive, but if the two atoms approach too close to one another, there is a repulsion. At some $r = R \equiv r_0$ there is equilibrium (minimum E).

The properties of a molecule are not obtained by the superposition of the properties of its individual atoms. A molecule has new properties (*i.e.*, a compound has different properties than those of the chemical elements it consists of). For example, H_2O is not the same body as either of its constituents, hydrogen or oxygen, taken alone.

When the atoms approach each other, the force in question is the *cohesion* or *chemical affinity* or *binding force* whose range is $R \sim$ a few Å. With E_0

being the molecule's energy (minimum energy), we have $E_{binding} = E_A + E_B - E_0$. If $E_{binding} > 0$, the reaction $A + B \to AB$ releases energy as the molecule establishes.[2]

In a free atom the nucleus is at rest and the electron revolves around it. But in a molecule the nuclei are not at rest; they move, because the molecule vibrates (and this requires an averaging on the part of the forces, equivalent to the equilibrium position of the nuclei). This nuclear harmonic motion is quantized ($\sim 10^{13}$ Hz) and exists even at $T = 0°K$. Equilibrium position means $R = \overline{R} = $ constant.

A molecule can be thought (as a model) as consisting of a rigid frame, at the corners of which atoms (nuclei) are positioned. The rods of the frame are the bonds. The electrons of each atom move within these bond distances and about the nuclei they belong to. But valence electrons can be shared by two or more nuclei. A charge density can be defined for a molecule which can have a dipole moment and hence electrical properties macroscopically manifested in the material. The chemical properties of the molecule depend on the nuclei present and the charge distribution about them.

In liquids, molecules execute Brownian motion. In crystalline solids, molecules (or atoms) are arranged to form lattices (frames); a crystal is an assembly (a collection) of molecules (or atoms) located at the frame points and vibrating thermally like springs, without distorting the shape of the solid. Such an ordering does not exist in liquids which have no definite shape, but they take the shape of their container. In gases the molecules are loose and free to fly. As we go from solids to gases, the intermolecular cohesion (bonds) becomes weaker and weaker.

Theoretically (and apart from the methods of a chemistry laboratory), how can we examine a molecule? The prescription is simple in principle: Write the Hamiltonian of the system, solve the Schrödinger equation, and find the energy of the system. In words it sounds easy. In actual practice it is terribly difficult, because — due to the explicit form of the wavefunctions and the potentials — the resulting Schrödinger differential equation is monstrous. What then, give up and go home and sit? No! To circumvent the difficulty, one can do some approximations. Idealizing the situation may cost us loss of exactitude,

---

[2]There is an interesting analogy between a physical system and an economic system in terms of stability. Parameters like energy (ability to do work), time, temperature, communication, transport logistics, dynamics (time-evolution of systems) etc. are common to physics, information theory, and economics.

but we cannot do any better, given that the situation is too complicated. (Physicists are notorious in rounding off things — the rough estimate counts in getting a feeling about what is going on — so that when asked to explain something about, say, a horse, they start with "Let us assume a cylindrical horse!".) But these tricks do not pertain to the kingdom of physics only. Approximations and simplifications are made in other sciences as well, like economics, sociology, political science etc., where direct experimentation with people and systems is not possible. In medicine, too, the prescription for a heart surgery is simple: Make the incisions, remove the saphenous vein from the leg, and transplant parts of it to the heart area to bypass the clogged coronary arteries. But when the heart surgeon sets off to apply this theoretical principle, the real situation he encounters in the patient's state may force him to alter the prescribed technique. No matter how well we plan things according to theoretical principles, practice will ultimately show us how to act.

Getting back to our case, we have to write the Hamiltonian of the system. The operator is

$$\mathcal{H} = \text{Kinetic energy} + \text{Potential energy}.$$

The first term is easy. Since we have a collection of nuclei and electrons in the molecule (Fig. 4.5), we just add up their kinetic energies. The potential energy will include three parts (for all the particles in pairs): The forces between the nuclei and the electrons, the repulsive forces between the electrons themselves, and the repulsive forces between the nuclei. We have

$$\mathcal{H} = -\frac{\hbar^2}{2}\sum_j \frac{1}{M_j}\nabla_j^2 - \frac{\hbar^2}{2m}\sum_i \nabla_i^2 - \frac{1}{2}\sum_{i,j}\frac{Z_j e^2}{r_{ij}} + \frac{e^2}{2}\sum_{i,k}\frac{1}{r_{ik}}$$

$$+ \frac{e^2}{2}\sum_{j,k}\frac{Z_j Z_k}{R_{jk}}$$

where the first term is the total kinetic energy of j nuclei (each of different mass M_j in the most general case); the second term is the total kinetic energy of i electrons (each of the same mass m — out of the sum); the third term is the Coulombic attraction potential energy between nuclei and electrons (with r_{ij} being the distances between nuclei and electrons, Ze being the nuclear charge, and e the electronic charge); the fourth term is the repulsion between electrons (and hence positive, with a plus sign in front); the last term is the repulsion between nuclei (where R_{jk} are the internuclear distances). The factors $(\frac{1}{2})$

come from the proper counting of pairs. (To avoid overcounting, *i.e.*, counting each particle twice, we divide by two.)

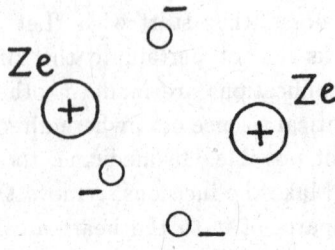

Fig. 4.5.

We can do the following simplifications: If we assume the nuclei fixed, the first term is zero. Also, considering center-of-mass coordinates, under the condition that R_{jk} = constant, the last term can be taken as constant and left out of this business, because it does nothing but shifting the total energy by a constant amount.[3] So we ignore the nuclear translation (motion of the center of mass) and the internuclear repulsion terms.

With ψ being the total (molecular) wavefunction of the system (molecule), we write the Schrödinger equation $\mathcal{H}\psi = E\psi$ (where the left-hand side is the operator $\mathcal{H}$ *operating on* ψ, and the right-hand side is the *total* energy E of the system *multiplied by* the same ψ, because E is the eigenvalue of $\mathcal{H}$). The equation becomes

$$\frac{\hbar^2}{2m}\nabla_i^2\psi + \left[E + \frac{e^2}{2}\sum_{i,j}\frac{Z_j}{r_{ij}} - \frac{e^2}{2}\sum_{i,k}\frac{1}{r_{ik}} \right]\psi = 0$$

where only the electronic kinetic energy is retained (first term). For H_2^+ this equation becomes

$$\nabla^2\psi + \frac{2m}{\hbar^2}\left(E + \frac{e^2}{r_A} + \frac{e^2}{r_B} \right)\psi = 0\,.$$

No exact solution can be found, even for H_2^+. Only approximations are possible. An approximation method has been tried, called *variational method* (see below).

---

[3]Later however, when the total energy is found, this (additive) constant must be added, because the potential is taken as zero at infinity, and we should have quantitative consistency.

The total molecular wavefunction ψ is a function of the coordinates of all the electrons and nuclei:

$$\psi = \psi(x_i, y_i, z_i, X_j, Y_j, Z_j)$$

where lower-case letters are the electronic coordinates, and upper-case letters are the nuclear coordinates.

The particular molecular wavefunctions ψ_A (molecular) and ψ_B (molecular) are obviously functions of (x_i, y_i, z_i) only.

4.2. Variational Method

Every wavefunction ψ (here the total molecular wavefunction) satisfies the operational Schrödinger equation

$$\mathcal{H}\psi = E\psi$$

which is a fundamental equation in quantum mechanics. (Here E is the total energy of the molecule.) We multiply both sides by ψ^* (the complex conjugate of ψ) and integrate over all space. Since the stationary values of the energy are constant (because of stability), E gets out of the integral:

$$E = \frac{\int \psi^* \mathcal{H}\psi dV}{\int \psi^* \psi dV} \tag{4.1}$$

where E represents the molecular energy states. If the function ψ is normalized, the denominator is equal to unity.

We can always write the Hamiltonian $\mathcal{H}$ of the system and the Schrödinger equation, but we must know ψ explicitly in order to solve it. Which wavefunctions ψ are appropriate (or at least fit) for this case? We do not know exactly. It is a question requiring much thought. But if we do not know ψ, we can approximate: Since E is minimum (stability of molecule), we can minimize Eq. (4.1), *i.e.*, we can take the derivative of E with respect to some parameter, and equate it to zero. But to do that, we must have approximate expressions for the molecular ψ. There are two methods to find such expressions:

(1) The *atomic orbitals method* (Heitler-London).[4]

(2) The *molecular orbitals method* (Hund-Mulliken-Herzberg).

The principle described above (of minimizing Eq. (4.1)) is just the variational principle used in mechanics and optics.

4.3. Method of Atomic Orbitals

Let there be n atoms labeled a, b, ..., i, ..., n. Let their valence electrons be specified as 1, 2, ..., k, ..., n (where we considered monovalent atoms so that the number of electrons be equal to the number of atoms, for simplicity). When the atoms are far apart from each other, they behave as independent, noninteracting entities, each with a noninteracting, unperturbed atomic wavefunction ψ_i. We thus initially have $\psi_a(1), \psi_b(2), \ldots, \psi_k(k), \ldots, \psi_n(n)$. The total wavefunction of their assembly is

$$\psi = \psi_a(1)\psi_b(2)\cdots\psi_k(k)\cdots\psi_n(n) = \prod_{i=1}^{n} \psi_i(i),$$

i.e., a product of all the initial atomic wavefunctions. Why a product? Because when we have a collection of independent degrees of freedom, the total wavefunction (or probability) is equal to the product of the individual wavefunctions (or probabilities of the individual events), and the total energy is equal to the sum of the individual energies. (That is, for the unperturbed initial situation we have $\mathcal{H} = \mathcal{H}_a + \mathcal{H}_b + \cdots + \mathcal{H}_n$ and $E = E_a + E_b + \cdots + E_n$.)

Now when the atoms approach each other to form a molecule, they share their valence electrons, and we have a new behavior: An electron may leave its atom — temporarily — and move on to the next atom and then to the next. This implies a redistribution of the electronic charge density, *i.e.*, a new wavefunction Ψ for the system (molecule), given by

$$\Psi = \psi_a(2)\psi_b(1)\cdots\psi_k(j)\cdots\psi_n(k) = \prod_{i=1}^{n} \psi_i(j),$$

with all the possible permutations thereof. As the atoms exchange electrons (like in garage sales swappings), electrons move around and a new wavefunction, Ψ, is obtained. But for every separate redistribution of valence electrons there is a new Ψ, so that a whole set of molecular functions $\Psi_1, \Psi_2, \ldots, \Psi_m$ is possible and fit for the molecular Schrödinger equation. These wavefunctions

---

[4]See *Zeitschrift für Physik*, **44**, 455 (1927).

are *approximate*, and represent the *perturbed* atomic wavefunctions (though they include the unperturbed ones as well). Thus as soon as the electron exchange (interaction) starts, the initial wavefunctions get perturbed. The overall (total) molecular situation is described by the overall molecular function

$$\Psi_{mol} = c_1\Psi_1 + c_2\Psi_2 + \cdots + c_m\Psi_m$$

where the constants c determine the mixing of the states (wavefunctions), that is, with what percentage each one participates and how much it contributes to Ψ_{mol}. These constants c can be determined by the variational method. Once they are found, Ψ_{mol} can be found. This in turn can be put into the Schrödinger equation whence the energy of the molecule E can be found. Ψ_{mol} is the general solution of the Schrödinger equation for the molecule. The set $\Psi_1, \Psi_2, \ldots$ consists of functions which are approximate particular solutions of the same equation.

This is actually a many-body problem circumvented cleverly. For a detailed discussion see the References by Fermi (p. 140) and Schiff (pp. 445–455) listed in the end of this book. See also C. Herring, "*Direct Exchange Between Well-Separated Atoms*" in *Magnetism*, vol. 2B, G.T. Rado and H. Suhl, eds., Academic Press, New York (1965).

4.4. Method of Molecular Orbitals

Consider a diatomic molecule whose atoms are described by the atomic wavefunctions ψ_A and ψ_B (if free). Either of these satisfy the Schrödinger equation

$$\mathcal{H}_{at}\psi_A = E_A\psi_A \quad \text{and} \quad \mathcal{H}_{at}\psi_B = E_B\psi_B$$

where $\mathcal{H}_{at}$ is the *atomic* Hamiltonian operator.

Let ψ be the *molecular orbital* (wavefunction) which describes the motion of a valence electron between atoms A and B. When the electron gets close to A (Fig. 4.6), it is influenced by (the nucleus and the electron cloud of) A, so that $\psi \rightarrow \psi_A$ (atomic). Similarly, when the electron approaches B, $\psi \rightarrow \psi_B$ (in the limit). Therefore, ψ can be written as the superposition (combination) of the two atomic functions:

$$\psi = c_A\psi_A + c_B\psi_B. \tag{4.2}$$

When the electron is close to A, we have $c_B = 0$. The constants c measure the contribution of each atomic wavefunction.

Fig. 4.6.

When the atoms recede from each other, E_A and E_B approach their unperturbed values $E_A^{(0)}$ and $E_B^{(0)}$.

With $\mathcal{H}$ being the molecular Hamiltonian and E the molecular energy, we have

$$\mathcal{H}\psi = E\psi.$$

We multiply by ψ^* to obtain

$$\psi^*\mathcal{H}\psi = \psi^* E\psi = E\psi^*\psi = E|\psi|^2$$

whence

$$E = \frac{\int \psi^* \mathcal{H}\psi dV}{\int |\psi|^2 dV}.$$

The condition for the coefficients is

$$c_A^2 + 2c_A c_B \int \psi_A \psi_B dV + c_B^2 = 1$$

where the *overlap integral* is defined as

$$S = \int \psi_A \psi_B dV.$$

It is taken over all space and corresponds to the overlap of the atomic wavefunctions (Fig. 4.7).

The transition from A to B (and vice versa) is represented by the *resonance integral*

$$\beta = \int \psi_A \mathcal{H}\psi_B dV$$

where ψ_A and ψ_B are the atomic wavefunctions (where they are close together), but $\mathcal{H}$ is the *molecular* Hamiltonian. The resonance integral does not depend only on one atom; it depends on both A and B. It represents the state of pertinence to both atoms at the same time.

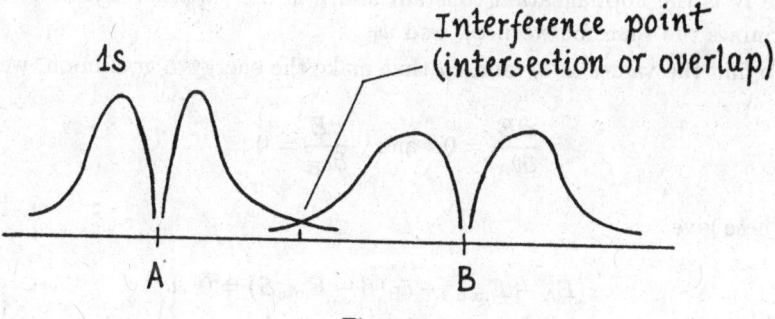

Fig. 4.7.

Then E becomes

$$E = \frac{c_A^2 E_A + 2c_A c_B \beta + c_B^2 E_B}{c_A^2 + 2c_A c_B S + c_B^2}$$

where the denominator is equal to unity, if normalization is provided. We also have

$$\left. \begin{array}{l} E_A = \displaystyle\int \psi_A \mathcal{H} \psi_A dV \\[2mm] E_B = \displaystyle\int \psi_B \mathcal{H} \psi_B dV \end{array} \right\} \tag{4.3}$$

where E_A and E_B are the two parts of the molecular energy that correspond to the (perturbed) atomic wavefunctions ψ_A and ψ_B, as the two atoms are close together. They are *not* (and they should not be confused with) the unperturbed atomic energies $E_A^{(0)}$ and $E_B^{(0)}$ when the two atoms are free. Anyway, notice that $\mathcal{H}$ is the molecular Hamiltonian acting on ψ_A and ψ_B in Eqs. (4.3).

When the two atoms are close together, their atomic functions overlap and $\beta \neq 0$, $S \neq 0$, so that their atomic energy levels split. Since $\beta < 0$, the lower-energy level is represented by the symmetric combination (mixing of states) $\psi_A + \psi_B$, while the upper-energy level is expressed by the antisymmetric combination $\psi_A - \psi_B$. As the two atoms approach each other, there is an attractive force (S and β increase, and as one level increases the other one decreases). When they get too close to each other, an internuclear repulsion sets up and the molecular energy increases again.

Equation (4.2) can be written as

$$\psi = N(\psi_A + \lambda \psi_B) \tag{4.4}$$

where N is the normalization constant and λ is the *partition constant* which determines the share between ψ_A and ψ_B.

To find the values of c_A and c_B that make the energy E minimum, we set

$$\frac{\partial E}{\partial c_A} = 0 \quad \text{and} \quad \frac{\partial E}{\partial c_B} = 0.$$

These give

$$\left.\begin{array}{l} c_A(E_A - E_{min}) + c_B(\beta - E_{min}S) = 0 \\ c_A(\beta - E_{min}S) + c_B(E_B - E'_{min}) = 0. \end{array}\right\} \tag{4.5}$$

We set the determinant of the coefficients equal to zero:

$$\begin{vmatrix} E_A - E_{min} & \beta - E_{min}S \\ \beta - E_{min}S & E_B - E_{min} \end{vmatrix} = 0,$$

$$(E_A - E_{min})(E_B - E_{min}) - (\beta - E_{min}S)^2 = 0.$$

For an A_2 type molecule (Fig. 4.8), $\psi_A = \psi_B$, $E_A = E_B$, and $\lambda = \pm 1$. It is a molecule consisting of two identical atoms; two atoms of the same kind come next to each other, *e.g.*, Cl_2, H_2, O_2. In case of A_2, two different energy states appear for a single electron, that is, the atomic energy level splits into two (Fig. 4.9), where

$$E_1 = \frac{E_A - \beta}{1 + S}$$

$$E_2 = \frac{E_A - \beta}{1 - S},$$

representing the molecular energy situation. The situation is not stable; instead of two energy levels one on top of the other, we obtain two different levels E_1 and E_2. The energy of one is greater, and that of the other is lower. Since $S < 1$, we have $E_A - E_1 < E_2 - E_A$.

Fig. 4.8.

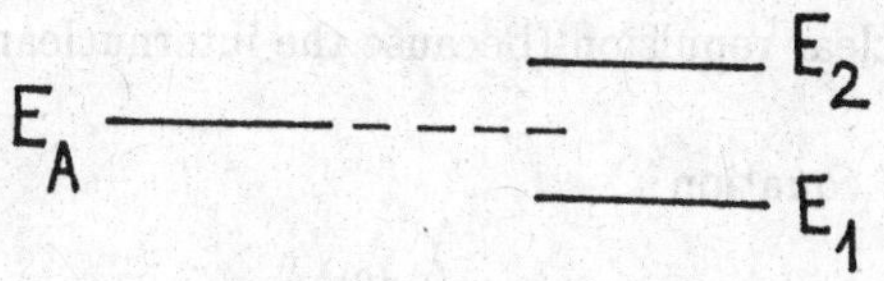

$$E_A \underline{\quad\quad\quad} \quad - - - - \underline{\quad\quad} \quad \begin{matrix} \underline{\quad\quad} E_2 \\ \\ \underline{\quad\quad} E_1 \end{matrix}$$

Fig. 4.9.

From Eq. (4.4), the wave functions that correspond to the two levels are

$$\psi_1 = N(\psi_A + \psi_B)$$

$$\psi_2 = N(\psi_A - \psi_B).$$

4.5. Examples

(1) For an A_2 type molecule: If we take H_2 we have two electrons, and the treatment is difficult. So, let us take the ion H_2^+ instead. It has two protons and one electron (Fig. 4.10). It is hard to solve the Schrödinger equation for a three-body system and find exact solutions. But we can use the method instead.

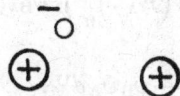

Fig. 4.10.

The Hamiltonian operator is

$$\mathcal{H} = -\frac{\hbar^2}{2m}\nabla^2 - \frac{e^2}{r_A} - \frac{e^2}{r_B}$$

where m is the electronic mass, and r_A and r_B are operators. In the center-of-mass coordinates (Fig. 4.11) we ignore the translational motion. We also

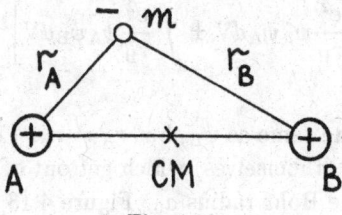

Fig. 4.11.

neglect the internuclear repulsion (because the internuclear separation is large enough).

The operational equation

$$\mathcal{H}_{at}\psi_A = \left(E_A^{(0)}\right)_0 \psi_A$$

is satisfied, where $\left(E_A^{(0)}\right)_0$ is the unperturbed ground-state energy of atom A, since the atomic Hamiltonian is acting. Explicitly, we have

$$\left(-\frac{\hbar^2}{2m}\nabla^2 - \frac{e^2}{r_A}\right)\psi_A = \left(E_A^{(0)}\right)_0 \psi_A.$$

The perturbed energy of A is

$$E_A = \int \psi_A^* \mathcal{H}\psi_A dV = \int \psi_A^* \left(-\frac{\hbar^2}{2m}\nabla^2 - \frac{e^2}{r_A} - \frac{e^2}{r_B}\right)\psi_A dV$$

$$= \left(E_A^{(0)}\right)_0 - \int \frac{e^2}{r_B}\psi_A^*\psi_A dV.$$

$$\beta = \int \psi_B^* \mathcal{H}\psi_A dV = \left(E_A^{(0)}\right)_0 \int \psi_B\psi_A dV - \int \frac{e^2}{r_B}\psi_B\psi_A dV$$

$$= \left(E_A^{(0)}\right)_0 S - \int \frac{e^2}{r_B}\psi_B\psi_A dV.$$

Notice that $\left(E_A^{(0)}\right)_0$ is decreased by some amount, to give E_A.

$$E_{1,2} = \frac{\left(E_A^{(0)}\right)_0 - \int \frac{e^2}{r_B}\psi_A\psi_B dV \pm \left[\left(E_A^{(0)}\right)_0 S - \int \frac{e^2}{r_B}\psi_B\psi_A dV\right]}{1 \pm S}$$

where the plus (minus) sign belongs to $E_1(E_2)$.

$$E_{1,2} = \left[\left(E_A^{(0)}\right)_0 - \int \frac{e^2}{r_B}\psi_A^*\psi_A dV \pm \int \frac{e^2}{r_B}\psi_A\psi_B dV\right]\left(\frac{1}{1 \pm \int \psi_A\psi_B dV}\right).$$

The function ψ_A is the same as ψ_B, but $r_A \neq r_B$. (The values of the radius operators are r_A and r_B (themselves) which get out of the integrals.)

Figure 4.12 shows the Bohr radius a_0. Figure 4.13 shows the situation for H_2^+, where energy is plotted as a function of radial distance r, in terms of e^2/a_0.

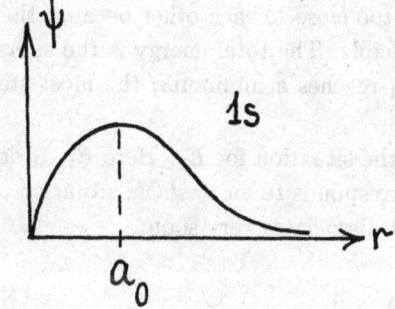

Fig. 4.12.

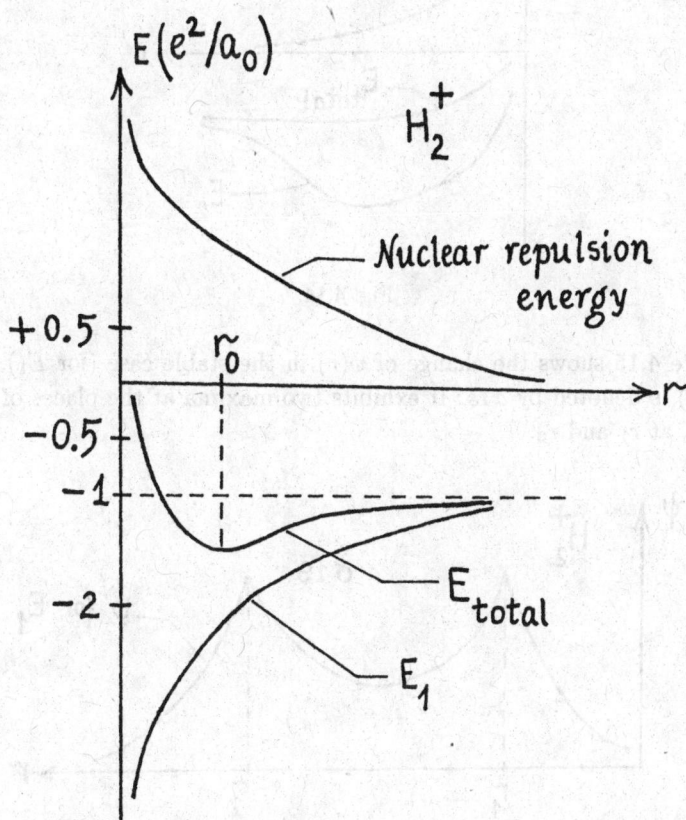

Fig. 4.13.

The atoms cannot get too close to each other because there is a repulsion and because of Pauli's principle. The total energy is the sum of the repulsion and E_1. For $r = r_0$, E_{total} reaches a minimum, the most stationary state of the molecule.

Figure 4.14 shows the situation for E_2. Here E_{total} does not show a minimum at all. So, E_2 corresponds to an *unstable* situation. For E_2 the molecule cannot stay for long; it dissociates very soon.

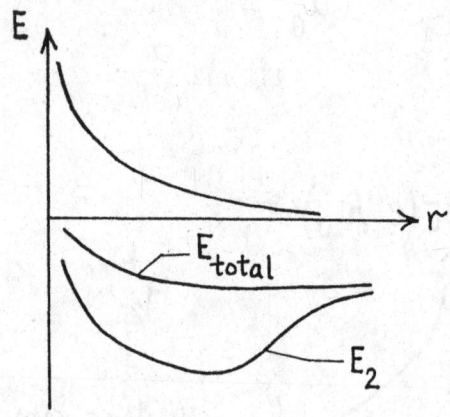

Fig. 4.14.

Figure 4.15 shows the change of $\psi(r)$ in the stable case (for E_1). In this case $\psi(r)$ is denoted by $\sigma 1s$. It exhibits two maxima at the places of the two atoms A, at r_1 and r_2.

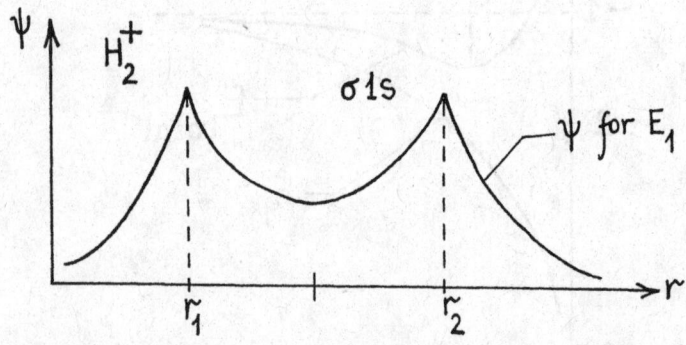

Fig. 4.15.

Figure 4.16 shows $r^2\psi^2$, the probability to find an electron somewhere in space. In the inner contour, the electron behaves as if it belonged to a single atom (its own atom); in the outer contour, the electron belongs to both atoms.

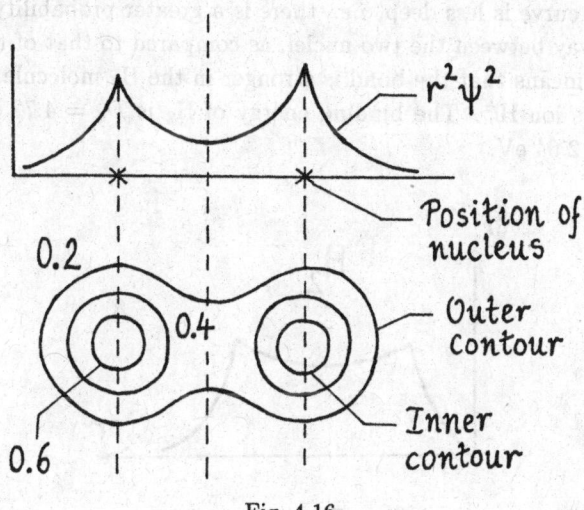

Fig. 4.16.

Figure 4.17 shows the probability that corresponds to the state of E_2. This is an *excited state*, denoted by σ^*1s. At the positions of the two nuclei ψ goes to infinity asymptotically. The situation is indeterminate there. It is as if the probability for the electron to be at the position of the nuclei were infinite, as if the electron were captured by the nuclei.

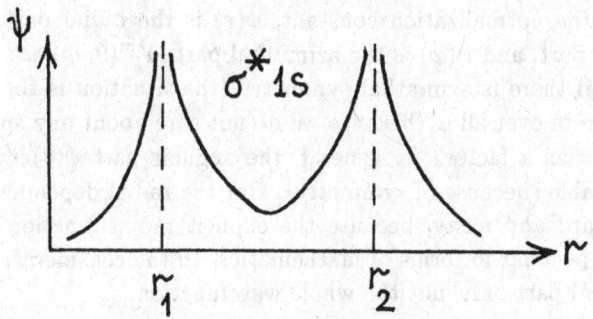

Fig. 4.17.

To summarize, there are two energy states for $1s$-electrons (where $n = 1$, $l = 0$): One is stable and denoted by $\sigma 1s$, and the other one is excited and unstable, denoted by $\sigma^* 1s$.

Figure 4.18 shows the situation for the nonionized H_2 molecule. The central trough of the curve is less deep, *i.e.*, there is a greater probability to find the electron midway between the two nuclei, as compared to that of the situation for H_2^+. This means that the bond is stronger in the H_2 molecule. H_2 is more stable than its ion H_2^+. The binding energy of H_2 is $D_e = 4.72$ eV, whereas that of H_2^+ is 2.64 eV.

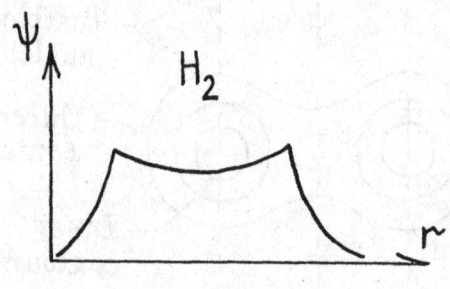

Fig. 4.18.

In general, the wavefunction of an electron can be written as a product of three factors (parts), where each factor depends only on one polar coordinate:

$$\psi_{nlm} = \psi(r, \theta, \varphi) = N\psi(r)\psi(\theta)\psi(\varphi) = N\psi(r)\psi(\theta)e^{-im_l\varphi}$$
$$= N\psi(r)Y_l^m(\theta, \varphi)$$

where N is the normalization constant, $\psi(r)$ is the radial part, $\psi(\theta)$ is the θ-dependent part, and $\psi(\varphi)$ is the azimuthal part. $Y_l^m(\theta, \varphi)$ are the spherical harmonics. If there is azimuthal symmetry (the situation is the same for all φ), we integrate over all φ (because we do not care about any specific φ) and we obtain 4π as a factor. In general, the angular part $\psi(\theta)\psi(\varphi)$ is elegant and manageable (because of symmetry). But the radial dependence may turn out to be hard and nasty, because the explicit radial function $\psi(r)$ can be capriciously peculiar in terms of mathematics. In the considerations above we took this $\psi(r)$ part only, not the whole wavefunction.

(2) For an AB type (asymmetric molecule): The molecule consists of two atoms of different kind. The minimum energy can be found from

$$(E_A - E_{min})(E_B - E_{min}) - (\beta - E_{min}S)^2 = 0 . \qquad (4.6)$$

The curve is a parabola and has two roots: E_1 and E_2, where the parabola intersects the r-axis (Fig. 4.19). For $E_A < E < E_B$, the left-hand side of Eq. (4.6) is negative, and the two roots lie outside E_A and E_B, as shown.

$$E_1 = E_A - \frac{(\beta - E_{min}S)^2}{E_A - E_B}$$

$$E_2 = E_B + \frac{(\beta - E_{min}S)^2}{E_A - E_B} .$$

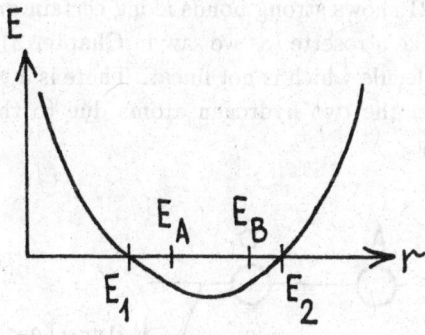

Fig. 4.19.

If ψ_A and ψ_B overlap, two molecular wavefunctions appear, giving different molecular energy levels. One level is of lower energy compared to the atomic E_A, and the other one is of higher energy. If $E_A \ll E_B$, the two roots $E_{1,2}$ differ too much from $E_{A,B}$.

In Eqs. (4.5), if we let $E = E_B$, we find $E_A = 0$; and if we let $E = E_A$, we find $E_B = 0$. Then the molecular functions are degenerate and become identical with $\psi_{A,B}$. This means that there is no bond between A and B or, equivalently, the two atoms do not react chemically.

If ψ_A and ψ_B do not overlap, then $S = 0$ and $\beta = 0$, i.e., $E_1 = E_A$ and $E_2 = E_B$, and there is no difference in energy.

Molecular energy levels fill with electrons like the atomic ones. Pauli's principle states that each level has two electrons with antiparallel spins. (These pairs of electrons form the covalent bonds.) The lower-energy levels fill first, and then the upper-energy ones. If the lower energy fills completely, binding electrons come forth to give stable molecules. If the total energy of these electrons is greater than that of the upper level, we have a stable molecule.

4.6. Molecular Stability

There are three conditions for a stable molecule with strong bonds:

(1) The two atomic wavefunctions ψ_A and ψ_B must overlap as much as possible. Since the square of a wavefunction gives the charge density in space, a large overlap of two wavefunctions means that the probability of finding the electron between the two atoms is high, and hence the bond is strong.

(2) The two wavefunctions ψ_A and ψ_B must be symmetric in the direction of the molecular axis (Fig. 4.20).

Figure 4.21 shows strong bonds along certain directions only, forming a shape like a rosette (as we saw in Chapter 3). Figure 4.22 shows the water molecule which is not linear. There is a wide angle (of about 120°) between the two hydrogen atoms due to the mutual repulsion between them.

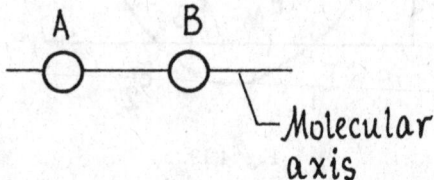

Fig. 4.20.

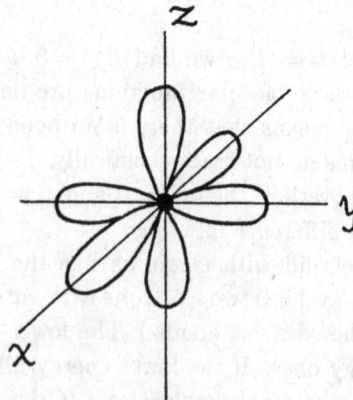

Fig. 4.21.

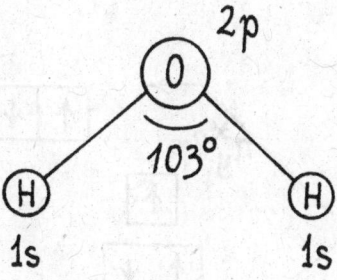

Fig. 4.22.

(3) The atomic energy levels E_A and E_B must not be too different from each other in value.

When two atoms are next to each other, there is a splitting in energy which depends on how much the atomic wavefunctions overlap. If we take the energy state of $1s$, we find two electrons with opposite spins in it (Fig. 4.23). This state splits into two: $\sigma 1s$ and $\sigma^* 1s$ (Fig.4.24, lower part). Each of them has two electrons with opposite spins. The $2s$ state splits into two states as well (Fig. 4.24). The $2p$ state has six electrons. The starred splittings are excited states. The $\pi_x 2p$ state can have two electrons of the same energy, because the energy state that corresponds to the x-direction (Fig. 4.25) is not split. In Fig. 4.24 the energy increases as we go upwards from the bottom to the top.

Figure 4.26 shows the case of hydrogen. The electron must absorb the energy $E = h\nu$ in order to go to the excited state $2s^*$. This state has a certain lifetime τ. After τ seconds the electron comes down to the ground state again and emits the energy quantum it absorbed. We thus have emission.

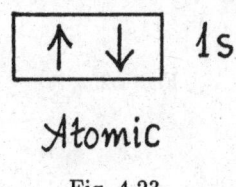

Atomic

Fig. 4.23.

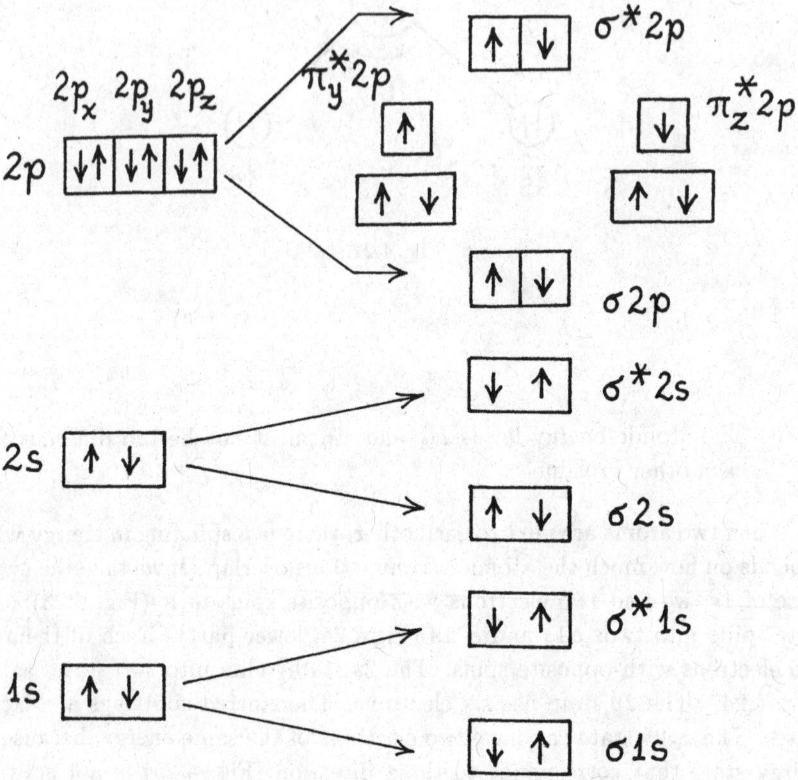

$2p_x$ $2p_y$ $2p_z$

$2p$ ↓↑ ↓↑ ↓↑

π_y^*2p

↑↓ σ^*2p

↑ π_z^*2p ↓

↑ ↓

↑ ↓

↑ ↓ $\sigma 2p$

↓ ↑ σ^*2s

$2s$ ↑ ↓

↑ ↓ $\sigma 2s$

↓ ↑ σ^*1s

$1s$ ↑ ↓

↓ ↑ $\sigma 1s$

Fig. 4.24.

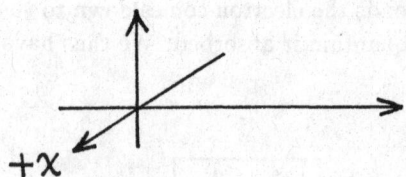

$+x$

Fig. 4.25.

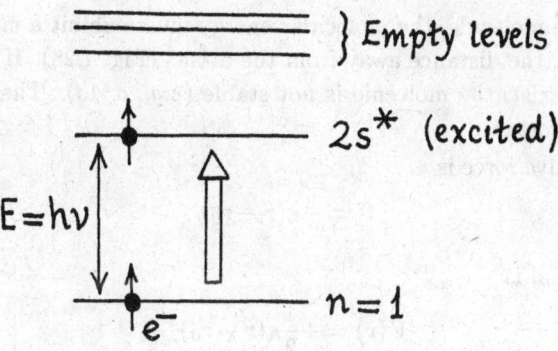

Fig. 4.26.

If the electron cannot stay continuously at the same energy level and makes a transition (Fig. 4.27), the molecule dissociates. In other words, the electron is not shared by the two atoms any more.

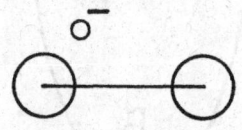

Fig. 4.27.

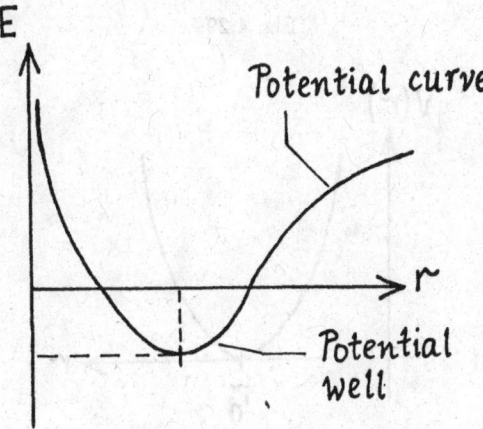

Fig. 4.28.

For a stable molecule, the molecular energy must exhibit a minimum, as it changes with r, the distance away from the atoms (Fig. 4.28). If this *potential well* does not exist, the molecule is not stable (*e.g.*, σ^*1s). The shape of the well is also important.

The attractive force is

$$F = -\kappa(r - r_0)$$

and the potential is

$$V(r) = +\frac{1}{2}\kappa(r - r_0)^2$$

which is a parabola (parabolic potential well) (Fig. 4.29 or 4.30). The minimum occurs at $r = r_0$.

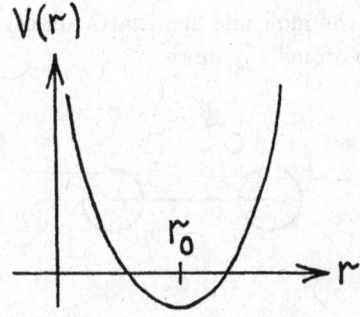

Fig. 4.29.

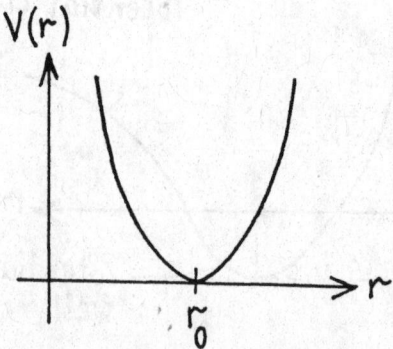

Fig. 4.30.

4.7. Molecular Vibration and Morse Potential

The energy of the molecular vibration is

$$E_{\text{vibr}} = h\nu_0 \left(\text{v} + \frac{1}{2} \right).$$

As the vibrational quantum number v increases, the energy intervals of the molecular energy levels decrease, and a correction term (nonlinear in v) is added:

$$E_{\text{vibr}} = h\nu_0 \left(\text{v} + \frac{1}{2} \right) - h\nu_0 \alpha \left(\text{v} + \frac{1}{2} \right)^2$$

where α is a constant that causes anharmonicity. The molecular potential energy curve is shown in Fig. 4.31. Its part which is below the r-axis can be taken as parabolic. The same figure shows that as we get away from the minimum energy, the point r_0 corresponds to large energies and shifts to the right, away from its place.

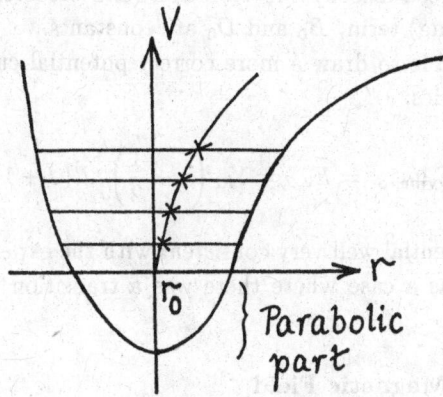

Fig. 4.31.

Since r_0 changes upon vibration, another potential was proposed as a correction, the *Morse potential*, given by

$$V(r) = D_e \left(1 - e^{-(r-r_0)/d} \right)^2$$

where d is a constant, r_0 is the equilibrium position, and D_e is the dissociation (or binding) energy. This potential is slightly different only for very large values of r, and gives better results with the parabolic approach.

If we adopt the Morse potential, we take the vibration and rotation energies together: $E_{\text{vibr,rot}} \equiv E_{\text{v},J}$. By using the Morse potential we obtain the following relation:

$$E_{\text{vibr,rot}} = h\nu_0 \left(\text{v} + \frac{1}{2}\right) - h\nu_0\alpha \left(\text{v} + \frac{1}{2}\right)^2 + B_0 h J(J+1)$$

$$- D_0 h J^2(J+1)^2 - a\left(\text{v} + \frac{1}{2}\right)(J+1)J \qquad (4.7)$$

where the first term is the first-order vibration energy; the second term is the second-order vibration energy; the third term is the first-order rotation energy; the fourth term is a second-order correction to the rotation; and the last term is a cross term that includes both quantum numbers v and J (vibration and rotation), to show the interaction energy between vibration and rotation. Since a is a small coefficient (compared to others), we can treat this last term as a small correction (perturbation) term. It signifies the fact that a fast rotating molecule is influenced also by the vibration (and vice versa), and hence the interaction (coupling) term. B_0 and D_0 are constants.

Dunham was able to draw a more correct potential curve.[5] He expressed the energy as a series:

$$E_{\text{vibr,rot}} = E_{\text{v},J} = Y_{l,\text{i}} \left(\text{v} + \frac{1}{2}\right)^l J^{\text{i}}(J+1)^{\text{i}}.$$

This way a potential well very consistent with the experimental results was achieved. This was a case where there was a transition from experiment to theory.

4.8. Effect of a Magnetic Field

Consider a monoelectronic (or monovalent) atom whose (valence) electron has mass m. If there is no external magnetic field ($H_0 = 0$), the Hamiltonian operator is

$$\mathcal{H} = -\frac{\hbar^2}{2m}\nabla^2 + V_0(r)$$

where ∇^2 and V_0 are operators. There is only a Coulombic radial potential V_0. The first term is the kinetic energy of the electron. (When the operator V_0 acts on ψ it just gives the value V_0 multiplied by ψ, where V_0 is a function

---

[5]See *Phys. Rev.*, **41**, 721 (1932).

of r. If V_0 is not radial, it is a function of $\mathbf{r}$, *i.e.*, it has θ and φ-dependence as well.)

Now when we turn on an externally applied field $H_0 \neq 0$, the momentum of the electron changes from $\mathbf{p}$ to $\mathbf{p} - \frac{q}{c}\mathbf{A}$, where q is the charge, c is the speed of light, and $\mathbf{A}$ is the vector potential that defines the magnetic field through $\mathbf{H}_0 = \vec{\nabla} \times \mathbf{A}$.

When the field in on, $\mathcal{H}$ changes drastically, to become

$$\mathcal{H} = \frac{1}{2m}\left(\mathbf{p} - \frac{q}{c}\mathbf{A}\right)^2 + V_0 + V_1$$
$$+ \mathcal{H}_1 + \mathcal{H}_2 + \mathcal{H}_3 + \mathcal{H}_4 + \mathcal{H}_5 + \mathcal{H}_6 + \mathcal{H}_7 + \mathcal{H}_8 \qquad (4.8)$$

where V_0 is the Coulombic potential; V_1 is the crystal potential; and the rest are correction (perturbation) terms, usually small. Whichever of them is small in strength (compared to others) can be treated as a perturbation to the total $\mathcal{H}$.

Notice that the full Eq. (4.8) holds for atoms in a molecule, so that there is a nonzero V_1 term.

Let us examine the terms of $\mathcal{H}$ beyond the Coulombic potential term.

As the electron moves in the electric field of the crystal, there is a crystal potential V_1 (or V_{cryst}). The electron is electrically influenced by the ions of the crystal (or molecule). The ions create commonly an electric field which acts on the electron. So, the electron has an electrical potential energy

$$V_1 = -eV = -e\sum_i V(x_i, y_i, z_i)$$

where e is the electronic charge, and the sum runs over all the coordinates of all the ions.

Since there is a charge distribution (Fig. 4.32), the potential V in spherical coordinates is

$$V = \sum \mathcal{A}_n^m r^n Y_n^m(\theta, \varphi)$$

which is a mathematical series. (Here m is a superscript, not a power.) The sum is over n and m. If m = 0, there is axial symmetry (and vice versa). If the potential is symmetric with respect to an axis in the crystal lattice, then m = 0. The case m = ± 2 occurs in rhombic crystals. To find the rhombic crystal potential, we let m = ± 2 and take the sum. For m = ± 3 we have trigonal crystals (threefold symmetry); for m = ± 4 we have tetragonal symmetry; for m = ±6, the symmetry is hexagonal. If we know the crystal symmetry, we know V_1. The spherical harmonics give V.

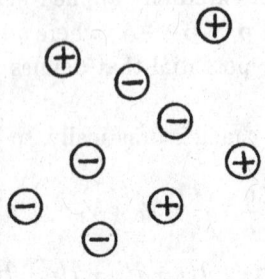

Fig. 4.32.

Since the charge distribution is discrete (distinct charges, Fig. 4.32), we take a sum. (For a continuous charge density, *i.e.*, a distribution $\rho(\mathbf{r})$, we take an integral — a continuous sum, or the limit of a sum. We can always go from a sum to an integral, provided that we write down the density of states of the integrand, or we multiply the discrete values by the appropriate Dirac's delta functions.)

In crystals, V_1 may be too large. We then have electronic energy splittings.

The term $\mathcal{H}_1$ represents the interaction between the orbiting electrons of the atoms of the molecule and the magnetic field. That is, the orbital angular momentum of the molecular electron interacts with the external magnetic field $\mathbf{H}_0$. As a result, the energy is split (Fig. 4.33). This is the *normal Zeeman effect*, and $\mathcal{H}_1$ is the energy term of this effect, given by

$$\mathcal{H}_1 = \mu_B \hbar H_0 (\Delta m_l)$$

where μ_B is the Bohr magneton, and Δm_l is the change in the orbital magnetic quantum number, which can be $\Delta m_l = 0, \pm 1$ (*i.e.*, m_l can change by 0 or ± 1) *if* $l = 1$. One observes energy splittings.

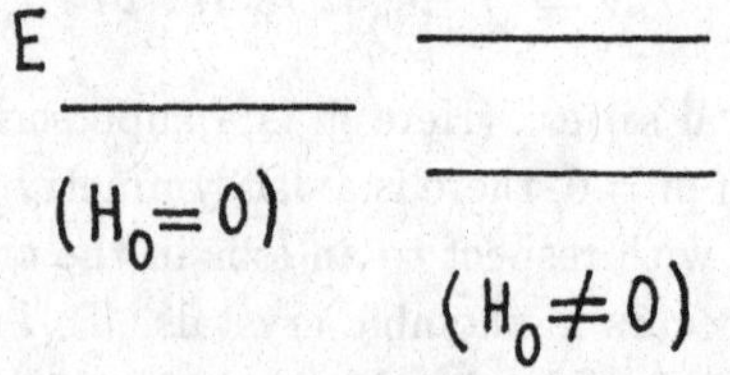

$$E \quad \underline{\qquad} \qquad \overline{\qquad}$$

$$(H_0 = 0) \qquad \underline{\qquad}$$

$$(H_0 \neq 0)$$

Fig. 4.33.

The term $\mathcal{H}_2$ is the spin-orbit interaction energy, given by

$$\mathcal{H}_2 = \lambda \mathbf{L} \cdot \mathbf{S}$$

where $\mathbf{L}$ is the (total) orbital angular momentum, and $\mathbf{S}$ is the (total) spin (which is interacting with $\mathbf{L}$). The scalar coefficient λ is the interaction strength, giving the importance of the interaction (energy), *i.e.*, how large it is. The two angular momenta ($\mathbf{L}$ has a classical analogue but $\mathbf{S}$ does not) mutually interact.

$\mathcal{H}_3$ is the interaction between the magnetic moment of the electron (that corresponds to its spin angular momentum) and the field $\mathbf{H}_0$. This is the spin Zeeman effect, where the electronic spin vector $\mathbf{S}$ interacts with the external field:

$$\mathcal{H}_3 = 2\mu_B \mathbf{H}_0 \cdot \mathbf{S}.$$

This term gives ESR (the electron spin resonance) because it shows the interaction between the electron spin and the external field.[6] ESR is observed in paramagnetic materials only. When we apply an external magnetic field $\mathbf{H}_0$ (Fig. 4.34), an internal field $\mathbf{H}_{int}$ sets up inside the body which is present around. We may have, in general, $H_{int} \neq H_0$. But in paramagnetic materials we have $H_{int} = H_0$.

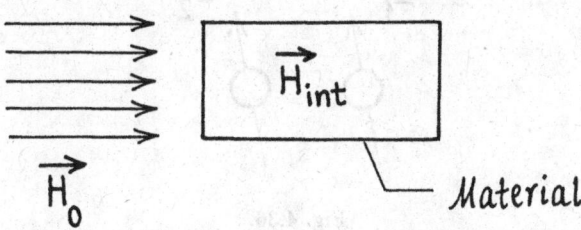

Fig. 4.34.

The term $\mathcal{H}_4$ is the interaction between the electron spin $\mathbf{S}$ and the nuclear spin $\mathbf{I}$ of the ion.

$\mathcal{H}_5$ is the quadrupole term, *i.e.*, the Hamiltonian arising from the electric quadrupole moment of the nucleus (of the ion). A nucleus may have a quadruple moment Q. If $Q \neq 0$ and if $I > \frac{1}{2}$, then this term gains importance, because NQR (the phenomenon of nuclear quadrupole resonance) is observed, *alias* PQR (pure quadrupole resonance).

---

[6]The value of the field at the position of the electron.

$\mathcal{H}_6$ is the term that gives the interaction between the nuclear spin **I** and the external field $\mathbf{H}_0$. This is NMR (the phenomenon of nuclear magnetic resonance).

$\mathcal{H}_7$ is the interaction between the nuclear spins $\mathbf{I}_1$ and $\mathbf{I}_2$, given by $\mathbf{I}_1 \cdot \mathbf{I}_2$, and in general,

$$\mathcal{H}_7 = \sum_{i,j} \mathbf{I}_i \cdot \mathbf{I}_j \, ,$$

shown in Fig. 4.35 and Fig. 4.36.

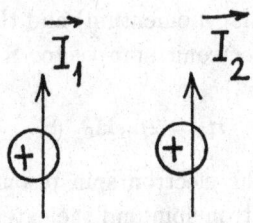

Fig. 4.35.

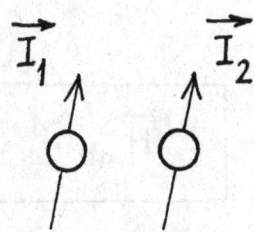

Fig. 4.36.

Finally, $\mathcal{H}_8$ is the smallest term which gives the *exchange interaction*. It is too small, and so it is difficult to measure it directly.

Notice that all the terms beyond V_0 exist only if there is an external $\mathbf{H}_0$ acting, except V_1, $\mathcal{H}_5$, $\mathcal{H}_7$ and $\mathcal{H}_8$, which can exist even if there is no $\mathbf{H}_0$.

4.9. Dunham's Calculations

Refer to Section 4.7. Dunham was able to use the Wentzel-Kramers-Brillouin method and calculate the energy levels of a vibrating rotor. He did it for any potential that can be expanded in powers of $r - r_0$ in the vicinity of the potential well, that is, for

$$V(r) = \sum_n \mathcal{A}_n (r - r_0)^n \qquad (n = 1, 2, \ldots)$$

where $\mathcal{A}_n$ are the expansion coefficients, and r_0 is the equilibrium position. Dunham's result is

$$E_{v,J} = \sum_{l,i} Y_{l,i} \left(v + \frac{1}{2} \right)^l J^i (J + 1)^i$$

where v and J are the rotational and vibrational quantum numbers respectively, and l and i are the summation indices. Y_l are the spherical harmonics, coefficients that depend on the molecular parameters.

Dunham performed these refined calculations in 1932. He calculated the first fifteen $Y_{l,i}$. The effective potential of the rotor is

$$V = \mathcal{A}_0 \xi^2 (1 + \mathcal{A}_1 \xi + \mathcal{A}_2 \xi^2 + \ldots) + B_0 J(J+1)(1 - 2\xi + 3\xi^2 - 4\xi^3 + \ldots)$$

where $\xi \equiv (r - r_0)/r_0$ and $B_0 = h/8\pi^2 \bar{\mu} r_0^2$. Dunham's first few harmonics are

$$Y_{00} = B_0 \bigg/ \left(3\mathcal{A}_2 - \frac{7}{4}\mathcal{A}_1^2 \right)$$

$$Y_{10} = W \left[1 + (B_0^2/4W^2)\left(25\mathcal{A}_4 - \frac{95}{2}\mathcal{A}_1\mathcal{A}_2 - \frac{135}{2}\mathcal{A}_1\mathcal{A}_5 \right. \right.$$
$$\left. \left. + \frac{453}{8}\mathcal{A}_1^2\mathcal{A}_2 - \frac{1155}{64}\mathcal{A}_1^4 \right) \right]$$

$$Y_{12} = -(12B_0^4/W^3)\left(\frac{19}{2} + 9\mathcal{A}_1 + \frac{9}{2}\mathcal{A}_1^2 - 4\mathcal{A}_2 \right)$$

$$Y_{13} = 16B_0^5(3 + \mathcal{A}_1)/W^4 \, .$$

The ratio $(B_0/W)^2$ is usually too small: For H_2, it is $\sim 10^{-3}$; for heavier molecules, it is $\sim 10^{-6}$. If it is small, then Eq. (4.7) which is

$$E = E_{\text{vibr}} + E_{\text{rot}} = \underbrace{h\nu_0 \left(v + \frac{1}{2} \right) - h\nu_0 x_0 \left(v + \frac{1}{2} \right)^2}_{\text{Vibration terms}}$$

$$+ \underbrace{hcB_0 J(J+1) + hcD_e J^2(J+1)^2}_{\text{Rotation terms}} - \underbrace{a_0 \left(v + \frac{1}{2} \right) J(J+1)}_{\substack{\text{Vibration-rotation coupling} \\ \text{due to mutual interaction}}}$$

now becomes

$$E_{v,J} = W\left(v+\frac{1}{2}\right) - Wx_0\left(v+\frac{1}{2}\right)^2 + Wy_0\left(v+\frac{1}{3}\right)^3 + Wz_0\left(v+\frac{1}{4}\right)^4$$

$$+ B_0 J(J+1) - D_e J^2(J+1)^2 + H_0 J^3(J+1)^3 + \dots \text{h.o.t.}$$

where $\nu_0 = \frac{1}{2\pi}\sqrt{\frac{\kappa}{\mu}}$, $W = h\nu_0$, $x_0 = \frac{h\nu_0}{4D_e} \simeq 0.01$ to 0.05, $B_0 = \frac{h}{8\pi^2 c \mathcal{I}_0}$, $\mathcal{I}_0 = \tilde{\mu} r_0^2$, $D_e = -\frac{h^3}{128\pi^6 r_0^6 c \nu_0^2}$, and $a_0 = \frac{3h^3 \nu_0}{16\pi^2 \mathcal{I}_0 D_e}\left(\frac{d}{r_0} - \frac{d^2}{r_0^2}\right)$, with d being the parameter of the Morse potential $V(r) = D_e(1 - e^{-(r-r_0)/d})^2$.

Here $W \simeq Y_{10}$, $B_0 \simeq Y_{01}$, $D_e \simeq -Y_{02}$, $H_0 \simeq Y_{03}$, $Wx_0 \simeq -Y_{20}$, $Wy_0 = Y_{30}$, $Wz_0 \simeq Y_{20}$, $\alpha \simeq -Y_{11}$, $\beta \simeq Y_{12}$, $\gamma \simeq Y_{21}$. Dunham proved that for the Morse potential, all $Y_{l0} = 0$ except Y_{10} and Y_{20}. But this is not the end. Sandemann extended Dunham's calculations to higher terms.

Notice that in the expression for E above the vibration part consists of the basic first-order term plus the second-order contribution (correction) term. The same thing holds for the rotation part.

Chapter 5

REPULSIVE FORCES

5.1. Classical Repulsive Potential

When two atoms get close together, they repel each other because their nuclei repel each other electrostatically, as being both of positive charge (Fig. 5.1). So, when the atoms approach each other, repulsive forces arise after a certain separation distance between them.

Fig. 5.1.

Consider an atom A with an outermost electron orbiting at a radius r_A, and an atom B approaching atom A from the right (Fig. 5.2). (Actually, A could be a hydrogen atom with its single electron, and B could be a proton approaching, or being close to, the hydrogen atom (Fig. 5.3).)

The potential of the repulsive force between A and B is

$$V_{rep} = e^2/r \, ,$$

while the total (Coulombic) potential is

$$V = V_{rep} + V_{attr} = \frac{e^2}{r} - \frac{e^2}{r_B} \, ,$$

179

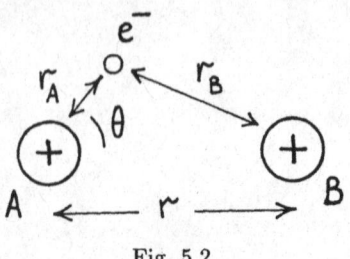

Fig. 5.2.

Fig. 5.3.

consisting of two terms, that is, the repulsive part between the positive nuclei (first term) and the attractive part between B and the electron (second term). From the geometry,

$$V = \frac{e^2}{r} - \frac{e^2}{\sqrt{r^2 + r_A^2 - 2r_A r \cos\theta}}.$$

So far the consideration has been classical. Now we will continue discussing the problem quantum mechanically, *i.e.*, we will discuss the calculation of the repulsive forces, using quantum mechanics. (See next section.)

5.2. Quantum Mechanical Repulsive Potential

Since the electron is not nailed in place, but moves around atom A, the potential changes and has an average value $\overline{V}$. Quantum mechanically,

$$\overline{V} = \int \psi_{1s} V \psi_{1s}^* d^3r \qquad (5.1)$$

where ψ_{1s} is a real function, the wavefunction of the 1s-electron, given by

$$\psi_{1s} = \left(\frac{1}{\pi a_0^3}\right)^{1/2} e^{-r/a_0}$$

where a_0 is the Bohr radius. Notice that ψ_{1s} has only a radial part. It does not have an angular part depending on θ. It is symmetric regarding to its φ-dependent part.

Thus one has to evaluate the integral in Eq. (5.1). But we can do some simplifications: We can expand V to a series. We change variable by introducing new variables, *i.e.*, we let $\xi \equiv 2r_A/a_0$ and $R = 2r/a_0$.

For $\xi/R < 1$, the expansion is

$$V = -\frac{2e^2}{a_0}\left[\xi\frac{\cos\theta}{R} + \frac{\xi^2}{R^3}\left(\frac{3}{2}\cos^2\theta - \frac{1}{2}\right) + \cdots\right].$$

For $\xi/R > 1$, the expansion is

$$V = -\frac{2e^2}{a_0}\left[\left(\frac{1}{\xi} - \frac{1}{R}\right) + \frac{R\cos\theta}{\xi^2} + \frac{R^2}{\xi^3}\left(\frac{3}{2}\cos^2\theta - \frac{1}{2}\right) + \cdots\right].$$

So the mean repulsive potential, when a proton approaches a hydrogen atom, is

$$\overline{V}_{rep} = \int_{\xi=R}^{\infty} -\frac{e^2}{a_0}\left(\frac{1}{\xi} - \frac{1}{R}\right)e^{-\xi}\xi^2 d\xi = -\frac{e^2}{a_0}\left(1 + \frac{a_0}{r}\right)e^{-2r/a_0}. \qquad (5.2)$$

As $r \to 0$, $\overline{V}_{rep} \to \infty$ (grows too large). In the vicinity of the equilibrium distance the repulsive potential is weak, *i.e.*, it has a small value. When $r = a_0/2$, the repulsive potential is 29.8 eV. For $r = a_0$, the value is 3.5 eV; and for $r = 2a_0$, it is 0.1 eV which is a small value, saying that the repulsive force is too small there and can be neglected.

Here $r_A \simeq a_0$ is the distance between the nucleus and the electron. (Actually, a_0 is the place where the probability of finding a 1s-electron peaks, or $\langle r \rangle = a_0$.)

The first result, that is, Eq. (5.2) was reached by Unsold, a theoretical physicist, in 1927. The exponential decrease in $\overline{V}_{rep}$ is due to the exponential decrease in the charge density of the atom as we get away from the nucleus (Fig. 5.4). Experimental and theoretical results are in good agreement.

$\overline{V}_{rep}$ is spherically symmetric because ψ_{1s} is so. As $r \to 0$, $\overline{V}_{rep}$ increases sharply; this can help us to define the radius of the atom.

For the K-shell of an atom with nuclear charge Ze, Eq. (5.2) becomes

$$\overline{V}_{rep}(1s^2) = -\frac{2e^2}{a_0}Z\left(1 + \frac{a_0}{Zr}\right)e^{-2Zr/a_0}.$$

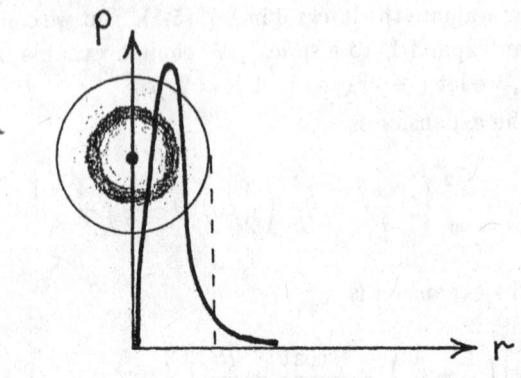

Fig. 5.4.

$\overline{V}_{rep}$ for atoms with filled shells can be calculated. For example, physicist Slater calculated it for He:

$$\overline{V}_{rep}(\text{He}) = 481 \exp(-r/0.412).$$

Mayer calculated it for Ne:

$$\overline{V}_{rep}(\text{Ne}) = 1.18 \exp(-r/0.395).$$

As Z decreases, the numerical factor is front of the exponential increases.

One can compare $\overline{V}_{rep}$ with the Coulombic potential which is just $V(\text{Coulombic}) \sim 1/r$.

The repulsive force depends on the screening of the nucleus and its charge by the favorable attractive charge of the electron cloud around it. If two atoms approach each other so that their electron cloud interpenetrate each other (overlap of ψ's), the repulsive force increases drastically because the two nuclei are too close to each other.

5.3. Total Potential

When two atoms mutually interact, we speak of two potentials: V_{rep} and V_{attr} (Fig. 5.5(a)). The former is positive and strongly repulsive for small r values, and then it decreases exponentially. The latter is negative and blows up at the origin (position of either nucleus). The combination of the two is the total interaction potential $V = V_{attr} + V_{rep}$, plotted in Fig. 5.5(b). It exhibits a negative potential well with a minimum for $r = r_0$. This minimum means stability, and r_0 is the interatomic separation where (for which) stability

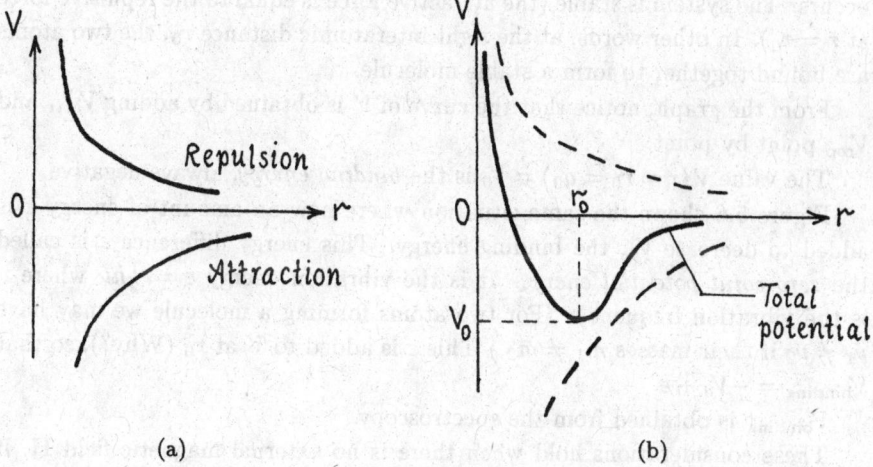

(a)   (b)

Fig. 5.5.

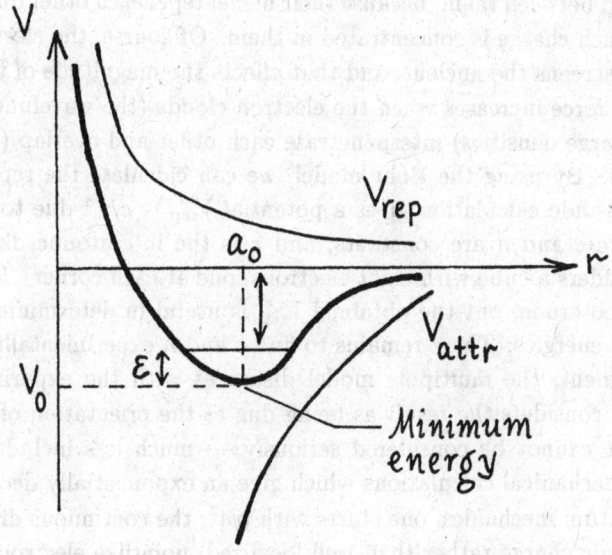

Fig. 5.6.

occurs: The system is stable (the attractive force is equal to the repulsive force at $r = r_0$). In other words, at the right interatomic distance r_0, the two atoms are bound together to form a stable molecule.

From the graph, notice that the curve of V is obtained by adding V_{attr} and V_{rep} point by point.

The value $V(r = r_0 = a_0) = V_0$ is the *binding energy*, always negative.

Figure 5.6 shows the same situation where now an amount of energy ε is added to decrease V_0, the binding energy. This energy difference ε is called the *zero-point potential energy*. It is the vibration energy $\varepsilon = \frac{1}{2}h\nu$ where ν is the vibration frequency. (For two atoms forming a molecule we may have $\nu_1 \neq \nu_2$ if their masses $m_1 \neq m_2$.) This ε is added to V at r_0 (Why?), so that $V_{\text{binding}} = -V_0 + \varepsilon$.

V_{binding} is obtained from the spectroscopy.

These considerations hold when there is no external magnetic field $\mathbf{H}$. If we apply an external field, the energy of the system changes (as we saw in the last section of the previous chapter).

5.4. Comments on the Repulsive Potential

Classically, when atoms (or molecules) approach one another, repulsive forces set up between them, because their nuclei repel each other due to the fact that too much charge is concentrated in them. Of course, the electrons form a cloud that screens the nucleus, and that affects the magnitude of the repulsive force. This force increases when the electron clouds (the wavefunctions of the electron charge densities) interpenetrate each other and overlap (between the two atoms). By using the Bohr model, we can calculate the repulsive force. The Born-Lande calculation gives a potential $V_{\text{rep}} \sim c/r^n$ due to a multipole force, where c and n are constants, and r is the interatomic distance. The model considers a cube with eight electrons, one at each corner. This model is certainly too crude, but the obtained V_{rep} is useful in determining molecular and crystal energies. There remains to find c and n experimentally, but to our disappointment, the multipole model disagrees with the experimental facts (because it considers the result as being due to the orientation of the atoms), and thus it cannot be considered seriously — much less included — in the quantum mechanical calculations which give an exponentially decreasing V_{rep}.

In quantum mechanics, one starts with $|\psi|^2$, the continuous distribution of the electronic charge rather than well-localized, pointlike electrons. Then one performs the quantum mechanical calculation — it is done in Section 5.2. The result is Eq. (5.2), the *Unsold equation*. Notice that V_{rep} decreases *exponen-*

tially with r, not as an inverse power r^{-n}. This is so because the electronic charge density in atoms decreases exponentially; there is no abrupt cut-off of it beyond the atomic radius along r, nor an r^{-n} dependence. (Even at infinity, there is a fantastically small probability to find a bound electron — that's quantum mechanics!)

V_{rep} for atoms with filled shells can be calculated, too (He, Ne etc.).

As the atoms approach each other, the repulsion energy increases sharply and steeply (which explains the hardness and macroscopic impermeability of solids). As $r \to 0$, this steepness of the curve can help us determine the atomic radius.

In molecules the repulsion and attraction (two opposite forces) are in equilibrium, and that is why we have the molecule in the first place. The interatomic equilibrium separation depends also on the attraction (and that is why atoms, ions, and molecules have different radii).

Chapter 6

INTERATOMIC MOLECULAR BONDS

6.1. Kinds of Bonds

The functionality of the bonds between atoms is not dissimilar to that of the bonds between people or organizations. We certainly do not intend to discuss the similarity here, because the reader is intelligent enough to make the parallelisms.

All the kinds of bonds owe their existence to electrostatic forces. (Of course, kinetic effects due to the electronic motion and feeble magnetic forces can contribute to the binding energy as well, but these do not alter the mainly Coulombic nature of the bonds.) The distribution of electrons around atoms and molecules is the first and foremost factor that determines the kind of the binding.

The interatomic electrostatic interaction (whose range is ~ 3 Å) is responsible for chemically binding the atoms together in molecules. The chemical bonds are effected by the valence electrons of atoms that come to a binding. The valence of an atom tells us about its tendency to form a bond.

The chemically active elements are either electropositive metals or electronegative metalloids. According to the electronegativity, there are three kinds of bonds: *Covalent, metallic* and *ionic*. They are all electrostatic, but since the electron clouds are different, we have different types of bonds (and different physical and chemical properties of the compounds).

The bond types are listed in the following table:

Name of bond	Participants	Example
Homopolar covalent (− −)	Metalloid-metalloid	Cl_2
Homopolar metallic (+ +)	Metal-metal	Na
Heteropolar ionic (+ −)	Metal-metalloid	NaCl

These are not the only types of bonds. There are other bonds, too, intermediate ones, depending on the electronegativity. *Homopolar* means of the same polarity: Valence electrons are shared by, and execute orbits around *several* nuclei. That is, the electronic charge density smears in space around neighboring nuclei, just like the way a national cause binds the members of a nation together and causes political covalence. *Heteropolar* means of different (opposite) polarity: The charge density gravitates near the more electronegative atom. And if there is a big difference in electronegativity, the valence electron of the electropositive atom jumps over to the electronegative atom. Thus a union of an ionic bond is formed. Example: NaCl, where the valence electron of Na gets to Cl and causes a union between two ions: Na^+ and Cl^-.

The full list of the bonds is as follows:

(1) *Ionic bonds*: One kind of atom gives an electron to another kind of atom. The resulting two opposite ions attract each other electrostatically. These Coulombic forces are the source of ionic bonds.

(2) *Covalent bonds*: Electron pairs play an important role here. Each of the electrons behaves as if it belonged to both atoms which get bound together. (This resembles the collaboration between two entities with the same common interests.) Example: The H_2 molecule.

(3) *Metallic bonds*: The crystal binding in metals can be explained by the fact that the positive metal ions are held together by free electrons. Example: Alkali metals. In transition metals (*e.g.*, Fe, W), covalent bonds between inner electron shells play a role, too, because these shells are not filled. The bonds effected by free electrons are not so strong; for example, the binding energy of Na is 26 kcal/mol, whereas in W (where covalent bonds of inner electrons participate as well) it is 210 kcal/mol.

These three kinds of bonds in molecules are the most important ones. But there are also other bonds which also come from molecular forces. These are the following:

(4) *Dispersive bonds* or *Van der Waals bonds*: They are due to the dispersive Van der Waals forces, to which the next chapter is devoted.

Whereas the other bonds have a range of 1 to 3 Å, the Van der Waals forces are considered at a range of 5 Å. These bonds are explained by the deformation that atoms cause to each other. This effect happens in two ways: (a) The molecules (or atoms) have a dipole moment which induces another dipole moment in other molecules (or atoms). So, the molecules (or atom) attract each other. (b) The molecule (or atom) does not have a dipole moment (its electric charge is spherically symmetric), but since the location of the valence electron changes continuously, there is a changing instantaneous dipole moment which induces other dipole moments in nearby molecules (or atoms). As a result, there is an attraction that binds the molecules (or atoms).

(5) *Hydrogen bond*: It is observed in organic molecules. In this case a hydrogen atom can bind two atoms together with a strong bond which in fact is of ionic character. Since hydrogen is a single-electron atom, it can form a covalent bond only with one atom. However, it can also be connected to two atoms simultaneously! The mechanism is as follows: The hydrogen atom gives its $1s$ electron to another atom. The created H^+ ion is just a proton, *i.e.*, a charged particle of small size. So, a second negative ion can approach and be connected to this same proton. (A third negative ion cannot approach, because of its large size.) Thus the H^+ ion is a mediator which connects two negative ions together. Such a bond is not strong. (Its binding energy is 2 to 10 kcal/mol.) The hydrogen bond explains the behavior of water molecules and the crystal structure of the protein molecules.

(6) Bonds between donor and acceptor levels in semiconductors.

6.2. Bond Structure

The chemical, electrical, and magnetic properties of compounds depend on the bonds, and specifically on the following:

(1) The number of valence electrons that form the bond.[1]
(2) The multiplicity and the number of bonds between neighboring atoms (or molecules).
(3) The configuration and distribution of bonds.
(4) The polarizability of bonds.

---

[1] For example, the shining of metals is due to their outermost valence electron which scatters the light.

(5) The new distribution of the electronic charge density after the bond has been formed.

These facts plus the knowledge of the interatomic distance (which is determined via diffraction, spectroscopic, or nuclear resonance methods) help us deduce molecular structures, angles, and symmetries.

Since the kind of the bond depends on the electronic charge density distribution between the atoms of the molecule, one cannot distinguish between the kinds of binding in every crystal. Moreover, the same crystal may sometimes exhibit different bonds along its different crystallographic axes. Diffraction methods are used to determine the electronic charge density in a molecule, and that is important, because the distribution of this density may give rise to polar bonds. The polarity in turn can be determined via ESR or NQR, to which we devote separate chapters in this book.

There are four basic methods to determine interatomic distances:

(1) Structural methods.
(2) Optical spectroscopy (with visible light).
(3) Radio-frequency spectroscopy.
(4) Resonance methods.

6.3. Polyatomic Molecules

In polyatomic molecules the electrons behave as if they belonged to many nuclei. The molecular wavefunction can be written as a combination of the atomic wavefunctions:

$$\psi_{\text{molecular}} = c_1\psi_A + c_2\psi_B + c_3\psi_C + \cdots$$

The strongest bond occurs between neighboring atoms because the overlap of the atomic wavefunctions is maximum there. Schematically, the molecule is shown by drawing a short straight line between the atoms (represented by their symbol letters), to show that the probability of finding a shuttling electron is large along this straight line. But we should remember that this is just convenient schematics.

In the water molecule, H_2O, the distribution of the p-electrons is as in Fig. 6.1. The distribution maxima (lobes) are along the three mutually perpendicular axes (six electrons). Figure 6.2 shows three of them. Out of these three, the p_x and p_y-electrons participate in the two bonds with the hydrogen atoms (Fig 6.3) where the electron clouds overlap. But due to the repulsion between the two hydrogen nuclei (protons), the angle between the two bonds widens from 90° to 103° or even 104.5°, and thus the molecule takes a triangular shape (Fig. 6.4). So in H_2O, the bond is effected by the two O–H connections, and with the participation of the two valence electrons of oxygen, $2p_x$ and $2p_y$. The atomic orbitals of these two electrons overlap with those of the single valence electrons of the two hydrogen atoms. It should be noted that the H–H distance in the water molecule is larger than the covalent H–H distance in the H_2 molecule (which is 0.74 Å).

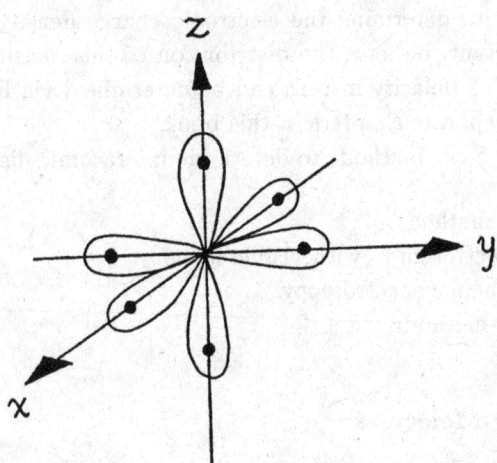

Fig. 6.1.

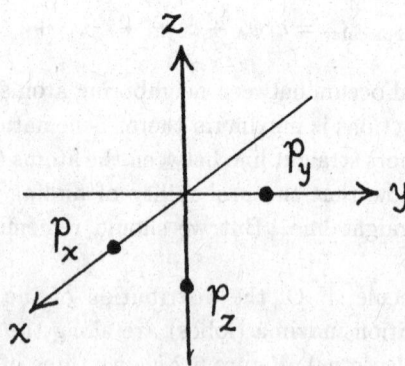

Fig. 6.2.

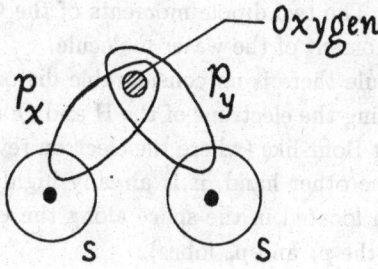

Fig. 6.3.

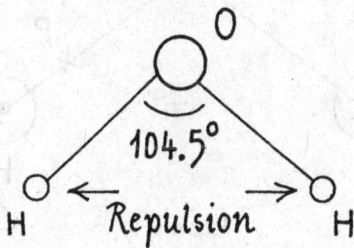

Fig. 6.4.

How long is the O–H distance in the water molecule? It is just the sum of the radii of the two atoms: 0.66 Å + 0.37 Å = 1.03 Å (Fig. 6.5). Here 0.66 Å is the mean Bohr radius of H, and 0.37 Å is the distance where the electronic distribution of O is maximum.

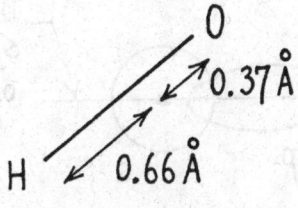

Fig. 6.5.

The water molecule is a polar molecule. It has a dipole moment. The O atom is in a significant position, and the two H atoms stay bound and loyal to it. The O–H bond has a dipole moment whose negative tip is at the O atom (which is more electronegative than the H atom). The positive tip is at the proton of the H atom. The two dipole moments of the O–H bonds add up to give the total dipole moment of the water molecule.

In the water molecule there is no considerable distortion of the electronic charge density concerning the electrons of the H and O atoms, because on one hand the H atom is not Bohr-like (where the electron revolves smoothly about the proton), and on the other hand, it is already highly probable to find an electron of the O atom located in the space along the O–H bond (at point P of Fig. 6.6, because of the p_x and p_y lobes).

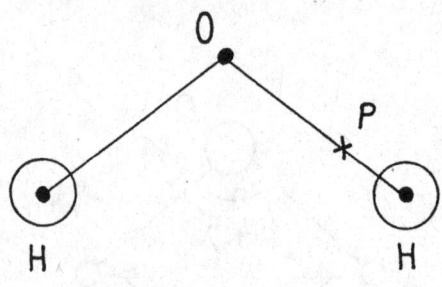

Fig. 6.6.

Molecules with the –OH radical are formed as follows: The configuration of O is $1s^2\,2s^2\,2p^4$. One out of the four $2p$-electrons forms the bond (Fig. 6.7), and thus an –OH ion is formed. The p-electron has a symmetry in a certain direction, but the molecule formed with the –OH radical is not of course linear along that direction.

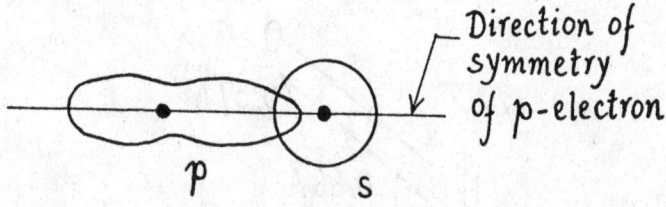

Fig. 6.7.

The bonds between two electrons can be of three types: $s - s$, $s - p$, and $p - p$. That is, $s - s$ means a bond between two s-electrons, and so on. The s-electron exhibits spherical symmetry (Fig. 6.8). If a molecule is formed with the participation of two electrons of the same kind and with spherical symmetry

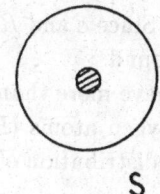

Fig. 6.8.

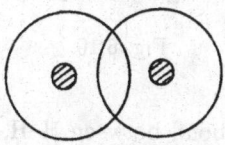

Fig. 6.9.

(Fig. 6.9), the resulting bond is covalent. If the electrons are of different type, the bond is ionic.

If the participating electrons are of different type, say s and p, then an electric dipole moment results. The molecular wavefunction is

$$\psi = \psi_s + \lambda \psi_p$$

where $|\lambda| \neq 1$. From the normalization condition and the fact that the wavefunctions are orthogonal to each other, we find that the electronic charge distribution is shared by the two atoms at the ratio of $1/(1+\lambda^2)$ and $\lambda^2/(1+\lambda^2)$, so that the sum is equal to unity:

$$\frac{1}{1+\lambda^2} + \frac{\lambda^2}{1+\lambda^2} = 1.$$

The charge is distributed this way. The difference between the two gives the electric dipole moment:

$$\mathfrak{p} = \left(\frac{\lambda^2 - 1}{\lambda^2 + 1}\right) eR \tag{6.1}$$

where R is the separation between the two nuclei.

If $|\lambda| > 1$ (as in the water molecule), ψ_p plays a bigger role in the formation of the dipole moment.

The distribution ratios by which the two atoms share the charge are given above. The center of the charge distribution coincides with the overall molecular center of mass.

One can measure p for water. Since e and R are known, one can determine λ. And if λ is known, ψ can be found.

Between two atoms, we may have more than one bond. For example, there is a *double bond* between two oxygen atoms (Fig. 6.10). It is shared by two p-electrons, p_x and p_y, where the distribution of the p_y-electron is far from the molecular axis.

$$O =\!\!= O$$

Fig. 6.10.

In H_2, if there is a single bond between H–H, we call it a σ-*bond*. If there are two bonds, we call them σ and π-*bonds* (Fig. 6.11).

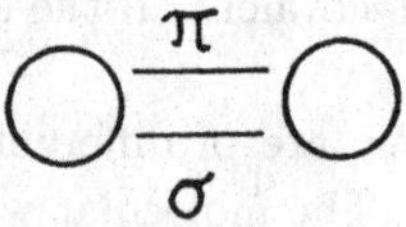

$$O \overset{\pi}{\underset{\sigma}{=\!\!=}} O$$

Fig. 6.11.

The binding energy of a π-bond is less than the binding energy of a σ-bond. For example, in the ethane molecule, the bond between the two carbon atoms is a σ-bond, with binding energy 71 kcal/mol (Fig. 6.12). This distance between the two C atoms in ethane is 1.55 Å. In acetylene (Fig. 6.13) the binding energy is 119 kcal/mol, about twice as large as that of the ethane. The separation between C–C in acetylene is 1.33 Å.

If we know the distance between atoms in a molecule, we can get an idea about the kind of the bond. That is why measuring this distance is important.

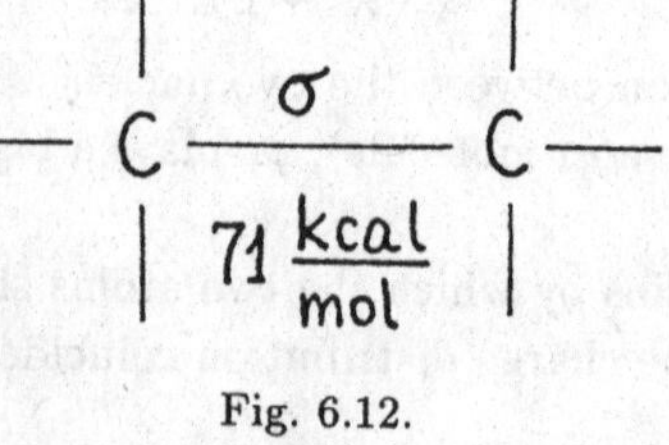

$$-\,\overset{|}{\underset{|}{C}}\,\overset{\sigma}{\underset{71\,\frac{kcal}{mol}}{-\!\!\!-\!\!\!-\!\!\!-}}\,\overset{|}{\underset{|}{C}}\,-$$

Fig. 6.12.

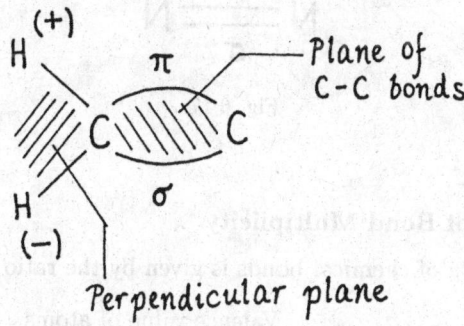

Fig. 6.13.

In the case of a σ-bond the molecule is perfectly symmetric; in the case of a double bond (σ and π), we take the plane in which the two bonds (σ and π) lie. Then the atoms connected to C form a plane which is perpendicular to the bond plane (Fig. 6.14). One atom is above (+) the bond plane, and the other one below (−). This is better shown in Fig. 6.15.

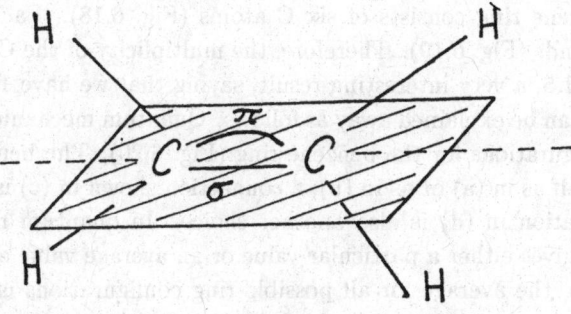

Fig. 6.14.

Fig. 6.15.

Notice that while the σ-bond is along the C–C axis, the π-bond curves itself around (Fig. 6.16).

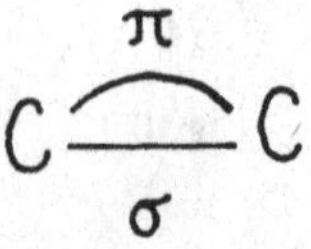

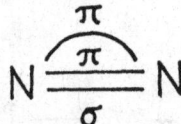

Fig. 6.16.

In nitrogen, N_2, we have three bonds: σ, π, π (Fig. 6.17).

$$N \overset{\overset{\displaystyle \pi}{\overset{\frown}{\pi}}}{\underset{\sigma}{=\!=\!=}} N$$

Fig. 6.17.

6.4. Definition of Bond Multiplicity

The *multiplicity* of chemical bonds is given by the ratio

$$\text{Multiplicity} = \frac{\text{Valence value of atoms}}{\text{Number of bonds formed}}.$$

In acetylene the multiplicity is $4/2 = 2$ (because the valence of C is 4, and two bonds are formed).

The benzene ring consists of six C atoms (Fig. 6.18). Each C atom can form four bonds (Fig. 6.19). Therefore, the multiplicity of the C atoms here is $(4-1)/2 = 1.5$, a very interesting result, saying that we have 1.5 bond per C atom! This can be explained away as follows: Quantum mechanically, there are several configurations for the benzene ring (Fig. 6.20). The benzene molecule can form itself as in (a) or as in (b); a connection shown in (c) is also possible; the configuration in (d) is also another choice. In quantum mechanics, the observation gives either a particular value or an average value of the observed quantity. So, the average for all possible ring configurations is 1.5 for the C atoms. In other words, one should find the atomic wavefunction of each C atom and then take the average of them.

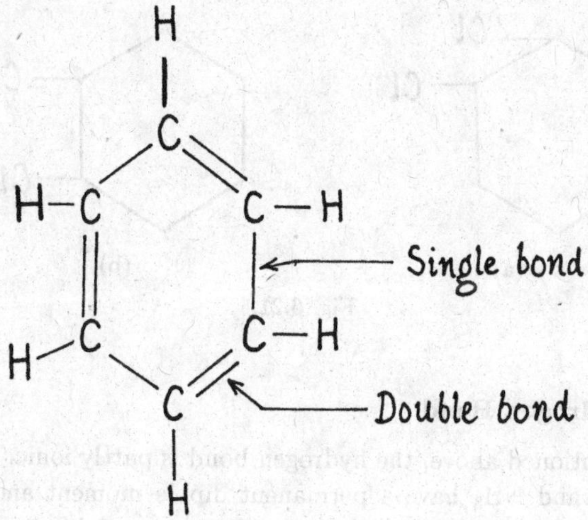

H

C

H—C   C—H   ←— Single bond

H—C   C—H

H   C   ←— Double bond

H

Fig. 6.18.

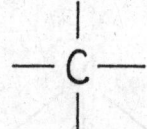

Fig. 6.19.

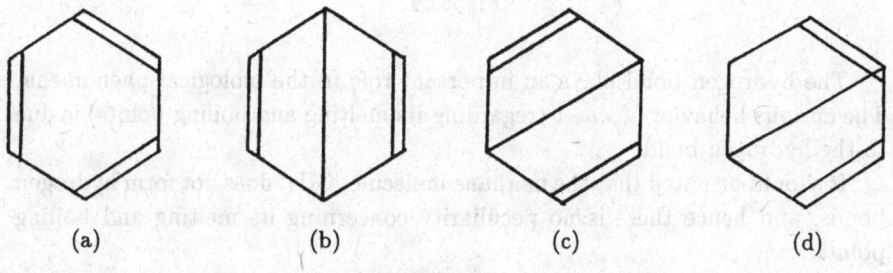

(a)   (b)   (c)   (d)

Fig. 6.20.

Similarly, there are two different dichlorobenzenes (Fig. 6.21), but there is no chemical difference between them. The Cl atoms form single bonds. There is no orthodichlorobenzene.

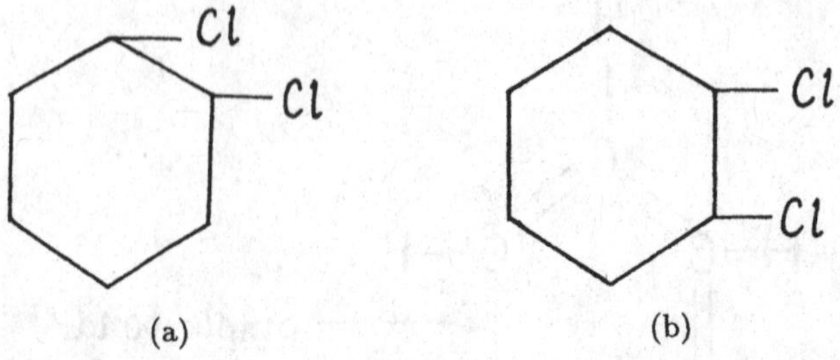

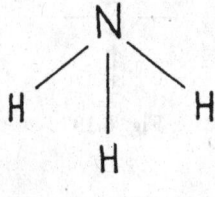

(a)   (b)

Fig. 6.21.

6.5. The Hydrogen Bond

As it is mentioned above, the hydrogen bond is partly ionic. So, molecules like H_2O, HF, and NH_3 have a permanent dipole moment and a dipole interaction between them. Thus they can form crystal binding at the right temperature, because the dipole interaction is attractive. The dipole moment of each molecule attracts and orients its counterparts of neighboring molecules.

Figure 6.22 shows the NH_3 molecule.

$$N$$

H   H

H

Fig. 6.22.

The hydrogen bond plays an important role in the biological phenomena. The curious behavior of water (regarding its melting and boiling points) is due to the hydrogen bond.

It should be noted that the methane molecule, CH_4, does not form hydrogen bonds, and hence there is no peculiarity concerning its melting and boiling points.

6.6. Metallic Bonds

Metallic bonds are not formed when two atoms come close together. They are formed only when many metal atoms get together. Thus we can consider a metal as a single molecule consisting of an infinite number of atoms.

So, when we bring two Cu atoms close together, no bond is formed. For the atomic wavefunctions to overlap, many atoms must come close together, not just two. As the wavefunctions overlap the energy increases, and hence each atom gets more and more able to form bonds (*i.e.*, draw energy and store it as order, just like the stability of an economic system as it is fed with more and more capital). As the atom extends its influence over further neighboring atoms, we observe a gathering of atoms, giving rise to a molecular structure, *i.e.*, the metallic crystal. The form of the structure depends on the less costly way of formation, that is, the state which requires the minimum energy.

Metallic bonds are formed between electropositive metal atoms. There are free electrons that travel between the atoms throughout the material, giving rise to the electrical and thermal conductivity. (These two conductivities exist together in metals.)

6.7. Van der Waals Bonds

The Van der Waals bonds are related to the polarization of the atoms. They are discussed sufficiently in Section 6.1 of this chapter.

6.8. Covalent Bonds

Covalent bonds are formed between two electronegative[2] metalloids, or even between two atoms of the same kind, *e.g.*, as in the H_2 molecule (Fig. 6.23), where two H atoms set up a bond. They are mentioned in Section 6.1 above.

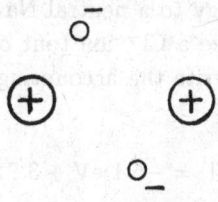

Fig. 6.23.

6.9. Heteropolar Ionic Bonds

Ionic bonds in molecules can be explained via the Coulomb potential. They are formed between two ions: A positive metal and a negative metalloid that attract each other. In other words, the two atoms that form the bond are of different electronegativity.

---

[2]The elements are electropositive, electronegative, or inert in terms of valence.

The conditions for strong bonds mentioned in Chapter 4 hold valid here. Also, the partitioning into σ and π-bonds occurs here, but the molecule is not symmetric.

For an AB type heteropolar ionic molecule, the molecular wavefunction is

$$\psi = c_A \psi_A + c_B \psi_B = N(\psi_A + \lambda \psi_B),$$

as discussed in Chapter 4.

The main difference between heteropolar and homopolar molecules lies in the formation of their dipole moments. As mentioned above, if $|\lambda| > 1$, then ψ_B plays a more important role in the bond, *i.e.*, the valence electrons spend more time next to atom B than next to A. In other words, after the formation of the molecule, the electronic charge density is redistributed so as to increase in the vicinity of atom B.

Equation (6.1) above shows that the dipole moment of the molecule is proportional to the *polarity factor* λ.

An heteropolar molecular bond is an admixture of ionic and covalent characters. Each character contributes by a percentage. The polarity of the bond depends on the difference between the electronegativities of the two atoms.[3]

Alkali halides (NaCl, KCl, etc.) are typically ionic compounds.

6.10. Ionic Bond Energy

How much energy is needed to form the bond? We should first make a Na^+ ion. So, we supply energy to a neutral Na atom, equal to its ionization energy $E(Na)$. Then we make a Cl^- ion (out of a neutral Cl atom) and we gain an energy $E(Cl)$. We write the accounting equation for the total (net) energy needed:

$$-E(Na) + E(Cl) = -5.1 \text{ eV} + 3.7 \text{ eV} = -1.4 \text{ eV}.$$

But this is not the answer, because we still have an energy gain due to the Coulombic interaction between the ions, given by the potential

$$V = \frac{e}{R} = \frac{4.8 \times 10^{-10} \text{ statCb}}{2.5 \times 10^{-8} \text{ cm}} = 1.9 \times 10^{-2} \text{ statvolt}$$

---

[3]The word *heteropolar* consists of two words of Greek origin: *Heteros* ('ἕτερος) = another, different, and *polos* (πόλος) = pole. The word *ionic* comes from the Greek word *ion* (ἰών, ἰοῦσα, ἰόν) which is the active participle of the irregular verb *eimi* (εἶμι) = to proceed, πορεύομαι. So, *ion* means "the proceeding one".

where $R = 2.5$ Å in NaCl, and the atoms are monovalent; so the interaction energy is

$$E = eV = \frac{e^2}{R} = 5.7 \text{ eV},$$

and the overall net energy is

$$E_{\text{binding}} = 5.7 \text{ eV} - 1.4 \text{ eV} = 4.3 \text{ eV}$$

which is the energy needed to form a NaCl molecule. It corresponds to 101 kcal/mol which is in close agreement with the experimentally found dissociation energy of 98 kcal/mol.

The calculation above is classical. A quantum mechanical calculation (outlined in Section 4.4 of Chapter 4) is also possible.

The ionic character of the NaCl bond can be proven by two methods:

(1) X-ray diffraction off NaCl crystals: The scattering intensity gives valuable data.

(2) Observation of the dipole moments of gaseous NaCl (gas molecules).

6.11. Summary

A molecule takes in energy from the environment and stores it as order and organization, just like a business must extract capital from somewhere to build its stable existence. This is not against the flow of entropy, because a molecule is not a closed system; it is an open system, interacting with its environment.

A molecular bond means emanation of binding forces between the atoms of the molecule. These forces bind the atoms together. Bonds can be formed between molecules, too; the treatment is the same.

As far as the interatomic forces are concerned, there are three kinds of them, giving rise to the three most important types of bonds:

(1) Covalent bonds (due to covalent binding forces).

(2) Ionic bonds (due to Coulombic interactions).

(3) Metallic bonds (due to forces between metal atoms).

Inert gases can form molecules, too (by the mediation of Van der Waals forces). Inert gases can become liquids at low temperatures, and hence attractive forces arise.

One should bear in mind that the ultimate source of any bond is electrostatic. Further, by the notion of "bond" we also mean the energy necessary to break the bond. Finally, there is no simple way to describe a molecule.

6.12. The H_2^+ (Hydrogen Molecule Ion) and the H_2 Molecule

Molecules like O_2, Cl_2, H_2, and N_2 are diatomic of type A_2 (or dumb-bell), where two atoms of the same kind are connected together by a covalent bond. An *ionized* H_2 molecule is an assembly of two protons and an electron, constituting a sovereign system. Actually, it is a three-body problem with no exact solution in general. The approximate assumption is that the protons are fixed at rest, while the electron moves; this is sensible, because $v(\text{electron}) \gg v(\text{proton})$ due to $m(\text{proton}) = 1836m(\text{electron})$. If the line between the protons is along x and the origin is in the middle, the Schrödinger equation — for the electron (written for some particular separation r between the protons) — is, in three dimensions,

$$-\frac{\hbar^2}{2m}\nabla^2\psi - \frac{e^2}{4\pi\varepsilon_0}\left[\frac{1}{\left(x-\frac{r}{2}\right)^2 + y^2 + z^2} + \frac{1}{\left(x+\frac{r}{2}\right)^2 + y^2 + z^2}\right]\psi = E\psi$$

where $\nabla^2 = \partial^2/\partial x^2 + \partial^2/\partial y^2 + \partial^2/\partial z^2$. For large r we obtain the known eigenenergies of the H atom, as if the proton were tied and belonged to either proton only, *i.e.*, the case of an H atom and a proton accompanying it, off at r. For small r (a few Å), a (nondegenerate) ψ should be of either even or odd parity (because V is symmetric). As r varies from large to very small values, the ground-state ψ changes shape. Figure 6.24 shows $\psi(x)$, *i.e.*, ψ evaluated along x for different values of the ion separation r (notice the symmetric ψ_0^s and the antisymmetric ψ_0^a where the subscript indicates ground state). The odd parity is the case of the He^+ ion whose E_0 is $4 \times (-13.6) = -54.4$ eV. The even parity is just the case of $E_{\text{binding}} = -13.6$ eV for an electron in an H atom.

The system's energy (*not* that of the electron only!) becomes minimum at the equilibrium separation R. Upon mutual approach (of the protons), the electron's energy goes down, and the electrostatic repulsion energy (between protons) goes up. Again, the zero energy is conventionally chosen when the protons are far away from each other at infinity ($r \to \infty$), with the electron attached to either proton, *i.e.*, an H atom and a proton far away. The system's

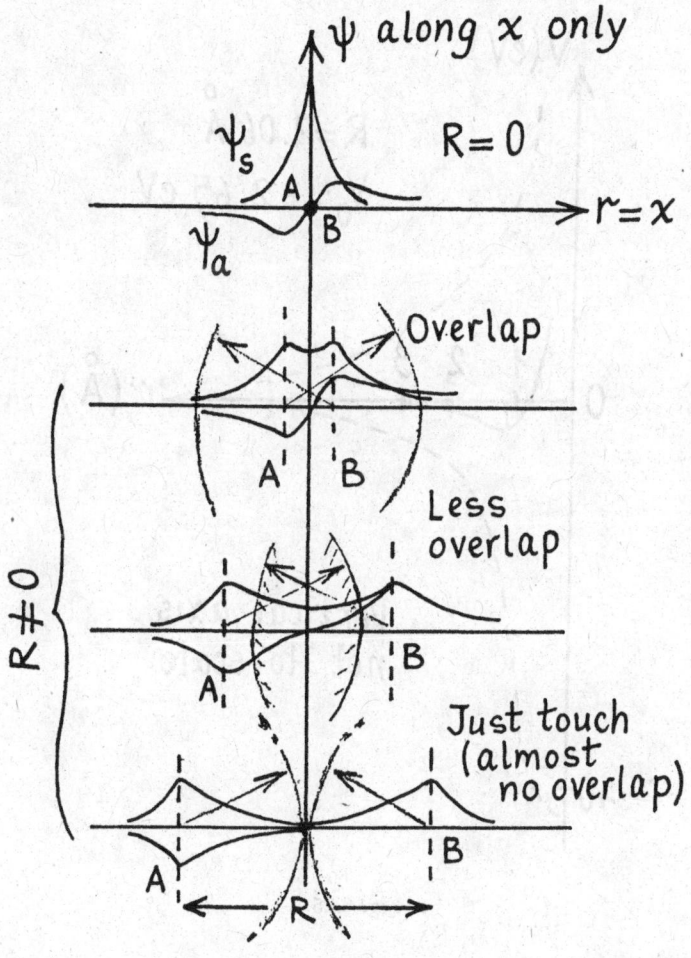

Fig. 6.24.

energy is found if we form the difference of the two previous energies, with a minimum of -2.65 eV at $R = 1.06$ Å (Fig. 6.25 where the upper dashed line is the proton repulsion energy, $+1.44/r$ eV, the lower dashed line is the energy reduction due to the resonant exchange of electrons between protons, and the solid line is the potential energy of H_2^+ vs. the interproton separation r), and refers to a zero-energy choice when a neutral H atom and a p^+ are at $r = \infty$ one relative to the other.

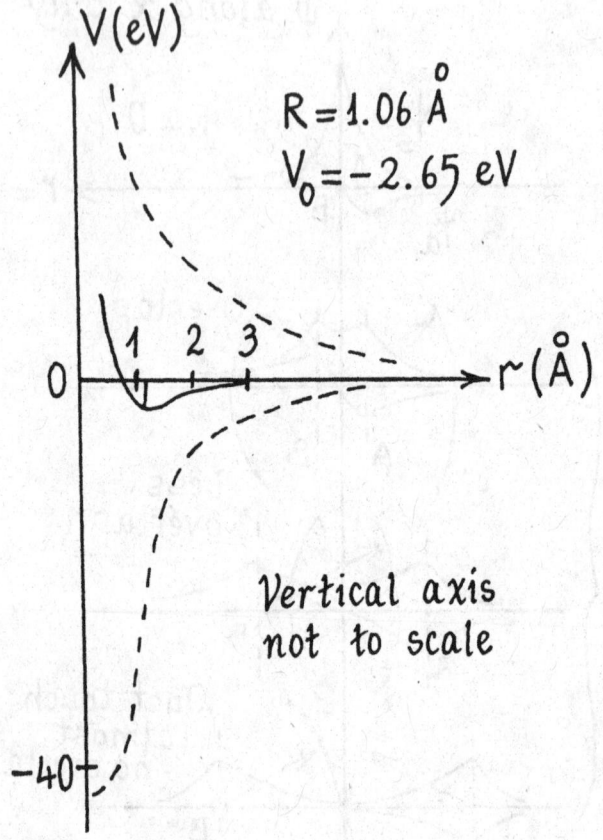

V(eV)

$R = 1.06 \overset{\circ}{A}$
$V_0 = -2.65 \ eV$

0

1   2   3

$r(\overset{\circ}{A})$

Vertical axis
not to scale

−40

Fig. 6.25.

Figure 6.26 shows the electronic charge distribution $|\psi|^2 = \psi^*\psi$ along x (for the equilibrium $r = R$) and the isoprobable curves of equal $|\psi|^2$ as drawn by Burrau, for $0.9|\psi|^2_{max}$, $0.8|\psi|^2_{max}$ etc., up to $0.1|\psi|^2_{max}$. Notice that it is too likely (highly probable) to find the electron between the protons — and it makes sense, because the bond there in between is just an electron shared by the two protons (nuclei). Again, we take $\psi = \psi(x,0,0)$.

Here only the even-parity ψ provides binding; the odd-parity ψ^a_0 gives an increasing E as r decreases, in other words, an overall repulsion only (between the protons), and thus no binding. Further, this single-electron bond is strong here, whereas if the two ions (nuclei) have opposite charge, it is feeble.

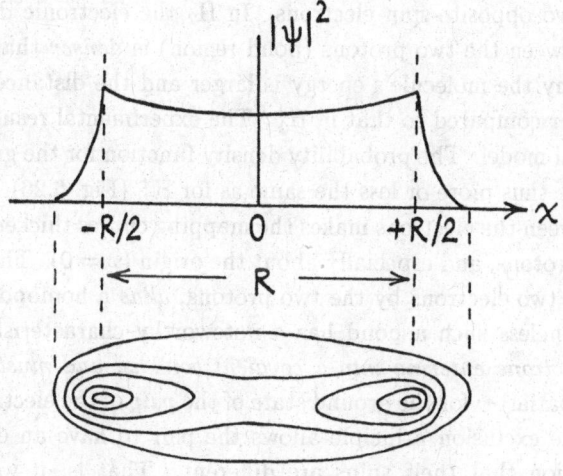

Fig. 6.26.

The H_2 molecule can come into being if an H_2^+ catches a free electron (*electron capture*). The first (approximate) calculations were made by Heitler and London in 1927. They wrote down the Schrödinger equation of the system and solved it approximately, using a method that followed the reasoning stated above. If r is large, the ground state of the system is that of two free, independent H atoms very far from each other. If, on the other hand, the separation r of the protons tends to nothing, we have a He atom. These are the two extreme limits. For r values in between, however, the case is not obvious; on the contrary, it is troublesome because of complications. Both the potential function V and the Hamiltonian (operator) $\mathcal{H}$ are again symmetric (in x), and hence the ψ functions are either of even or odd parity (in x), that is, the same situation as in H_2^+. The theoretical computation gives the ground-state energy of the system; it is a function of r and its variation can be represented as well by the solid line in Fig. 6.25. The equilibrium occurs at $R = 0.74$ Å, and we have $E_{binding} = 4.48$ eV which means that a neutral H_2 molecule has a stronger and tighter bond than H_2^+, and the two protons are closer to each other.

H_2 is a simple molecule and *different* than H_2^+ (its molecule ion) because it has two electrons. Its solution resembles the method of molecular orbitals. The bond in H_2 in covalent.[4] The ground state of H_2 corresponds to the state

---

[4]The literal meaning of the word *covalent* in Latin is "of the same strength".

of H_2^+ with two opposite-spin electrons. In H_2 the electronic distribution in the region between the two protons (bond region) is *denser* than that in H_2^+, and that is why the molecule's energy is larger and the distance between the protons shorter compared to that in H_2^+. The experimental results agree with this theoretical model. The probability density function for the ground state of (neutral) H_2 is thus more or less the same as for H_2^+ (Fig. 6.26), save that the repulsion between the electrons makes the mapping curves thicker in the region between the protons, and especially about the origin ($x = 0$). The bond is just the sharing of two electrons by the two protons, *alias* a homopolar (covalent) bond. Nevertheless such a bond has a noteworthy characteristic: *The spin of the two electrons entering into a covalent bond is, and must be, opposite.* Why? The (spatial) ψ for the ground state of the *pair* of the electrons is of even parity, and the exclusion principle allows the pair to have an even function, on the condition that their spins are different. That is, if we interchange the spatial and spin coordinates (of the two electrons), the result has to be antisymmetric. Therefore, if the spins are the same (parallel), one of the two electrons has necessarily to change state and get excited to a higher energy, because it cannot stay there otherwise, according to the principle. But such a change would not ensure any binding at all — no H_2 would be produced. Therefore, the only possibility of forming H_2 out of two H atoms is that the two electrons have antiparallel (opposite) spins.

To give some numerical data, we have (in equilibrium) $R(H_2) = 1.38a_0$ $= 0.74$ Å, $R(H_2^+) = 2a_0 = 1.06$ Å, and $E_{\text{dissociation}}(H_2) = 4.72$ eV $\sim 2E_{\text{dissociation}}(H_2^+)$. The internuclear Coulombic repulsion (in equilibrium) is 19.3 eV for $H_2 > 13.5$ eV for H_2^+, because the nuclei are closer together. There must be added a force of 17.8 eV due to the repulsion of the electrons. The dissociation energy of the molecule is equal to the difference between two energies:

$$E_{\text{dissociation}}(\text{molecule}) = E(\text{molecule}) - E(\text{of two H atoms})$$

$$= -(2 \times 34) + 17.8 + 19.3 - (-27.2) = -3.7 \text{ eV}$$

where 34 eV is the electron energy in molecular orbital (taken twice); 17.8 eV is the electronic and 19.3 eV is the nuclear repulsion. The result is less than the experimental value by 1 eV, due to the peculiarity of the theoretical computational methods used for H_2, for which corrections are possible to improve the theoretical result.

6.13. Electrostatic Energy of Ionic Crystals (NaCl)

To directly measure interatomic forces is not a trivial job. We measure energy changes instead, as a crystal goes from an atomic array to another (chemical change), where such energies are Coulombic, because the atomic forces are so.

Consider the ionic NaCl lattice (two ions, one positive and the other negative, attract each other electrostatically, and a bond is formed). We think of Na^+ and Cl^- as hard spheres touching each other (due to Coulombic attraction) and arranged in a cubical structure whose section is like a checkerboard, with Na^+ and Cl^- alternating in position. Figure 6.27 shows the arrangement which is the same in the other two dimensions, too. The spacing of ions (lattice constant) is $d = 2.81$ Å.

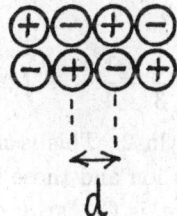

Fig. 6.27.

If the ions are pushed together too much, a repulsive force arises that balances the attraction, and the bond is hence stable and steady.

This model of rigid spheres forming a three-dimensional cubical lattice structure was set up by the help of X-ray diffraction.

How much energy is needed to rip the lattice apart and destroy the crystal order entirely, to get separate ions? This energy will be equal to the energy that dissociates the NaCl molecules into Na^+ and Cl^- ions, plus the vaporization heat of NaCl.

The total energy of separation of NaCl to ions is (experimentally) 7.92 eV/molecule. Since there are $N_A = 6.02 \times 10^{23}$ molecules/mol (Avogadro number), the vaporization energy is 7.64×10^5 Joules/mol. And since 1 eV/molecule = 23 kcal/mol, the *chemical dissociation energy* of table salt is 183 kcal/mol. This can be figured out also theoretically, by finding the amount of energy needed to pull NaCl apart, *i.e.*, the work needed to break all the ion pairs, in other words, the sum of the potentials of the pairs. That is, we choose any ion and calculate its potential energy with each of the other ions around, one by one, and then sum up. To ensure correct counting and not overcounting, we divide by 2, because the sum is double the energy/ion, as it is shared by pairs of charges. So, half the sum is the energy/ion. However, we are looking for the energy/molecule, and since a molecule has *two* ions, we multiply back by 2. Therefore the sum is just what we want.

Any ion interacts electrostatically with (each of) its nearest neighbor(s) as $\sim e^2/d$ (where the lattice constant d is the distance between the centers of the spherical ions), if each ion is monovalent (one-electron). Evaluating, we have $e^2/d = 5.12$ eV. Now we sum, considering ion pairs (interaction in pairs) along a straight line on either side of the chosen ion. If the latter is Na^+, it has two Cl^- ions (each at d) on either side, two Na^+ ions at $2d$ on either side, and so on, so that the sum along the line is

$$V_{partial} = \frac{e^2}{d} \left(-\frac{2}{1} + \frac{2}{2} - \frac{2}{3} + \frac{2}{4} + \cdots \right)$$

$$= -\frac{2e^2}{d} \left(1 - \frac{1}{2} + \frac{1}{3} - \frac{1}{4} + \cdots \right) \simeq -\frac{2e^2}{d} \ln 2 = -1.386 \frac{e^2}{d} ,$$

because the series converges to $\ln 2$. This is not the end. We still have the interactions between the chosen ion and those belonging to the next adjacent line. The nearest neighbor there, is Cl^- at d; on its either side there are two Na^+ at $\sqrt{2}d$; the next closest pair (two Cl^-) is at $\sqrt{5}d$ etc. For the whole chain of that line we have

$$V_{next\ line} = \frac{e^2}{d} \left(-\frac{1}{1} + \frac{2}{\sqrt{2}} - \frac{2}{\sqrt{5}} + \frac{2}{\sqrt{10}} - \cdots \right) ,$$

and since the structure is three-dimensional, there are four such lines. Of course, there are also the next to the next lines, but as we move away from the chosen ion, the strength of V diminishes as $\sim (distance)^{-1}$, and becomes unimportant after some extent (distant neighbors do not contribute significantly). If we include all the important lines and sum, the final result becomes $V = (-1.747)\ e^2/d > V_{partial}$. Letting $e^2/d = 5.12$ eV, we obtain $V = -8.94$ eV, a bit higher than the experimental result. Why? Because we did not include the repulsion between ions when close. Notice that the ions are not in reality hard spheres; their silhouette is made of electron clouds which are compressed a bit as the ions come close together (like rubber balls), and some energy goes to their deformation (when they are pulled apart, this energy is freed back). So actually, the energy required to pull the ions apart is (a bit) less than the theoretical result (which ignores the actual fact that repulsion favors our effort to beat the Coulombic attraction and break the bond).

If we know the repulsive force, we can correct our theoretical model and result. For example, if we measure the macroscopic compressibility of the bulk, we can get an idea about the dependence of the repulsive force between ions. Such a measurement reveals that the repulsion is equal to minus 1/9 of the Coulombic attraction. Subtracting this contribution, we find V(dissociation energy) ~ 8 eV, closer to the experimental value of 7.92 eV. It is still at variance with it, because we have not included the kinetic energy of crystal vibrations.

This model is powerful because it reveals the following:

(1) The greatest part of the crystal energy is electrostatic: The lattice stands and is kept together by electrical Coulombic forces.

(2) A phenomenological property of the macroscropic bulk is obtained from atomic-level facts: We understand substance behavior in terms of atomic properties.

6.14. Energy Levels of Diatomic Molecules and Level Diagrams

The diagram of the energy levels of a molecule must include the split atomic energy levels. This splitting depends on how much the wavefunctions overlap. (Does the overlap increase or decrease as the interatomic separation r decreases?) Nevertheless one expects to have a molecular energy-level diagram which is the same as the atomic one, *both* for $r \to 0$ and $r \to \infty$. The former situation means that we have two atoms one atop of the other, with coinciding nuclei, giving a combined atom A + B. The latter case corresponds to two free atoms far apart. Somewhere in between (for $r = r_0$) we have the molecule (AB).

Figure 6.28 shows the states of an A_2 type molecule (middle column) and the transition from the extreme case of a unified (merged) two-atom system (right) to a single atom (left). In the middle we have the diatomic molecule made of two identical atoms. In this case we took O_2. Notice that the lowest molecular-energy state is $\sigma 1s$ which goes over to $1s$ which is the lowest level of the unified system. The next molecular state is $\sigma^* 1s$ which goes over to $2p$ (as the corresponding wavefunctions have mutually similar symmetry). The vertical separation between the levels depends on the interatomic distance. For example, energy levels lying too close to each other (like $\sigma 2p$ and $\pi 2p$) can overlap (especially for high principal quantum numbers). If we know the ordering of the levels, we can obtain valuable information about certain properties of the molecule (electronic structure, binding force, magnetic properties etc.).

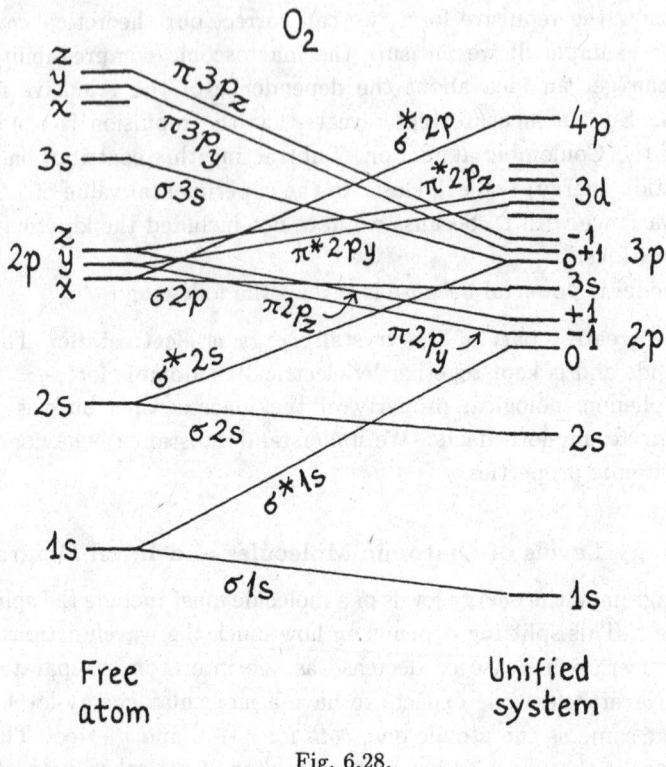

Fig. 6.28.

Figure 6.29 shows the ordering of the energy levels of an A_2 type molecule. The column on the left presents the atomic states, and the one on the right the molecular states.

The table next page gives the electron distribution in some diatomic molecules and their ions (A_2^+ type ions). The single bond is taken as unit. So, the multiplicity is calculated as the difference between the numbers of electrons. The molecular state electrons lie in states above and below the unsplit atomic states. In the last column, the index on the left shows the spectral-line multiplicity (1 = singlet, 2 = doublet, 3 = triplet).

The Greek capital letter denotes the total angular momentum of the electrons with respect to the axis, and the designation "even" or "odd" refers to the molecular symmetry.

The data referring the simple molecules and listed in the table are experimental. We immediately see that the formation of Be_2 (and inert gas molecules) is not possible ($E_{\text{dissociation}} = 0$ eV). The bond multiplicity

Molecule or ion	Molecular states										Number of valence electrons		Bond multiplicity	$E_{dissociation}$ (in eV)	Spectroscopic symbol
	$\sigma 1s$	$\sigma^* 1s$	$\sigma 2s$	$\sigma^* 2s$	$\sigma 2p$	$\pi_y 2p$	$\pi_x 2p$	$\pi_z^* 2p$	$\pi_y^* 2p$	$\sigma^* 2p$	Bonding	Antibonding			
H_2^+	1										1		1/2	2.64	$^2\Sigma$ even
H_2	2										2		1	4.72	$^1\Sigma$ even
He_2^+	2	1									2	1	1/2	2.5	$^2\Sigma$ odd
He_2	2	2									2	2	0	0	$^1\Sigma$ even
Li_2	2	2	2								2		1	1.4	$^1\Sigma$ even
Be_2	2	2	2	2							2	2	0	0	$^1\Sigma$ even
N_2^+	2	2	2	1	2	2	2				7	2	2.5	6.35	$^2\Sigma$ even
N_2	2	2	2	2	2	2	2				8	2	3	7.38	$^1\Sigma$ even
O_2^+	2	2	2	2	2	2	2	1			8	3	2.5	6.48	$^2\Sigma$ even
O_2	2	2	2	2	2	2	2	1	1		8	4	2	5.08	$^3\Sigma$ even
F_2^+	2	2	2	2	2	2	2	2	1		8	5	1.5	0	$^2\Sigma$ even
F_2	2	2	2	2	2	2	2	2	2		8	6	1	2.8	$^1\Sigma$ even
Ne_2	2	2	2	2	2	2	2	2	2	2	8	8	0	0	$^1\Sigma$ even

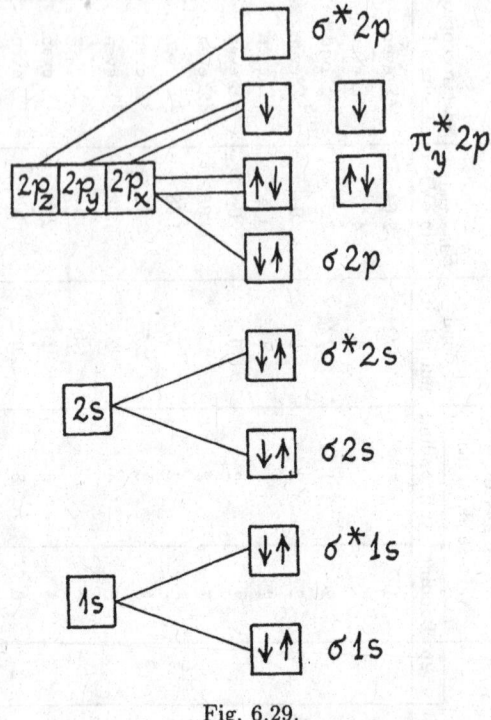

Fig. 6.29.

depends on the number of the available binding electrons in forming molecules out of atoms. In molecules with single bonds (H_2, Li_2), the binding energy decreases with increasing r, the interatomic separation. (Remember that as r increases, the value of the overlap integral S decreases.)

The molecules H_2, Li_2, N_2, and F_2 are diamagnetic: The molecular-energy levels are filled with electrons in pairs each, a situation that affects the spin and orbital angular momenta. But O_2 — which is paramagnetic — is an exception: It has a magnetic moment twice the value of μ_B. All its levels are filled up to π^*2p which has only two electrons, and the total number of electrons is 16. So, π^*2p has two cells and is doubly degenerate. (See Fig. 6.29.) According to a principle analogous to Hund's rules, these two cells contain one electron each, of parallel spins. These two electrons of π^*2p prevent binding, *i.e.*, they are *antibonding* electrons. Quantum mechanics has thus succeeded to fully explain the fact that O_2 is paramagnetic[5] although it is diatomic.

---

[5]Another success is the exact calculation of the energy levels of, and interatomic separation in H_2.

Consequently, if one of these two electrons leaves the molecule, E_{binding} increases. Therefore, $E_{\text{binding}}(O_2^+) > E_{\text{binding}}(O_2)$, *i.e.*, we need more energy to make the ion than to make the neutral molecule.

The number of valence electrons in N_2 are two less than those in O_2, and hence, if one electron is detached from N_2, E_{binding} *decreases*: $E_{\text{binding}}(N_2^+) < E_{\text{binding}}(N_2)$. All the $2p$-electrons of N_2 are *bonding* electrons, *i.e.*, willing to form a bond. That is why if one of them leaves, $E_{\text{dissociation}}(N_2^+)$ becomes less than that of N_2.

What happens when many-electron atoms combine to form a molecule? Naturally, the overlap of their inner electrons is not as much as that of their outer electrons. If we consider K-electrons, and when the molecular states $(\sigma 1s)^2$ and $(\sigma^* 1s)^2$ are all full, the total electron density in these states will be

$$\frac{2(\psi_A + \psi_B)^2}{2(1+S)} + \frac{2(\psi_A - \psi_B)^2}{2(1-S)} = \left(\frac{2}{1-S^2}\right)(\psi_A^2 + \psi_B^2 - 2\psi_A\psi_B S)$$

where the denominators can be evaluated from the normalization condition

$$N^2 \int \psi^2 \, dV = 1$$

over all space. For K-shells, the amplitude of S drops fast, as the number of electrons increases. For example, in H_2 we have $S = 0.68$ (or 68%), in Li_2 $S = 0.01$, and in C (single-bonded diamond atom where $r_0 = 1.54$ Å) $S = 10^{-5}$. If $S = 0$, the equation above lapses to $2(\psi_A^2 + \psi_B^2)$, the sum of the unperturbed electron clouds in the K-shells. That is why one can conveniently separate the electrons into two categories: Those partaking of the bond (outer electrons) and those unaffected by the chemical binding (inner electrons). The first group is just the valence electrons. The second group's energies are just the atomic energies of the atom they belong to. This is equivalent to saying that in molecules (and crystals) the energy levels of the atomic inner electrons remain unchanged and unperturbed, and indeed, this is the case. (It is nothing else than the physical model about the linear combination of the atomic orbitals, a method so useful in quantum physics.)

6.15. Aspects of the Theory of Binding (Chemical Bonds)

Bonds involving electron pairs are almost equally strong whether the bound atoms are of the same kind (A_2) or of different kind (AB). In the case of one electron between two different atoms there is a greater probability to find the electron near the atom where the electron minimizes its energy. That is, as nature always tends toward configurations of the least energy, the electron spends most of its time in the vicinity of whichever atom offers greater tranquility (= lowest energy), and hence $|\psi|^2$ of the electron varies accordingly. If the bond involves two electrons, there is a Coulombic repulsion between them, forcing the electrons to stay apart. However, if either atom's electron affinity is strong enough, it can overcome the Coulombic repulsion and give rise to an ionic bond. An ordinary covalent bond involves two electrons only. In the event that there are three electrons on the scene, the third one has to have higher energy; such an excitation, however, weakens the binding, and that is indeed what we observe in H_2^-. In other cases the excitation lets no bond formation at all, *e.g.*, in HeH. Finally, we *can* have several covalent bonds in a molecule between its atoms, like in the case of O_2 (two bonds), N_2 (three bonds), or CH_4 (methane, where the C atom forms four covalent bonds, one with each H).

It is interesting to note that the Li_2 molecule is possible (spectroscopy attests to its existence). This diatomic molecule has $E_{dissociation} = 1.14$ eV and $r_0 = 2.67$ Å. The theoretical calculations concerning Li_2 are similar to those concerning H_2, and that is only understandable, because they cannot be otherwise. In Li_2, however, the atomic $1s$ states are full, and the valence electrons belong to $2s$. If we calculate $E_{dissociation}$ this way, we find 1.09 eV, a result in agreement with the experimental one. Since the repulsive force is stronger, the internuclear separation is larger in Li_2; that is why it becomes easier to dissociate Li_2, and hence its $E_{dissociation}$ is less than $E_{dissociation}$ of H_2, as less costly.

Molecules are three-dimensional and this complicates the mathematics involved in their description. We know that Beethoven was able to divinely transsubstantiate human feelings or natural beauties into music, thereby creating immortal monuments of art; similarly, a scientist has to mathematically express what is going on in molecules and atomic binding, *i.e.*, he or she has to transfer real physics into mathematics, so that the phenomena can be described and written down not in tedious words, but in an artful, elegant, brief, and consistent mathematical language that carries the beauty of a minor art. In the light of this thought, we can attempt to describe atomic binding (of two atoms) mathematically, by the use of simple square potential wells in one dimension. Consider two square wells of depth V_0 each, and an electron contained in each. The problem is simple, quantum mechanical. In Fig. 6.30(a) the wells are well-separated (two free atoms); in (b) they are close together, and in (c) almost merged (without a barrier in between — one single well). The symmetric ψ (for the ground state of the electron) is shown above, and the antisymmetric below. Each well is an atom having a bound electron with its ψ. The plot in (b) is the approximate description of a diatomic molecule where the two atoms (wells) are close together at some separation distance (with a narrow wall between them), *i.e.*, the wells are located at the positions of the nuclei.

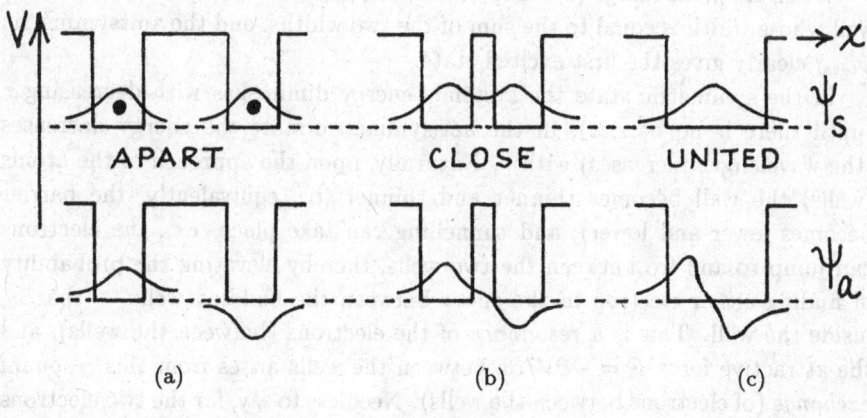

Fig. 6.30.

The wavefunction diminishes at the walls, as expected (the particle is mostly at the center of the well), and there is some nonzero part (tail, probability) outside the two walls. Notice that the situation resembles the case of the modes of a vibrating string fixed at two ends, with its lowest and its excited modes.

Both ψ and $-\psi$ are eigenstates here. In plot (a) the atoms do not interact, because they are considerably separated far apart, and no matter what the sign of ψ is, the *system*'s energy is the same. Interaction sets on when the wells get closer and closer; then the sign of ψ_{total} is immaterial. However, since the symmetric ψ_{total} gives lower energy, different eigenenergies are obtained, depending on whether ψ_{total} is symmetric (same ψ in both wells) or antisymmetric (opposite ψ). Why? Because the quantity

$$\langle E \rangle = \int_{-\infty}^{\infty} \psi^*(x, t) \left(-\frac{\hbar^2}{2m} \frac{\partial^2}{\partial x^2} \right) \psi(x, t) dx = \text{integrate by parts}$$

$$= -\frac{\hbar^2}{2m}\left\{\left[\psi^*\frac{\partial\psi}{\partial x}\right]_{-\infty}^{\infty} - \int_{-\infty}^{\infty}\frac{\partial\psi}{\partial x}\frac{\partial\psi^*}{\partial x}dx\right\} = \text{from the normalization}$$

$$= \frac{\hbar^2}{2m}\int_{-\infty}^{\infty}\frac{\partial\psi}{\partial x}\frac{\partial\psi^*}{\partial x}dx$$

where $E(\text{operator}) = (-\hbar^2/2m)\nabla^2$, depends on $|d\psi/dx|^2$ where $|\psi'| = |d\psi/dx|$ is less in the symmetric choice.

When the wells merge (c), the symmetric ψ_{total} gives the ground state for a well whose width is equal to the sum of the two widths, and the anitsymmetric ψ_{total} clearly gives the first excited state.

In the symmetric state the system's energy diminishes with decreasing r (until there is no barrier); in the antisymmetric state the energy decreases (the wavelength increases) with r. Naturally, upon the approach of the atoms (wells) the wall becomes thinner and thinner (or, equivalently, the barrier becomes lower and lower), and tunnelling can take place, *i.e.*, the electrons can jump to and fro between the two wells, thereby elevating the probability of finding either electron in the space between the huddled wells — that is, inside the wall. This is a *resonance* of the electrons (between the wells), and the attractive force $F = -\partial V/\partial r$ between the wells arises from this *resonant exchange* (of electrons between the wells). Needless to say, for the two electrons to be in the ground state (and have ground-state spatial ψ), the Pauli principle requires that their spin be opposite (antiparallel). (Otherwise no binding is possible, as one of them will be forced to go to an excited state, because it will not be able to coexist with the other in the same state without violating the Pauli principle.) That is quantum mechanics.

This theoretical analysis (model) works well for the case of making a diatomic molecule by letting two atoms approach each other (or bringing them together on purpose). Interestingly enough, a general physical principle is involved here, too: *Any two identical physical systems* (be they pendulums, resonant coil-capacitance circuits interacting via mutual inductance, or any other systems) *getting so close as to resonate and interact, the individual energy of each splits into two levels, and the splitting is proportional to the interaction strength.* For example, two coupled AC circuits have *two* (resonant) frequencies (one slightly above and one slightly below the frequency of either circuit when alone. In dynamics, too, two interacting pendula have *two* normal mode frequencies (greater and less than the natural frequency of either pendulum).

In case of three potential wells (atoms) interacting, one energy splits into *three* levels. (The same for circuits and pendula.) In case of N wells in interaction, the splitting is N-fold, and each level has its own frequency. In solids N is usually large, because we have a collection of many building blocks (say, atoms) interacting and vibrating; thus an energy level of an atom splits into a whole bunch of closely spaced levels which is just a *band* of energies (energy band).

We have the following types of bonds:

(1) Heteropolar ionic bonds: AB type bonds are discussed in Sections 4.4 and 6.9. The molecular orbital is $\psi_A + \lambda\psi_B$, and $|\lambda| \neq 1$. (See Eq. (4.4).) The conditions for bond formation are the usual (see Section 4.6) and stay the same: The energies must be almost equal, the overlap must be as maximum as possible, and the initial atomic orbitals must be symmetric along the bond direction (*i.e.*, orbitals ψ_A and ψ_B must be selected so). If $\lambda > 1$, ψ_B gains importance (atom B attracts the electron and keeps it busy near itself more than A, *i.e.*, the electron density (or cloud) increases in the vicinity of B and decreases near A — this is the distribution of the electron upon the redistribution of its density). Thus a $\mathfrak{p} \propto \lambda$ is produced, and the electron density of ψ_A and ψ_B is proportional to λ^{-2} (approximately). The electron charges of A and B go as $(1+\lambda^2)^{-1}$ and $\lambda^2/(1+\lambda^2)$ and coincide with the (positions of the) nuclei. As the valence electrons get distributed about the nuclei, the molecule develops a

$$\mathfrak{p} \simeq \left(\frac{\lambda^2 - 1}{\lambda^2 + 1}\right) eR$$

where the moleculear $\mathfrak{p}$ can be found if λ is known. Further, if ψ_A and ψ_B are known, $\mathfrak{p}$ can be improved. However, λ cannot be determined accurately, and hence we go backwards and find λ from $\mathfrak{p}$. The molecular dipole moment $\mathfrak{p}$ is proportional to the difference in electronegativity between the two atoms.

In hydrogen-halogen molecules the bond ionicity increases as we go from HI to HF. The ionic bond percentage is related to the effective charges of the dipole.

In the table below polar molecules are listed, where $\mathfrak{p}$ is experimental, and λ is calculated from it. The values are taken with respect to the halogen atom (negative side of the dipole). The effective charge is $\mathfrak{p}/eR$, and the charges are thought as if condensed at the nuclei and expressed as percentages of (the electron charge) e (so that for a fully covalent bond the value is 0, while for a fully ionic monovalent bond it is $\pm 1e$). The molecular dipole moments are as follows:

Molecule	$R(\text{Å})$	$\mathfrak{p}$ (in CGS units)	λ	Effective charge	Difference in electronegativity
KCl	2.79	6.3 ⎫	2	0.47	2.2
HF	0.92	1.91 ⎪	1.9	0.43	1.9
HCl	1.27	1.03 ⎬ $\times 10^{18}$	1.3	0.17	0.9
HBr	1.41	0.78 ⎪	1.2	0.11	0.7
HI	1.61	0.38 ⎭	1.06	0.05	0.4

$E_{\text{binding}}(\text{NaCl})$ is given in Section 6.10. If for a moment we take the bond as covalent — though incorrect — we should take the average binding energy of the Na crystal and of the molecular Cl_2 in order to calculate the energy of the NaCl molecule, to find

$$E(\text{NaCl}) = \frac{1}{2}\left(V(\text{Na}) + \frac{1}{2}E(Cl_2)\right) \simeq 37 \text{ kcal/mol}$$

which is way below (much less than) 101 kcal/mol.

The existence of ions can be proved by the fact that electrolytes are produced when we dissolve salts in liquid solvents. The ionic character can be proven as follows:

(a) In solid NaCl, from the confirmation obtained from the X-ray diffraction intensities and the data thereof.

(b) In alkali halide gas molecules, by observing and measuring $\mathfrak{p}$.

(2) Metallic bonds: To have an overlap[6] of atomic wavefunctions, two atoms must come mutually close together. If *many* atoms do so, the

---

[6]For the sake of expressional purism, do not confuse the *overlap* of two state functions with the *mixing* of two state functions (or *coupling* of two states). The former refers to the interpenetration of the ψ functions (or electron clouds) of two *different* atoms to form a chemical bond — and this is what we mean here. The latter refers to the communication of two states *of the same atom* (e.g., ground state and excited state, or two excited states) that couple (coexist) for a while as the electron of the atom is getting from one to the other. During this transition, the situation is described by a composite wavefunction consisting of a percentage (coefficient) of one state and a bit of the other state in an admixture, until the electron settles in a definite pure state. In other words, the electron has one foot on one state and the other foot on the other state, so to say. The electron cloud stretches and is for a while shared by both states, and the situation is intermediate, as one end of the cloud is in the initial state, and the other end is reaching out to the final state. The coefficients in front of the two states are percentages of the share (participation), or give the (partial) probabilities of finding the electron in either state in a measurement during the transition.

overlap happens again, and when this overlap increases the energy gain, it makes it possible for each atom to favor the formation of as many bonds as possible. Each atom wants to form bonds with as many neighboring atoms as possible. As a result, a structure can be formed where atoms whose number extends boundlessly in space are densely concentrated in a certain region of space. In such a structure the packing (coordination) number becomes maximum. Experiments prove that metals crystallize usually producing various structures which have a high symmetry, a high coordination number, and a simple appearance.

If we let many atoms come mutually close, will they form molecules (of type A_2 or A_3), or will they form crystals (with atomic building blocks)? It will depend on which of the two choices costs less energy (and corresponds to the least energy). Anyway, a metallic crystal is already a huge, single polyatomic molecule to some extent.

Figure 6.31 shows the overlap of the wavefunctions of $2s$-electrons in the (metallic) Li crystal.

Fig. 6.31.

(3) Covalent bonds: The (ionized) H_2^+ molecule has one electron. To solve this problem, we write $\mathcal{H}\psi_A = E_0\psi_A$ where

$$\mathcal{H} = -\frac{\hbar^2}{2m}\nabla^2 - \frac{e^2}{r_A} - \frac{e^2}{r_B},$$

if we exclude the term that corresponds to the repulsion of the nuclei. The ground-state energy of the atom is $E_0 = -e^2/2a_0$. Hence

$$\mathcal{H}\psi_A = E\psi_A - \left(\frac{e^2}{r_B}\right)\psi_A,$$

$$E_A = \int \psi_A \mathcal{H}\psi_A \, dV = E_0 - \int \frac{e^2}{r_B}\psi_A \, dV,$$

$$\beta \equiv \psi_B \mathcal{H}\psi_A \, dV = E_0 S - \int \frac{e^2}{r_B}\psi_A\psi_B \, dV.$$

And the part of the energy of the ion that belongs to the electron becomes

$$E_{1,2} = E_0 - \frac{\int \left(\frac{e^2}{r_B}\right) \psi_A^2 \, dV \pm \int \left(\frac{e^2}{r_B}\right) \psi_A \psi_B \, dV}{1 \pm \int \psi_A \psi_B \, dV}$$

where we can evaluate the integrals by using hydrogenic (atomic) wavefunctions. Figure 6.32 shows the plots for H_2^+. Notice that the minimum total energy occurs at $r = 2a_0 = 1.06$ Å for $\psi_A + \psi_B$, which is the equilibrium separation of the two nuclei. For $\psi_A - \psi_B$ there is no minimum for the total energy — no equilibrium for this choice. Figure 6.33 shows the electron density plot (in contours) for H_2^+. The plotted ψ^2 is for the stable $\sigma 1s$ (attraction) and for the excited $\sigma^* 1s$ (repulsion) state. The electron density (along the molecular axis) is plotted in Fig. 6.34 for the two cases. The dashed lines show the free atomic ψ_A^2 and ψ_B^2. The total curve for H_2^+ (for the case $\psi_A + \psi_B$) is the mean $(1/2)(\psi_A^2 + \psi_B^2)$. Note that the electron density increases between the nuclei in $\sigma 1s$, and hence strong attractive forces arise, ensuring a stable bond and hence a molecule. In $\sigma^* 1s$ the electron density drops to zero between the nuclei. This means that it is not likely to find an electron there, that is, the two nuclei are probably bare and unscreened, and hence their mutual Coulomb repulsion wins, allowing no possibility for a molecule to form, as the two nuclei repel each other.

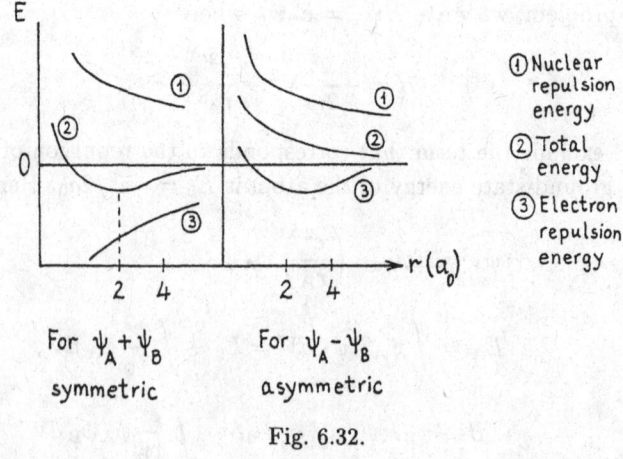

Fig. 6.32.

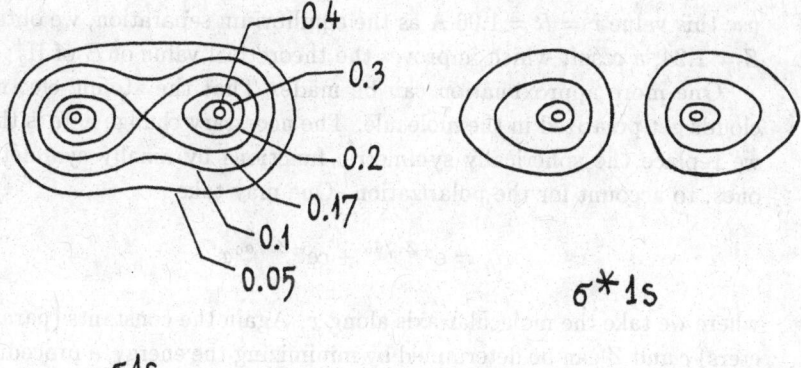

Fig. 6.33.

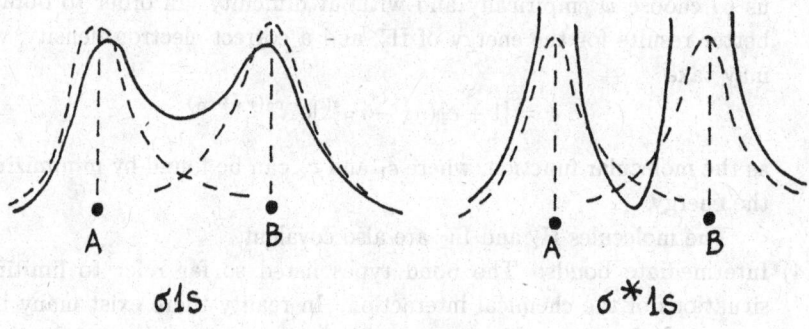

Fig. 6.34.

Notice that the electron distribution ψ^2 in H_2^+ is quite different than what it would be if we took the free isolated atoms for comparison; in other words, if we set off to find the molecular wavefunction by taking the unperturbed functions ψ_A and ψ_B and graphically superimposing them, the result will be in error. Thus the perturbed wavefunctions must be chosen — and properly. One can take the atomic (hydrogenic) wavefunctions where the nuclear charge is written as Z:

$$\psi_A = \sqrt{\frac{Z^3}{\pi a_0^3}} e^{-Zr/a_0}$$

where Z, the effective charge, can be found variationally, by requiring that the energy E becomes minimum for a certain (value of) r. If we use this value $r = R = 1.06$ Å as the equilibrium separation, we obtain $Z = 1.24$, a result which improves the theoretical value of E of H_2^+.

One more approximation can be made: That the atomic electron clouds get polarized in the molecule. The necessary change here is that we replace the spherically symmetric functions by axially symmetric ones, to account for the polarization. One may take

$$\psi_A = e^{-Zr/a_0} + ce^{-Zr/a_0}x$$

where we take the molecular axis along x. Again the constants (parameters) c and Z can be determined by minimizing the energy, a procedure which corrects the result for the molecular energy and brings the theoretical and experimental results to a better agreement.

A dumbbell molecule has an ellipsoidal electron cloud which allows us to choose ψ empirically and without difficulty. In order to obtain better results for the energy of H_2^+ and a correct electron density, we may take

$$\psi = [1 + c_1(r_A - r_B)^2]e^{-c_2(r_A + r_B)}$$

as the molecular function, where c_1 and c_2 can be found by minimizing the energy.

The molecules H_2 and Li_2 are also covalent.

(4) Intermediate bonds: The bond types listed so far refer to limiting situations of the chemical interaction. In reality there exist many intermediate bonds — all quite possible.

A_2 type molecules (or crystals made of the same atoms) form covalent bonds where the valence electrons are shared and move in orbits that enclose several nuclei, resulting in a symmetric electron density between neighboring nuclei. The bonds are homopolar. If the electronegativity is different (heteropolar bonds in molecules or crystals), the center of mass of the electron distribution shifts towards the atom whose electronegativity is greater. Now if there is a big difference between the electronegativities, the valence electron of the electropositive atom gets fully over to the electronegative atom. In that case the binding is fully ionic.

So, the electronegativity difference (between the interacting atoms) and the nature of the atoms play an important role here.

One-electron bonds are weak, except if the atoms are alike.

In all kinds of bonds the type of interaction is the same: Coulombic. However, since the electron cloud differs along the periodic table, the *character* of the interaction (binding) is different. The physical and chemical properties of the bodies are due to this difference in the nature of the chemical bonds.

In metals the outer valence band is partially filled. So, in alkali metals the atomic valence shell has one s-electron, while the neighboring p-states are empty. In heavy metals at the end of the periodic table, the energy states s, p, d, and f are occupied, but some of the lower states are empty. In metalloids most of the electronic structure is filled, except some states which are empty. In C the s–p shells are full in half and each atom has four electrons plus four places empty. In metalloids the empty closest higher-energy level lies too high compared to the binding energy, and hence even if atoms come close together, it seems energetically unlikely for them to share electrons.

6.16. Bond Multiplicity

We use the linear combination $\psi_A \pm \psi_B$ to calculate the molecular ψ, but we do not have to use the $1s$ electronic wavefunctions (of atoms A and B) only. We can equally use $2s$, $2p_x$, $2p_y$ etc. functions, too. The density distribution of valence electrons in molecules depends on the chosen functions. For example, molecular wavefunctions consisting of two s or p_z functions are axially symmetric, *i.e.*, the valence-electron density along the molecular axis is maximum. The electron cloud that is formed by the interaction of two p_y or p_z-functions is different: It is oval and has its axes perpendicular to the molecular axis. The combination $\psi_A(p_z) + \psi_B(p_z)$ increases the electron density, giving a stable molecule whose state is $\pi_z 2p$, but the molecular ψ does not have a symmetry center; the distribution of the electron density extends over two regions which are parallel to the molecular axis. Between these regions, the molecular ψ has a nodal plane which includes the molecular axis and the z-axis along which $\psi = 0$. The electron cloud exhibits no axial symmetry. Such a binding occurs with two p_y-electrons, and the cloud lies in a plane which is rotated by 90° and includes the y-axis (along which it is maximum).

Figure 6.35 shows molecular functions coming from the binding of s and p-electrons. A symmetric ψ has a symmetry axis; an asymmetric ψ has no such axis, *i.e.*, $\psi_s = \psi_A(s) + \psi_B(s)$ is symmetric, and so is $\psi_s = \psi_A(p_z) - \psi_B(p_z)$. The former refers to the molecular orbital symbol $\sigma 1s$, and the latter to $\pi_z^* 2p$. Both are even functions. Functions $\psi_a = \psi_A(s) - \psi_B(s)$ and $\psi_a = \psi_A(p_z) + \psi_B(p_z)$ are asymmetric and odd; the former refers to $\sigma^* 1s$ and the latter to $\pi_z 2p$.

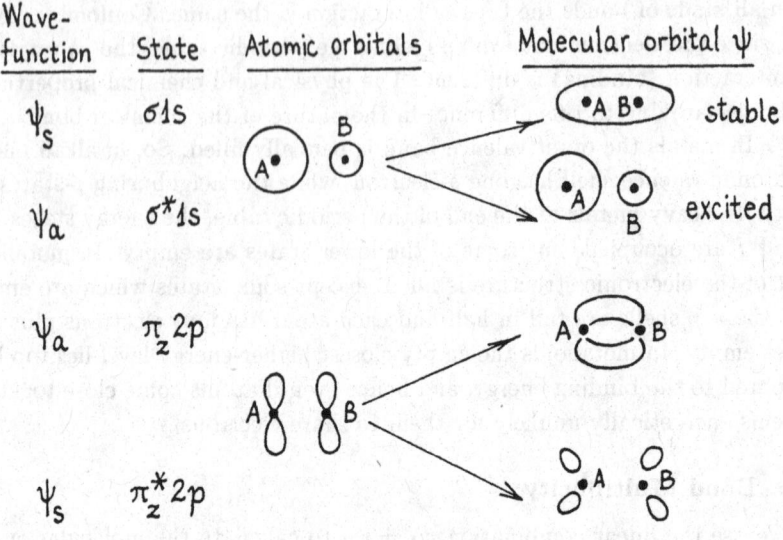

Fig. 6.35.

The tetravalent C atom forms double and triple bonds. The structure of ethane ($r_0 = 1.55$ Å), ethylene ($r_0 = 1.33$ Å), and acetylene ($r_0 = 1.2$ Å) is shown in Fig. 6.36 where C atoms get bound together by multiple bonds. The separation between the atoms decreases with the C–C distance as the C–C bond multiplicity rises. (Notice the different lengths of the bond lines in Fig. 6.36(a), (b), and (c).) For diamond ($r_0 = 1.54$ Å) and graphite ($r_0 = 1.41$ Å) we have fourfold and threefold bonds respectively (for C–C), and the multiplicity is 1 and $4/3 = 1.33$ respectively. So, the difference between the two r_0 values for C–C corresponds to the difference between the multiplicities.

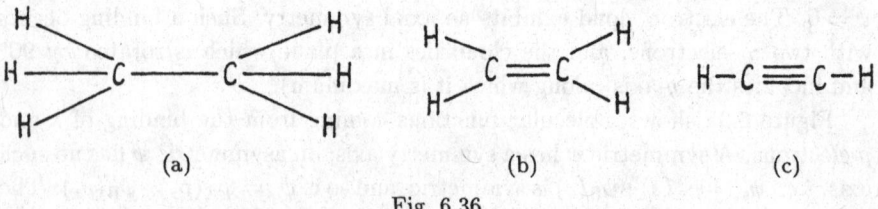

(a)   (b)   (c)

Fig. 6.36.

The benzene molecule (C_6H_6) is hexagonal. Each C atom forms a single or a double bond with another C atom. Hence the C–C bond multiplicity is $(4 - 1)/2 = 1.5$ and $r_0 = 1.39$ Å. Figure 6.37 shows the r_0(C–C) distance plotted against the bond multiplicity; so, when the multiplicity rises from 1 to 3, r_0 drops by 22%. In carbon carbides the structure is cubic (like in NaCl) and the coordination number is 6. We assume that C maintains its valence (4) in carbides. Then the multiplicity is $4/6 = 0.66$. By using the metal-carbon separation distance and the coordination number, we can find the radius of the C atom (0.8 Å) corrected according to the radii of the metal atoms. This result is greater than 0.77 Å, the radius of the single-bonded C atom.

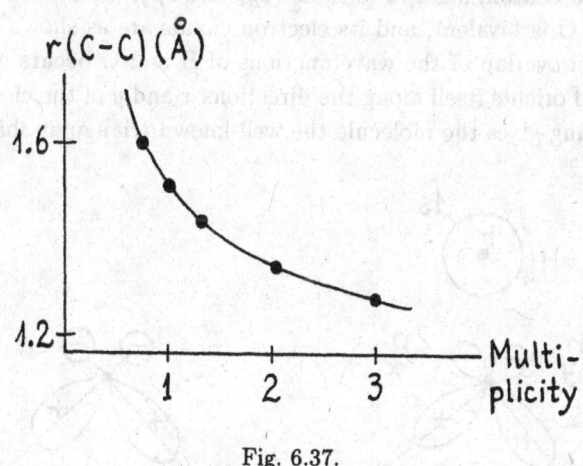

Fig. 6.37.

The valence of an atom in different compounds is independent of the coordination number. The multiplicity is either an integer or the ratio of two integers. For metallic and ionic bonds, it is < 1, and the interatomic distance — even in these cases — depends on the bond multiplicity, and this dependence constitutes a correction to the atomic and ionic radius (according to various coordination numbers). Any change depends on the changing valence of the atom. For example, in CO_2 and CO the distance C–O is 1.15 and 1.13 Å respectively, almost the same, corresponding to the double bond C=O. The slight decrease in the distance in CO is due to the fact that the valence is slightly less. That is why CO is more reactive than CO_2 which is chemically inert: Not all the valence value is used.

6.17. Details on Polyatomic Molecules

Each electron can move following an orbit that goes around several nuclei in a polyatomic molecule. Thus a molecular orbital ψ is the linear combination of the atomic orbtials ψ_A, ψ_B, ..., etc., and can be expressed as

$$\psi = c_1\psi_A + c_2\psi_B + c_3\psi_C + \cdots$$

where c_i are constant coefficients. In order to have a strong bond, the overlap of the atomic wavefunctions must be maximum, *i.e.*, all the atomic wavefunctions must be equally important. The most important part is the linear combination of the wavefunctions of the nearest-neighbour atoms.

As an example, consider H_2O. Each of the H atoms has one valence electron, while the O atom has two of them ($2p_x$ and $2p_y$, which contribute to the binding). So, O is bivalent, and its electron clouds are as shown in Fig. 6.38. The maximum overlap of the wavefunctions of H and O occurs when the hydrogenic cloud orients itself along the directions x and y of the cloud of O, and thus the binding gives the molecule the well-known triangular shape.

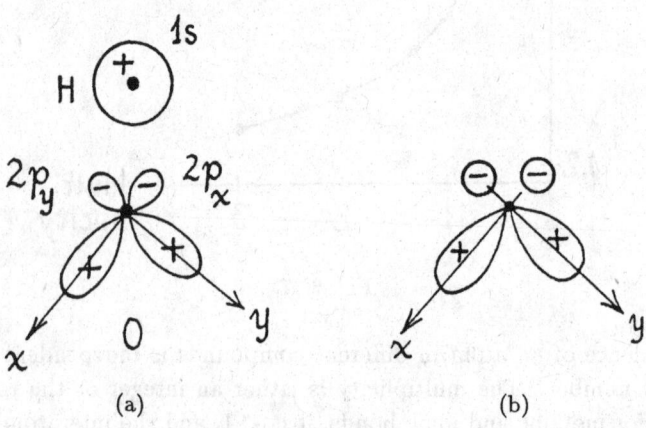

(a)   (b)

Fig. 6.38.

Since O is more electronegative than H (atomically), an O–H bond possesses an electric dipole moment (whose negative tip is at O). The *vector* sum of the dipoles of the (two) O–H bonds gives the (total) dipole moment of H_2O, which is large. Hence the polar character of H_2O. The positive tips (protons) of the two dipoles repel each other, and the initial right angle between x and y becomes wider (109.5°) in the H_2O molecule. This happens as the s-states are shared in the bonds and a tetrahedral structure of sp^3-hybridization comes about. The two H atoms stay tied to O where $r_0(O–H) = 1.07$ Å, being roughly equal to the sum of the radii of O and H: $r_0(O–H) = r(O) + r(H) = (0.66 + 0.37)$ Å.

We should mention that $r_0(H–H) = 1.6$ Å in H_2O, greater than $r_0(H–H) = 0.74$ Å in H_2. Why? Because there is some interaction between the H atoms

— of course much weaker than the interaction between H and O.

In H_2O the protagonistic role is played by the O–H bond, and the triangular shape is due to the p^2-electrons (valence) of O. Hence we can discriminate the molecular atoms as those which are directly connected (bound) to each other and those not connected by bonds (just lying side by side). Between the atoms of the first category we can schematically draw lines to represent the bond. Figure 6.39 shows this simple (but satisfactory) device for three molecules.

Bonds are characterized by the following:

(1) *Structural characteristics*: Interatomic distances, coordination numbers, valence angles etc.
(2) *Binding forces*: Energy and vibration frequencies.
(3) *Macroscopic molecular properties*: Optical, electrical, magnetic and other properties.

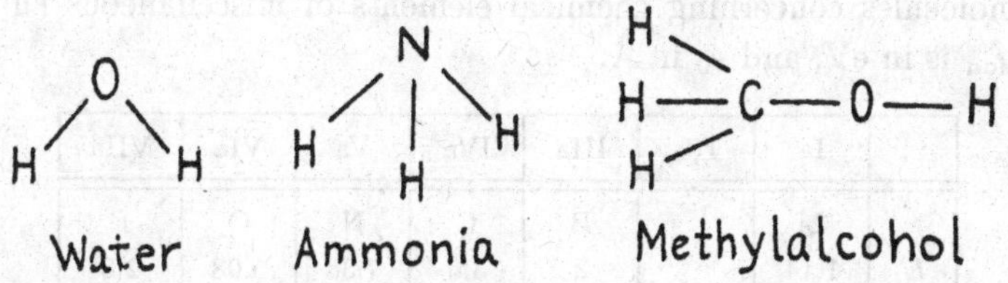

Water   Ammonia   Methylalcohol

Fig. 6.39.

We should not make the mistake of thinking that every molecule can be analyzed and examined by reducing it to its initial constituent atoms. This is not always possible, because molecular properties are results of the *whole picture* of the valence electrons, especially in molecules that have bonds of multiple coupling (benzene, semiconductors, and electrically conductive crystals, like graphite and metals). However, for H_2O such an analysis is possible and immediate:

$$\psi \equiv \psi(H_2O) = c_1\psi_1(H, 1s) + c_2\psi_2(H, 1s) + c_3\psi_3(O, 2p_x) + c_4\psi_4(O, 2p_y)$$

is the molecular orbital, consisting of the proper linear combination of four wavefunctions, one for each electron entering into the bond. But instead of this, we could take two other wavefunctions and form their mixture:

$$\left. \begin{array}{l} \psi(\text{first O–H bond}) = \psi_1(H, 1s) + \lambda\psi_3(O, 2p_x) \\ \psi(\text{second O–H bond}) = \psi_2(H, 1s) + \lambda\psi_4(O, 2p_y) \end{array} \right\} \qquad (6.2)$$

where λ is a multiplicative factor (parameter) yet to be determined. Since p_x and p_y-functions are mutually orthogonal, their overlap integral vanishes:

$$\int \psi(p_x)\psi(p_y)\,dV = 0.$$

Of course, $\psi_1(H, 1s)$ and $\psi_2(H, 1s)$ are identical hydrogenic (atomic) orbitals (wavefunctions). The two wavefunctions (orbitals) in Eq. (6.2) correspond to the two O–H bonds.

Consider the elements of subgroup Vb in the periodic table. Their ground state is $3p$. Hence the electrons form three bonds perpendicular to each other, and this explains why NH_3 has a pyramidal (three-dimensional) shape and structure. Here, too, the repulsion between the protons (H nuclei) widens the angles between the bonds (just as in the case of H_2O).

Now consider the special case of diatomic homopolar molecules. The table below gives the interatomic distances and the dissociation energies of diatomic molecules concerning chemical elements of miscellaneous subgroups. $E_{\text{dissociation}}$ is in eV, and r_0 in Å.

	Ia	IIb	IIIa	IVa	Va	VIa	VIIa
	Li		B	C	N	O	F
E	1.14		3	3.6	7.38	5.08	2.8
r_0	2.67		2.59	1.31	1.09	1.21	1.45
	Na		Al	Si	P	S	Cl
E	0.76				5.03	3.6	2.48
r_0	3.08				1.89	1.9	2.01
	K	Zn	Ga	Ge	As	Se	Br
E	0.51				3.96	2.7	1.97
r_0	3.92					2.19	2.28
	Rb	Cd	In	Sn	Sb	Te	I
E	0.49	0.09			3.7	2.3	1.54
r_0						2.59	2.66
	Cs	Hg	Tl	Pb	Bi	Po	At
E	0.45	0.08		0.7	3.34		
r_0							

Finally, let us discuss bonds in biology, *i.e.*, of very large molecules. Quantum mechanics proved to be very beneficial when it comes to its practical application in explaining chemical bonds and their nature.

In short, atoms donate or share electrons and thus form stable molecules where the chemical binding is effected by Coulombic forces, and where the strong bonds are ionic (if the electrons are shared unevenly) or covalent (if they are shared equally). A bond is studied with regard to its strength and length. The former is just $E_{binding}$ (or energy required to break the bond), *i.e.*, the depth of the potential well; the latter varies between 0.5 and 1.5 Å, and is expressed as the separation r_0 between the two atoms at their equilibrium position (at the bottom of the well). $E_{binding}$ for ionic and covalent bonds varies between 1 and 5 eV, typically. To convert it into kcal/mol, remember that 1 kcal = 4180 Joules = 2.61×10^{22} eV, and 1 mol = 6.022×10^{23} atoms, so that 1 kcal/mol = 0.0433 eV per bond.

In describing the bonds schematically, we use a single line (between atoms) for a single bond, a double line for a double bond etc. Some bond energies are given below:

Bond	$E_{binding}$(eV)	Bond	$E_{binding}$(eV)
H–H	4.48	C=N	5.85
C–C	3.53	N=N	4.23
N–N	1.66	C=O	7.49
O–O	1.51	O=O	4.16
C–N	3	C≡C	8.31
O–H	4.77	C≡O	11.08
C–O	3.53	N≡N	9.74
C=C	6.32		

The *valence* of a molecular bond is the largest number of bonds that an atom can form. The *bond angle* is just the spatial stereometrical angle between two bonds in a molecule where the two bonds belong to the same atom and originate from it. A pair of shared electrons gives a single bond, and atoms whose valence is greater than 1 are able to make single, double, or triple bonds. Since oxygen's valence is 2 (because two electrons are needed to complete the — outer — $2p$-shell and get it filled), O can make two (covalent) bonds (at most), like in H_2O. Carbon's valence is 4 (because four electrons are needed for $2p$), and so it can form four single (covalent) bonds in CH_4 (methane) where we have a pyramidal tetrahedron whose bond angles are 109° (Fig. 6.40).

Bond angles determine the way clusters of molecules get grouped into combinations. For example, a double or triple bond is usually very stiff and rigid; so, the groups of atoms can hardly rotate around the bond axis.

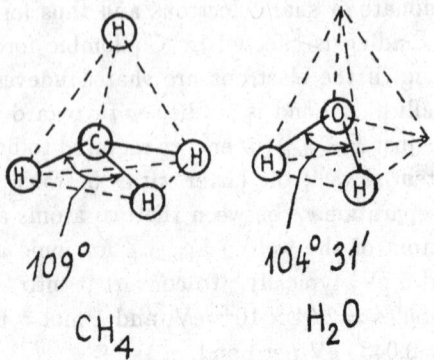

Fig. 6.40.

Apart from ionic and covalent bonds, there are weak bonds (Van der Waals bonds) resulting from the attraction between molecular dipole moments which are either permanent, or results of charge separations that happen if the electron cloud of a polar (by structure) molecule finds itself in close proximity to a nonpolar (electrically balanced) molecule. The table below gives the permanent dipoles of some molecules:

Molecule	$p \times 10^{-30}$ (Cb m)
CO_2	0
CO	0.37
H_2O	6.2
HCl	3.6
NH_3	4.9
HNO_3	7.2

If either of the atoms involved is H, the weak bond is called *hydrogen bond.*

Organic molecules are those containing C; they are called so because organic matter consists of them. Biology and biochemistry deal with large organic molecules. Organic chemistry deals with both small and large organic molecules. Biochemistry says that weak bonds hold together the long chains of molecules in hemoglobin, the O_2 carrier in blood and valuable oxygenation agent of the tissues and cells. So, hydrogen bonds cross-link the pairs in DNA, a double-helix molecule related to the reproduction of the genetic code and heritage. We should note that these large molecules, called *macromolecules*, are actually chains of molecules with molecular weight 10^5 to 10^6. Biology deals mostly with hydrogen and covalent bonds, structurally. Almost all the molecules life is based on are made of C, N, H, P, O, and S. Many enzymes (ferments) are made of proteins, and so are cells and DNA. Proteins are macromolecules, too, consisting of smaller molecules or units that contain N and that are called *aminoacids*. Figure 6.41 shows the chemical structure of an aminoacid where the basic amino-group is NH_2 which is linked to a carboxyl group (COOH). The side group is R which can be a combination that sticks to the C atom from the left side. It could be just one H atom bound to C, or a methyl group (CH_3), or any other group, quite complex one. The figure shows one of the existing twenty variations (aminoacid molecules) named *ala* (alanine), *gly* (glycine) etc. Aminoacids can cluster together. This condensation occurs in the form of long chains connected by covalent bonds between the −COOH radical of one aminoacid and the −NH_2 group of the other, giving rise to a *peptide link*. Hundreds of such links between aminoacids form proteins or *polypeptide chains*. Figure 6.42 shows the formation of a peptide link: A H_2O molecule parts with the rest and gets off. Immediately after this split-away, the NH_2 group of one aminoacid associates with the COOH group of the other one (a).

Proteins are different because the number and arrangement of the aminoacid links differ, *i.e.*, the order is different. Primarily the structure consists of a linear succession of C, N, or O atoms, which is the main trunk of the polypeptide chain. If we find and map this aminoacid sequence (of the protein molecules)

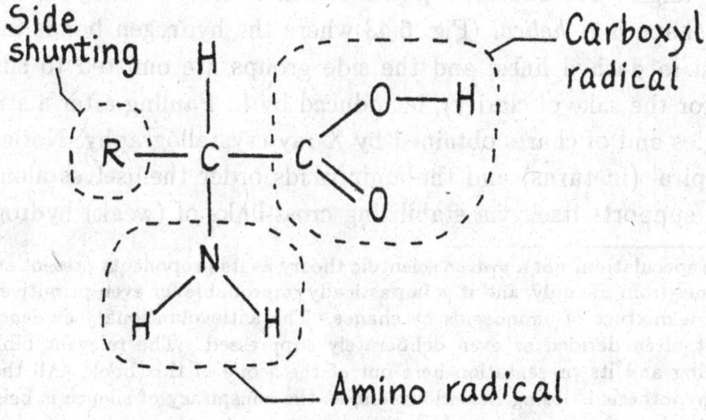

Fig. 6.41.

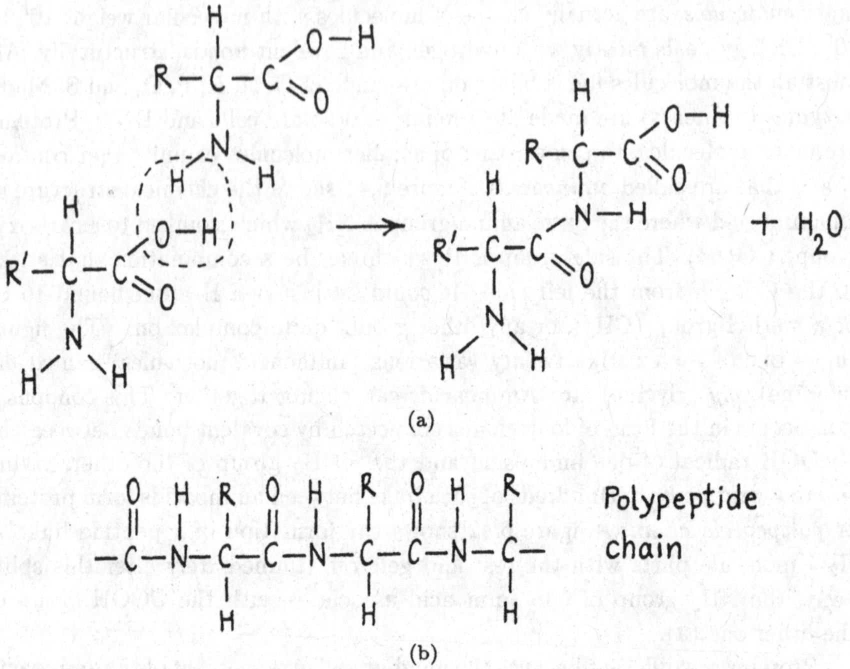

(a)

(b)

Fig. 6.42.

in tissue cells, we can check if it hints at the "theory" of evolution[7] of species, but no such clues have been found. The secondary structure of a protein depends on the shape of the polypeptide chain, and that in turn depends on the bond angles between C, N, and O which are on the main trunk. An ordinary shape is *α-helical* (Fig. 6.43 where the hydrogen bonds in the helix are shown in dashed lines, and the side groups are omitted to simplify the drawing for the sake of clarity), introduced by L. Pauling after a study of the bond angles and of charts obtained by X-ray crystallography. Notice that the helix is spiral (in turns) and the aminoacids order themselves along it. The structure supports itself via stabilizing cross-links of (weak) hydrogen bonds

---

[7]It is just a speculation, *not a proven* scientific theory as its proponents present and advertise it. Life comes from life only, and it is fantastically improbable for even primitive life to have come from a mixture of aminoacids by chance. The antievoluntionary evidence is equally strong, but often derided or even deliberately suppressed. The relevant bibliography is overwhelming and its presentation here out of the scope of this book. All the same, the evolution hypothesis is losing its seriousness, as the conspiracy of silence is being revealed. And despite the existing strong evidence, it is only an *hypothesis*, because an established theory has a *proof*.

(dashed lines in the figure) between N, H, or O atoms of the peptide groups. This same arrangement is repeated as a constant pattern every five turns along the spiral, that is, over a length of 27 Å. We have 3.6 R-groups per turn (totally 18 of them per pattern).

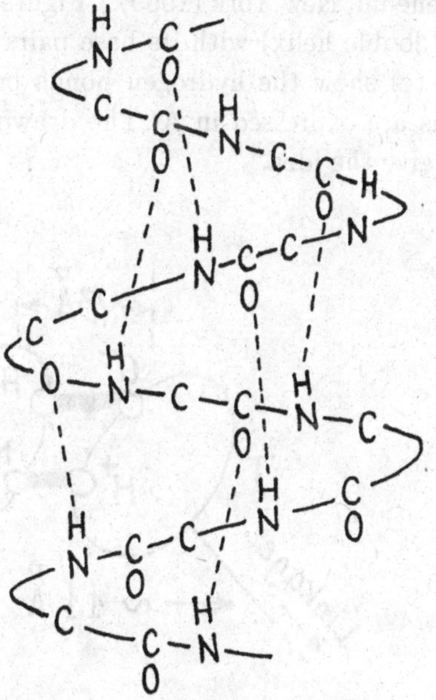

Fig. 6.43.

The DNA molecule carries our genetic code. Briefly told, our cells should be able to perform certain tasks so that life and its continuation be possible. Further, this ability must be reproduced in passing from a generation to the next. Mother nature provides for this ceaseless process and ensures its safeguarded continuation, by building in information about reproduction in chromosomes which consist of genes. In other words, an encoding is being made. Genetics tries to map the genes and see if it is possible to influence them in order to eliminate hereditary diseases (like coronary atherosclerosis which leads to fatal heart attacks). In each gene there is inscribed and contained enough information to produce a particular protein molecule. The main molecular stuff in the chromosomes is DNA (de-oxyribonucleic acid); it is made of a double helix of sub-unit (smaller) molecules which are cross-linked by hydrogen bonds.[8]

---

[8]This was proven by J.D. Watson and F. Crick in 1953.

The smaller molecules are *nucleotide base pairs*, and there are four of them: Adenine (A), cytocine (C), guanine (G), and thymine (T). Given their shapes, we find out that A can be hydrogen-bound to T, and G to C only. So, as we move along the spiral, A pairs with T, and G is across of C. (See J.D. Watson, *The Double Helix*, Atheneum, New York (1969).) Figure 6.44 shows a segment of the DNA molecule (double helix) with its base pairs (A–T and G–C) that match. Parts (b) and (c) show the hydrogen bonds between the base pairs where the bond lengths are expressed in Å. The drawings are oversimplified for clarity, and just to give the idea.

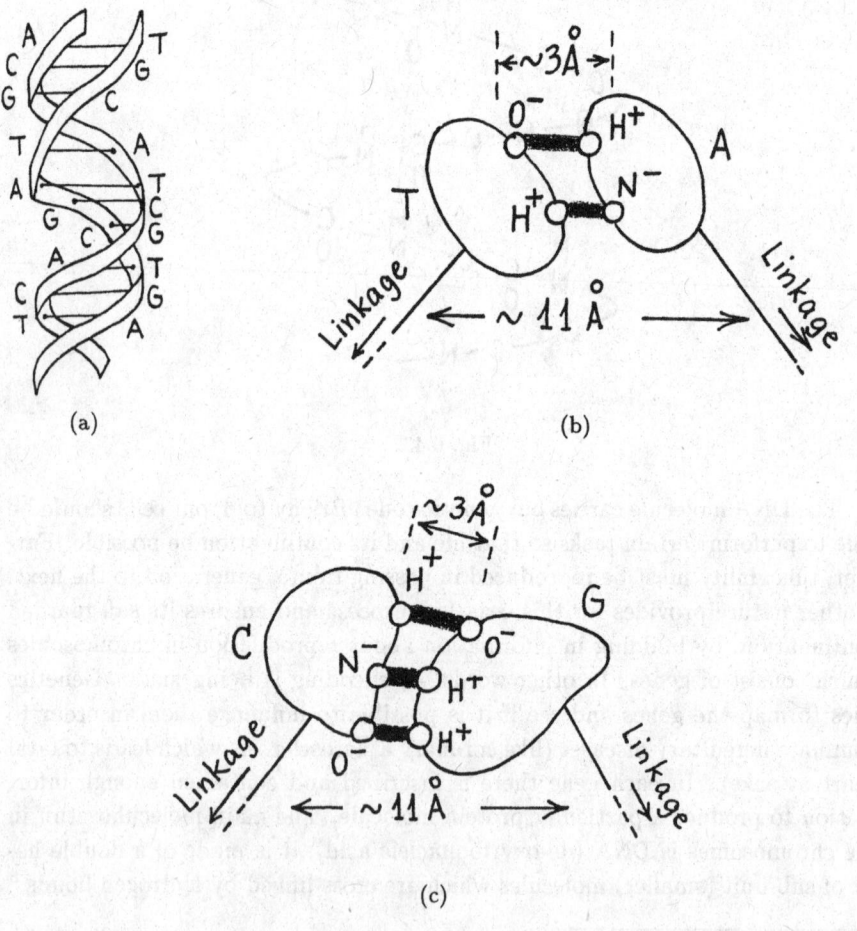

Fig. 6.44.

In the process of the cell reproduction the double helix opens at one end by disentanglement. The unraveled main tapes allow their base pairs to find new mates to get linked with. Again, only A–T and G–C linkings are possible. Since the cellular fluid medium contains many stray base pairs freely floating around, a new double helix can set up from them, and thus the genetic code (information) is both reproduced by duplication and passed on from the parent generation to the next. The chance of error here (by which a monstrous freak could be formed) is too slim, though not zero; an improper mismatch would be due to either a mutation all by itself (spontaneous deformation) or the effect of external radiation which causes mutations (*e.g.*, cancer) by inflicting chemical damage (whereby the cells start illogical activities instead of performing their regular natural duties).

Molecular biology is almost the sum of molecular physics, biochemistry, and genetics. It is a science that thrives in the United States and Germany, both theoretically and experimentally, at an advanced level and with full-steam research. Topics like genetic engineering, genetic mapping, genetic code, protein synthesis, the importance of chemical bonds etc. are relevant, but very specialized. Nice pertinent literature exists to satisfy all levels of curiosity.

Chapter 7

DISPERSIVE FORCES

7.1. Dipole Moment

An *electric dipole* is a pair of two equal and opposite point charges fixed in place and separated by a distance l (Fig. 7.1(a)). If we draw a vector from the negative charge to the positive charge (in the direction opposite to the field line), we can describe the dipole by this vector, denoted by $\vec{p}$ (Fig. 7.1(b)). This same vector represents also the *dipole moment*[1] defined as

$$\vec{p} = q\,\mathbf{l} = q\,l\,\hat{\mathbf{l}}$$

where q is either of the two charges, and $\hat{\mathbf{l}}$ is a unit vector along l. Obviously, $|\vec{p}| = \mathfrak{p} = ql$.

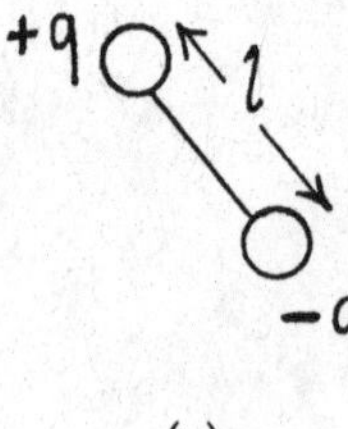

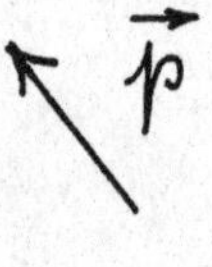

(a)   (b)

Fig. 7.1.

---

[1] The dipole moment is the first moment of a multipole expansion where the zeroth moment is the charge q itself, the second moment is the quadrupole moment Q, and so on.

An *external* electric field can redistribute charges in a material and thus create induced dipoles, or it can orient already existing dipoles. The relation between $\vec{p}$ and the applied field $\vec{\mathcal{E}}$ is

$$\vec{p} = \alpha \vec{\mathcal{E}} \tag{7.1}$$

where α, the proportionality constant, is the *polarizability* of the medium where the dipole resides. If the medium is not electrically isotropic, then α is a tensor (matrix) that couples the various directions. So, in the most general case we have

$$\begin{pmatrix} p_x \\ p_y \\ p_z \end{pmatrix} = \begin{pmatrix} \alpha_{xx} & \alpha_{xy} & \alpha_{xz} \\ \alpha_{yx} & \alpha_{yy} & \alpha_{yz} \\ \alpha_{zx} & \alpha_{zy} & \alpha_{zz} \end{pmatrix} \begin{pmatrix} \mathcal{E}_x \\ \mathcal{E}_y \\ \mathcal{E}_z \end{pmatrix}.$$

Physically, α is a measure of the ease with which an atom or a molecule can be deformed by $\vec{\mathcal{E}}$. Figure 7.2 shows a molecule distorted by an external field. A dipole is produced. The molecules of a dielectric medium can thus be polarized to produce a polarization in the medium, *i.e.*, an assembly of produced (induced) dipoles.

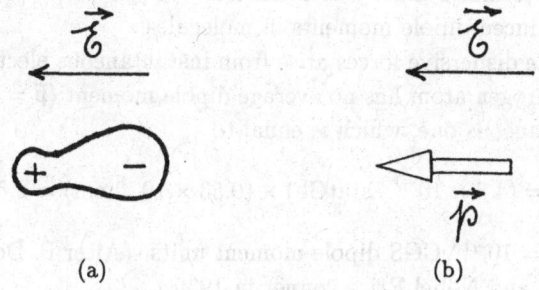

(a)   (b)

Fig. 7.2.

Molecules with no permanent dipole moment are called *nonpolar*. But if brought into an electric field, they may exhibit a dipole moment. Molecules having a permanent dipole moment are called *polar*. The water molecule is polar because of its shape that gives rise to a dipole moment (Fig. 7.3).

For quantum mechanical reasons, atoms have no permanent dipole moment in the absence of $\vec{\mathcal{E}}$, whereas molecules might.[2] Some molecules have degenerate

---

[2]However, it is possible (like in the hydrogen atom) that unperturbed degenerate states of opposite parities can produce an induced permanent dipole moment in atoms. But this happens only in hydrogen because of a special dynamical symmetry, and even not in the ground state.

rotational states of both parities, and if the corresponding energy levels are closely spaced as compared to the thermal energy, they can produce a molecular permanent electric dipole. Also, molecules are usually not spherical. So, they can have a permanent dipole moment.

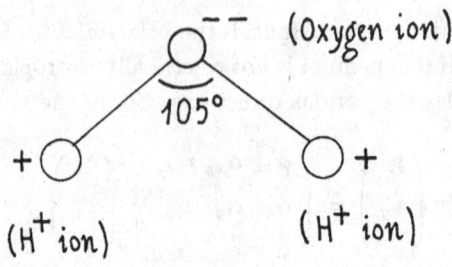

Fig. 7.3.

7.2. Dispersive Forces

Dispersive forces are derived from a potential. It is due to these forces that we have intermolecular bonds. The attractive Van der Waals forces arise from permanently induced dipole moments in molecules.

In atoms, the dispersive forces arise from instantaneous electric dipole moments. The hydrogen atom has no average dipole moment ($\bar{\mathfrak{p}} = 0$), but it does have an instantaneous one, which is equal to

$$\mathfrak{p} = ea_0 = (4.8 \times 10^{-10} \text{ statCb}) \times (0.53 \times 10^{-8} \text{ cm}) = 2.5 \text{ Debye}$$

where 1 Debye = 10^{-18} CGS dipole moment units. (After P. Debye, a famous Dutch physicist and Nobel Prize winner in 1936.)

Atoms whose valence shells are filled have a spherically symmetric charge distribution, and hence no permanent dipole. Nor can they awake induced dipole moments in each other. The Van der Waals forces between atoms are due to instantaneous dipole moments only. The average atomic dipole moment is zero, but there may be an instantaneous one in atoms.

At a distance where the atomic wavefunctions do not overlap, the instantaneous atomic dipoles interact mutually and give rise to attraction. If the electronic motion in atoms is so synchronized as to align these dipoles (and thus reduce the energy), a mutual attraction sets up. The first quantum mechanical calculation of these forces was made by London in 1930. So, the forces were called *London forces*, and the corresponding potential, *London potential*.

7.3. The Charged Oscillator Model[3]

When a molecule enters a region of an electric field and acquires a dipole moment, that means that it has a polarizability α which can be measured, by measuring $\vec{\mathfrak{p}}$ and employing the relation in Eq. (7.1).

Consider now two identical charged oscillators. (A charged oscillator is a small mass m with a charge q, connected to a spring of constant κ, as shown in Fig. 7.4.) They could be two simple electric dipoles (*e.g.*, two hydrogen atoms). For simplicity, let us take them in one dimension (Fig. 7.5). Let their positive charges (nuclei) be kept fixed at distance R from each other, and let their negative charges (electrons) move (vibrate) along the x-direction with displacements x_1 and x_2 respectively. For each dipole we can write

$$\mathfrak{p} = ex = \alpha \mathcal{E}$$

where x is the one-dimensional vibration displacement of the negative charge. If this charge moves away from its equilibrium position by x, a restoring force F acts on it, given by Hooke's law (about first-order elasticity)

$$F = -\kappa x = e\mathcal{E}$$

whence the spring elasticity constant is

$$\kappa = \frac{e^2}{\alpha}.$$

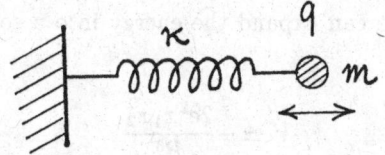

Fig. 7.4.

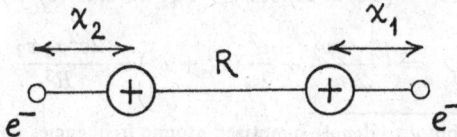

Fig. 7.5.

---

[3]The harmonic oscillator is a good model for the molecular vibrations.

An oscillator (spring) of constant κ vibrates with a frequency[4]

$$f = \frac{1}{2\pi}\sqrt{\frac{\kappa}{m}} = \frac{1}{2\pi}\sqrt{\frac{e^2}{\alpha m}}. \tag{5}$$

The energy of the oscillator is

$$E = \text{(Kinetic energy)} + \text{(Elastic potential energy)}$$

$$= \frac{p^2}{2m} + \frac{1}{2}\kappa x^2 = \frac{p^2}{2m} + \frac{e^2 x^2}{2\alpha}$$

where p is the momentum. So, when there is no mutual interaction, the energies of the two oscillators are

$$E_1 = \frac{p_1^2}{2m} + \frac{e^2 x_1^2}{2\alpha} \quad \text{and} \quad E_2 = \frac{p_2^2}{2m} + \frac{e^2 x_2^2}{2\alpha}.$$

The total energy is $E = E_1 + E_2$.

With R being the separation between the oscillators along the x-direction, the electrostatic (Coulombic) energy, as they interact, is

$$V = \text{Potential energy} = \frac{e^2}{R} - \frac{e^2}{R + x_1 + x_2} - \frac{e^2}{R + x_1} - \frac{e^2}{R + x_2}$$

where the first term is the Coulombic repulsion between the positive (nuclear) charges; the second term is the same between the negative (electron) charges; and the other two terms are the attractions between opposite charges.

If $R \gg x_1, x_2$, we can expand the energy into a series and keep the first term only, to find

$$V \simeq -\frac{2e^2 x_1 x_2}{R^3}. \tag{6}$$

So, in case of interaction, the total energy of the system is

$$E_t = \text{(Total kinetic)} + \text{(Total elastic)} + V$$

$$= \frac{p_1^2 + p_2^2}{2m} + \frac{e^2}{2\alpha}(x_1^2 + x_2^2) - \frac{2e^2 x_1 x_2}{R^3}$$

---

[4]We reserve the symbol ν to denote quantized atomic frequencies only.
[5]If both the proton and the electron vibrate with a *relative* displacement x, then m is their total effective mass.
[6]Notice that after the expansion (and the disposal of the insignificant terms), the remaining leading term is $\sim 1/R^3$.

where there is a cross-term that couples x_1 and x_2. (Here the mass and polarizability of the two oscillators are the same.)

The interaction of the oscillators lifts the degeneracy and we have two separate vibration frequencies. To decouple them, we use the normal (new) coordinates

$$X_1 = \frac{1}{2}(x_1 + x_2) \quad \text{and} \quad X_2 = \frac{1}{2}(x_1 - x_2)$$

whence (from $p_1 = m\dot{x}_1$ and $p_2 = m\dot{x}_2$) we find the momenta in the new coordinates:

$$P_1 = \frac{1}{\sqrt{2}}(p_1 + p_2) \quad \text{and} \quad P_2 = \frac{1}{\sqrt{2}}(p_1 - p_2).$$

Then E_t of the system becomes

$$E_t = \frac{1}{2m}(P_1^2 + P_2^2) + \frac{e^2}{2\alpha}(X_1^2 + X_2^2) + \frac{e^2}{R^3}(X_1^2 - X_2^2)$$

$$= \left[\frac{P_1^2}{2m} + \left(\frac{e^2}{2\alpha} + \frac{e^2}{R^3}\right)X_1^2\right] + \left[\frac{P_2^2}{2m} + \left(\frac{e^2}{2\alpha} - \frac{e^2}{R^3}\right)X_2^2\right]$$

where the two oscillators are now decoupled as though noninteracting, so that the total energy consists of the sum of two energies, each belonging to an independent oscillator (in the new, decoupling coordinates).

In the new coordinate system, the vibration eigenfrequencies corresponding to each oscillator are

$$f_1 = \frac{1}{2\pi}\sqrt{\frac{e^2}{m\alpha}\left(1 + \frac{2\alpha}{R^3}\right)}$$

$$f_2 = \frac{1}{2\pi}\sqrt{\frac{e^2}{m\alpha}\left(1 - \frac{2\alpha}{R^3}\right)}$$

where the correction term $2\alpha/R^3$ comes from the mutual interaction.

The total oscillator energy $E_0 = hf_0$ must now be modified by a factor. Indeed, if we write the zero-point energy

$$E_0 = \frac{1}{2}hf_1 + \frac{1}{2}hf_2 = \frac{1}{2}hf_0\left(\sqrt{1 + \frac{2\alpha}{R^3}} + \sqrt{1 - \frac{2\alpha}{R^3}}\right)$$

and expand it into a series, we find

$$E_0 = hf_0\left(1 - \frac{\alpha^2}{2R^6} + \dots \text{h.o.t.}\right) < hf_0 \tag{7.2}$$

where f_0 is the frequency of either oscillator when there is no interaction:

$$f_0 = \frac{1}{2\pi}\sqrt{\frac{e^2}{m\alpha}}.$$

In Eq. (7.2) we see a decrease in the energy due to the interaction, and this gives rise to attraction. Therefore, attractive forces are at work due to the dipoles; since the dispersion of light in optics is related to these forces, they are called *dispersive forces*.

The frequency change in the hydrogen molecule is such that in one of the oscillators the frequency increases (f_1), and in the other decreases (f_2).

If there is an interaction, the total energy decreases. The interaction energy (the term that decreases the total energy) is $\sim 1/R^6$, as we see in Eq. (7.2).

7.4. Dispersive Potential

The potential of the dispersive forces can be obtained by expanding the interaction (multipole expansion) and keeping higher-order terms as well. The expression is

$$V_{\text{disp}} = -\frac{e^2}{a_0}\left(\frac{1}{R^6} + \frac{c_1}{R^8} + \frac{c_2}{R^{10}} + \cdots\right)$$

where a_0 is the Bohr radius, and c_1, c_2 etc. are constants.

For hydrogen (if we ignore the zero-point energy) we have

$$V_{\text{disp}} \simeq -\frac{6e^2}{a_0}\left(\frac{1}{R^6} + \frac{22.5}{R^8} + \frac{236}{R^{10}} + \cdots\right)$$

where the term $22.5/R^8$ is equal to 10% of the preceding term, and hence too large to be neglected. At very low temperatures and if the body is still a liquid, this term can explain the interaction and the bonds, by taking into account these forces.

The interaction range of the Van der Waals forces is ~ 5 Å.

For He we have

$$V_{\text{disp}} = -1.62\frac{e^2}{a_0}\left(\frac{1}{R^6} + \frac{7.8}{R^8} + \frac{30}{R^{10}} + \cdots\right) \tag{7.3}$$

which is enough to explain the interatomic attractive forces in liquid He at low temperatures. If we take into account the repulsive forces as well, the total interaction curve reaches a minimum at $R = 5.5a_0$, *i.e.*, the Bohr radius for

He is 5.5 times the radius of the hydrogen atom. If we let $R = 5.5a_0$ and $a_0 = 0.53$ Å in Eq. (7.3), we find the numerical value

$$V_{\text{disp}}(\text{He}) = 0.74 \times 10^{-3} \text{ eV}$$

which explains the binding energy of liquid He. In liquid He the interatomic binding forces are of this order.

The dispersive forces hold for all atoms, not only for inert gas atoms with filled shells. They also hold for ions and between molecules. Of course, if there is another strong and quantitatively important interaction around to steal the performance, the dispersive forces fade away and remain in the shadow as a small perturbation. Otherwise they gain importance in the interaction between molecules. After all, this is the only interaction in inert gases, since there is no chemical affinity between their atoms.

7.5. Total Molecular Interaction

If the molecules have a permanent or an induced dipole, an electrostatic interaction appears *in addition to* the dispersion interaction stated above.

If the dipole is permanent, the molecules try to orient themselves so as to minimize their mutual interaction energy — though thermal motions compete against to spoil this a bit (unless the temperature is very low).

At very low temperatures, the dipole interaction energy of the orderly oriented dipoles is

$$V_{\text{orientation}} = -\frac{2p^2}{R^3} .$$

If α is large, there is also an induced dipole whose energy is independent of temperature:

$$V_{\text{ind}} = -\frac{2\alpha p^2}{R^6} .$$

In general, when two molecules interact, we have for the total molecular interaction

$$V_{\text{total}} = V_{\text{orientation}} + V_{\text{ind}} + V_{\text{disp}} .$$

Further interesting details can be discussed here about this particular topic, but we will omit them for space reasons.

7.6. Summary

In molecular physics, we are interested in molecules that are of the size of 1 to 3 Å. (This is the typical range of the bonds.) Since the Van der Waals forces have a range of 5 Å (they appear when the atoms come so close to each other), these forces are in fact not considered in molecular physics.

The dispersive (Van der Waals) forces are derivable from a potential and give rise to a bond of the same name. They are due to permanent or induced dipole moments in atoms.

The dispersive potential is in general proportional to $1/R^6$, but the consideration of only one term in the expansion may not be sufficient, because the next term is not small — it is comparable to $1/R^6$.

At very low temperatures the dispersive forces (which in fact are too weak) play a significant role, and must be considered and included in the calculations. For example, for He we should retain the first three terms of the expansion in Eq. (7.3).

The attractive Van der Waals forces are responsible for the cohesion between free atoms, but they are not capable of forming bonds and binding the atoms into molecules. They do not play such a role.[7] This phenomenon resembles the cohesion within EU; the member countries do not fuse — each one maintains her national identity, consciousness, and culture. The politico-economical cohesion respects the individuality of the members and is not able to form out of them a conglomerate with strong bonds.

The dispersion forces are universal, acting between any atoms. If the existing interaction is strong, they remain as a small correction term. In inert gases they are the only interatomic attraction forces, accounting for any attraction at all.

7.7. Van Der Waals Interaction in H_2

It is a long-range interaction between two H atoms, both in their ground states. The interaction energy can be found by two ways:

(1) Solution of the problem by *perturbation theory*.
(2) Application of the *variational principle*.

Refer to Fig. 8.14. The two H nuclei are labelled as A and B at a constant separation $\mathbf{R}$ between them, along $\mathbf{x}$. The position vectors of electrons #1 and 2 (with respect to their nuclei) are $\mathbf{r}_1$ and $\mathbf{r}_2$, measuring their displacements away from the nuclei. We have

---

[7]For chemical forces to appear and bind the atoms, the latter should come too close together so that their electronic wavefunctions (charge clouds) overlap. Then an electron exchange starts, and the atoms share each other's electrons (in covalent bonds). In noble gases, since no such phenomenon occurs, the Van der Waals forces are the only ones acting there, providing the cohesion.

$\mathcal{H}$(of electrons #1 and 2) = Unperturbed part + Perturbation = $\mathcal{H}^{(0)} + \mathcal{H}'$

where

$$\mathcal{H}^{(0)} = \frac{p_1^2}{2m} + \frac{p_2^2}{2m} + V_1 + V_2 = -\frac{\hbar^2}{2m}(\nabla_1^2 + \nabla_2^2) - \frac{e^2}{r_1} - \frac{e^2}{r_2}$$

$$\mathcal{H}' = \text{(Mixed states)} = \begin{pmatrix} \mathcal{H}'_{11} & \mathcal{H}'_{12} \\ \mathcal{H}'_{21} & \mathcal{H}'_{22} \end{pmatrix} = \frac{e^2}{R} + \frac{e^2}{r_{12}} - \frac{e^2}{r_{1B}} - \frac{e^2}{r_{2A}}. \tag{8}$$

Notice that the interactions are electrostatic. The solution of $\mathcal{H}^{(0)}$ is $\psi^{(0)}(\mathbf{r}_1, \mathbf{r}_2) = \psi_{100}(\mathbf{r}_1)\psi_{100}(\mathbf{r}_2)$ because we have two noninteracting H atoms in their ground states $|100\rangle$. If $R \gg a_0$, then the interaction part $\mathcal{H}'$ can be taken as perturbative. Since R is large (compared to the atomic radius), we can expand $\mathcal{H}'$ (in powers of R^{-1}) and retain the lowest-order terms, because we want the leading term only:

$$\mathcal{H}' = \frac{e^2}{R}\left\{ 1 + \left[1 + \frac{2(x_2 - x_1)}{R} + \frac{(x_2 - x_1)^2 + (y_2 - y_1)^2 + (z_2 - z_1)^2}{R^2} \right]^{-1/2} \right.$$

$$\left. - \left[1 - \frac{2x_1}{R} + \frac{r_1^2}{R^2} \right]^{-1/2} - \left[1 + \frac{2x_2}{R} + \frac{r_2^2}{R^2} \right]^{-1/2} \right\}$$

$$\simeq \frac{e^2}{R^3}(y_1 y_2 + z_1 z_2 - 2x_1 x_2)$$

where the last term represents the dipole-dipole interaction due to the instantaneous configuration of electrons #1 and 2 in their corresponding atoms (that form two dipoles for a moment). Here we neglect terms $\sim 1/R^4$ (dipole-quadrupole interaction) and $\sim 1/R^5$ (interaction between quadrupoles), and also h.o.t.

The fact that $\psi^{(0)}$ is even (in $\mathbf{r}_1$ and $\mathbf{r}_2$) and $\mathcal{H}'$ is odd (in $\mathbf{r}_1$ and $\mathbf{r}_2$) reduces the average value of the leading term of $\mathcal{H}'$ (for $\psi^{(0)}$) to zero. Also, all the neglected h.o.t. of $\mathcal{H}'$ have zero mean value (for $\psi^{(0)}$) (because they are expressed in terms of Y functions of nonzero order). What is left? The leading term of the interaction (energy). It is actually the second-order perturbation (of the term of the interaction between dipoles) which is $\propto (\mathcal{H}')^2$, or $\sim R^{-6}$. We

---

[8]A matrix means interaction (coupling of states). Its off-diagonal elements give the value of the coupling (cross-terms). Its diagonal elements are pure states. If there is no interaction (no mixed states), the off-diagonal elements are zero.

conclude that the leading term of the energy varies as $1/R^6$, if the separation distance R is large.[9]

The perturbation theory gives the same result. The second-order energy change in the two H atoms is

$$\Delta E(R) = \sum_n{}' \frac{|\langle 0|\mathcal{H}'|n\rangle|^2}{E_0 - E_n} \tag{7.4}$$

where the sum is restricted, to exclude the case $n = 0$ which makes the denominator zero and blows up the result. Here we actually sum over discrete states (of the pair of the unperturbed H atoms) *and* integrate over continuous states. Continuous? Yes, because we include dissociated states as well! So, index n runs over *all* states, except the ground state $\psi^{(0)}$ which is excluded. The numerator is a square, and therefore, positive; the denominator is negative ($E_n > E_0$). Thus ΔE is negative, *i.e.*, the force is attractive, and the potential is $\sim R^{-6}$ (for large R).[10]

The quantity $-\Delta E(R)$ is positive and has an upper limit which can be found.[11] We let $E_n = E^*$ where E^* is the lowest possible excited state (of the two H atoms) which has nonvanishing $\langle 0|\mathcal{H}'|*\rangle$. The denominator becomes a constant that can be factored out and taken outside of $\sum'$. There remains to be found

$$\sum_n{}' |\langle 0|\mathcal{H}'|n\rangle|^2 = \sum_n{}' |\langle 0|\mathcal{H}'|n\rangle\langle n|\mathcal{H}'|0\rangle - (\langle 0|\mathcal{H}'|0\rangle)^2$$

$$= \langle 0|\mathcal{H}'^2|0\rangle - (\langle 0|\mathcal{H}'|0\rangle)^2$$

where we inserted $|n\rangle\langle n|$ which is just 1, but constitutes a trick that facilitates the calculation. But we have $\langle 0|\mathcal{H}'|0\rangle = 0$ and hence

$$-\Delta E(R) \le \frac{\langle 0|\mathcal{H}'^2|0\rangle}{E^* - E_0}.$$

For H atoms, $|*\rangle$ corresponds to the state whose principal quantum number is 2. Therefore, we can evaluate the result as $E^* = -2(e^2/8a_0)$, $E_0 = -2(e^2/2a_0)$, and $E^* - E_0 = 3e^2/4a_0$. So,

$$\mathcal{H}'^2 = \frac{e^4}{R^6}(y_1^2 y_2^2 + z_1^2 z_2^2 + 4x_1^2 x_2^2 + 2y_1 y_2 z_1 z_2 - \ldots)$$

[9]F. London, *Z. Physik*, **63**, 245 (1930).
[10]This holds for *any* two atoms together in ground states that are spherically symmetric and nondegenerate.
[11]A. Unsöld, *Z. Physik*, **43**, 563 (1927).

from above. But $\langle y_1 y_2 z_1 z_2 \rangle = 0$ because we have a cross product of odd Cartesian components of $\mathbf{r}_1$ and $\mathbf{r}_2$. The first three terms are of the form

$$\int |\psi_{100}(\mathbf{r})|^2 x^2 d^3r = \frac{1}{3} \int |\psi_{100}(\mathbf{r})|^2 r^2 d^3r$$

$$= \frac{1}{3\pi a_0^3} \int_0^\infty e^{-2r/a_0} r^2 4\pi r^2 dr = a_0^2$$

where $r^2 d^3r = r^2 d\Omega dr = r^2 d\varphi \sin\theta d\theta\, dr$, and integration over angles gives 4π. We have $\langle 0|\mathcal{H}'^2|0\rangle = 6e^4 a_0^4/R^6$ and $\Delta E(R) \geq -8e^2 a_0^5/R^6$.

The perturbation theory and the variational method give the two limits (upper and lower) of the coefficient of the leading term $1/R^6$. Now we will see the upper limit given by the variational calculation.

How should we choose the trial ψ? Suppose that it is independent of R. Then $\langle E \rangle = \langle \mathcal{H}' \rangle \sim R^{-3}$. The upper limit of this dependence on R is not what we want, because we are looking for the coefficient of R^{-6}, not R^{-3}.

Therefore, ψ must contain a term $\propto \mathcal{H}'$, so that $\langle E \rangle \propto \mathcal{H}'^2$ and hence $\sim R^{-6}$ (the desired interaction). The choice is

$$\psi(\mathbf{r}_1, \mathbf{r}_2) = \psi_{100}(\mathbf{r}_1)\psi_{100}(\mathbf{r}_2)(1 + \lambda\mathcal{H}')\,,$$

with λ being the variational parameter that is to be allowed to vary. Caution! ψ is *not* normalized. Therefore, we cannot use $E_0 \leq \int \psi^* \mathcal{H}\psi\, dV$. We use

$$E_0 \leq \frac{\displaystyle\int \psi^* \mathcal{H}\psi\, dV}{\displaystyle\int |\psi|^2 \, dV}$$

where the integration is over all space. This gives

$$E_0 + \Delta E(R) \leq \frac{\displaystyle\iint \psi^{(0)}(1 + \lambda\mathcal{H}')(\mathcal{H}^{(0)} + \mathcal{H}')\psi^{(0)}(1 + \lambda\mathcal{H}')d^3r_1\, d^3r_2}{\displaystyle\iint \psi^{(0)^2}(1 + \lambda\mathcal{H}')^2 d^3r_1\, d^3r_2}$$

where λ is (taken) real, and $\psi^{(0)}$ is the product of $\psi_{100}(\mathbf{r}_1)$ and $\psi_{100}(\mathbf{r}_2)$, the two ground-state wavefunctions of the two H atoms together. But $\psi^{(0)}$ *is* normalized (because it is an eigenstate of $\mathcal{H}^{(0)}$), having the eigenvalue $E_0 = -e^2/a_0$. Therefore,

$$E_0 + \Delta E(R) \leq \frac{E_0 + 2\lambda\langle 0|\mathcal{H}'^2|0\rangle + \lambda^2\langle 0|\mathcal{H}'\mathcal{H}^{(0)}\mathcal{H}'|0\rangle}{1 + \lambda^2\langle 0|\mathcal{H}'^2|0\rangle}$$

where $\langle 0|\mathcal{H}'|0\rangle = \langle 0|\mathcal{H}'^3|0\rangle = 0$ and $\langle 0|\mathcal{H}'\mathcal{H}^{(0)}\mathcal{H}'|0\rangle = \sum(\text{Squares}) = 0$ because the squares are like

$$\int \psi_{100}(\mathbf{r}) x \mathcal{H}^{(0)} x \psi_{100}(\mathbf{r}) d^3 r \, .$$

We will stop at the order $\mathcal{H}'^2$. We write the expansion

$$\frac{E_0 + 2\lambda\langle 0|\mathcal{H}'^2|0\rangle}{1 + \lambda^2\langle 0|\mathcal{H}'^2|0\rangle} = \left[E_0 + 2\lambda\langle 0|\mathcal{H}'^2|0\rangle\right]\left[1 + \lambda^2\langle 0|\mathcal{H}'^2|0\rangle\right]^{-1}$$

$$= E_0 + (2\lambda - E_0\lambda^2)\langle 0|\mathcal{H}'^2|0\rangle + \text{h.o.t.}$$

Since $E_0 < 0$, the expansion passes through a minimum upon variation of λ, at $\lambda = 1/E_0$. Then

$$E_0 + \Delta E(R) \le E_0 + \frac{\langle 0|\mathcal{H}'^2|0\rangle}{E_0} = E_0 - \frac{6e^2 a_0^5}{R^6} \, .$$

The two limits (upper and lower) on the interaction energy are

$$-\frac{8e^2 a_0^5}{R^6} \le \Delta E(R) \le -\frac{6e^2 a_0^5}{R^6}$$

where the coefficients of $1/R^6$ can be numerically evaluated. However, this result is in fact *wrong*, because $\mathcal{H}'$ contains only the static interaction between two dipoles (atoms), ignoring the detail of the *retardation* in the information transmission from a dipole to the other, as the propagation of the electromagnetic interaction has a finite speed, the speed of light c. Therefore, $\Delta E \sim -R^{-7}$ in fact, for large R. (When we say "large" we compare it with the wavelength of the electromagnetic radiative transition: $R \gg \hbar a_0 c/e^2 = 137a_0$.) But since R is large, the Van der Waals interaction becomes unimportant, and hence the two limits obtained above are practically correct for our needs.[12]

Notice that in writing $\mathcal{H}'$ above, the quantities $1/R$ and $1/r$ are operators, as $\mathcal{H}'$ is an operator, that is, the Hamiltonian. But in the expressions for energy (such as Eq. (7.4)) the right-hand side consists of numbers.

The Van der Waals case is best treated in Park (pp. 507–511) as forces between atoms in molecules. See also Kittel, p. 78.

Example problem: (a) Consider the internuclear potential for H_2 obtained in Section 8.10. Prove that, for large R, it is not $\sim -R^{-6}$. Give reason.

---

[12] H.B.G. Casimir and D. Polder, *Phys. Rev.*, **73**, 360 (1948).
L. Pauling and E.B. Wilson, Jr., *Introduction to Quantum Mechanics*, McGraw-Hill, New York (1935).

(b) What are the next terms ($\sim 1/R^4$) in the expansion of $\mathcal{H}'$? Prove that the Van der Waals interaction has no contribution of R^{-4}, because the matrix elements $\langle \psi^{(0)}|\mathcal{H}'|\psi^{(0)}\rangle_{\text{diagonal}} = 0$. Here $\psi^{(0)}$ is the unperturbed ground state. (c) Expand the series in Eq. (7.4) until the first nonzero term. That will give you a lower limit on $-\Delta E(R)$. Is it the same as that obtained by using the variational principle? Explain. (Note that the perturbation theory and the variational method give two opposite limits for the coefficient of the Van der Waals term.) (d) In the perturbation theory, the leading term in energy varies as $\sim R^{-6}$ at large separations R. How do you reconcile that with the statement of part (a)?

Solution: It is left to the reader as an exercise.

Chapter 8

THE DIATOMIC MOLECULE

8.1. The Spectrum

The spectrum of a diatomic molecule consists of

(1) Electronic energies.
(2) Vibrational states (due to the vibration of the two atoms).
(3) Rotational states (due to the rotation of the whole molecule).

We will discuss these states separately below.

8.2. Electronic Energies

The (mechanical) stability of a molecule is due to the atomic electrons. It can be proven quantum mechanically (by the double potential-well theory) that when two atoms share electrons, an attractive force is formed between them (for certain electronic energy levels). Thus the electron clouds play the role of a cohesive tissue that holds the molecule together. Mainly the ground state of the electrons binds the molecule, and there is no guarantee that the excited states will contribute to the binding. Hence if we excite the electrons to higher states we run the risk of breaking the molecule apart to a complete dissociation.

As long as the molecule keeps stable, transitions between electronic energy states give a spectrum pretty much the same as that of an atom, in the ultra-violet and the visible region of the electromagnetic spectrum. The energies of these transitions are given by

$$E_{\text{el}} \sim \frac{\hbar^2}{md^2} \tag{8.1}$$

where m is the mass of the electron, and d is the size of the molecule. Equation (8.1) is approximate and can be derived from the uncertainty principle.

8.3. Vibrational States

If the amplitude of the vibration stays small (compared to the molecular size), the molecular vibration is almost that of a simple harmonic oscillator of force constant κ, vibrating according to Hooke's law.[1] The energy of the vibrational states is

$$E_{\text{vib}} = \left(\text{v} + \frac{1}{2}\right)\hbar\omega = \left(\text{v} + \frac{1}{2}\right)\hbar\sqrt{\frac{\kappa}{\tilde{\mu}}}$$

where v is the *vibrational quantum number* (for the quantization of the vibrational energy) and $\tilde{\mu}$ is the reduced mass (of the two atoms in the molecule).

Here κ is unknown, but it can be found roughly: As the molecule can be stretched at most to a length equal to its size d, its elastic potential energy becomes $(1/2)\kappa d^2$. But this is about equal to E_{el}, because the electronic states are the ones which ensure the binding (up to a distance d). Hence $\kappa \sim E_{\text{el}}/d^2$, so that

$$E_{\text{vib}} \sim \hbar\sqrt{\frac{\kappa}{\tilde{\mu}}} \sim \hbar\sqrt{\frac{\hbar^2}{md^4\tilde{\mu}}} \sim E_{\text{el}}\sqrt{\frac{m}{\tilde{\mu}}}\,.$$

The vibrational states are evenly spaced. Given that we have typically $\tilde{\mu} \sim 10^4 m$, the result is

$$E_{\text{vib}} = 10^{-2} E_{\text{el}}$$

which means that the energies of the transitions between vibrational states are in the infrared region.

8.4. Rotational States

The Hamiltonian for the rotational kinetic energy of a rigid rotor is

$$\mathcal{H} = \frac{P^2}{2\mathcal{I}}$$

where P is the angular momentum, and $\mathcal{I}$ is the moment of inertia of the rotor (here, molecule). (Note that $\mathcal{H}$ and P are operators in this expression.) If we

---

[1] The classical harmonic oscillator is a good model for the molecular vibrations.

solve the equation we find the eigenenergies:

$$E_{\text{rot}} = \frac{\hbar^2}{2\mathcal{I}}k(k+1) \tag{8.2}$$

which are quantized according to the *rotational quantum number k*. (Note that for molecules the notation is P and k, whereas for revolving atomic electrons we have L and l respectively.) We have $k = 0, 1, 2, \ldots$. Since $\mathcal{I} = \tilde{\mu}d^2$, we have

$$E_{\text{rot}} \sim \frac{\hbar^2}{2\mathcal{I}} \sim \frac{\hbar^2}{2\tilde{\mu}d^2} = E_{\text{el}}\frac{m}{\tilde{\mu}}$$

which is down by another factor of 10^2 compared to E_{vib}. This means that the emitted energies during the transitions between rotational states are in the far infrared region.

The selection rule for rotational transitions is $\Delta k = \pm 1$. So, the interstate transition energy (from a k value to its immediately next, since k changes by one unit only) is

$$\Delta E = \frac{\hbar^2}{2\mathcal{I}}[k(k+1) - (k-1)(k-1+1)] = \frac{k\hbar^2}{\mathcal{I}}$$

with $k = 1, 2, 3, \ldots$. So, k can take all integer values until the value where the molecule dissociates. This gives a spectrum of equally spaced energies (spectral lines) which characterize a typical molecular spectrum.

One may ask: How many rotational states can a diatomic molecule have? What is the maximum possible value of k? To answer this question, one refers to the centripetal force, that is one finds how fast the molecule can turn before it flies apart. Since we know the masses of the two atoms and their separation, we calculate the rotational energy and set it equal to E_{rot} to find (the maximum) k via Eq. (8.2).

When one expands the energy of a molecule to a series, *i.e.*, to the powers of $\sqrt{m/\tilde{\mu}}$, the first three terms are E_{el}, E_{vib}, and E_{rot}. The translational energy of the molecule as a whole is ignored because the motional velocity of the atomic electrons in their orbits is much higher than the translational velocity of the molecule, so that the inclusion of the latter (vector addition of velocities) does not affect the energy appreciably.

8.5. Vibration-Rotation Spectrum

In a diatomic molecule, the vibrational and rotational transitions occur together and simultaneously. This gives an infrared molecular spectrum.

Emission spectroscopy is not possible with molecules, because in a discharge tube the electrons have usually high energies and thus they make the molecules come apart. For this reason, absorption spectroscopy is done.

The molecule is thought as an harmonic oscillator which absorbs an energy quantum and raises its vibrational quantum number by one, according to the selection rule for an oscillator. Also, it changes its angular momentum by one unit.

In going from k to $k + 1$, the absorbed energy (quantum) is

$$\Delta E = \Delta E_{\text{vib}} + \frac{\hbar^2}{2\mathcal{I}}[k(k+1) - (k+1)(k+2)] = \Delta E_{\text{vib}} + \frac{\hbar^2}{2\mathcal{I}}[-2(k+1)]$$

with $k = 0, 1, 2, \dots$.

In going from k to $k - 1$, the absorbed energy is

$$\Delta E = \Delta E_{\text{vib}} + \frac{\hbar^2}{2\mathcal{I}}[k(k+1) - (k-1)k] = \Delta E_{\text{vib}} + \frac{\hbar^2}{2\mathcal{I}}(2k)$$

with $k = 1, 2, 3, \dots$. (No $k = 0$, because if it is already in the lowest rotational state; it cannot decrease further.)

In both cases, $\Delta E_{\text{vib}} = \hbar\omega$.

For the diatomic HCl molecule, $\Delta E_{\text{vib}} \sim 0.4$ eV and $(\hbar^2/2\mathcal{I}) \sim 10^{-3}$ eV. This gives $d \sim 1$ Å. For the same molecule we have

$$\frac{\text{Lowest rotational energy}}{\text{Lowest vibrational energy}} = \frac{2\hbar^2/2\mathcal{I}}{(1/2)\hbar\omega} \sim \frac{1}{80}$$

which almost agrees with $\sqrt{m/\bar{\mu}}$.

The amount $\hbar^2/2\mathcal{I}$ is so small,[2] that many rotational states have the chance to be excited before the molecule dissociates, and hence the large number of k values observed in the spectrum of HCl as distinct spectral lines.

If we take the HCl spectrum (plot of intensity vs. wavelength), we see that the spectral lines (intensity peaks) are not equally spaced, which means that there must be something wrong with the theory. But the theory is *not* wrong; it is only oversimplified by assumptions, namely the following ones:

(1) Upon rotation, the molecule gets strained and hence its moment of inertia changes. The theory did not allow for it.

(2) Hooke's law is for perfect, ideal systems, not for vibrating real molecules; that is, the molecule is not a perfect oscillator (whereas in the theory we took it as such).

---

[2]The lowest rotational energy is way below the thermal energy at room temperature.

(3) The theory neglects the Coriolis force on a rotating molecule.

It is left to the enthusiasts to perform the calculations including these contributions, and find the potential from the spacings of the spectral lines. Of course, one need not live with diatomic molecules only; one can extend one's curiosity to polyatomic molecules, too, to find vibration frequencies of bonds and rotation energies of radicals. But we should warn the volunteers that the mathematics involved is messy and complicated. However, the reward is great, because this way one becomes able to look at the spectra and recognize materials of large molecular weight. A computer is necessary to expedite such calculations.

8.6. Classical Treatment of Vibrations

In classical dynamics, one can study the vibrations of two oscillating bodies coupled together by an intermediate massless spring. This is the model of a diatomic molecule of linear structure (where the two oscillating masses represent the atoms and the spring represents the bond). Finding the frequencies and modes (shapes, fashions) of the vibration is an eigenvalue problem where one determines the eigenfrequencies (of the vibrating system) and the *normal coordinates* (of the system) that *decouple the motion*, that is, coordinates in which when the system is expressed, it appears to be consisting of two separate and independent (decoupled) oscillators.

The two masses vibrate, being coupled by the same common spring: They are not independent in reality. However, we can find (artificial) normal coordinates that decouple them.

When we study the linear vibrations of a system of two masses connected by a spring, we take the motion along the line joining the masses. The force constant of the spring is κ, and its unstretched length d. The generalized coordinates of the two masses are just their one-dimensional Cartesian coordinates x_1 and x_2 along x, from some origin O (Fig. 8.1).

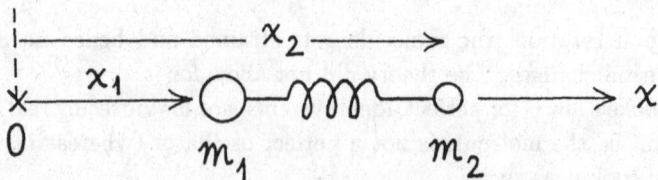

Fig. 8.1.

The kinetic and potential energy of the system are given by

$$T = \frac{1}{2}(m_1\dot{x}_1^2 + m_2\dot{x}_2^2)$$

$$V = \frac{1}{2}\kappa(x_2 - x_1 - d)^2$$

where $x_2 - x_1 - d$ is the *net* stretch. V is not a homogeneous quadratic function of x_1 and x_2. So, there is no linear transformation leading us to the normal coordinates. For this reason, we use a mathematical trick: We let $s = x_2 - d$ and use the new coordinate s to write

$$T = \frac{1}{2}(m_1\dot{x}_1^2 + m_2\dot{s}^2)$$

$$V = \frac{1}{2}\kappa(s - x_1)^2 . \tag{8.3}$$

The Lagrangian of the system is

$$\mathcal{L} = T - V = \frac{1}{2}(m_1\dot{x}_1^2 + m_2\dot{s}^2) - \frac{1}{2}\kappa(s - x_1)^2$$

whence the differential equations of motion (for x_1 and s) are

$$\left.\begin{array}{r} m_1\ddot{x}_1 + \kappa x_1 - \kappa s = 0 \\ m_2\ddot{s} + \kappa s - \kappa x_1 = 0 \end{array}\right\} \tag{8.4}$$

which are coupled. We try solutions of the form $x_1 = \mathcal{A}e^{i\omega t}$ and $s = \mathcal{B}e^{i\omega t}$ (which are oscillatory, with amplitudes $\mathcal{A}$ and $\mathcal{B}$, and frequency ω).[3] We substitute these solutions into Eqs. (8.4) and obtain the algrebraic equations

$$\left.\begin{array}{r} (-m_1\omega^2 + \kappa)\mathcal{A} - \kappa\mathcal{B} = 0 \\ -\kappa\mathcal{A} + (-m_2\omega^2 + \kappa)\mathcal{B} = 0 \end{array}\right\} \tag{8.5}$$

where we have an eigenvalue problem, with ω yet to be found. Equations (8.5) form a system of simultaneous equations. For a nontrivial solution, the determinant of the coefficients should vanish:

$$\begin{vmatrix} m_1\omega^2 - \kappa & \kappa \\ \kappa & m_2\omega^2 - \kappa \end{vmatrix} = 0$$

whence we have

---

[3]These solutions are just guesses that we hope that they will work, and indeed they do.

$$(m_1\omega^2 - \kappa)(m_2\omega^2 - \kappa) - \kappa^2 = 0$$

and $\omega_1 = 0$, $\omega_2 = \pm\sqrt{\kappa(m_1 + m_2)/m_1 m_2}$. (In ω_2 we reject the negative result because it although exists mathematically, it has no physical meaning, as a frequency cannot be a negative quantity). The eigenfrequencies of the system are ω_1 and ω_2.

For $\omega = \omega_1$ we have $\mathcal{A} = \mathcal{B}$, and for $\omega = \omega_2$ we have $\mathcal{A} = -(m_2/m_1)\mathcal{B}$. (See below for the physical interpretation of these results concerning the vibration amplitudes.)

For $\omega_2 \neq 0$ we obtain the solution

$$x_1 = \mathcal{A}_1 e^{i\omega_2 t} + \mathcal{A}_1' e^{-i\omega_2 t}$$

$$s = \mathcal{A}_2 e^{i\omega_2 t} + \mathcal{A}_2' e^{-i\omega_2 t}.$$

For $\omega_1 = 0$ the solution is

$$x_1 = \mathcal{B}_1 + \mathcal{B}_1' t$$

$$s = \mathcal{B}_2 + \mathcal{B}_2' t.$$

The general solution is

$$x_1 = \mathcal{B}_1' t + \mathcal{B}_1 + \mathcal{A}_1 e^{i\omega_2 t} + \mathcal{A}_1' e^{-i\omega_2 t}$$

$$s = \mathcal{B}_2' t + \mathcal{B}_2 - \frac{m_1}{m_2}(\mathcal{A}_2 e^{i\omega_2 t} + \mathcal{A}_2' e^{-i\omega_2 t})$$

and the normal coordinates are

$$X_1 = \frac{m_1}{m_2}x_1 + s = \frac{m_1 + m_2}{m_2}(\mathcal{B}_1' t + \mathcal{B}_1)$$

$$X_2 = x_1 - s = \frac{m_1 + m_2}{m_2}(\mathcal{A}_1 e^{i\omega_2 t} + \mathcal{A}_1' e^{-i\omega_2 t}).$$

Hoping that we will not confuse the reader, we would do better if we associated the subscript 1 with ω_1, and 2 with ω_2. Then the notation would be more consistent. We would have

$$x_1 = \mathcal{A}_1' t + \mathcal{A}_1 + \mathcal{A}_2 e^{i\omega_2 t} + \mathcal{A}_2' e^{-i\omega_2 t}$$

$$s = \mathcal{A}_1' t + \mathcal{A}_1 - \frac{m_1}{m_2}(\mathcal{A}_2 e^{i\omega_2 t} + \mathcal{A}_2' e^{-i\omega_2 t}).$$

For X_1 we have $x_1 = s = x_2 - d$ which corresponds to a uniform translational motion of the two masses together in one direction (translation of their center of mass) (Fig. 8.2). This is the $\omega = \omega_1 = 0$ mode (no vibration).

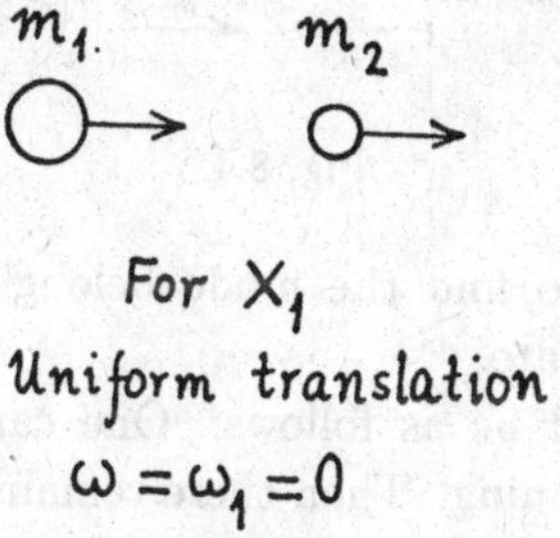

For X_1

Uniform translation

$\omega = \omega_1 = 0$

Fig. 8.2.

For X_2 we have

$$x_1 = -\frac{m_2}{m_1}s = -\frac{m_2}{m_1}(x_2 - d)$$

which corresponds to an óscillation (relative to the center of mass) *alias* a "breathing mode" (Fig. 8.3). This is the $\omega = \omega_2 \neq 0$ mode, and the "breathing" takes place with different amplitudes, if $m_1 \neq m_2$. The larger the mass the smaller the amplitude (motion away from the equilibrium position), and that is understandable even intuitively (as a large mass is more reluctant to move). If $m_1 = m_2 \equiv m$, the two amplitudes are equal, as expected, and the vibration ("breathing") of the two masses is perfectly symmetric (Fig. 8.4). In Fig. 8.2, Fig. 8.3, and Fig. 8.4 the size of the arrows indicates the size of the amplitudes. In Fig. 8.3 we have $m_1 > m_2$. By inertial sluggishness, the larger mass has a shorter amplitude.

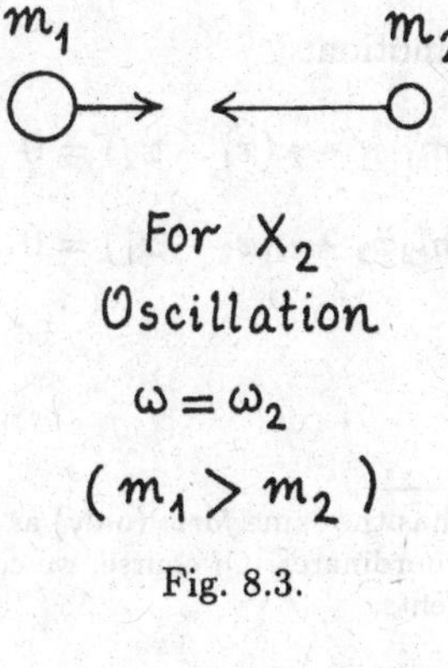

For X_2

Oscillation

$\omega = \omega_2$

$(m_1 > m_2)$

Fig. 8.3.

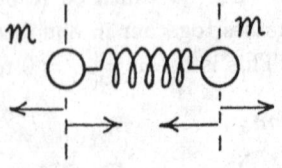

Fig. 8.4.

Notice that in order to find the mode belonging to one coordinate, we suppress the other coordinate.

Another solution could be as follows: One can foresee X_1 without any calculation, from the beginning. Then there remains only the relative motion for which we need just one equation with the reduced mass $\bar{\mu}$ of the system, whence we obtain $\omega_2 = \sqrt{\kappa/\bar{\mu}}$.

The equations of motion can be reached by still another way: Instead of taking x_1 and x_2 as the Cartesian coordinates of the masses, we can consider x_1 and x_2 as being the *displacements* (from equilibrium) of the two masses (Fig. 8.5). Then we have

$$T = \frac{1}{2}(m_1 \dot{x}_1^2 + m_2 \dot{x}_2^2)$$

$$V = \frac{1}{2}\kappa(x_2 - x_1)^2$$

with $x_2 - x_1$ being the net displacement.[4] By writing the Lagrangian $\mathcal{L} = T - V$ and the equation for the generalized coordinates

$$\frac{d}{dt}\left(\frac{\partial T}{\partial \dot{q}_i}\right) - \frac{\partial T}{\partial q_i} + \frac{\partial V}{\partial q_i} = 0,$$

we obtain the equations of motion:

$$m_1 \ddot{x}_1 + \kappa(x_1 - x_2) = 0$$

$$m_2 \ddot{x}_2 + \kappa(x_2 - x_1) = 0.$$

---

[4] Notice that in this approach T has the same *form* (only) as Eq. (8.3), but now the variables are the displacements, not the coordinates. Of course, we could choose other letters, say ξ_1 and ξ_2, to denote the displacements.

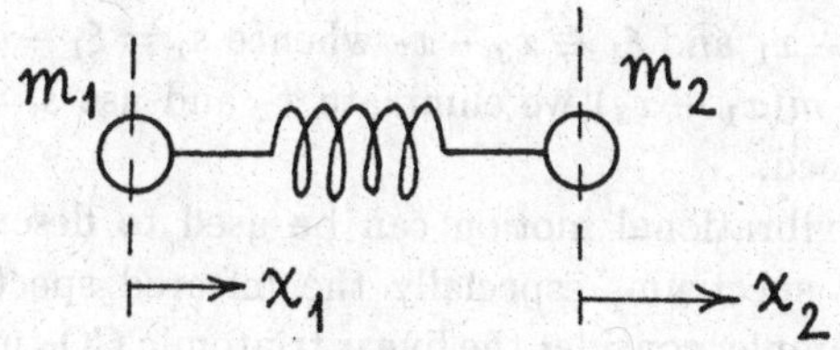

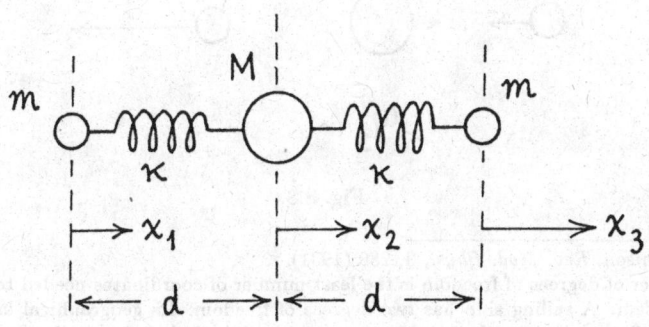

Fig. 8.5.

But by inspection, one realizes that these two equations can be written down directly, without any need of Lagrangian formalism. They are nothing else than Hooke's law for the restoring force of a spring stretched by x:

$$F = ma = m\ddot{x} = -\kappa x,$$

written for x_1 and x_2. Here κ is the stiffness of the spring.

8.7. The Triatomic Molecule

Consider three masses, m, M, and m, connected by two identical springs (of constant κ each). Their unstretched length is d. The system is and moves on a straight line (Fig. 8.6). The displacements (*not* the coordinates) of the three masses (in a molecule, atoms) are x_1, x_2, x_3, as shown. We have

$$T = \frac{1}{2}(m\dot{x}_1^2 + M\dot{x}_2^2 + m\dot{x}_3^2)$$

$$V = \frac{1}{2}\kappa(x_2 - x_1 - d)^2 + \frac{1}{2}\kappa(x_3 - x_2 - d)^2 .$$

Fig. 8.6.

We let $\xi_1 \equiv x_2 - x_1$ and $\xi_1 \equiv x_3 - x_2$ whence $s_1 = \xi_1 - d$ and $s_2 = \xi_2 - d$.

From $Mx_2 = -m(x_1 + x_3)$ we eliminate x_2 and use $s_i = Ae^{i\omega t}$ and $s_2 = Be^{i\omega t}$, and we proceed.

This model of vibrational motion can be used to describe molecular vibrations and their spectrum, especially the infrared spectra of polyatomic molecules.[5] For example, consider the linear triatomic CO_2 molecule (Fig. 8.7). The motion has three degrees of freedom[6] (as it is along the molecular axis, a straight line). So, we will obtain three normal coordinates. One of them has $\omega_1 = 0$ (simple translation of the center of mass, as before). The other two have modes shown in Fig. 8.8. The amplitudes are so arranged (in size and direction)

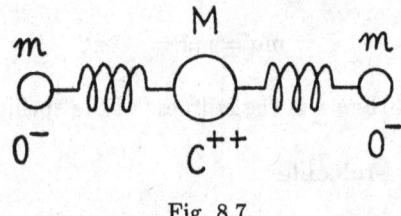

Fig. 8.7.

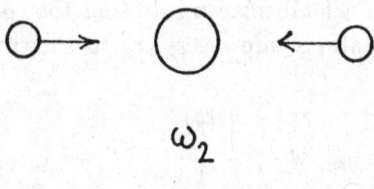

ω_2

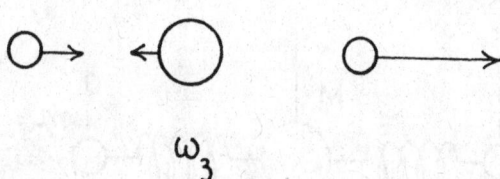

ω_3

Fig. 8.8.

---

[5]D.M. Dennison, *Rev. Mod. Phys.*, **3**, 280 (1931).

[6]The number of degrees of freedom is the least number of coordinates needed to fully determine a system. A sailing ship has two degrees of freedom: Its geographical longitude and latitude. A flying plane needs a third one, the altitude. We can thus determine its position fully. To locate the house of a person in a city we need *at least* two postal coordinates (bits of information or degrees of freedom): The name of the street and the number of the building.

as to keep the center of mass fixed. In the ω_2 mode the C atom does not move, and the two oxygen atoms vibrate in opposite phase, with equal amplitudes ("breathing" about C). In the ω_3 mode the C atom moves towards one of the oxygen atoms which in turn approaches it, while the other oxygen atom moves away. The system oscillates as shown.

It is important to note that only ω_3 is observed spectroscopically. The frequency ω_2 is not observed because in this mode the electrical center of the molecule coincides with the center of mass, and on the overall, there is no oscillating dipole

$$\vec{p} = \sum q\mathbf{r},$$

and hence no dipole radiation is emitted by this mode.

Let us now solve the problem of the triatomic molecule rigorously, by employing the method used in classical mechanics for small oscillations. If the displacements of the three masses are x_1, x_2, x_3, we have

$$V = \frac{1}{2}\kappa(x_2 - x_1 - d)^2 + \frac{1}{2}\kappa(x_3 - x_2 - d)^2.$$

Let $\xi_i = x_i - (x_0)_i$ be coordinates relative to the equilibrium positions $(x_0)_i$. Then $(x_0)_2 - (x_0)_1 = (x_0)_3 - (x_0)_2 = d$. We have then

$$V = \frac{1}{2}\kappa(\xi_2 - \xi_1)^2 + \frac{1}{2}\kappa(\xi_3 - \xi_2)^2$$

$$= \frac{1}{2}\kappa(\xi_1^2 + 2\xi_2^2 + \xi_3^2 - 2\xi_1\xi_2 - 2\xi_2\xi_3)$$

where the cross terms indicate coupling (interaction or mixing of states). In classical and quantum mechanics interactions between states can be described by matrices. If the matrix has nonzero off-diagonal elements, this means that states mix (one affects the other, or the entity in question — whether an oscillating mass or an electron making a transition — partakes of both states for a while). So, the $\mathbb{V}$ matrix is

$$\mathbb{V} = \begin{pmatrix} \kappa & -\kappa & 0 \\ -\kappa & 2\kappa & -\kappa \\ 0 & -\kappa & \kappa \end{pmatrix}.$$

But the kinetic energy involves no coupling (no cross terms), corresponding to a pure state:

$$T = \frac{1}{2}m\dot{\xi}_1^2 + \frac{1}{2}m\dot{\xi}_3^2 + \frac{1}{2}M\dot{\xi}_2^2$$

whence the $\mathbb{T}$ matrix is diagonal (the nonzero elements are along the main diagonal only, *i.e.*, every state interacts only with *itself*, not with others):

$$\mathbb{T} = \begin{pmatrix} m & 0 & 0 \\ 0 & M & 0 \\ 0 & 0 & m \end{pmatrix}.$$

The secular determinant becomes

$$|\mathbb{V} - \omega^2\mathbb{T}| = \begin{vmatrix} \kappa - \omega^2 m & -\kappa & 0 \\ -\kappa & 2\kappa - \omega^2 M & -\kappa \\ 0 & -\kappa & \kappa - \omega^2 m \end{vmatrix} = 0$$

which when expanded gives the equation

$$\omega^2(\kappa - \omega^2 m)[\kappa(M + 2m) - \omega^2 mM] = 0$$

whose three solutions are

$$\omega_1 = 0$$

$$\omega_2 = \sqrt{\kappa/m}$$

$$\omega_3 = \sqrt{\frac{\kappa}{m}\left(1 + \frac{2}{M}\right)}.$$

That one of the eigenfrequencies turns out to be zero means that this solution does not correspond to any oscillation; it has to do just with a uniform translation, without a relative motion between the three masses. The molecule moves rigidly altogether along its axis, and V does not change (indifferent equilibrium). There is no restoring force, and so there is no ω. (Although there are three degrees of freedom, one of them is translational, not vibrational.) We can fix matters so that ω_1 is disposed of from the beginning: We can write $m(x_1 + x_3) + Mx_2 = 0$ which means that the center of mass is held fixed at rest. Thus one coordinate is eliminated, and there remain only the other two.

If we constrain the molecule to move along its axis only, there is one rigid translational motion possible. If we allow it to vibrate in three dimensions, we will have six translations. The molecule will translate (along three axes) or rotate. So, if the system has N degrees of freedom, there will be six frequencies equal to zero and $N - 6$ vibrational frequencies. We can always cut down the degrees of freedom initially, by putting constraints on the coordinates, arising from the conservation of linear and angular momentum.[7]

The frequency ω_2 corresponds to the simple harmonic motion of a mass m, that is, to the vibration of the two masses m while M stays at rest. In the ω_3 mode M moves (the quantity κ/m is modified by a multiplying factor in the parentheses).

The amplitudes (eigenvectors) can be found as follows:

$$(\kappa - \omega_i^2 m)\mathcal{A}_{1i} - \kappa\mathcal{A}_{2i} = 0$$

$$-\kappa\mathcal{A}_{1i} + (2\kappa - \omega_i^2 M)\mathcal{A}_{2i} - \kappa\mathcal{A}_{3i} = 0$$

$$-\kappa\mathcal{A}_{2i} + (\kappa - \omega_i^2 m)\mathcal{A}_{3i} = 0\,.$$

We also have the normalization condition

$$m(\mathcal{A}_{1i}^2 + \mathcal{A}_{3i}^2) + M\mathcal{A}_{2i}^2 = 1\,.$$

If we set $\omega_i = \omega_1 = 0$, from these equations we obtain $\mathcal{A}_{11} = \mathcal{A}_{21} = \mathcal{A}_{31}$. (The first subscript denotes the label of the mass, and the second subscript the mode in question.) This is the translational motion of the molecule as a whole. From the normalization condition we find that the size of the three amplitudes is $1/\sqrt{2m + M}$.

For $\omega_i = \omega_2$ we obtain $\mathcal{A}_{22} = 0$ and $\mathcal{A}_{12} = -\mathcal{A}_{32}$. The center mass is stationary (its amplitude is zero), while the two end masses vibrate out of phase (to conserve momentum). The value of the amplitude is $\mathcal{A}_{12} = 1/\sqrt{2m}$.

For $\omega_i = \omega_3$ we obtain

$$\mathcal{A}_{13} = \mathcal{A}_{33} = \frac{1}{\sqrt{2m\left(1 + \dfrac{2m}{M}\right)}} \quad \text{and} \quad \mathcal{A}_{23} = \frac{-2}{\sqrt{2M\left(2 + \dfrac{M}{m}\right)}}$$

where the two end masses oscillate with the same amplitude and phase, while the center one is out of phase and has a different amplitude.

---

[7]Apart from the translation, vanishing frequencies may show up if the first and second derivatives of V at equilibrium are zero. Since also a vanishing third derivative is necessary for equilibrium stability, oscillations may still occur if the fourth derivative is nonzero. However, the forces are nonlinear and hence the oscillations are anharmonic in this case. They deserve no treatment in this book, because they do not pertain to the formalism of small oscillations, and furthermore, they seldom occur.

The three normal modes are shown in Fig. 8.9.[8] In the next section we calculate the general (longitudinal) motion of the triatomic molecule.

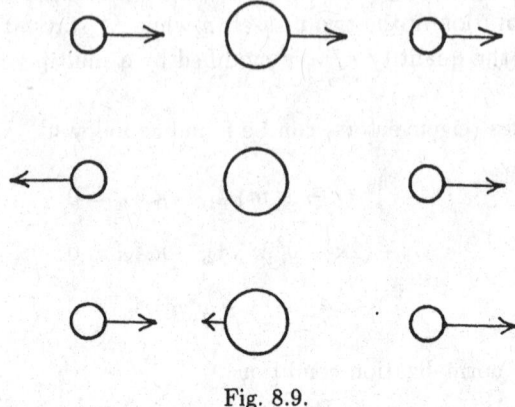

Fig. 8.9.

The foregoing treatment concerns longitudinal vibrations only. What if the molecule vibrates in a direction normal to its axis? We will have to find the lateral modes, considering a three-dimensional case (with nine degrees of freedom). The way is in principle a kid's stuff, but the mathematics involved in a three-dimensional motion is so convolute that we leave it for the reader, following the formal recipe of authors and professors to leave for the students every calculation that starts rapidly smelling trouble and complication. We will give only a description of the result in words here:

The three-dimensional case (with vibrations in all directions) has some vanishing frequencies corresponding to translations. If the molecule is *linear*, there are three degrees of freedom (x, y, z) for the translational motion, and *two* degrees of freedom for the rotational motion. But since rotation about the molecular axis is not a mode, we have four modes of vibration, all told. Of these two are longitudinal, discussed above. The other two are lateral, normal to the molecular axis. But since the molecule is symmetric, these two lateral modes are degenerate (because the vibration along the y-axis cannot be distinguished from the vibration along the z-axis, and hence the two frequencies are the same). In other words, all directions normal to the molecular axis are indistinguishable, and any two of them (in a plane perpendicular to the molecular axis) would do for the two degenerate modes of vibration. If these two

---

[8] A general vibrational motion is usually a linear combination of the two normal modes (ω_2 and ω_3). If we know the initial conditions, we can find the amplitudes and phases.

modes are cophasic, the atoms follow a straight line; if they are out of phase, the atoms describe an ellipse (Fig. 8.10), and the two modes correspond to the *rotation* of a dumbbell (if the center atom were absent) rather than a vibration in the strict sense of the word. Figure 8.10 shows the degenerate modes of a symmetric triatomic molecule. Notice that the two outer atoms have equal amplitudes (because of symmetry); as they move, say, counterclockwise, the middle atom moves clockwise because angular momentum must be conserved.

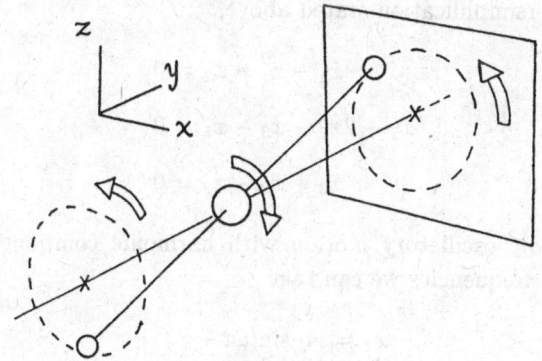

Fig. 8.10.

We should state again that by a convenient reformulation of the problem, it is intelligent to preclude from the beginning the involvement of the vanishing frequencies (that correspond to the translational motion) and thus lessen the strain put on us by the mathematics.

8.8. The Triatomic Molecule — General Motion

In this section we will mathematically obtain the solutions to the equations of motion of a triatomic molecule for *longitudinal* vibrations only. Refer to Fig. 8.11. For simplicity, we will take $m_1 = m_2 = m_3 \equiv m$. Although it is illegitimate, we will take $m = 1$ and $\kappa = 1$ for simplicity, in order not to carry along coefficients that dangle around. (We concentrate on mathematics here, not on physics.)

The differential equations of motion of the system are

$$m_1 \ddot{x}_1 + \kappa(x_1 - x_2) = 0$$

$$m_2 \ddot{x}_2 + \kappa(x_2 - x_1) + \kappa(x_2 - x_3) = 0$$

$$m_3 \ddot{x}_3 + \kappa(x_3 - x_2) = 0$$

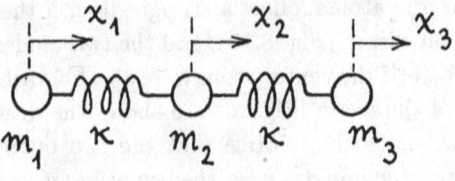

Fig. 8.11.

or, with the oversimplification stated above,

$$\ddot{x}_1 + x_1 - x_2 = 0$$

$$\ddot{x}_2 + 2x_2 - x_3 - x_1 = 0$$

$$\ddot{x}_3 + x_3 - x_2 = 0.$$

For a periodic oscillatory motion with harmonic components of various amplitudes and frequencies we can take

$$x_1 = A_1 \sin(\omega t + \varphi)$$

$$x_2 = A_2 \sin(\omega t + \varphi)$$

$$x_3 = A_3 \sin(\omega t + \varphi)$$

where φ is some phase which we have the freedom to include in our trial solutions. The eigenvalue problem here is to find the frequencies ω. We substitute the trial solutions into the equations of motion to have

$$\left.\begin{array}{r} (1 - \omega^2)A_1 - A_2 = 0 \\ -A_1 + (2 - \omega^2)A_2 - A_3 = 0 \\ -A_2 + (1 - \omega^2)A_3 = 0 \end{array}\right\} \tag{8.6}$$

whose solution is reached via the relation

$$\begin{vmatrix} (1 - \omega^2) & -1 & 0 \\ -1 & (2 - \omega^2) & -1 \\ 0 & -1 & (1 - \omega^2) \end{vmatrix} = 0$$

which yields, after expansion,

$$\omega^2(1 - \omega^2)(\omega^2 - 3) = 0$$

whence $\omega_1 = 0$, $\omega_2 = 1$, $\omega_3 = \sqrt{3}$ (only positive values are physically admissible).

The system has three degrees of freedom. The general solution describing the motion is

$$x_1 = A_{11} \sin(\omega_1 t + \varphi_1) + A_{12} \sin(\omega_2 t + \varphi_2) + A_{13} \sin(\omega_3 t + \varphi_3)$$

$$x_2 = A_{21} \sin(\omega_1 t + \varphi_1) + A_{22} \sin(\omega_2 t + \varphi_2) + A_{23} \sin(\omega_3 t + \varphi_3)$$

$$x_3 = A_{31} \sin(\omega_1 t + \varphi_1) + A_{32} \sin(\omega_2 t + \varphi_2) + A_{33} \sin(\omega_3 t + \varphi_3)$$

where some amplitudes are expressible in terms of others. From Eqs. (8.6) we have $A_2 = (1 - \omega^2)A_1$ whence $A_{21} = A_{11}$, $A_{22} = A_{12} = 0$, and $A_{23} = -2A_{13}$. (We substituted ω_1, ω_2, ω_3 into the equation for ω.) Again from Eqs. (8.6) we have $A_3 = A_1$ whence $A_{31} = A_{11}$ and $A_{33} = A_{13}$. The second equation of Eqs. (8.6) gives

$$-A_{12} + (2 - \omega_2^2)A_{22} - A_{32} = 0 ,$$

$$-A_{12} + A_{22} - A_{32} = 0$$

because $\omega_2 = 1$. This is the same as

$$-1 + \frac{A_{22}}{A_{12}} - \frac{A_{32}}{A_{12}} = 0 .$$

Since $A_{22} = A_{12} = 0$, we get $A_{32} = -A_{12}$.

Now the general solution of the motion becomes

$$x_1 = A_1 \sin(\omega_1 t + \varphi_1) + A_2 \sin(\omega_2 t + \varphi_2) + A_3 \sin(\omega_3 t + \varphi_3)$$

$$x_2 = A_1 \sin(\omega_1 t + \varphi_1) - 2A_3 \sin(\omega_3 t + \varphi_3)$$

$$x_3 = A_1 \sin(\omega_1 t + \varphi_1) - A_2 \sin(\omega_2 t + \varphi_2) + A_3 \sin(\omega_3 t + \varphi_3)$$

where the second subscript in the amplitudes is now redundant.

If we know the initial conditions, we can describe the motion for any later time t. Just for the purpose of illustration, suppose that the system is initially at rest, and the rightmost mass m_3 has an initial displacement to the right, equal to unity. Then the initial conditions are $x_1(0) = x_2(0) = 0$, $x_3(0) = 1$, $\dot{x}_1(0) = \dot{x}_2(0) = \dot{x}_3(0) = 0$, whence we have

$$\left. \begin{array}{r} A_1 \sin \varphi_1 + A_2 \sin \varphi_2 + A_3 \sin \varphi_3 = 0 \\ A_1 \sin \varphi_1 - 2A_3 \sin \varphi_3 = 0 \\ A_1 \sin \varphi_1 - A_2 \sin \varphi_2 + A_3 \sin \varphi_3 = 1 \end{array} \right\} \tag{8.7}$$

and

$$\left.\begin{array}{c} A_1 \cos\varphi_1 + A_2 \cos\varphi_2 + \sqrt{3}A_3 \cos\varphi_3 = 0 \\ A_1 \cos\varphi_1 - 2\sqrt{3}A_3 \cos\varphi_3 = 0 \\ A_1 \cos\varphi_1 - A_2 \cos\varphi_2 + \sqrt{3}A_3 \cos\varphi_3 = 0 \,. \end{array}\right\} \tag{8.8}$$

From Eqs. (8.8) we obtain $3A_1 \cos\varphi_1 = 0$, *i.e.*, $\varphi_1 = \pi/2$, and $\varphi_2 = \varphi_3 = \pi/2$, too. From Eqs. (8.7) we obtain $A_1 = 1/3$, $A_2 = -1/2$, and $A_3 = 1/6$. For the given initial conditions, the most general solution at any time t becomes

$$x_1 = x_1(t) = \frac{1}{3} - \frac{1}{2}\sin\left(t + \frac{\pi}{2}\right) + \frac{1}{6}\sin\left(\sqrt{3}t + \frac{\pi}{2}\right)$$

$$x_2 = x_2(t) = \frac{1}{3} - \frac{1}{3}\sin\left(\sqrt{3}t + \frac{\pi}{2}\right)$$

$$x_3 = x_3(t) = \frac{1}{3} + \frac{1}{2}\sin\left(t + \frac{\pi}{2}\right) + \frac{1}{6}\sin\left(\sqrt{3}t + \frac{\pi}{2}\right)\,.$$

As the reader notices, the treatment is purely classical. The mathematical reasoning is rather straightforward, but the algebra is tedious. We apologize to the reader for this mathematical extravagance: It is boring inherently, not that we made it so.

8.9. Rigid Rotor and Reduced Mass

In studying the pure rotation spectrum, we consider the model of the *rigid rotor*, *i.e.*, two masses m_1 and m_2 joined by a rigid massless rod. This is a mechanical model for an AB type diatomic molecule (*e.g.*, HCl) where the rod plays roughly the role of the bond between the atoms. If the two masses are equal (same kind of atoms), we have a *dumbbell molecule* of type A_2 (*e.g.*, O_2, H_2, N_2, Cl_2).

In general, a *diatomic molecule* consists of two atoms and a bond between them. We have $m_1 \equiv m_A$ and $m_2 \equiv m_B$. We assume for a moment that the interatomic distance (between centers of nuclei) R (or r_0, or d) is constant in the rigid rotor model (*i.e.*, we have no vibrations, but pure rotations only). If the rotor rotates about an axis perpendicular to R and passing through the center of mass of the molecule (Fig. 8.12), the total moment of inertia (about the center of mass) is

$$\mathcal{I} = m_1(R - x)^2 + m_2 x^2\,. \tag{8.9}$$

If we take moments with respect to the center of mass, we have

$$m_1(R - x) = m_2 x$$

whence we read off x and substitute it into Eq. (8.9) to have

$$\mathcal{I} = m_1 \left(R - \frac{m_1 R}{m_1 + m_2} \right)^2 + m_2 \left(\frac{m_1 R}{m_1 + m_2} \right)^2 = \frac{m_1 m_2}{m_1 + m_2} R^2 \equiv \tilde{\mu} R^2$$

where the quantity $\tilde{\mu} \equiv m_1 m_2 / (m_1 + m_2)$ is the *reduced mass* of the rotor (diatomic molecule), otherwise given as

$$\frac{1}{\tilde{\mu}} = \frac{1}{m_1} + \frac{1}{m_2}.$$

Thus $\mathcal{I} = \tilde{\mu} R^2$. The physical meaning of this is the following: The rotating rotor (diatomic molecule) in Fig. 8.12 is equivalent to *a single mass* $\tilde{\mu}$ (shown in Fig. 8.13) rotating at a distance R from (and about) an axis. Then the two configurations in Fig. 8.12 and Fig. 8.13 have the same $\mathcal{I}$. If $m_1 = m_2 \equiv m$ (dumbbell), then $\tilde{\mu} = m/2$, as expected.

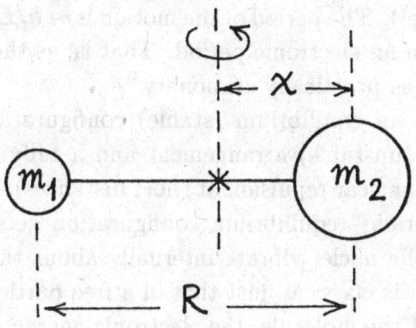

Fig. 8.12.

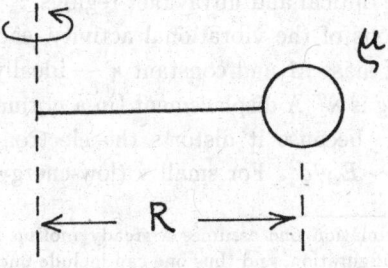

Fig. 8.13.

When a mass rotates in a circle, the quantum mechanical solution (of the Schrödinger equation) gives the (quantized) energies of this rotator mass:

$$(E_{\text{rot}})_J = \frac{h^2 J(J+1)}{8\pi^2 \tilde{\mu} R^2} = \frac{h^2 J(J+1)}{8\pi^2 \mathcal{I}} = Bh\, J(J+1) \qquad (8.10)$$

where J is the same as k used in Eq. (8.2) above. These are the energy levels of a quantum mechanical rigid rotator. Its energy is quantized (by steps in J), unlike that of a classical rotor (dumbbell) whose rotational energy is continuous in value (takes continuous values).

The energy of the rotor is $E = L^2/2\mathcal{I}$ where L is the angular momentum and $\mathcal{I}$ is the moment of inertia. In classical dynamics, E and L are just ordinary quantities; in quantum mechanics they are operators (respectively, $\mathcal{H}$ — Hamiltionian operator — and L_{op}).

8.10. Molecular Energies

In molecules $m_{\text{nucleus}} \gg m_{\text{electron}}$. Thus E(nuclear motion) $\ll E$(electronic motion around nuclei). The period of the motion is $\sim \hbar/E$, and hence a nuclear period is longer than an electronic period. That is, as the electrons move, the nuclei can be taken as practically stationary.[9]

The nuclei have an equilibrium (stable) configuration that lies between a fully swollen out (unstable) arrangement and a fully contracted one (also unstable because of nuclear repulsion at short distances). As Schiff ingeniously puts it, the "quasi-rigid" equilibrium configuration gets translated in space and also rotates. The nuclei vibrate internally about this equilibrium. (The translational motion is classical, just that of a free particle.)

If d is the size of the molecule, the electronic energy (of the motion of the valence electron[10]) is $E_{\text{el}} \sim \hbar/md^2$ (where m is the mass of the electron). Its momentum uncertainty is $\Delta p \sim \hbar/d$. So, E_{el} is the least kinetic energy. Since $d \sim$ Å, E_{el} falls in the optical and ultraviolet regions.

In the considerations of the vibrational activity, each normal mode is an harmonic oscillator of mass M and constant κ — ideally so. Obviously, $M \sim$ nuclear mass. How big is κ? A displacement (in a normal mode) by d changes the energy by $\sim E_{\text{el}}$, because it distorts the electron could (electronic ψ) appreciably. Thus $\kappa \sim E_{\text{el}}/d^2$. For small v (low-energy vibrations), we have

---

[9]In the adiabatic approximation one assumes a steady motion for the electrons at each instantaneous nuclear configuration, and thus one can include nuclear motion, too.

[10]Which spans the entire volume of the molecule by its motion. We do not consider an inner-shell bound electron close to a nucleus.

$$E_v \sim \hbar \sqrt{\frac{\kappa}{M}} \sim \frac{\hbar^2}{\sqrt{mM}d^2} \sim E_{el}\sqrt{\frac{m}{M}} .$$

Thus $E_v \sim \frac{1}{100}E_{el}$ refers to transitions in the near infrared region.

In the rotational energy considerations the moment of inertia of a molecule is $\sim Md^2$. For low-energy rotational modes, the angular momentum is just $\hbar$. Hence

$$E_{rot} \sim \frac{\hbar^2}{Md^2} \sim \frac{m}{M}E_{el} ,$$

i.e., $E_{rot} \sim \frac{1}{100}E_v$. It lies in the far infrared.

Since m/M is small ($\sim 10^{-3}$), Born and Oppenheimer[11] proved that the three kinds of energy are just higher orders of an approximation made in an expansion in terms of m/M, the expansion parameter, or of the ratio

$$\frac{\text{Nuclear vibrational displacement (typical)}}{\text{Internuclear separation } (\sim d)} .$$

The displacement x of an oscillator is $\sim \sqrt{E_v/\kappa} \sim \sqrt{E_v/E_{el}}\,d$, whence the expansion parameter becomes $\sqrt{E_v/E_{el}} \sim \sqrt[4]{m/M}$. Thus the zeroth-order approximation is E_{el}. The second-order step is E_v, and the fourth-order step is E_{rot}. The other orders in between are zero.

The Schrödinger wave equation of a molecule is

$$\left[-\frac{\hbar^2}{2m}\sum_{i=1}^{n}\nabla_i^2 - \sum_{j=1}^{N}\frac{\hbar^2}{2M_j}\nabla_j^2 + V \right]\psi = E\psi$$

where there is no time dependence, and we sum over n electrons (all of the same mass m) and N nuclei (of different — in general — mass M_j each). All the Coulombic interactions between pairs of charges are included in V (as additive scalars). The kinetic energies of the nuclei are of the fourth order (in the expansion). If we neglect them, the nuclear coordinates $\mathbf{R}_j$ in ψ become just parameters, and the wave equation involves only electronic coordinates $\mathbf{r}_i$, *i.e.*, describes the motion of electrons relative to *fixed* nuclei. Then $\psi \rightarrow \psi_j(\mathbf{r}_i)$ with eigenvalue $E(\mathbf{R}_j)$. Only after this step can we find the nuclear motion. How? By taking $E(\mathbf{R}_j)$ as a potential and finding a nuclear wavefunction $u(\mathbf{R}_j)$. Then the whole wavefunction is $\psi(\mathbf{r}_i, \mathbf{R}_j) = \psi_j(\mathbf{r}_i)u(\mathbf{R}_j)$ where ψ_j regards the equation

---

[11]M. Born and J.R. Oppenheimer, *Ann. Physik*, **84**, 457 (1927).

$$\left[\frac{-\hbar^2}{2m} \sum_{i=1}^{n} \nabla_i^2 + V \right] \psi_j = E(\mathbf{R}_j)\psi_j$$

where for each different nuclear configuration, $E(\mathbf{R}_j)$ is an eigenenergy of the equation. Of course, different electronic states give different solutions, but ψ_j and $E(\mathbf{R}_j)$ must be continuous functions of $\mathbf{R}_j$ in the molecule.

Combining all these equations, we have

$$\left[-\sum_{j=1}^{N} \frac{\hbar^2}{2M_j} \nabla_j^2 + E(\mathbf{R}_j) \right] \psi = E\psi,$$

$$\psi_j \left[-\sum_{j=1}^{N} \frac{\hbar^2}{2M_j} \nabla_j^2 + E(\mathbf{R}_j) - E \right] u = \sum_{j=1}^{N} \frac{\hbar^2}{2M_j} \left[u\nabla_j^2 \psi_j + 2\vec{\nabla}_j u \cdot \vec{\nabla}_j \psi_j \right].$$

We can now neglect the fact that ψ_j depends on $\mathbf{R}_j$, so that the right-hand side of the last equation becomes zero. We are left with the part that represents the nuclear motion only:

$$\left[-\sum_{j=1}^{N} \frac{\hbar^2}{2M_j} \nabla_j^2 + E(\mathbf{R}_j) \right] u = Eu.$$

How can we neglect $\vec{\nabla}_j \psi_j$? We can, because the extent (amplitude) of nuclear motion is too small compared to the internuclear separation (at equilibrium). The expansion parameter is small, and the electronic ψ_j is not subject to appreciable changes during the nuclear motion. Needless to say, this approximation holds for low vibrational and rotational modes (a few excited ones).

We thus obtain electronic ψ_j functions *and* $E(\mathbf{R}_j)$ which is a potential for the nuclear motion. The solution for complex molecules is very difficult. Only simple cases can be considered. For example, for the special case of the H_2 molecule the problem can be solved. Heitler and London[12] made an approximation for H_2. (Then one can make simplifying assumptions about $E(\mathbf{R}_j)$ and find the nuclear motion for any diatomic molecule in general.)

In H_2, $|\mathbf{R}_j| = R =$ distance between two H nuclei. Here R cannot be taken large compared to $a_0 = \hbar^2/me^2$, *i.e.*, the approximations made in

---

[12]W. Heitler and F. London, *Z. Physik*, **44**, 455 (1927).

calculating the Van der Waals interaction cannot be made here. But we have (see Fig. 8.14)

$$\psi(\mathbf{r}_1, \mathbf{r}_2) = \psi(\mathbf{r}_1, \text{ground state})\psi(\mathbf{r}_2, \text{ground state}),$$

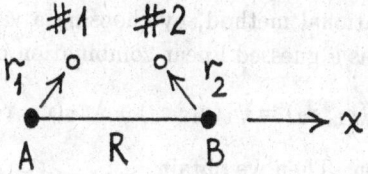

Fig. 8.14.

a product of two hydrogen-atom wavefunctions, with exchange degeneracy. That is, two degenerate wavefunctions are taken together:

(1) Electron #1 belongs to atom A, and #2 to B.
(2) Electron #1 belongs to atom B, and #2 to A.

The linear combination of the (unperturbed) wavefunctions (homopolar binding) brings us to less energy than the separate wavefunctions, and the phenomenon is called *resonance degeneracy* (not to be confused with *resonance scattering* directly). As we saw in the text, when two classical oscillators interact at resonance (the unperturbed frequency is the same), the produced two normal modes are of lower and higher frequency respectively. Analogously, two interacting quantum mechanical levels which are resonant, and therefore, degenerate, produce two eigenvalues, one of lower and one of higher energy.[13]

For H_2 we have

$$[\mathcal{H} - E(R)]\psi(\mathbf{r}_1, \mathbf{r}_2) = 0$$

where

$$\mathcal{H} = -\frac{\hbar^2}{2m}(\nabla_1^2 + \nabla_2^2) + e^2\left(\frac{1}{r_{12}} + \frac{1}{r_{1A}} - \frac{1}{r_{2B}} - \frac{1}{r_{1B}} - \frac{1}{r_{2A}}\right)$$

with

$$\psi_1(\mathbf{r}_1, \mathbf{r}_2) = \psi_A(\mathbf{r}_1)\psi_B(\mathbf{r}_2)$$

$$\psi_2(\mathbf{r}_1, \mathbf{r}_2) = \psi_A(\mathbf{r}_2)\psi_B(\mathbf{r}_1)$$

---

[13]In general, there may exist *many* degenerate (unperturbed) states where their degeneracy may not be due to exchange.

where $E(R)$ is to be calculated approximately, and ψ_A and ψ_B are ground-state hydrogen wavefunctions. (Here ψ_1 and ψ_2 are eigenfunctions of *different* unperturbed $\mathcal{H}$ operators, because we are not doing degenerate perturbation, whereas in He we do. Here the exchange-degenerate wavefunctions are not solutions of the same $\mathcal{H}$.) $E(R)$ is the internuclear potential for H_2.

We apply the variational method, by choosing a variational parameter λ, and writing $\psi(\mathbf{r}_1, \mathbf{r}_2)$ as a guessed linear combination of the form

$$\psi(\mathbf{r}_1, \mathbf{r}_2) = \psi_1(\mathbf{r}_1, \mathbf{r}_2) + \lambda\psi_2(\mathbf{r}_1, \mathbf{r}_2)$$

which is a trial function. Then we obtain

$$E(R) \leq \frac{(1+\lambda^2)\langle 1|\mathcal{H}|1\rangle + 2\lambda\langle 1|\mathcal{H}|2\rangle}{1+\lambda^2+2\lambda\xi}$$

where the matrix elements (expectation values) are

$$\mathcal{H}_{11} = \mathcal{H}_{22} = \iint \psi_1\mathcal{H}\psi_1\,d^3r_1\,d^3r_2$$

$$\mathcal{H}_{12} = \mathcal{H}_{21} = \iint \psi_1\mathcal{H}\psi_2\,d^3r_1\,d^3r_2$$

and

$$\xi = \iint \psi_1\psi_2\,d^3r_1\,d^3r_2$$

where the wavefunctions are real, and $\mathcal{H}$ is Hermitean (and symmetric with respect electrons #1 and 2). All these expressions depend on R (which varies as the molecule vibrates). Thus

$$\frac{dE(R)}{d\lambda} = \frac{2(1-\lambda^2)(\mathcal{H}_{12}-\xi\mathcal{H}_{11})}{(1+\lambda^2+2\lambda\xi)^2}$$

which vanishes if $\lambda = \pm 1$. $E(R) = \mathcal{H}_{11}$ if $\lambda = 0$ or $\pm\infty$. Therefore, $\lambda = \pm 1$ are the two extrema (one minimum and one maximum). For $\lambda = +1$ we have $\langle\mathcal{H}\rangle_{min}$ with $\psi = \psi_1 + \psi_2$ and $E(R) \leq (\mathcal{H}_{11}+\mathcal{H}_{12})/(1+\xi)$. Equality holds for the upper limit on $E(R)$ which is just the internuclear (and intramolecular) potential (in a diatomic molecule). Notice that ψ is symmetric upon the interchange of spatial coordinates of the electrons #1 and 2 ($\mathbf{r}_1 \to \mathbf{r}_2$ and $\mathbf{r}_2 \to \mathbf{r}_1$). So, it should be multiplied by the singlet (antisymmetric) spin state $\chi = \frac{1}{\sqrt{2}}[|\uparrow\downarrow\rangle - |\downarrow\uparrow\rangle]$.

The Pauli principle states that the spatial (configurational) states of the electrons must be different *if* they have parallel spins (and thus they repel each other). In the ground state of H_2 the lowest energy level is attained (and hence the strongest bond) when the two electrons are found between the two H nuclei (because although there is a repulsion between the electrons, it is obscured by the attraction of *two* nuclei per each electron). To attain this, the two electrons must have the same spatial wavefunction, but antiparallel spins. Therefore, molecular stability requires a singlet state.[14]

Now let us consider the nuclear motion for a general diatomic molecule of nuclear masses M_1 and M_2, reduced mass $\tilde{\mu} = M_1 M_2 / (M_1 + M_2)$, and internuclear (relative) position vector $\mathbf{R}$ where polar coordinates will be used. The relative motion is described by

$$\left[-\frac{\hbar^2}{2\tilde{\mu}} \nabla^2 + E(R) \right] u(R, \theta, \varphi) = Eu(R, \theta, \varphi)$$

where we do not sum for two nuclei, because we use $\tilde{\mu}$ (we reduced the system to one equivalent body of effective mass $\tilde{\mu}$).

$E(R)$ for low E_{el} states in diatomic molecules can be represented in a very satisfactory way by the Morse potential (which has three adjustable parameters). The case is treated in the text and we will not repeat it here.

To find the rotations and vibrations, we can separate the last equation above into radial and angular parts to obtain

$$u(R, \theta, \varphi) = \frac{f(R)}{R} Y_{K, \mathsf{m}_K}(\theta, \varphi)$$

where $f(R)$ is a radial function (depending on R only), and K and m_K are angular momentum quantum numbers (like L and m_L in atoms in a central potential). The radial part is

$$-\frac{\hbar^2}{2\tilde{\mu}} + \frac{\partial^2 f}{\partial R^2} + V'(R)f = Ef$$

where

$$V'(R) = E(R) + \frac{\hbar^2 K(K+1)}{2\tilde{\mu}R^2}$$

is the *effective potential* (with $K = $ integer $= 0, 1, 2, 3, \ldots$). This is the case of one-dimensional motion for a particle of mass $\tilde{\mu}$ moving in $V'(R)$. The boundary condition is $f = 0$ for $R \to 0$. For small K, V' is the Morse potential.

---

[14]In He, the excited $1s2s$ level has a diminished Coulombic repulsion between the electrons (lower energy) because of the exclusion principle. The triplet states in He are lower than the singlet state (for the same spatial configuration). In the ground state, the $1s^2$ configuration has only a singlet state. Other states are not viable there.

If vibrations of small amplitude (about the minimum R_0) are important only, one can expand V' about R_0[15]:

$$V'(R) = V_0' + \frac{1}{2}\kappa(R - R_0)^2 + c_1(R - R_0)^3 + c_2(R - R_0)^4 + \text{h.o.t}.$$

If we neglect c_1, c_2, and higher-order terms, and let $R \to -\infty$, the eigenenergies to be obtained will be harmonic oscillator eigenvalues shifted by an additive term V_0'. So, E will be energies with V_0'. Here c_1 and c_2 are constant coefficients of the expansion. The motion is along R only.

This is an approximation, valid for reasonable K and v values (where K is the rotational quantum number). To improve the approximation, we consider c_1 and c_2 as perturbations (to the oscillator). The expectation values of these terms are to be computed by matrix elements. Both c_1 and c_2 contribute to E with comparable orders of magnitude. (The c_1 term causes a second-order effect; the c_2 term turns up as first-order effect.)[16] We have the eigenenergies

$$E = V_0' + \hbar\sqrt{\frac{\kappa}{\tilde{\mu}}}\left(\text{v} + \frac{1}{2}\right) - \frac{\hbar^2 c_1^2}{\tilde{\mu}\kappa^2}\left[\frac{15}{4}\left(\text{v} + \frac{1}{2}\right)^2 + \frac{7}{16}\right] + \frac{3\hbar^2 c_2}{2\tilde{\mu}\kappa}\left[\left(\text{v} + \frac{1}{2}\right)^2 + \frac{1}{4}\right]$$

to the lowest order in c_1 and c_2 (where v = integer = 0, 1, 2, 3, ...).

Now we expand all the appearing constants in powers of the quantity $K(K + 1)$ to obtain

$$R_{\min} = R_0 + \frac{\hbar^2 K(K+1)d^2}{2\tilde{\mu}R_0^3 E_0} = \text{Equilibrium} + \text{Stretch}$$

$$V_0' = -E_0 + \frac{\hbar^2 K(K+1)}{2\tilde{\mu}R_0^2} - \frac{\hbar^2 K^2(K+1)^2 d^2}{4\tilde{\mu}^2 R_0^6 E_0}$$

$$\kappa = \frac{2E_0}{d^2} - \frac{3\hbar^2 K(K+1)}{\tilde{\mu}R_0^2 d^2}\frac{d}{R_0}\left(1 - \frac{d}{R_0}\right)$$

$$c_1 = -E_0/d^3 \quad \text{and} \quad c_2 = 7E_0/12d^4$$

where $E(R)$ is the Morse potential

$$E(R) = E_0\left(e^{-2(R-R_0)/d} - 2e^{-(R-R_0)/d}\right),$$

---

[15] In fact $R_{\min} = R_0$ only if $K = 0$.
[16] See Schiff, Chapter 8, Problem 2.

with d being the width of the attractive region $\lesssim R_0$ (Fig. 8.15). The coefficients depend on the parameters of $E(R)$. Here we kept terms up to second order in $v + \frac{1}{2}$ and $K(K+1)$.

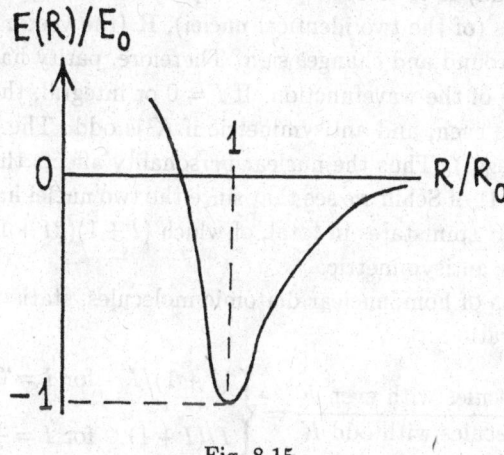

Fig. 8.15.

$R_{\min}$ shows that the molecule swells out upon rotation. $V_0' = $ Equilibrium energy $(-E_0)$ + Rotational energy (second-order). (Notice that the rigid rotor first-order rotational energy is $\hbar^2 K(K+1)/2\mathcal{I}$ where $\mathcal{I} = \tilde{\mu}R_0^2 = $ molecular moment of inertia.) $\kappa = $ Stiffness + Change because of the elongation. The anharmonic terms c_1 and c_2 have also corrective terms (for stretching), but they are neglected to this second order.

Consider the result for E above. We expand the second term to obtain

$$\sqrt{\frac{2E_0\hbar^2}{\tilde{\mu}d^2}}\left(v + \frac{1}{2}\right)\left[1 - \frac{3\hbar^2 K(K+1)}{4\tilde{\mu}R_0^2 E_0}\frac{d}{R_0}\left(1 - \frac{d}{R_0}\right)\right].$$

The last two terms merge to form

$$\left(-\frac{15}{16} + \frac{7}{16}\right)\frac{\hbar^2}{\tilde{\mu}d^2}\left(v + \frac{1}{2}\right)^2 = -\frac{\hbar^2}{2\tilde{\mu}d^2}\left(v + \frac{1}{2}\right)^2$$

which the reader can easily recognize as the vibrational energy to the second order. Notice that as v or K increases, the level spacings decrease to lesser values than those for an harmonic oscillator and a rigid rotor.

In an A_2 type molecule ψ is symmetric upon interchange of the spatial and spin coordinates of the two atoms (whose nuclei are the same). This is so if the nuclear spin is $I = 0$ or integral. If $I = \frac{1}{2}$(Odd integer), then ψ is antisymmetric. From Section 14 of Schiff one sees that the angular part $Y_{K,m_K}(\theta, \varphi)$ (spherical harmonics) determines the parity of the nuclear wavefunction. The parity is even (odd) if K is even (odd). Upon interchange of the configurational coordinates (of the two identical nuclei), $\mathbf{R}$ (the vector of their relative position) turns around and changes sign. Therefore, parity has to do with the spatial symmetry of the wavefunction. If $I = 0$ or integral, the spin state χ is symmetric if K is even, and antisymmetric if K is odd. The reverse happens if $I = \frac{1}{2}$(Odd integer). Thus the nuclear personality affects the situation.

From Section 41 of Schiff we see that since the two nuclei have spin $\hbar I$ each, there are $(2I + 1)^2$ spin states in total, of which $(I + 1)(2I + 1)$ are symmetric and $I(2I + 1)$ are antisymmetric.

Consider a gas of homonuclear diatomic molecules, statistically in equilibrium. Then the ratio

$$\frac{\text{Number of molecules with even } K}{\text{Number of molecules with odd } K} = \begin{cases} (I+1)/I & \text{for } I = 0 \text{ or integral} \\ I/(I+1) & \text{for } I = \frac{1}{2}(\text{Odd integer}) \end{cases}$$

times the Boltzmann exponential factor, if the ΔE_{rot} spacing is *not* small compared to thermal energies ($\sim kT$). Notice that these are just the statistical weights (that modify the Boltzmann factor). So, the appearance of alternating intensities in the rotational spectrum of an A_2 type molecule is thus explained.[17]

As an exercise, the reader might consider the vibrational energy of a diatomic molecule, and compute the contribution to it by the fifth and sixth-power terms of the expansion of V'. (Use the matrix elements of x^5 and x^6 for an harmonic oscillator.) The reader might prove that these contributions are indeed negligible.

8.11. The Sum-up of the Diatomic Molecule

The electron configuration in an atom explains the chemical behavior of the atom and the nature of the chemical bond it forms with other atoms. A bond is the result of an interaction (or reaction) that finally keeps the atoms together by forming stable structures (molecules and crystalline solids).

The *electrovalent* or *ionic* bond is a simple case, with the common example of NaCl where Na gives up its $3s$-electron to Cl which in turn completes its vacancy in the $3p$ subshell. If the two atoms are initially separated, energy is

---

[17]This is a good means to determine statistics and spins of nuclei.

needed to realize the electron transfer. We obtain two ions which attract each other and keep together because of their equal and opposite charge: One is positive and the other negative. When they approach each other enough, their potential energy decreases, and the eventual bound state (of Na^+Cl^-) has lower energy than the state in which the two atoms are free apart and neutral.

To remove the electron of Na, we should give an energy of 5.1 eV which is the *ionization energy* (or *potential*) of Na. On the other hand, Cl has an electron *affinity* (willingness to receive) of 3.8 eV, which means that a neutral Cl atom is capable of attracting a free electron as extra charge, and when such an electron comes and occupies the empty place in $3p$, it needs 3.8 eV to be removed from there. Therefore, the net cost that incurs us in making the Na^+ and Cl^- ions (separately) is $5.1 - 3.8 = 1.3$ eV, an expense of energy on our part. Since the electrostatic potential of attraction (of the ions mutually) is -5 eV, this amount of energy is ample to reimburse us for our expense in creating the Na^+ and Cl^- ions. In other words, the linkage is possible, because our investment to create the necessary conditions for the compound can be paid back (several times over) from the energy fund of the compound, once it forms. The potential energy of mutual attraction depends of course on how close the ions approach one another before the Pauli principle forbids and stops any further overlap of their electron clouds.

Instead of one electron, an ionic bond can be formed by the transfer of two electrons (per atom). For example, ionic bonds can be formed by alkaline earth elements where two electrons per atom are involved, *i.e.*, to form the compound, each atom gives up two electrons (*e.g.*, $Mg^{++}Cl_2^-$). Loss of three or more electrons is not so common, and anyway, the kind of the bond ceases being ionic then.

In a *covalent homopolar* bond, two free atoms participate *symmetrically* (whereas in the ionic bond the process of electron transfer is asymmetric). H_2 is a case of covalent bond (where the structure consists of two protons and two electrons in cohabitation). How does it work? Consider two dipoles in interaction (Fig. 8.16). If they are far away from each other (a), they do not interact appreciably. If they get close, with their like charges farther apart than their unlike charges (b), the net result is a mutual attractive force (anti-parallel dipoles). Otherwise the interaction is repulsive, as in (c) (parallel dipoles) and (d) (opposite dipoles). Since the charges of a molecule move, being not at rest, whenever the electrons happen to find themselves in the region between the two protons, the attractive force exerted by the electrons on the protons sometimes by far exceeds the repulsive force between the two protons. The net (attractive) force is more than enough to make the cohabitation possible. Hence in H_2 the attraction is provided by two electrons (one per atom). This pair of electrons (due to the contribution of one electron by each atom) has its charge cloud concentrated mostly and primarily between the two protons, and that is exactly what the bond is all about: Shared electrons, *i.e.*, an electron-pair bond. For such a bond it is necessary to have two electrons of *opposite spins* (and no more than *two* electrons), because the Pauli principle says that two electrons can come together and be in the same region of space in close proximity, only if their spins are antiparallel; otherwise the state is forbidden by the exclusion principle, in spite of the fact that it may be energetically possible and favorable. If the spins are parallel, the lowest state that is possible is one where the electron clouds take position anywhere else *but* the spatial region between the protons. Then no bond is possible between them; the protons repel each other, and since this repulsion wins, the net resultant interaction is repulsive, favoring no molecule formation at all.

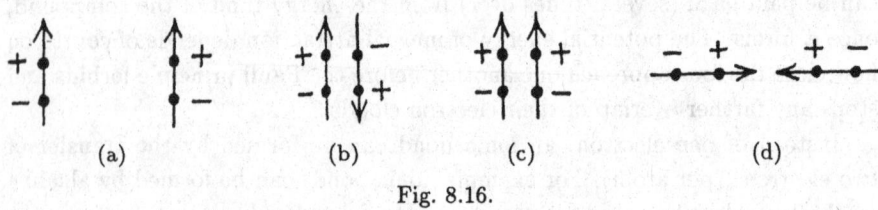

(a)   (b)   (c)   (d)

Fig. 8.16.

By stating that only two electrons are involved (*per bond*, that is), we do not mean that an atom has only one bond (of electron pair). In fact an atom can form *several* such bonds, because it may have several electrons in its outer shell to contribute, and thus form covalent bonds with several other atoms, *e.g.*, C has four such electrons and can form four covalent bonds (at most), *i.e.*, electron-pair bonds with other atoms. Example: CH_4 (methane) (Fig. 8.17). (The C–H bonding is important in organic chemistry.) In CH_4, the C atom is at the center of a regular tetrahedral pyramid where the H atoms are at the corners. Since C has four electrons in the L-shell, each of them forms a covalent bond with each of the H atoms. Similarly, in complex organic molecules we have covalent bonds of the same pattern. In Fig. 8.17(b) the electron cloud (shaded region) stretches between the central C atom and each of the four H atoms, representing the two electrons that set up the covalent bond (one from C and one from H, for each bond or cloud).

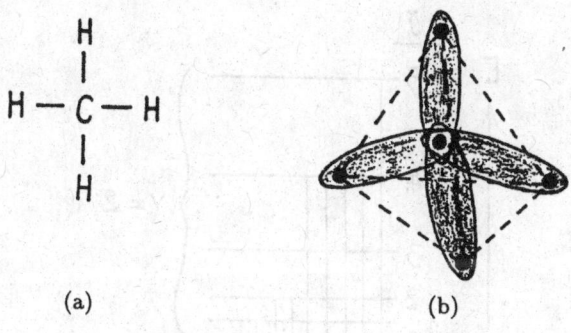

(a)   (b)

Fig. 8.17.

The *metallic* bond is not so of chemical character as the two other bonds. It is a structure where the outermost electrons are not localized at definite lattice sites. They are rather detached. This secession from the parent atom makes them free to travel and move throughout the bulk of the crystal (metal). That is, the charge clouds extend to cover many atoms of the metallic object, as if we had an arrangement of positive ions (lacking electrons by removal, one or more than one) immersed in a sea of free electrons whose swift motion provides the good thermal and electrical conductivity of metals, and whose attraction to the positive ions provides the bonding that keeps the system (crystal) together. (Since the electron sea looks like a gas with respect to its properties, we call it *electron gas* and we talk of this model of metallic solids.) So, in metallic crystals the atoms want to form covalent bonds by sharing electrons, but there are no sufficient valence electrons, and hence the available electrons are necessarily shared by many atoms collectively. That is why this bond is not directional; the shape of the crystal lattice is determined by *close packing* (putting as many atoms as they can be accommodated in a certain volume — the maximum number). So, we have the face-centered cubic and the hexagonal packing in metallic crystal lattices. The rest pertains to solid state physics.

These three types of bonding are the most important and common in the structure of solids. We can say that a crystalline solid is a huge single molecule made of atoms kept together by chemical bonds.

Figure 8.18 shows the energy-level diagram of a diatomic molecule. For each ΔE_v there can be many combinations of J, giving a series of closely spaced levels. Different pairs of v correspond to different series, so that the outcome spectrum consists of bands. Each band belongs to a particular vibrational

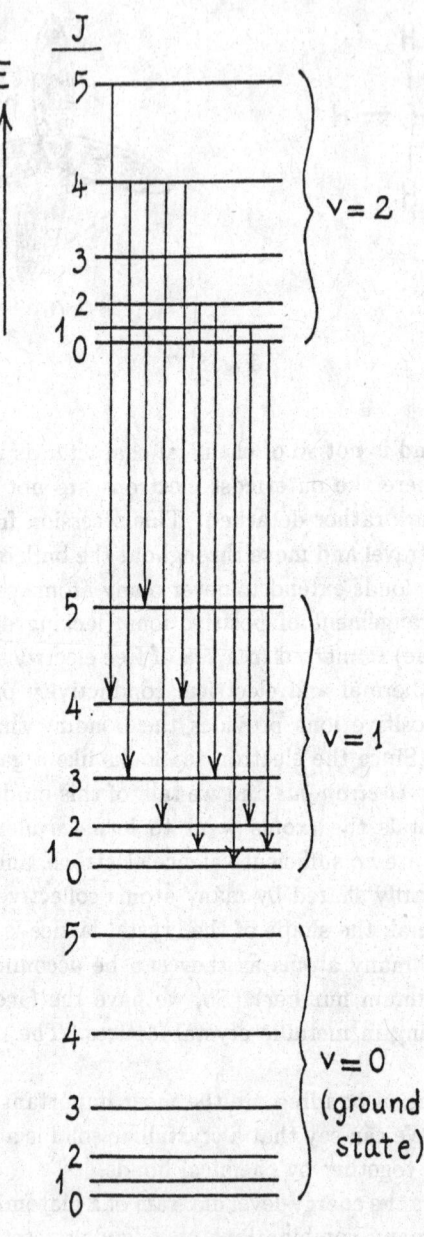

Fig. 8.18.

transition $v \rightleftarrows v'$. Each individual line within a band belongs to a particular rotational transition $J \rightleftarrows J'$. The diagram is for vibrational and rotational levels. For each level of v there is a series of closely spaced levels of J. The figure shows the possible transitions that correspond to the band $v = 2 \rightarrow v' = 1$ in the band spectrum.

Figure 8.19 shows a typical band spectrum.

The energy levels of atoms are consequences of the kinetic and potential energies (with respect to the nucleus) of their electrons. In molecules the energy levels are not that simple; they have additional features because of additional degrees of freedom that have to do with the relative motion of the nuclei and the rotation of the molecule as a whole. The energies of these motions are quantized, as the Schrödinger equation dictates. The same analysis and considerations are applicable to complex molecules, too. For example, a molecule with many atoms has many different fashions (shapes, forms) of vibrational motion, called *modes*. Each mode has its own set of energy levels E_v (and corresponding frequencies via $E_v = h\nu(v + \frac{1}{2})$). Needless to say, the energy level diagram (and the spectrum) will be complicated. However, what holds true for simple diatomic molecules in general, holds also true for molecules with three, four or more atoms — that is a principle. By and large, almost all the cases that find their way to the laboratory give infrared radiation, and the analysis of the associated molecular spectra is indeed a powerful means — actually a tool — that yields analytical information — and in fact a valuable one — concerning the molecular structure of complex cases, the strength of the bonds, and the rigidity of the molecular structural frame.

Fig. 8.19.

Notice that on one hand we have chemistry which is directly the science of tangible matter, and on the other hand — going down to the atomic sizes and levels — we have quantum physics where matter eludes us to appear as condensed energy or waves of probabilities of events in space. And interestingly enough, molecular physics lies right in the middle between these two realms.

Chapter 9

MOLECULAR SPECTRA

9.1. Formation of Molecular Spectra

The molecular energy E consists of the following terms (parts):

$$E = \text{Electronic energy} + \text{Vibrational energy}$$
$$+ \text{Rotational energy} + \text{Other terms}$$

where the electronic energy is just the binding energy of the atomic electrons (*i.e.*, electron energy of the valence electrons bound to the atoms of the molecule). The "other terms" are small and include spin interactions, quadrupole effects etc. If we ignore the small terms, we have

$$E = E_{\text{el}} + E_{\text{vibr}} + E_{\text{rot}} \qquad (9.1)$$

where $E_{\text{el}} > E_{\text{vibr}} > E_{\text{rot}}$. The last two terms are molecular energies due to *molecular* vibrations and rotations, not existing in free atoms. E_{el} binds the electron to the atom and hence to the molecule (*e.g.*, H_2 molecule). For a single, isolated atom, we have E_{el} levels only. When two atoms come close together and interact, each E_{el} level splits into two. That is, if E_{el} is for a single atom, for two interacting atoms it becomes $(E_{\text{el}})_1$ *and* $(E_{\text{el}})_2$.

E_{el} depends on the *angular momentum quantum number l*. For a single electron bound to a single atom we have

l	Specification of state
0	s (ground state)
1	p
2	d
3	f
4	g } (excited states)
5	h
$\vdots$	$\vdots$

As l takes integer values, we call the corresponding states s, p, d etc., and the corresponding electrons s-electron, p-electron and so on, to indicate which orbit the electron is in.[1] If the atom has many electrons, we consider the valence electron. According to the energy of the electron, the atomic (actually electronic) states are specified by the *total* orbital angular momentum of the electrons L, *i.e.*, by the *spectral terms* (state designations) S, P, D, F, G, ...

In the case of *molecules* the spectral terms (states) are Σ, Π, Δ, Φ, Γ, ... , *i.e.*, we revert to the corresponding Greek capital letters (instead of using the Latin letters which we reserve for atoms). This notation for the spectral terms (and term schemes) is widely used in spectroscopy.[2]

The molecular excitations (and spectra) are the following:

(1) *Electronic excitations* (which regard the atomic electrons in the molecule).

(2) *Vibrational excitations.*

(3) *Rotational excitations* (which already exist thermally).

The first two can be caused by irradiating the molecule by electromagnetic waves (ordinary energy in the visible and near infrared region). These two are of higher energy relatively to the third.

---

[1] The symbol s stands for *sharp*, p for *principal*, d for *diffused*, and f for *fundamental*, referring to the appearance of the *atomic* spectral lines. After the letter f the state specification follows the regular order of the letters of the alphabet: g, h, i etc.

[2] *Atomic* spectroscopy can basically be done as follows: If, for example, one is to study the lines of Na, one uses a Na lamp, a diffraction grating (that diffracts the yellow light of the lamp), and a rotating spectrometer. At different grating angles one watches the spectral lines through the spectrometer. These lines correspond to transitions (de-excitations) of Na atoms. Besides the two most characteristic ones (D_1 and D_2), one gets other colored lines which are present, too, at various angles. If one uses a Hg lamp instead, one obtains the Hg spectrum. Down the way in this book we devote a separate chapter to *molecular* spectrometry.

The molecular energy levels are quantized (just like the atomic ones), with quantum jumps between them. The discrete energy changes in molecules can occur by the absorption or emission of quanta, as electrons get from an orbit (electronic energy state) to another in the atoms. This process gives E_{el}. There are two more ways by which energy can change *in molecules* (not in atoms): The molecule can absorb or emit a quantum and thus change its vibrational or rotational state. These three types of energy are independent of each other, and hence they can be written as a sum to give the total (molecular) E.[3] This is just Eq. (9.1).

The amounts of energy involved are typically as follows: $E_{el} \sim 2$ to 10 eV (100 kcal/mol), $E_{vib} \sim 0.2$ to 2 eV (5 to 50 kcal/mol), and $E_{rot} \sim 10^{-5}$ to 10^{-3} eV (10^{-2} kcal/mol).

At room temperature (thermal energies $\sim 1/40$ eV), the thermal kinetic energy of the molecules is $kT \sim 1$ kcal/mol. So, if we assume a Boltzmann distribution, the molecules are mostly in their ground electronic and vibrational state, but in excited rotational states (because the ground rotational state is too low and corresponds to temperatures below room temperature).

These numbers tell us that transitions ΔE_{el} correspond to the visible and ultraviolet radiation; transitions (emission or absorption) ΔE_{vibr} fall in the near infrared region of the electromagnetic spectrum; and transitions ΔE_{rot} fall in the far infrared and the microwave regions.

In fact, rotational and vibrational actions in molecules take place together, giving the *vibration-rotation spectra* (because the two types of transition between levels occur simultaneously together).

Each electronic level has vibrational (sub)levels specified by the vibration quantum number v, and each vibrational (sub)level in turn has rotational sub-sublevels specified by the rotational quantum number J.[4]

In this book we will go over these spectra *briefly* because we cannot afford a detailed discussion. The interested reader should consult books particularly written on molecular spectra and spectroscopy; fans of acute expertise who need further details should turn to specialized research literature. For the beginners, elementary but nevertheless good and instructive discussions can be found in the following sources (which are included in the list of References):

---

[3]Whenever we have *independent* degrees of freedom, we can write their total energy (or Hamiltonian) as a *sum* of the individual energies (or Hamiltonians) (they are additive), and their total wavefunction ψ as a *product* of the individual wavefunctions. This is consistent with the probability theory which states that the probability of many *independent* events to occur together equals the product of the probabilities of each separate event, since in quantum mechanics $|\psi|^2$ is the probability of finding something somewhere.

[4]The prefixes *sub-* are dropped in practice.

Richtmyer, pp. 489–525; Schiff, pp. 445–455; Fermi, pp. 140–144; Park, pp. 243–247; Sears, pp. 779–782.

Certainly, there are other good sources as well, but as it is mentioned in the Preface, when the space is limited it is hard to decide what to suggest among many equally interesting works.

9.2. Molecular Potential

Consider a two-atom molecule (two positive nuclei, and electron clouds around them mediating the binding). As long as the molecule is stable, its potential energy curve $V(r)$ has a minimum (because a minimum in energy indicates stability). From previous chapters we know that this minimum corresponds to an interatomic (internuclear) separation $r = R$. (Here r is the distance away from either nucleus). At $r = R$ we have stability, as the repulsion between the nuclei equals the attraction due to the electronic bonds, and the two forces are in equilibrium (the attraction is equal to the repulsion), so that R is the ideal size for the established molecule. For $r < R$, repulsive forces between the nuclei (ion cores) prevail. The attractive forces prevail for larger r values. Figure 9.1 shows the repulsive and attractive potential curves and their resultant (total) which gives $V(r)$ or (the molecular energy) E. Notice that as $r \to 0$, V (or E) is positive and strongly repulsive. As r increases, V becomes negative and forms an attractive potential well with a minimum (well depth) at R. For $r > R$, the potential energy is still attractive (negative) and goes asymptotically to zero as $r \to \infty$. The range of the attractive force is $\sim$ a few Å *effectively* (because *theoretically* it extends to infinite distances, though itself is weak). This effective range is just the size d of the molecule over which the attractive force is essentially and substantially valid.

The vibrational energy is quantized in levels specified by $v = 0, 1, 2, \ldots$. For small excursions (oscillations) away from equilibrium, the displacement is proportional to the restoring force, that is, the molecule vibrates (about the equilibrium position) like a simple harmonic oscillator, and the potential well is almost a parabola (Fig. 9.2). As v increases (we go to excited vibrational states), it becomes easier to rip the molecule apart than to unite the atoms by making them approach each other. Then the curve is not parabolic any more, and a new potential describes the situation (for the lowest electronic states): It is the *Morse potential*[5] given by

---

[5] P.M. Morse, *Phys. Rev.*, **34**, 57 (1929).

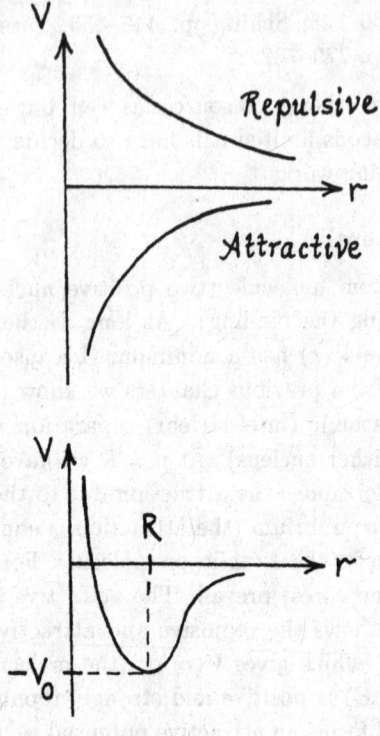

Fig. 9.1.

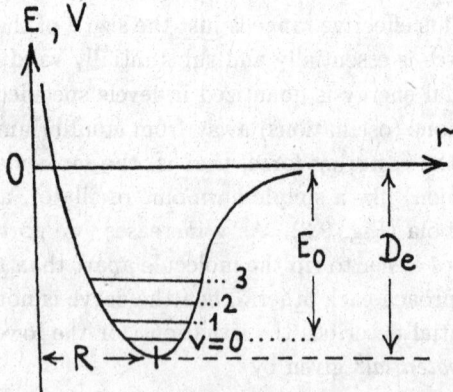

Fig. 9.2.

$$V(r) = V_0(e^{-2(r-R)/a} - 2e^{-(r-R)/a})$$

where $-V_0$ is the minimum value, and the parameter a is the width of the attractive region. For large r, $V \to 0$. V becomes minimum at $r = R$, and large as $r \to 0$, if $a \lesssim R$. If the energy is taken as zero when the atoms are at infinity, V becomes negative as r decreases (due to the Van der Waals attraction and the Heitler-London resonance). As r keeps decreasing, the internuclear repulsion dominates and V grows rapidly large and positive.

As v takes higher and higher values, the vibrations become nonlinear (*anharmonic*) until the molecule eventually dissociates.

Notice that the lowest vibrational state v = 0 does not coincide with the energy minimum (lowest point of the curve). At $T = 0°$ K the molecule still vibrates because it still keeps a nonzero energy for itself (a vibrational energy), although it gives off all its remaining energy. This is a law of nature, saying that we cannot extract *everything* from a system; nature retains some privacy. This is determined by Heisenberg's uncertainty principle: Since at $r = R$ (at the minimum energy point) the momentum is $p = 0$, we must have an indeterminacy $\Delta p = 0$. Given that we know the position for sure $(r = R)$, we should also have $\Delta x = 0$ which violates the principle. So, the molecule cannot reach this absolute lowest energy without violating nature.

Refer to Fig. 9.2. The difference $D_e - E_0$ is the *zero-point energy* (which is always nonzero). Here D_e is the dissociation (or binding) energy of the molecule, and E_0 is the lowest vibrational energy (corresponding to the level v = 0). So, the zero-point energy is the difference between the theoretical minimum energy and the energy of the lowest vibrational state.

To clarify things, we must state that R is the *theoretical* internuclear separation (position of equilibrium). Instead of R, experimentally we measure r_0, the *virtual internuclear distance*, which is different from R. Whatever the method of measuring r_0 is, the differences from R are slight, as expected. Here r_0 is the internuclear distance at the lowest vibrational state, whereas R corresponds to the lowest energy the molecule would have if it were not vibrating at all. All experimental results lead to r_0, not R. (Similarly, for the chemical bonds, we actually measure E_0, not D_e, *i.e.*, the energy between the zero level of the potential and the ground vibrational state.)

At room temperature most of the molecules are in the ground vibrational state (v = 0). If in this state the instantaneous value of the internuclear separation is r, then r_0 is actually $r_0 = \langle r \rangle$ where $\langle r \rangle$ is the average value of r. Each experimental method in molecular physics follows a different way to reach $\langle r \rangle$, and to our surprise each way gives different results! Nevertheless, as we said above, the difference $R - \langle r \rangle$ is too small at every turn.

Since the potentials vary as the inverse powers of r, what we measure spectroscopically is $\langle 1/r^2 \rangle$, not $\langle r \rangle$. Hence we have

$$\frac{1}{r_0} = \left[\left\langle \frac{1}{r^2} \right\rangle \right]^{1/2} .$$

The experimental procedures of NMR give $\langle 1/r^3 \rangle$ whence

$$\frac{1}{r_0} = \left[\left\langle \frac{1}{r^3} \right\rangle \right]^{1/3} .$$

Curiously, the method of X-ray diffraction does not give the internuclear separation; it gives instead the average separation between the centers of mass of the electronic clouds, but it is not a big sin to accept it as r_0.

Consider again a diatomic molecule (Fig. 9.3). It consists of two atoms bound by (the sharing of) an electron. This electron is common to the two atoms, and keeps them bound together. Let E_1 (E_2) be the binding energy of the electron to atom A_1 (A_2). The situation can be thought as electron sharing. Figure 9.4 shows the potential energy (its change versus distance)

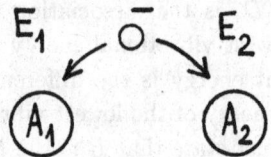

Fig. 9.3.

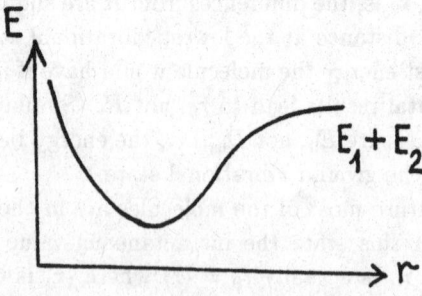

Fig. 9.4.

of the binding electron. The potential curve is actually $E_1 + E_2$ for large r, and exhibits a well (minimum), as explained above. The electron can have various energy states specified by the principal quantum number $n = 1, 2, 3, 4, \ldots$.[6] Figure 9.5 shows the potential curves (electronic energy levels) for several quantum numbers n and l. As the quantum numbers increase, the curves get more closely spaced. The dashed curve at the top ($l \to \infty$) is the potential well that corresponds to the ionization of the molecule. Notice that the minimum points of the curves do not lie on a vertical straight line (Fig. 9.6). If the dashed curve has still a minimum, it corresponds to the case where *the molecule is still stable upon ionization* (already ionized). But for some molecules the situation is different: When we excite the electron that

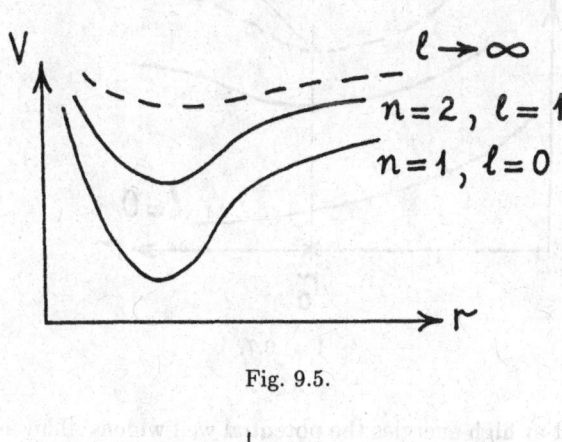

Fig. 9.5.

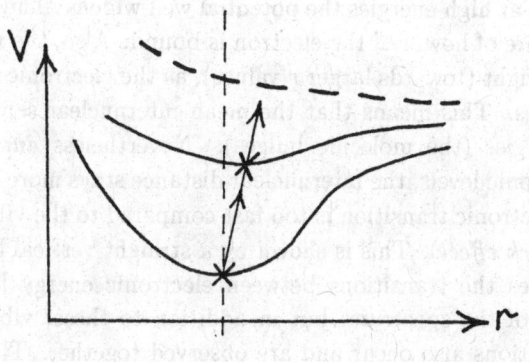

Fig. 9.6.

[6]For an electron, the lowest state is $n = 1$. For an oscillator, n (or v) $= 0$.

binds the atoms together, the molecule dissociates after some definite upper energy level. The potential curve corresponding to this state (plotted versus the distance) does *not* exhibit any minimum (Fig. 9.6, dashed line at the top). So, when E_{el} increases by excitation to a certain level (shown by the dashed line), there is no stability, and the molecule breaks up.

There is also another case (which is encountered rarely): The high-energy level does show a minimum while the low-energy state does not! (Fig. 9.7). Molecules of this type are weakly bound; they can be torn apart very easily. (The electron is excited.)

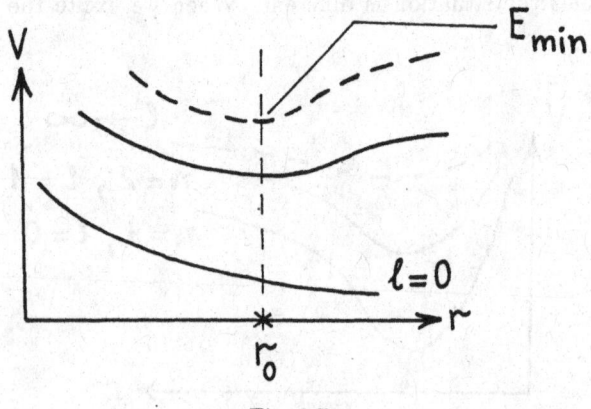

Fig. 9.7.

Notice that at high energies the potential well widens. The sharpness of the well is a measure of how well the electron is bound. Also, the well gets slightly shifted to the right (towards larger r values), as the electronic energy increases to higher states. This means that the mean internuclear separation is larger at higher energies (the molecule bulges). Nevertheless *during* a transition between electronic levels, the internuclear distance stays more or less the same, because an electronic transition is too fast compared to the vibrational motion (*Condon-Franck effect*). This is shown by a straight vertical line in Fig. 9.8.

In molecules the transitions between electronic energy levels fall in the visible region of the spectrum, but in addition to these, vibrational and rotational transitions also occur and are observed together. The latter ones do not fall in the optical region. One obtains a spectrum consisting of bands and lines (Fig. 9.9), a mixed and complicated spectrum. This is a *band spectrum*, belonging typically to molecules. Molecules give band spectra.

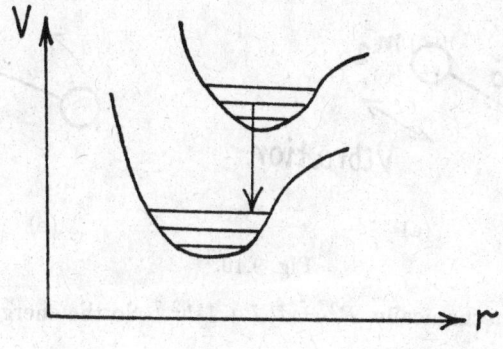

Fig. 9.8.

Fig. 9.9.

9.3. Vibration-Rotation Spectrum

Consider a molecule as a two-atom system. Its reduced mass $\tilde{\mu}$ is given by

$$\frac{1}{\tilde{\mu}} = \frac{1}{m_1} + \frac{1}{m_2}$$

where m_1 and m_2 are the masses of the individual atoms. The molecule can (a) vibrate (as the atoms oscillate with a very small vibration amplitude), and (b) rotate (Fig. 9.10(a) and (b)). In both cases the pertinent energy is quantized. The vibration is taken to be linear. If we assume that the electron is steady in a given electronic state, then the molecular energy is

$$E = E_{\text{rot}} + E_{\text{vibr}}.$$

The molecular angular momentum is **P** and its magnitude squared is $|\mathbf{P}|^2 = P^2$. The rotational energy of the molecule is

$$E_{\text{rot}} = \frac{P^2}{2\tilde{\mu}r_0^2}$$

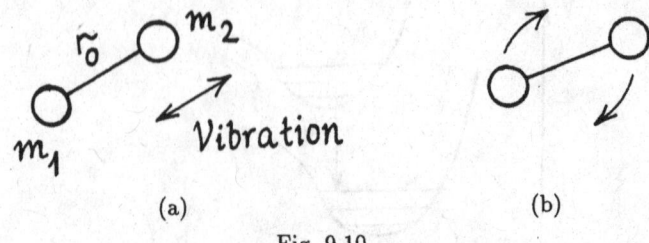

(a)   (b)

Fig. 9.10.

where quantum mechanically, $P^2 = J(J+1)\hbar^2$.[7] So the energy becomes

$$E_{\text{rot}} = \frac{J(J+1)\hbar^2}{2\tilde{\mu}r_0^2}$$

which is the quantum mechanical rotational kinetic energy of a rigid rotor. If we define the *rotation constant* as

$$B_0 \equiv \frac{h}{8\pi^2 \tilde{\mu} r_0^2}$$

(where $\hbar = h/2\pi$ is the Heisenberg constant), we obtain

$$E_{\text{rot}} = B_0 h\, J(J+1) \tag{9.2}$$

where J is the *total* angular momentum quantum number of the molecule. Equation (9.2) is very important in spectroscopy.

Now let us take Eq. (3.17) and write $E_{\text{rot}} + E_{\text{vibr}}$ instead of E:

$$\frac{\partial^2 f(r)}{\partial r^2} + \frac{2\tilde{\mu} f(r)}{\hbar^2}[E - V(r)] - J(J+1)\frac{f(r)}{r_0^2}\left(\frac{r_0}{r}\right)^2 = 0,$$

$$\frac{\partial^2 f(r)}{\partial r^2} + \frac{2\tilde{\mu} f(r)}{\hbar^2}[E - V(r)] - \frac{2\tilde{\mu} f(r)}{\hbar^2}E_{\text{rot}}\left(\frac{r_0}{r}\right)^2 = 0,$$

$$\frac{\partial^2 f(r)}{\partial r^2} + \frac{2\tilde{\mu} f(r)}{\hbar^2}\left\{E_{\text{vib}} - V(r) + E_{\text{rot}}\left[1 - \left(\frac{r_0}{r}\right)^2\right]\right\} = 0.$$

For small J values, $r \to r_0$ (*i.e.*, r is close to r_0: $r \simeq r_0$), and hence the last term can be neglected. Because if we say that $r = r_0$ (approximately), the E_{rot} term becomes zero. So, we have

$$\frac{\partial^2 f(r)}{\partial r^2} + \frac{2\tilde{\mu} f(r)}{\hbar^2}[E_{\text{vibr}} - V(r)] = 0,$$

---

[7]Notice that the expression for the rotational energy is classical *in form*, but P^2 is a quantum mechanical *operator*, not a number.

the Schrödinger equation for vibrational energies only.

As the molecule vibrates (Fig. 9.11(a)), the restoring force is $F = -\kappa(r-r_0)$ (where the minus sign indicates the fact that F is opposite to the displacement, in order to restore equilibrium: As the oscillator moves away, F tends to call it back; as it approaches equilibrium, F pushes it away). F is linear in r, so that the system behaves like a simple harmonic oscillator (Fig. 9.11(b)), obeying Hooke's law: A mass m tied to a spring whose length y changes by Δy, as m vibrates (oscillates) up and down or back and forth. Since the force is derived from a potential V, we have

$$F = -\frac{\partial V}{\partial r}$$

where $V(r) = \frac{1}{2}\kappa(r - r_0)^2$, the elastic potential of a spring. The change in length is $r - r_0 \equiv x$. We consider one-dimensional vibration (along x).

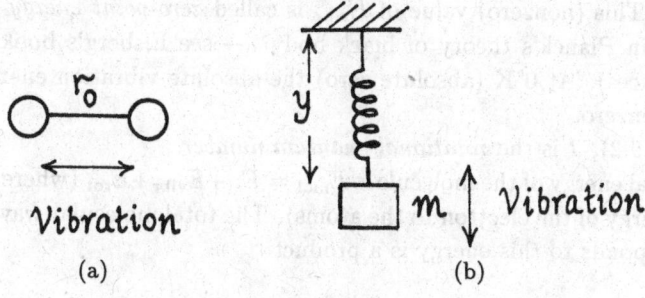

Fig. 9.11.

If the vibration amplitude is small, we can say that the force changes linearly with the distance away from the equilibrium position (Hooke's law). We have

$$\frac{\partial^2 f(x)}{\partial x^2} + \frac{2\tilde{\mu}f(x)}{\hbar^2}\left(E_{\text{vibr}} - \frac{1}{2}\kappa x^2\right) = 0$$

which is the linear harmonic oscillator equation. To solve it, we set

$$\alpha \equiv \frac{2\tilde{\mu}}{\hbar^2} - E_{\text{vibr}}, \quad \beta \equiv \frac{\sqrt{\tilde{\mu}\kappa}}{\hbar}, \quad \xi \equiv x\sqrt{\beta}$$

for simplicity, in order to make the equation concise, and not to carry along many constants as multipliers. In this mathematical trick α and β are constants. Thus we obtain the following differential equation with constant coefficients:

$$\frac{\partial^2 f(\xi)}{\partial \xi^2} + \beta\left(\frac{\alpha}{\beta} - \xi^2\right)f(\xi) = 0$$

which can be solved first by an approximation and then by an expansion into a series. We have

$$f(\xi) = V(\xi)e^{-\xi^2/2}$$

where $V(\xi) = \mathcal{A}_v\xi^v$, with v = finite. When one reaches the solution, one finds the condition $(\alpha/\beta) = 2v + 1$. If we substitute the values of α and β, we find

$$E_{\text{vibr}} = h\nu_0 \left(v + \frac{1}{2} \right) \tag{9.3}$$

where $\nu_0 = \sqrt{\kappa/\bar{\mu}}$. So, the energy of linear vibrations is quantized by the integer number v = 0, 1, 2, ... which is the *vibrational quantum number*. Notice that $E_{\text{vibr}} \neq 0$ even if v = 0 (whereas for $J = 0$ we have $E_{\text{rot}} = 0$, by virtue of Eq. (9.2)). E_{vibr} can never become zero (by virtue of Eq. (9.3)); the lowest vibration energy (for v = 0) is $E(v = 0) = (1/2)h\nu_0$ where ν_0 is the vibration frequency. This (nonzero) value of E_{vibr} is called *zero-point energy* (which is important in Planck's theory of black body[8] — see Eisberg's book listed in the References). At 0°K (absolute zero) the absolute vibration energy of the atoms is nonzero.

In Eq. (9.2), J is the *rotational quantum number*.

The total energy of the molecule is $E_{\text{mol}} = E_{\text{el}} + E_{\text{vibr}} + E_{\text{rot}}$ (where E_{el} is the binding energy of the electron to the atoms). The total molecular wavefunction that corresponds to this energy is a product:

$$\psi_{\text{mol}} = \psi_{\text{el}}\psi_{\text{vibr}}\psi_{\text{rot}} .$$

These results are obtained under the assumption that we take the molecule as a system consisting of two rigid masses (without paying any attention to the internal structure of the atoms). But since the atoms have wheels and gears, these results are approximate and rough; they just give a gross idea about the energy levels of the molecule.

Figure 9.12 illustrates schematically the energy states of a diatomic molecule. Every level of n has levels of v, and every level of v has levels of J whose spacing increases as J increases to higher values. If the potential were ideally parabolic, the vibrational levels would be all equally spaced (as expected

---

[8]A black body is not necessarily black in terms of color. For example, imagine a whitewashed house with an open window under a blazing sun in Mexico. As you look towards the window, you see it dark and black, because it traps and absorbs the incident radiation coming from outside. Any radiation that has a chance to escape out is masked by the outer ambient radiation (as the white surface reflects almost everything). The window is an example to a black body.

from Eq. (9.3)), but since for higher v values the curve departs from a parabolic shape, the v levels get closer together (Fig. 9.13). The diagram in Fig. 9.12 is not to scale, and the *J*-level spacings are exaggerated (with respect to the v-level spacings) for clarity. Only a few levels (for each quantum number) are shown. The spacing of the electronic levels gets smaller and smaller as n increases until it reaches the continuum where the levels are so close together as to give way to a continuous domain of accessible energy for the electron. Here the

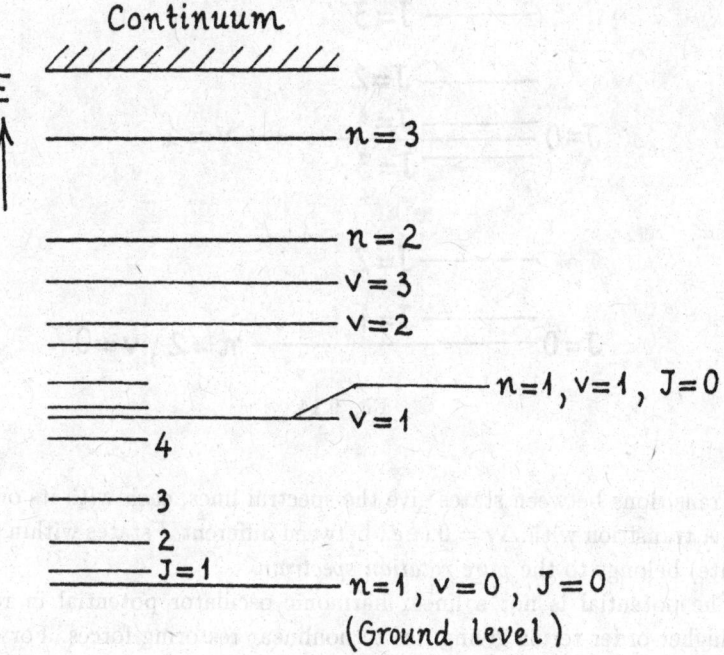

Fig. 9.12.

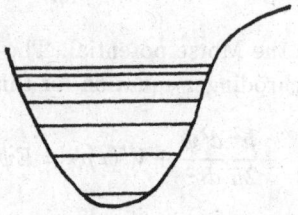

Fig. 9.13.

electron becomes free as it gets detached from the atom to emerge to freedom. A free electron can have any (kinetic) energy value in the energy continuum. The state $n = 1$ is the ground level of the electron bound to the atom.

Figure 9.14 shows the $n = 2$ level alone and a few of its v and J levels. (It is an excited electronic level.) Notice how the J levels are more and more spaced out as J increases for every v. This is expected from Eq. (9.2).

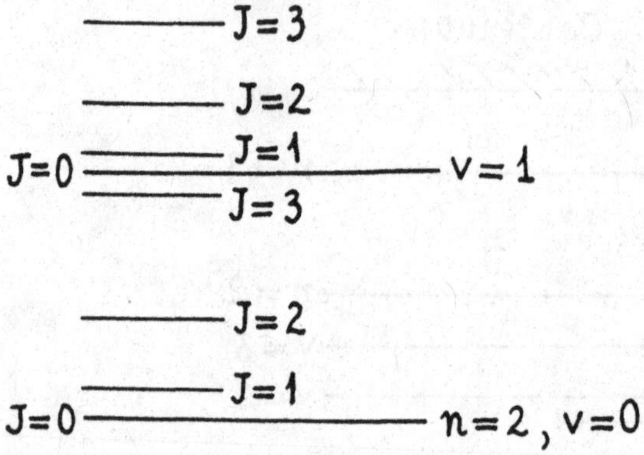

Fig. 9.14.

Transitions between states give the spectral lines, each with its own intensity. A transition with $\Delta v = 0$ (*i.e.*, between different J states within the same v state) belongs to the *pure rotation spectrum*.

The potential is not a linear harmonic oscillator potential in reality; it has higher-order terms giving rise to nonlinear restoring forces. For example, consider a particle of mass $\bar{\mu}$ moving in a potential

$$V'(r) = V(r) + \frac{\hbar^2 J(J+1)}{2\bar{\mu}r^2}$$

which for small J can be the Morse potential. The one-dimensional (with the motion along r) radial Schrödinger equation for this particle is

$$-\frac{\hbar^2}{2\bar{\mu}}\frac{d^2\psi}{dr^2} + V'(r)\psi = E\psi$$

where $\psi \to 0$ as $r \to 0$. For small vibrational departures from equilibrium, we can expand V' about its minimum at r' (where $r' = r_0$ only if $J = 0$):

$$V'(r) = V_0 + \frac{1}{2}\kappa(r - r')^2 + \alpha(r - r')^3 + \beta(r - r')^4 + \text{h.o.t.}$$

where α and β are constant coefficients determining the strength of their corresponding terms. If we neglect everything beyond the second term, we obviously have a linear harmonic oscillator (with an extra constant term V_0 that shifts the energy by a constant amount) and the solution is straightforward, as above (idealized situation). But if we want to include the contributions of α and β — however small — to the energy, we must use perturbation theory. A one-dimensional harmonic oscillator perturbed by an extra potential term x^3 is subject to a change in each of its energy levels to second order. So, the $(r - r')^3$ term will contribute a second-order effect. The $(r - r')^4$ term must be included, too, because its contribution is comparable to that of the previous term. The case is worked out in Schiff (pp. 453–454).

The spacing (interval) between adjacent vibrational energy levels is (from Eq. (9.3)) $\Delta E_{\text{vibr}} = (1/2)h\nu_0$. The rotational intervals are $\Delta E_{\text{rot}} = 2B_0 J$. We have $\Delta E_{\text{vibr}} \gg \Delta E_{\text{rot}}$. The electron energy is $E_n \propto 1/n^2$, and the levels get crowded as n increases. E_n is the energy with which the electron is bound to the atom, and consequently, to the molecule.

When the electron receives external energy (radiation), it absorbs a quantum $\Delta E_{\text{el}} = h\nu = E_2 - E_1$ and gets from $n = 1$ to $n = 2$ (it gets excited) (Fig. 9.15). As it gets to the upper level of higher energy, it makes an *absorption* (of the incident radiation). When it is in the upper level it can emit the difference ΔE_{el} back as a photon and return to the lower state (the ground state, in this case) making an *emission*. (If it is in a further excited state it can descend to an allowed orbit of lower energy other than the ground state, provided that the transition is allowed quantum mechanically.)

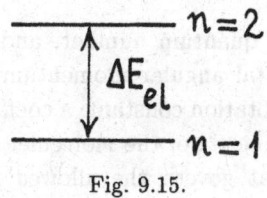

Fig. 9.15.

We never measure E_n itself; what we measure is the change or *difference* ΔE_n that corresponds to the transition. Also, we cannot measure directly the limit value $E_0 = (1/2)h\nu_0$.

9.4. Transitions

Transitions between levels give the spectral lines. Let us suppose that there is no transition between electronic energy levels, that is, $\Delta E_{\text{el}} = 0$ or $\Delta n = 0$. (The atoms stay at a given state n.) In other words, the transition shown by an arrow in Fig. 9.16 does not occur. Then any change ΔE in the molecular energy E will be due to vibrational and rotational transitions:

$$\Delta E = \Delta E_{\text{vibr}} + \Delta E_{\text{rot}} ,$$

since $\Delta E_{\text{el}} = 0$. This ΔE corresponds to an emission or an absorption. (For example, in microwave spectroscopy we have absorption; if we send microwaves to the molecule, it absorbs certain wavelengths corresponding to its rotational state intervals.)

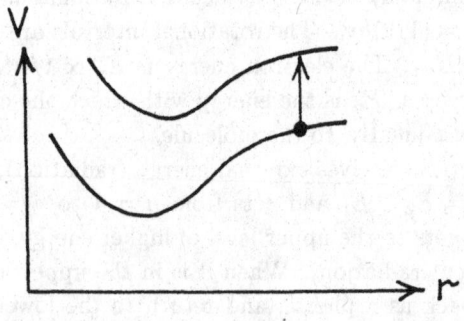

Fig. 9.16.

For a diatomic molecule we have

$$E_{\text{vibr,rot}} = \left(v + \frac{1}{2}\right) h\nu_0 + Bh \, J(J+1)$$

where v is the vibrational quantum number, and J (or k) is the rotational quantum number, *alias* total angular momentum quantum number. B (we drop the subscript) is the rotation constant, a coefficient given by $B = h/8\pi^2 \mathcal{I}$ where $\mathcal{I}$ is the moment of inertia of the molecule.

The selection rules that govern the allowed transitions are $\Delta v = \pm 1$, $\Delta J = \pm 1$ (both numbers can change by one), but for large J values the latter rule breaks down. There are exemptions for $\Delta J = 0$, *i.e.*, $\Delta J = 0$ is not always forbidden. That J changes by one is shown in Fig. 9.17 where we see a few radiative transitions between two vibration-rotation bands, and the corresponding spectrum with the lines (the intensities are not to scale). It is

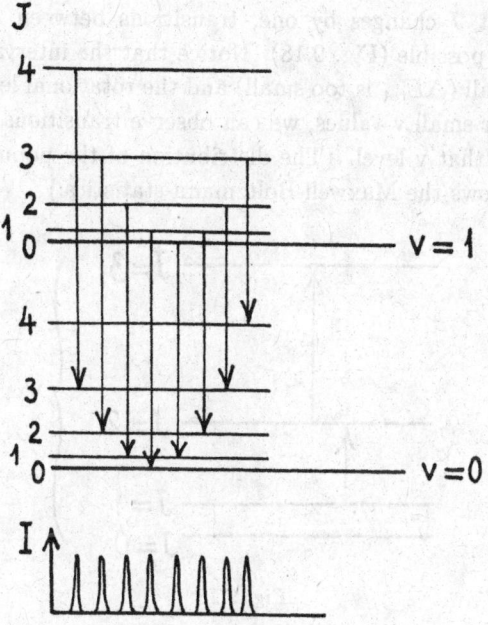

Fig. 9.17.

possible to also have a transition where only J changes (again from 1 to 2, from 2 to 3 etc. or vice versa).

We assume that B is constant. The situation in any transition is

$$\Delta E = h\nu_0(v_2 - v_1) + Bh[J_2(J_2 + 1) - J_1(J_1 + 1)] = h\nu$$

which is an energy *difference*. In case of emission we have $+h\nu$; in case of absorption (removal of radiation) we have $-h\nu$, just for correct bookkeeping purposes. Since the vibration energy is larger than the rotation energy, the latter shows up as fine structure.

If $J_2 = J_1 + 1$, we have $\Delta E = h\nu_0 + 2Bh\, J_1$. If $J_2 = J_1$, we have $\Delta E = h\nu_0$. If $J_2 = J_1 - 1$, we have $\Delta E = h\nu_0 - 2Bh\, J_1$. (Remember that as a result of the selection rule $\Delta J = \pm 1$, we can have $J_2 = J_1$, $J_2 = J_1 + 1$, $J_2 = J_1 - 1$ only.) Here J can be $J = 0, 1, 2, \dots$. The general expression is

$$\Delta E = h\nu_0 + 2Bh\, \mathrm{m}_J$$

where m_J is a number that can take the following values: $\mathrm{m}_J = 0, \pm 1, \pm 2, \dots,$ $\pm J$. The multiplicity is $2J + 1$, *i.e.*, m_J takes $2J + 1$ values.

Provided that J changes by one, transitions between J levels within a given v level are possible (Fig. 9.18). Notice that the intervals between the J levels are too small (ΔE_{rot} is too small) and the rotational levels are so closely crowded, that for small v values, we can observe transitions from a J level to the next, within that v level. (The distribution of the population among the energy levels follows the Maxwell-Boltzmann statistics.)

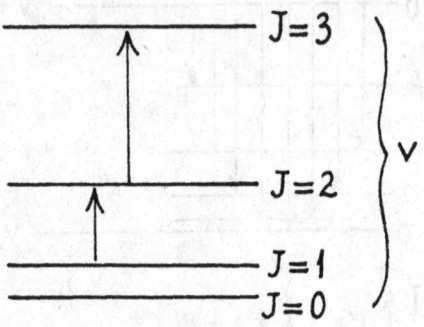

Fig. 9.18.

Not all lines have the same intensity in the spectrum. Figure 9.19 shows the absorption lines (intensity I vs. frequency). The middle line (Q-*line*) corresponds to $m_J = 0$. For $m_J > 0$ we have the P-*lines*, and for $m_J < 0$ we have the R-*lines*. The curve is symmetric about the Q-line (Fig. 9.20).

Except for the nitrogen monoxide molecule, many molecules have no permanent dipole moment, and hence the Q-line is not observed. That is, due to the fact that diatomic molecules have no permanent average dipole moment

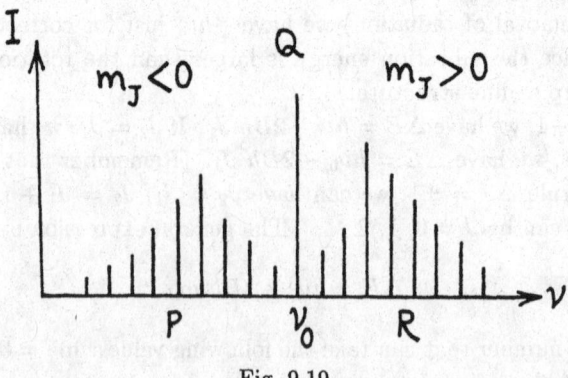

Fig. 9.19.

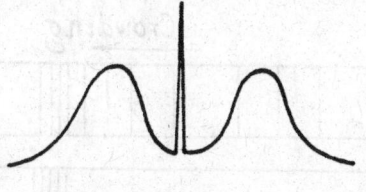

Fig. 9.20.

$(\vec{p} = 0)$, the Q-line does not appear. The exception to this case in NO.

Now let us assume that $B \neq$ constant. Then the Q-line is not a single line, but many lines instead:

$$\Delta E = h\nu_0 + B_2 h \, J_2(J_2 + 1) - B_1 h \, J_1(J_1 + 1) \,.$$

If $J_2 = J_1 - 1$, we have $\Delta E = h\nu_0 + B_2 h \, J_1(J_1 - 1) - B_1 h \, J_1(J_1 + 1)$ which is the P-*branch*. If $J_2 = J_1 \equiv J$, we have $\Delta E = h\nu_0 + B_2 h \, J(J+1) - B_1 h \, J(J+1)$ which is the Q-*branch*. And if $J_2 = J_1 + 1$, we have $\Delta E = h\nu_0 + B_2 h (J_1 + 1)(J_1 + 2) - B_1 h \, J_1(J_1 + 1)$ which is the R-*branch*.

If $B_1 < B_2$, we have $\mathcal{I}_1 > \mathcal{I}_2$, *i.e.*, the moment of inertia is larger in one state than in the other.

One can plot the three branches (P, Q, and R) as J vs. ν, and thus obtain a *Fortrat diagram* (Fig. 9.21). On the three branches, as we move down the curves, the points get closer together. Each point corresponds to a certain value of J. There is no point corresponding to $J = 0$.

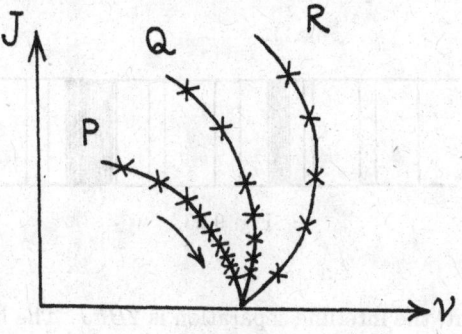

Fig. 9.21.

The spectrum thus obtained looks like as in Fig. 9.22. In the P-branch the lines get crowded together as one moves to the right. In the Q-branch we have lines at equal intervals but compact (very closely spaced together).

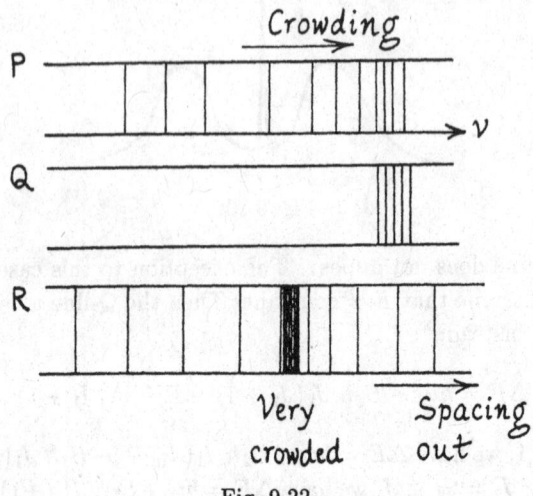

Fig. 9.22.

In the R-branch we see a bunch of very tightly squeezed lines, and as one moves to the right, they get wide apart and spaced out.

If these three spectra are superimposed, one obtains a complicated outcome consisting of lines and bands (bunches) (Fig. 9.23) This is a *band spectrum* given by a diatomic molecule. If the transition $v_2 \rightarrow v_1$ is specified and known, we have this outcome. But if our business involves transitions corresponding to $\Delta n = \pm 1$ as well, the spectrum becomes too complex and intricate.

Fig. 9.23.

If B is constant, the interline separation is $2BhJ$. The lines form a certain geometrical series.

If one has the (superimposed) spectrum in Fig. 9.23, one can find and tell apart the three branches (Fig. 9.24). The frequency difference $\Delta \nu$ is divided by two. One works on the right half of the spectrum (Fig. 9.25) and separates the three branches.

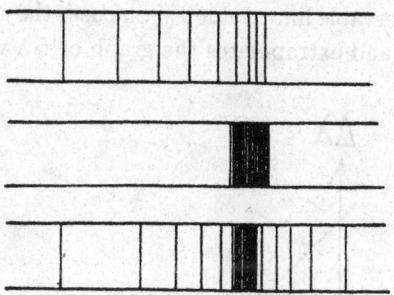

Fig. 9.24.

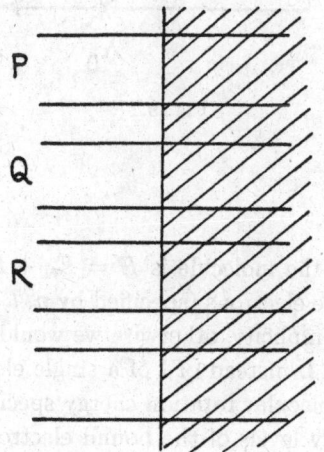

Fig. 9.25.

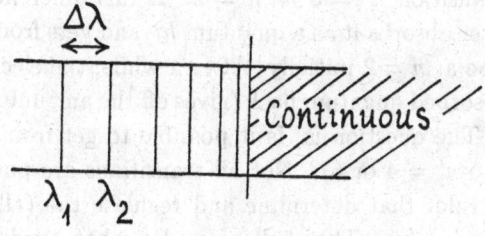

Fig. 9.26.

To find the unobservable limit value λ_0, one uses the graphical method: One takes $\Delta\lambda$ (Fig. 9.26) and extrapolates the graph of $\Delta\lambda$ vs. λ to λ_0 (Fig. 9.27).

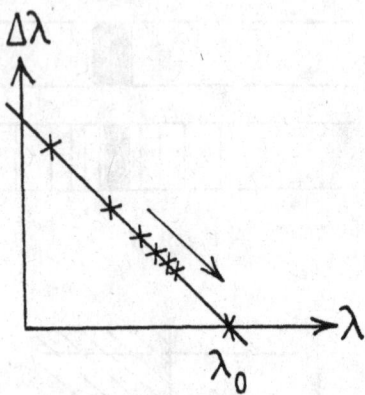

Fig. 9.27.

9.5. Selection Rules

The total energy of the molecule is $E = E_{\text{el}} + E_{\text{vibr}} + E_{\text{rot}}$ where E_{el} is the binding energy of the electrons (specified by n, l, m_l — a single electron is usually considered for simplicity, otherwise we would have n, L, m_L for many electrons, *i.e.*, the total $\mathbf{L}$ instead of l of a single electron), E_{vibr} is specified by v, and E_{rot} is the molecular rotation energy specified by J.

The electronic energy levels of the bound electron are shown in Fig. 9.28 where each state has a fine structure due to vibrations and rotations. So, E is specified by many energy levels at once. The transition between electronic energy levels (Fig. 9.29) is represented by an energy difference, say, $\Delta E_n = E_3 - E_2$ for a transition $n' = 3 \rightleftarrows n = 2$. If this difference is given to the molecule, the latter absorbs it as a quantum $h\nu$ and gets from $n = 2$ to $n' = 3$, if it happens to be at $n = 2$ initially. After a while, the electron at the upper level emits the absorbed quantum back (gives off the amount ΔE_n) and returns down to $n = 2$. The question is, Is it possible to get from $n = 1$ to $n' = 3$ (or from $n = 2$ to $n' = 4$ or 5)? Not all transitions are possible in quantum mechanics. The rules that determine and regulate the (allowed) transitions are called *selection rules*. They tell us under what conditions a transition is possible. Finding the selection rules means finding conditions that favor transitions.

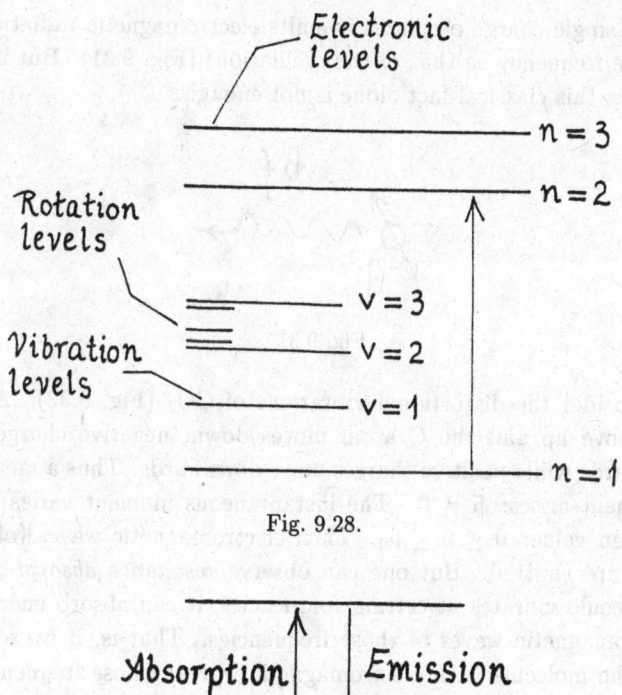

Fig. 9.28.

Fig. 9.29.

Consider the CO_2 molecule (Fig. 9.30). It is linear. The O atoms approach C and recede from it as the molecule vibrates in the direction shown. Does any emission or absorption of energy take place? The classical electromagnetic theory answers as follows: If the O atoms vibrate symmetrically and with the same speed along the molecular axis, the mean dipole moment is zero: $\bar{p} = 0$. Since the charges (ions) approach (or recede from) each other by the same amount, the instantaneous dipole moment is also zero. Therefore, the molecule neither emits nor absorbs electromagnetic energy.

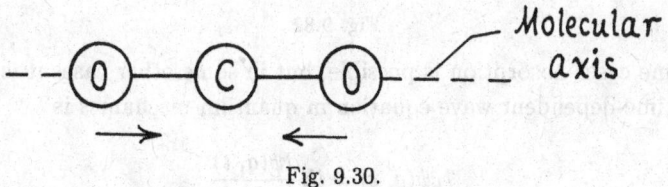

Fig. 9.30.

When a single charge oscillates it emits electromagnetic radiation around (of the same frequency as that of the oscillation) (Fig. 9.31). But in the case of a molecule this classical fact alone is not enough.

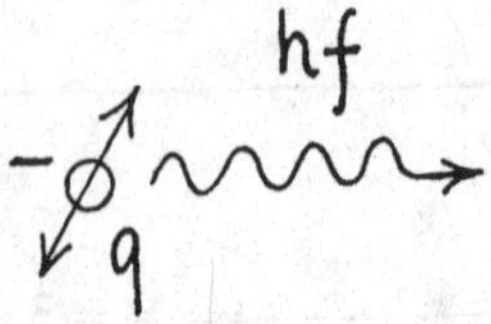

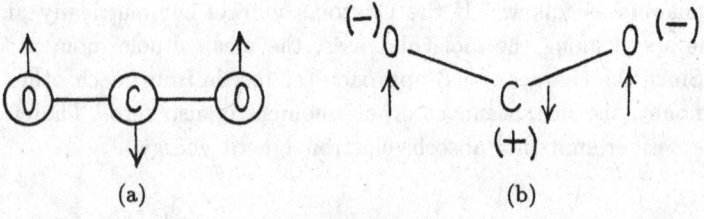

Fig. 9.31.

Now consider the distortional vibrations of CO_2 (Fig. 9.32). As the two O atoms move up and the C atom moves down, negative charges are displaced upwards while positive charges move downwards. Thus a mean *nonzero* dipole moment arises: $\bar{p} \neq 0$. The instantaneous moment varies between 0 and a certain value: $0 \leq p \leq p_0$. Then electromagnetic waves (of the same frequency) *are* emitted. But one can observe *resonance absorption* as well: As the molecule vibrates at certain frequencies, it can absorb energy of incident electromagnetic waves of those frequencies. That is, if we send energy (irradiate the molecule with electromagnetic waves) whose frequency ensures transitions between vibrational states, the system absorbs this energy because there is a resonance between the frequency of the field and the frequency that corresponds to the transition quantum $\Delta E = h\nu$. The incident frequency either happens to be in tune with one of the quantized and fixed frequencies of the transitions of the system, or we can choose it so by intention, in order to cause transitions.

(a)   (b)

Fig. 9.32.

In some cases absorption is possible, but in some other cases it is not. The time-dependent wave equation in quantum mechanics is

$$\mathcal{H}\psi(q,\,t) = i\hbar\frac{\partial\psi(q,\,t)}{\partial t}$$

whose solution is

$$\psi(q, t) = \psi(q)\psi(t) = \psi(q)e^{-2\pi i \frac{E}{h} t} \tag{9.4}$$

where $E = \hbar\omega = h\nu$ and $\omega = 2\pi\nu$. Here q is a generalized coordinate which may as well be $\mathbf{r}$. So, actually $\psi(q, t) = \psi(\mathbf{r}, t) = \psi(x, y, z, t)$. Notice that the solution (Eq. (9.4)) is separable to space-dependent and time-dependent parts which are written as a product of two factors:

$$\psi = (\text{Spatial part}) \times (\text{Temporal part}).$$

Let us suppose that two energies, $E_{n'}$ and E_n, are accessible to the molecule, *i.e.*, the system can take (or have) two energies specified by (the quantum numbers) n' and n. Then E in Eq. (9.4) is $E = E_{n'} - E_n$. The two wavefunctions that correspond to these two states are $\psi_{n'}$ and ψ_n respectively (two eigenfunctions associated with the two eigenvalues-eigenenergies). The *total molecular wavefunction* ψ is

$$\psi = c_n\psi_n(q, t) + c_{n'}\psi_{n'}(q, t),$$

the superposition[9] of the two states.[10,11] As a result of the transition (emission

---

[9]The principle of superposition (where the total outcome is the resultant of the contribution of the separate parts, each acting alone) is widely used in physics (*e.g.*, in vector analysis, wave theory, electric field mapping, magnetostatics, geometrical optics etc.).

[10]The two states are coupled (mixed) to form the total state ψ (which has a bit from one and a bit from the other). The coefficients c are the percentages (shares) of the contribution of each state to ψ. This happens whenever two states *interact*: The system partakes of both during the smooth transition (just like when one occupies two houses when he is moving out). As the system reaches n', it has not yet left n entirely, and is partly still anchored at n for a while, so that the states mix. This does *not* mean that upon an observation we will find the system in both states at the same time. Actually, each coefficient c gives the probability of finding the system in the associated state. When we make an experimental measurement we will find the system either in state n *or* in n' (with the associated likelihood). (And what we measure is the *eigenvalue* of the operator, *not* its expectation value.) When we leave a cat unfed for a long time in a locked room and then open the door to see what has happened, we will find the cat either dead *or* alive, not half-dead or half-alive! So we will have, according to Schrödinger,

$$\psi(\text{cat}) = c_1\psi(\text{cat dead}) + c_2\psi(\text{cat alive}).$$

[11]The phenomenon of coupling is not encountered in quantum-mechanical states only. In classical mechanics, for example, we have the mechanical coupling (when two connected pendula affect each other), or we can speak of decoupling an apparatus from the floor vibrations (by placing it on absorber springs or using other clever devices); we also have the electrical coupling between two oscillators (circuits), the acoustical coupling between two tuning forks etc. Similarly, when two atoms are brought together, they interact (get coupled). Upon every coupling, the energy splits into two. If we have N interacting particles (like in solids) we have an N-fold splitting: The degenerate energy of every particle (atom) splits into a band (bunch) of energies which spread — as we see in crystal (solid state) physics. Coupling (mixing) of states is treated by matrices.

or absorption, $n \rightleftarrows n'$), ψ changes in t. We say that the molecule is in an electromagnetic field. And conversely, if there is an electromagnetic field, ψ depends on time, and so do the coefficients c:

$$\psi(q, t) = c_n(t)\psi_n(q, t) + c_{n'}(t)\psi_{n'}(q, t). \tag{9.5}$$

The total energy changes in time, *i.e.*, $\mathcal{H}$ changes. If $\mathcal{H}^{(0)}$ is the operator corresponding to the (unperturbed) static case and $\mathcal{H}'$ is the small perturbation term (part), we have $\mathcal{H} = \mathcal{H}^{(0)} + \mathcal{H}'$ and

$$(\mathcal{H}^{(0)} + \mathcal{H}')\psi(q, t) = i\hbar \frac{\partial \psi(q, t)}{\partial t}$$

where $\mathcal{H}'$ includes all the time-dependent terms. Only this part is of interest:

$$\mathcal{H}'\psi(q, t) = i\hbar \left[\frac{\partial c_n}{\partial t}\psi(q, t) + \frac{\partial c_{n'}}{\partial t}\psi(q, t) \right] \tag{9.6}$$

where the amount of change in the coefficients c is also included. To solve this equation, we express the wavefunction in terms of the the spatial center-of-mass coordinates:

$$\psi_n(q, t) = \psi_n \underbrace{(X, Y, Z)}_{\substack{\text{Center-of-mass} \\ \text{coordinates}}} \; \psi_n \underbrace{(x, y, z)}_{\substack{\text{Intramolecular} \\ \text{(internal)} \\ \text{coordinates}}} \; e^{-2\pi i \frac{E_{n'} - E_n}{\hbar} t}. \tag{12}$$

The wavefunction is separated into three parts (factors): One depends on the center-of-mass coordinates; the other depends on internal coordinates; and the third one is time-dependent only.

We multiply Eq. (9.6) by ψ_n^* and integrate it over all space:

$$c_n \int \psi_n^* \mathcal{H}' \psi_n dV + c_{n'} \int \psi_n^* \mathcal{H}' \psi_{n'} dV = i\hbar \frac{\partial c_n}{\partial t} \times 1, \tag{9.7}$$

or in Dirac notation

$$c_n \langle n|\mathcal{H}'|n \rangle + c_{n'} \langle n|\mathcal{H}'|n' \rangle = i\hbar \dot{c}_n$$

where the coefficients c are left out of the integrals because they are time-dependent only, not space-dependent. The unity on the right-hand side of Eq. (9.7) comes from the normalization condition

---

[12]Notice that the imaginary unit i makes the exponential a propagating wave (as it behooves ψ); otherwise (without i) it would be a decaying exponential.

$$\int \psi_n^* \psi_n dV = \int |\psi_n|^2 dV \equiv \langle n|n\rangle = 1. \tag{13}$$

Now we consider the initial condition. We assume that initially $c_n = 0$ (the molecule is in state n' purely) and we are going to find how much it will change later on by smearing from a state to another. So, the first term of Eq. (9.7) becomes zero. We solve the remaining by imposing the following condition:

The potential energy (of the interaction between a dipole $\vec{p}$ and a field $\vec{\mathcal{E}}$) that corresponds to $\mathcal{H}'$ is

$$\mathcal{H}' = E' = \vec{p} \cdot \vec{\mathcal{E}} = p_x \mathcal{E}_X + p_y \mathcal{E}_Y + p_z \mathcal{E}_Z$$

where $\vec{p}$ depends on (is a function of) the intramolecular coordinates, and the electric field $\vec{\mathcal{E}}$ depends on the center-of-mass coordinates. The components (projections) of $\vec{p}$ are expressed in terms of the molecular coordinates. We have

$$p_x \mathcal{E}_X = c_{n'} \int \psi_n^* p_x(x,y,z) \mathcal{E}_X(X,Y,Z) \psi_{n'} dx\, dy\, dz\, dX\, dY\, dZ = i\hbar \frac{\partial c_n}{\partial t},$$

$$\frac{\partial c_n(t)}{\partial t} = \frac{c_{n'}(t)}{i\hbar} \exp\left[-2\pi i \left(\frac{E_{n'} - E_n}{\hbar}\right) t\right] \int \psi_n^*(X,Y,Z) \psi_n^*(x,y,z)$$

$$\times\; p_x(x,y,z) \mathcal{E}_X(X,Y,Z) \psi_{n'} dx\, dy\, dz\, dX\, dY\, dZ.$$

But we do not care about the external appearance of the molecule and its translation in space (after all, every molecule is translated in space). So, we integrate over whatever we are not interested in, that is, over $dX\,dY\,dZ$ which signifies translation and which is not important to the molecule and is out of interest. So, the matrix element (or expectation value, or average) of the operator that mixes the states n and n' and causes a transition (interaction) between them is

$$(p_x)_{nn'} \equiv \langle n|p_x|n'\rangle \equiv \langle p\rangle_{nn'} \equiv \overline{p}$$

$$= \int \psi_n^*(x,y,z) p_x(x,y,z) \psi_{n'}(x,y,z) dx\, dy\, dz.$$

[13]As $|\psi|^2$ is the probability of finding a particle somewhere, when integrated (summed continuously) over *all* space it gives unity (= certainty, 100%). If we are concerned about where Julie Andrews probably is, and include all the globe, we find 1, because she will certainly and inevitably be somewhere in the world. "All space" here means the entire universe, because in quantum mechanics there is a likelihood — however fantastically small — that the particle might be *anywhere* in space, even away from where it is expected to be found classically.

If this integral turns out to be zero, then $c_n =$ constant. Hence if $\bar{p}_x = 0$, c_n does not change. Therefore, no absorption or emission takes place, and this transition ($n \rightleftarrows n'$) is forbidden.

If it turns out to be $(p_x)_{nn'} \neq 0$, the emerging condition gives the pertinent selection rule, and the transition (absorption/emission) is possible (allowed), subject to the selection rule.

For the hydrogen atom, $\bar{p}_x = e\bar{x}$ (bars indicate mean values). The necessary condition for $\bar{x} \neq 0$ gives the selection rules for the hydrogen atom.

So, if $\Delta c_n(t) = 0$, we have $c_n =$ constant, and hence no coupling (transition) between the two states. The system remains in one state. If $\Delta c_n(t) \neq 0$, there is emission or absorption (transition from one state to the other).

Consider Eq. (9.5). The wavefunction that describes the initial situation is $\psi(q, 0)$. If initially $c_n = 0$, the situation is

$$\psi(q, 0) = c_{n'}\psi_{n'}(q, t) = \psi_{n'}(q, t)e^{-2\pi \frac{\Delta E}{h} t}$$

where the transition occurs during a time interval specified by the exponential factor (which is $c_{n'}(t)$). Notice that the exponential decay is dimensionless, as it should be.

9.6. Macroscopic Reasons of Selection Rules in Molecular Spectroscopy

Consider linear vibrations (along the molecular axis). As the molecule vibrates, its dipole moment must change, so that we obtain a spectrum. This happens in the distortional vibrations (off the molecular axis) of the CO_2 molecule, for example. If the dipole moment indeed changes, we observe a vibration spectrum. To observe a vibrational Raman spectrum, the selection rule is as follows: Consider the relation $\vec{p} = \alpha\vec{\mathcal{E}}$. If during the vibration the polarizability α of the molecule changes, we observe a vibrational Raman[14] spectrum.

To observe a pure rotational spectrum, the instantaneous dipole moment must be $p_0 \neq 0$. This is the selection rule of this spectrum. To observe a rotational Raman spectrum, the macroscopic selection rule is the following: The molecular polarizability α must be anisotropic in a direction perpendicular to the rotation axis of the molecule. As the molecule rotates, it is as if p changed

---

[14]In the 20th century, the Hindu physicist Sir C.V. Raman — among his other accomplishments — made a thorough investigation of the vibrations of bowed violin strings, to scientifically analyze what Hungarian gypsies had known all along practically and intuitively, with a strong innate feeling about it.

(anisotropy), and hence we have transitions (emission). And if there is a varying $\vec{\mathcal{E}}$ field around, there is absorption of energy from it, should its frequency be at resonance with an interlevel transition frequency. If there are many frequencies beating against each other (Fig. 9.33),[15] the system selects those frequencies which agree with the frequencies of allowed transitions (multichromatic field case), just like a company which recruits applicants whose talents agree with the requirements of the promotional vertical mobility within the ranks of the company.

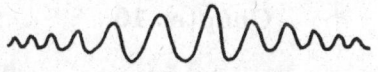

Fig. 9.33.

9.7. Remarks

The quantity J in Eq. (9.2) and in the discussion thereafter is the rotational quantum number of the *molecule* (as a whole), identified as k in Section 8.4, and should not be confused with the atomic J (total *electronic* angular momentum). For molecules J is what L is for atoms; it is the *molecular* rotation quantum number, to be distinguished from the atomic J of $\mathbf{J} = \mathbf{L} + \mathbf{S}$.

9.8. Summary

This chapter was only a general review regarding molecular spectra. Armed with this much knowledge, we can get into some more detail now, but this will necessitate a fresh chapter.

---

[15]In nature we usually have an infinite number of frequencies but with a definite average frequency and a small spread (width) about it.

Chapter 10

PURE ROTATION SPECTRUM

10.1. Rotation of a Diatomic Molecule

The spectroscopy of pure rotations is a good way of collecting information about the parameters of gaseous molecules, provided that the molecule in question satisfies the selection rules and gives a rotation spectrum.

The principle is as follows: If a charged rigid rotor were to rotate, it would classically emit electromagnetic radiation (just like a single charge would, revolving in a circle). The emitted radiation would have the frequency of rotation. By the same token, an incident sinusoidal radiation could set the rotor into rotation by imparting energy to it. In this case the rotor (actually, the hit molecule) absorbs energy from the incident radiation.

A rotating molecule can be thought as such a rigid rotor whose charges (ions) radiate upon rotational motion. Spectral lines corresponding to this picture form the *pure rotation spectrum* and fall in the far infrared region.

The quantum states of such a molecule are characterized by a quantum number k which has fixed integral values and which need not be anything else than the angular momentum quantum number J. In other words, the rotational states are quantized just like the electronic states in an atom, but they are determined by the angular momentum, with $J = 0, 1, 2, \ldots$. It is customary and pedagogical to start the analysis with a simple diatomic molecule and afterwards extend to polyatomic systems. The simplicity lies with the fact that for a diatomic molecule, the molecular axis (the line that connects the two atoms) is a symmetry axis. So, any rotation about an axis normal to the symmetry axis would have the same $\mathcal{I}$.

314

It all starts with the classical relations $\mathbf{P} = \mathcal{I}\vec{\omega}$ and $E = (1/2)\mathcal{I}\omega^2$ whence $E = |\mathbf{P}|^2/2\mathcal{I}$. Since $\mathbf{P}$ is an *operator* here (and not just a classical vector), quantum physics enters here via the substitution of the quantized value of $|\mathbf{P}|^2$. So, the energies are given by

$$E = \frac{\hbar^2 J(J+1)}{2\mathcal{I}} \tag{10.1}$$

where J is confined to $J = 0, 1, 2, \ldots$. For $J = 0$ there is no E_{rot} (while there is still E_{vibr} because $E_{\text{vibr}} \neq 0$ even for v = 0). So, the lowest *rotational* (zero-point rotational) energy is 0 (whereas the zero-point *vibrational* energy is $(1/2)h\nu_0 \neq 0$). Even if $E_{\text{rot}} = 0$, there is a nonzero E_{vibr}, and thus the uncertainty principle is not violated.

For real molecules, r_0 must be used instead of R in Eq. (8.10), where r_0 is the virtual internuclear separation which is not equal to R, the equilibrium separation.

If J_2 (J_1) indicates the upper (lower) rotational level, from the relation

$$\Delta E_{\text{rot}} = Bh[J_2(J_2+1) - J_1(J_1+1)] \tag{10.2}$$

concerning successive transitions (between $J_2 \rightleftarrows J_1$), we obtain the following table for the pure rotation (infrared) spectrum of a rigid rotor:

J	E_{rot}	ΔE_{rot}
0	0	.
1	$2Bh$ $>2Bh$	
2	$6Bh$ $>4Bh$	
3	$12Bh$ $>6Bh$	
4	$20Bh$ $>8Bh$	
$\vdots$	$\vdots$	$\vdots$

The energy levels are shown in Fig. 10.1. The spacing increases with J. Notice that $\Delta E \propto J$, but the spacings of the lines are all equal because they are proportional to the constant number B. This is explained as follows:

A dipole radiation is possible if the molecule has an electric dipole moment, *i.e.*, one end with excess positive charge and another end with excess negative (and equal) charge. Then the selection rule is $\Delta J = \pm 1$ (just like the case of l for a single electron in a field). Here $+1$ means absorption, and -1 means emission. In getting, say, from J to $J - 1$, the radiated frequency is

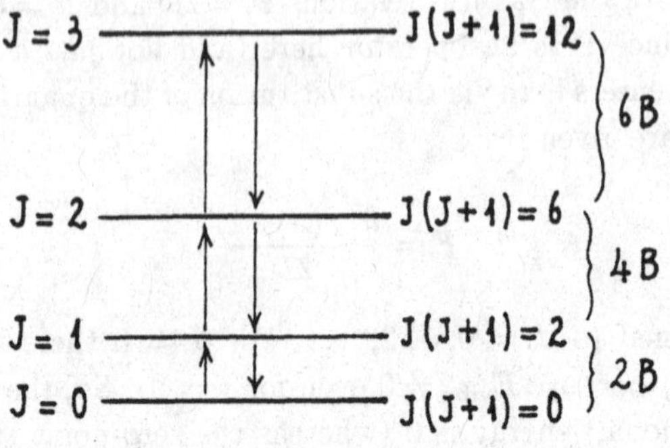

Fig. 10.1.

$$\Delta E = h\Delta\nu = 2BhJ,$$

$$\Delta\nu = \frac{\Delta E}{h} = 2BJ \tag{10.3}$$

from Eq. (10.2) where $J(J+1) - (J-1)(J-1+1) = 2J$. So, the produced spectrum has equally spaced lines whose frequencies are multiples of B.[1]

Gaseous hydrogen halides give infrared absorption spectra with equally spaced lines and wave numbers equal to multiples of B. These are pure rotation lines due to transitions between rotational states only (where J changes and hence the rotational state changes — without a change in the vibrational state necessarily). The following table shows the case of HCl[2] (where J is the greater value involved in the transition):

J	$\bar{\nu}(cm^{-1})$	$\Delta\bar{\nu}(cm^{-1})$
4	83.03	20.70
5	103.73	20.57
6	124.30	20.73
7	145.03	20.48
8	165.51	20.35
9	185.86	20.52
10	206.38	20.12
11	226.50	

---

[1]In case of absorption, J is that of the final state. In case of emission, of the initial state.
[2]M. Czerny, *Z. Physik*, **44**, 235 (1927).

Notice that in this table which gives the absorption spectrum of HCl in the far infrared, the spacings $\Delta\bar{\nu}$ are all nearly equal, but they tend to decrease as J increases. B also decreases with increasing J, *i.e.*, $\mathcal{I}$ increases. The increase in the molecular $\mathcal{I}$ means that the molecule swells as if less bound together, and hence pulled more and more apart as it rotates faster and faster. This is due to the centrifugal force. As J increases the molecule picks up angular speed and r_0 increases (we have elongation) giving rise to an increase in $\mathcal{I}$. This necessitates a correction term in E_{rot}, a term that need be considered only for high J values:

$$E_{\text{rot}} = BhJ(J+1) - D_0 h \, J^2(J+1)^2$$

where D_0 is the *centrifugal distortion* (or *elongation*) *coefficient*.[3] (See Eq. (4.7).) So, as J increases the spacings between the spectral lines decrease.

By measuring the spacing $\Delta\bar{\nu}$ between the lines, one can find $\Delta\nu$, and hence — via Eq. (10.3)— one can find $\mathcal{I}$ from the rotation constant B. For example, for the HCl molecule we have from the table above $J = 4$ corresponding to $\nu = 83.03 \times 3 \times 10^{10}$ Hz whence $\mathcal{I} = 2.7 \times 10^{-47}$ kg m^2. To illustrate the way better, let us work out an example problem explicitly.

Example problem: The infrared rotation spectrum of HI gives $\Delta\bar{\nu} = 12.8$ cm^{-1}. Calculate the separation (bond distance) H–I. The atomic masses of the hydrogen and iodine atoms are $A(\text{H}) = 1$ and $A(\text{I}) = 127$.

Solution: $\Delta\nu = c\Delta\bar{\nu} = 3 \times 10^{10} \times 12.8 = 3.84 \times 10^{11}$ Hz (where $\Delta\bar{\nu} = 1/\lambda$). The energy difference between two adjacent spectral lines is

$$\Delta E = h\Delta\nu = 6.62 \times 10^{-27} \times 3.84 \times 10^{11} = 2.54 \times 10^{-15} \text{ ergs}.$$

$$\Delta E = 2Bh = \frac{2h^2}{8\pi^2\mathcal{I}} = 2.54 \times 10^{-15} \text{ ergs}$$

whence $\mathcal{I} = 4.37 \times 10^{-40}$ gr cm^2.

The reduced mass of hydrogen iodide is

$$\tilde{\mu} = \frac{1}{N_A}\left(\frac{1 \times 127}{1 + 127}\right) = 1.65 \times 10^{-24} \text{ gr}$$

where N_A is Avogadro's number. We have now $r_0^2 = \frac{\mathcal{I}}{\tilde{\mu}} = 2.65 \times 10^{-16}$ cm^2, and hence $r_0 = 1.67$ Å.

---

[3]Caution: Do not confuse the elongation constant D_0 with D_e, the dissociation energy of the molecule. The accidental choice of notation is poor, but the subscripts make the difference.

10.2. Rotation of Linear Polyatomic Molecules

All the considerations above hold true for polyatomic molecules, too, provided that they are linear. Here, too, as the molecule gets to excited rotational states, it rotates faster and swings out, thereby increasing its I. By using the powerful spectroscopic method, one can measure molecular parameters. For example, from the microwave spectra of certain molecules we can calculate r_0 of their various bonds. Some results are as follows:

Molecule	Bond	$r_0(\text{Å})$	Molecule	Bond	$r_0(\text{Å})$
NNO	NN	1.12	ClCCN	ClC	1.63
	NO	1.19		CC	1.21
BrCN	BrC	1.79		CN	1.05
	CN	1.15	HC $\equiv$ C–CN	HC	1.05
HCN	HC	1.06		CC	1.38
	CN	1.15	C $\equiv$ C	1.20	
ClCN	ClC	1.62		CN	1.15
	CN	1.16			

The way is best illustrated by an example problem.

Example problem: Consider the (linear) O $=$ C $=$ S molecule where the atomic mass numbers are $A(\text{O}) = 16$, $A(\text{C}) = 12$, and $A(\text{S}) = 32$ and $A(\text{S}) = 34$ for the two isotopes of S. The microwave spectrum of the molecule is as follows:

$J_1 \rightleftarrows J_2$		$^{16}\text{O} = {}^{12}\text{C} = {}^{32}\text{S}$	$^{16}\text{O} = {}^{12}\text{C} = {}^{34}\text{S}$
1	2	24,325.92 MHz	23,732.33 MHz
2	3	36,488.82	
3	4	48,651.64	47,462.40
4	5	60,814.08	

(a) Calculate B and D_0. (b) Calculate r_0 between C–O and C–S.

Solution: From $\Delta E = h\Delta\nu$, Eq. (10.2), and the relation between B and I, one immediately finds the quantities tabulated below.

Molecule	B (MHz)	$\mathcal{I} \times 10^{-40}$ (gr cm^2)	D_0 (kHz)
$^{16}O = {}^{12}C = {}^{32}S$	6081.480	137.9	1600
$^{16}O = {}^{12}C = {}^{34}S$	5932.843	141.4	1400

Now refer to Fig. 10.2. Let $m(^{16}O) = m_1$, $m(^{12}C) = m_2$, $m(^{32}S) = m_3$, and $m(^{34}S) = m_4$. The center of mass of the first molecule is O. $\mathcal{I}$ about O for the first molecule is

$$\mathcal{I}_O(32) = m_1 r_1^2 + m_2 r_2^2 + m_3 r_3^2 .$$

The center of mass O′ of the second molecule is shifted to the right by x because ^{34}S is heavier than ^{32}S. So, by Steiner's theorem we have

$$\mathcal{I}_O(34) = \mathcal{I}_{O'}(34) + Mx^2$$

where M is the total mass of the second molecule, and $\mathcal{I}_{O'}$ is its moment of inertia about its center of mass O′. We have

$$\mathcal{I}_{O'}(34) + Mx^2 = m_1 r_1^2 + m_2 r_2^2 + m_4 r_3^2 . \qquad (10.4)$$

We take moments with respect to O and O′:

$$m_1 r_1 + m_2 r_2 = m_3 r_3$$

$$m_1(r_1 + x) + m_2(r_2 + x) = m_4(r_3 - x)$$

whence

$$x = \frac{m_4 - m_3}{m_4} .$$

From Eq. (10.4) we have

$$\mathcal{I}_{O'}(34) = m_1 r_1^2 + m_2 r_2^2 + \left[m_4 - \left(\frac{m_4 - m_3}{m_4} \right)^2 \right] r_3^2$$

where $m_1 = 16$, $m_2 = 12$, $m_3 = 31.98$, and $m_4 = 33.97$ which are to be converted to grams by using N_A.

From $\mathcal{I}_O(32) - \mathcal{I}_{O'}(34)$ we find r_3, and by successive substitutions we eventually find $r_3 = 1.03$ Å, $x = 0.03$ Å, $r_2 = 0.52$ Å, and $r_1 = 1.68$ Å. Hence $r_0(\text{C–O}) = r_1 - r_2 = 1.16$ Å and $r_0(\text{C–S}) = x + r_3 = 1.55$ Å.

Any error is due to the inaccurate m values (A values) and the inability to measure R exactly (zero-point energy uncertainty).

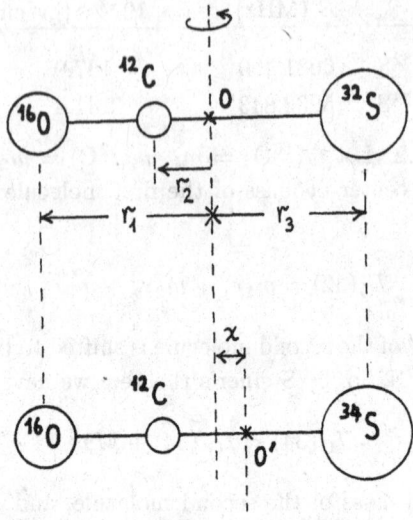

Fig. 10.2.

10.3. Rotation of Nonlinear Molecules

Nonlinear molecules can be either symmetric or asymmetric. The former have a threefold or a manifold rotational symmetry axis (symmetric tops), whereas the latter have three different moments of inertia (like, for example, CH_3ON which has no symmetry at all). Molecules having one symmetry plane (like CH_2BrCl) or two such planes (like H_2O) are also considered as symmetric tops.

With asymmetric molecules one is burdened by the trouble of having to find three moments of inertia. But if there is some symmetry, the burden is less. For example, H_2O is symmetric $(mm(C_{2v}))$ and hence only two moments of inertia are enough, if we want to find its two bond distances.

10.4. Rotation of Planar Molecules

The water molecule is planar. So, its three moments of inertia are not mutually independent. From classical mechanics we know that if I_1, I_2, and I_3, are the principal moments of inertia of any flat body (lamina), we have $I_1 + I_2 = I_3$ where these moments of inertia are taken with respect to the principal axes. The third principal axis is perpendicular to the other two which lie in the plane of the flat body (Fig. 10.3).

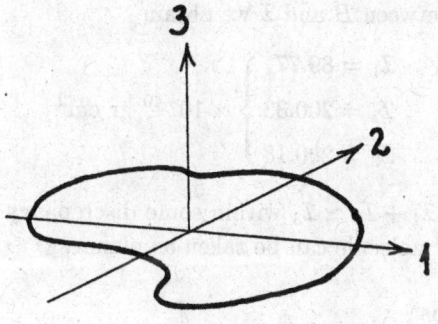

Fig. 10.3.

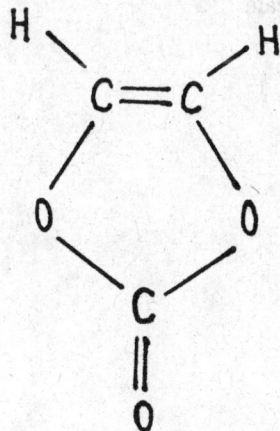

Fig. 10.4.

The microwave spectrum can help us determine whether the molecule examined is flat (planar) or not. For example, consider the vinylene carbonate molecule (Fig. 10.4). Is it flat? If we look at its spectrum we find

$$\left.\begin{array}{l} B_1 = 9346.70 \\ B_2 = 4188.46 \\ B_3 = 2891.54 \end{array}\right\} \text{MHz}.$$

From the relation between B and $\mathcal{I}$ we obtain

$$\left.\begin{array}{l} \mathcal{I}_1 = 89.77 \\ \mathcal{I}_2 = 200.33 \\ \mathcal{I}_3 = 290.18 \end{array}\right\} \times 10^{-40} \text{ gr cm}^2 .$$

We have indeed $\mathcal{I}_1 + \mathcal{I}_2 \simeq \mathcal{I}_3$ within some discrepancy (error). So, within that same error the molecule can be taken as planar.

10.5. Observations

One cannot go over this topic without remarking that the quantization of the rotational (and the other) levels reveals the existence of order and harmony in nature, just like in the case of the orbits of planets in the solar system or the intervals of a musical scale.

Chapter 11

VIBRATION-ROTATION SPECTRUM

11.1. General Theory

If the molecule did not rotate at all, we would obtain a pure vibration spectrum, because the molecule is not rigid: It pulsates as its atoms move back and forth under the regime of elastic restoring forces. During the vibration about the equilibrium position, the molecule can radiate (classically) if its one end is positively charged and the other end negatively charged. The frequency of the emitted radiation would be equal to the vibration frequency, but in reality this is not the case, because the molecule rotates at the same time. So, the total motional energy of the molecule is always $E_{\text{vibr}} + E_{\text{rot}}$, *i.e.*, the two motions occur and coexist together (if we ignore, of course, the translation of the molecule as a whole). This gives the *vibration-rotation spectrum*. The spectral line splits into two lines (the degeneracy is lifted) whose frequencies are slightly above and below the frequency of the pure vibration.[1] Rotation affects vibration and vice versa, and the two actions are coupled together.

Harmonicity breaks down if the vibration amplitude is large: Then the system is not a linear harmonic oscillator any more. (This is true also for the simple pendulum in freshman physics.) However, lack of harmonicity does not mean lack of periodicity. If the motion is still periodic, then one can do a Fourier analysis like in acoustics: One analyzes the emitted field into components (wavetrains) whose frequencies include the fundamental vibration

---

[1]This phenomenon happens whenever two systems or motions (or general actions) are coupled together (*e.g.*, in oscillators, tuning forks etc., even in the Zeeman effect where an external magnetic field changes the electronic energies in an atom).

frequency and its harmonics. Each component of this whole set is going to be further split by the rotation.

This situation gives the vibration-rotation spectrum which is in the infrared region. A representative example is the near-infrared absorption spectrum of HCl. Its shape (graph of the intensity of the lines vs. ν) allows for interesting interpretations and conclusions.[2]

As a simple linear harmonic oscillator to deal with, one considers a diatomic molecule[3] whose atoms can vibrate, *i.e.*, execute linear harmonic motion under the influence of a force that holds them and varies linearly with the separation distance. The force is derivable from a potential $V(r)$ whose plot is a parabola, if the motion is truly harmonic. This force is exerted by one atom on the other. The slope of the parabola at any point r gives the value of the force at that point. We can take $V(r \to \infty) \to 0$, as usual. At some r_0, $V(r = r_0)$ is minimum, and thus r_0 is the equilibrium separation of the atoms (at which the atoms *would* be at rest *if* there *were* no vibrations). For $r < r_0$ the force becomes strongly repulsive. For $r > r_0$ it is attractive. So, the force tries always to restore equilibrium by calling the atoms back as they depart from the equilibrium configuration either by stretching or contraction of the molecule along r. If the molecule somehow gains an energy larger than $V(r_0) \equiv V_0$, it dissociates. This simple description (model) satisfies the gross molecular behavior, and is depicted in Fig. 11.1 where the parabola is shown dashed.

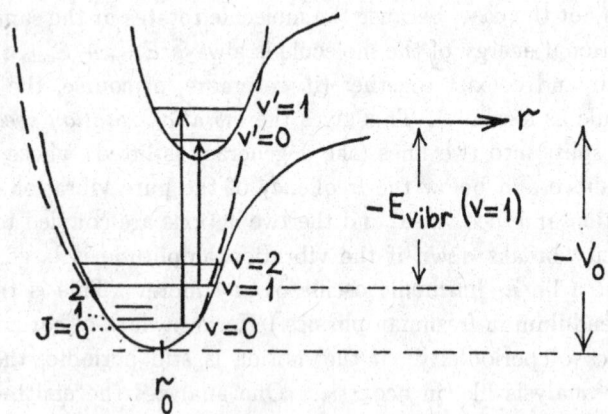

Fig. 11.1.

[2]E.S. Imes, *Astrophys. J.*, **50**, 251 (1919).
[3]It is represented by the model of the rigid rotor whose exact solution (for the energies) is known in quantum mechanics.

This description may make one think that the accessible energy to the molecule is continuous, but this is not the case. The vibrational energies are quantized according to the vibrational quantum number v, and their stepwise structure is shown in Fig. 11.1. Each vibrational level has a fine structure, *i.e.*, it consists of minute rotational levels. So, each vibrational state is in fact degenerate because of the (presence of the) rotation.

The lowest vibrational energy is nonzero; that is, the molecule still vibrates (and r is not exactly constant) in its ground vibrational state. So, r_0 is just the *mode* (most probable value of the trend) of r.[4]

The vibrational states are enumerated by v (v = 0, 1, 2, ...). These energies are negative. (The zero energy corresponds to a dissociated molecule with its atom at rest at $r = \infty$.) Therefore, the depth of the parabolic potential well, *i.e.*, $-V_0$, is just the *dissociation energy*. When the molecule is at some state v, its dissociation energy is then $-E_{\text{vibr}}(v)$, *i.e.*, the interval between that state and the zero level (see Fig. 11.1 for v = 1). Notice that for any v, $E_{\text{vibr}}(v) < V_0$.

Whether v takes on finite or infinite values depends on whether the shape of the curve of $V(r)$ allows room for it, *i.e.*, it depends on the curve itself.

Why do we not use *negative* values for v? Actually we could — there is no essential reason prohibiting that — but it would be very awkward and impractical to start enumerating the states, say, from −100 instead of 0. So, it depends on how we want to mark (label) the lowest-energy state and then continue upwards.

As long as v is small (first few vibrational levels), $V(r)$ is a parabolic curve, corresponding to a linear harmonic oscillator. This is apparent from the expansion of $V(r)$ about $r = r_0$:

$$V(r) = V_0 + \left.\frac{dV}{dr}\right|_{r_0}(r - r_0) + \frac{1}{2}\left.\frac{d^2V}{dr^2}\right|_{r_0}(r - r_0)^2 + \frac{1}{6}\left.\frac{d^3V}{dr^3}\right|_{r_0}(r - r_0)^3 + \text{h.o.t.}$$

where the second term is zero because $dV/dr = 0$ at $r = r_0$. Notice also that $V(r = r_0) = V_0$. The first and third terms together correspond to a linear harmonic oscillator where the quantity (coefficient)

---

[4]From the statistical theory we know that the mode is (necessarily) equal to the mean (average) value only if the distribution is too narrow (they coincide). For a wide distribution, mode $\neq$ mean. The mode is obtained by setting the derivative of the distribution equal to zero and obtaining the extremum, whereas the mean is the first moment of the distribution. The mode is the value which most of distributed elements prefer to take, and hence they throng to accumulate about it. This includes the fashion trends (in French, *mode*) that concern elegant ladies, too!

$$\kappa^* \equiv \frac{d^2 V}{dr^2}\bigg|_{r=r_0}$$

is called *effective spring constant* (*i.e.*, it corresponds to κ of a real spring as if it were actually present in the system).

In the vicinity of r_0 the curve is parabolic and the system behaves like a simple harmonic oscillator whose vibrational energies are given by Eq. (9.3) where ν_0 is the *ground-state vibration frequency* given by

$$\nu_0 = \frac{1}{2\pi}\sqrt{\frac{\kappa^*}{\tilde{\mu}}}$$

where $\tilde{\mu}$ is the reduced mass of the vibratory system, *i.e.*, the molecule.

The value v = 0 gives the zero-point energy. The relation between the *spectroscopic dissociation energy* D_e and the *chemical dissociation energy* E_0 is

$$D_e = E_0 + \frac{1}{2}h\nu_0$$

where the second term is the zero-point energy. Hence D_e is just V_0, the minimum chemical binding energy (*real* binding energy), and E_0 is $E(v = 0)$, the binding (or dissociation energy) due to linear vibrations (from the lowest vibrational state to $V = 0$).

To obtain a spectrum, the selection rule is that the vibration must change the dipole moment of the molecule, *i.e.*, $\Delta v = \pm 1$.

In the spectrum of a diatomic molecule rarely does one see only vibration frequencies (lines). The spectrum usually shows rotation lines, too, together with the vibration lines, because the molecule vibrates and rotates simultaneously, so that the two actions are not separated. We can tell one from the other from the values of the frequencies. At room temperature, the molecule is in excited rotational states. As the molecule rotates *and* vibrates, these two actions interact and are coupled together, one affecting the other in a vicious circle. (See Eq. (4.7), last term.)

As the molecule gets to higher v states, the $V(r)$ curve deviates from an ideal parabola and takes the shape shown in Fig. 11.1 by a solid line. So, the v states get closer and closer to each other, and their corresponding energies also deviate from those of an harmonic oscillator (because higher-order terms in the expansion of $V(r)$ gain importance and start making themselves felt). As a result of this, the selection rule $\Delta v = \pm 1$ becomes $\Delta v = 0, \pm 1$.[5]

---

[5]Transitions with $|\Delta v| > 1$ are possible, but not very probable.

If we disregard electronic transitions, the overall energy will be $E = E_{vibr} + E_{rot}$, *alias vibration-rotation levels.* Since $Bh \ll \Delta E_{vibr}$, the pure rotational levels appear as a dense bunch of lines for each vibrational level. (Fig. 11.1).

Molecular spectra consist of vibration-rotation *bands.* They are due to transitions between vibrational levels mainly, plus occasional transitions between (pure) rotational levels. Since B is small, the lines are too closely spaced, so as to make the spectrum appear continuous, and hence the name *band.* If we increase the resolving power of the spectrometer, we see that the bands are not continuous, but are rather bunches of lines.

The selection rule for rotational transitions is $\Delta J = \pm 1$ for a diatomic molecule, and $\Delta J = 0, \pm 1$ for polyatomic molecules.

If ν_0 is the pure axial vibration frequency and ν_{rot} the rotation frequency, one sees ν_0 and a series of lines $\nu_0 \pm \nu_{rot}$ on either side of ν_0 in the spectrum (Fig. 11.2). The P and R-branches look continuous if the spectrometer is crude, but they actually consist of many lines. (The continuity comes partly from an averaging over the frequencies of many molecules). At ν_0, the intensity $I \to 0$.

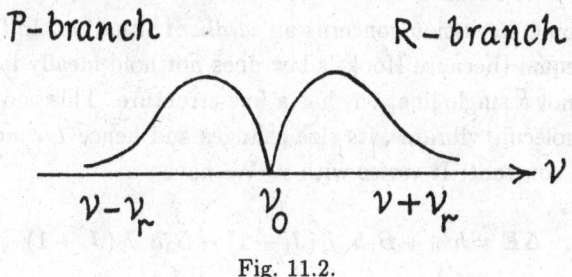

Fig. 11.2.

From the equation $E = E_{vibr} + E_{rot}$ we have for a transition between $v \rightleftarrows v'$ and $J \rightleftarrows J'$:

$$\Delta E = h\nu_0(v - v') + Bh[J(J+1) - J'(J'+1)]$$

where $\Delta v = v - v' = 1$ and $\Delta J = 0, \pm 1$ (selection rules). We encounter three possibilities:

(1) If $\Delta J = J - J' = 1$, we have $J' = J - 1$ and hence $\Delta E = h\nu_0 + 2BhJ$, with $J = 1, 2, 3, \ldots$ (but *not* 0).

(2) If $\Delta J = J - J' = -1$, we have $J = J' - 1$ and hence $\Delta E = h\nu_0 - 2BhJ'$, with $J' = 1, 2, 3, \ldots$.

(3) If $\Delta J = J - J' = 0$, we have $J = J'$ and hence $\Delta E = h\nu_0 \equiv \Delta E_{v,v'}$ (pure vibrational transition).

In general, $\Delta E = h\nu_0 + 2Bhm_J$, with $m_J = 0, \pm1, \pm2, \dots$. For $m_J = 0$ we have pure vibration (without any rotation at all) and $\Delta E = h\nu_0$, *i.e.*, ν_0 is observed in the spectrum as the Q-line in the middle (between the P and R-branches).

In diatomic molecules (except for NO), the Q-line at the center of the band is *not* observed, because for diatomic molecules we have $\Delta J = \pm1$ only, and hence, possibility (3) above cannot occur. Therefore, there is no line for pure vibration. The closest adjacent lines to it will be $\pm2Bh$. The vibration-rotation band of a diatomic molecule is thus equally spaced in frequency, and the spacing is $2B$.

To observe the Q-line, there must be a dipole moment along the molecular axis; and among the stable diatomic molecules only NO does have such a dipole. Since $\Delta J = 0$ is possible with polyatomic molecules, the Q-line is observed in their spectra.

For $m_J > 0$ we obtain a series of equally spaced lines (R-branch) above ν_0, and for $m_J < 0$ we obtain a symmetrical series (P-branch) below ν_0.

The foregoing discussion concerns an *idealized* situation. In fact, the spacings are *not* equal (because Hooke's law does not hold ideally in reality), and the Q-line is not a single line, but has a fine structure. This is due to the fact that as the molecule vibrates, its size changes and hence $\mathcal{I}$ changes. Thus B does not stay constant: It varies with v. We have

$$\Delta E = h\nu_0 + B_1 h \, J_1(J_1 + 1) - B_2 h \, J_2(J_2 + 1)$$

where B_1 and J_1 (B_2 and J_2) correspond to the upper (lower) state. If we let $J_1 - J_2 = -1$, we obtain for the P-branch:

$$\Delta E = h\nu_0 - h(B_1 + B_2)J_2 + h(B_1 - B_2)J_2^2 \qquad (11.1)$$

with $J_2 = 1, 2, \dots$. If we let $J_1 - J_2 = +1$, we obtain for the R-branch:

$$\Delta E = h\nu_0 + h(B_1 + B_2)J_1 + h(B_1 - B_2)J_1^2 \qquad (11.2)$$

with $J_1 = 1, 2, \dots$. And if we let $J_1 - J_2 = 0$, we obtain for the Q-branch:

$$\Delta E = h\nu_0 + (B_1 - B_2)J_1 + h(B_1 - B_2)J_1^2 \qquad (11.3)$$

where J_1 is just J, and $J_1 = J_2 \equiv J = 0, 1, 2, \ldots$.[6]

Since B_1 and B_2 belong to different electronic states, we have $B_1 \neq B_2$, so that $\mathcal{I}_1 \neq \mathcal{I}_2$ (and the wavefunctions and interatomic forces are different in the two states, too). In the vibration-rotation bands the variation of B from a level to the next is usually slight, but even there (*i.e.*, even within the same electronic level, with $\Delta n = 0$), excited rotational states sometimes have considerably larger $\mathcal{I}$ (and hence smaller B) values (because the molecule stretches out as it rotates faster). So, the last terms in Eqs. (11.1), (11.2), (11.3) grow significant, and as J increases, the trend of $\bar{\nu}$ in a branch gets reversed, thereby showing a *head*. This happens in the P-branch for $B_1 > B_2$, and in the R-branch for $B_1 < B_2$. So, one branch folds back upon itself, and its tail overlaps with the other branches, and hence its lines mix with those of the other branches in the spectrogram. The position in the spectrogram where a branch turns back is called *band head*. Here there is a crowding of lines, and the rest of them shade away from this point, towards the high or low frequencies, depending on the band. So, the lines of the R-branch may get on the top of the lines of the P-branch. Surprisingly, a Q-branch (if present) may also show a head (though Eq. (11.3) does not seem to mathematically provide for such a turnout).

The situation is shown in Fig. 11.3 where we have two Fortrat diagrams, *i.e.*, we plot the discrete values of J (on the ordinate axis) versus ν (on the abscissa axis). If $B_2 > B_1$, the last term in Eqs. (11.1), (11.2), (11.3) is negative. In the R-branch this term overtakes the second term (because $J_1^2 > J_1$) as J_1 increases, and hence ΔE shows a trend to decrease (whereas it was increasing). The R-branch shows a head towards higher frequencies. As the rotational quantum number increases, the lines of the R-branch intermingle with those of the other two branches towards lower frequencies. As the rotational

---

[6]This analysis may at a first glance look tiring and apt to confuse and discourage people, making them think that a physical theory is messy and junky. Although, admittedly, there is some vanity of physicists involved here, to make others say "How can you understand these things?", we should make it clear that things in nature are so hopelessly complex, that there is no any easier way to describe them on a piece of paper. One has to devise and set up such delicate methods, in order to express complex entities like the structure of matter or how nature works. For this reason, there is precious wisdom here as much as there is in the Holy Scriptures or in ancient Greek tragedies. Further, similar scholastic and minute analyses are made in all sciences to explain details as better as possible and in a gnat-refining way, like in political analysis, strategic studies, art criticism, economics, sociology, psychology, medicine, theology, military intelligence, policy planning, computer science, theory of action, athletism, *savoir vivre*, hobbies etc.

quantum number increases, the lines of the R and Q-branches (and a bit of the
P-branch) spread out in the direction of lower frequencies (Fig. 11.3(a)).

All these considerations hold true also for linear polyatomic molecules.

It may sound strange, but the case $\mathcal{I}_1 < \mathcal{I}_2$ ($B_1 > B_2$) is also possible,
because if the transition is electronic ($\Delta n \neq 0$), an upper electronic state
may have a *lower* $\mathcal{I}$. In that case, the appropriate Fortrat diagram is that in
Fig. 11.3(b). Here the P-branch shows a head towards lower frequencies. The
head lines are on the low-frequency side of Q, and can be easily distinguished.

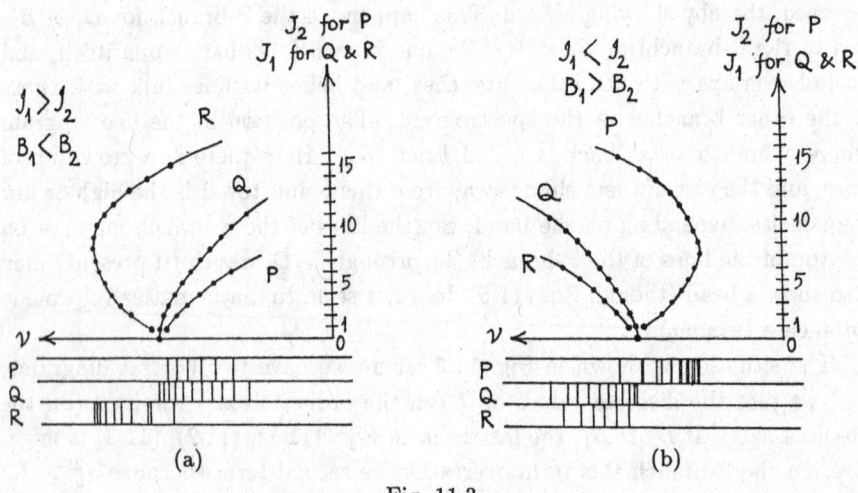

(a)   (b)

Fig. 11.3.

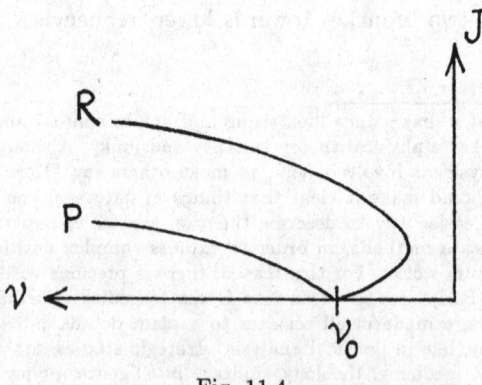

Fig. 11.4.

Towards higher frequencies the lines get weaker in intensity and spaced out. The Q and R-branches have lines which get scarce towards higher frequencies.

What we get in the spectrogram is the superposition of the three branches, one atop of the other. The resulting spectrum is thus complicated, and it takes an experienced spectroscopist to interpret it.

Figure 11.4 shows another Fortrat diagram where the Q-branch is missing.

11.2. The Most General Case

In the most general case, a transition involves a change in the electronic state *and* a change in the vibration-rotation state. Actually, this is the case in most molecular band spectra. (The case of $\Delta n = 0$ is special.) Hence the electronic bands are observed in the visible and ultraviolet regions.

In a transition $n \rightleftarrows n'$, $v \rightleftarrows v'$, if these quantum numbers are fixed, J may change to give rise to a *single band*. If we consider all the changes of v within a fixed and given $n \rightleftarrows n'$, then we obtain a *band system*. And if we consider all possible $n \rightleftarrows n'$ transitions, we obtain a *band spectrum* (for a molecule) which is formed by many band systems.

Sometimes we use K instead of J for the rotations, where K is the projection of J on the molecular axis. We may have $K \neq J$. In any case, the selection rule is $\Delta K = 0, \pm 1$ *and* $\Delta J = 0, \pm 1$. If $K = J$, the former relation lapses to the latter. In the most general case we have $K \neq J$ and hence a fine structure. Apart from that, the full bands (that include electronic jumps, too) are still more complex compared to plain vibration-rotation bands, because $\Delta K = 0$ (and $\Delta J = 0$) is also possible.

Again we have three branches:

(1) The P-branch: ΔK and Δv have opposite signs.
(2) The R-branch: ΔK and Δv have the same sign.
(3) The Q-branch: $\Delta K = 0$ (or $\Delta J = 0$, if $K = J$). This branch does not exist if the transition is within the same molecular electronic level (*e.g.*, within the Σ state).

If K and J are not identical quantities, we have a fine structure, *i.e.*, a single band consists of many branches of each category.

If we include electronic transitions, we should rewrite Eqs. (11.1), (11.2), (11.3) as follows: The total molecular energy is

$$E = E_{\text{el}} + E_{\text{vibr}} + BJ(J+1).$$

In getting from the unprimed electronic state to the primed electronic state we have

$$E = E_{\text{el}} + E_{\text{vibr}} + BJ(J+1) \rightarrow E' = E'_{\text{el}} + E'_{\text{vibr}} + B'J(J+1).$$

For the P-branch we have (v, $J-1 \rightarrow$ v', J) and

$$\Delta E = \Delta E_{\text{el}} + \Delta E_{\text{vibr}}(\text{v, v}') - (B+B')J + (B-B')J^2. \qquad (11.4)$$

For the R-branch we have (v, $J \rightarrow$ v', $J-1$) and

$$\Delta E = \Delta E_{\text{el}} + \Delta E_{\text{vibr}}(\text{v, v}') + (B+B')J + (B-B')J^2. \qquad (11.5)$$

For the Q-branch we have (v, $J \rightarrow$ v', J) and

$$\Delta E = E - E' = \Delta E_{\text{el}} + \Delta E_{\text{vibr}}(\text{v, v}') + (B-B')J(J+1). \qquad (11.6)$$

J belongs to the larger of the two values involved in the transition. If $K \neq J$, then we must write K instead of J in these equations (unless, of course, the levels have no K, in which case we write J).

The corresponding frequency for the transition is $\nu = \Delta E/h$.

How large is the number of single bands included in a band system? It may be quite large. The selection rule of an harmonic oscillator, $\Delta \text{v} = \text{v} - \text{v}' = \pm 1$ breaks down and includes higher values than $|\pm 1|$, but not too high (because the probability of such transitions is low).

We suggest the book by G. Herzberg, *Spectra of Diatomic Molecules*, Van Nostrand, Princeton, N.J. (1950), where the reader will find the Swan bands of C_2 (of the electronic molecular transition $^3\Pi \rightarrow {}^3\Pi$) and the bands of CN (violet, $^2\Sigma \rightarrow {}^2\Sigma$; and red, $^2\Pi \rightarrow {}^2\Sigma$).

11.3. Anharmonicity

Real molecules are not simple harmonic oscillators. So, one must include a correction term in E_{vibr}, coming from the perturbation theory. As we move to higher v values (close to dissociation), the spacing of the vibrational levels decreases (something that can be seen experimentally). The perturbative correction term is quadratic in v and negative, so as to decrease E_{vibr}. We have

$$E_{\text{vibr}} = \left(\text{v} + \frac{1}{2}\right)h\nu_0 - \left(\text{v} + \frac{1}{2}\right)^2 h\nu_0 \alpha$$

where the second term is the correction to E_{vibr}, and α is a positive constant called *anharmonicity coefficient*, namely the perturbation strength. (See Eq. (4.7).) Anharmonicity is represented by this quadratic term and causes deviation from the behavior of an ideal harmonic spring.

At high (excited) vibration energies (where the levels are closely spaced) the selection rule $\Delta v = \pm 1$ becomes $\Delta v = \pm 1, \pm 2, \pm 3, \ldots$ and harmonics (overtones) of vibrations are included in the spectrum.

From the measurement of $\Delta \bar{\nu}$ in the spectrum, one can determine α.

With polyatomic molecules we obtain spectral bands which — apart from the overtones — include many different vibrations (and combinations thereof), making the spectrogram too complicated and perplexing. This is inevitable and maybe frustrating, but the advantage of a thorough analysis is the fact that one gets a better picture of what is going on.

11.4. Vibration of Polyatomic Molecules

A polyatomic molecule can be represented by an (ideal) model of point masses (or small balls) coupled by flexible springs. The vibrations of such a system are too complex to be written down and described mathematically (even by Lissajous figures). Fortunately, one can resort to the analysis of the motion into *normal modes* (which decouple the motion and present the moving masses as though independent). That is, one regards the motion as consisting of simple components which are few in number, but form the resultant motion consistently. Each such mode of vibration is expressible in *normal coordinates* that describe the vibrational motion of the balls (atoms).

A polyatomic molecule with N atoms has three degrees of freedom *for each atom* (because every atom can vibrate along the x, y, and z-axis). So, the total number of the molecular degrees of freedom is $3N$. Three out of them belong to the global translation of the molecule in space, along the x, y, and z-direction (and hence are not vibration-related). Another three belong to the rotation of the molecule about the three Cartesian axes. So, there remain $3N - 6$ vibration modes (per molecule). Now if the molecule is *linear*, there are $3N - 5$ modes because the motions along the two axes perpendicular to the molecular axis are identical. (Vibration off the axis is a separate mode and means bending.) So, if the molecule is triatomic, it has $3 \times 3 - 5 = 4$ modes (shapes, ways) of vibration (of which two are off its axis and identical). If it is diatomic, it has $3 \times 2 - 5 = 1$ mode of vibration (just stretching and contracting along its axis).

Sometimes we may have degenerate (identical) modes which reduce the total number of the observed modes. For example, in CH_4 (five atoms) there must be $3 \times 5 - 6 = 9$ modes, but only four are observed, because the molecular symmetry degenerates the rest (and so we get two triply degenerate, one doubly degenerate, and one nondegenerate modes).

In fact, a molecule with a rotational symmetry (threefold or more) has degenerate modes. For instance, AB_2 type bent molecules like H_2O belong to the symmetry group $mm(C_{2v})$, while CAB_3 type molecules (like CH_3Cl) which belong to $3m(C_{3v})$ and are tetrahedral, have three doubly degenerate and three nondegenerate modes. So, symmetry puts a limit on the observable modes. This is beneficial to us because we can go the other way around and find the symmetry by tracing down the different vibration modes. Here is where the merit of spectroscopy comes into play. However, one should be very cautious because sometimes some frequencies are missing (because of their weak intensity or their position in the electromagnetic spectrum outside the region examined). Also, sometimes some other frequencies are by chance degenerate (accidental degeneracy) and overlapping, so that two frequencies may be counted as one by deception.

11.5. General Remarks

(1) The number of normal modes in a molecule depends on the number of its atoms.

(2) The degeneracy depends on symmetry (which is described by the symmetry groups. (See Chapter 2.))

(3) All vibrations causing a change in the electric dipole moment of the molecule give infrared spectra. However, the infrared region is difficult to work in. Spectroscopists prefer the microwave region for pure rotational spectra (and thus they sensitively measure molecular parameters and obtain information about dipole moments, quadrupole moments, and nuclear spins). Since it is possible to work with microwaves coming from monoenergetic sources (microwave generators), the resolution is good and hence the measurements are accurate.

(4) All vibrations causing a change in the polarizability of the molecule give Raman spectra.

(5) Vibrations that are not characterized by (3) and (4) above, do not give infrared or Raman spectra.[7]

(6) If a molecule has different isotopes of the same atom, it gives a slightly different band spectrum for each case with a different isotope. So, if the specimen consists of a mixture of molecules with different isotopes, the band

---

[7]Many known molecules are tabulated in terms of their normal modes, degeneracies, and spectra. If one knows the chemical composition (formula) of a molecule and if the molecule gives an observable infrared or Raman spectrum with detectable fundamental vibration frequencies, one can find the symmetry by just looking it up from the tables.

spectrum presents extra lines which can help us discover rare isotopes.

(7) When we have a band of frequencies, we find out the ones that correspond to the normal frequencies (fundamental vibration frequencies) by following the criteria stated below:

(a) Knowledge of modes beforehand (theoretically) and their activity in the infrared or Raman region.

(b) The fundamental frequencies are more intense than the harmonics, combinations, or differences.

(c) The bending vibration frequencies are lower than the stretching vibration frequencies.

(d) Band contour and general trends and appearance of the band (*if* it is not complicated by the appearance of rotation fine-structure lines).

(8) If we have a spectrometer with a low dispersion, we can achieve the inclusion of a wide spectral range in the spectrogram, but not without a cost: A band may get so much squeezed as to appear as a single line.

11.6. The CO_2 Molecule

The CO_2 molecule is linear, with four normal modes. The corresponding frequencies are ν_1, ν_2, ν_3 (Fig. 11.5). Notice that ν_2 is degenerate (the same motion in two directions, both lateral and perpendicular to the molecular axis — and to each other). The molecule has a symmetry and belongs to $\infty/mm(D_{\infty h})$. Purely speaking, the word *lineal* would be more advisable than the word *linear*.

The measured normal (fundamental vibration) frequencies of CO_2 are $\nu_1 = 1340$ Hz, $\nu_2 = 667$ Hz, and $\nu_3 = 2349$ Hz. In the infrared region there appears a whole bunch of vibration bands. Of the many frequencies present these three values are found to be the normal-mode frequencies. (See (7) in the previous section.) The main mission of the spectroscopist here is to find which vibration corresponds to which modes. We first find out the number of vibration modes. We next predetermine or guess which of them are infrared-active and which ones Raman-active. Then we proceed experimentally (see (7a) and (7b) above) to observe the fundamental vibrations.

The ν_1 mode is symmetric, and hence it does not change $\mathfrak{p}$ (because a stretch is compensated by a contraction, and the net result is nil). So, ν_1 does not appear in infrared spectra, nor do *any other frequencies of such symmetric modes*.

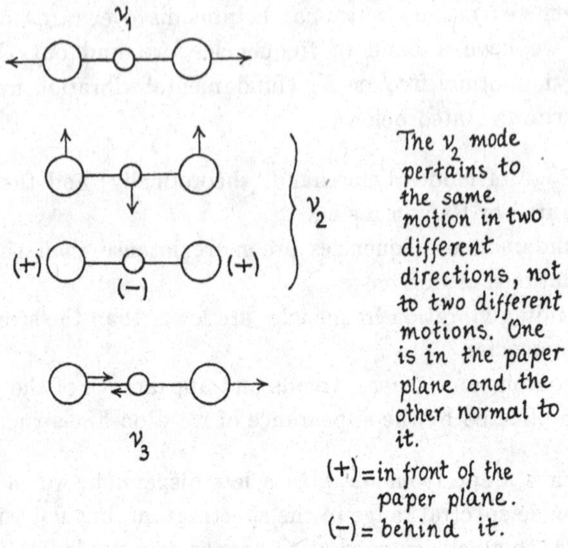

ν_1

ν_2 — The ν_2 mode pertains to the same motion in two different directions, not to two different motions. One is in the paper plane and the other normal to it.

(+) = in front of the paper plane.
(−) = behind it.

ν_3

Fig. 11.5.

The ν_2 mode describes a bending during vibration, and is degenerate in two directions normal to each other. One motion is in the plane of the paper, and the other perpendicular to it, but identical. In this case, as the molecule bends, we do have formation of a net p that changes between two values, from $-p_0$ to $+p_0$, and passes through zero for a moment. This changing p can interact with an external (incident) varying $\mathcal{E}$ of the same frequency. Thus we have an absorption that appears as ν_2 in the infrared spectrum. We say that ν_2 is *active* in the infrared region (whereas ν_1 is *inactive* in the same region).

The ν_3 mode is a linear asymmetric vibration that produces a p (at the two extreme phases of the vibrational motion). Thus ν_3 participates (and is active) in the infrared spectrum.

How about Raman spectra? Does the polarizability α change? The ν_1 mode does change it (by the stretch and the contraction).[8] So, ν_1 is active in the Raman spectrum. Since α is a scalar (matrix element), ν_2 does not change α at the two extreme phases of the motion. The same is true for ν_3. So, ν_2 and ν_3 are inactive in the Raman spectrum.

This is an example for the *principle of mutual exclusion* stated as follows: *If a molecule is symmetric about an axis* (has a symmetry axis), *whichever*

---

[8] A long molecule has a greater α.

vibration frequency is active in the infrared region is inactive in the Raman region, and vice versa.

In the spectrum of CO_2 no Q-line is observed in the ν_3 mode because the change of $\mathfrak{p}$ is along the molecular axis (Fig. 11.6(a)). J and v do not change simultaneously. In the ν_2 mode, the change of $\mathfrak{p}$ is perpendicular to the molecular axis as the molecule vibrates. So, $\Delta J = 0, \pm 1$, and a Q-line appears (Fig. 11.6(b)).

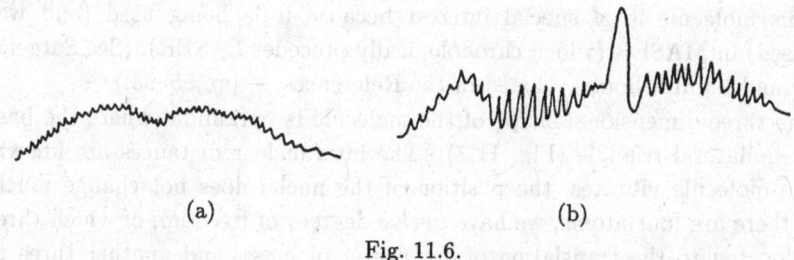

(a)   (b)

Fig. 11.6.

It should be mentioned that ν_1 consists of *two* Raman lines in fact, because the first harmonic of ν_2 is almost equal to ν_1: $2\nu_2 \simeq \nu_1$. This phenomenon is called *Fermi resonance*: The line splits into two (1286 Hz and 1388 Hz). This may confuse the observer, and the production of such additional lines corresponding to unexpected normal (fundamental) modes may be held responsible for the overcounting of the normal frequencies. So, the spectroscopist must be careful. He or she should check for any Fermi resonances.

The simultaneous appearance of the vibration-rotation spectrum, *i.e.*, the appearance of the rotational fine structure as well, mixes up things. A quick glance at the spectrum is enough to observe that the spacings of the fine rotational lines (in the P and R-branches) are all equal, which means that the molecule has only one $\mathcal{I}$ — and it should, since it is linear. A closer inspection reveals the fact that these spacings are wider — actually twice as wide — than the expected. Why? Because of the differing statistical weights of odd and even J in molecules having the ∞/mm symmetry. That is, in such molecules only even J values are observed. Whereas in a common rotation spectrum the selection rule is $\Delta J = \pm 1$, for Raman lines of pure rotation we have $\Delta J = \pm 2$, and so the lines are spaced twice as wide apart as are the lines of the rotation spectrum itself.[9] (This is in contrast to the case where the statistical weights are the same in molecules having the $\infty m (C_{\infty v})$ symmetry.)

---

[9]Since the statistical weights depend on spin, this situation can help us find the symmetry. For example, we can explain why the structure of the N_2O molecule is N-N-O instead of N-O-N.

We realize that the band spectrum of a polyatomic molecule is not simple, because several vibrational modes can be excited, and if oscillating electric dipoles are involved, one obtains vibration-rotation bands.[10] As an example we have just considered the CO_2 spectrum. Let us now discuss another example, the ammonia molecule in gaseous state.

11.7. The NH_3 Molecule

This molecule is of special interest because it is being used (and was first used) in MASER (which chronologically precedes LASER). (See Sargent, Scully and Lamb's book — listed in the References — pp. 55–63.)

The three-dimensional shape of the molecule is pyramidal where the base is an equilateral triangle (Fig. 11.7). The internuclear distances are known. As the molecule vibrates, the position of the nuclei does not change much. Since there are four atoms, we have twelve degrees of freedom, of which three are allocated to the translation of the center of mass, and another three to rotational motion. Out of the remaining six, only four are obtained (because of the symmetry degeneracy): ν_1, ν_2, ν_3, ν_4. Out of these, ν_1 and ν_3 correspond to the symmetric motion of the nuclei with respect to the axis of the pyramid. There are two modes for ν_2 *and* two for ν_4. The total number of modes is six.

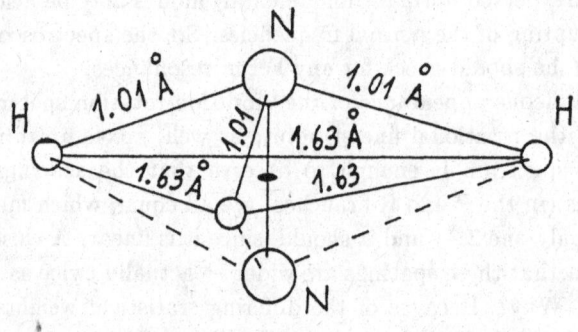

Fig. 11.7.

As the system vibrates, the N atom can be found on either side of the triangular base of the pyramid, giving rise to a mirror symmetry with respect to this base. Classically, one cannot tell one configuration from the other, because the two situations are indistinguishable. But in quantum mechanics they can be distinguished from each other. If z is the distance of the N nucleus

---

[10]The first description was made by Hund in 1927.

from the base of the pyramid, then the potential $V(z)$ of the molecule has two minima corresponding to the two possible stable positions of N, as the latter vibrates along z. Between the two potential wells there is a barrier (Fig. 11.8). The two minima are the equilibrium positions of N (with respect to the base). This consideration is good so far, provided that the H atoms (actually, protons) do not move. But in fact they do move, lowering the potential barrier between the two minima. Classically, the N nucleus is inside either of the two wells (and stays there forever, as long as the barrier is rigid, just like a marbleball in an unshaken cup — where the wall of the cup constitutes a gravitational potential barrier for the marbleball which does not have enough kinetic energy to overcome it and get out). But in quantum mechanics it is *possible* to find N tunneling through the barrier. In other words, there is a nonzero chance $|\psi|^2$ — however slim — that N is, say, at $z = 0$, *i.e.*, in the plane of the base or slightly off it.

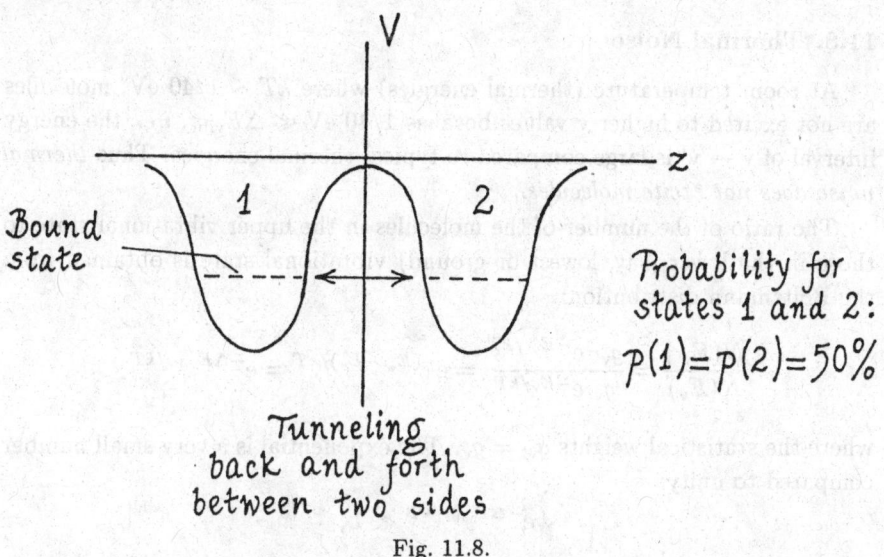

Fig. 11.8.

If the barrier tends to disappear, $V(r)$ tends to take the shape of an harmonic oscillator potential, and the energies become more and more distinct (like the energies of the oscillator). But if the barrier is high and rigid, they pair off without exact correspondence, because a small nonzero part of ψ gets into the barrier to give a nonzero penetration probability in there.

This phenomenon is called *ammonia inversion*. It gives a special spectrum where every vibration-rotation state is doubled, so that we have double vibration-rotation lines in the band. The spacing $\Delta\bar{\nu}$ of the doublet is equal to the spacing of the initial doublet state plus the spacing of the final doublet state. This doubling occurs also in the pure rotational states. Relevant measurements were taken by Dennison in 1932, and later by Randall and Wright.

The transition between the lowest two levels is called *inversion line* ($\bar{\nu} = 0.66$ cm^{-1}). One of them is the ground level. The line was studied by Williams and Cleeton in 1934. Twelve years later, Good found that the inversion line had a fine structure (30 lines over a width of 0.2 cm^{-1}) due to pure rotations which produced radiative transitions. Good was able to further discover that there was a hyperfine structure, too, as a result of a nonzero quadrupole moment in the ^{14}N nucleus, changing the nuclear electric field. (The quadrupole moment is zero in the isotope ^{15}N. So, NH$_3$ molecules with that isotope do not exhibit hyperfine structure.) This is a case where experimental measurements and theoretical calculations go together.[11]

11.8. Thermal Noise

At room temperature (thermal energies) where $kT \sim 1/40$ eV, molecules are not excited to higher v values because $1/40$ eV $\ll \Delta E_{\text{vibr}}$, *i.e.*, the energy interval of v $\rightarrow$ v$'$ is large compared to typical thermal energies. Thus *thermal noise does not excite molecules.*

The ratio of the number of the molecules in the upper vibrational state to those in the lower (say, lowest or ground) vibrational state is obtained from the Boltzmann distribution:

$$\frac{N(E_{\text{v}'})}{N(E_{\text{v}})} = \frac{g_{\text{v}'}\, e^{-E_{\text{v}'}/kT}}{g_{\text{v}}\, e^{-E_{\text{v}}/kT}} = e^{-(E_{\text{v}'} - E_{\text{v}})/kT} = e^{-\Delta E_{\text{vibr}}/kT}$$

where the statistical weights $g_{\text{v}'} = g_{\text{v}}$. The exponential is a very small number compared to unity:

$$e^{-\Delta E_{\text{vibr}}/kT} \ll 1,$$

so that almost all molecules are in v $= 0$ at room temperature. This means that at room temperature any transition is within the ground vibrational state (v $= 0 \rightarrow$ v$' = 0$), namely purely rotational.

---

[11]The quadrupole moment splits the energies, but the splitting is different for every different interaction of the nuclear spin with the spins of other nuclei. Also, note that a quadrupole effect occurs only if $I > 1/2$ for the nucleus. For N-14 we have $I = 1$ and so we have splitting. For N-15 we have $I = 1/2$ where the two opposite possibilities (orientations) cancel each other (in terms of Q).

Now if we compare ΔE_{rot} with kT we see the following: Each rotational state has a statistical weight $2J + 1$ (because $-J \leq m_J \leq J$). So, the ratio of the molecules in J' to those in J is

$$\frac{N(J')}{N(J)} = \frac{2J' + 1}{2J + 1}\, e^{-\Delta E_{\text{rot}}/kT}$$

where $\Delta E_{\text{rot}} = Bh[J'(J' + 1) - J(J + 1)]$. This exponential is again less than unity; it means that states with J' are encountered (occur) $N(J')/N(J)$ times as often as states with J, and hence the spectral line intensities are weaker by this proportion, for transitions $J \rightarrow J'$.

There are, however, exceptions. For example, in I_2 we find many molecules in excited vibrational states at room temperature. This affects the infrared bonds.

11.9. Dissociation Energy

The molecular energy when the molecule is in a bound state is lower than the energy of the separate atoms resting at infinity; otherwise the molecule would not exist. So, if the molecule acquires enough energy somehow from the environment, it dissociates all by itself. The energy that must be given to the molecule (when it is in its ground state) so that it dismisses its atoms and sends them to rest at infinity, is the *dissociation energy* (or *heat of dissociation*) of the molecule, denoted by D_e.

This automatically means that any transitional ΔE in the spectrum is less than D_e, *i.e.*, any quantum corresponding to the difference between two states, namely to an interstate transition, cannot exceed D_e. For example, for HCl we have the following accounting arithmetic:

Action	Heat needed
Dissociation $2HCl \rightarrow H_2 + Cl_2$	92×10^3 Joules
Combination $H + H \rightarrow H_2$	211×10^3
$Cl + Cl \rightarrow Cl_2$	120×10^3
	$D_e = 423 \times 10^3$ Joules

or 4.4 eV per molecule, greater enough than the highest energy of the vibration band which is $\Delta E_{\text{vibr}} = 1.4$ eV.

11.10. Conclusions

(1) The pure rotational spectrum is just a simple band because it corresponds to transitions within a given vibrational state, *i.e.*, to $\Delta v = 0$. In that case, $\Delta E_{vibr} = 0$ in Eqs. (11.4), (11.5), (11.6).

(2) A vibration-rotation (and a pure rotation) spectrum is given by a molecule only if the molecule has a p. A_2 type molecules have no p, *i.e.*, no agent to interact with incident radiation and cause absorption, and hence they do not give such spectra. A gaseous medium consisting of such molecules allow infrared radiation to pass through with a perfect transparency. Such *homonuclear* molecules are O_2, N_2, H_2, Cl_2 etc.

(3) Transitions with $|\Delta v| > 1$ are possible but not likely, and hence their intensity is low. Had the vibrational levels been equally spaced (harmonic oscillator model), the transition frequencies for $|\Delta v| > 1$ (higher-order bands) would have followed the ratio 1:2:3:4 ... (like in the case of classical vibrational harmonics, *e.g.*, a musical tone or a violin string). However, if v is small, the discrepancy (from the model of a classical vibratory system) is not big. Indeed, for HCl we obtain a band at 3.46 μm for $v = 0 \rightleftarrows v' = 1$, and the next bands strike at 1.76 μm, 1.20 μm, and 0.91 μm (in decreasing intensity, too), pretty much in agreement with the classical proportions stated above.

(4) The structure of the vibration-rotation spectra of real molecules is quite complex, actually more complicated than what is discussed in the previous sections. Experimentally obtained and known vibration-rotation spectra of many molecules exist, but their qualitative analysis is difficult. They veil "mysteries" and it is worth searching and interpreting them. For example, one can even see transition lines with $\Delta J = 0$ (hyperfine structure, *i.e.*, extra lines), especially if $\Delta E(J \rightleftarrows J')$ between the sublevels is too small an interval. That is, although $\Delta J = 0$, the quantum gap between the sublevels is so small, that a jump through is possible and permitted, because a very small ΔE gap implies a high probability for a favored transition. This series of additional lines causes no gap at the center of the spectral band in the spectrogram. Spectral complexity is encountered also in triatomic molecules, *e.g.*, in CO_2, where many (more than forty) different infrared vibration-rotation bands are observed in the infrared region of 1 to 15 μm. (See Section 11.6.)[12]

---

[12]The water molecule has also many bands in the infrared region. That is why the atmosphere of our planet (containing H_2O and CO_2) absorbs infrared radiation. Namely, the water vapor spectrum exhibits many vibration-rotation bands in the region of 0.69 to 6.26 μm, in addition to the pure rotation bands.

11.11. Quantum States and Coupling Schemes in Molecules

In molecular physics, one usually sweeps the electrons under the carpet, despite the fact that they are present and contribute to the energy. A conscientious treatment would consider a molecular wavefunction that includes the configurational coordinates and spins of the electrons, too, in addition to those of the ions (nuclei). Because although the inner electronic core belongs closely to the ion, the valence electrons behave as if they belonged commonly to the whole molecule instead of only to the nuclei of their origin, just like a federal army which guards and belongs to the whole confederation rather than to an individual state only, albeit it might have been raised by that state. In short, the electrons feel "cosmopolitan" in identity in their molecular world.

For simplicity, and lest we make the reader feel all at sea, we will consider a diatomic molecule in this section.

Fortunately, the electronic and the rest of the molecular behavior are separable in terms of quantum mechanical calculations. That is, one can write the molecular wavefunction ψ as a product:

$$\psi = \psi(\text{electronic}) \, \psi(\text{vibrations of ions}) \, \psi(\text{rotations of molecule}) \qquad (11.7)$$

where ψ(electronic) gives E_n, the quantum mechanical energy of the electron in some (electronic) state n. It is interesting to state that E_n is influenced by (actually depends on) where the ions are, and hence E_n in a molecule is *not* the same as its counterpart in a free atom. For instance, the Coulombic energy of the nuclei is added to the electronic energy to shift it by some amount, and make it E_n(molecular), *alias* E_{el}, as denoted above.[13] So, E_n(molecular) has a minimum for a given nuclear configuration in space (*i.e.*, for a given internuclear separation, if the molecule is diatomic). In other words, E_n(molecular) is actually a *potential function* (of the instantaneous internuclear separation r) corresponding to a potential energy that produces the confederation of the atoms which are held together in the molecule. This function is shown in Fig. 11.1.

The other two factors of the product in Eq. (11.7) give the rest of the molecular energy, *i.e.*, $E_{\text{vibr}} + E_{\text{rot}}$. So, the *total* molecular energy E (that corresponds to ψ) is

$$E = E_{\text{el}} + E_{\text{vibr}} + E_{\text{rot}}$$

---

[13]In fact, the situation is more compound than that, but any further involvement in detail would be outside the scope of this section.

where

$$E_{el} \equiv E_n(\text{molecular}) = E_n(\text{free atom}) + E(\text{Coulombic between nuclei})$$

where we see that $E_n(\text{free atom})$ is shifted by the second term.

In a free atom the electronic spin **S** couples with the orbital angular momentum **L**. This *LS*-coupling occurs when the spin-orbit interaction is small. An analogous phenomenon occurs in molecules, too. Let Λ be the component of the orbital angular momentum of the electrons along the molecular axis (which in the case of a diatomic molecule is the fictional line that joins the two nuclei center to center[14]). *To a first approximation*, Λ is constant. We have $|\Lambda| = \Lambda \hbar$, with $\Lambda = 0$ or positive integer. In molecular physics Λ (actually m_Λ, but we may as well call it Λ) is the counterpart of m_l in atomic physics (or of M_L in multielectronic atoms). The *actual* angular momentum L of the electrons *is not constant in a molecule*, and hence a quantum number L does not exist (it is meaningless) for a molecule. In molecules Λ is the corresponding counterpart of L, because states that correspond to different Λ values have different energies, and hence the molecular states are specified as follows:

Λ	State
0	Σ
1	Π
2	Δ
3	Φ
$\vdots$	$\vdots$

i.e., in analogy with the atomic case. The analogue of *LS*-coupling is the ΛS-*coupling* where S is the total spin of the electrons.[15] The multiplicity is $2S+1$, and a general state in the ΛS-coupling is represented by the term scheme

$$^{2S+1}\Lambda$$

e.g., $^1\Sigma$, $^3\Sigma$, $^1\Pi$, $^3\Pi$ etc. The selection rules for transitions are $\Delta\Lambda = 0, \pm 1$ and $\Delta S = 0$.

In analogy with the coupling schemes for atoms (*LS* and *jj*-couplings), there are two possibilities concerning the *molecular* electrons:

---

[14]This is also a symmetry axis along which we consider angular momentum components in molecular physics.

[15]We should keep in mind that for a one-electron atom we have **l**, **s**, l, m_l, and m_s. For a many-electron atom we have **L**, **S**, L, M_L, and M_S.

(1) Spin-orbit effects are smaller (and hence less important) than extra-electronic effects (like nuclear motion and molecular rotation), in which case the ΛS-coupling described above is a better scheme to specify an electronic state. In that case, the state of an (*one*) electron is specified as $|n \text{ v } K J \Lambda S\rangle$ in Dirac notion where the quantum numbers in the ket are the so-called good quantum numbers for this case.

(2) The spin-orbit interaction is more important than the extraelectronic effects, in which case we should go to another coupling scheme — more convenient for this case — analogous to the jj-coupling in the atomic case. In this case we have other quantum numbers, *i.e.*, a different set of them to label the states and use it as a basis. In this case, the state is specified as $|n \text{ v } J \Omega\rangle$, and the (new set of) good quantum numbers are those in the state ket (eigenvector): N, v, J, Ω.

Let us explain these two cases:

We suppose that nuclear vibrations and rotations affect the energy more than the spin-orbit effects do. The electronic state couples then with the vibration/rotation states of the nuclei first. To describe the situation, we add the orbital angular momentum of the electrons to the rotational angular momentum of the nuclei — vectorially, of course. But we must be careful here: The axis to which the electronic orbital angular momentum is referred *is not the same* with the axis about which nuclear rotation takes place (the molecule rotates)! The former is just the molecular axis, whereas the latter is an axis *perpendicular* to it. So, we need another quantum number, K, for the rotation. This is an integer where $K \geq \Lambda$ always, because we always have $\mathbf{K} \perp \Lambda$ for diatomic molecules. The selection rule for K is $\Delta K = 0, \pm 1$. Therefore, the good quantum numbers for this case are n, v, K, Λ, S. Now we can introduce the (less important) spin-orbit interaction. Since we have obtained the total orbital angular momentum of the electrons *and* the nuclei, $\mathbf{K}$, we can add it to $\mathbf{S}$ (the electronic spin) to obtain the *final* resultant angular momentum $\mathbf{J}$ whose quantum numbers are J and M_J, as usual. We have $|K - S| \leq J \leq K + S$ where J is an integer. Like in the atomic LS-coupling, the electronic spin-order interaction here splits states of different J, giving rise to a fine structure in the electronic-vibration-rotation levels.

If $S = 0$ and $J = K$, we have a singlet molecular state (*e.g.*, $^1\Sigma$) because $2 \times 0 + 1 = 1$.

The scheme just described is the ΛS-coupling, or more accurately, the $\Lambda K S$-coupling (for a *molecular* electron). Now if the electronic spin-orbit effects are large and dominant over the molecular vibrations/rotations, then the scheme described above paralyzes and becomes incapable of specifying the state (because its so established selection rules fail and become inappropriate, if not incorrect). This new case is analogous to the jj-coupling of electrons in free atoms. We first combine the electronic orbital angular momenta and spins (about the molecular axis) together to find a vector whose quantum number is Ω.[16] This vector specifies the *electronic* state which in turn is coupled with (interacts with) a vibration-rotation state specified by v and J (where J represents the rotational states, just like the case when $S = 0$).

We can summarize the analysis as follows: The molecular energy E has a state expressible either as $|n$ v K J Λ $S\rangle$ (ΛS-coupling) or $|n$ v Ω $J\rangle$. The change of E with K and J is small, and hence lines corresponding to $\Delta E_{\text{rot}}(K \rightleftarrows K')$ are too closely spaced together in the spectral band (as described in the previous sections of this chapter). In the general case where $J \neq K$ (because $J = K$ is a special case), ΔE is still less in its J-dependence (*i.e.*, E differs only by a small bit in the J values for the initial and final states) so that it contributes only a fine structure to the spectral band lines. In the Ω-*coupling* we obtain lines due to ΔJ, and ΔE depends more on v (*i.e.*, Δv $=$ v$'$ $-$ v gives a band) while its K or J-dependence is smaller.

Before leaving the topic, we should point out that E is *negative* because we have (arbitrarily) chosen $V = 0$ when the atoms are free, and the $V(r)$ curve is below the r-axis. In the relation $E = E_{\text{el}} + E_{\text{vibr}} + E_{\text{rot}}$ the *average* electronic energy E_{el} is *negative* (because it is a binding energy, too — of the electrons to the atom, that is). The part $E_{\text{vibr}} + E_{\text{rot}}$ is positive but much less than E_{el} (so that the outcome is still negative).

A last word of comment here. Notice that as in the case of additive independent events where we have the product of the probabilities, here for the total molecular energy we likewise have the product of wavefunctions (Eq. (11.7)) of the independent degrees of freedom, giving the sum of the particular energies (Eq. (9.1)) of which the Hamiltonian consists.

11.12. Thermal Expansion of a Molecular Structure

When a molecule receives the proper[17] energy, it expands. Its size gets larger and larger, until the added energy is consumed in breaking the bond,

---

[16]Ω is an integer if the number of electrons in the molecule is even, and half integer if the number is odd.

[17]We say proper, because it should correspond to interstate vibrational transitions between vibrational levels.

and then the molecule is ripped apart. The enlargement of the molecular dimensions is due to the fact that the vibration amplitude increases, as energy keeps being supplied to the system. Similarly, when a crystal is heated, it expands. The incoming heat energy is spent in loosening and finally breaking the interatomic or intermolecular bonds in the crystal, so that we have the transition from the solid state (with fixed atomic or molecular positions about which the atoms or molecules vibrate in the rigid lattice) to the liquid state (where the molecules enjoy some freeom, but not as much as that of the gaseous state in which they fly completely free).

We can study the case of a diatomic molecule, say, KCl. Consider the pair of neighboring K and Cl ions in the molecule (or crystal) of KCl. The mutual interaction potential energy of this ion pair (and in general, any polar crystal such as KCl) is, approximately,

$$V(r) \simeq -C_1 \frac{e^2}{r} + \frac{C_2}{r^n}$$

where r is the separation distance between ions, n is a constant exponent (approximately equal to 9), and C_1 and C_2 are positive constants (of proportionality). The first term is just the Coulomb attraction, and the second term is the potential energy (part) that corresponds to the repulsive force. In this expression we have $C_1 < 1$ because neighbor ions are present as well. (For KCl, $C_1 \sim 0.29$.) The constant C_2 is determinable in terms of C_1 and the equilibrium separation r_0 of the pair, because at equilibrium,

$$\frac{dV}{dr}\bigg|_{r=r_0} = 0 = \frac{C_1 e^2}{r_0^2} - \frac{C_2 n}{r_0^{n+1}}$$

whence $C_2 = C_1 e^2 r_0^{n-1}/n$. Since r_0 is known from X-ray scattering data (for KCl, $r_0 \simeq 3.136$ Å), C_2 can be evaluated. Note that $C_1 \sim 0.29$ for all crystals of the type of KCl.

The potential energy of the ion pair plotted against the separation r of the constituents is seen in Fig. 11.9. The resulant potential energy is represented by the solid curve, and is the algebraic sum of the Coulombic potential (which gives the electrostatic attractive force) and the repulsive part which is $\sim 1/r^{10}$. (The two parts plotted as a function of r each, are shown with dashed curves.) The minimum of the total potential occurs at r_0, the equilibrium separation (or size). In molecules $r_0 \sim 1$ to 3 Å.

Viewing the system as an idealized theoretical (mathematical) model (with simplified features and without complications), one can consider small displacements away from r_0. Then one can expand the function V into a Taylor

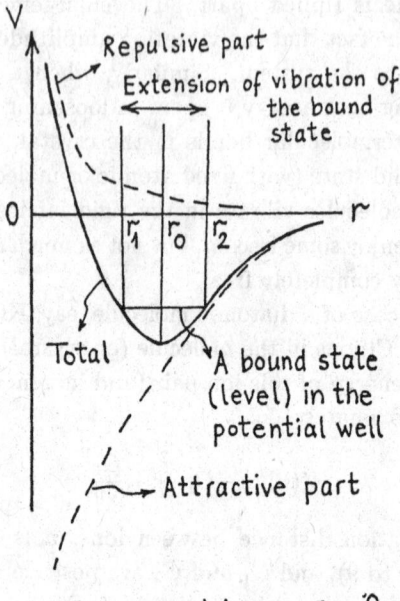

Fig. 11.9.

series about the value $r = r_0$. For small $|r - r_0|$, with r being close to r_0, *i.e.*, as $r \to r_0$, the mathematical expansion is

$$V(r) = V(r_0) + (r - r_0)\frac{dV}{dr}\bigg|_{r_0} + (r - r_0)^2\frac{1}{2!}\frac{d^2V}{dr^2}\bigg|_{r_0}$$

$$+(r - r_0)^3\frac{1}{3!}\frac{d^3V}{dr^3}\bigg|_{r_0} + \text{h.o.t.}$$

where all derivatives are evaluated at $r = r_0$. Since the force is zero at exactly $r = r_0$, the second term of the expansion vanishes. Then we have

$$V(x) = A\frac{x^2}{2!} + B\frac{x^3}{3!} + \text{h.o.t.} \tag{11.8}$$

where $x \equiv r - r_0$, $V(x) \equiv V(r) - V(r_0)$, and A and B are differential coefficients. The first term is quadratic and gives the elastic force, so that if the displacement is small, the potential energy curve has a parabolic shape, because the first term dominates, corresponding to a linear harmonic oscillator.

That is, if a little amount of energy is given to the pair, the molecule will execute small oscillations symmetrically (of symmetrical amplitude) about (the point) $r = r_0$. But if a large amount of energy is supplied, then the second term in the expansion gains importance. As this term becomes appreciable, the motion deviates from being harmonic, and the oscillation is not symmetric about r_0 any more; the molecule does not behave like a simple harmonic oscillator any longer. The presence of the higher-order terms distorts the parabolic shape of the curve as r gets larger and larger. The curve is nearly parabolic only for $r \to r_0$ (Fig. 11.9). For large r the motion is not harmonic, and as $V \to 0$, no bound state (no potential well or net attraction able to sustain a bound state) remains: The system gets out of the well and is no longer a bound molecule. There is no force that holds the ions together.

The amplitude $r_2 - r_0$ (in the direction of increasing r) is larger than the amplitude $r_0 - r_1$ on the other side (see Fig. 11.9). That is why the time average of the absolute deviation $|r - r_0|$ is *not* zero (as it would be if the force were purely elastic and the motion simply harmonic).

Thus $\langle r \rangle > r_0$ because $\langle r - r_0 \rangle \gtrsim 0$. The time average is slightly larger than zero. This means that if all the ion pairs in the crystal receive this amount of energy by being heated, the crystal expands, due to the favored asymmetry on one side. This is the net reponse of the crystal.

From Eq. (11.8) we can derive the force between the ions, by taking the derivative of the potential which creates the force. Thus we have

$$F = -\frac{dV}{dx} = -Ax - \frac{B}{2}x^2 - \text{h.o.t.} \tag{11.9}$$

where the first term represents the elastic force and is actually just the harmonic restoring force itself. The term is linear and leads to the elastic coefficient (Young's modulus) of the crystal under a tensile or compressive strain. The second term is quadratic and not symmetric for positive and negative values of x, because the power of x is even. (Whereas an odd power is symmetric for positive and negative displacements.) This asymmetry arising from the quadratic terms, contributes unequal amplitudes in the two directions on either side of r_0.

The common force law is just $F = -Ax$.

Since F is given by Eq. (11.9), one can write and solve the equation of motion of this ion pair, but one should remember to use the reduced (effective) mass of the system, not the mass of either ion. The solution can be reached approximately.

This model is simplified; in reality the situation is more complex. The idealization of the problem may be interpreted as dodging the issue, but it nevertheless is useful and better than nothing. As one may easily realize, the situation with complex molecules is hopelessly difficult.

The quadratic term in the expansion of V contains information about the thermal expansion coefficient of the crystal. Any further study thereof concerns solid state physics.

11.13. Quantum Mechanical Observations

Consider Eq. (11.7). Notice that as in the case of additive independent events (or degrees of freedom) we have the product of probabilities, here for the total molecular energy we likewise have the product of wavefunctions of independent degrees of freedom, giving the sum of the particular energies of which the Hamiltonian (of the system) consists. The Hamiltonian is the sum of independent terms (Eq. (9.1)).

impurity — to calculate the Lagrangian ... Which is only ... but in (1s), and in
at the ... The other modes ... and their corresponding eigenfrequencies are present.
... If the molecule is not linear, then it ... does not make any sense to
... of ... fundamental modes ... These coordinates that all ... be in equation motion for
the modes.

Chapter 12

SPECTRA OF POLYATOMIC MOLECULES

12.1. General

In this chapter, by "polyatomic molecules" we mean molecules consisting of more than two (but not more than five or six) atoms. We will discuss, separately, their vibrational and rotational spectra.

12.2. Linear Vibrations of Polyatomic Molecules

The study of polyatomic molecules can be thoroughly achieved by using *group theory* which can be applied to molecules. Although the application of the theory would give us more latitude for comfortable molecular studies, in this book we will not follow it, because it would be too scholastic for our purposes. Perhaps in this business we will pay a cost by excluding a group theoretical treatment, but we can live with it to a first-order approximation, keeping always in mind that our substitute way can never unseat or supersede group theory in terms of rigor and precision.

A molecule that consists of N atoms has $3N$ degrees of freedom, because a body of mass m can move in three directions in space and hence has three degrees of freedom (Fig. 12.1) (Actually, it can move towards any direction, but the motion can be eventually referred and projected to the three Cartesian axes.) Three out of these $3N$ degree belong to the translation of the molecule as a whole. Another three belong to the rotation energies of the molecule. So, a molecule with N atoms has (is left with) $3N - 6$ degrees of freedom for vibration. These are called *normal modes of vibration* or *vibrational modes* (shapes, fashions). If the molecule is of linear structure, it is customary — for simplicity — to analyze the longitudinal vibrations only, but in reality (and in spectroscopy) the other modes and their corresponding frequencies are present, too. If the molecule is not linear, then it already does not make any sense to speak of longitudinal modes; one considers just all the designated modes for the molecule.

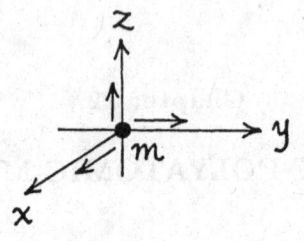

Fig. 12.1.

We will consider four examples:

(1) The H_2O (water) molecule: It has $3 \times 3 - 6 = 3$ normal modes of vibration shown in Fig. 12.2, with corresponding frequencies ν_1, ν_2, ν_3. In the first normal mode (a), as the atoms vibrate, the two H atoms approach the O atom while the latter moves in a direction opposite to the resultant of the motion of the two H atoms (to conserve momentum), and then the atoms move away from each other. The second vibration mode is shown in (b). In the third mode (c), the O atom oscillates sideways.

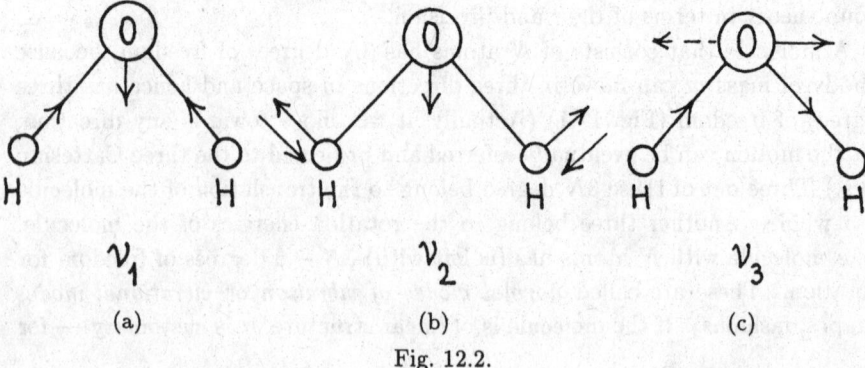

Fig. 12.2.

When one watches the microwave and infrared absorption bands of water vapor, one observes three spectral lines corresponding to ν_1, ν_2, ν_3. The corresponding wave numbers ($\bar{\nu} = 1/\lambda$) are $\bar{\nu}_1 = 3652$ cm^{-1}, $\bar{\nu}_2 = 1595$ cm^{-1}, and $\bar{\nu}_3 = 3756$ cm^{-1}.

(2) The CO_2 (carbon dioxide) molecule: One can observe and measure three bands and their corresponding spectral lines, but in practice only two bands (frequencies) can be actually observed and measured: $\bar{\nu}_2 = 667$ cm^{-1} and $\bar{\nu}_3 = 2349$ cm^{-1}. There is also an observable fundamental vibration of $\bar{\nu}_1 = 1340$ cm^{-1}, but it is very weak. The reason rests with the fact that CO_2 is a linear molecule (Fig. 12.3(a)). When it vibrates in the mode specified by ν_1 (at ν_1), there is no change in the dipole moment. So, the corresponding vibration frequency ν_1 cannot be observed, that is, the absorption lines belonging to ν_1 are too weak. So, the first vibration mode is too weak to be observed.

The line that corresponds to ν_2 actually consists of two lines very closely spaced (very close to each other): ν_{2a} and ν_{2b} (Fig. 12.3(b) and (c)). In both of these there is a change in the dipole moment, and hence the corresponding vibration mode of ν_2 can be observed. In ν_{2a}

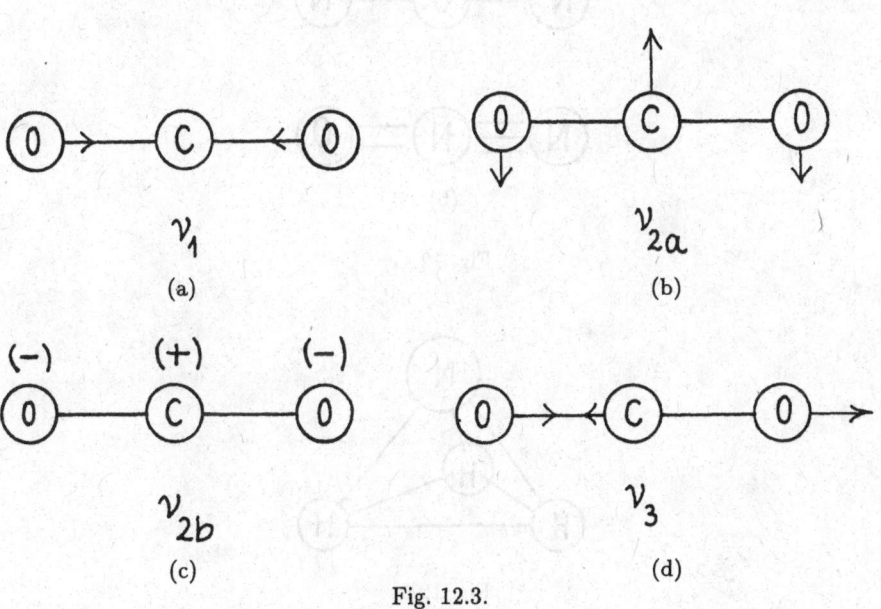

Fig. 12.3.

the vibration occurs in the plane of the paper; in ν_{2b} the vibration is in a direction perpendicular to the plane of the paper (as the C atom comes out of the plane towards the reader, the two O atoms move backwards and away from the reader). So, we have two fundamental lines, ν_{2a} and ν_{2b}, very close to each other, because the value of the dipole moment is about the same in both cases.

A third mode is shown in (d). Initially the dipole moment is zero, but it changes afterwards as the rightmost O atom recedes from the other two atoms which approach each other.

Frequencies ν_2 and ν_3 are observed in the infrared region (in terms of their value). We say that modes ν_2 and ν_3 are *active* in the infrared region.

(3) The N_2O molecule: There are two possible formations of the chemical bonds in this molecule (Fig. 12.4(a) and (b)). In case (a) we should not observe ν_1 which ought to be very weak. However, when we obtain the spectrum we see that ν_1, ν_{2a}, ν_{2b}, and ν_3 are all present and strong. Since we observe them all, we conclude that the structure shown in (b) is the correct one for this molecule.

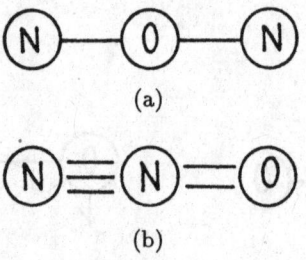

(a)

(b)

Fig. 12.4.

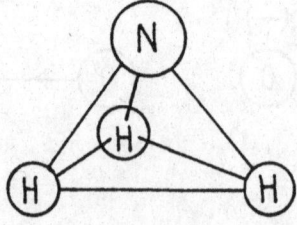

Fig. 12.5.

(4) The NH_3 (ammonia) molecule: It has four atoms. One expects to observe $3 \times 4 - 6 = 12 - 6 = 6$ vibration modes. The molecule is like a symmetric top. The three H atoms form an equilateral triangle which in turn forms the base of a pyramid at the apex of which the N atom is located (Fig. 12.5). Modes ν_1 and ν_2 are shown in (a) and (b) respectively, of Fig. 12.6. We could have called the next two cases ν_3 and ν_4 respectively ((c) and (d)), but we realize that they are in fact one case whose energy is twofold degenerate, and two corresponding vibration frequencies show up: $2\nu_3$. In (d) the shaded triangular plane bends, *i.e.*, the vibration plane rotates, and hence the corresponding energy is twofold degenerate.[1] This double degeneracy is shown by $2\nu_3$, because the triangular plane formed by the H atoms rotates.

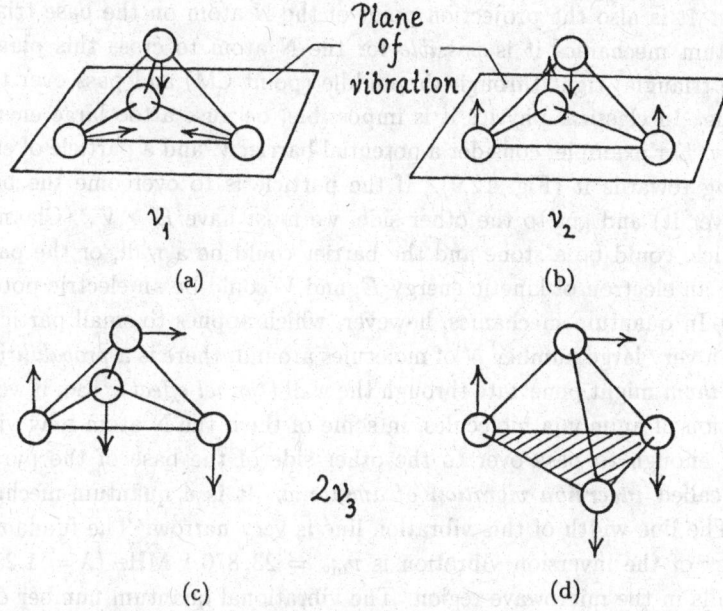

Fig. 12.6.

Figure 12.7 shows the distortional vibrations: We can have (a) or (b), again a doubly degenerate case denoted by $2\nu_4$.

---

[1]If two different systems have the same energy, we say that this energy value is twofold (doubly) degenerate.

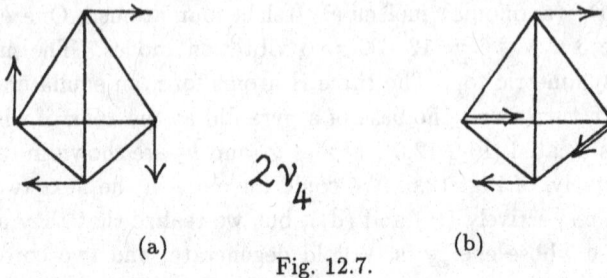

(a)   (b)

Fig. 12.7.

As the electric dipole moment changes, all these cases are observed in the infrared region where they are active. We have $\bar{\nu}_1 = 3336$ cm^{-1}, $\bar{\nu}_2 = 933$ cm^{-1}, $2\bar{\nu}_3 = 3410$ cm^{-1}, and $2\bar{\nu}_4 = 1630$ cm^{-1}.

In Fig. 12.8 CM is the center of mass of the base triangle formed by the H atoms. It is also the projection point of the N atom on the base triangle. In quantum mechanics it is *possible* for the N atom to cross this plane (of the base triangle) right through the middle (point CM) and pass over to the other side. In classical physics it is impossible, because a too large energy is necessary. For example, consider a potential barrier V and a particle of energy E coming towards it (Fig. 12.9). If the particle is to overcome the barrier (jump over it) and get to the other side, we must have $E \geq V$. (Classically, the particle could be a stone and the barrier could be a wall, or the particle could be an electron of kinetic energy E, and V could be an electric potential barrier.) In quantum mechanics, however, which applies to small particles, if we have a very large number N of molecules around, there is a *probability* that *some of them* might penetrate through the wall (*tunnel effect*).[2] So, if we have a collection of ammonia molecules, in some of them the N atom may vibrate strongly enough to pass over to the other side of the base of the pyramid. This is called *inversion vibration of ammonia*. It is a quantum-mechanical effect. The line width of this vibration line is very narrow. The fundamental frequency of the inversion vibration is $\nu_{\text{inv}} = 23{,}870.1$ MHz ($\lambda = 1.25$ cm) which falls in the microwave region. The vibrational quantum number of ν_{inv} is v $= 0$, and the transition selection rules are Δv $= 0$, $\Delta J = 0$.

The ammonia inversion is used in making atomic clocks. One fills a microwave cavity with ammonia at low pressure and compares the absorption spectrum with an electronic oscillator. If one makes the frequency difference $\Delta f = 0$, one has an atomic (or molecular) clock (Fig. 12.10). This is the

---

[2]Even if we send one ($N = 1$) particle, it is *likely* that is passes through.

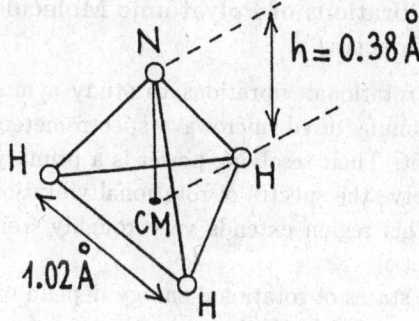

Fig. 12.8.

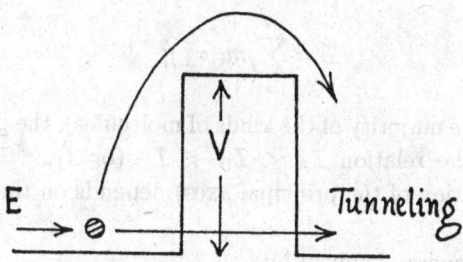

Fig. 12.9.

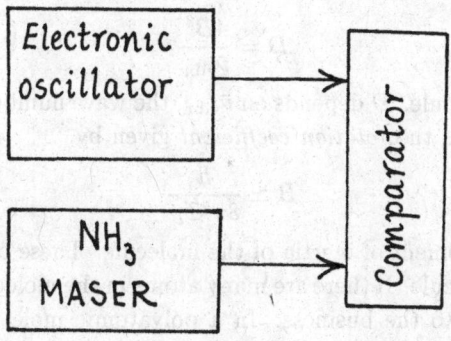

Fig. 12.10.

principle. This vibration frequency is used as a standard frequency, because it is constant, as it is taken from nature. It is of high accuracy because molecular vibrations are subject to very little changes.

12.3. Rotational Vibrations of Polyatomic Molecules (Microwave Spectra)

One can use the rotational vibrations to study symmetry properties in molecules. Ready, manufactured microwave spectrometers are commercially available in the market. Their resolving power is a primary factor of concern. They are used to observe the spectra of rotational vibrations which fall in the microwave region. This region extends very roughly from 10^3 to 10^5 MHz (10 to 10^{-1} cm).

In a molecule, the states of rotational energy depend on the principal moments of inertia of the molecule. If the molecule consists of N atoms — each of mass m_i — located at a perpendicular distance $(r_\perp)_i$ from the molecular axis, then the *moment of inertia* is defined as

$$I = \sum_{i=1}^{N} m_i (r_\perp)_i^2 .$$

In general (in the majority of the kinds of molecules), the principal moments of inertia satisfy the relation $I_A < I_B < I_C$ (or $I_1 < I_2 < I_3$). Their value (and the location of the principal axes) depends on the symmetry of the molecule.[3]

The rotational energy is given by

$$E_{\text{rot}} = Bh \, J(J+1) - Dh \, J^2(J+1)^2$$

where D is the *centrifugal distortion coefficient* given by

$$D = \frac{4B^3}{\bar{\nu}_{\text{vibr}}} \tag{12.1}$$

for a diatomic molecule. D depends on $\bar{\nu}_{\text{vibr}}$, the wave number of the molecular vibration lines. B is the *rotation coefficient* given by

$$B \simeq \frac{h}{8\pi^2 I_B}$$

where I_B is the moment of inertia of the molecule. These considerations hold for a diatomic molecule. If there are many atoms in the molecule, the symmetry properties come into the business. In a polyatomic molecule, one has three moments of inertia: I_A, I_B, and I_C, and three corresponding coefficients, called *rotation constants*:

$$A = \frac{h}{8\pi^2 I_A}, \quad B = \frac{h}{8\pi^2 I_B}, \quad C = \frac{h}{8\pi^2 I_C}$$

---

[3]A symmetry plane is in general perpendicular to one of the principal axes.

where, in general, $A \geq B \geq C$. Thus we can find $\mathcal{I}$ and get an idea about molecular symmetry. The molecular symmetry and the rotation constants are related as the following table summarizes:

Molecular structure (symmetry)	Conditions on the moments of inertia	Rotation coefficients	Example
Single atom	$\mathcal{I}_A = \mathcal{I}_B = \mathcal{I}_C = 0$	$A = B = C = \infty$	Na
Linear molecule	$\mathcal{I}_A = 0 < \mathcal{I}_B = \mathcal{I}_C$	$A = \infty,\ B = C > 0$	CO_2
Planar molecule	$\mathcal{I}_A + \mathcal{I}_B = \mathcal{I}_C$	$(1/A) + (1/B) = 1/C$	Vinylene carbonate
Symmetric top:			
(a) Prolate (elongated) molecule	$0 < \mathcal{I}_A < \mathcal{I}_B = \mathcal{I}_C$	$A > B = C > 0$	CH_3Cl
(b) Oblate (pancake) molecule	$0 < \mathcal{I}_A = \mathcal{I}_B < \mathcal{I}_C$	$A = B > C > 0$	NH_3
Asymmetric top	$0 < \mathcal{I}_A < \mathcal{I}_B < \mathcal{I}_C$	$A > B > C > 0$	CH_3CN
Spherical symmetry	$0 < \mathcal{I}_A = \mathcal{I}_B = \mathcal{I}_C$	$A = B = C > 0$	CH_4

Remarks regarding the table above:

(1) In general, $\mathcal{I}_A < \mathcal{I}_B < \mathcal{I}_C$.

(2) In linear molecules, $\mathcal{I}_A$ is with respect to the molecular axis. Example: CO_2 (Fig. 12.11).

(3) For any planar (flat) molecule, $\mathcal{I}_A + \mathcal{I}_B = \mathcal{I}_C$. Vinylene carbonate (Fig. 12.12) was found to be planar, by microwave measurements.

(4) The monochloromethane molecule (Fig. 12.13) is an example to a long symmetric top molecule.

(5) The ammonia molecule (Fig. 12.14) is an example to a short symmetric top molecule.

(6) The ethane cyanide molecule (Fig. 12.15) is an example to an asymmetric top molecule.

(7) The methane molecule (Fig. 12.16) has spherical symmetry. No rotation spectrum can be observed for such molecules. The reason can be explained by the selection rules.

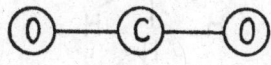

Fig. 12.11.

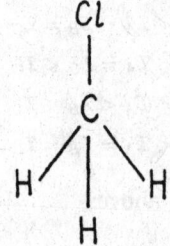

Fig. 12.12.

Fig. 12.13.

Fig. 12.14.

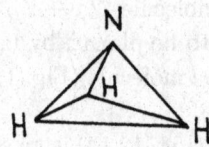

Fig. 12.15.

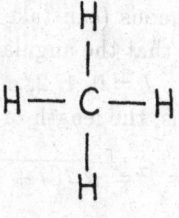

Fig. 12.16.

For a rigid linear molecule we have

$$E_{\text{rot}} = \frac{h^2}{8\pi^2 \mathcal{I}_B} J(J+1) = Bh\, J(J+1)$$

where J is the rotational quantum number than can take the values $J = 0, 1, 2, \ldots$ (integers).

If the molecule is linear, but not a rigid rotor, we have

$$E_{\text{rot}} = Bh\, J(J+1) - Dh\, J^2(J+1)^2$$

where the second term is corrective, but has a small contribution. D depends on the molecule. For a diatomic molecule D is given by Eq. (12.1) where $\bar{\nu}_{\text{vibr}}$ is the wave number of vibrations.

To visualize a rotation, consider a linear molecule rotating in a plane which is perpendicular to the axis zz' about which the molecule rotates (Fig. 12.17). Let the axis pass through the center of mass of the molecule. The angular momentum (classically) is

$$P = \sum_i m_i v_i r_i = m_1 v_1 r_1 + m_2 v_2 r_2 + \ldots = \mathcal{I}\omega .$$

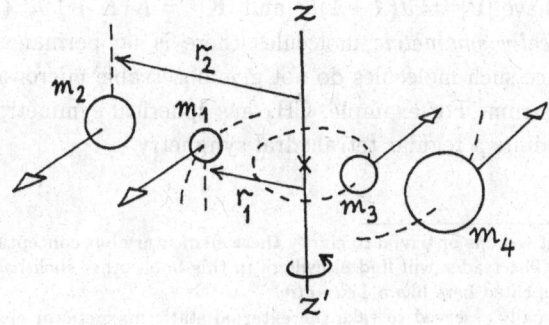

Fig. 12.17.

Classically, **P** can be continuous (can take any value), but quantum mechanically is limited by the fact that the angular momentum quantum number J can take integral values only: $J = 0, 1, 2, \ldots$. Thus **P** is quantized (takes discrete values in jumps), that is, the length of the vector operator **P** is

$$|\mathbf{P}| = P = \sqrt{J(J+1)}\hbar. \tag{12.2}$$

Quantum mechanics textbooks call P also J, but this may confuse the reader. Do *not* confuse P with J. The former (left-hand side of Eq. (12.2)) is the *length of the vector operator of angular momentum* given by the right-hand side of Eq. (12.2) which is the *eigenvalue* of the vector operator **P**. The latter, *i.e.*, J, is the *quantum number* of the angular momentum operator **P**.[4]

If we apply an external electric field, say, in the z-direction, the projection of **P** on the z-axis (along the field) is also quantized by the magnetic quantum number m_J, and can take $2J + 1$ values, because m_J takes the values $m_J = J, J - 1, \ldots, -J$, *i.e.*, $2J + 1$ of them altogether.

For *symmetric top* molecules, the rotational energy states are given by

$$E(J, K) = Bh\, J(J+1) \pm (A - C)K^2 h$$

where the plus (minus) sign is for rigid prolate (oblate) symmetric top molecules. The rotational levels depend on the two rotational quantum numbers J and K where $J = 0, 1, 2, \ldots$, and $K = 0, \pm 1, \pm 2, \ldots, \pm J$. As **P** is quantized in space, it has a component (projection) on the z-axis[5] given by (and equal to) m_J, and a projection *along the symmetry axis of the molecule* which is also quantized and can take the value $K\hbar$, specified by the quantum number K. So, K is a measure of the quantum-mechanical angle between **P** and the molecular axis. (If the two axes happen to coincide, then obviously, K lapses to J.) So, we have $|\mathbf{P}|^2 = J(J+1)\hbar^2$ and $|\mathbf{K}|^2 = K(K+1)\hbar^2$ (Fig. 12.18).

For *spherically symmetric* molecules there is no permanent dipole moment, and hence such molecules do not give observable microwave absorption (rotation) spectrum. For example, CH_4 has spherical symmetry (Fig. 12.19); actually, it exhibits a regular tetrahedral symmetry.

---

[4] We do not find it tedious or trivial to clarify these elementary but conceptual facts of quantum mechanics. The reader will find elsewhere in this book other such basic clarifications, too, a practice repeated here like a *Leitmotiv*.

[5] The z-axis is usually reserved to take the external static magnetic or electric field along. The field direction is usually the z-direction. Since we deal with electric dipoles here, the field to be externally applied in this case is electric, not magnetic.

Molecular axis

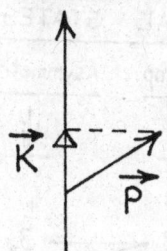

Fig. 12.18.

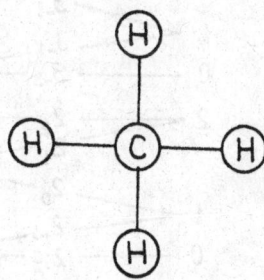

Fig. 12.19.

For *asymmetric top* molecules, the energy levels unfortunately cannot be expressed by an explicit expression. Here again, $P = \sqrt{J(J+1)}\hbar$, and each level is $2J + 1$ times degenerate in the absence of any external field.

The E_{rot} levels and the rotation constants are given in the diagram that appears on the next page.

Notice that as J increases, the spacing of the energy levels of a linear molecule increases, too. For $J = 1$ we have two different states: $K = +1$ and $K = -1$. That is, the energy level is split into two. For $J \geq K$, for example, if $J = 3$, we have $K = 3, 2, 1$. For $J = 1$ in J_j, we have a splitting into three: $K = 0, -1, +1$. In J_j (for example, $J_j = 2_{-2}$) J is the rotational quantum number, and j is a second quantum number that gives the substates of energy of J.

The prolate-top and oblate-top symmetries are the limit cases. One defines a parameter (coefficient) $\mathfrak{h}$, called the *asymmetry parameter*, as

ROTATIONAL STATES

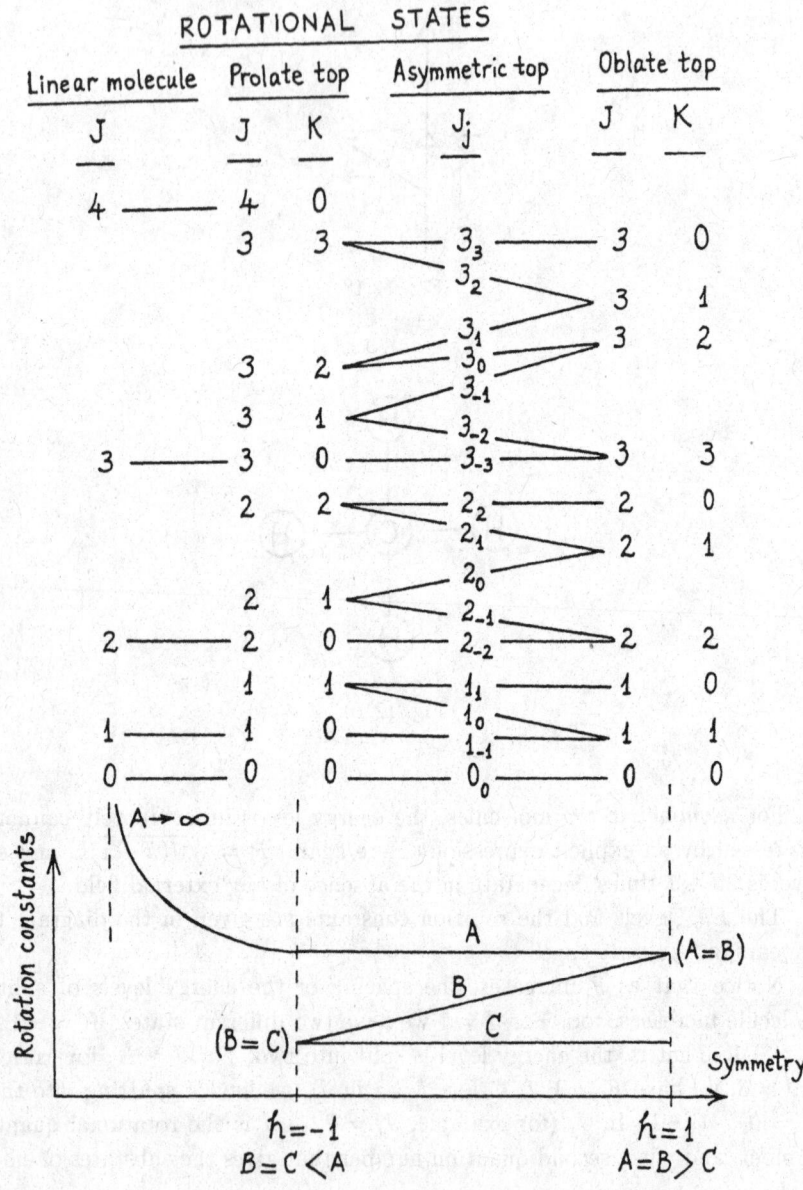

$$\mathfrak{h} \equiv \frac{2[B - \frac{1}{2}(A+C)]}{A - C}$$

which is used to determine and specify the energy states of the symmetric and asymmetric tops. This parameter varies between -1 and $+1$: $-1 \leq \mathfrak{h} \leq +1$. Its value tells us whether the molecule in question is closer to the symmetric prolate top, to the asymmetric top, or to the symmetric oblate top. In other words, at the two extremes $\mathfrak{h}$ is equal to ± 1 where for $\mathfrak{h} = -1$ we have $B = C < A$ (prolate), and for $\mathfrak{h} = +1$ we have $A = B > C$ (oblate). The asymmetric top case is intermediate in between. For linear molecules, $A \to \infty$.

For a symmetric top we have degeneracy: $E_{rot}(J, K) = E_{rot}(J, -K)$. The two energy states of K are degenerate and coincide. So, each state is $2(2J+1)$ times degenerate altogether.

For an asymmetric top the energy states split into two (except for the $K = 0$ state). The state change is continuous as A, B, and C change. If for a given J one connects smoothly the prolate side with the oblate side by straight lines, one can find the energy levels for the asymmetric top which is in between. This is done in the diagram.

For $K \neq 0$, the energy states of a symmetric top is doubly degenerate.

For an asymmetric top the highest state is J_j, and the next lower ones follow as J_{J-1}, $J_{J-2}, \ldots, J_{-J}$.

An explicit formula for the rotational states of an asymmetric rotor was attempted by King *et al.*[6]:

$$E_{rot}(\text{asymm.}) = \frac{h}{2}(A + C)J(J + 1) + \frac{h}{2}(A - C)E_j(\mathfrak{h})$$

where the values of $E_j(\mathfrak{h})$ are tabulated depending on $\mathfrak{h}$ and for different values of J and j.

12.4. Selection Rules

There are selection rules for the rotation absorption spectrum, *i.e.*, for J and K. The selection rule for symmetric tops states that transitions corresponding to $\Delta J = 1$ and $\Delta K = 0$ are allowed and possible. The selection rule for asymmetric tops is $\Delta J = 0, \pm 1$. The transitions are subject to the limitations posed by the selection rules. Transitions violating the selection rules are forbidden in quantum mechanics.

---

[6]G.W. King, R.M. Hainer, and P.C. Cross, *J. Chem. Phys.*, **11**, 27 (1943).

The full table of the selection rules is given below (where $\mathcal{E}$ is the applied electric field, and $\mathcal{E}_m$ is the microwave field for the spectroscopy):

Molecule	$\mathcal{E} = 0$	$\mathcal{E} \| \mathcal{E}_m$	$\mathcal{E} \perp \mathcal{E}_m$
Linear	$\Delta J = 1$	$\Delta J = 1$	$\Delta J = 1$
		$\Delta m_J = 0$	$\Delta m_J \pm 1$
Symmetric top	$\Delta J = 1$	$\Delta J = 1$	$\Delta J = 1$
	$\Delta K = 0$	$\Delta K = 0$	$\Delta K = 0$
		$\Delta m_J = 0$	$\Delta m_J \pm 1$
Asymmetric top	$\Delta J = 1$	$\Delta J = 0, \pm 1$	$\Delta J = 0, \pm 1$
		$\Delta m_J = 0$	$\Delta m_J \pm 1$

These selection rules allow the following transitions:

For linear molecules and symmetric tops:

$$\Delta E_{\text{rot}} = 2Bh(J+1).$$

For asymmetric tops:

$$\Delta E_{\text{rot}} = h(A+C)(J+1) + \frac{h}{2}(A-C)[E_j(J+1) - E_j(J)]$$

for $\Delta J = 1$, and

$$\Delta E_{\text{rot}} = \frac{h}{2}(A-C)[E_{j'}(J) - E_j(J)] \quad \text{for} \quad \Delta J = 0.$$

Since A, B, $C \sim 5000$ MHz, for linear and symmetric top molecules we obtain microwave spectra where the interval between two successive spectral lines is typically $\Delta \nu \sim 10^4$ MHz, a too wide separation. Hence we observe only a few transitions (and maybe the fine structure of their lines, if the spectrometer is sensitive enough).

As far as asymmetric tops are concerned, the line separation depends primarily on $A - C$, but there is something else competing with that: The difference ΔE_j (that appears inside the brackets above) is continuous. Hence in an interval of 10^3 MHz in the spectrum of an asymmetric-top molecule, one can usually observe hundreds of lines!

12.5. Comments

Line spectra can be observed with gases. In condensed specimens, since the molecules are quite close to each other, the rotational spectral lines get broadened to an almost continuous spectrum, ruining the discrete line structure of the spectrum. However, since now the spectrum is almost continuous, one can observe absorption lines manifested as dark bands on the continuous spectrum (which serves as background).

Besides gases, one can comfortably study also liquids and solids whose vapor pressure is of the order of 10^{-1} to 10^{-3} Torr, though C.H. Townes was able to work at higher pressures.

One should not forget the limitation that *molecules with no dipole moment* (or with a small one) *cannot give observable microwave (rotation) spectra* (because of the selection rules).

If the axis about which the moment of inertia I is taken passes through the center of mass of the molecule, one obtains the inertia ellipsoid (Fig. 12.20). The axes of this ellipsoid are called *principal axes* and their lengths are determined by principal moments of inertia. (The inertia tensor is diagonal along the principal axes, with the products of inertia vanishing.) Refer to Fig. 12.20. If the cross-section is a straight line, the molecule is linear; if it is an ellipse, we have a top; and if it is a circle, we have spherical symmetry.

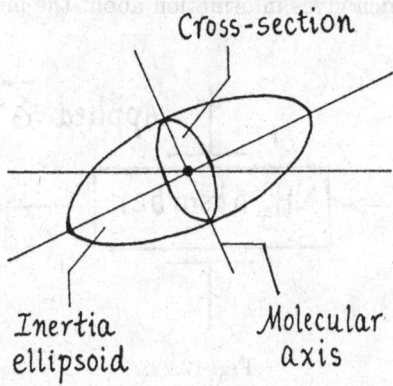

Fig. 12.20.

12.6. Fine Structure of the Molecular Spectrum

The fine structure is due to energy splittings because of the following reasons:

(1) *External magnetic field*: Its effect is to split the energy levels (and hence give more lines; actually, a line that appears single at the absence of the field shows structure when the field is on).

(2) *External electric field*: It causes splitting, too, due to the *Stark effect* (Fig. 12.21).

(3) *Nuclear quadrupole moment*: It causes splittings that are seen as fine lines.

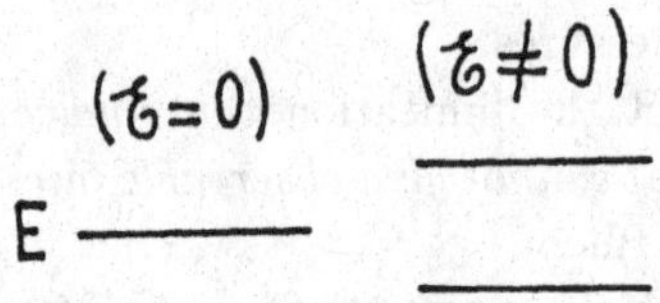

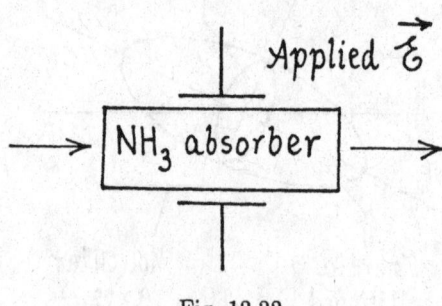

Fig. 12.21.

The Stark effect and the nuclear quadrupole moments provide us with information about the number of molecular energy splittings. To produce the Stark effect, we put our specimen (*e.g.*, NH_3) inside a microwave cavity and we apply an electric field (Fig. 12.22). The spectrum then exhibits line splittings, *i.e.*, fine structure, which gives information about the molecular structure.

$$\text{Applied } \vec{\mathcal{E}}$$

$$\longrightarrow \boxed{NH_3 \text{ absorber}} \longrightarrow$$

Fig. 12.22.

12.7. Thermodynamic Properties

If we want to calculate a thermodynamic quantity, *e.g.*, the enthalpy, of a specimen (compound), the matter comes down to knowing the principal moments of inertia $\mathcal{I}_A$, $\mathcal{I}_B$, and $\mathcal{I}_C$ of the molecule of that compound. This is a point where macroscopic thermodynamics meets microscopic quantities,[7] and is derived from them. So, if we want to calculate enthalpy and entropy, we must first find $\mathcal{I}_A$, $\mathcal{I}_B$, and $\mathcal{I}_C$ which can be obtained from microwave (rotation) spectra.

If one measures the principal moments of inertia, one can calculate the thermodynamic functions of the specimen. These functions determine its macroscopic thermodynamics, that is, its thermodynamic properties.

For example, the enthalpy is

$$H = E_0 + \frac{5}{2}\mathcal{R}T + \mathcal{R}T^2\frac{d(\ln Z)}{dT}.$$

The entropy is

$$S = \frac{3}{2}\mathcal{R}\ln M + \frac{5}{2}\ln T - \mathcal{R}\ln P - 7.26 + \frac{5}{2}\mathcal{R} + \mathcal{R}\left(\ln Z + T\frac{d(\ln Z)}{dT}\right)$$

where E_0 is the vibrational zero-point energy (per mol of an ideal gas); M is the molecular weight; T is the absolute temperature; P is the pressure, $\mathcal{R}$ is the universal gas constant; and Z is the partition function.

Finding H and S means finding E_0 and Z first. Actually, all thermodynamics follow from the knowledge of Z. If we know the partition function, we can calculate the thermodynamic functions (enthalpy, entropy, free energy, internal energy, pressure, specific heat) because they are all related to it explicitly. The partition function is (in its concise form)

$$Z = \sum_i g_i e^{-(\Delta E)_i/kT}$$

where we sum over all states. Here $(\Delta E)_i$ is the energy difference between the i-th state the the zero-point state; g_i is the statistical weight (degeneracy) of the i-th state (for example, a rotational state of an asymmetric top is $2J + 1$ times degenerate; so, for that case, $g = 2J + 1$); and k is the Boltzmann constant.

---

[7]A *phenomenological theory* explains the "how" of the phenomena in a descriptive way, without getting too much into reasons. A *fundamental theory* explains the "why"; it studies the phenomena esoterically with reasons, searching for what is going on in the depths of them, in their foundations. Thermodynamics is a (macroscopic) phenomenological theory, whereas statistical physics is a (microscopic) fundamental theory (that reaches microstates). Quantities like entropy and enthalpy connect the two theories as they partake of both, under different capacity. (For example, in thermodynamics the internal energy U is a state function which in the language of statistical mechanics becomes the mean energy $\overline{E}$.) Thermodynamics has its own history, tradition, and methods, and may not need statistical physics for its progress, but nevertheless it is derived from it. Similarly, physical optics is a phenomenological theory, whereas the electromagnetic theory is its fundamental theory.

In fact, Z involves $\mathcal{I}_A$, $\mathcal{I}_B$, and $\mathcal{I}_C$, if written explicitly. So, to calculate Z, one must know the principal moments of inertia first. And these can be found from the microwave spectra.

12.8. Microwave Spectroscopy

An ordinary microwave spectroscope gives the intense absorption lines only, because cylstron noise sometimes outgrows and obscures the absorption signals. To cure this, we insert an electrode into the waveguide of the apparatus, to increase its sensitivity via Stark modulation and phase detection.

If the external electric field is zero, a molecule absorbs energy and gets from state n to state n'. But if a field is on, n shifts to m, and n' to m'. As a consequence, the microwave absorption frequency shifts from ν to ν'. If the field is sinusoidal, with frequency f, the absorption (at ν') varies with the same frequency, *i.e.*, the absorption gets modulated by the frequency f. Thus, one can connect to the detector a narrow-band amplifier (or phase detector) adjusted to f, to increase the sensitivity of measurement after detection. That is, the electronic noise is filtered and left out without being amplified along, and the signal/noise ratio goes up.

The line at ν' is the *Stark component* of the (original) line at ν. An original (unshifted) line ν may have many such components, used in identifying microwave spectral lines.

Microwave spectroscopy gives information about molecular structure and thermodynamic properties. Molecules are stable and equilibrated systems made of positive nuclei and electrons (around the nuclei). One equilibrates the other, that is, quantum mechanics dictates that the nuclei cannot devour up the electrons, and the electrons, on their part, keep the nuclei bound together. Molecular spectroscopy is a beneficial means to obtain information on such complex systems. Moreover, if instead of one single molecule there is a collection of molecules, their internal properties are influenced by this group coexistence.

In stereochemistry we take the molecule as a number of nuclei in a stable structure, and we describe the structure, ignoring the electrons. But quantum physics holds out a better consideration: It tries to determine ψ, the probability function of the distribution of the (negative) electronic charge density within the molecular volume.[8]

---

[8]The two different kinds of charge behave differently because the mass difference is large: $m_p/m_e \sim 1836$.

Microwave spectroscopy deals with the distribution of nuclei inside the molecule (intramolecular order) and also with the state of the electric field gradient at different positions inside the molecule, and tries to pump out information. The topic is too broad nowadays, much broader than what is merely mentioned here.

Infrared spectroscopy deals with the energy changes when the nuclei change position inside the molecule, in which case the force fields (valence forces) change. In short, nature is at work, and pertinent spectra are observed. When a quantum is absorbed the force field changes. By studying the resulting spectrum, we can derive information regarding molecular dynamics.

To date, many chemical compounds have been studied by using microwave spectroscopy. The rotational spectrum gives one single constant B. In order to determine the correct molecular stereometry, we should know which atomic isotopes are lodged in the molecule. Because although the theory says that a given molecule has the same angles and bond lengths for all isotopes, surprisingly enough, experiments show that this is *not* so!

12.9. Stark Effect in Molecules

In the previous section we saw that we can obtain the Stark components (shifted lines) and follow them by instruments connected to the output of the spectrometer (microamperemeter and recorder).

Let us consider the Stark effect in a linear molecule that has a permanent dipole moment $\mathfrak{p}$. In a uniform field $\vec{\mathcal{E}}$, the change in the rotational energy is

$$E_{\text{rot}}(J, m_J) = \frac{(\mathfrak{p}\mathcal{E})^2}{2Bhc^2} \frac{J(J+1) - 3m_J^2}{J(J+1)(2J-1)(2J+3)},$$

so that

$$E_{\text{rot}}(0, 0) = -\frac{(\mathfrak{p}\mathcal{E})^2}{6Bhc^2}.$$

Numerical example: For $\mathcal{E} = 300$ Volt/cm, $B = 0.5$ cm^{-1}, and $\mathfrak{p} = 1$ Debye $= 10^{-18}$ CGS unit, a molecule has its levels shifted by

$$E = 25 \times 10^{-6} \times \frac{J(J+1) - 3m_J^2}{J(J+1)(2J-1)(2J+3)} \text{ cm}^{-1}$$

$$= 0.75 \times \frac{J(J+1) - 3m_J^2}{J(J+1)(2J-1)(2J+3)} \text{ MHz}.$$

The selection rules are $\Delta J = 1$, $\Delta m_J = 0$. The following table shows the Stark shift (difference between the unshifted line and the shifted component) for a linear molecule:

$J \to J'$	$m_J =$	0	1	2	3
$0 \to 1$		+0.5333	—	—	—
$1 \to 2$		−0.1524	+0.1238	—	—
$2 \to 3$		−0.0254	−0.0071	+0.0476	—
$3 \to 4$		−0.0092	−0.0056	+0.0052	+0.0288

This is a quadratic Stark effect: There are components (shifted lines) on either side of the actual (unshifted) line, towards the low and high frequencies. The diagram is shown in Fig. 12.23, qualitatively. The unshifted line is in the middle, drawn thicker. It is flanked by the components (only one in the case $J = 0 \to 1$). The effect gets weaker and the shift decreases as J increases. For example, for $J = 3 \to 4$ the shift is 10^{-2} MHz, but if we intensify the field ten times, the shift becomes 1 MHz.

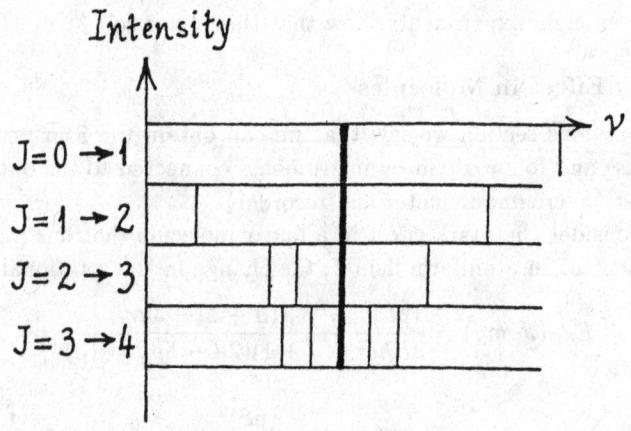

Fig. 12.23.

Chapter 13

MOLECULAR SPECTROSCOPY

13.1. Introduction

If we observe the hydrogen spectrum emitted by a discharge tube filled with hydrogen gas, we see an atomic spectrum given by H atoms (dissociated H_2 molecules in the tube) *and* other spectral lines given by H_2 molecules not yet dissociated. The same thing happens in the spectrum of a C arc: Bands are seen in the violet region of the optical spectrum (which form a *head*, *i.e.*, are sharper, towards lower frequencies and shade away towards higher frequencies, fading away gradually). If we use a spectrometer of a higher resolving power, we see that these bands (which belong to CN molecules) consist of many lines accumulated together on the lower-frequency side and spread out towards the higher-frequency side. If the resolving power is low, the lines may appear as a single line, or at most as a narrow continuous spectrum (band).

Band spectra are of molecular origin. Every band in the spectrogram is a bunch of many spectral lines.

In terms of the *method* used (and the action that the incident radiation is subject to upon meeting the molecule), the spectra are of three categories:

(1) *Absorption spectra.*
(2) *Emission spectra.*
(3) *Raman spectra.*

In terms of the microscropic *mechanism* that produces them, molecular spectra are of the three kinds:

(1) *Electronic spectra.*

373

(2) *Vibration spectra.*

(3) *Rotation spectra.*

Usually, the last two coexist together, forming the *vibration-rotation spectra.*

In terms of the *wavelength* involved (and consequently, the technique used), molecular spectra are classified as follows:

(1) *Optical* (visible region) *spectra.*

(2) *Infrared spectra.*

(3) *Microwave spectra.*

There is a relation between the second and third grouping in terms of the range they cover in the electromagnetic spectrum. For example, electronic spectra are in the visible and ultraviolet region, while vibration spectra are in the infrared. The second grouping gives the three spectral ranges of the molecular spectra, corresponding to three types of transitions. The corresponding energies vary and fall in different regions in the electromagnetic spectrum. Quantum mechanics explains the reason, and a theoretical analysis thereof is given in the previous chapters.

Emission spectra are obtained mostly from free atoms. An atom emits radiation after being excited by it first. So, bright lines appear in the spectrogram, corresponding to the transition frequencies. Absorption spectra are obtained from molecules. A molecule absorbs incident radiation at a resonant frequency equal to that of an interlevel transition $\Delta E = h\nu$. The absorbed frequencies appear as dark lines over a background of a continuous spectrum, right at the places of the corresponding frequencies of the latter.

Molecular spectroscopy is done mostly by absorption.

Electronic spectra belong to excited molecules (actually, excited electrons in molecules) and are in the visible and ultraviolet range. Microwave spectra are due to transitions between pure rotational levels. Infrared spectra are due to vibrations-rotations. The Raman spectra can be vibrational Raman or rotational Raman, and will be discussed below.

13.2. Utility of the Emission and Absorption Spectroscopy

Information about molecular parameters and symmetry can be obtained from molecular spectra. Certain molecular properties and quantities can be measured. Electronic spectra, too, can help us draw some conclusions about molecules.

13.3. Electronic Spectra

Classically, any vibrating charge (and consequently, an electron) radiates electromagnetic waves (energy). In quantum mechanics (and atomic physics) we have electronic "vibrations" in an atom, where the word "vibrations" must be taken in its loose meaning, which is to be construed as jumps (of the electron) between electronic levels, from orbit to orbit (from n to n') in the atom. When n increases (decreases) we have absorption (emission) of quantic energy.

So, in quantum mechanics electronic radiation (and its spectrum) comes from transitions between electronic states, but it is influenced by the vibrations of the nuclei and the rotations of the molecule, *i.e.*, there is a coupling between a pure electronic state and a vibration-rotation state. In other words, an outer electron emits visible radiation at a wavelength which is perturbed by the motion of the nuclei and the rotation of the molecule as a whole. The latter always splits the energies (and hence the lines), as in the case of CN stated above.

The electronic bonds are discussed in Section 11.2 where we saw that when the electronic state changes in a transition, the resulting energy change is large enough, so that the band falls in the visible or ultraviolet region.

Notice that the electronic states are labeled by (the quantum number) n where $n = 1, 2, 3, \ldots$ (is an integer starting with 1), whereas the states of an harmonic oscillator are labeled by (the quantum number) v where v $= 0, 1, 2, 3, \ldots$ (is an integer starting with 0). In quantum mechanics texts one finds the formula for the energies of the harmonic oscillator given as

$$E_n = h\nu \left(n + \frac{1}{2} \right),$$

but in our book we used v instead of n, lest it be confused with the electronic state label n.

13.4. Infrared Spectroscopy

When a body is heated it radiates heat (*i.e.*, infrared rays). If these rays are collimated towards a sample (whose absorption is to be studied) and then are dispersed by a prism or a grating, an infrared spectrum is obtained.[1] Of course, the prism must be transparent to infrared rays (NaCl or KBr). The spectrum is viewed through a detector which scans all narrow parts of the

---

[1] Do not forget that *radiated* (not conducted or convected) heat constitutes electromagnetic *rays* or *waves*, requiring no carrier medium for its propagation.

range of the wavelengths present. The resolution of a detector depends on the width of the narrowest part (of the spectrum) that the instrument can discern. During the scan of the continuous spectrum one sees dips (intensity drops) corresponding to wavelengths absorbed by the molecules of the sample. These wavelengths correspond to absorptive transitions in the molecules.

13.5. Resolution

The *resolution* — or *resolving power*, we are not going to play with the words[2] — of the spectroscope is the difference $\Delta\lambda$ between the two closest lines (wavelengths) that the instrument can discriminate (tell apart).

The difference between high and low resolution resembles the difference between a coarse and a fine sieve. We will illustrate the point by another example: Let us suppose that the state of some country passes a law that prohibits private production of spirits and alcoholic beverages, both on a large scale (industrially, for commercial use) and individually, for personal use. A factory of the private sector cannot get on with its operation because it will be detected immediately. Nevertheless an individual citizen can be engaged in this activity at home, for private consumption. This individual can go unde-tected because the resolution of the law enforcement is low. His/her activity will be too fine for the coarse state control. However, the state may dispatch patrols to search houses throughout the territory, thereby increasing the res-olution of the detection process. Consequently, the law enforcement becomes more penetrating and of a higher resolution, being able to spot illegal alcohol production on a smaller scale. The better the patrols work the higher the resolution (more incidents are detected).

13.6. Raman Spectroscopy

Monochromatic rays are sent over to a gas. The observer looks into the gas at right angles to the incident rays and sees the scattered rays (by the molecules of the gas). These rays are resolved to their wavelengths (by a prism) and thus discrete spectral lines are obtained. If the incident radiation frequency is ν_0, a quantum $h\nu_0$ can elastically collide with the molecules and get scattered (with the same ν_0) (*Rayleigh scattering*). However, the hit molecule can get excited to an upper vibrational level (*e.g.*, $v = 0 \rightarrow v' = 1$) and gain energy

---

[2]Purists distinguish between the two terms with respect to their definition, but we do not intend to enter into demagogy here. We may use the two terminologies interchangeably as long as we stick to the canonical definition given in this section. It is the conventional notion that counts, not the bureaucratic nomenclature.

by the absorption of a quantum $h\nu$ (which is equal to ΔE_{vibr}). In that case the scattered ray will have an energy equal to $h(\nu_0 - \nu)$. This gives rise to a Raman line (whose frequency is $\nu_0 - \nu < \nu_0$), namely a *Stokes line*. But the opposite can also happen: The molecule may initially be in state v = 1 and get de-excited to state v = 0 *upon collision* (not spontaneously), in which case the scattered photon frequency will be $\nu_0 + \nu$ (of higher frequency than the incident ν_0), giving rise to an *anti-Stokes line* of the same (Raman) spectrum. The Raman spectrum contains both vibration and rotation lines (of $\nu_0 \pm \nu$). Obviously, since most of the molecular population is in the ground vibrational state, the Stokes lines are more intense than the anti-Stokes lines. The rotational Raman lines are not that simple, because at room temperature the molecules are in high J states.

Experimentally, we keep the sample gas in a transparent container and we observe emission and absorption at a certain temperature, by sending monochromatic light (of frequency ν_0) to the gas. We watch at right angles to the incident rays and see and examine the produced spectrum (Fig. 13.1) which is due to absorptions and emissions effected by the gas molecules. What we see is actually the spectrum obtained by scattering, namely by the scattered photons.

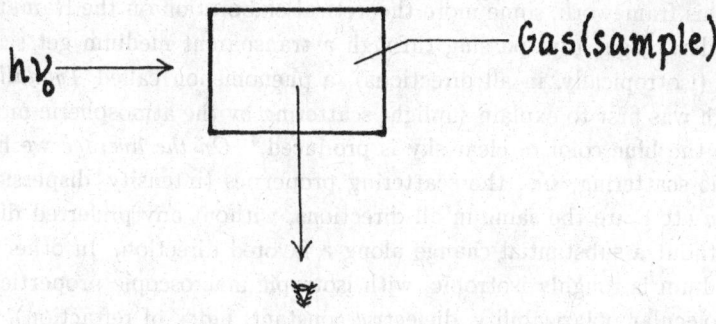

Fig. 13.1.

Figure 13.2 shows the *Raman spectrum* $(\nu_0 \pm \nu)$, a picture[3] named after the Hindu physicist Raman. The central line ν_0 is the frequency of the incident light. Besides ν_0 one observes (on its either side) other lines, too, the Stokes and anti-Stokes lines. The scattering of ν_0 is just the Rayleigh scattering. From

---

[3]The literal definition of *spectrum* is as follows: Spectrum is the picture of the analysis of light into its component wavelengths.

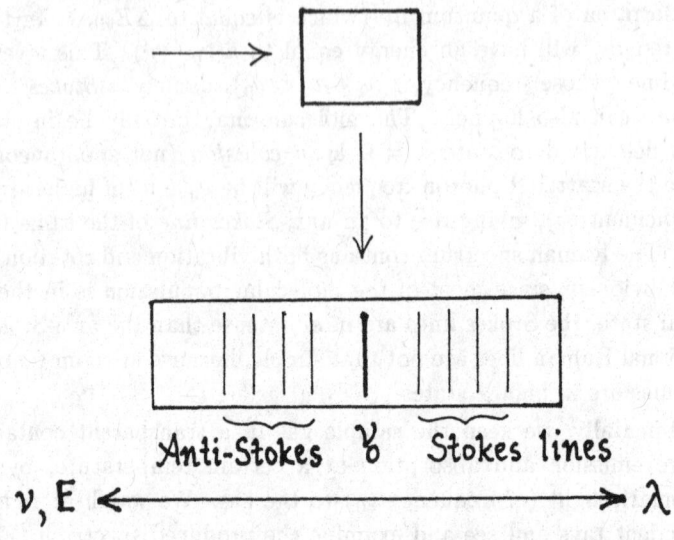

Fig. 13.2.

$E = h\nu$ we conclude that the energy of the anti-Stokes lines is higher than that of the Stokes lines.

In this framework, some more theoretical elaboration on the Raman effect is of order. Light rays passing through a transparent medium get scattered around (isotropically, in all directions), a phenomenon called *Tyndall effect*. Rayleigh was first to explain sunlight scattering by the atmospheric molecules whence the blue color of clear sky is produced.[4] *On the average* we have an isotropic scattering, *i.e.*, the scattering properties (intensity, dispersion, absorption etc.) are the same in all directions, without any preferred direction and without a substantial change along a favored direction. In other words, the medium is roughly isotropic, with isotropic macroscopic properties (density, molecular polarizability, dielectric constant, index of refraction). (Local anisotropies, peculiarities, asymmetries, and other minor anomalies and irregularities do not count.)

Individual molecules (or groups thereof) whose sizes are smaller than the incident wavelength λ_0 scatter the light. An elastic scattering will not change λ_0, as Rayleigh put it. But this does not happen in reality, because the molecules

---

[4]It is due to the scattering off the air molecules that we are able to see a light beam from a projector in the night from a sideways direction perpendicular to the beam, without directly seeing the projector (light source) itself. So, in vacuum we cannot see the beam, except, of course, if the projector is directed towards us.

move, and hence we obtain the Stokes and anti-Stokes lines *in addition to* the scattered λ_0. In other words, λ_0 suffers a change (shift) upon scattering, *i.e.*, *part of* the incident quanta may get absorbed by the molecules and re-emitted later. Whatever ν_0 is, the Stokes and anti-Stokes lines exist flanking it in the spectrogram (even if it changes). This fact was experimentally established by Raman in 1928.

A word of caution here is proper, concerning the difference between the fluorescence (radiative transition) lines and Raman lines. In fluorescence, the obtained lines correspond to naturally fixed transitions with which the incident radiation frequencies happen to be resonant. That is, we spectroscopically recognize the sample material from the characteristic lines it gives. In the Raman scattering — as we watch the spectrum — we have *frequency shifts* produced by the (character of) the sample substance, rather than the frequencies *per se*.

As mentioned above, in an elastic collision, ν_0 does not change, *i.e.*, if the hit molecule stays in the same state. But if the molecule gets to another state, say, from $E(v)$ to $E(v')$, the bookkeeping equation of the energy conservation reads

$$h\nu_0 + E(v) = E(v') + h\nu_{sc}$$

where ν_{sc} is the scattered frequency, and $E(v) - E(v') = \Delta E_{vibr}$. So, the frequency of the obtained line will be

$$\nu_{sc} = \nu_0 + \frac{\Delta E_{vibr}}{h} = \nu_0 \pm \nu$$

where ΔE_{vibr} is positive (negative) when we have emission (absorption) by the molecule. In other words, as the molecule goes from state v to state v', we have

$$\nu_{sc} = \nu_0 + [E(v) - E(v')]/h$$

where if the second term in the brackets is positive (v > v'), we have emission ($\nu_{sc} > \nu_0$); if it is negative (v < v'), we have absorption ($\nu_{sc} < \nu_0$). So, the molecule makes absorption and emission of quanta. The frequency of the emitted/absorbed quantum is equal to the shift between the incident frequency ν_0 and the scattered frequency which otherwise is called *Raman line frequency*.

Although a Raman line is related to the molecular absorption/emission line as described above, its spectral intensity is *not* related to that of the latter, because the selection rules are not the same. In the Raman spectroscopy we can see transitions otherwise forbidden! This confused the minds of scientists at first as something incomprehensible, but its importance is nevertheless obvious. What makes the effect interesting is that *an interlevel transition is allowed* (say, from state 1 to state 2) *if and only if there is another level* (say, state 3) (or more than one) *where usual transitions are possible* (*allowed*) *between this third level and the other two separately* (*i.e.*, between state 1 and state 3 *and* between state 2 and state 3). This can be interpreted as a stepwise *two-stage transition* (first from 1 to 3, and next from 3 to 2). In the case of the usual radiative transition (fluorescence), any state has a lifetime (*i.e.*, the molecule stays in it for a while); in this interpretation, however, we cannot apply this to state 3.

We considered ΔE_{vibr} above, but it does not have to be ΔE_{vibr} only; transitions between electronic, rotational, or vibration-rotation levels are all possible in the Raman effect, if allowed. So, the Raman lines of polyatomic molecules that scatter light may cover all the types of spectra, regardless if these Raman spectra are indeed observable or are inhibited by other selection rules.

The selection rule for the usual rotation spectrum is $\Delta J = \pm 1$. But for the rotational *Raman* spectrum (*i.e.*, the lines associated with it) we have $\Delta J = 0, \pm 2$. Here $\Delta J = 0$ corresponds to a transition from state 1 to state 3 ($J \rightarrow J \pm 1$), and then form state 3 to state 2 ($J \pm 1 \rightarrow J$). This selection rule brings no change to the rotational energy of the molecule, and hence it corresponds to the elastic (Rayleigh) scattering where the incident exciting frequency ν_0 remains unmodified. So, for the observed rotational Raman lines the selection rule must be $\Delta J = \pm 2$ which spaces the spectral lines two times as much as does the selection rule $\Delta J = \pm 1$ for the lines of the usual (ordinary) rotation spectrum. This can be found from

$$E \propto J(J+1),$$

$$\Delta E \propto J_1(J_1 + 1) - J_2(J_2 + 1) = J(J+1) - (J-2)(J-1) = 2(2J-1)$$

where J is the higher level of rotation. So, the shifts follow this spacing for every J. The spacing between the ν_0 line and the first Raman line (on either side of ν_0) is larger than the (equal) spacings between the rest of the lines.

Now let us consider the vibration-rotation Raman spectrum. In addition to the rotational Raman lines (and mostly on the low-frequency side only), a band of vibration-rotation lines appears in the spectrogram, with selection rule $\Delta J = 0, \pm 2$ where $\Delta J = 0$ again corresponds to the unchanged rotational energy whose corresponding line comes to appear slightly off the position of the absent Q-line. On its either side there come the lines of $\Delta J = \pm 2$. This vibration-rotation Raman band lies far away from the pure rotational Raman band.

If we have a vibrational transition from v = 0 to v = 1, and the v = 1 level is so sufficiently low that at room temperature (in thermal noise) many molecules are in v = 1, another Raman band (albeit less intense) is observed on the high-frequency side of ν_0, due to *spontaneous* de-excitations from v = 1 to v = 0.

A_2 type molecules give Raman spectra (rotational and vibration-rotation, but the former are not easily seen). Such molecules exhibit peculiarities and a special behavior, which will not be discussed here, because the case is treated well in Section 20.9 of Richtmyer's book (listed in References). The ortho and para-states and the wavefunctions of these molecules are discussed there, along with the effects on their band spectra.

As far as heteronuclear (AB type) molecules are concerned, their Raman spectra are observed directly. For example, when we irradiate HCl (in gaseous state) with light from a Hg lamp, we obtain a Raman spectrum where the $\lambda_0 = 4047$ Å of Hg excites a vibration-rotation transition at 4581 Å. The spacing between 4047 Å and 4581 Å agrees with the location of the absent Q-line of HCl. On either side of λ_0 one observes Raman lines corresponding to the pure rotational Raman spectrum. Another Hg line ($\lambda_0 = 4358$ Å) is also observed with a bunch of Raman lines flanking it, whose average spacing is about 41.5 cm^{-1} and thus agrees with *double the spacing* $\Delta\bar{\nu}$ of the usual rotation spectrum of HCl shown in the table in Section 10.1.

In short, (a) the selection rule for Raman spectra is $\Delta J = 0, \pm 2$ (where due to the possibility of ± 2 the spacings are twice as wide as the ones of pure rotation spectra), and (b) the Raman spectra are of great value and strong interest, because they give us appreciable information about molecules and their quantum mechanical states.

13.7. Selection Rules

Like in atoms, molecular transitions, too, are subject to selection rules (limitations on quantum numbers).[5]

Every kind of spectrum has its own selection rules, but there are also *general selection rules* determining the possibility of a transition, *i.e.*, whether a molecule is allowed to make certain transitions and give rotation or vibration spectra. The selection rules determine forbidden transitions; without the selection rules, the possibility of a transition is doubtful.

---

[5]For example, in atoms an $s \rightarrow p$ or a $p \rightarrow d$ transition is possible, but not an $s \rightarrow d$ transition. Without a selection rule at hand (which nature puts), one may contest the possibility of a transition.

As far as the electronic transitions are concerned, the same selection rules that hold for an atom hold also for a molecule. The existing molecular selection rules forbid A_2 type molecules (*e.g.*, Cl_2) to give infrared vibrational spectra, because of symmetry. That is, as the molecule vibrates along its axis (symmetrically about its center of mass), the charges move symmetrically back and forth, and the charge vibrations on the two sides cancel the effect of each other. This amounts to no *net* displacement of charge, and hence the molecule cannot resonate with a varying field sent to it. We say that the molecule is *inactive* in the infrared region and has forbidden (by nature and construction) vibrational transitions. (However, if the symmetry of the molecule is somehow distributed — *e.g.*, by a collision — the selection rule may break down, but such a probability is usually low, and the resulting — then allowed — lines are of extremely weak intensity.) Contrary to the A_2 type molecules, an AB type molecule (*e.g.*, HCl) can be viewed as a vibrating *single* charge (net effect because of the asymmetry), and hence it is allowed to interact with an incident radiation field and thus give an infrared vibrational spectrum.

All these considerations are valid for polyatomic molecules, too, but the possibility of certain vibrations is debatable there; because certain vibrations are allowed while others are not. Each vibrational frequency corresponds to a different fashion (mode) of vibration. To have an infrared vibrational spectrum, the *instantaneous* dipole moment must be $\mathfrak{p}_0 \neq 0$ (though the permanent — average — dipole moment *may* be zero). In other words, we must have $d\mathfrak{p}/dq \neq 0$ where q is any generalized coordinate that describes the molecular vibrational motion. *The vibration must change the dipole moment of the molecule.* Then an infrared vibrational spectrum can be observed.

To have a microwave rotational spectrum, the dipole moment of the molecule must change in the direction of the incident radiation, as the molecule rotates. This necessitates the existence of a permanent $\mathfrak{p}$ already. So, to observe a pure rotation spectrum, we should have $\mathfrak{p} \neq 0$.

The general selection rule for Raman spectra says that the molecule must be polarizable as it moves: $d\alpha/dq \neq 0$. That is, an α must be formed by the molecular motion, and furthermore, α must change during the vibrations. So, the selection rule can be stated as follows: To have a Raman spectrum, *the vibration must change* α. An A_2 type molecule gives a vibrational Raman spectrum, because when the bond length changes, α changes. A vibrational Raman spectrum is thus observed. (However, molecules with threefold and manifold rotational symmetry axes do *not* give Raman spectra. For example, the benzene molecule has a sixfold axis; so, it does not give a rotational Raman spectrum as it rotates.) Here α is the *polarizability*.

The selection rule for pure rotational spectra is that the instantaneous dipole of the molecule must be $\mathfrak{p}_0 \neq 0$. The selection rule for a rotational Raman spectrum is that α must be anisotropic in a direction perpendicular to the rotation axis of the molecule, so that as the molecule rotates, it presents a changing $\mathfrak{p}$ (which can interact with the incident varying field and make an absorption, or it can make a dipole emission).

13.8. The Electromagnetic Spectrum

The continuous electromagnetic spectrum extends from very high to very low frequencies (very short to very long wavelengths, *i.e.*, from γ and X-rays to radio and TV waves). The optical (visible) part occupies a very narrow region of the spectrum (roughly, from 4000 Å to 7000 Å) (Fig. 13.3).

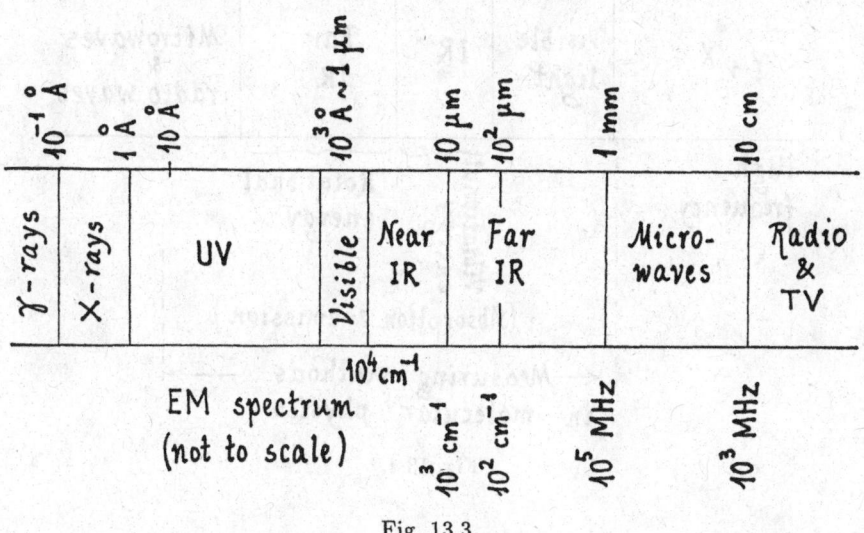

Fig. 13.3.

Notice that in the high-frequency region spectroscopists use Å (as the unit of λ). In the middle region they use cm^{-1} (unit of $\bar{\nu}$, the wave number). In the low-frequency region they use MHz (unit of ν). To the dismay of the purists, energy is also expressed in these units!

The dimension of $\bar{\nu}$ is one over length (and hence its unit is cm^{-1}), because $\bar{\nu}$ is the number of waves per unit length (or number of wavelengths included in an interval of 1 cm):

$$\bar{\nu} = \frac{1}{\lambda} = \frac{\nu}{c} \quad \text{waves/cm, or } cm^{-1} .$$

Molecular spectroscopy has to do with three regions of the spectrum (visible, infrared, microwaves) (Fig. 13.4), with different spectrometers for each. The relevant intervals of energies and the corresponding measurements are shown in Fig. 13.4 and Fig. 13.5. The parts of γ and X-rays correspond to high energies (of electromagnetic waves) and lie outside the spectroscopic interest of molecular physics (because no molecular spectrum is observed there). The ultraviolet region has to do mostly with atoms (*e.g.*, Hg emits partly in this region). The molecular E_{vibr} lies in the infrared, and E_{rot} in the microwave region.

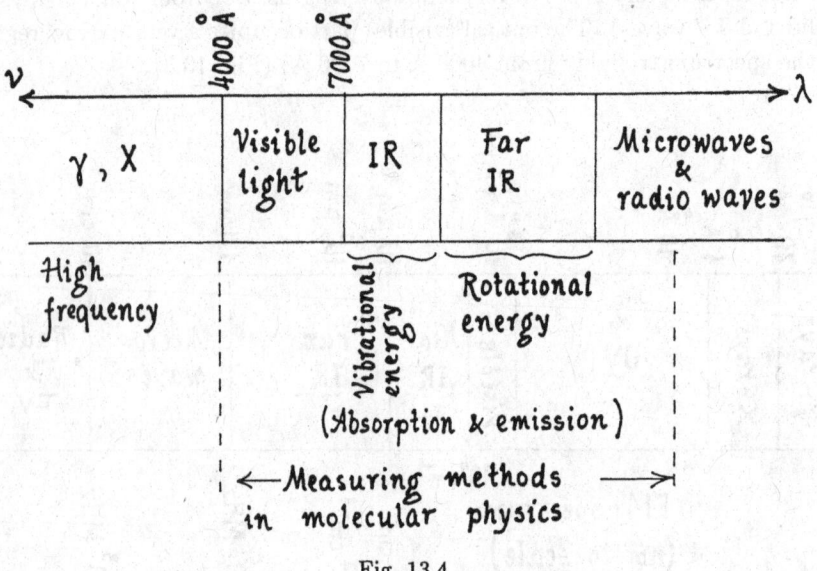

Fig. 13.4.

Fig. 13.5.

Classically, energy is continuous and can take any value. An emitted (radiated) classical field is also continuous (in terms of frequency and energy). But microscopically, energy is quantized, *i.e.*, exchanged in quanta (photons), just like economic values exchanged in money which in fact consists of multiples of a definite currency unit which is the economic quantum.[6]

To recapitulate what has been said in the previous chapters, the molecular energy E is

$$E = E_{el} + E_{vibr} + E_{rot} + \text{Smaller terms}$$

where E_{el} is the term of electronic binding energies (states of the electrons bound to the atoms of the molecule) specified by n, l, m_l, and m_s. E_{vibr} is the term of the energies of the molecular vibrations (vibrating nuclei) specified by v. And E_{rot} are the energies specified by J. The other terms include spin interactions, quadrupole effects etc.

13.9. Experimental Spectroscopy

Spectroscopy is done by setting up and operating the following arrangement:

(1) Source of incident radiation.
(2) Specimen, *i.e.*, substance of interest whose spectrum (absorption/emission) is to be studied (irradiated by the radiation of the source).
(3) Dispersion system (prism or diffraction grating appropriate for the radiation used).
(4) Detector (a system that gives the intensity of the spectral lines on a screen (oscilloscope) or a dial (electrometer), or as a plotted curve (recorder)).

This arrangement is called *spectroscope* or *spectrometer*, depending on the purpose of the observation.

In the case of emission spectroscopy with atoms, the source and specimen are usually merged into one entity, *i.e.*, they are the same thing (*e.g.*, a discharge tube whose light is to be analyzed). In other cases the source can be a lamp (ordinary light source) or a LASER, as appropriate. For example, ordinary lamps (lightbulbs) used in household lighting give a continuous spectrum more or less like that of the sun, with all the visible wavelengths present.

---

[6]There is a direct similarity and correspondence between physical and economic quantities. For example, energy corresponds to capital; time to time; entropy to entropy (or informatic ignorance — disorder); heat to market mobility; temperature to communications and transport; force to force (and pressure); elasticity to elasticity etc.

(There are also solar simulation lamps, used, for instance, in solariums and winter tanning in Swiss health resorts.)

In this section we will briefly say a few words about light sources in spectroscopy.

In the *emission* spectroscopy one can use as a source the following:

(1) A discharge tube: For gaseous samples only.

(2) Flames: For example, city gas, oxygen, acetylene.

(3) Arc lamps: DC arcs give neutral atomic spectra, while AC arcs give neutral atomic *and* ion spectra.

In the *absorption* spectroscopy, tungsten-filament lamps are used. Ordinary lamps will do.

In *infrared* spectroscopy we use Nernst lamps.

In *Raman* spectroscopy we can use Hg arc lamps, and also He, Cd, and Ar lamps. Toronto type low-pressure Hg lamps (in spiral shape) are widely used. LASER (*alias* optical MASER) can also be used if we want an intense source.

In *microwave* spectroscopy we use clystrons (special electronic tubes that generate microwaves).

The interested parties can buy ready, manufactured spectroscopes that are commercially available. They are sensitive and expensive, and come with all their accessories, for spectral analysis and professional laboratory use. But one can make a spectroscope of one's own — evidently, less sensitive — by building it part by part, for home use and hobby purposes. In that case the lamps (light sources) that can be used for molecular spectroscopy are of common kinds, easily found in the market. For example, we suggest the following kinds of lamps:

Group A: INCANDESCENT LAMPS

(1) *Ordinary incandescent lamps*: Used commonly in households and public places.

(2) *Tungsten halogen lamps*: Their Wattage range is 15 to 2000 Watts. There are also "cool lamps" of this kind, for applications where the beam heat may damage the specimen. These lamps are used in indoor, accent, and task lighting.

(3) *Special lamps*: Daylight-blue lamps, roadstand lamps, airfield lamps, garden party and decorative lamps etc.

Group B: FLUORESCENT LAMPS

Used commonly in household, public and commercial lighting. There are versions with no stroboscopic effect, suitable for light regulation and dimming. They are economical.

Group C: GAS DISCHARGE LAMPS

They are high-intensity, high or low-pressure vapor lamps of several kinds:

(1) *High-pressure Hg lamps*: They have high light output and give accurately directed light. They are discharge lamps with mercury vapor, used in photochemical processes, outdoor lighting, decorative floodlights etc.

(2) *Low-pressure Hg lamps*: They produce ozone and ultraviolet radiation (dangerous to the eyes and skin upon direct exposure). They are used for sterilization and odor elimination.

(3) *Compact source Hg lamps*: Their maximum arc stability is unaffected by voltage fluctuations. They supply high energy concentration in a small space and high luminance in the visible and ultraviolet spectral regions. They are designed for sophisticated optical systems used in projection and instrumentation (slide projectors, microscope lighting etc.).

(4) *Blacklight blue lamps*: They are fluorescent and high-pressure Hg lamps used in testing, inspection, and analysis (in chemistry, metallurgy, food processing, medicine, criminology, archaeology, public signs, theaters, night clubs, circuses, spectacular effects etc.).

(5) *High-pressure metal halide lamps*: They are gas discharge lamps of stable lumen output and high illuminance levels, producing accurately controlled light. They are used in indoors industrial and public lighting (CTV broadcasts, color filming, shop windows, office lighting, open-air performances, spotlight projectors, skating rinks, stadiums, plant irradiation etc.).

(6) *Low-pressure Na lamps*: They are Na vapor discharge lamps of high visual acuity, low luminosity, and little glare, good for humid and dusty places and chemically aggressive and corrosive atmospheres.

(7) *High-pressure Na lamps*: They have high luminous efficacy and are used in road lighting, optical systems, spectroscopy, security and marine lighting etc.

Group D: SPECIAL LAMPS

They are designed for special purposes. They produce a powerful and accurately controlled light beam, and are of high quality standards. They are used in TV studios, movie-shooting sets, discos, opera-house stages, search-lights, photography, spectral analysis, cosmetic treatment, agriculture, over-head projectors, photocopiers, music halls etc. This category includes the quartz halogen (incandescent) lamps (used in microscopes, image projectors, microreaders, aviation etc.), low-pressure halogen lamps, arc lamps, gas-discharge tin halide lamps, gas-discharge xenon lamps, ultraviolet lamps (used in germicide, pharmaceutical manufacturing, bacteriological research, water sterilization, curing processes, cabarets etc.), ultraviolet actinic lamps (used in insect trapping, skin-disease treatment, photoprinting), standard spectral lamps (emitting monochromatic radiation for spectral analysis), infrared lamps (used in physics laboratories, drying processes, animal rearing, health care, portable furnaces etc.), and quartz infrared lamps.

Technical characteristics and other particulars about all these groups of lamps can be found in lighting and lamp catalogues published by Philips, Siemens, Westinghouse, General Electric etc.

13.10. Philosophical Implications

A spectrum reveals the beauty (and functioning) of nature as excitingly as a painting or a piece of poetry would, refering to natural beauties.[7] But the merit is not only of esthetic or axiological significance. It goes far beyond, into philosophical considerations. For example, one cannot help making the following parallelism: An institution has a legal status on the basis of which it functions. Similarly, a molecule has rules put by nature (*e.g.*, selection rules put by quantum mechanics) in accordance with which it exists and behaves. One cannot bypass indifferently these built-in philosophical facts on one's way through the discipline of physics — facts that make one stop and think.

Another striking analogy is the following: In economics, prices are determined by the interaction of the market forces and by some government intervention and adjustment. In physics, energies are determined by the interaction of physical forces and the intervention of quantum mechanics (*e.g.*, uncertainty principle, Pauli principle etc.). There are also similarities in absorption processes. In other words, one realizes that there are invariant mechanisms whose essence is the same everywhere.

In classical ballet one speaks of physical energy or philosophy of motion (where feelings are expressed by movements and musical support); but these philosophical concepts can be found in physics, too. What we are trying to

---

[7]For this reason, it ought to be unfair to call physics "hard stuff".

say — and what reinforces the merit of physics — is that physics cannot be separated or alienated from philosophy and other subjects — including daily life (where we always bump to it) — because philosophy constitutes an inevitable substratum that keeps all knowledge united in common grounds or overlapping intersections with an organic cohesion.

In politics, the stability of a country is determined by the interaction of political forces (and counterforces, according to Newton's third law) and the intervention of state laws and government policy. In physics, molecular stability is determined by the interaction of binding forces, centripetal forces (responsible for rotations), and restoring forces (that account for vibrations). The concept of natural rotation can be philosophically extended to the notion of periodicity and repeatedness, and hence to the sequence of events, the seasonal events in nature, or the philosophical circularity of time. On the other hand, the sense of vibrations and rhythm can be associated with a whole bunch of phenomena, from clockmaking to eating habits, from dance steps to the rhythmicality of poetic verses, and so on. There is a remarkably close affinity and intermingling relationship between philosophical concepts and common patterns that pervade all disciplines (especially sciences and fine arts[8]), as if they were all parts (members) of an harmonious unity analogous to what is called *homeostasis* (state of perfect health by the perfect cooperation and coexistence of all parts to make one feel as a unity) in medicine and information theory.

These considerations are highly philosophical. Philosophy enters physics by right. Those physicists who are afraid of that, should become engineers (just like economists who turn to business administration). Physics without philosophy is good for *technicians of science*, not physicists. When a physicist extends his/her interest to the epistemology, ontology, and metaphysics (in short, to the philosophy) of his/her scientific subject, he/she may be contemptuously accused by science technicians of dealing with "cheap eclecticism" or "cheap syncretism"; such subjective reactions are, unfortunately, not rare.

---

[8]There is a mutual approach between sciences and arts: A sunset is a topic common to both, for example. Sciences and arts serve each other (including the mediation of the technological means) and sometimes coexist in substance or in the means of performance, something that can be seen in a Cousteau documentary, a Jules Verne novel, a Da Vinci painting, an Homeric saga, a Rachmaninov piano concerto, a Walt Disney movie, a Dostoevsky work, a Eugene O'Neal play, or a Gothic cathedral.

Chapter 14

MOLECULAR SPECTROMETRY

14.1. Kinds of Molecular Spectroscopy (Spectroscopic Methods)

Classical spectroscopy is one of the experimental methods in molecular physics. We have the following branches of it:

(1) *Optical spectroscopy* (with visible light).
(2) *Infrared spectroscopy* (which falls in the infrared region of the electro-magnetic spectrum and regards the atomic vibrations and the duet of rotations-vibrations).
(3) *Raman spectroscopy*.
(4) *Microwave spectroscopy* (which regards the molecular rotations).

Since no molecular action emits (or absorbs) ultraviolet waves (rays), the ultraviolet region of the spectrum does not concern molecular physics. (Some electronic transitions belong to this region though.)

The optical spectroscopy is of two kinds:

(1) *Absorption spectroscopy*: We send electromagnetic radiation to a sample which absorbs certain wavelengths, and the resulting spectrum has dark lines corresponding to the absorbed wavelengths which are hence missing.
(2) *Emission spectroscopy*: We heat the sample (or supply energy to it via electrical discharges). Its atoms get excited to higher energy states, and as they return back to lower states (or the ground state) they emit the absorbed energy quantum back, giving off spectral lines (of various intensities that are related to transition probabilities) which are of the wavelengths of the emitted quanta.

Emission spectroscopy with discharge tubes cannot be done with molecules because the molecules dissociate in the tube. This method is good for atoms only.

We can do four actions concerning the spectra:

(1) *Spectroscopy*: Watching the spectrum (from the Greek verb *scopein* = to watch).
(2) *Spectrometry*: Measuring the spectrum (from the Greek verb *metrein* = to measure).
(3) *Spectrography*: Recording the spectrum (from the Greek verb *graphein* = to record, to write down).
(4) *Spectrophotography*: Taking the picture of the spectrum (from the Greek word *photographia* = picture, recording through light).

Accordingly, the instruments are called *spectroscope, spectrometer, spectrograph*, and *spectrophotograph*.

14.2. Absorption Spectrometry

The spectrometer works as follows: The sample is put on the way of the incident light (electromagnetic radiation), between the light source and the prism (beyond which there is a detector which could be a meter, a recorder, a photographic plate, or any similar device which can record intensity). We first place an empty (transparent) container between the beam chopper (flag) and the prism (Fig. 14.1), and thus we do the zero calibration (0%) of the detector. We next place a standard sample whose concentration (dissolved matter per unit volume) is known and we do the 100% adjustment with it. We divide the interval into a hundred equal parts, and thus the recorder has now a scale of divisions from 0 to 100. We finally place the sample for which the measurement is to be made. As light passes through it, there will be absorption. The absorbed wavelengths correspond to the vibrations and rotations of the sample molecules. This is a quantitative spectral analysis with a dispersive system. One can find the concentration of the unknown sample and identify the molecules it contains (from the absorbed wavelengths).

Figure 14.2 shows the plot of the intensity measured by the recorder. The upper plot is the intensity (versus λ) when there is no sample in between.[1]

---

[1]Actually, we place the empty container so that the *optical path* be accounted for, both in the absence and presence of the sample. That is, since the sample is kept in a container, when we remove it we should leave there an empty (identical) container, not a void space.

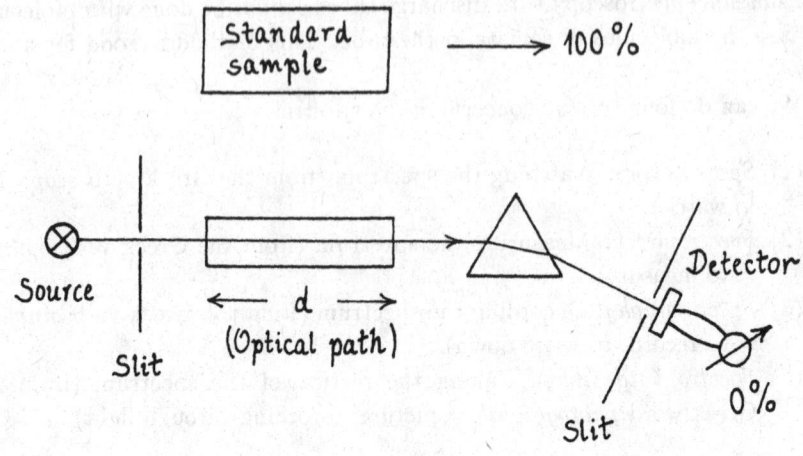

Fig. 14.1.

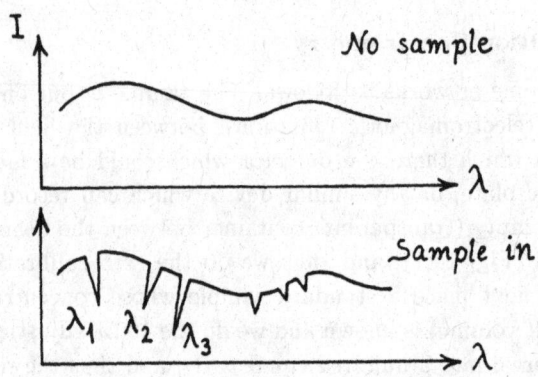

Fig. 14.2.

The lower graph represents the same measurement when we place the sample.

In the spectrum we observe dips (absorption minima) at certain wavelengths. These λ_1, λ_2, λ_3,... correspond to the quanta absorbed by the sample. Thus the spectrum gives us information about the structure of the sample and its molecular situation.

14.3. Emission Spectrometry

The idea here is that the sample to be studied becomes a source of electromagnetic waves (visible light) whose emitted light is studied, and hence the source is identified. Some materials give off very characteristic lights (spectra), *e.g.*, the yellow light of Na.

A primitive and very crude spectroscopy is as follows: We heat the sample and thus obtain a special source of electromagnetic waves (Fig. 14.3). The radiated light is observed. For example, we have a burning Bunsen burner and sprinkle table salt (NaCl) on it from a saltshaker. We immediately see a yellow flame, the characteristic yellow color of Na. Also, if we touch the tip of a wire and hold it in the Bunsen flame, we see again a slight yellow flame, because as we touched the wire, traces of NaCl from the perspiration of our fingers were deposited on it (Fig. 14.4). In short, as we hold the sample in the flame (whose temperature is high), the sample atoms get excited (or even ionized). As they return to their previous state they emit electromagnetic radiation.

As far as emission spectroscopy for atoms is concerned, we can use discharge tubes, a diffraction grating, and a spectrometer for observation (to actually see

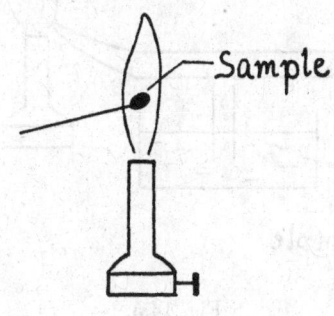

Fig. 14.3.

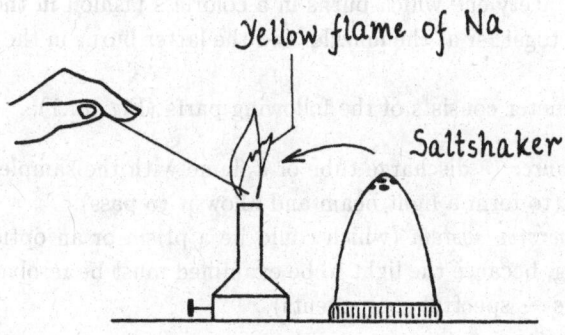

Fig. 14.4.

the colored spectral lines of various intensities at various angles, as we rotate the spectrometer with respect to the grating). The source can be a sodium lamp or a mercury lamp or a neon lamp whose spectrum is of interest and to be observed. (In the case of Na we observe the two main yellow lines $\lambda_2 = 5890$ Å and $\lambda_1 = 5896$ Å — the famous D_2 and D_1 lines of Na respectively — and in addition to them, more colored lines of weaker intensities, corresponding to other transitions, as the excited Na atoms get de-excited to lower states by the emission of visible quanta.)

One can use the *flame spectrophotometer* whose source works as follows: A reverse vacuum pump blows air through a pipe which communicates with a tank where we hold the sample material (Fig. 14.5). The sample can be in the form of a solid dissolved in a liquid, or a molten material. The other end of the pipe forms a muzzle that meets the Bunsen flame.

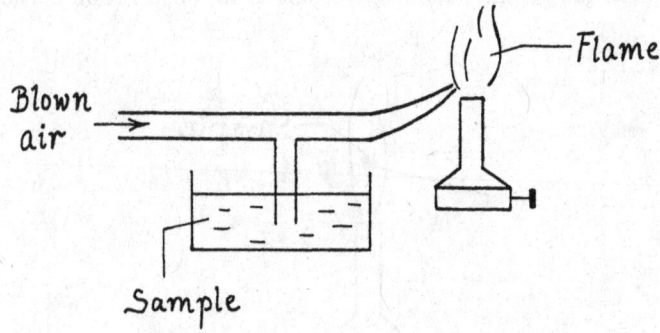

Fig. 14.5.

Still another arrangement is that shown in Fig. 14.6. The Bunsen burner is replaced by acetylene which burns in a colorless fashion in the air. Air and acetylene mix together at the muzzle, and the latter burns in the air, providing the flame.

A spectrometer consists of the following parts (Fig. 14.7):

(1) The *source* (a discharge tube or a flame with the sample).
(2) A *slit* (to form a light beam and allow it to pass).
(3) A *dispersive system* (which could be a prism or an optical diffraction grating, because the light to be examined must be resolved to its wavelengths — spectral components).
(4) A *detector*.

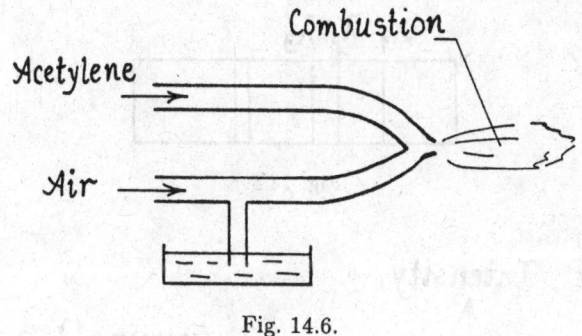

Fig. 14.6.

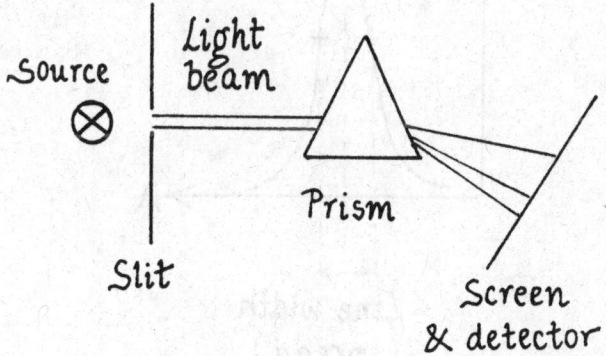

Fig. 14.7.

The light that comes in the form of a thin beam is resolved to wavelengths. The detector can be a photographic plate on which bright spectral lines corresponding to the present distinct wavelengths appear. The picture of the spectrum obtained looks like that in Fig. 14.8: A dark background and a succession of bright lines (just the opposite of that of the absorption spectroscopy where we have dark lines of missing — absorbed — wavelengths on a bright background of light). Of course each line is not ideally sharp (purely monochromatic). It has a line width (Fig. 14.9) where we have the nominal (main) wavelength λ_0 in the middle, and some weak dosages of immediately adjacent (slightly off) wavelengths on its either side. (The line widening and its reasons are not to be discussed here.) Thus we obtain a succession of lines: $\lambda_1, \lambda_2, \lambda_3, \ldots$. The detector can also be a photocell (whose exposed area is very small) or a photomultiplier tube, or even a photosensitive semiconductor (Fig. 14.10). We can also have an intensity recorder to measure and record

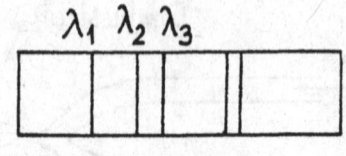

Fig. 14.8.

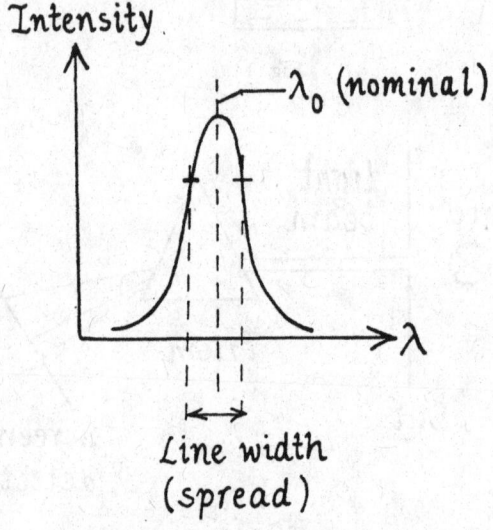

Fig. 14.9.

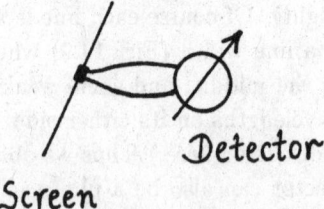

Fig. 14.10.

(plot) intensity (Fig. 14.11). Thus we can read the intensity from the plot. If we want an increased sensitivity and precision, we can connect the detector output to an amplifier and a phase detection arrangement whose output is in turn connected to an oscilloscope (on the screen of which we observe the outcome) (Fig. 14.12). Such a detection system has an augmented sensitivity.

In front of the source slit one places a beam chopper (flag) which can be opened and closed by an electromagnet (coil) modulated at 50 or 20 Hz

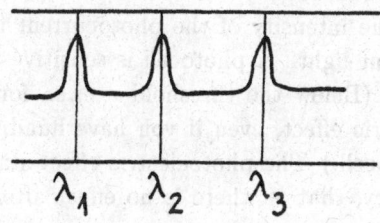

λ_1 λ_2 λ_3

Fig. 14.11.

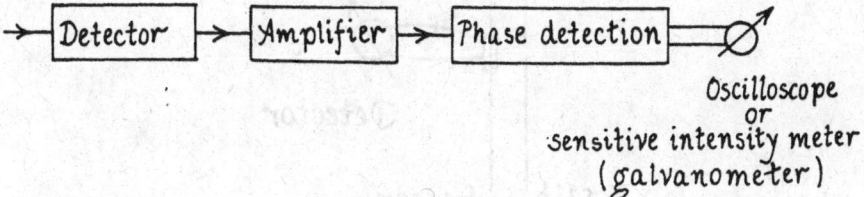

Oscilloscope
or
sensitive intensity meter
(galvanometer)

Fig. 14.12.

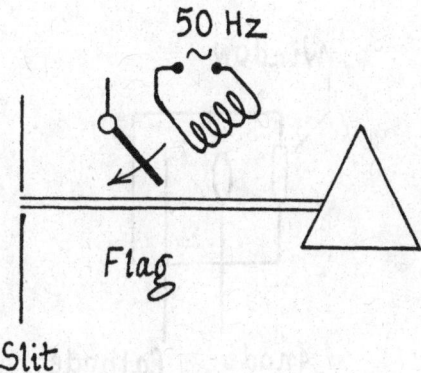

50 Hz

Flag

Slit

Fig. 14.13.

(Fig. 14.13). (This cannot be done manually because the frequency is too high.) Thus the beam passage is modulated by opening or intercepting the light track. This is a mechanical-electric-optical chopping system.

To make the detector photocell area small, one places a slit before it (Fig. 14.14). We operate this second slit by a mechanical system and thus we change λ. That is, if d_0 is the slit aperture at standard calibration and d_λ is the measurement aperture (in divisions), we can evaluate $d_\lambda - d_0$.

A photocell is a vacuum tube whose operation is based on the photoelectric effect (Fig. 14.15). The intensity of the photocurrent is proportional to the intensity of the incident light. A photocell is sensitive to certain frequencies after some threshold. (Below the threshold — *i.e.*, for lower frequencies — there is no photoelectric effect, even if you have hundreds of lamps around, illuminating the photocell.) The photoelectric effect starts taking place after the threshold frequency, that is, there is no effect after a certain maximum

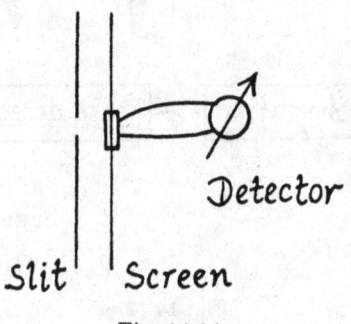

Detector

Slit *Screen*

Fig. 14.14.

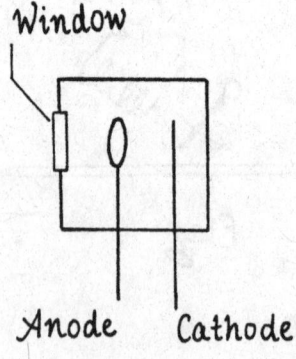

Window

Anode *Cathode*

Fig. 14.15.

λ. Hence the photocell is useless beyond that maximum λ (or for frequencies less than the threshold frequency). Every photocell must be used up to the maximum λ to which it is sensitive. Also, its performance depends on its light window.

The dependence of a photocell (or a photomultiplier) on λ has two components:

(1) It depends on the threshold photoelectric frequency of the cathode.
(2) It depends on the interval $\Delta\lambda$ transmitted by the window material.

Sometimes researchers place filters in front of the source. A filter transmits one wavelength (or a few ones), and thus the measurement sensitivity increases (because we have only a definite λ). In its simplest form, a filter can be a colored glass (which allows one λ — the one of its own color — plus certain others, while it trims the rest).

14.4. Prism Material

The prism of a spectrometer is not always made of glass. Its material depends on the wavelength we work with. (For example, if we work with ultraviolet light, we should use quartz instead of glass, because the latter does not transmit ultraviolet rays.[2]) The material chosen in each case is that which transmits best the used λ. For example, if we work in the ultraviolet region with $\lambda = 1200$ to 2000 Å, we use CaF_2 or LiF prisms. In the same region, for $\lambda = 2000$ to 4000 Å we use quartz as the best transmitter. In the optical (visible light) region, for $\lambda = 4000$ to 7000 Å we use miscellaneous glasses. Glass can be used up to $\lambda = 20,000$ Å. Beyond $\lambda = 2$ μm glass is no good because the absorption is large. So, for $\lambda = 2$ to 6 μm we use LiF; for $\lambda = 5$ to 15 μm we use NaCl; for $\lambda = 15$ to 25 μm we use KBr; and for $\lambda = 25$ to 40 μm (microwave region) we use TlBrI. The last three are highly hygroscopic materials: They retain humidity (water) from the air. Therefore, it is necessary to clean the prism, cover it well, and keep it in a dry place. Otherwise it retains moisture and the transparency of its surface disappears.

14.5. Raman Spectrometry

In Raman spectroscopy, the source must be monochromatic and very intense. The observation is made at right angles to the incident light beam

---

[2]Since ordinary glass absorbs ultraviolet radiation, it does not make sense to sunbathe behind glass panels. For the same reason, the glass of the ultraviolet lamps is quartz, not ordinary glass. Quartz glass allows the passage of ultraviolet rays.

(Fig. 14.16). If we watch at right angles to the beam, we observe Rayleigh scattering[3] (which gives polarized light) and Raman scattering. The latter is useful in studying and investigating the molecular linear vibrations and the rotation-vibration.

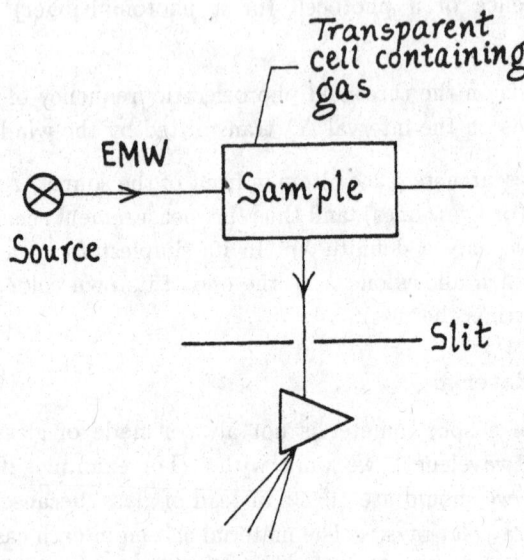

Fig. 14.16.

As it is mentioned above, the source is monochromatic, of a single wavelength λ_0 and frequency ν_0. On either side of the ν_0 line we obtain (there appear) two lines at equal intervals $\pm\Delta\nu$ (if the molecule has a single line). The line at $\nu_0 - \Delta\nu$ is called *Stokes line*; the one at $\nu_0 + \Delta\nu$ is called *anti-Stokes line* (Fig. 14.17). Stokes lines are too intense, whereas anti-Stokes lines are weaker.

If the difference $\Delta\nu$ corresponds to a definite vibration frequency of the molecule, and if the source frequency is chosen in the visible region, we can examine the molecular vibrations under visible light.

In Raman spectroscopy today, we use LASER rays as a source, because they are highly nonchromatic (pure). Any other source of visible light would have

---

[3]Rayleigh scattering explains why the sky is blue. The source is the Sun, and the sample just the molecules of the atmosphere which scatter and diffuse the blue frequency the most. At sunrise and sunset, as the angle changes and the sunbeams have to pass through a thicker layer of air before they reach us, they lose most of their blue color via scattering, and hence we see beautiful red and yellow tints which constitute the remaining complementary of blue.

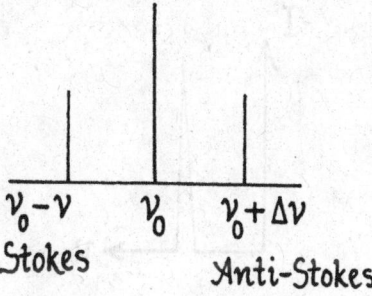

Fig. 14.17.

a *spectral line width* $\Delta\nu$ (Fig. 14.18) which would be too wide for our purposes (as we need a monochromatic ν_0 as pure as possible, corresponding to the quantum to be absorbed by the sample molecules, *i.e.*, we need a resonant beam — resonant to molecular states). (Note that $\Delta\nu$ corresponds to $0.7 I_{max}$ where I_{max} is the intensity maximum at the peak.) So, ordinary sources are not purely monochromatic (they do not emit one, single and exact ν_0): They have a natural width due to the uncertainty principle, further widened by electric field influences and collisions of atoms (collision broadening). But in a LASER source the line (distribution of intensity versus frequency) is very narrow and sharp, almost a delta function $\delta(\nu-\nu_0)$ (Fig. 14.19), because it corresponds to a definite electronic transition. The sharpness is due to the coherence of the source, and the narrow line is due to the fact that it corresponds to a transition between energy levels. (And if we tune the LASER frequency so that it is exactly the same as, and at resonance with an interlevel

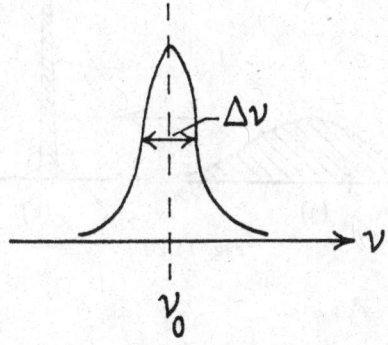

Fig. 14.18.

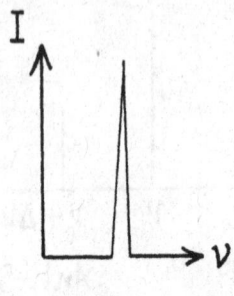

Fig. 14.19.

transition of the sample, we can achieve excitations in the sample. This tunability of LASER that gives resonance is a great advantage.)

The spectrum of an ordinary white light source (*e.g.*, the Sun or the lightbulb of a desk lamp) is a broad and dull curve that contains all frequencies. Figure 14.20(a) shows the distribution of a lightbulb, ranging from the infrared (or even from the microwave) to the ultraviolet region.[4] Figure 14.18 shows the case of an ordinary monochromatic source (where a narrow band of frequencies about a nominal and dominant one — which has the highest intensity — is emitted, or allowed to pass by a selective filter). Figure 14.20(b) shows the LASER case. Notice that in going from (a) to (b), the energy (area under the curve) stays the same; so, the intensity has necessarily to increase as the distribution (frequency interval) gets narrow (and the area remains constant).

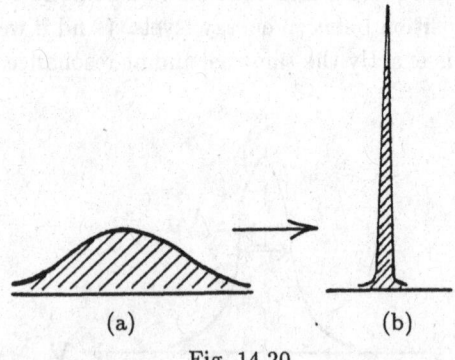

(a)   (b)

Fig. 14.20.

---

[4]Where the infrared radiation appears as radiated heat, and the fraction of the included ultraviolet part is weak and, of course, totally invisible.

The relevant aspects of LASER theory and technology constitute a topic too vast to be condensed to a section and included here — it would be like presenting a drop from an ocean; it would also be out of the scope of this book anyway. The only thing we have to state is that as a light source, the LASER is highly monochromatic (pure), very intense, coherent, and unidirectional. Hence it is greatly advantageous.

The Raman spectrum is obtained as follows: When light is incident on a molecule, the electrons vibrate. For example, let a beam of quanta come over an atom (Fig. 14.21). The incoming ray (electromagnetic energy) is the sum of quanta $h\nu$ (it consists of them). The electrons vibrate in this electromagnetic field. If the molecule has a permanent dipole moment, an electromagnetic wave is emitted.

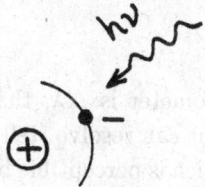

Fig. 14.21.

A molecule may or may not have an electric dipole moment $\mathbf{p}$. That depends on the *polarizability* α of the molecule, *i.e.*, its ability or tendency (or willingness) to become electrically polarized. In other words, α is a measure of the inclination of the molecule to exhibit a dipole moment inside an electric field $\vec{\mathcal{E}}$. We have $\vec{\mathbf{p}} = \alpha\vec{\mathcal{E}}$.[5]

If the incident ray is an electromagnetic wave, then we have

$$\mathbf{p} = \alpha\mathcal{E}_0 \sin(\omega_a t)$$

where $\mathcal{E}_0$ is the amplitude (maximum value) of the incoming electromagnetic wave, ν_0 is the frequency of the incident vibration, and $\omega_0 = 2\pi\nu_0$.

If this electromagnetic wave touches the molecule, the electrons vibrate at this ν_0. But the molecule performs also linear and rotational vibrations (in which the electrons participate, too). If the vibrational angular frequency is ω_v, then α is a function of it for certain directions: $\alpha = f(\omega_v)$. We have explicitly

$$\alpha = \alpha_0 + \alpha' q_i \cos(\omega_v t + \delta)$$

---

[5]The external field does not have to be always static.

where α_0 is the static value of the polarizability, and q_i is the displacement (amount of change of place) of the electron upon vibration at ω_v. There is a different q_i value for every different ω_v. We further have

$$\mathfrak{p} = \alpha_0 \mathcal{E}_0 \sin(\omega_0 t) + \alpha' \mathcal{E}_0 q_i \sin(\omega_0 t) \cos(\omega_v t + \delta)$$

$$= \alpha_0 \mathcal{E}_0 \sin(\omega_0 t) + \frac{1}{2}\alpha' \mathcal{E}_0 q_i \{\sin[(\omega_0 - \omega_v)t - \delta] + \sin[(\omega_0 + \omega_v)t + \delta]\}.$$

Whether the phenomenon occurs or not depends on the coefficient α'. Different molecules vibrate in different directions, and q_i are different. If the molecule executes linear or rotational vibrations and if an electric dipole moment is produced ($\alpha' \neq 0$), one can observe a Raman spectrum; otherwise one cannot.

14.6. Resolution

The *resolution* of a spectrometer is $\Delta\lambda$, that is, the interval of the two closest lines that the instrument can resolve (tell one from the other), *i.e.*, the smallest distinguishable $\Delta\lambda$ which is perceptible by the instrument (Fig. 14.22). If we take two spectral lines too close to each other, there is a limit beyond which the separation $\Delta\lambda$ is not clear any more to the spectrometer. The smallest separation of two lines that the instrument can "see" as distinct, is its resolution limit. Anything beyond that (smaller than $\Delta\lambda$) is too fine for the instrument. The smaller the $\Delta\lambda$ the better (the higher the resolution). In the region of visible light, $\Delta\lambda$ is in general good (whereas in the infrared region it is too large).

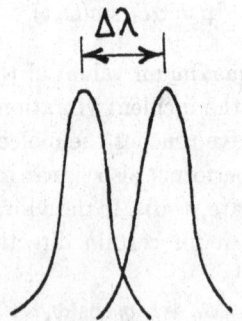

Fig. 14.22.

14.7. Microwave Spectrometry

The spectrum of the rotational energy transitions of the molecules falls in the microwave region. This region extends from 3000 MHz ($\lambda \simeq 10$ cm) to 10^5 MHz ($\lambda \simeq 0.3$ cm). Microwaves have centimetric and millimetric wavelengths. This electromagnetic radiation is produced by electronic methods. Microwave sources are the following:

(1) *Crystrons.*[6]
(2) *Tunnel diods.*

Microwave spectrometry can be done by absorption. The microwave absorption spectrometer is shown in Fig. 14.23. The waves have the property of (invisible) light.[7] The electromagnetic waves that emerge from the clystron travel down a pipe which directs them towards a special waveguide. The system is symmetric, that is, the waveguide propagates the waves in two opposite directions, as shown. Two detectors (D_1 and D_2) and the comparison method are used for sensitive measurements. The currents produced at D_1 and D_2 are the same. We next place the *absorption cell* as shown. It contains the sample

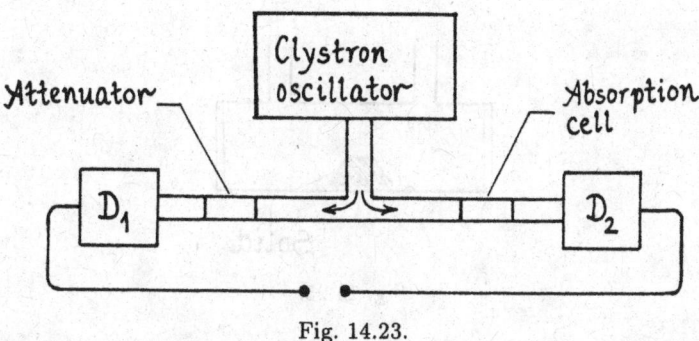

Fig. 14.23.

---

[6]The Greek word *clystron* ($\kappa\lambda\widehat{\upsilon\sigma}\tau\rho o\nu$) comes from the verb $\kappa\lambda\acute{\upsilon}\zeta\epsilon\iota\nu$ (= to flood) whence the words *cataclysm* and *clyster wash* ($\kappa\lambda\widehat{\upsilon\sigma}\mu\alpha$). It should not be confused with the similar word $\kappa\lambda\epsilon\widehat{\iota\sigma}\tau\rho o\nu$ which comes from the verb $\kappa\lambda\epsilon\acute{\iota}\epsilon\iota\nu$ (= to shut) and means shutter, a part of the photographic camera.

[7]Microwaves were discovered by the end of World War I. They are extensively used in radars, space telecommunications, espionage, physics research, cooking ovens etc. In the 1970s, the U.S. Embassy building in Moscow was scanned by a microwave irradiation by the Russians for espionage reasons, an action that allegedly caused cancer incidents. As far as their fast cooking ability is concerned, when a spectator in an oven demonstration asked the showgirl how the waves cooked the meal, she simply reminded him of the heat he felt when he rubbed his hands.

gas at low pressure (Fig. 14.24). The cell has windows made of thin mica (so that the absorption is little). If the sample is a solid, we use a resonator *cavity* (Fig. 14.25). Since there is absorption, the current I_2 gets weaker. For this reason, we place an attenuator on the other side (the side of D_1) which weakens I_1, too, and brings it down to the same level as I_2. The interpolation of the attenuator results in a weakening of the electromagnetic wave equal to the weakening due to the absorption on the other side or otherwise.

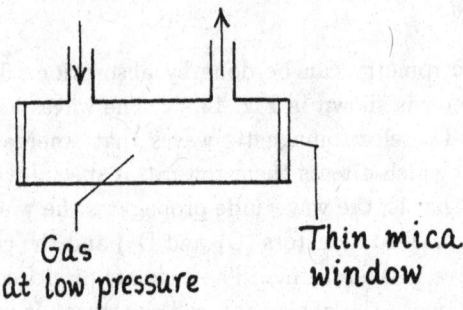

Gas
at low pressure

Thin mica
window

Fig. 14.24.

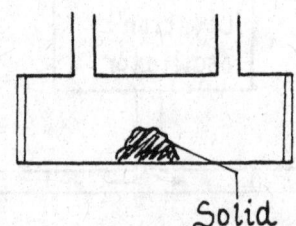

Solid

Fig. 14.25.

The potential difference $\Delta V = V_2 - V_1$ is amplified (Fig. 14.26) and sent over to an oscilloscope. A sweeping voltage oscillator provides a sawtooth potential (Fig. 14.27(a)). If the sweeping voltage oscillator is connected to the clystron (Fig. 14.28), the clystron frequency increases and decreases (fluctuates) as $f_0 \pm nt$, with a period t where $0 < t < \tau$, and n is a constant (Fig. 14.27(b)). The spot on the screen of the oscilloscope moves and the frequency changes (Fig. 14.29). The horizontal axis is the frequency axis (Fig. 14.30) where we observe absorptions corresponding to certain frequencies and appearing as dips. The vertical axis of the oscilloscope is the intensity axis (Fig. 14.31).

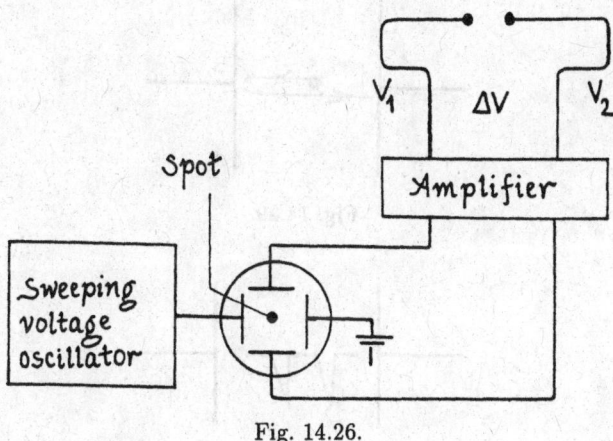

Fig. 14.26.

Fig. 14.27.

(a) (b)

Fig. 14.28.

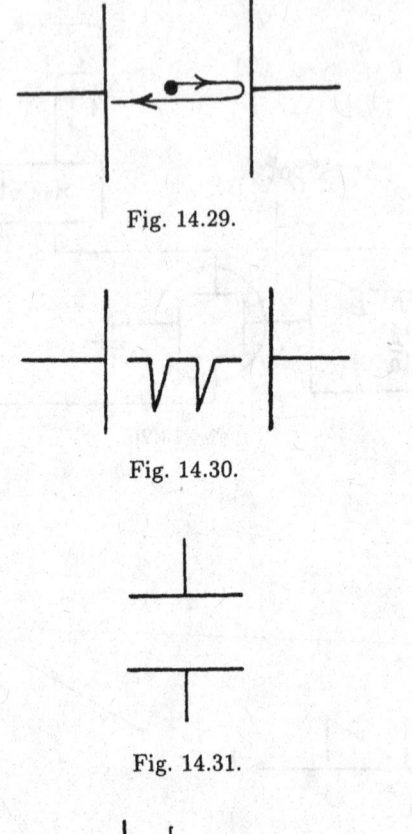

Fig. 14.29.

Fig. 14.30.

Fig. 14.31.

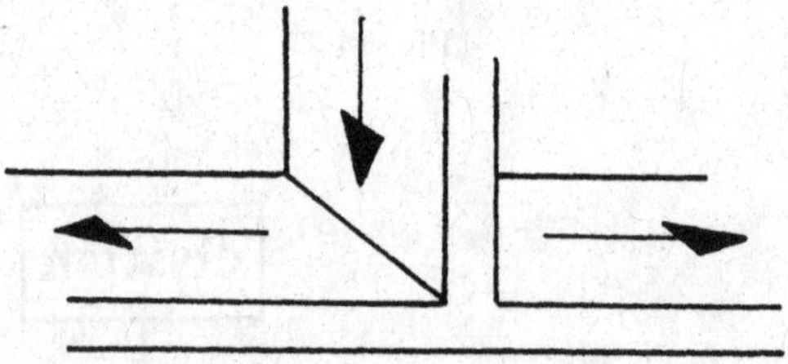

Fig. 14.32.

The waveguide used here is called *"magic tee"* (Fig. 14.32), because it consists of rectangular pipes at right angles to each other, forming an inverted T. As we dispatch and drive the electromagnetic waves through the vertical pipe, they get separated in two directions at the intersection: One part reaches the attenuator and D_1, and the other one is incident on the sample and then on D_2.

The microwave absorption spectrometer is used in the investigation of the rotation spectrum. Since microwaves have the properties of light, they get absorbed, reflected, diffracted, and transmitted by waveguides.

Notice that the waves propagate in two directions in this system, and if the system is symmetric, the potentials at D_1 and D_2 are equal. The absorption cell is placed inside (or on the way of) one of the waveguides, and filled with the gas to be examined, which is sent in through pipes at low pressure.

If a solid sample is of question, the solid material is placed at the points of the cavity that correspond to maximum current. The body absorbs the incoming radiation, and the intensity drops. To compensate this weakening, we place an attenuator on the other side.

To summarize the detection system described above, we repeat that the potential difference is fed to an amplifier whose output is handed over to an oscilloscope. A sweeping voltage is applied to the horizontal deflection. The same sweeping voltage taken from the oscillator is fed to the clystron, too. The frequency of the clystron changes in the form of $f_0 \pm nt$, with a period t. As the spot moves horizontally, the frequency changes linearly. When the sample makes absorptions at certain frequencies, these absorptions appear on the screen of the oscilloscope.

14.8. Conclusion

It is unfortunate that we cannot analyze here in detail the experimental spectroscopic and spectrometric techniques used in laboratories, although the matter is very interesting. The foregoing discussion was so crude that we have not even scratched the surface of the topic. A thorough treatment would require volumes. The topic is exciting not only from a purely physical point of view, but also because it is a discipline that professionally interests, and is used by other sciences like chemistry, biology, astrophysics, geology, engineering, archaeology, hygienic and forensic medicine, criminology, and other branches and scientific activities that use spectral analysis. The versatility of the utility of spectroscopy and the interesting aspects of its techniques (laboratory methods, instrumentation and machinery, technological sophistication, scientific deontology, standardized procedures and rules of practice) endow it with a grandeur of a minor art. The colorful image of spectroscopy as a science reminds one of the elegance and kaleidoscopic splendor of a Viennese operetta, actually more than that: Spectroscopy possesses a meritorious profile that can be appreciated by insightful physicists.

Chapter 15

DIFFRACTION METHODS

15.1. Diffraction Methods

To study crystal and molecular structure we use three diffraction methods:

(1) *X-ray diffraction*: It is used in solid state physics to study crystals and molecular structure. We benefit from the high penetration ability of X-rays into materials which are opaque to other less energetic radiations. X-rays can be used for media consisting of atoms of large atomic number Z, because atoms of small Z do not diffract X-rays; they just let them pass through. (The absortivity of X-rays is proportional to Z^3.)

(2) *Electron diffraction*: It is a difficult method but gives good results. It is used where X-rays are useless. However, this method has limitations, too, because electrons are not highly penetrating particles. (Although they may zip by fast, they interact electrostatically and thus get their penetration depth reduced.) Since their penetration is not deep, the method is used for thin layers, films and foils.

(3) *Neutron diffraction*: We employ neutrons to study molecular properties because neutrons leave off energy around as they pass through a medium. The disadvantage here is that the resolution of neutrons is low, and that we need a nuclear reactor to get them from.

We will now discuss these three methods briefly.

410

15.2. X-ray Diffraction

The diffraction of X-rays was first studied by von Laue in 1912 (whence the order of the X-ray wavelength, 10^{-8} cm, was determined). X-rays were scattered off a crystal and thus gave diffraction patterns on a photographic film. The patterns suggested that the molecules (or atoms) of the crystal were orderly arranged. By studying these patterns one can infer (and build models of) the crystal structure and symmetries. The interpretation of the diffraction patterns was first made by Bragg.

The principle is as follows: Consider a crystal with layers of orderly arranged atoms (or molecules) where the separation between layers (lattice constant) is d. Two parallel X-rays are incident on the crystal (Fig. 15.1). One is scattered by an atom (or molecule) of the first (upper) layer, and the other one by an atom of the second (lower) layer. The path difference between the two is $\overline{CB} + \overline{BD}$. If this is an integral multiple of the wavelength λ of the X-rays, constructive interference takes place between the two rays, and we obtain a bright spot on the screen (photographic plate):

$$\overline{CB} + \overline{BD} = n\lambda \tag{15.1}$$

where n is an integer. But since $\overline{CB} = \overline{BD} = d\sin\theta$, we have for Eq. (15.1)

$$2d\sin\theta = n\lambda$$

which is *Bragg's law*. Here n is the order of diffraction, an integer. If we know λ, we can measure θ and n, and thus find d, the interatomic separation, which is a parameter of the crystal. Also, by studying the shape of the diffraction pattern on the film, we can mentally reconstruct the stereoscopic arrangement of the atoms in the crystal, and hence deduce the symmetries.

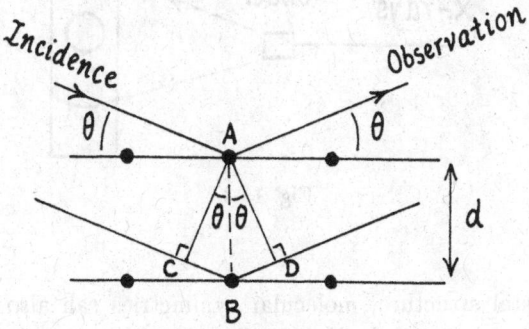

Fig. 15.1.

In the *rotating-crystal method*, a photographic film is cylindrically wrapped around a crystal to which X-rays are sent. The crystal is rotated around, and a diffraction pattern is obtained on the film in the radial directions. This pattern is then interpreted and the structure of the crystal is deduced (Fig. 15.2).

In the *powder method* our sample is the powder of a solid where micro-crystals are present, scattering off the incident X-rays and giving diffraction patterns (Fig. 15.3).

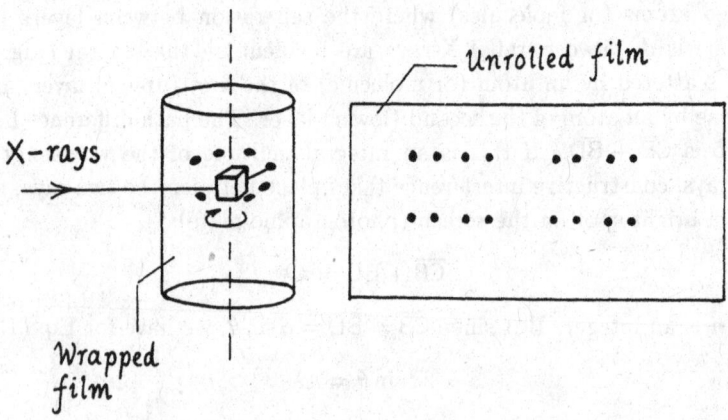

Fig. 15.2.

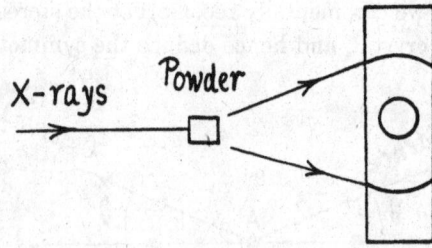

Fig. 15.3.

Besides crystal structure, molecular symmetries can also be studied via X-ray diffraction, but in this book we will not step into details. The interested reader should consult books and references on solid state physics.

15.3. Electron Diffraction

The diffraction of electrons is used to obtain information about the molecules of gases, or at most thin films, because electrons cannot penetrate too much into solids.[1] They get captured, *i.e.*, absorbed soon in solids, because they are charged, and so, they interact electrostatically with the ions in the solid.

The de Broglie wavelength $\lambda = h/mv$ of the accelerated electrons is of the order of the internuclear distances in molecules, and hence they can be diffracted by scattering. That is, they get scattered by molecules.

An electron gun is used as electron source. Electrons emitted by the heated filament of the gun are accelerated under a potential difference of the order of 10^4 Volts and sent into a gaseous sample. (Monochromaticity, *i.e.*, a single λ of the electron beam can be achieved by a stabilized potential.) Nonrelativistically, from the energy accounting equation

$$\frac{1}{2}mv^2 = eV$$

we obtain the electron de Broglie wavelength

$$\lambda = \frac{h}{m}\sqrt{\frac{m}{2eV}} = \frac{h}{\sqrt{2meV}} \, .$$

Magnets can be used to focus the beam, that is, to adjust the electron optics. After being scattered by the gas molecules, the electrons fall onto a detector which can be just a photographic film. Of course, to prevent diffusion, the whole process is allowed to take place in vacuum (of the order of 10^{-7} Torr, achieved by mechanical *and* diffusion pumps). Otherwise the beam will be diffused before hitting the detector.

It is obvious that large-size molecules scatter the electrons more than small-size ones. By observing and studying the diffraction patterns (on the photographic film), one can make deductions about molecular symmetries concerning the molecules of the sample.

This method was first tried by Davisson and Germer in 1927. Their sample, however, was a Ni crystal, not a gas. Their detector was a galvanometer that measured the current intensity of the electrons that were scattered by the crystal.

---

[1]Since electrons cannot travel in solid bulks for long, the sample materials are gases. But some diffraction experiments with solids are possible, too: We can study surface effects in thin layers of solids. We can also determine the molecular structure of very thin crystals.

By rotating the detector around the crystal and measuring the intensity of the scattered beam in different directions, one can plot the scattered intensity as a function of the scattering angle θ, in polar coordinates. The curve (for Ni) is shown in Fig. 15.4 where the radial vector is the intensity of the scattered

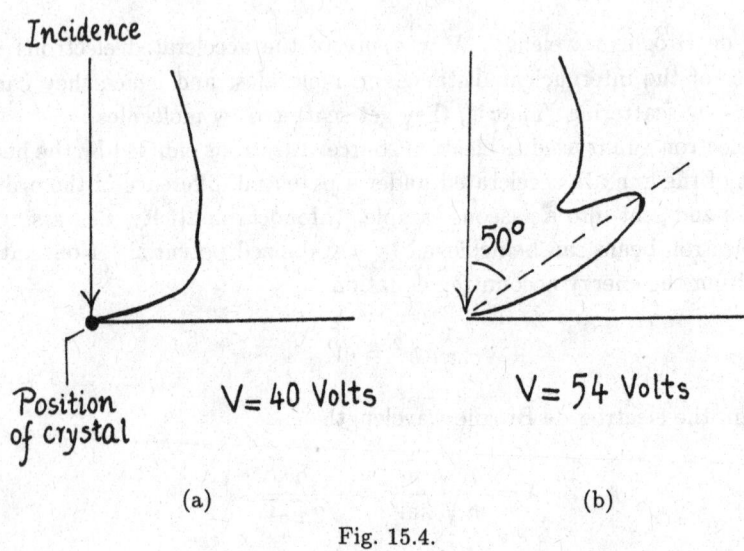

Incidence

Position of crystal

$V = 40$ Volts

(a)

$50°$

$V = 54$ Volts

(b)

Fig. 15.4.

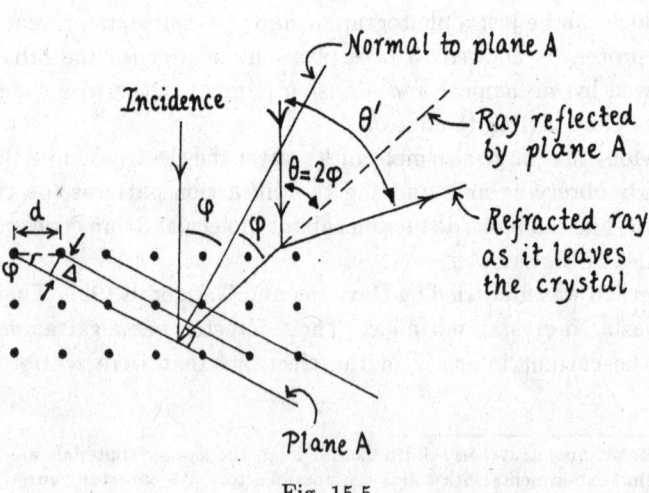

Normal to plane A

Incidence

θ'

Ray reflected by plane A

$\theta = 2\varphi$

φ

φ

d

φ

Δ

Refracted ray as it leaves the crystal

Plane A

Fig. 15.5.

beam. The scattering angle is measured between the direction of incidence and the radial vector. For V = 40 Volts, the curve is smooth (Fig. 15.4(a)). As the applied voltage increases, a protrusion starts becoming more and more evident, and at V = 54 Volts a salient peak of the scattering intensity can be seen at $\theta = 50°$ with respect to the incident beam (Fig. 15.4(b)). This is due to the crystalline structure of Ni (with regular interatomic separations and layers) which produces constructive interference. As the voltage increases further, the peak disappears because of destructive interference.

Figure 15.5 shows the reflection and refraction of the incident beam together. The beam is reflected by the slanted plane (of atoms) A, so that $\theta = 2\varphi$ is the angle between the incident and the reflected (scattered) beam, if we assume that there is no refraction. (In fact, the beam leaves the crystal surface of an angle θ' with respect to the incidence because of the refraction.) In the case of Ni, for 54 Volts, the selective reflection (direction of the salient peak) occurs at $\theta = 50°$. That is, θ is the angle between the incident beam and the direction of the maximum of the diffracted part of the beam.

The Bragg diffraction condition is

$$n\lambda = 2\Delta \cos \varphi$$

where λ is the de Broglie wavelength of the electron, n is the order of the diffraction, and Δ is the separation between the slanted layers.

Since $\Delta = d \sin \varphi$, we obtain

$$n\lambda = d \sin \varphi \cos \varphi = d \sin 2\varphi = d \sin \theta$$

where $\theta = 2\varphi$ and d is the lattice constant. If we know λ, we can measure n and θ, and thus find the crystal parameter d.

In a gaseous sample the molecules are not fixed in place, nor orderly arranged; they fly around randomly. When the sample is a gas, the obtained total scattering intensity is equal to the sum of three intensities: The incoherent atomic scattering plus the coherent atomic scattering plus the coherent molecular scattering. The first component is due to the excitation of atoms to any of their higher levels and their subsequent de-excitation followed by the emission of the absorbed energy. This is incoherent because one atom does not know or care about what the other does. The scattered electrons of this incoherent component can be of a whole bunch of energies, giving a wide spectrum. This component does not exhibit any maxima. The second and third components together (coherent parts) depend on the geometry of the apparatus.

The full formula of the total scattering intensity is

$$I_{\text{total}} = I_{\text{atomic excitation}} + I_{\text{atomic coherent}} + I_{\text{molecular coherent}}$$

$$= I_{\text{atomic excitation}} + C \left\{ \sum_{i=1}^{N} \frac{\mathcal{F}_i}{s^4} + \sum_{i=1}^{N} \sum_{j=1}^{N} \mathcal{A}_{ij} \frac{\mathcal{F}_i \mathcal{F}_j}{s^4} \frac{\sin(sr_{ij})}{sr_{ij}} \right\} \quad (15.2)$$

where C is a constant that includes the geometry, N is the total number of atoms in the molecule, $\mathcal{F}_i$ is the scattering factor of the i-th atom, r_{ij} is the distance between the i-th and j-th atom, $\mathcal{A}_{ij}$ is an exponential factor depending on the vibration of the atoms in the molecule, and s is a parameter given by

$$s = \frac{4\pi \sin(\theta/2)}{\lambda} \quad (15.3)$$

with θ being the scattering angle, and λ the de Broglie wavelength of the electrons.

$\mathcal{F}_i$ depends on the atomic number Z_i of the i-th atom. Actually, $\mathcal{F}_i = Z_i - f_i$ where Z_i pertains to the positive nuclear charge. Since this is screened by the negative charge of the cloud of the electrons around it, we subtract f_i to account for it.

$\mathcal{A}_{ij}$ is given by $\mathcal{A}_{ij} = \exp(-\bar{x}_{ij} s^2)$ where $\bar{x}_{ij}$ is the mean value of the *relative* vibration amplitude of the i-th and j-th atoms.

After some simplifications, $I_{\text{coherent molecular}}$ (which is the part we are interested in) in Eq. (15.2) can be reduced to

$$I_{\text{coherent molecular}} = C \left\{ \sum_{i=1}^{N} \sum_{j=1}^{N} Z_i Z_j \frac{\sin(sr_{ij})}{sr_{ij}} \right\}$$

where Z_i and Z_j are the atomic numbers of the i-th and j-th atom respectively. This function exhibits maxima and minima. It is the coherent molecular part, that is, the molecules agree to give intensity peaks at certain θ values, where θ is included in s through Eq. (15.3).

The point here is to experimentally find the parameters s and r_{ij} via the electron diffraction method, by measuring the intensities. One can also proceed theoretically, by setting up a model for the structure of the molecule in question and selecting proper values for the interatomic separations r_{ij}. Then one can compare the theoretical and experimental results. It is interesting to note that one can work experimentally and theoretically at the same time in such a research.

The total diffraction pattern obtained is always I_{total} as expressed above, but one tries to isolate the $I_{\text{coherent molecular}}$ part and amplify it, in order to obtain information about the molecular parameters.

There are two good methods that work here: The *visual method* and the *sector method*, both useful in molecular structure research. For space reasons we will not discuss them here, because they both involve a whole ritual of elaborate experimental actions.

The finally obtained radial function that represents the molecular scattering curve is

$$f(r) = \int_{s_1}^{s_2} I_{\text{molecular}} s e^{-as^2} \sin(sr) ds$$

where a is a constant (to be chosen conveniently), and s_1 and s_2 are the limits of the scattering angle within which scattering intensity measurements are possible (for a given λ). Computing this integral is not a trivial job. One needs a computer to do it.

This function $f(r)$ represents the probability that a certain r value (interatomic distance) exists inside the molecule in question.[2] The function may peak for several values of r. For example, for CCl_2F_2 the plot is as in Fig. 15.6. It has four salient maxima because there are four possible interatomic distances for five atoms present in the molecule.

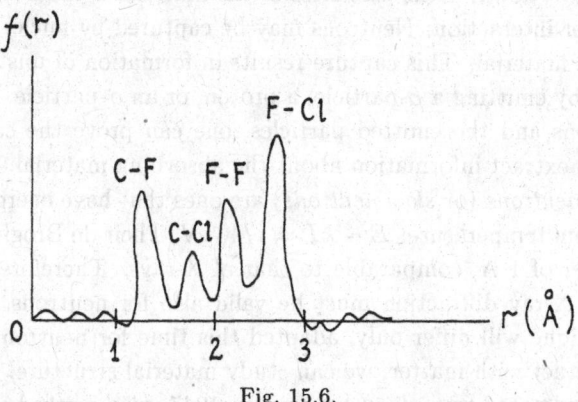

Fig. 15.6.

The function sometimes does not give salient maxima. The situation then becomes troublesome (because the maxima overlap), but there are remedial approximations through which the maxima are taken as Gaussian curves and thus their positions are located.

The method of electron diffraction works only if the molecular symmetries are more or less already known beforehand. If they are, we obtain better

---

[2] All the painful toils were just for this!

results for the molecular parameters. If they are unknown, the results are vague. The method is liable to give quite unsatisfactory results if the molecular shape is completely unknown. Also, the method works best if the molecular symmetries are as many as possible (because this reduces the number of the parameters to be determined). The more the symmetries the less the number of the parameters, and then $f(r)$ gives more precise results. Of course, one can always assume reasonable values for some parameters and hence obtain other parameters, but such a practice gives unreliable results.

15.4. Neutron Diffraction

Neutrons are electrically neutral particles, thus exhibiting no Coulombic interaction with charges. This means that they can penetrate deep into matter (and namely into the sample material to be studied) without being affected much. From this aspect they are advantageous to use in the study of solids.

As they cross a medium, they do not interact with ions or with the electrostatic potentials of the electrons and nuclei in the bulk. Their main interaction with matter is through the elastic collisions with nuclei, whereby they lose some energy and slow down, if the thickness of the material is large enough. There is also another interaction: Neutrons may be captured by nuclei and thus absorbed in the material. This capture results in formation of unstable isotopes which decay by emitting a β-particle, a proton, or an α-particle. By detecting these emissions and the emitted particles, one can prove the capture of the neutrons and extract information about the absorbing material.

Thermal neutrons (or *slow neutrons*) are ones that have energy that corresponds to room temperature: $E \sim kT = 1/40$ eV. Their de Broglie wavelength is of the order of 1 Å, comparable to that of X-rays. Therefore, whatever is valid for the X-ray diffraction must be valid also for neutrons. The experimental technique will differ only, adapted this time for neutrons. By letting neutrons interact with matter, we can study material structure.

Such experiments were done by Zinn in 1947, who worked with a calcite crystal and a beam of thermalized neutrons taken from a reactor.[3] The beam was scattered by the crystal and the intensity was measured. The plot is shown in Fig. 15.7, with a prominent maximum at a certain angle.

Wollan and Schull[4] worked with diamond powder and a monoenergetic neutron beam in 1948. Detection can be made by using either a BF_3 proportional counter, or an indium plate that captures the neutrons of the scattered beam to

---

[3]W.H. Zinn, *Phys. Rev.*, **71**, 755 (1947).
[4]E.O. Wollan and C.G. Shull, *Phys. Rev.*, **73**, 834 (1948).

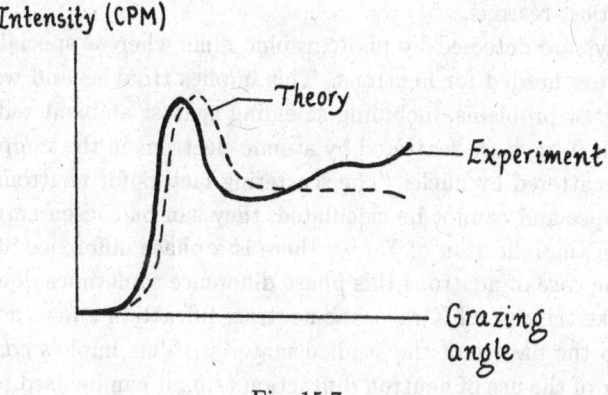

Fig. 15.7.

form unstable nuclei which disintegrate by the emission of β-particles. These particles are in turn detected by a photographic plate.

Today the neutron-diffraction method is being used to study powders and crystalline solids, including single crystals. It is the complementary of the X-ray diffraction.

15.5. X-ray Diffraction vs. Neutron Diffraction

Although the neutron diffraction resembles X-ray diffraction, there are the following main differences:

(1) The technology of a neutron spectrometer differs than that for X-rays.

(2) The neutron-diffraction method sometimes is not preferred because neutron sources are not easily and readily available next to molecular physics laboratories. (Necessarily, the opposite happens: The molecular physics laboratory moves next to a reactor or an accelerator.) The neutrons obtained from a reactor are fast (too energetic), and so they must be thermalized (brought to $\lambda \sim 1$ Å) via elastic collisions in carbon moderators during their passage through them after their emergence from the reactor. They should also be monoenergetic, and so they are first let get scattered off a NaCl crystal before they are used. Still, their monochromaticity is not as good as that of X-rays.

(3) A neutron beam is weaker than an X-ray. To compensate this, a wide neutron beam can be taken, but this requires a large size on the part of the sample. The researcher aims at getting more and more neutrons into the incident beam and simultaneously narrowing the beam for

practical reasons.

(4) X-rays are detected by photographic films whereas special BF_3 detectors are needed for neutrons. This implies troubles and worries about detector problems, including shielding against ambient radiation.

(5) While X-rays are scattered by atomic electrons in the sample, neutrons are scattered by nuclei. The scattering factors for neutrons depend on isotopes and cannot be calculated; they can be chosen empirically.

(6) Upon the reflection of X-rays there is a phase difference of π, whereas in the case of neutrons this phase difference sometimes does not occur.

(7) Unlike the case of X-rays, the neutron diffraction knows no limitations as to the nature of the studied material. This implies convenience in favor of the use of neutron diffraction (*e.g.*, it can be used to sensitively determine the positions of hydrogen atoms in a crystal).

(8) When we want to determine the positions of the H atoms in a material, it is easier to use X-rays in examining a material containing uranium atoms compared to a material containing carbon atoms.

15.6. Neutron Sources

Neutrons can be obtained by two processes: (α, n) or (γ, n), though they are weak sources compared to others. Actually, there are three ways to generate neutrons:

(1) Radioactive materials (*e.g.*, beryllium bombarded by α particles).
(2) Particle accelerators.
(3) Nuclear reactors.

Let us briefly discuss these three sources:

(1) Neutrons are obtained via the exothermic reaction

$$^9_4\text{Be} + {}^4_2\text{He} \rightarrow (^{13}_6\text{C}) \rightarrow {}^{12}_6\text{C} + {}^1_0 n + 5 \text{ MeV},$$

but they are not monoenergetic. Here an alpha-beryllium source (Be bombarded by 5-MeV α-particles) emits a beam of neutrons (which were discovered by Chadwick via this reaction). Since the reaction is exothermic, the energy of the emitted neutrons is larger than that of the incoming α-particles. Here we need a radioactive substance emitting α-particles in the first place. This substance is mixed with Be so that the collision cross-section is large (the probability that alphas hit Be nuclei is high). Neutrons thus obtained are not monochromatic, because ^{12}C can get into several excited states, robbing energy from the neutron. (If after the reaction ^{12}C stays excited, this amount of energy stays with it.) Secondly, the incoming alphas may not be monoenergetic, because of collisions. (Before reaching Be, alphas pass through matter and lose energy, so that these incident alphas have an energy spread, resulting in an energy spread in the emitted neutrons.) Another disadvantage is that this source produces also undesirable γ and X-rays which can show up in the detector as noise.

(2) Protons can be accelerated up to 10 MeV, to achieve the endoenergetic reactions

$$^{7}_{1}Li + p^{+} \rightarrow {}^{7}Be + n - 1.647 \text{ MeV}$$

$$^{9}Be + p^{+} \rightarrow {}^{9}B + n - 1.85 \text{ MeV}$$

$$^{3}H + p^{+} \rightarrow {}^{3}H + n - 0.765 \text{ MeV}$$

where the emitted neutron has less energy than the initial accelerated proton. Here the neutron energy is not so spread, because by making the protons monoenergetic, we can have monoenergetic neutrons.

(3) Fusion reactions can be used to produce free neutrons:

$$^{2}H + {}^{2}H \rightarrow {}^{3}He + {}_{0}n + 3.27 \text{ MeV}$$

$$^{2}H + {}^{3}H \rightarrow {}^{4}He + {}_{0}n + 17.6 \text{ MeV}$$

where the apparatus, called *omnitron*, is relatively simple. The deuterons here are of 50 keV, which is sufficient.

The omnitron is a device that produces neutrons for neutron diffraction, shown in Fig. 15.8. Deuterons of energy of 50 keV are used for the fusion reactions

$$^{2}H + {}^{2}H \rightarrow {}^{3}He + n + \text{Energy}$$

$$^{2}H + {}^{3}H \rightarrow {}^{4}He + n + \text{Energy}$$

which are easily obtained (and important, because they release a great deal of energy). The deuterons with a kinetic energy given above are able to make these reactions occur with a high probability.

Another possible reaction with deuterons is

$$^{2}H + {}^{9}Be \rightarrow {}^{10}B + n + 4.35 \text{ MeV} .$$

Further, a fission reaction is possible (with ^{235}U) whence neutron beams can be taken out from the fission reactor and thermalized down to 1/40 eV. Such a reactor gives on the average 2.5 neutrons per fission, of mean energy of 2 MeV (to be thermalized afterwards). The typical flux of the beam is 10^{13} neutrons/cm^2 sec.

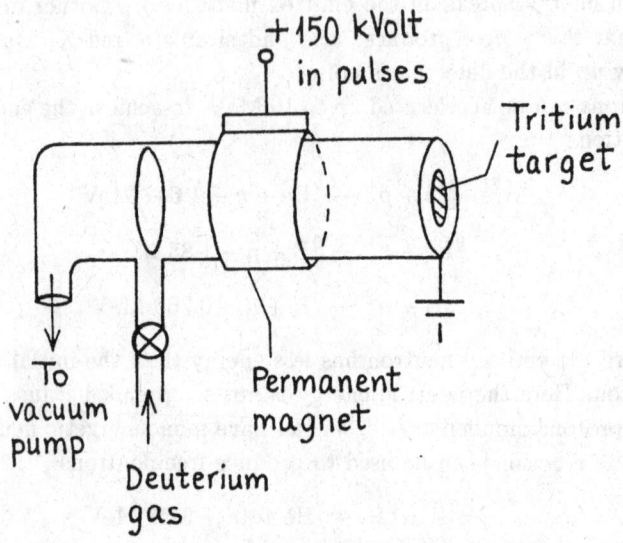

Fig. 15.8.

15.7. Neutron Detectors

It is difficult to detect particles that carry no charge. For this reason, exoenergetic reactions are used as an indirect means of detecting neutrons:

$$^{6}\text{Li} + n \rightarrow {}^{3}\text{H} + \alpha + 4.8 \text{ MeV}$$

$$^{10}\text{B} + n \rightarrow {}^{7}\text{Li} + \alpha + 2.3 \text{ MeV}.$$

A proportional counter can be used where the produced α-particles ionize the gas in the detector chamber, thereby producing signal pulses. For slow neutrons, BF$_3$ gas is used where boron nuclei decay after the neutron bombardment, emitting detectable α-particles. (This is just the second reaction written above. The proportional counter contains BF$_3$ gas at 40 cm Hg, enriched with ^{10}B isotopes. The neutrons are swiftly caught by the ^{10}B nuclei.

As a result, an alpha is emitted which ionizes the gas, and this gives a count whose intensity is proportional to the energy of the α-particle.)

The reactions (n, α) occur with thermal neutrons. The produced alphas are of MeV.

This counter may affect the resolving power.

A scintillation counter is another choice, where scintillations are triggered by the produced alphas. By counting the scintillations, we obtain a measure of the neutron intensity. A scintillation counter is better to use because instead of a gas, it employs a solid crystal which has more atoms per unit volume.

A difficulty arising here is that the energy of the produced alphas is greater ($\sim$ MeV) than that of the neutrons, and the relation between the two is not trivial.

15.8. Molecular Diffraction

Since a de Broglie wavelength escorts any particle moving with a speed v, what holds true for electrons and neutrons can hold true for molecules (and atoms), too. Molecules themselves can be diffracted by crystals. This is a way through which molecules are used as probes to decipher other molecules. Such a diffraction was first achieved by Stern *et al.* who let H_2 molecules, and, separately, He atoms, scatter off LiF and NaCl crystals. The molecular beam was taken from a hot oven (at a given temperature). The beam hit the crystal and got scattered. For a given de Broglie λ, a diffraction pattern with intensity maxima was observed.

So, apart from the three diffraction methods we discussed above, we have the molecular diffraction, too.

15.9. Visual Method in Electron Diffraction

A diffraction pattern has intensity maxima and minima (I vs. s). An experimental scattering curve can be set up by measuring the relative intensity values of the pattern on a photographic film (by following the light intensity at different points of the pattern). The positions of the maxima and minima are fixed by eye estimation, *i.e.*, by just watching the intensity variation.

The last term in Eq. (15.2) is $I_{molecular}$ which is of interest to us, and which passes through maxima and minima, as θ varies.[5] The molecular scattering

---

[5]Since Total scattering = Molecular scattering + Atomic coherent and incoherent scattering (by "scattering" here we mean scattered intensity), the overall diffraction pattern consists of a fast falling part plus a superimposed part that varies only a little and that depends on the molecular parameters. We try to separate it in the electron-diffraction experiments, and amplify it, to obtain information about molecular parameters.

is of the form $(\sin x)/x$. Theoretically, one has to simplify $I_{molecular}$ by letting $\mathcal{F}_i\mathcal{F}_j/s^4 \to Z_iZ_j.$, and abandoning the exponential decrease factor $\mathcal{A}_{ij}$ (see Section 15.3). These simplications do no big harm to the curves; specifically, they do not move the positions of the minima. One then imagines a molecular model and chooses convenient r_{ij} values. Finally, one compares the so obtained scattering curves (calculated theoretically through $I_{coherent\ molecular}$) to those found experimentally.

One adopts the theoretical curve that agrees the most with the experimental one. Thus the r_{ij} values picked for the calculation of that curve are considered as the molecular parameters.

As a simple example, consider SiF_4 (silisium tetrafluoride). Simple, because only one parameter is to be determined here. There are four equal Si–F distances and six F–F distances. So,

$$I_{molecular} = 6CZ_F^2 \left\{ \frac{\sin[sr_0(FF)]}{sr_0(FF)} \right\} + 4CZ_FZ_{Si} \left\{ \frac{\sin[sr_0(SiF)]}{sr_0(SiF)} \right\}.$$

Since the molecule is tetrahedral, the two r_0 values are related as $r_0(FF) = \sqrt{8/3}r_0(SiF)$. Also, $Z_F = 9$ and $Z_{Si} = 4$.

$$I_{molecular} = C \left\{ \frac{486 \sin\left[\sqrt{\frac{8}{3}}sr_0(SiF)\right]}{\sqrt{\frac{8}{3}}sr_0(SiF)} + \frac{504 \sin[sr_0(SiF)]}{sr_0(SiF)} \right\}.$$

Now the theoretician must have Jobian patience in this championship, or a fast computer. Because he has to calculate $I_{molecular}$ for different values of $sr_0(SiF)$ (between 0 and 50). Then he plots the curves of the results, and fixes the positions of the maxima and minima. The calculated curve for SiF_4 is shown in Fig. 15.9.

Now the experimentalist takes over. He photographs the electron diffraction and finds the following:

(1) The position of the diffraction rings.
(2) The distance between the sample and the photographic film (or the fluorescent screen).
(3) The s values for each maximum and minimum, using the known λ of the electrons.

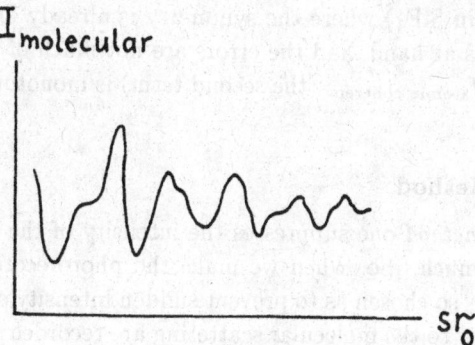

Fig. 15.9.

For each maximum and minimum, there is a value for sr_0(SiF):

Maximum	Minimum	sr_0(SiF)	$\lambda = \dfrac{4\pi \sin(\theta/2)}{s}$	r_0
1	—	8.30	5.406 Å	1.535 Å
—	2	10.69	6.932 Å	1.542 Å
2	—	13.09	8.266 Å	1.584 Å
—	3	17.82	11.59 Å	1.538 Å
3	—	20.25	13.02 Å	1.555 Å
—	4	22.67	15.08 Å	1.503 Å
4	—	27.42	17.27 Å	1.588 Å
—	5	29.80	19.32 Å	1.542 Å
5	—	32.20	21.34 Å	1.509 Å

$$(\bar{r}_0 = 1.544 \text{ Å})$$

In the visual method the experimentalist should be very careful, because errors creep in, rendering the results unacceptably wrong. For example, caution is due for the possibility of secondary scattering. Also, sometimes people ignore scattering by H atoms which is obscured by scattering by heavy atoms. This gives rise to errors in the (determination of the) molecular parameters. Furthermore, we may have double rings of diffraction, reducing the number of the extrema. Other error sources exist, too.

These errors become significant if the molecular structure in question is completely unknown. It is like walking in the darkness, without hints. But in simple cases (like in SiF_4) where the symmetry is already known beforehand, one has some leads at hand, and the errors are not much of importance.

In Eq. (15.2), $I_{atomic\ coherent}$ (the second term) is monotonic and drops fast with θ.

15.10. Sector Method

In the *sector method* one suppresses the intensity of the atomic scattering which varies too much. So, when we make the photorecording, we rotate a sector whose size is so chosen as to prevent sudden intensity changes. Thus the variations pertinent to the molecular scattering are recorded better. A curve is obtained where the intensity is measured with a microphotometer. One rotates the photographic film fast to reduce the intensity, and makes a recording with the microphotometer. Such a recording is seen in Fig. 15.10. To obtain the molecular-scattering curve, one draws a smooth optimum curve through the middle of the ups and downs (oscillatory variations) of the recorded curve, so that equal amounts of areas remain on either side of the drawn best-fitting curve. This last curve is the molecular-scattering plot. It gives the internuclear distances directly, and is used to calculate the radial distribution function $f(r)$ which is mentioned in Section 15.3.

Figure 15.6 shows the curve obtained (calculated) from the experimental molecular scattering curve for CCl_2F_2. The four maxima belong to the existing four different internuclear distances in the molecule. Of these, $r_0(CCl)$ and $r_0(CF)$ can be obtained from the curve directly; the rest can be calculated from these two. The following table shows the results for certain molecules:

Molecule	Bond distance	Angle	$r_0(\text{Å})$
CCl_2F_2	C–F	109.5° ± 3° (F–C–F)	1.33 ± 0.02
	C–Cl	108.5° ± 2° (Cl–C–Cl)	1.77 ± 0.02
CF_3Cl	C–F	108.6° ± 0.4° (F–C–F)	1.328 ± 0.002
	C–Cl		1.751 ± 0.004
CH_3Cl	C–H	110° ± 2° (H–C–H)	1.11 ± 0.01
	C–Cl		1.784 ± 0.003
CCl_4	C–Cl	Tetrahedral	1.769 ± 0.005

Consider the curve produced by two atoms. The product of the area under this curve times the internuclear distance is proportional to the *scattering power* of the two atoms.

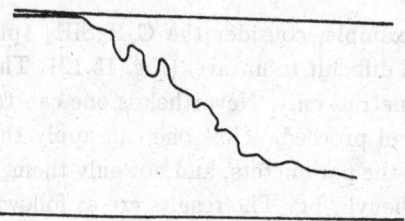

Fig. 15.10.

As another example, consider the benzene molecule. The three C–C maxima obtained are 1.393 Å (ortho-), 2.410 Å (para-), and 2.786 Å (meta-carbon atom). Also, there are four peaks for the four C–H bonds: 1.08 Å, 2.13 Å, 3.40 Å, and 3.89 Å. Of all these, the most fitting ones for an hexagonal molecule are $r_0(CC) = 1.393 \pm 0.005$ Å and $r_0(CH) = 1.08 \pm 0.02$ Å, together with their uncertainties.

The limitations on the method of electron diffraction are mentioned in Section 15.3. The method is no good if the shape and symmetry are unknown, if one does not have an idea at all about them beforehand. Even an intuitional guess might help. The method works (gives satisfactory results) even if the benzene molecule, for example, is distorted a bit, but not too much, say, within a limit where the C atoms are off the hexagonal plane by 0.10 Å at most. That is, the results are correct if the benzene molecule has $6/mmm(D_{6h})$ symmetry. Consider SiF_4 which is tetrahedral; its symmetry is $\bar{4}3m(T_d)$. But if it is squatty (pressed down low) or almost flattened (with symmetry $\bar{4}2m(D_{2d})$), the results will be wrong (Fig. 15.11).

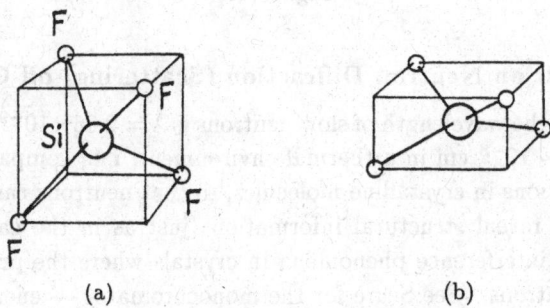

(a)   (b)

Fig. 15.11.

The molecular-scattering curve is not sensitive for light atoms. So, one first assumes the hydrogen-bond parameters as known beforehand, and then one proceeds. For example, consider the $C_6H_5SiH_3$ (phenylsilane) molecule which is complex and difficult to unravel (Fig. 15.12). There are many parameters and a few symmetries only. Nevertheless one can take $r_0(SiH) = 1.42$ Å (tetrahedral bond) and proceed. Thus one can apply the electron-diffraction method to determine the parameters, and not only them, but also measure the deformation of the phenyl ring. The results are as follows:

$$r_0(SiC_I) = 1.84 \pm 0.005 \text{ Å}$$

$$r_0(C_I C_{II} \text{ and all C--C}) = 1.39 \pm 0.005 \text{ Å}$$

$$r_0(C_{II}H_{II} \text{ and all C--H}) = 1.10 \pm 0.02 \text{ Å}$$

$$\text{Angle } C_{VI}C_IC_{II} = 117.4°$$

$$\text{Angle } C_{III}C_{IV}C_V = 120.8° .$$

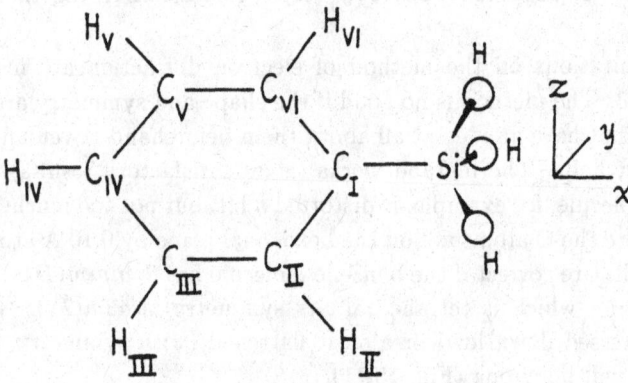

Fig. 15.12.

15.11. Details on Neutron Diffraction (Scattering) off Crystals

The de Broglie wavelength of slow neutrons is $\lambda = 2.86 \times 10^{-9}/\sqrt{E(eV)}$ cm, and becomes $\sim 10^{-8}$ cm in a thermal environment, *i.e.*, comparable to interatomic separations in crystalline molecules, so that neutrons can be diffracted by crystals to reveal structural information (just as in the case of X-rays). Thus we have interference phenomena in crystals where the probe is a beam of thermal neutrons. (See Segrè for the monochromator — energy selector — for monoenergetic slow neutrons, p. 573.)

Monoenergetic neutron beams are obtained by mechanical velocity selectors. The rest of the experimental technique resembles that for X-rays. Hydrogen, which is hard to detect by X-ray diffraction, can be studied by neutron diffraction. Thus neutron diffraction is a supplement of X-ray diffraction wherever the latter is not efficient.

Consider Bragg reflections from a crystal lattice consisting of parallel planes separated by a distance d and containing indistinguishable (spinless) scatterer nuclei of scattering length a. The neutrons scattered at a depth nd down the layers have a scattering amplitude

$$\mathcal{A} = \sum_{n=1}^{N}(-a e^{2iknd\cos\theta}) = -a\left[\frac{1 - e^{2iNkd\cos\theta}}{1 - e^{2ikd\cos\theta}}\right] e^{2ikd\cos\theta}$$

where the sum is a geometrical progress, and N is the total number of planes the neutrons penetrate through. The scattering intensity is

$$\text{Intensity} \sim |\mathcal{A}|^2 = |a|^2 \frac{\sin^2(Nkd\cos\theta)}{\sin^2(kd\cos\theta)} \tag{15.4}$$

which is large if the denominator becomes zero, that is, for values of θ where $kd\cos\theta = $ (Integer multiple)π, or $2d\cos\theta = $ (Integer multiple)λ which is Bragg's law.

The sample crystal may contain several isotopes of the same nucleus; then in Eq. (15.4) one uses the weighted sum of the different scattering lengths, which gives the *effective* scattering length, if the scattering is to be coherent. In reality there are no perfect crystals: The various isotopes have a random distribution throughout the lattice. This gives rise to some incoherent scattering besides the coherent one.

Consider the NaCl crystal where the order is simple: d is constant, and Na atoms alternate with Cl atoms (Fig. 15.13(a)). The structure is cubic. In the first-order reflection the optical path for reflection from Na planes is different by $\lambda/2$ from the path for reflection from Cl planes, as the two kinds of planes alternate. So, if Na and Cl atoms effect the same phase change in the scattered neutron wave, the two effects are subtractive and the intensity is low. The opposite happens when Na and Cl scatter the beam with opposite phase change.

In the second order we have the opposite situation: When Na and Cl cause the same phase change, the intensity is strong. The neutron diffraction by a cubic crystal is shown in Fig. 15.13(b).

If the structure is complicated, the change in the intensity from order to order can be found from the *form factor*, *i.e.*, the effective coherent scattering amplitude (per unit cell)

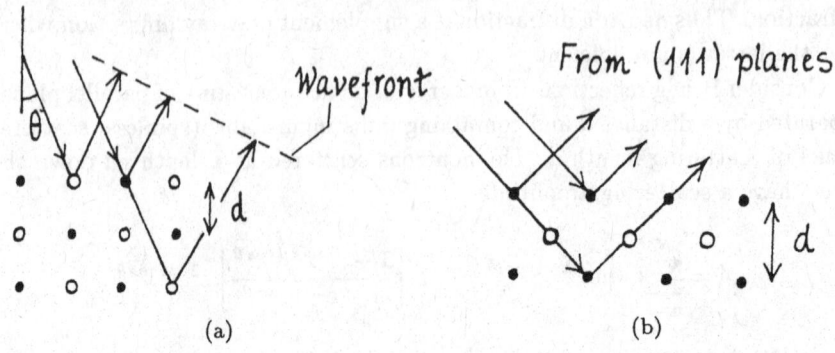

Fig. 15.13.

$$f = \left| \sum_i a_i e^{2\pi i n r / d} \right| \tag{15.5}$$

where d is the spacing between the lattice planes, n here is the order of the Bragg reflection, and r is the normal distance of the i-th nucleus to a reflection plane which is taken as reference. We sum over all nuclei in the unit cell. Notice that f depends on the signs of a_i. From here one can find the intensity for different orders, including the case of NaCl, because Eq. (15.5) is the general case. Needless to say, the intensity decreases from order to order (like in X-rays), and this is superimposed on the effects discussed above, that is, the phase changes in the first and second order.

How can we have slow neutrons for our experiment? We let the beam pass through a polycrystalline material. Those neutrons satisfying Bragg's law are removed from the beam by reflection. Those with $\lambda > 2d$ continue their way through. High velocities are thus cut off. Figure 15.14 shows the distribution (intensity versus λ) of very slow (cold) neutrons. The filter material is usually beryllium oxide. (Do not confuse this arrangement with the *mechanical* monochromator mentioned above, where velocity-selecting disks are used.)

Another use of Bragg scattering is the investigation of magnetic properties. Do not forget that a, the scattering length, depends on the orientation of a neutron spin with respect to the magnetic field produced by atomic electrons. If the sample is ferromagnetic, the reflected intensity depends on the neutron spin polarization. Thus neutrons can be polarized by the Bragg technique.

The neutron diffraction (Laue) patterns obtained with ferromagnetic materials depend on the scatterer's magnetization.

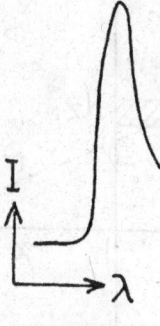

Fig. 15.14.

In antiferromagnetic materials the atomic (nuclear) spins are oriented (aligned) in pairs with opposite senses, just like two (saturated) ferromagnetic media (lattices) of equal and opposite magnetizations, taken together in a superposition. Macroscopically, these materials do not exhibit net ferromagnetism; but if probed by neutron diffraction, this internal situation comes out. For example, manganese oxide is antiferromagnetic. The powder diffraction patterns for crystalline manganese oxide reveal its antiferromagnetic structure (see *Phys. Rev.* **76**, 1256 (1949)).

Neutron scattering explains the index of refraction at a molecular level. The forward neutron scattering is coherent (and so is in X-rays and visible light); when it interferes with the incident beam, it produces what we macroscopically call index of refraction (of the sample material). We will connect here a to the refractive index n.

Assume a spherically symmetric scattering. Consider a sample layer of thickness h, containing N nuclei/cm^3. A field (plane wave) $\psi = e^{ikx}$ falls perpendicularly on the slab (Fig. 15.15). The wave at point P in the forward direction is

$$\psi(\mathsf{P}) = \text{(Incoming stuff)} - \text{(Scattered stuff)}$$

$$= e^{ikx} - 2\pi Nh \int_0^\infty \frac{a}{r}(e^{ikr})z\,dz$$

where the amount absorbed by the slab is negligible. From $r^2 = x^2 + z^2$ we have $zdz = r\,dr$. Thus

$$\psi(\mathsf{P}) = \text{(Leftover stuff)} = e^{ikx} - 2\pi Nha \int_x^\infty e^{ikr}dr\,.$$

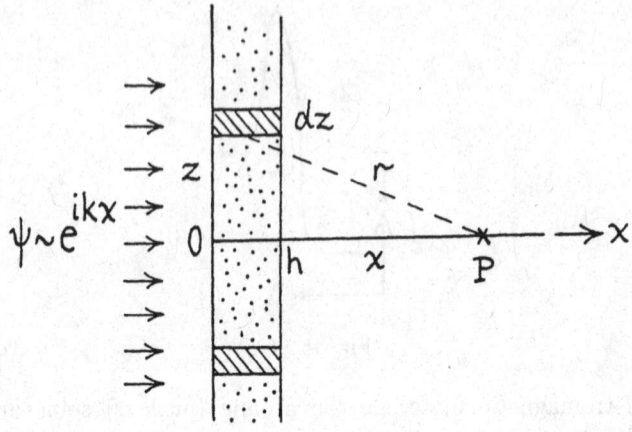

Fig. 15.15.

By considering Fresnel zones, we can calculate the integral. Or we can let $k \to k + ik'$ and take the limit as $k' \to 0$ (very small). Then the integral becomes zero for $r \to \infty$, the upper limit. We have then

$$\psi(P) = e^{ikx}\left(1 - \frac{2\pi i N h a}{k}\right).$$

The propagation vector in vacuum is k, and in the medium, nk where n is the refractive index of the medium. Then

$$\psi(P) \simeq e^{ik(x-h)+inkh} = e^{ikx}e^{ikh(n-1)}.$$

If $kh(n-1) \ll 1$, we have

$$\psi(P) \simeq e^{ikx}[1 + ikh(n-1)]$$

and from above,

$$n = 1 - \frac{2\pi N a}{k^2} = 1 - \frac{\lambda^2 N a}{2\pi}$$

where $n < 1$ if $a > 0$, and $n \simeq 1$ for slow neutrons. (Typically, $N \sim 10^{23}/\text{cm}^3$, $\lambda \sim 1$ Å, $a \sim 10^{-12}$ cm, and $n = 1 \pm (2 \times 10^{-6})$.) If $n < 1$, we can have total reflection from a surface, which happens if $\sin\theta_{\text{inc}} \geq n$. But since $n \simeq 1$, $\theta_{\text{inc}} \simeq 90°$ (tangential incidence).[6] The complementary angle is $\theta' = 90° - \theta_{\text{inc}}$.

---

[6]That is why you cannot see your face mirrored when you look normally to the glass panel of a window. The reflected intensity is the least (because $\theta_{\text{inc}} = 0°$). But if you look at the glass panel at a shallow complementary angle ($\theta_{\text{inc}} \simeq 90°$), it becomes as reflective as a mirror (the reflected intensity is strong) and you can see the images of objects or places.

Then $\sin\theta_{\text{inc}} = \cos\theta' \simeq 1 - \frac{\theta'^2}{2}$, or $n - 1 \leq -\theta'^2/2$. The limiting angle is

$$\theta' = \sqrt{\frac{Na\lambda^2}{\pi}} = \lambda\sqrt{\frac{N}{\pi}}\sqrt[4]{\frac{\sigma}{4\pi}} \tag{15.6}$$

where σ is the *total* scattering cross-section. Total reflection occurs at a given critical angle $(\theta_{\text{inc}})_{\text{crit}}$ for $\lambda > \lambda_{\text{crit}}$ defined by Eq. (15.6). This is a way of making a monochromator (actually, velocity selector) for neutrons.

In the case of reflection from magnetized surfaces (mirrors), n of a ferromagnetic material (for neutron beams) depends on the orientation between the magnetization of the material and the neutron spin, namely,

$$n_{\pm} = 1 - \frac{Na\lambda^2}{2\pi} \mp \frac{\mu_n H}{2E_n}$$

where μ_n is the magnetic moment of the neutron (*not* the nuclear magneton!), E_n is the neutron (kinetic) energy, H is the magnetic field in the mirror material (in the direction of the mirror plane), a is the scattering length of the mirror medium, and $\pm$ have to do with the orientation between the field and the neutron spin. Neutrons having opposite spins may have different θ' angles. Thus the total reflection from a magnetized surface can polarize neutrons (we mean, their spins).

How can we polarize neutrons? For a nice treatment, see Segrè, p. 579. We will not discuss it here.

Consider now total reflection from hydrocarbons. Since the C/H ratio may differ, the neutron reflection can measure the scattering amplitude of C divided by that of H. The measurements yield for the scattering lengths: $a(\text{H}) = 3.78 \times 10^{-13}$ cm and $a(^{12}\text{C}) = -6.63 \times 10^{-13}$ cm (for mono-isotopic C, without spin).

The solution of the following problems is left to the reader:

Example problem: What is the probability that in a scattering of a neutron off a proton, the neutron's spin flips? (See Segrè, pp. 564–572 on neutron scattering by hydrogen scatterers and complex nuclei, effect of chemical binding etc.)

Example problem: Consider neutron diffraction from NaCl. Find the scattered λ at 1°, 5°, 10°, 20°, and 40° in the first and second order. Find the relative intensity $(I_{\text{scattered}}/I_{\text{incident}})$, if the distribution of neutrons is $v^3 \exp(-mv^2/2kT)$.

Example problem: Relate the complex index of refraction to the complex scattering length a. (b) Relate the index of refraction to the scattering where the scattered wave amplitude is $\mathfrak{f}(\theta)$.

15.12. Neutron Scattering by Molecules

Slow neutrons can be used to probe molecules and crystals and study structure. A neutron of (kinetic) energy larger than the molecular binding energy hits a H nucleus (proton) in a molecule and dislodges it out of the molecule (molecular smash). Half of the neutron's energy is transferred to the proton in an elastic s-wave collision and scattering. If the neutron's energy is less than the molecular vibration energy $h\nu_0$, the neutron *cannot* lose any amount of energy to vibration or the knocking out of the H nucleus into freedom. Thus it is as if the proton had a mass equal to that of the whole molecule. Consequently, a slow neutron cannot lose energy so easily. In a thermal environment neutrons hardly lose any energy. This means that if the energy is less than $h\nu_0$, the reduced mass of the neutron + proton system is $\tilde{\mu} = M$, not $M/2$. Therefore, the scattering cross-section gets four times as big when we pass from large energies to energies less than that of a chemical bond.

If $\mathcal{H}'$ is the perturbation (interaction energy) operator (and $\langle f|\mathcal{H}'|i\rangle$ its matrix element between initial and final states), the differential scattering cross-section is (as the Born approximation provides)

$$\sigma(\theta) = \frac{\tilde{\mu}^2}{4\pi^2\hbar^4}|\langle f|\mathcal{H}'|i\rangle|^2$$

where

$$\langle f|\mathcal{H}'|i\rangle = \int \psi_f^* \mathcal{H}' \psi_i \, dV$$

over all space. Here ψ_i and ψ_f are plane waves and normalized. Since $\tilde{\mu}(\text{bound}) = 2\tilde{\mu}(\text{free})$, we have $\sigma(\text{bound}) = 4\sigma(\text{free})$.[7] This can be proven by another way, too, proposed by Weisskopf and Blatt. The incoming slow neutron has a de Broglie λ larger than the molecular size. Thus only s-wave scattering (spherically symmetric) need be considered. The center-of-mass system merges with the system in which the molecule is at rest. In this system, $d\sigma/d\Omega =$ constant. The scattering by a free proton is also spherically symmetric in this system (but there the system moves at half the speed of neutron with respect to the lab system). In the center-of-mass system we have $d\sigma/d\Omega' =$

---

[7]The Born approximation does not work at low thermal energies, but nevertheless it is not a big mistake to assume it as valid and standing for a moment in this case.

(differential cross-section) $= a^2$ where $d\Omega'$ is the solid angle element in the center-of-mass system (versus $d\Omega$ in the lab system), and a is the neutron-proton incoherent scattering length (see Segrè, Sections 10.2, 10.3, and 12.7 where the treatment includes singlet and triplet states and virtual ones).

If θ is the scattering angle in the lab frame, we have

$$\frac{d\sigma}{d\Omega} = \frac{d\sigma}{d\Omega'}\frac{d\Omega'}{d\Omega} = a^2\frac{d(\cos 2\theta)}{d(\cos\theta)} = 4a^2\cos\theta\,.$$

But for $\cos\theta = 1$ (in the forward direction) the two cross-sections have to be equal (because the proton does not gain momentum by the collision, and hence the scattering in the forward direction does not know whether the proton was free or bound to a molecule), that is, $4a^2 = d\sigma/d\Omega = $ constant. Totally,

$$\sigma_{\text{total}}(\text{bound}) = \int\frac{d\sigma}{d\Omega}d\Omega = 4\pi \times (\text{Constant}) = 16\pi a^2$$

which is four times as big as $\sigma_{\text{total}}(\text{free}) = 4\pi a^2$.

The same considerations hold for crystal lattices, too, not only for loose molecules.

Now one can roughly make an estimation of the energy where the cross-section increases due to bonds. For example, the C–H chemical bond in paraffin has a longitudinal vibration energy of $1/3$ eV (3000 cm^{-1}) and a transverse vibration of $1/15$ eV (600 cm^{-1}). (Segrè has graphs of the scattering cross-section curve for cetane, methane, H_2, ethylene, ethane, propane, and 1,3-butadiene, where σ — in barns — is plotted vs. the neutron time of flight — in μsec/m, and vs. energy in eV and λ in Å, p. 567.)

Now let us consider scattering of low-energy (slow) neutrons by complex nuclei in molecules. For simplicity, consider a free and loose nucleus. In low-energy scattering only s-wave scattering need be considered, because again, the neutron's λ is stronger than the range of the strong nuclear forces, *i.e.*, the dimensions of the nucleus. Asymptotically, the wavefunction of the system is

$$\psi \to e^{ikx} + c\frac{e^{ikr}}{r}$$

$$= (\text{Unscattered part — plane wave — in the forward direction})$$
$$+(\text{Scattered part, spherical wave, see Fig. 15.16})\,.$$

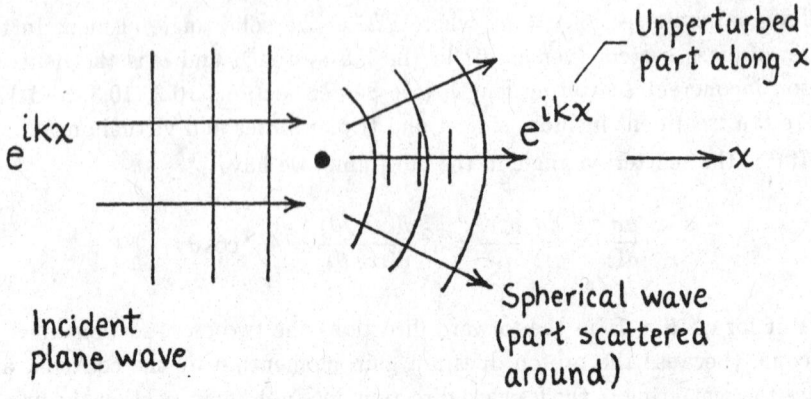

The unperturbed part is of smaller amplitude compared to the incident because part of the energy goes to the scattered part.

Fig. 15.16.

Here $k = 1/\bar{\lambda} = p/\hbar$, and c is a constant: $c = (1 - \eta_0)\frac{i}{2k}$ where for elastic scattering, $\eta_l = e^{2i\delta_l}$, and δ_l is the phase shift. For s-waves ($l = 0$) we have

$$c = \frac{\bar{\lambda}}{2i}(e^{2i\delta_0} - 1) = \bar{\lambda}e^{i\delta_0}\sin\delta_0 \simeq \bar{\lambda}\delta_0 \,,$$

because δ_0 is taken small. It is real, because the scattering is elastic. Totally,

$$\sigma_{\text{total}} = 4\pi\bar{\lambda}^2\sin^2\delta_0 \,.$$

The scattering length is defined as $a = -\bar{\lambda}\sin\delta_0$. If $|a/\bar{\lambda}| = |\sin\delta_0| \ll 1$, then a is the slope of $r\psi$ at $r = R \equiv$ nuclear radius, in the plot of $r\psi$ vs. r. Then elastically, $\sigma_{\text{total}} = 4\pi|a|^2$ where $a \sim 10^{-12}$ cm, typically.

In case we have a reaction alongside with scattering, a and δ_0 are complex numbers. Nevertheless for thermal neutrons, $\text{Re}\{a\} \gg \text{Im}\{a\}$. If a is real, the scattered part (wave) is $-ae^{ikr}/r$. The scattered part is not phase shifted with respect to the incoming wave when $a < 0$. If $a > 0$, then the phase shift is $\delta_0 = \pi$. Therefore, $\delta_0 = 0$ means no scattering at all. But mostly, $a > 0$ except for some cases, *e.g.*, neutron-proton collision in the singlet state.

The quantity a depends on the scatterer if the neutrons are slow, and on the relative orientation angle between the spins of the neutron and of the scatterer, if the latter has a nonzero spin. We can find a experimentally, if we allow scattering by materials with many nuclei. That is, we use bulk matter. The scattering depends on a of the nuclei in the material and on the way of their arrangement in the bulk, because waves scattered by different nuclei mutually interfere. As we saw above, this interference can explain phenomena like index of refraction, Bragg reflection, and scattering by polyatomic complex molecules.

Theoretically, we take an incident plane wave moving towards $+x$ and many pointlike scatterers (nuclei, scattering centers), each having a scattering length a_i. We work in the asymptopia, $i.e.$, the distance r_i from the (i-th) scatterer to the observation point P is large (Fig. 15.17), and the straight line from the scatterer to P touches the curved trajectory of the scattered neutron asymptotically at P. (This is plausible, because our detector can approach the

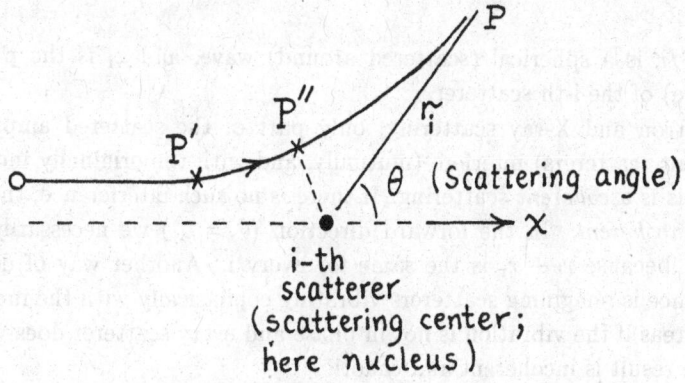

θ = angle between initial & final direction of bullet.

At P′ the bullet starts feeling the repulsive scattering potential of scatterer, and scattering deflection starts (the trajectory curves away; the scatterer's influence range extends appreciably until P′). P″ marks the distance of closest approach to scatterer. At P ($\to\infty$), r_i touches the trajectory asymptotically.

Fig. 15.17.

scatterer at most to within a few cm, whereas the size of the scatterer is ~Å. So, $r_i \sim$ cm $\gg$ Å, *i.e.*, the point of observation is asymptotically far away from the scatterer.) We ignore repeated scattering, concentrating on a single one. Then after scattering, the wave amplitude $\mathcal{A}$ consists of the amplitude of the unaffected (unscattered) part (which is still a plane wave) plus that of the spherically scattered part, subtracted from the original incident wave:

$$\mathcal{A} = e^{ikx} - \sum_i a_i \frac{\exp[ik(r_i + x_i)]}{r_i} \tag{15.7}$$

where we sum over i (all scatterers or nuclei). The *scatttered* intensity is

$$I = |\mathcal{A}_{\text{scattered}}|^2 = \left| \sum_i a_i \frac{\exp[ik(r_i + x_i)]}{r_i} \right|^2$$

where e^{ikr}/r is a spherical (scattered around) wave, and x_i is the position (coordinate) of the i-th scatterer.

In neutron and X-ray scattering, only part of the scattered amplitudes (by different scatterers) interfere (mutually, and with the originally incoming wave). This is a *coherent* scattering. If there is no such interference, the scattering is *incoherent*. In the forward direction ($\theta = 0°$) we necessarily have coherence, because $r_i + x_i$ is the same for every i. Another way of describing coherence is imagining scatterers vibrating cophasically with the incoming wave, whereas if the vibration is not in phase and every scatterer does what it wants, the result is incoherent and chaotic.

In a gaseous sample, as the molecules fly and move around randomly, the interference contributions of waves scattered by different molecules either cancel out or average to zero. Then

$$I_{\text{scattered}} = \sum_i I_i \equiv \text{Sum of individual intensities scattered by the molecules}.$$

If the gas is monatomic and of a single isotope, we have spherical symmetry in the scattering (in the center-of-mass system). Then the scattering cross-section is $\sigma = 4\pi|a|^2$ where the sign of a cannot be found experimentally.

Of course, things are different, depending on whether the scatterer nucleus is free or bound. If the hit nucleus is free, then $1/\tilde{\mu} = 1 + (1/A)$ where A is the atomic mass. If the target nucleus is bound to a molecule, then $1/\tilde{\mu} = 1 + (1/M)$ where $\tilde{\mu}$ is the reduced mass of the system (bullet + target), and M is the mass of the *molecule*. Then again from the Born approximation one obtains

$$\sigma(\text{bound}) = \left(\frac{1 + (1/A)}{1 + (1/M)} \right)^2 \sigma(\text{free})$$

and

$$a(\text{bound}) = \left(\frac{A+1}{M+1} \right) \left(\frac{M}{A} \right) a(\text{free}).$$

Also, a nucleus can be viewed as bound if the energy transfer (in the collision) is too small compared to the molecular binding energy. On the other hand, nuclei bound to solids behave very stiffly, that is, $M \to \infty$.

If the gas is of diatomic molecules and the nuclear species has two isotopes (without spin, or with spin, but ignored), and further, if we can admit the fact that λ of the neutrons is way larger than the interatomic distances in the molecule, then we can add up the amplitudes scattered by the two scatterers (nuclei) per molecule. Let the scattering lengths of the two nuclei be a_1 and a_2. Let the abundances of the two isotopes be a_1 and a_2 (where $a_1 + a_2 = 1$, *i.e.*, we consider relative abundances). We may have three situations for the gas:

(1) Molecules with both nuclei of isotope 1 and abundance a_1^2.
(2) Molecules with both nuclei of isotope 2 and abundance a_2^2.
(3) Mixed-marriage nuclei with abundance $2a_1 a_2$.

On the average, the scattering cross-section (per molecule) is

$$\bar{\sigma} = 4\pi [a_1^2 (2a_1)^2 + a_2^2 (2a_2)^2 + 2a_1 a_2 (a_1 + a_2)^2]$$

$$= 4\pi [4(a_1 a_1 + a_2 a_2)^2 + 2a_1 a_2 (a_1 - a_2)^2]$$

which is equivalent to the scattering by two equal nuclei which scatter coherently, each having a scattering length $a = a_1 a_1 + a_2 a_2$, and simultaneously also incoherently, where the coherent part (cross-section) is

$$\sigma(\text{coherent}) = 4\pi (a_1 a_1 + a_2 a_2)^2,$$

and the incoherent part is

$$\sigma(\text{incoherent}) = 4\pi a_1 a_2 (a_1 - a_2)^2$$

where the cross product means incoherence. Therefore, the overall molecular scattering is $\sigma = 4\sigma(\text{coherent}) + 2\sigma(\text{incoherent})$ (where for coherent scattering we add up amplitudes, whereas for incoherent scattering we sum intensities). Thus $\sigma_{\text{total}} = \sigma(\text{coherent}) + \sigma(\text{incoherent}) = 4\pi(\mathsf{a}_1 a_1^2 + \mathsf{a}_2 a_2^2)$.

If there are N isotopes in a random mixture, each with abundance a_i, then

$$a = \sum_{i=1}^{N} \mathsf{a}_i a_i$$

is responsible for interference in that material.

What happens if the nuclei do have spin I but are not polarized? Then one isotope (with spin I) behaves like a coexistence of two isotopes where one has abundance $(I+1)/(2I+1)$ and scattering length $a_\uparrow$ (parallel neutron and nuclear spin case), and the other has abundance $I/(2I+1)$ and scattering length $a_\downarrow$ (antiparallel case). Here the relative abundances measure also the multiplicities (statistical probabilities) of the two spin orientations respectively.

When we make a measurement in the case where the scattered waves (by the two scattering centers) mutually interfere, what we measure is $(a_1 + a_2)^2$. But if we measure the scattering by the two isotopes separately, then we measure a_1^2 and a_2^2. Hence we can have the (relative) signs of the two scattering lengths.

A word of caution is proper here: If the neutron spin flips upon scattering, the neutron waves do not interfere, because of the incoherence.

Suppose that we have scattering with interference. Since we have a collection of nuclei (in the sample medium), we cannot tell which nucleus caused the scattering; the only thing we know is that the interference is due to scattering by at least two (unspecifiable) nuclei. But if the neutron spin flips, we can detect the particular nucleus which scattered the neutron, because I_x of that nucleus has changed. Therefore, we should include the scatterer's wavefunction, too, in Eq. (15.7). For example, let a neutron with spin $|\uparrow\rangle$ be incident towards $+x$. Let the scatterer's wavefunction be u (whose explicit form does not matter here). Quantum mechanically, the whole system is initially described by the total wavefunction

$$\Psi = e^{ikx} \begin{pmatrix} 1 \\ 0 \end{pmatrix} u \equiv e^{ikx} |\uparrow\rangle u \equiv \psi_{\text{neutron}} \psi_{\text{nucleus}}$$

$$\equiv (\psi_{\text{neutron}})_{\text{space}} \times (\chi_{\text{neutron}})_{\text{spin}} \times u$$

where the matrix is the spin part of the neutron's ψ, i.e., $|\uparrow\rangle$, and e^{ikx} is its spatial part, in a product (because space and spin coordinates are independent). Let u_i = wavefunction of the i-th nucleus whose spin flips, a' = scattering length when there is a spin flip, and a = scattering length when there is no flip. Asymptotically, after the scattering, Ψ becomes

$$\Psi \rightarrow e^{ikx} \begin{pmatrix} 1 \\ 0 \end{pmatrix} u - \sum_i a_i \frac{e^{ik(r_i + x_i)}}{r_i} \begin{pmatrix} 1 \\ 0 \end{pmatrix} u - \sum_i a_i' \frac{e^{ik(r_i + x_i)}}{r_i} \begin{pmatrix} 0 \\ 1 \end{pmatrix} u_i$$

which is the solution that describes the event of scattering, including the possiblility for spin flip (as a correction term).[8] The intensity is

$$I_{\text{scattered}} = \int_{\substack{\text{Scatterer} \\ \text{coordinates}}} |\text{Modulus of the scattered part of the wave}|^2 dx_i \,.$$

There will be cross terms uu_i^*, but they go away, because

$$\int u_i^* u_j \, dV = \delta_{ij} \qquad \text{(Krönecker delta)}$$

(by definition). Only terms with $|u|^2$ and $|u_i|^2$ remain in the sum. Hence

$$I_{\text{scattered}} = (\text{Scattering without flip}) + (\text{Scattering with flip})$$

$$= \left| \sum_i a_i \frac{e^{ik(r_i + x_i)}}{r_i} \right|^2 + \sum_i \frac{|a_i'|^2}{r_i^2}$$

where the first term represents interference (between scatterings from many scatterers), and the second term means incoherence. And incoherence means that one wave does not know what the other wave does.

Readers interested in scattering in ortho- and parahydrogen should read Segrè, p. 571.

Note that the total probability of *independent* events occurring together is the product of the individual probabilities for each event. So, in quantum physics, ψ_{neutron} and ψ_{nucleus} above were taken in a product together for total $\Psi = \psi_{\text{neutron}} u$, because the spatial part of ψ_{neutron}, its spin part, and the nuclear u are independent degrees of freedom in our system considered (neutron plus nucleus).

We should state that the application of neutron scattering to problems of molecular physics and research therein is currently fashionable, with many possibilities. The method itself is an efficient tool to investigate molecules.

---

[8]To have consistency, it is advisable to replace x by z, unless $|\uparrow\rangle$ means spin "up" along x.

Chapter 16

NUCLEAR MAGNETIC RESONANCE

16.1. Resonance Methods

To study molecules, one uses experimental resonance methods which apart from their experimental substance and value, they are endowed with a rich theoretical background. These methods are the following:

(1) *Nuclear magnetic resonance* (NMR).
(2) *Electron paramagnetic (spin) resonance* (EPR or ESR).
(3) *Electron-nuclear double resonance* (ENDOR) which is the combination of NMR and EPR.
(4) *Nuclear quadrupole resonance* (NQR).

We will devote some of the chapters that follow to the analysis of these methods.

16.2. Perturbation Theory

Consider an electron that belongs to an atom or to a molecule. We can describe it by a wavefunction. Its evolution can be described by the Schrödinger wave equation. We first have to write the Hamiltonian for the electron, *i.e.*, its total energy:

$$\mathcal{H} = \text{(Kinetic energy)} + \text{(Electrostatic potential energy)},$$

but this is not the whole story. This is not the full (total, complete) Hamiltonian: There are terms to be added beyond the electrostatic potential energy,

442

i.e., energy correction terms of the form $\mathcal{H}_1 + \mathcal{H}_2 + \ldots$ whose explicit form must be found.

For atoms with many electrons, or for molecules with two or more atoms, we cannot solve the Schrödinger equation exactly in practice, because the differential equation is extremely complicated.[1] But if the correction terms are small (of lesser importance compared to the dominant first two terms), we can treat them as a perturbation to the dominant part, that is, we can apply the *perturbation theory* to the system (atom or molecule) and find the energy terms.

The first noteworthy perturbation term is $\lambda \mathbf{L} \cdot \mathbf{S}$ where λ is a constant (called the *interaction strength*). This term comes from the interaction between the electron's orbital angular momentum $\mathbf{L}$ and the electron spin $\mathbf{S}$. It has the dimension of energy (and so do the other terms, too).

The second term is the interaction between the strong external magnetic field $\mathbf{H}_0$ (if there exists) and $\mathbf{S}$.

The third term is the Zeeman energy, *i.e.*, the interaction of $\mathbf{H}_0$ with $\mathbf{L}$.

The fourth term is due to the interaction of the nuclear spin $\mathbf{I}$ with the orbital angular momentum $\mathbf{L}$.

The fifth term is the interaction between $\mathbf{I}$ and $\mathbf{S}$.

Next comes the sixth term, $\mathcal{H}_6$, which is the quadrupole interaction. This gives the energy of (the phenomenon of) NQR.

The next term gives the energy states that give rise to NMR, *i.e.*, the interaction of the nuclear spin $\mathbf{I}$ with $\mathbf{H}_0$.

Finally, we have the interaction between the electron spin $\mathbf{S}$ and the external field $\mathbf{H}_0$; this is the EPR term.

In the NQR we have energy states coming from the interaction between the nuclear quadrupole moment and the electric field gradient $\vec{\nabla}\vec{\mathcal{E}}$.

In the NMR, in addition to the main phenomenon, we also observe the interaction between the nuclear spin (of the nucleus whose resonance is being observed) and the nuclear spins of neighboring nuclei in the molecule.

In the EPR the biggest (perturbative) potential energy is the *crystal potential* V_{cryst} caused by the electric field of the crystal. In terms of size (quantitative importance), it is hierarchically placed after the electrostatic potential term in the total Hamiltonian.

---

[1]In principle we lightheartedly say: "Just solve the Schrödinger equation and find the quantized energies (eigenenergies)", thinking that we can get away by this simple command. In reality things are not so simple. Sometimes we have to approximate, *e.g.*, split the potentials into manageable parts (terms). Approximations and approximation methods are ways of getting around difficulties in quantum physics.

EPR is observed as a fine structure over the potential of the crystal electric field. Further, the interaction between the neighboring nuclear spins is observed as a hyperfine structure.

16.3. NMR Theory

NMR is a vast topic with a vast library of research papers and literature. The phenomenon comes from the interaction of the magnetic moment of the nucleus with an externally applied strong magnetic field H_0.

An atom (say, a H atom) can be thought of as a tiny, Lilliputian magnet: The electron revolving about the nucleus (Fig. 16.1) is a moving charge (closed current loop) that produces a magnetic field at the nucleus. This is equivalent to a tiny magnetic moment μ located at the nucleus, that is, the current loop is equivalent to a *magnetic dipole*, *alias* a tiny rod magnet (because a magnet has a μ) (Fig. 16.2 and Fig. 16.3). But to speak of a magnetic moment, there is no need to always have an atom around; every elementary particle (whether charged or uncharged) has a magnetic moment if it has a *nonzero* spin. For example, consider the electron. We think it as a sphere carrying a charge e and rotating about its own axis, just like the planets do as at the same time revolve about the Sun (so that the atom is a miniature planetary system in micrography).[2]

If P is the spin angular momentum of this rotating sphere (Fig. 16.4), its magnetic moment μ is proportional to it and equal to

$$\mu = \frac{e g_e}{2mc} P \tag{16.1}$$

where m is the mass of the electron and g_e is the *Lande factor* for the electron, a constant number. The expression that appears in front of P in Eq. (16.1) is the constant of proportionality. Equation (16.1) is for a single electron rotating about itself.

If we have a proton instead, then

$$\mu = \frac{e g_p}{2 m_p c} P$$

where g_p is the Lande factor for the proton, m_p is the mass of the proton, and c is the speed of light.

---

[2]Actually, the spin of the electron has no classical analogue. It is an intrinsic quantum-mechanical property associated with the particle, and should not be construed as a classical spin angular momentum, strictly speaking. The quantum-mechanical nomenclature "spin" is perhaps a poor choice, and the notion of an actual classical spinning should not be taken seriously, although some physicists think otherwise. In his famous *Lectures*, Feynman calls it "quantangular momentum".

In general, for any spinning system, the magnetic moment is $\mu \propto \mathbf{P}$ and

$$\mu = \frac{eg}{2Mc}\mathbf{P}$$

where M is the mass of the system. In case of a nucleus, M is the mass of the nucleus (protons + neutrons), $\mathbf{P}$ is the total angular momentum of the nucleus, and μ is the nuclear magnetic moment (equal to the vector sum of all the moments of the individual nuclear particles suitably added).

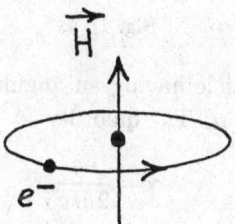

Fig. 16.1.

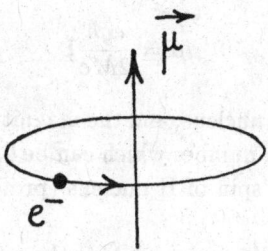

Fig. 16.2.

(a)

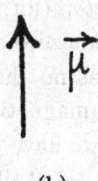

(b)

Fig. 16.3.

Fig. 16.4.

Therefore, to every particle having an angular momentum **P** there corresponds a magnetic moment $\boldsymbol{\mu}$. The quantity

$$\gamma = \frac{eg}{2Mc}$$

is called *gyromagnetic ratio*.

Since $\mathbf{P} = \hbar\mathbf{I}$ (where **I** is the nuclear spin vector) for a nucleus, we have for the nuclear magnetic moment

$$\boldsymbol{\mu} = \frac{eg\hbar}{2Mc}\mathbf{I}$$

where M is the mass of the nucleus, and the magnitude of **P** is $P = \sqrt{I(I+1)}\hbar$. Here I is the spin quantum number which can be $0, \pm 1/2, \ldots$ and the multiples thereof. For example, the spin of H nucleus (proton) is $1/2$, that of deuteron is 1 etc.

If m is the mass of the electron and M is the mass of nucleus, the quantities

$$\mu_B = \frac{e\hbar}{2mc} \quad \text{and} \quad \mu_n = \frac{e\hbar}{2Mc}$$

are called *Bohr magneton* (for the electron) and *nuclear magneton* respectively.

The number g is nonzero and independent of the electric charge; that is, even if the particle has no charge, it still has a μ (for example, the neutron does have a μ). The magnetic moment is zero *only if* the spin number I is zero. For the proton we have $g_p = +5.585$. For the neutron, $g_{\text{neutr}} = -3.826$. For the electron, $g_e = -1/2$ (if there is no interaction).

To cut the long word short, μ corresponds to an elementary magnet, and $\mu = g\mu_n\mathbf{I}$.

Since the nuclear μ is equivalent to a rod magnet, if it is put in an external (laboratory) field $\mathbf{H}_0$ (say, in the z-direction), it tries to orient itself along (or opposite to) the field, that is, it wants to take the direction of the field (or the opposite direction) (Fig. 16.5).

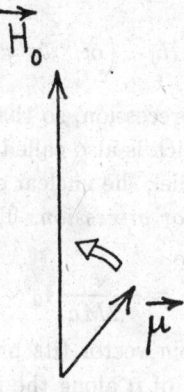

Fig. 16.5.

A torque (turning moment or couple) acts on μ, given by

$$\vec{T} = \mu \times \mathbf{H}_0.$$

Newtonian physics says that the torque is equal to the rate of change of the angular momentum. Therefore,

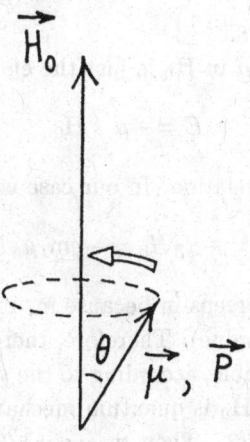

Fig. 16.6.

$$\vec{T} = \frac{d\mathbf{P}}{dt} = \mu \times \mathbf{H}_0 = \frac{eg}{2Mc}\mathbf{P} \times \mathbf{H}_0$$

which tells us that $\mathbf{P}$ (or the moment μ) *precesses* like a top about $\mathbf{H}_0$. The direction of $\mathbf{H}_0$ is the precession axis.

As μ precesses about $\mathbf{H}_0$ at a constant angle θ (Fig. 16.6), the coefficient

$$\omega_0 \equiv \frac{eg}{2Mc}H_0 \quad \left(\text{or} \quad \vec{\omega}_0 = \frac{eg}{2Mc}\mathbf{H}_0\right)$$

is the angular velocity of the precession, so that the relation $\omega_0 = 2\pi\nu_0$ gives the *precession frequency* ν_0 which is also called *Larmor frequency*.

If vectors $\mathbf{P}$ and μ are parallel, the nuclear elementary magnet precesses in $\mathbf{H}_0$. Its motion is called *Larmor precession*. The equation that describes the precession of μ is

$$\frac{d\mu}{dt} = -\frac{e}{2Mc}\mathbf{H}_0 \times \mu.$$

The z-component of the spin vector (its projection on the z-axis) is $m_I \hbar$ (Fig. 16.7). So, the component of μ along the field is

$$\mu_z = gm_I\mu_n \equiv \mu_\parallel$$

where m_I is the magnetic quantum number which can take the values $0, \pm 1, \ldots, \pm I$, *i.e.*, $2I + 1$ values.

The component perpendicular to the field is

$$\mu_\perp = g\mu_n\sqrt{I(I+1) - m_I^2}.$$

This is shown in Fig. 16.8.

The potential energy of μ in $\mathbf{H}_0$ is just the energy of a magnet in a field:

$$E = -\mu \cdot \mathbf{H}_0$$

which is a classical physics relation. In our case we have

$$E = -\mu_Z H_0 = -gm_I\mu_n H_0$$

where quantum mechanics creeps in because m_I can take discrete (quantized) values (not any continuous value). Therefore, there are certain discrete energy values (levels or states) possible, according to the values of m_I. In other words, the angle θ between μ and $\mathbf{H}_0$ is quantum mechanical.

For example, let $I = 3/2$. Then $m_I = \pm 1/2, \pm 3/2$ only. The positive (negative) values of m_I correspond to the negative (positive) energies. The

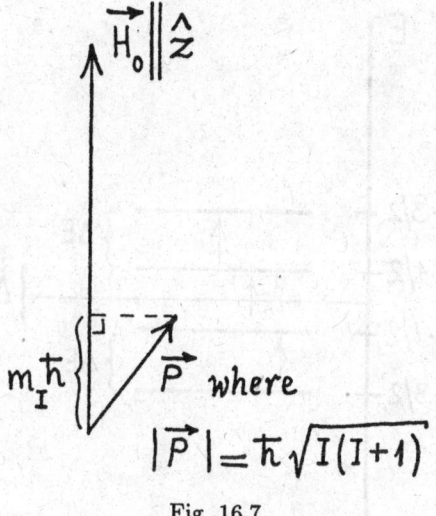

Fig. 16.7.

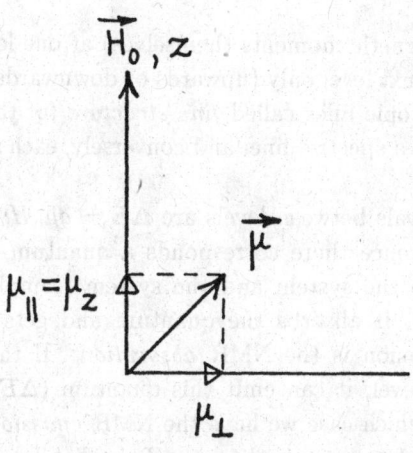

Fig. 16.8.

four energy states are shown in Fig. 16.9. They are the only possible ones for $I = 3/2$. They are equally spaced (the three ΔE differences are equal). The original level of the potential energy is arbitrary. For $I = 3/2$, the energy level is split into four.

Since the selection rule is $\Delta m_I = \pm 1$, transitions only between adjacent levels are possible and allowed by quantum mechanics; particles with magnetic

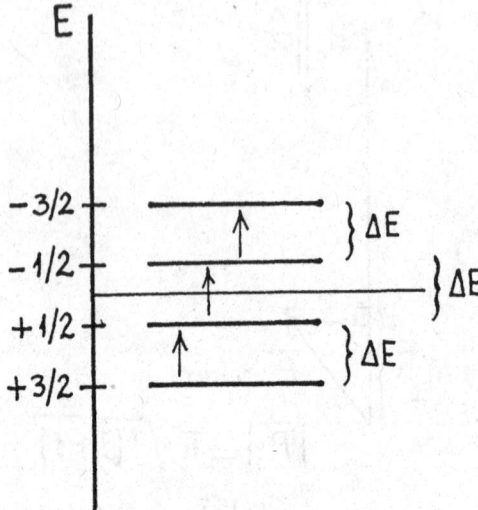

Fig. 16.9.

moments (or the magnetic moments themselves) at one level can pass (jump) to the immediately next level only (upwards or downwards). These transitions are seen as spectroscopic lines called *fine structure* (of the main spectrum).[3] Each transition gives a spectral line, and conversely, each spectral line belongs to a transition.

The energy intervals between levels are $\Delta E = g\mu_n H_0$ (since $\Delta m_I = \pm 1$). To this energy difference there corresponds a quantum $\Delta E = h\nu_0$. If this quantum is given to the system and the system (sample with nuclei) is at a lower energy level, it absorbs the quantum and gets to a higher energy level. This phenomenon is the NMR *absorption*. If the system is already at a higher energy level, it can emit this quantum (ΔE difference) and get to a lower level, in which case we have the NMR *emission*. So, depending on whether an amount of energy equal to $g\mu_n H_0$ is absorbed or emitted, we have NMR absorption or emission. Since this happens at the frequency ν_0 (*i.e.*, at the Larmor frequency), we have *nuclear magnetic resonance* between the electromagnetic wave (of frequency ν_0 at resonance) that we send to the system from outside, and the system itself that absorbs it (because the frequency of

---

[3]The main spectrum has a degenerate energy level unsplit, *i.e.*, several substates have the same energy in the absence of the field. The presence of the magnetic field removes the degeneracy and splits the level into four. The single energy takes now discrete and different values according to the spin orientations.

the sent radiation matches the transition frequency between levels, so that there is a resonance).

Of course, if $H_0 = 0$, there is no splitting: $\Delta E = 0$. When we turn off the field there is no NMR. There is only one single energy (degenerate state) for the nuclear moment.

Upon the application of $\mathbf{H_0}$, we get $2I + 1$ splittings (levels) and the degeneracy is lifted.

The phenomenon of NMR was first put forth theoretically by C.J. Gorter in 1942, who suggested that NMR can also be observed experimentally. He introduced NMR as a new concept and then the experimentalists took over, but they could not observe the phenomenon in the first trials. The reason is as follows: The distribution of the magnetic moments to the energy levels is uniform, that is, the moments are distributed in equal numbers per level. But when $\mathbf{H_0}$ is on, the energy level of the dipoles along the field is lower, and hence it invites more dipoles. The difference of the number of the dipoles can be found by assuming a Boltzmann distribution for the distribution among the four energy levels. This number difference turns out to be too small to give a noticeable observation.

For a Boltzmann distribution, the ratio of the numbers of nuclei in states $+1/2$ and $-1/2$ (Fig. 16.10), *i.e.*, those parallel to those antiparallel to the field, is

$$\frac{N(+1/2)}{N(-1/2)} \simeq 1 + \frac{\Delta E}{kT} = 1 + \frac{g\mu_n H_0}{kT}.$$

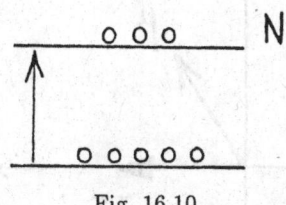

Fig. 16.10.

At room temperature, and for $I = 1/2$, $H_0 = 10$ kGauss, $g = g_{\text{proton}} = 5.585$, $k = 1.38 \times 10^{-16}$ erg/°K, and $\mu_n = 5.049 \times 10^{-24}$ erg/Gauss, we obtain

$$\frac{N(+1/2)}{N(-1/2)} = 1 + 14 \times 10^{-6}.$$

This means that if there are one million moments (nuclei) in one state, there are one million plus fourteen of them in the other state.[4] Since the difference is too small, the transition cannot be observed experimentally, in spite of the availability of sensitive methods. So, in order to observe NMR, the following trick is necessary:

As the nuclear moments precess about the applied H_0, we apply another field H_1 perpendicular to H_0, where H_1 is sinusoidal of the form $2H_1 \cos(\omega t)$. This can be a rotating field whose amplitude rotates in the xy-plane, so that its projection along the x-axis shrinks and grows sinusoidally (Fig. 16.11). As H_1 changes continuously direction, it exerts a torque on μ. The situation at any instant is shown in Fig. 16.12. If H_1 has the direction shown (at some instant), it tries to pull the magnet μ to its own direction. Half a period later, the situation is as in Fig. 16.13. But since the angular frequency (pulsation) ω of H_1 can be chosen and adjusted arbitrarily, we can see it $\omega \simeq \omega_0$, *i.e.*, equal to the Larmor precession frequency of μ. Then the rotation of H_1 and the precession of μ are synchronized about H_0. The two rotate together: μ precesses about H_0, and H_1 rotates simultaneously, staying in the xy-plane. So, if at $t = 0$ the situation is as in Fig. 16.12, half a period later, it is as in

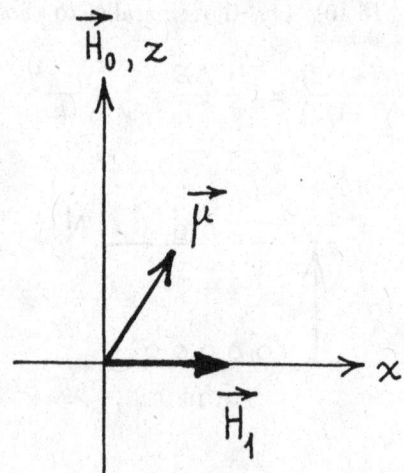

Fig. 16.11.

---

[4]This difference gives the paramagnetic (Curie) susceptibility χ of the dipoles, that is, the relation $\mathcal{M} = \chi H_0$. At room temperature χ is too low, and one has to go to low temperatures to measure it directly. Here $\mathcal{M}$ is the magnetization.

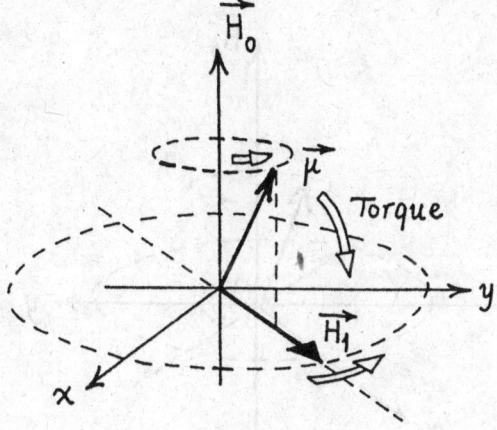

Fig. 16.12.

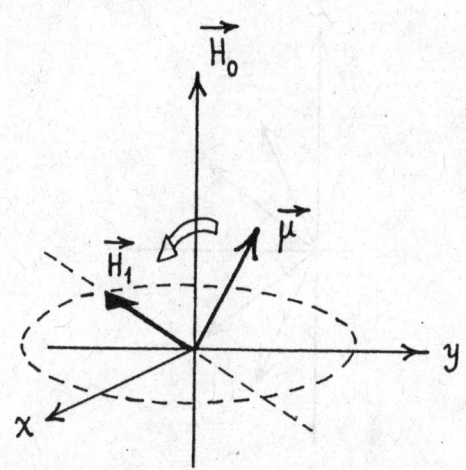

Fig. 16.13.

Fig. 16.14. Thus when $\omega \simeq \omega_0$, the motions of μ and $\mathbf{H}_1$ are synchronized and μ is always pulled down and absorbs energy. Now if at $t = 0$ we start with a situation as in Fig. 16.13 and maintain the simultaneous phase, then μ is always pulled up this time.

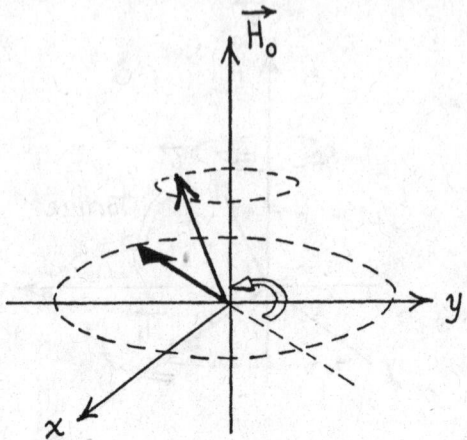

Fig. 16.14.

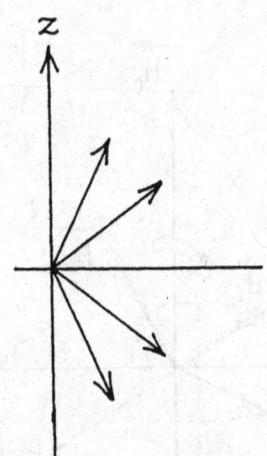

Fig. 16.15.

Since for $I = 3/2$ there are four states (Fig. 16.15), when μ is pulled down it can switch to another state of higher energy; and when it is pulled up (by $\mathbf{H}_1$) it can flip to a state of lower energy. (When a magnet orients itself with $\mathbf{H}_0$ the energy decreases; a full orientation means stabilization or minimum energy.) So, the moments μ in the sample material make transitions between energy levels. We thus need an alternating field applied with $\omega \simeq \omega_0$ and compatible with the value of H_0. If we apply this $\mathbf{H}_1$, we can measure absorption and emission.

16.4. Utility of NMR

NMR is extensively used in biochemistry, molecular biology, and pharmacy. The specimen (material whose nuclei are to be tested for nuclear resonance) is put into a test tube which in turn is inserted into the solenoid of a resonance circuit (Fig. 16.16). We intend to measure the energy absorbed by the material (in fact by the magnetic moments of its nuclei). The observed absorption is proportional to the magnetization $\mathcal{M}$ given by

$$\vec{\mathcal{M}} = \sum_{i=1}^{N} \mu_i$$

which is a *vector sum* of all the moments present in the specimen (there are many of them). We sum up to N, the number of nuclei participating in the absorption. Further we have

$$\vec{\mathcal{M}} = \chi \mathbf{H}_0$$

where χ is the *magnetic susceptibility*, the proportionality coefficient. If the specimen is paramagnetic, we talk about paramagnetic susceptibility (which is our case); if it is a ferromagnetic material, we talk about *permeability* which is outside this subject.

To observe absorption, we place the specimen in a series resonance circuit (Fig. 16.16(b)). The impedance of the resonance circuit is

$$\mathcal{Z} = R + i\left(\omega L - \frac{1}{\omega C}\right)$$

where R is the Ohmic resistance of the circuit, C is the capacitance, and L is the inductance of the solenoid. We have

$$L = L_0(1 + 4\pi\chi\eta)$$

where L_0 is the inductance when the solenoid contains vacuum instead of the specimen. This L_0 is modified by the multiplying parenthesis when we insert the specimen. Here η is the *filling factor*, due to the fact that the specimen does not fill the whole volume inside the solenoid. Therefore, η is a fraction: $0 < \eta < 1$.

If there is both absorption and emission in the medium, the expression for χ is complex, given by

$$\chi = \chi' - i\chi'',$$

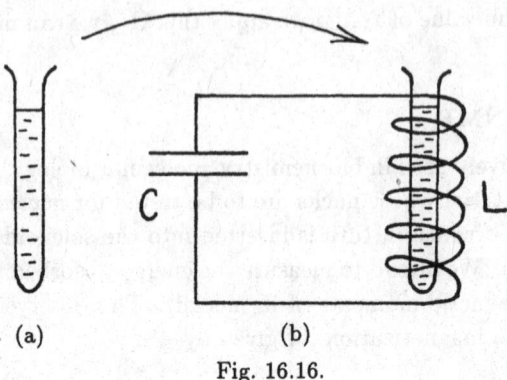

Fig. 16.16.

in general taken with the minus sign. Here χ' (the real part) refers to dispersion, and χ'' (the imaginary part) refers to absorption. So, χ is separable to real and imaginary parts.

The impedance of the resonance circuit becomes

$$Z = R + i\left(\omega L - \frac{1}{\omega C}\right) = R + 4\pi\omega L_0\eta\chi'' + i\left[\omega L + 4\pi\eta\omega L\chi' - \frac{1}{\omega C}\right]$$

where $\Delta R \equiv 4\pi\omega L_0\eta\chi''$ is in the real part, and $\Delta L \equiv 4\pi\eta\omega L\chi'$ is in the imaginary part. The changes ΔR and ΔL in the resonance circuit give the absorption and emission.

The change in the quality factor, dQ, is equal to the increase in absorption in the resonance circuit, which in turn is proportional to χ'', the imaginary part of χ:

$$dQ = -4\pi\eta Q\chi''(\omega)$$

where Q is the *quality factor* initially: $Q = \omega^2/R$.

The change in resonance frequency, $d\omega$, is the dispersion:

$$d\omega = -2\pi\eta Q\omega_s^2\chi'(\omega)$$

which is related to χ'. Here ω_s is the resonance frequency of the specimen.

The mean energy absorbed is

$$\overline{E} = \frac{1}{2}\omega H_0^2\chi''\eta V \tag{16.2}$$

where V is the volume. To convince the reader that the right-hand side of this equation has the dimension of energy, let us make a dimensional analysis: η

is a dimensionless number. $\vec{\mathcal{M}} \cdot \mathbf{H}_0$ is an energy (where $\vec{\mathcal{M}} = \chi \mathbf{H}_0$, so that $\vec{\mathcal{M}} \cdot \mathbf{H}_0$ becomes $\chi'' H_0^2$). Energy multiplied by ω (one over time) becomes power. Power multiplied by volume becomes energy.

Equation (16.2) shows that when H_0 is large, $\overline{E}$ is large, and hence easily measurable. Therefore, H_0 should be large.

The obtained signals are usually weak. A special electronic equipment is needed for the observations. $\mathbf{H}_0$ must be homogeneous, and the whole apparatus must be shielded and protected from external ambient fields. Also, the power supplies must be of good quality in the detection device.

16.5. Relaxation Times

When we insert a hot body into cold water (Fig. 16.17), it takes some time for the heat to pass from the body to the water. The body cools down exponentially in time. Similarly, when $\mathbf{H}_0$ is applied, the nuclear spins do not take the quantized states abruptly and immediately: The moments need a time τ_1 to orient themselves along $\mathbf{H}_0$. This τ_1 is the *spin-lattice relaxation time*. The absorbed energy depends on the relaxation time of the system; that is, the time that elapses until the energy is absorbed (or emitted) depends on the structure of the material.

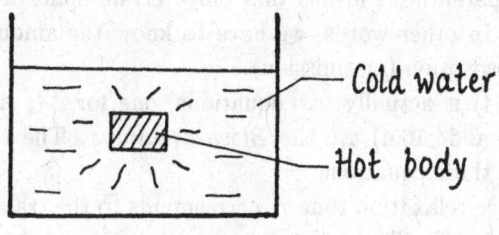

Cold water

Hot body

Fig. 16.17.

During the orientation of the dipoles, the component of the magnetization along the field, $\mathcal{M}_z$, reaches its maximum value exponentially in time, so that

$$\frac{d\mathcal{M}_z}{dt} \propto \frac{1}{\tau_1},$$

i.e., the time rate of change of the magnetization is inversely proportional to τ_1. (The total magnetization vector is $\vec{\mathcal{M}}$ which has a component $\mathcal{M}_z$ along the direction of $\mathbf{H}_0$.) If there is absorption in the phenomenon of resonance, the magnetization $\mathcal{M}_z$ of the moments μ increases in value up to a final value $\mathcal{M}_0$. If there is emission (or induced energy release outwards), $\mathcal{M}_z$ decreases.

We have

$$\left(\frac{d\mathcal{M}_z}{dt}\right) = \frac{\mathcal{M}_0 - \mathcal{M}_z}{\tau_1} \tag{16.3}$$

which can be obtained from the precessional equation of motion

$$\frac{d\mu}{dt} = \mu \times \mathbf{H}_0 .$$

We multiply both sides by N, the number of spins, to have

$$\left(\frac{d\vec{\mathcal{M}}}{dt}\right) = \vec{\mathcal{M}} \times \mathbf{H}_0$$

where the parenthesis on the left-hand side means that only certain spins will be taken into account (those affecting the change).

The other two components of $\vec{\mathcal{M}}$ undergo a change which is inversely proportional to another time τ_2, the *spin-spin relaxation time*:

$$\left(\frac{d\mathcal{M}_{x,y}}{dt}\right) = -\frac{\mathcal{M}_{x,y}}{\tau_2} \tag{16.4}$$

where again the parenthesis means that only certain spins are to be included in the equation. In other words, we have to know the amount of spins that partake of the absorption (or emission).

Equation (16.4) is actually two equations, one for $\mathcal{M}_x$ and one for $\mathcal{M}_y$. Equations (16.3) and (16.4) are the *Bloch equations*. The rotating field $\mathbf{H}_1$ should also enter these equations.

The spin-lattice relaxation time τ_1 corresponds to the transfer of heat from the spin to the crystal. The spin-spin relaxation time τ_2 corresponds to the transfer of energy from a spin to neighboring spins of the same kind. The spin-spin relaxation time cannot be observed from outside; it can be "seen" from the width of the resonance curve. Actually, τ_2 cannot be measured directly. What we measure experimentally is τ_2^* where

$$\frac{1}{\tau_2^*} = \frac{1}{\tau_2} - \frac{1}{2\tau_1} . \tag{16.5}$$

This equation was found theoretically. From the widening of the resonance curve, one calculates τ_2^*, and thus via Eq. (16.5) one measures indirectly the actual time τ_2.

We have always $\tau_2 < \tau_1$.

16.6. Bloch Equations

In thermal equilibrium (at temperature T), as the nuclei are in the field $\mathbf{H}_0$, the total magnetization is fully along $\mathbf{H}_0$: $\mathcal{M}_x = \mathcal{M}_y = 0$ and $\mathcal{M}_z = \mathcal{M}_0 = \chi_0 H_0 = (N\mu^2/3k)H_0/T$ where the quantity in parenthesis is the *Curie constant*, and k is the Boltzmann constant.

If $I = 1/2$, the magnetization depends on the population difference (per unit volume) $\Delta N = N_1 - N_2$ between the two states (levels): $\mathcal{M}_z = \mu\Delta N$. In thermal equilibrium, the population ratio is given by the Boltzmann factor (for the energy difference $\Delta E = 2\mu H_0$)

$$\left(\frac{N_2}{N_1}\right)_{\text{equil}} = e^{-2\mu H_0/kT},$$

and the equilibrium magnetization is $\mathcal{M} = N\mu \tanh(\mu H_0/kT)$.

But if $\mathcal{M}_z$ is not in thermal equilibrium, we assert that it can reach equilibrium at a rate proportional to the departure from the equilibrium value $\mathcal{M}_0$, i.e.,

$$\frac{d\mathcal{M}_z}{dt} = \frac{\mathcal{M}_0 - \mathcal{M}_z}{\tau_1}. \tag{16.6}$$

So, if at $t = 0$ the unmagnetized sample is brought inside $\mathbf{H}_0$ then $\mathcal{M}_z$ increases from 0 (initial value) to $\mathcal{M}_0$ (final value). Before turning on $\mathbf{H}_0$ (that is, when $H_0 = 0$), we have $N_1 = N_2$ (thermal equilibrium). But to reach the new equilibrium, some spins flip over when $H_0 \neq 0$. We integrate Eq. (16.6) to find

$$\int_0^{\mathcal{M}_z} \frac{d\mathcal{M}_z}{\mathcal{M}_0 - \mathcal{M}_z} = \frac{1}{\tau_1}\int_0^t dt,$$

$$\ln\left(\frac{\mathcal{M}_0}{\mathcal{M}_0 - \mathcal{M}_z}\right) = \frac{t}{\tau_1} \quad \text{and} \quad \mathcal{M}_z(t) = \mathcal{M}_0(1 - e^{-t/\tau_1}),$$

and the magnetic potential energy $-\vec{\mathcal{M}} \cdot \mathbf{H}_0$ decreases as $\mathcal{M}_z$ reaches its new equilibrium value.

The Bloch equations (of motion) are

$$\frac{d\mathcal{M}_z}{dt} = \gamma(\vec{\mathcal{M}} \times \mathbf{H}_0)_z + \frac{\mathcal{M}_0 - \mathcal{M}_z}{\tau_1}$$

$$\frac{d\mathcal{M}_{x,y}}{dt} = \gamma(\vec{\mathcal{M}} \times \mathbf{H}_0)_{x,y} - \frac{\mathcal{M}_{x,y}}{\tau_2}$$

where the rightmost extra terms are due to effects not included in $\mathbf{H}_0$: Alongside with the precession, $\vec{\mathcal{M}}$ reaches (relaxes to) its equilibrium value $\vec{\mathcal{M}}_0$. If in H_0 we have $\mathcal{M}_x \neq 0$ and $\mathcal{M}_y \neq 0$, these two components reach zero exponentially (because in thermal equilibrium there is no transverse component). So, the relaxation is included in the Bloch equations above.

It should be noted that the magnetic energy remains the same as the transverse components decay to zero (because no energy leaves the spin system as $\mathcal{M}_x$ and $\mathcal{M}_y$ relax to zero). If initially the spins are cophasic in their precession, after a time τ_2 they will be dephased to a randomness (as $\mathcal{M}_x$ and $\mathcal{M}_y \to 0$) because of the the fact the different local fields make the spins precess at different frequencies eventually. So, τ_2 is a measure of the time for which the spins (that contribute to $\mathcal{M}_x$ and $\mathcal{M}_y$) stay all cophasic.

Notice that the Bloch equations are not symmetric, because we necessarily have H_0 in a preferred direction. So, we apply a sinusoidal field H_1 (which stays in the xy-plane) and study the magnetization in the two combined fields H_0 and H_1.

16.7. Precession Frequency and Power Absorbed

Consider a spin precessing about H_0, with $\mathcal{M}_z = \mathcal{M}_0$. Then we have

$$\frac{d\mathcal{M}_z}{dt} = 0 \quad \text{and} \quad \frac{d\mathcal{M}_x}{dt} = \gamma H_0 \mathcal{M}_y - \frac{\mathcal{M}_x}{\tau_2}$$

$$\frac{d\mathcal{M}_y}{dt} = -\gamma H_0 \mathcal{M}_x - \frac{\mathcal{M}_y}{\tau_2}.$$

This is the case of a two-dimensional damped oscillator with solutions

$$\mathcal{M}_x = A e^{-t/\tau} \cos(\omega t)$$

$$\mathcal{M}_y = -A e^{-t/\tau} \sin(\omega t),$$

so that upon substitution, one of the equations becomes

$$-\omega \sin(\omega t) - \frac{\cos(\omega t)}{\tau} = -\gamma H_0 \sin(\omega t) - \frac{\cos(\omega t)}{\tau_2},$$

and the free precession frequency becomes $\omega_0 = \gamma H_0$, while τ is identified as $\tau = \tau_2$. The precessing spin can make absorption at resonance from a driving field of frequency $\omega_0 = \gamma H_0$ (or close to it). The response of the spin to the driving field will be characterized by the absorption curve (Fig. 16.18) where the full width at half maximum (FWHM) will be $\Delta\omega \simeq 1/\tau_2$.

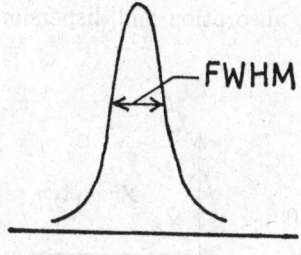

Fig. 16.18.

If H_1 is the amplitude of the sinusoidal rotating field, we have

$$H_x = H_1 \cos(\omega t) \quad \text{and} \quad H_y = -H_1 = \sin(\omega t).$$

The power absorption is made from this driving field. If we solve the Bloch equations, we find the absorbed power:

$$P(\omega) = \frac{\mathcal{M}_z \tau_2 \omega \gamma H_1^2}{1 + (\omega_0 - \omega)^2 \tau_2^2}$$

where again the FWHM of the resonance curve is $\Delta\omega = 1/\tau_2$.

To summarize, we realize that with the applied static $\mathbf{H}_0$ alone, the change in the magnetization in the sample material is too small. To be able to measure it, we also apply a weak field $\mathbf{H}_1$ perpendicular to $\mathbf{H}_0$, varying with a frequency ν. If this ν is set equal to the Larmor frequency, many of the dipoles orient themselves along $\mathbf{H}_0$ and a resonance (absorption at $\nu = \nu_0$) is observed. $\mathbf{H}_1(\nu)$ is a rotating field.

The sample in the solenoid has a Curie susceptibility χ expressed as $\chi = \chi' - i\chi''$. Provided that there is no saturation and that $H_1^2 \tau_1 \tau_2 \ll 1$, the solution of Bloch equations gives

$$\chi' = \frac{1}{2}\chi_0 \nu_0 \tau_2 \frac{\Delta\omega}{1 + \tau_2^2 \Delta\omega}$$

$$\chi'' = \frac{1}{2}\chi_0 \nu_0 \tau_2 \frac{1}{1 + \tau_2^2 \Delta\omega}$$

where χ_0 is the susceptibility in the equilibrium, ν_0 is the resonance frequency, and $\Delta\omega$ is the frequency mismatch (width):

$$\Delta\omega = 2\pi(\nu_0 - \nu).$$

Figure 16.19 shows the absorption and dispersion curves. The absorption is equal to $2H_1^2\chi''$.

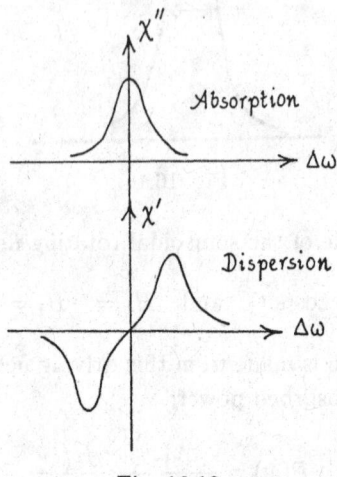

Fig. 16.19.

16.8. Nuclear Polarization

The orientation of nuclei with nonzero I can be such that samples with different m_I can have different populations. For $I = 1/2$, let us have $N_\uparrow$ nuclei with spin pointing up, and $N_\downarrow$ nuclei down (with respect to the external field). Then $m_I = \pm 1/2$. The *nuclear polarization* (of this system of nuclei) is

$$\text{Polarization} = \frac{N_\uparrow - N_\downarrow}{N_\uparrow + N_\downarrow}.$$

We can define the following functions, called *degrees of orientation* (of order i):

$$f_1 = \frac{1}{I} \sum_{m_I} m_I N_{m_I}$$

$$f_2 = \frac{1}{I^2} \left[\sum_{m_I} m_I^2 N_{m_I} - \frac{1}{3} I(I+1) \right]$$

$$f_3 = \frac{1}{I^3} \left[\sum_{m_I} m_I^3 N_{m_I} - \frac{1}{5}(3I^2 + 3I - 1) \sum_{m_I} m_I N_{m_I} \right]$$

and so on, where $-I \leq m_I \leq I$, and the normalization is

$$\sum_{m_I} N_{m_I} = 1.$$

Notice that if all N_{m_I} are the same, then all f_i vanish. In general, only the f_i where $i \leq 2I$ may be nonzero. For $I = 1/2$, f_1 is just the polarization. For $I > 1/2$ (and for every m_I) we may have $N_{m_I} = N_{-m_I}$ and $N_{m_I} \neq N_{m'_I}$ (because $|m_I| \neq |m'_I|$). Then those f_i with odd i are zero, and some of f_i with even i are nonzero. That is, the system is *aligned*, but not polarized (like a group of deuterons with zero m_I).

We know that the energy differences that have to do with nuclear orientation are $\sim \mu_n H_0$. The thermal energy fluctuations (thermal noise or agitation) are $\sim kT$. If $\mu_n H_0 \sim kT$ (they are of comparable order of magnitude), thermal motions break down the nuclear orientation and order. For example, if $H_0 \sim 10^5$ Gauss, the exponent is $\mu_n H_0/kT \sim 10^{-3}/T$ whence we have to use extremely low temperatures, next to absolute zero ($\sim 10^{-3}$ °K), reached by adiabatic demagnetization. In 1934, Gorter cooled down the sample in a magnetic field.

If $I = 1/2$, the ratio of the number of nuclei pointing (oriented) up (parallel to the external field H_0) to those pointing down (at T) is

$$\frac{N_\uparrow}{N_\downarrow} \sim e^{g_I \mu_n H_0/kT}$$

where a Boltzmann distribution is assumed, and the ratio is equal to the exponential Boltzmann factor. The polarization is then $f_1 = g_I \mu_n H_0/2kT$.

The disadvantage of this technique is the difficulty of the necessary very low temperature (and also, the impractically long relaxation time). So, Gorter used an indirect method (in 1948). He polarized paramagnetic salts which are more easily polarizable. Interestingly, they generate magnetic fields ($\sim 10^6$ Gauss) at the positions of the nuclei; these fields give rise to the nuclear orientation.

Alignment (but not polarization) can be achieved in crystals (which are anisotropic) where at low temperatures the nuclear moments align themselves parallel to the crystal axis (without overall polarization). This was done cryogenically in the laboratory by Bleaney, in 1951.

Another method of producing orientation dynamically is sending electromagnetic waves (radiation) over the sample. Then the population of one of the hyperfine-structure substates increases, and nuclear polarization is attained.

If we have oriented radioactive nuclei, we can follow the anisotropy of the gamma emissions to check the alignment or polarization. This method is technically feasible, and currently in use.

16.9. Line Width in NMR

A nucleus has a magnetic moment μ and a spin angular momentum $\hbar I$. Since the two are parallel, we have $\mu = \gamma \hbar I$ where γ is a constant. The energy (of interaction) with an external field H_0 is $E = -\mu \cdot H_0 = -\mu_z H_0 = -\gamma \hbar I_z H_0$, since H_0 is along z. Since the allowed values of I_z are $m_I = I, I - 1, \ldots, -I$, we have $E = -\gamma \hbar H_0 m_I$, and $\Delta E \propto \Delta m_I$ for transitions.

In a H_0 a nucleus with $I = 1/2$ has two levels: $m_I = \pm 1/2$ (Fig. 16.20). The energy difference is $\Delta E = \hbar \omega_0 = \gamma \hbar H_0$, or $\omega_0 = \gamma H_0$ which is the magnetic resonance absorption condition. For a proton, $\gamma = 2.675 \times 10^4$/sec Gauss and $\nu_0(\text{MHz}) = 4.258 H_0(\text{kGauss}) = 42.58 H_0(\text{Tesla})$. For an electron, $\nu_0(\text{GHz}) = 2.80 H_0(\text{kGauss})$.

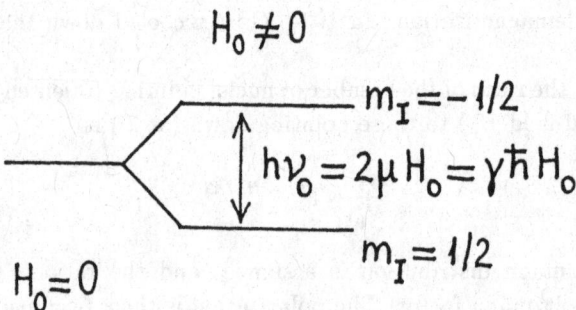

Fig. 16.20.

The equations of motion can be derived from the fact that the rate of change of the angular momentum (of a system) is just the torque acting on the system. The torque on μ in H_0 is $\mu \times H_0$. The gyroscopic equation is $\hbar dI/dt = \mu \times H_0$ or $d\mu/dt = \gamma \mu \times H_0$. The nuclear magnetization is $\vec{M} = \sum \mu / V$ where the sum is over all nuclei (in the volume considered). If there is only one isotope (of importance), one single γ value is taken. Then we have $d\vec{M}/dt = \gamma \vec{M} \times H_0$, and the rest is treated in Section 16.6. At $t = 0$ the unmagnetized ($\mathcal{M}_z(t = 0) = 0$) sample is placed in a static field H_0. The magnetization increases as time passes and tends to its (new) equilibrium value $\mathcal{M}_0 = \chi H_0$ (Fig. 16.21) whence we obtain the longitudinal relaxation time τ_1. The magnetic energy density $-\vec{M} \cdot H_0$ decreases, because some spins (part of the

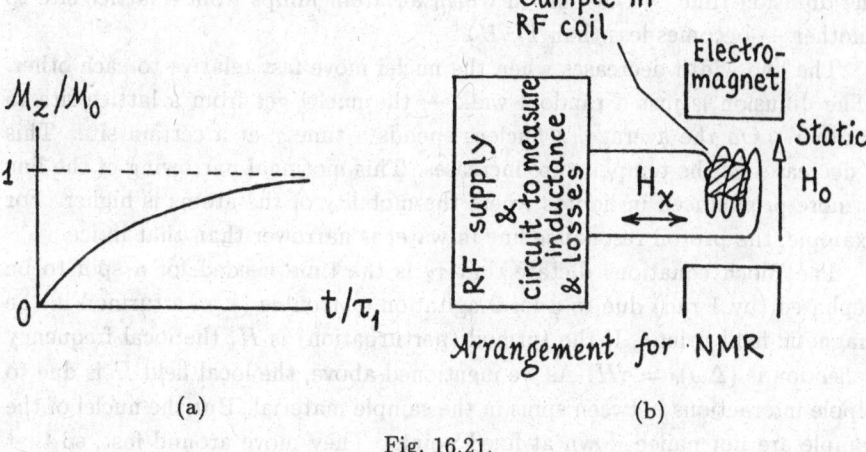

(a)   (b)

Fig. 16.21.

population) get to the lower energy level, and the asymptotic value is $-\mathcal{M}_0 H_0$, reached at $t \gg \tau_1$. Energy is transferred from the spin system to the lattice (system of vibrations), whence τ_1 is the spin-lattice relaxation time. Kittel (listed in the References) treats the case of the longitudinal magnetization relaxation (in metals and insulators), followed by processes of phonon emission, absorption, and scattering (p. 503).

In a stiff lattice of magnetic dipoles the line broadens because of magnetic dipole interaction. The field **H** experienced by a (test) dipole μ_1 as a result of another dipole μ_2 (source dipole) located at **r** from μ_1 is

$$\mathbf{H} = \frac{3(\mu_2 \cdot \mathbf{r})\mathbf{r} - \mu_2 r^2}{r^5},$$

according to magnetostatics. Therefore, the dependence (order of magnitude) is $H \sim \mu/r^3$ where only nearby neighbors have pronounced contributions and need be considered. For two protons separated by 2Å, the local field is

$$H = \frac{1.4 \times 10^{-23} \text{ Gauss cm}^3}{(2 \times 10^{-8} \text{ cm})^3} \simeq 2 \text{ Gauss}.$$

Consider the NMR line of a metallic ^{7}Li sample. Does the diffusion of nuclei affect the line? Yes. At low temperatures the experimental and theoretical values (for a stiff lattice) are the same, in agreement. But as the temperature rises, the line width decreases because of the increasing diffusion rate. At $T = 230°$K we observe a sudden and sharp decrease in the line width (because the diffusion time — the time in which an atom jumps from a lattice site to another — becomes less than $1/\gamma H$).[5]

The line width decreases when the nuclei move fast relative to each other. (The diffusion is just a random walk — the nuclei get from a lattice site to another.) On the average, a nucleus spends a time τ at a certain site. This τ decreases as the temperature increases. This motional narrowing of the line is more pronounced in liquids where the mobility of the atoms is higher. For example, the proton resonance line in water is narrower than that in ice.

The Bloch equations dictate that τ_2 is the time needed for a spin to be dephased (by 1 rad) due to a local agitation or trouble (*e.g.*, a turmoil in the magnetic field value). If the turmoil (perturbation) is H, the local frequency deflection is $(\Delta\omega)_0 = \gamma H$. As we mentioned above, the local field H is due to dipole interactions between spins in the sample material. But the nuclei of the sample are not nailed down at fixed points. They move around fast, so that H felt by a certain spin (nucleus) fluctuates in time and changes randomly, because the angle between μ and r changes (Fig. 16.22). For example, H can have a positive value $+H$ for a time τ, and then change to $-H$. During τ, the

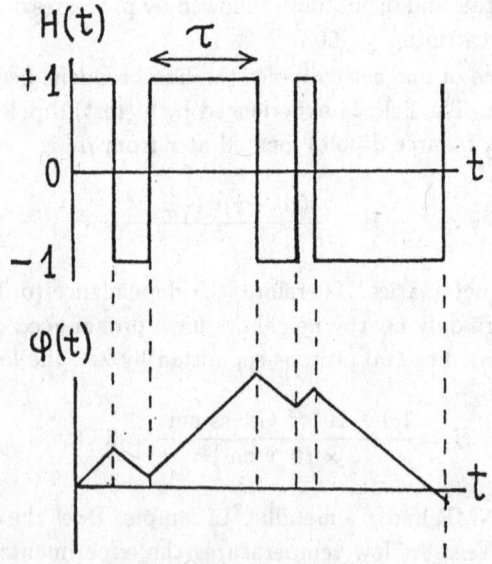

Fig. 16.22.

[5]H.S. Gutowsky and B.R. McGarvey, *J. Chem. Phys.*, **20**, 1472 (1952).

spin precesses by an additional phase $\Delta\varphi = \pm\gamma H\tau$ with respect to the phase of the steady precession about the constant external H_0. If τ is brief enough ($\Delta\varphi \ll 1$), we have *motional narrowing*. If we wait for N durations of τ each (in the applied field H_0), we will have for the dephasing angle φ

$$\overline{\varphi^2} = N(\Delta\varphi)^2 = N\gamma^2 H^2 \tau^2$$

which is nothing else than the random-walk principle. That is, if x is the net displacement from the initial position, after N random steps of length d each to the right or to the left (or to random directions, in three dimensions), we have $\overline{x^2} = Nd^2$.

How many steps are needed to dephase a spin by 1 rad? The answer is $N = 1/\gamma^2 H^2 \tau^2$. Why 1 rad? Because those dephased more than that do not give absorption signals. N steps are made in $\tau_2 = N\tau = 1\gamma^2 H^2 \tau$ sec. Whereas if the lattice were rigid, the answer would be $\tau_2 = 1/\gamma H$.

Therefore, in the case of fast nuclear motion (where the characteristic time is τ) the line width is $\Delta\omega = 1/\tau_2 = \gamma^2 H^2 \tau = (\Delta\omega)_0^2 \tau$ where $(\Delta\omega)_0$ is the line width for a stiff lattice.

Notice that $\Delta\omega \ll (\Delta\omega)_0$ because (by assumption) $(\Delta\omega)_0 \tau \ll 1$, so that we have $\Delta\varphi \ll 1$ (too small). The line gets narrow as τ shrinks, *i.e.*, we have a motional narrowing.[6]

Example problem: Compare $(\Delta\omega)_0$ and $\Delta\omega$ for water.

Solution: The rotational relaxation time for water molecules (at room temperature) is $\sim 10^{-10}$ sec, given by measurements made for the electric permittivity. So, $(\Delta\omega)_0 \sim 10^5$ sec^{-1} and $(\Delta\omega)_0 \tau \sim 10^{-5}$. Thus $\Delta\omega = (\Delta\omega)_0^2 \tau \sim 1$ sec^{-1} where we see that the nuclear motion narrows the proton resonance line in *liquid* water to 10^{-5} of the static width obtained in rigid frozen water.

Notice that the theory of motional narrowing is different than the considerations about the optical line width which is due to strong interatomic collisions (*e.g.*, in a gas discharge). There the line is broad as τ is short. But in the case of nuclear spins we have weak collisions. Secondly, in the optical case the strong interatomic collisions disturb the oscillation phase, whereas in NMR

---

[6]N. Bloembergen, E.M. Purcell, and R.V. Pound, *Phys. Rev.*, **73**, 679 (1948).

the phase is not interrupted: It can change gradually in a collision, despite the fact that the frequency may change abruptly.

16.10. Actual Magnetic Moment

A rotating sphere of mass m and charge e has a magnetic moment $\mu = e\mathbf{P}/2mc$ where $\mathbf{P}$ is the angular momentum and c is the speed of light. For the atomic nucleus, $\mu(\text{actual}) = \gamma e\mathbf{P}/2m_\mathrm{p}c$ where m_p is the mass of proton, and γ is the gyromagnetic ratio (for proton, $\gamma_\mathrm{p} = +5.585$; for neutron, $\gamma_\mathrm{n} = -3.826$ where the negativity comes from the fact that the dipole moment of neutron behaves like that of the electron, which is negative). This formula for $\mu(\text{actual})$ is empirical, not fully explained by nuclear theory.

Vector $\mu(\text{actual})$ acts in a direction called *magnetic axis*. The magnetic moments tabulated below are not $\mu(\text{actual})$, but they depend on it, as we will see below.

Quantum mechanically, the nuclear angular momentum is $|\mathbf{P}| = P = \hbar\sqrt{I(I+1)}$, taking discrete, quantized values, where I is the value of nuclear spin (measured units of $\hbar$). If the atomic number is even, I is an integer; if it is odd, I is equal to the multiples of $1/2$.

Many particles in a nucleus have their spins paired off oppositely; there remains only a small nuclear spin. So, all atoms whose atomic number is even and their Z is even, have zero magnetic moment, so that NMR is no good for studying their nuclei. (Two thirds of all the elements are in this category.) $\mu(\text{actual}) = \frac{\gamma e\hbar}{2m_\mathrm{p}c}\sqrt{I(I+1)} = \gamma\mu_\mathrm{n}\sqrt{I(I+1)}$ where the nuclear magneton is $\mu_\mathrm{n} = e\hbar/2m_\mathrm{p}c = 5.05038 \times 10^{-24}$ erg/Gauss.

The magnetic moment $\mu(\text{actual})$ of a nucleus in a field $\mathbf{H}_0$ has a potential energy $V = -\mu(\text{actual})H_0\cos\theta = -\mu_\parallel H_0$ where θ is the angle between μ and $\mathbf{H}_0$, a quantum mechanical angle that takes certain values only. Here $\mu_\parallel$ is the component of μ along $\mathbf{H}_0$, the external field. The quantization condition says that $\mu_\parallel$ can be $m\hbar$ only, where m can be $-I, I+1, \ldots, 0, \ldots, I$, that is, it can take $2I + 1$ values (orientations) altogether in space. The component is

$$|\mu_\parallel| = \frac{\gamma e}{2m_\mathrm{p}}\frac{mh}{2\pi} = \gamma m\mu_\mathrm{n}\,.$$

The maximum value of $\mu(\text{actual})$ along the field occurs for m $= I$: $\mu_\parallel(\text{max}) = \gamma\mu_\mathrm{n}I = \mu_\mathrm{n}\mu$ where $\mu = \gamma I$, and μ is the nuclear magnetic moment listed in tables. The table on the following page lists the γ coefficient, the spin, and the magnetic moment of certain nuclei.

Nucleus	γ	Spin I	Magnetic moment μ (in units of μ_n)
^{1}n	-3.826	1/2	-1.913
^{1}H	+5.585	1/2	+2.793
^{4}He	0	0	0
^{7}Li	+2.171	3/2	+3.257
^{9}Be	-0.785	3/2	-1.178
^{10}B	+0.600	3	+1.801
^{12}C	0	0	0
^{13}C	+1.405	1/2	+0.702
^{14}N	+0.404	1	+0.404
^{15}N	-0.566	1/2	-0.283
^{16}O	0	0	0
^{19}F	+5.256	1/2	+2.628

16.11. Utility of Magnetic Resonance in General

Zavoisky[7] was the first scientist to experimentally deal with magnetic resonance (with solid samples). The first samples were paramagnetic salts. Liquids were tried later by Purcell and Bloch. The resolution is high in diamagnetic liquids.

The following information can be obtained:

(1) Excitations *en bloc* and collectively in spin systems (assemblies of spins).
(2) Line-width observations, effect of spin motions (or of the surrounding lattice) on the line width, and pertinent changes.
(3) Internal local magnetic fields felt by spins and interpreted via the resonance line.
(4) Crystal defects and electronic structure (revealed by the resonance absorption). Study of color centers.

We have the following kinds of resonance: NMR, NQR, EPR (or ESR), ENDOR, FMR (ferromagnetic resonance), SWR (spin-wave resonance in ferromagnetic layers), AFMR (antiferromagnetic resonance), and CESR (conduction-electron spin resonance).

---

[7]E. Zavoisky, *J. Phys. USSR*, **9**, 211, 245, 447 (1945).

We recommend Kittel where AFMR is treated concisely (p. 522).

The most important of all is NMR, used in organic chemistry (as opposed to the rest which are used in solid state physics) as a remarkable means and tool to obtain information about the structure of complex molecules. Specifically, NMR is used in molecular physics to study the following:

(1) As a result of the chemical structure, a slight shift (chemical shift) is observed in the nuclear resonance frequency. By measuring these shifts, we can study molecular structure, especially those radicals (and the molecules to which they are bound) which cannot be studied by X-ray diffraction (H_2O, ethylalcohol, CH_2, H_2 etc.). Thus we can examine the molecular structure of derived structures, isomers, chemical bonds and their electronegativity, chemical exchange, resonance bonds etc.

(2) Knight-shift study. This phenomenon occurs in conductors (due to the electrons close to the Fermi surface) and gives rise to paramagnetism. Of course, we do not observe it in every metal. The study of Knight shift enables us to learn many things about alloys, ferromagnetic and antiferromagnetic substances, electron densities in valence bands etc. Pertinent measurements and academic work have been in progress since long. This shift: (a) Is proportional to H_0. (b) Depends on temperature. (c) Is more in heavy elements. (d) Is pertinent to the change in the relaxation time τ_1.

(3) Measurement of the relaxation times. This scientific activity has gained importance in molecular physics, because it provides information about the structure of liquids and internal rotations that may take place in molecules of solids at given temperatures. Moreover, many molecules containing hydrogen (*e.g.*, $CaSO_4.6H_2O$) can be studied concerning the measurement of the separation between protons and the bond orientations.

(4) Study of the fine structure of spectral lines. This is also a source of information on molecular structure and nuclear physics phenomena.

(5) Sensitive measurement of nuclear magnetic moments.

(6) Measurement of the shape, width, and area of the resonance (absorption) curve. This helps us to study plastics and polymers, and many ionic, organic, and biological molecules.

16.12. Further Discussion on NMR

Spectroscopy and nuclear physics experiments reveal that nuclei have a magnetic moment. A magnet has a magnetic moment μ by which it can be represented (Fig. 16.23). An atom with a revolving electron has also a μ at the nucleus. Therefore, an atom can be thought as a microscopic, tiny magnet. Even a naked nucleus has a μ due to its intrinsic spin **I**.

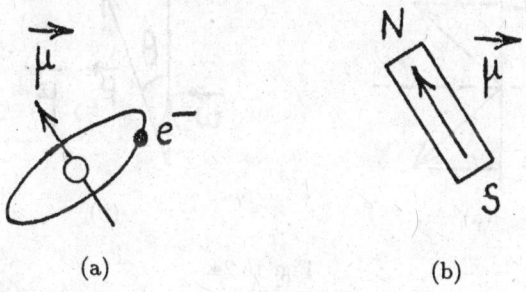

(a)   (b)

Fig. 16.23.

A current loop is equivalent to a magnetic dipole. Such a dipole has a magnetic moment which is proportional to the angular momentum **P**. For example, a proton has

$$\mu = \frac{eg}{2mc}\mathbf{P}$$

where m is the proton mass. We have $g_{\mathrm{p}} = 5.585$ and $g_{\mathrm{n}} = -3.826$. We also have $P = \sqrt{I(I+1)}\hbar$ where for proton and neutron, $I = 1/2$.

The spin **I** of a nucleus is the vector sum of the spins of all the protons and neutrons in the nucleus. If the spins are opposite, then they add up to zero, and the total nuclear $I = 0$ (*e.g.*, if Z, the atomic number, is even and A, the mass number, is a multiple of 4, then the magnetic moment is zero: $I = 0$). (Actually, when the opposite spins are paired, the leftover I is a very small number.)

The nuclear magneton is

$$\mu_{\mathrm{n}} = \frac{e\hbar}{2Mc}$$

where M is the mass of the nucleus and e is the electron charge. The nuclear magnetic moment μ behaves like a bar magnet in an external field H_0. The interaction energy is

$$E = -\boldsymbol{\mu} \cdot \mathbf{H}_0 = -\mu H_0 \cos\theta$$

which is quantized so that for $I = 1/2$ it can take two different values, because μ takes two different positions, with $m_I = \pm 1/2$ (Fig. 16.24). The energies are $E_{1,2} = \mp g\mu_{\mathrm{n}}H_0$.

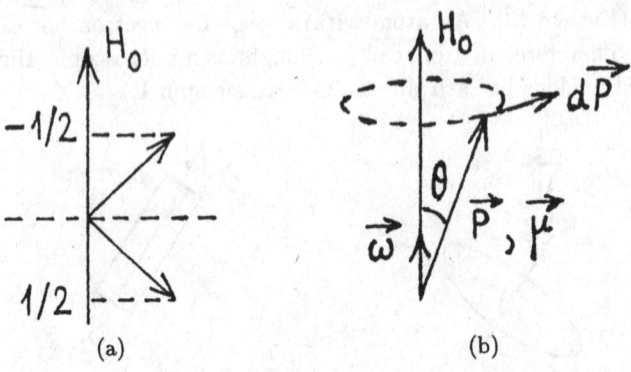

Fig. 16.24.

If $\Delta E = g\mu_n H_0$ (an amount of energy equal to this difference) is given to the system as a quantum, some of the nuclear moments (nuclei) of the sample in the lower-energy state absorb this $\Delta E = h\nu_0$ at resonance and get to the upper level. This is the NMR phenomenon, whose possibility was first claimed by C.J. Gorter in *Physica*, **3**, 503, 995 (1936) and **9**, 591 (1942).

If $I = I$ (*i.e.*, for a given I, any I), there are $2I + 1$ energy levels. The selection rule is $\Delta m_I = \pm 1$. The resonance frequency is $\nu_0 = g\mu_n H_0/h$.

A moment μ in a constant $\mathbf{H}_0$ precesses because the dipole feels a torque

$$\vec{T} = \mu \times \mathbf{H}_0 = \frac{d\mathbf{P}}{dt} = -\frac{eg}{2Mc}\mathbf{H}_0 \times \mathbf{P}$$

where $\mathbf{P}$ precesses with the Larmor precession frequency ω.

To explain away the hyperfine structure of the spectral lines, we have to accept that a nucleus possesses an intrinsic spin angular momentum associated with it (causing nuclear paramagnetism). We have $g_n < 0$, *i.e.*, the neutron dipole moment behaves like that of the electron which is negative.

Nuclei with $I = 0$ cannot be studied by NMR, because

$$\mu = \frac{eg\hbar}{2Mc}\sqrt{I(I+1)} = g\mu_n\sqrt{I(I+1)}.$$

The nuclear paramagnetic susceptibility is

$$\chi_n = N_0 g^2 \mu_n^2 I(I+1)/3kT.$$

The nuclei can be considered as bar magnets with μ each; they feel a torque in a uniform H_0. Each nuclear spin executes a gyroscopic motion in H_0 (precession about H_0) at the Larmor frequency ν_0.

The angle θ can take certain (quantized) values; this means that the potential energy is quantized and takes certain values, actually $2I + 1$ of them. The magnetic quantum number (projection of **I** along $H_0\hat{z}$) can take the values $m_I = -I, -I - 1, \ldots, -1, 0, 1, \ldots, I$, *i.e.*, $2I + 1$ in all.

Consider an ^{1}H nucleus (spin $= 1/2$) in a constant H_0.

Two different states for μ are possible (because $I = 1/2$ and $m_I = \pm 1/2$), one of lower, and one of higher energy. We have $\mu_\parallel = \pm \frac{1}{2} g\mu_n$ (Fig. 16.25).

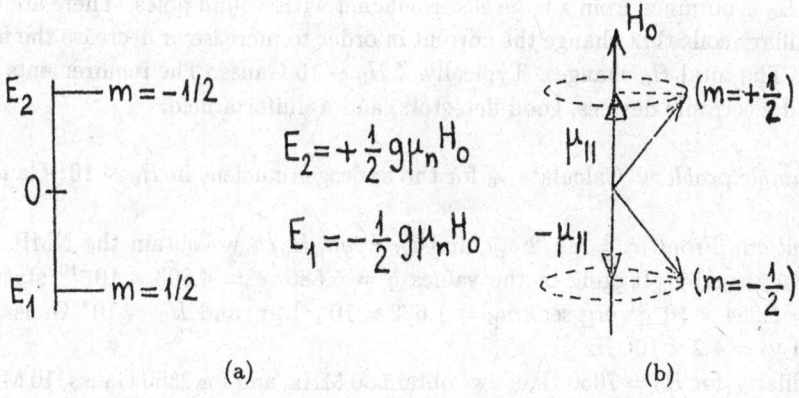

(a)   (b)

Fig. 16.25.

An energy difference $\Delta = g\mu_n H_0$ must be spent for μ to get from the $+1/2$ state to $-1/2$. If Δ is absorbed (or emitted) as a quantum, we have

$$\nu_0 = \frac{\Delta E}{h} = \frac{g\mu_n H_0}{h} = \text{Larmor frequency}.$$

At thermal equilibrium the difference in the number of nuclei between these two levels is $\sim 10^{-6}$, a difference too small to be observed spectroscopically (because the electromagnetic wave sent to the sample has to cover hundreds of *kilometers* before it is absorbed by a noticeable quantity!). For this reason we resort to the resonance method. A system of nuclear moments is resonanted by cophasic stimulations. We irradiate the sample with electromagnetic waves picked at the Larmor ν_0.[8] The magnetic moments jump from an energy level to another, and these changes are noticeable.

---

[8]We can shift it a bit and watch the results.

NMR can be observed in two ways:

(1) Keep H_0 constant and apply a sinusoidal H_1 perpendicular to it. The frequency of H_1 is smoothly shifted to the Larmor ν_0. Resonance is observed at this frequency, *i.e.*, the sample nuclei absorb energy (from the field) at this ν_0.

(2) Hold the sinusoidal high-frequency field H_1 constant and apply a field H_0, shifted to obtain resonance, and measure $\Delta\chi$.

H_0 is obtained from a large electromagnet with wound poles. There are two auxiliary coils that change the current in order to increase or decrease the field H_0. The total H_0 changes. Typically, $\Delta H_0 \sim 10$ Gauss. The requirements are good electronic devices, good detectors, and a uniform field.

Example problem: Calculate ν_0 for the hydrogen nucleus in $H_0 \sim 10^4$ Gauss.

Solution: From $\mu_n = e\hbar/2m_p c$ and $\nu_0 = g\mu_n H_0/h$ we obtain the NMR frequency ν_0, by plugging in the values $g = 5.585$, $e = 4.803 \times 10^{-10}$ statCb, $\hbar = 1.054 \times 10^{-27}$ erg sec, $m_p = 1.672 \times 10^{-24}$ gr, and $H_0 = 10^4$ Gauss, to find $\nu_0 = 4.2 \times 10^7$ Hz.

Similarly, for $H_0 = 7050$ Gauss we obtain 30 MHz, and for 2350 Gauss, 10 MHz.

Whereas the study of the hyperfine structure gives us information about the nuclear spin, the study the nuclear magnetic moment requires a different method, the *magnetic resonance*. The method is based on the fact that two magnetic fields act simultaneously on the nucleus. One of them is constant (and of relatively large intensity), and the other one alternating (varying with a high frequency) and normal to the first. If a nucleus of spin I and magnetic moment μ is acted upon by the two fields (Fig. 16.26), the constant field H_0 exerts a mechanical moment μH_0 on it, which tends to orient μ along $\mathbf{H_0}$. But the nucleus — which is like a top with angular momentum I — instead of orienting itself, precesses with the *Larmor pulsation*

$$\omega_0 = \frac{\text{Mechanical moment}}{\text{Spin}} = \frac{\mu H_0}{I}. \tag{16.7}$$

As the nucleus moves so, the weak alternating field acts on it, resulting in nothing. But if the frequency of the alternating field happens to be (or is made) equal to the Larmor precession frequency (magnetic resonance effect), then the orientation is achieved. Thus if I is known, and ω_0 and H_0 are measured, the only remaining unknown is μ which can be determined through Eq. (16.7).

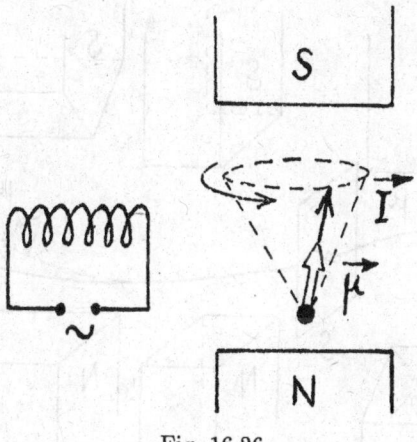

Fig. 16.26.

Here ω_0 is the circular frequency of resonance (or resonance frequency), H_0 is the field intensity, I is the nuclear spin (quantum number), and μ is the magnetic moment of the nucleus.

The resonance can be achieved by several methods. In the *method of molecular beams* (developed by I.I. Rabi[9]), a beam of molecules (or atoms) passes successively through an inhomogeneous magnetic field (analogous to that used in the Stern-Gerlach experiment and caused by the pole arrangement M_1 in Fig. 16.27), then through a magnetic resonance set-up C, and finally through another inhomogeneous field generated by arrangement M_2 before it hits detector D. (M_1, M_2, and C are electromagnets.) In all three fields the sense of the field is the same (in Fig. 16.27, vertically upwards). As the particle moves down the apparatus within inhomogeneous magnetic fields, it is acted upon by forces which curve its trajectory around. Since the sense of the gradient dH/dz of the first field (in M_1) is, say, downwards, while that in M_2 is opposite (upwards), a particle (which initially moves towards the appropriate direction) can pass through slit S and reach D, in spite of the curvatures of its trajectory. Figure 16.27 shows such a trajectory of a particle whose μ forms an angle with the direction of the field in M_1. Vector μ describes a cone by precession. If during the passage this angle decreases (because of the inserted resonance arrangement C), as the particle passes through M_2 it will feel a larger force, and hence the curvature of its trajectory will change (dashed line). As a result, the particle may miss D and go undetected. So, the resonance phenomenon is

---

[9]American physicist, Nobel laureate (1944) for his work on nuclear magnetic moments. He raised an incorruptible monument of science that honored his homeland.

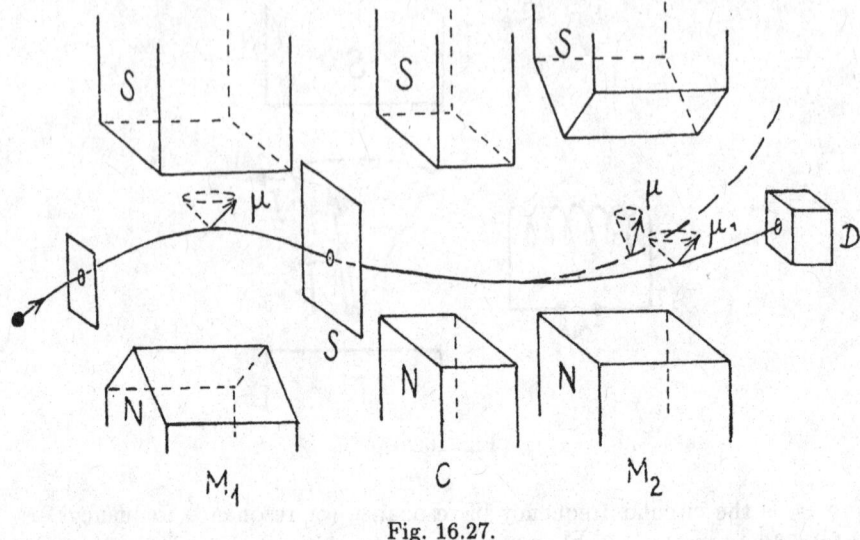

Fig. 16.27.

observed and verified by the decrease in the number of particles reaching (and counted by) D.

A variation of this method is the *method of nuclear induction* developed by Felix Bloch.[10] An amount of matter is subjected to magnetic resonance. As a result of the induced orientations of the magnetic moments, a rotating magnetic field inside the material is created, as it can be shown. This field induces an electromotor force (over a properly placed coil) whence the resonance can be detected.

Another variation is the *method of absorption due to resonance.*[11] The material is placed in an electromagnetic resonator (cavity) which in turn is placed between the poles of a strong magnet. By the help of a high-frequency electric source (usually microwaves), the resonator is excited to an oscillation. On the nuclei of the material there act both the constant magnetic field and that of the high-frequency electromagnetic field. If the frequency of the latter and the intensity of the former are properly chosen, resonance can be obtained, where the material absorbs energy (from the field) and the resonator oscillation diminishes.

Such measurements conducted on water, paraffin, oil, and other substances containing hydrogen, made it possible to find the magnetic moment of proton and deuteron. The same thing was achieved for other elements, too.

---

[10]German scientist, Nobel laureate (1952).
[11]It is developed by E.M. Purcell, American Nobel Prize winner (1952).

Consider now a substance containing N noninteracting nuclei of spin 1/2 and magnetic moment μ each. The substance is in a field H_0. Each spin points either "up" (parallel to $\mathbf{H}_0$) or "down" (antiparallel to $\mathbf{H}_0$). The two possible energies (energy states) (of each nucleus) are $E_\pm = \mp\mu H_0$. If $N_\uparrow$ is the number of spins pointing up, and $N_\downarrow$ is the number of those pointing down, we have $N_\uparrow + N_\downarrow = N$. The energy levels of a nucleus of spin 1/2 in an external field are shown in Fig. 16.28 (where $\mu > 0$).

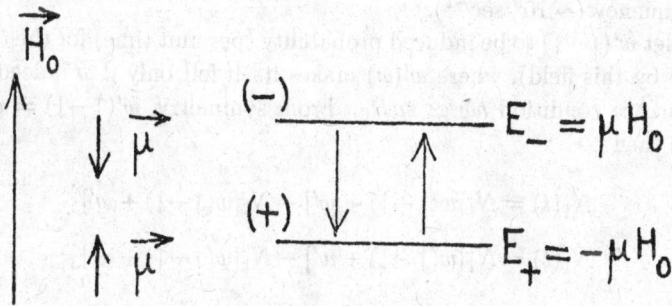

Fig. 16.28.

The overall Hamiltonian of this system is

$$\mathcal{H} = \mathcal{H}_1 + \mathcal{H}_2 + \mathcal{H}_3$$

where $\mathcal{H}_1$ is the interation between all μ and $\mathbf{H}_0$, $\mathcal{H}_2$ is the lattice part (whatever has nothing to do with spin, and the rest of the atoms in the system), and $\mathcal{H}_3$ is the interaction between the nuclear spins and the lattice (which may change the nuclear spin states by transitions, as it happens in the case of a moving nucleus which generates a variable magnetic field at the locations of other nuclei, giving rise to transtions). The nucleus-lattice interaction may make the spin turn from up to down. We call the probability (per unit time) of this event $w(\uparrow\to\downarrow)$. Then we have

$$\frac{w(\downarrow\to\uparrow)}{w(\uparrow\to\downarrow)} = \frac{e^{-E_+/kT}}{e^{-E_-/kT}} = e^{(E_- - E_+)/kT} = e^{2\mu H_0/kT}$$

where the lattice is taken as a big system (in internal equilibrium) at temperature T and in contact with the spins. Typically, $\mu \sim 5 \times 10^{-24}$ erg/Gauss and H_0(lab field) $\sim 10^4$ Gauss. Hence the exponent $\mu H_0/kT \sim 5 \times 10^{-4}/T \ll 1$, except at very low T. (This is true also for electrons, although their μ is 10^3 times as big.) The lattice acts as a heat reservoir for the spins.

If we expand the exponential, we find

$$w(\downarrow\to\uparrow) = w(\uparrow\to\downarrow)(1 + 2\mu H_0/kT)$$

where $\mu H_0/kT \ll 1$.

If there is also an external alternating field whose angular frequency ω satisfies the relation $\hbar\omega \simeq E_- - E_+ = 2\mu H_0$, then this variable field causes resonant transitions between nuclear spin states. For $H_0 = 10^4$ Gauss, ω is a radio frequency ($\sim 10^8$ sec^{-1}).

Now let $w'(\uparrow\to\downarrow)$ to be induced probability (per unit time) for the transition (induced by this field), where $w'(\omega)$ makes itself felt only if ω indeed satisfies the resonance condition $\hbar\omega \simeq 2\mu H_0$. From symmetry, $w'(\uparrow\to\downarrow) = w'(\downarrow\to\uparrow)$. We have then

$$\dot{N}_\uparrow(t) = N_\downarrow[w(\downarrow\to\uparrow) + w'] - N_\uparrow[w(\uparrow\to\downarrow) + w']$$

$$\dot{N}_\downarrow(t) = N_\uparrow[w(\uparrow\to\downarrow) + w'] - N_\downarrow[w(\downarrow\to\uparrow) + w'],$$

$$\frac{d}{dt}(N_\uparrow - N_\downarrow) = -2N_\uparrow[w(\uparrow\to\downarrow) + w'] + 2N_\downarrow[w(\downarrow\to\uparrow) + w']$$

$$= -2[w(\uparrow\to\downarrow) + w']n + 2\mu H_0 N w(\uparrow\to\downarrow)/kT \equiv \dot{n}$$

where $n = N_\uparrow - N_\downarrow$ is the population difference, and $4\mu H_0 N_\downarrow w(\uparrow\to\downarrow)/kT = 4\mu H_0 w(\uparrow\to\downarrow)(\frac{1}{2}N - n)/kT \simeq 2\mu H_0 N w(\uparrow\to\downarrow)/kT$, because $n \ll N$ at laboratory temperatures.

If there is no radio-frequency field and there is equilibrium ($w' = 0$), then $\dot{n} = 0$, and the equation above gives the extra number of spins in equilibrium, $n_{eq} = \mu H_0 N/kT$. This follows from the canonical distribution (in equilibrium)

$$N_{\uparrow\downarrow} = N\left(\frac{e^{\pm\mu H_0/kT}}{e^{\mu H_0/kT} + e^{-\mu H_0/kT}}\right) \simeq N\left(\frac{1 \pm \mu H_0/kT}{2}\right).$$

So, we have

$$\dot{n} = -2w(\uparrow\to\downarrow)(n - n_{eq}) - 2nw'.$$

Set $w' = 0$ and integrate this to obtain

$$n(t) = n_{eq} + (n_0 - n_{eq})e^{-2w(\uparrow\to\downarrow)t}$$

where $n_0 = n(t = 0)$. Notice that $n(t) \to n_{eq}$ exponentially, with a characteristic *relaxation time* $\tau = 1/2w(\uparrow\to\downarrow)$. If w, the spin-lattice interaction is strong and large, τ shrinks short. If the interaction between the spins and the heat reservoir (lattice) is feeble, then $w \to 0$, and one feels the need of a radio-frequency field to save the situation and have resonance. In that case we have $\dot{n} = -2nw'$, and hence $n(t) = n_0 \exp(-2w't)$. Then $n \to 0$ exponentially, with a characteristic time $1/2w'$ which is inversely proportional to the interaction strength with the alternating field. There are more nuclei in the lower state than in the upper. So, the induced transition (by the alternating field) causes absorption of energy (or power) which is $(N_\uparrow - N_\downarrow)(2\mu H_0 w')$. Since the system of nuclear spins is isolated ($w = 0$), the absorbed energy (from the alternating field) increases the energy (or spin temperature) of the spins. When the spin temperature becomes infinite, we have no population difference, because we have $N_\uparrow = N_\downarrow$ and hence $n = 0$. At that point the absorption of radio-frequency power stops, because the spin system gets *saturated*.

However, we can have a steady-state absorption of radio-frequency power, if we provide for a good communication between the spin system and the lattice. Then the power absorbed by the spins is given back to the lattice, and since the latter is large, it withstands this wave of return without suffering appreciable rises of its temperature, *i.e.*, its heat capacity is large enough. Since in the steady state $\dot{n} = 0$, we have $w(n - n_{eq}) = -w'n$, that is,

$$ n = n_{eq} \left(\frac{1}{1 + \dfrac{w'}{w}} \right) . $$

This means that n departs from (and is less than) its equilibrium value by some amount that depends on w'/w, the transition probabilities compared. For $w \gg w'$ we have $n \to n_{eq}$, and for $w \ll w'$ we have saturation ($n \to 0$).

16.13. NMR Methods

A nucleus has an angular momentum due to rotation and also a magnetic moment. NMR methods are used to measure nuclear magnetic moments.

The nuclear magneton is $\mu_n = eh/4\pi m_p c$. The rotational angular momentum is $I\hbar$. The nuclear magnetic moment is $\mu = gI\mu_n$ where the Lande factor is $g = \mu/I\mu_n$, so that $\mu = gIeh/4\pi m_p c$. A nucleus with μ precesses about the direction of an external field H_0 with the Larmor frequency given by $\nu = \mu H_0/Ih$.

In the beams technique mentioned in the previous section, the molecule feels a force in the inhomogeneous field, given by

$$ F = \mu_z \left(\frac{\partial H}{\partial z} \right)_{\text{magnet}} . $$

If we know the velocity of the molecule (determined by the source temperature and the apparatus geometry), we can find the deflection as it passes through the inhomogeneous fields. If μ_z stays the same, the two deflections are opposite to each other. To make them equal and opposite (and thus lead the molecule right to the detector), *i.e.*, to refocus the beam, one has to adjust the field gradients in the two magnets, which is an easy job.

In Fig. 16.27 magnet C produces the constant field H_0. At the same place there is another arrangement (not shown in the figure) producing the high-frequency oscillatory field H_1 (which is normal to H_0). A molecule (with a nucleus having a μ) precesses about H_0 with ν. The interaction with H_1 changes the angle between μ and $\mathbf{H}_0$ (by creating a force moment that increases or decreases the angle). If the frequency f of H_1 is not equal to ν, the created force moment is soon out of phase with the precession, and the net result is nil. If $f = \nu$ (chosen so), the change in angle is appreciable. The detected beam intensity vs. H_0 for the ^{7}LiCl molecule is a typical curve shown in Fig. 16.29, where the dip shows the resonance absorption at $f = \nu = 5585$ Hz occurring at a given field strength. Here f is kept constant and H_0 is varied. The curve as a function of H_0 is obtained. Beams of LiF or Li_2 can also be used. The curve concerns the ^{7}Li nucleus, for which g(measured) $= 2.1688$ and $I = 3/2$, so that $\mu = 3.2532\mu_n$ can be found.

Millman and Kusch measured μ of proton in 1941. Kellogg, Rabi, Ramsey, and Zacharias measured μ_p/μ_d in 1939. Hence one finds that $\mu_p = 2.7896\mu_n$

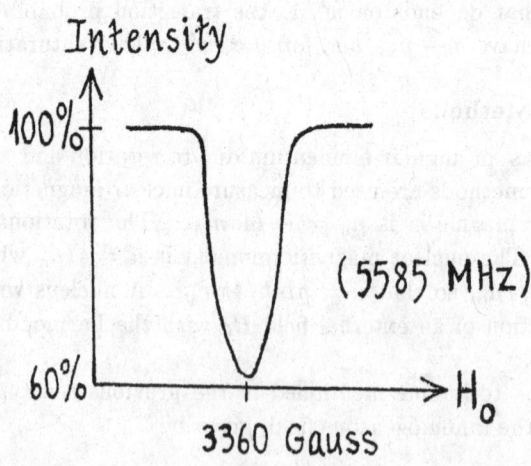

Fig. 16.29.

and μ_d(deuteron) $= 0.8565\mu_n$. Since a deuteron is a protron plus a neutron (and $\mu_d = \mu_p + \mu$(neutron)), one finds μ(neutron) $= -1.933\mu_n$. But it is also possible to determine μ(neutron) by letting a slow neutron beam pass through the magnetic resonance molecular beam apparatus. Since for neutrons $I = 1/2$, we find for free neutrons $\mu = -1.935\mu_n$. This was done in 1940 by Bloch and Alvarez.

In 1946, Bloch measured the frequency of the alternating field at which the molecular beam intensity became minimum. This frequency equals the Larmor frequency of the nucleus in a constant field. Bloch was able to measure the induced electromotor force due to the orientation of the spins in the solid (or liquid) sample.

The experimental *nuclear induction* apparatus used by Bloch, Hansen, and Packard[12] is shown in Fig. 16.30. The sample is surrounded by an input coil whose axis is along the x-direction. A high-frequency generator supplies an alternating field H_x along x at a frequency f. Another coil (output coil) surrounds the sample, having its axis along y. Whenever the flux along y changes, an induced electromotor force is produced in the output coil which is connected to a circuit to measure the electromotor force. Of course, this arrangement is placed inside a constant H_0 along z, *i.e.*, between the poles of an electromagnet which are above and below the sample. The duty (and effect) of H_0 is to make the spins precess about it.

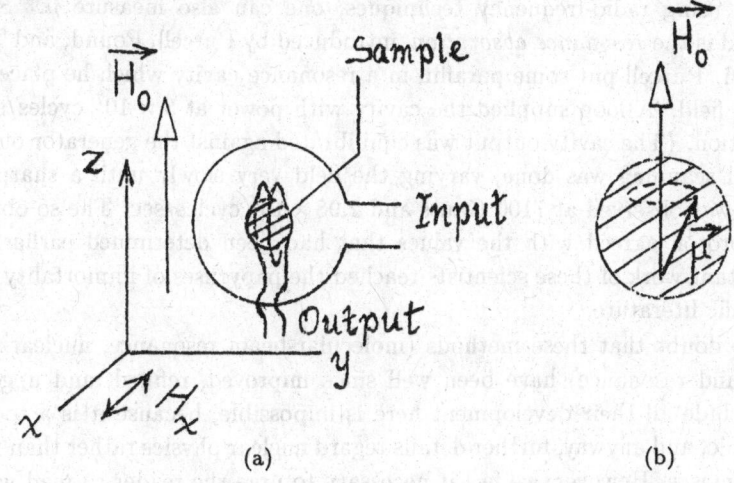

Fig. 16.30.

[12]See *Phys. Rev.*, **70**, 475 (1946).

Since the sample contains many spins (because it is bulky), one speaks of the paramagnetic magnetization (magnetic moment per unit volume). As one works with a bulky sample, a certain amount of time passes before the magnetization sets up fully. This is the *relaxation time*, depending on temperature (*i.e.*, on the interactions between the spins and the electronic motions and distributions).

The magnetization of the material produces a magnetic flux. Since the alternating field changes, the flux (along y) changes, and the output coil records an electromotor force. As f gets close to the Larmor precession ν, the flux change increases.

Bloch used water as sample first. The observed relaxation time was a few seconds. A solution of $Fe(NO_3)_3$ in water gave a relaxation time of 10^{-4} sec at $f = \nu = 7.765 \times 10^6$ sec^{-1} and for $H_0 = 1826$ Gauss. By improving the machine, Bloch and his collaborators were able to measure magnetic-moment ratios concerning the moments of the deuteron, the triton, and the neutron. Later, Rogers and Staub (1949) used a high-frequency rotating field to prove that μ(neutron) and μ_p have opposite signs.

The essence here is that in order to determine μ well, we should know the constant field H_0 well at the resonance frequency. Interestingly enough, one can do the experiment backwards, too: If μ of the sample is known, one can use this method to precisely measure magnetic field intensities (*e.g.*, in cyclotrons)!

By using radio-frequency techniques, one can also measure μ. Such a method is the *resonance absorption*, introduced by Purcell, Pound, and Torrey in 1946. Purcell put some paraffin in a resonance cavity which he placed in a strong field. A loop supplied the cavity with power at 3×10^7 cycles/sec for excitation. (The cavity output was equilibrated against the generator output.) A field scanning was done, varying the field very slowly until a sharp resonance was observed at 7100 Gauss and 2.98×10^7 cycles/sec. The so obtained μ of proton agreed with the values that had been determined earlier. The important work of these scientists reached the papyruses of immortality in the scientific literature.

No doubt that these methods (molecular-beam resonance, nuclear induction, and resonance) have been well since improved, refined, and upgraded. To include all their development here is impossible, because it is a too wide subtopic, and anyway, further details regard nuclear physics rather than molecular phyiscs. However, we feel it necessary to urge the reader to read without fail pp. 217–235 and 239–244 of Segrè (listed in the References) where the pertinent treatment is very important, and in our opinion, the best and most straightforward. We do not think that the reader will have a complete knowledge of the topic of NMR without reading the text fragments indicated. The treatment includes magnetic moments, methods of measurement etc.

The nuclear magnetic moment values today are

$$\mu(\text{proton}) = 2.79270\mu_n$$

$$\mu(\text{neutron}) = -1.91316\mu_n$$

$$\mu(\text{deuteron}) = 0.85738\mu_n$$

where the sum of the first two is *almost* equal to the third, being off by $0.022\mu_n$.

To recapitulate, the pertinent relations are

$$\mu = g_I \mu_n \mathbf{I}$$

$$\mu = \gamma_I \hbar I$$

(Fig. 16.31). The force in the inhomogeneous field (in the beams set-up) is

$$F = \mu_{\text{eff}} \frac{\partial H_z}{\partial z}$$

where the gradient is just the field inhomogeneity. The deflection is

$$\ddot{z} = \mu_{\text{eff}} \frac{\partial H_z}{\partial z} \frac{1}{m}$$

where m is the mass of the molecule. Hence

$$z = \frac{1}{2} \mu_{\text{eff}} \frac{\partial H_z}{\partial z} \frac{1}{m} \left(\frac{x}{v}\right)^2$$

where v is the speed of the molecule, and x is the length over which the field extends. If ω is the angular frequency of the variable field, the *resonance condition* is

$$\hbar\omega = g_I \mu_n H_z \Delta m_I .$$

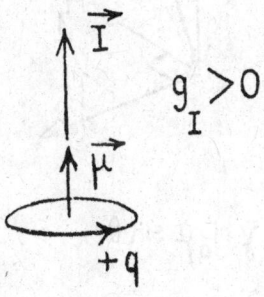

Fig. 16.31.

When molecules are deflected from the beam, the (undeflected) beam intensity drops, and this is due to the inhomogeneity of the field (except if H_z is such that $\mu_{\text{eff}}(H_z) = 0$, in which case the intensity does not decrease as a function of the field — it reaches its maximum and the beam is fully undeflected).

Magnet C (Fig. 16.27) produces a field $H_z \sin(\omega t)$ where H_z is constant. This field induces transitions between various hyperfine-structure levels (according to the selection rules). When ω is such that resonance occurs (the resonance condition is satisfied), the energy difference between the levels equals the field quanta energy $\hbar\omega$. Now when these transitions occur, magnet M_2 does not refocus (because the magnetic moment of the molecule has changed) and thus the beam reaching D is weaker now. (The factor g_I is determined from H_z and ω.)

In Bloch's method one works with bulks of matter (*e.g.*, water, that is, hydrogen nuclei) in a constant external field $\mathbf{H}_0$ along z (Fig. 16.32). The Boltzmann statistics says that there are more protons whose magnetic moment is oriented antiparallel to the field. So, the slight average magnetization of the bulk matter is (according to Langevin)

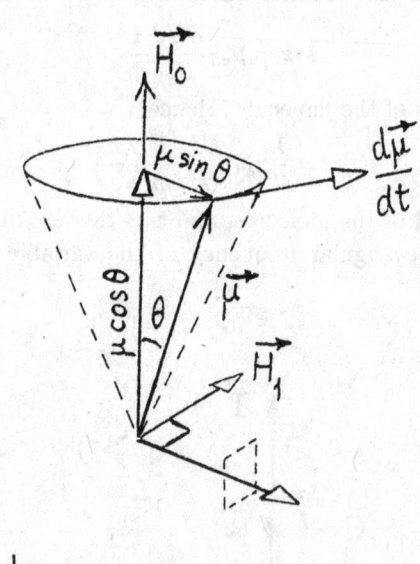

$$\frac{d\mu}{dt} = \gamma H_0 \mu \sin\theta$$

Fig. 16.32.

$$\mathcal{M} = N\mu^2 \frac{H_0}{kT}\frac{I+1}{I} = \chi H_0$$

where N is the number of nuclei per unit volume (cm^3), and $\mathcal{M}$ is the magnetic moment per unit volume: $\vec{\mathcal{M}} = \mu/V = \chi H_0$. This $\mathcal{M}$ is feeble and imperceptible in the environment of the static field H_0. That is why we apply a rotating field H_1 in a plane normal to H_0: To make $\vec{\mathcal{M}}$ observable.

If the classical (mechanical) angular momentum (per unit volume) associated with $\mathcal{M}$ is P, then $\vec{\mathcal{M}} = \gamma P$ where γ is the gyromagnetic ratio, and the equation of motion is

$$\text{Torque} = \frac{dP}{dt} = \gamma P \times H_0.$$

This equation holds also quantum mechanically, if we replace the classical quantities by the expectation (or mean) values of their corresponding operators, that is, if we interpret the situations as $\langle P \rangle$ = average angular moment, and $\langle \gamma P \rangle$ = average magnetic moment.

Since H_0 is constant, we dot both sides by P to obtain $dP^2/dt = 0$ or P^2 = constant. Dotting by H_0 we obtain $P \cdot H_0$ = constant. Therefore, μ precesses about H_0 with a period τ where $\tau\gamma\mu H_0 \sin\theta = 2\pi\mu \sin\theta$, and the circular frequency is $\omega = \gamma H_0 \equiv \omega_{\text{Larmor}}$.

Larmor's theorem says that if H_0 is along z where the reference frame rotates about H_0 with ω, the motion is the same as if H_0 were absent and the frame were stationary. If H_1 exists (where $H_1 \perp H_0$), rotating with ω, for $\omega \neq \omega_{\text{Larmor}}$, H_1 tends to once increase and once decrease θ, so that the average effect washes out to zero. But if $\omega = \omega_{\text{Larmor}}$, the field pursues μ and acts on it as a constant field, varying the angle θ which soon increases, and μ precesses, describing cones of growing apex, and rotating about z until it becomes antiparallel to $+z$. Then by precession it again becomes parallel to $+z$, and so on. So, effectively, the rotating field is actually just a *nonrotating* alternating field along y, of magnitude $H_1 \cos(\omega t)$ where H_1 is its amplitude. Figure 16.33 shows the trace of the endpoint of μ (or of the magnetic polarization vector) of hydrogen nuclei in an alternating field at resonance frequency. The strong field is along z, and the small alternating field along x or y. This field can be thought as two fields of amplitude $H_1/2$, rotating at frequencies $\pm\omega$ around z, where only one contributes if $\omega = \omega_{\text{Larmor}}$ (because the other one has $\omega = -\omega_{\text{Larmor}}$ and is off resonance). When the angle between μ and H_0 becomes large, an electromotor force (of frequency ω_{Larmor}) is induced in a coil fixed in a plane parallel to H_0 and normal to H_1. We can detect this electromotor force and find γ from H_0 and ω_{Larmor}.

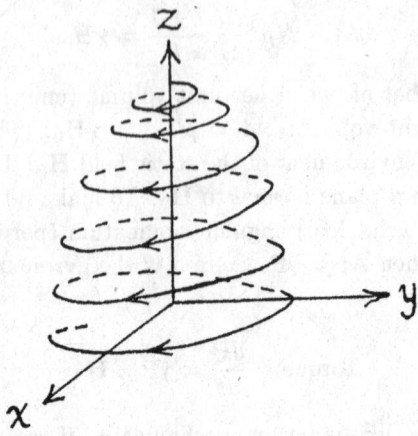

Fig. 16.33.

Of course, the sample material has some relaxation time which is important in studying molecular structure in chemistry, molecular physics, and solid state physics. The relaxation time can be infinite (as in Fig. 16.33).

In the Purcell method we detect the resonance by measuring the radio-frequency energy absorbed by the sample in the field. When the radio-frequency equals ω_{Larmor}, the absorption curve flicks to a maximum.

Figure 16.34 shows the experimental apparatus for measuring magnetic moments by using Purcell's method.

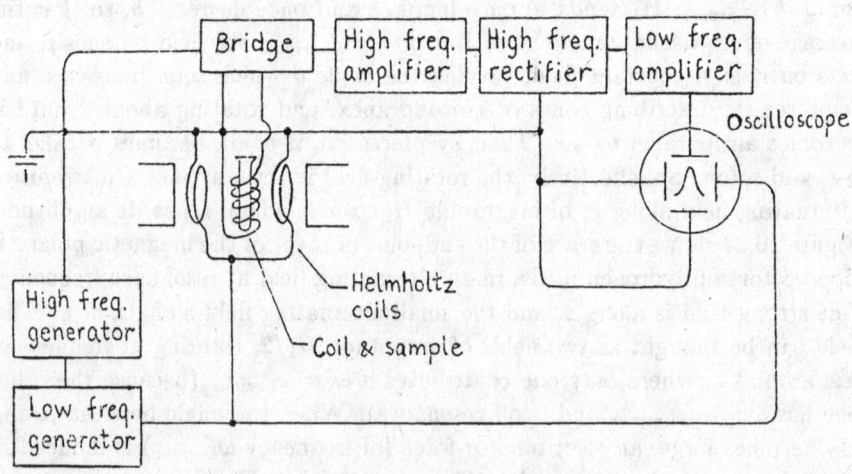

Fig. 16.34.

In this section we considered methods of measuring spins and magnetic moments both of stable and radioactive nuclei. There are other methods, too, good for radioactive nuclei. They use the angular correlation effect, as the gamma-ray quanta emitted by oriented nuclei are anisotropic, and give information about the nuclear orientation. The case is explained in Segrè, p. 234. Also, Chapter 8 of the same reference treats the Zeeman effect for nuclei, as it is directly observed by the method of nuclear recoilless emission.

Getting back to the molecular beams technique, the reader should note that the quantity (field gradient) $\partial H_z / \partial z$ is the *inhomogeneity* of the magnetic field in the beams set-up (Fig. 16.27). The nuclei of the sample material are subject to an alternating field at resonance frequency *and* another field (a strong one) perpendicular to the first.

The technique deserves further detailed discussion, but since we cannot afford the space here, we will have to dispense with it. Good experimental work is going on in the beams laboratories of the physics department of New York University, but it concentrates itself on beam scattering experiments, not on resonance methods.

Chapter 17

EXPERIMENTAL NMR TECHNIQUES

17.1. The NMR Method

An atomic nucleus (Fig. 17.1) whose spin $\mathbf{I}$ is nonzero, possesses a magnetic moment μ which is proportional to the angular momentum $\mathbf{P}$:

$$\mu = \gamma \mathbf{P}$$

where γ is the *gyromagnetic constant* of proportionality, given by

$$\gamma = \frac{e}{2Mc}g$$

with M being the mass of the nucleus, c the speed of light, and e the electronic charge. The factor g is a dimensionless number that serves as scaling and is called *Lande factor*.

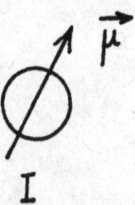

Fig. 17.1.

Since $\mathbf{P} = \hbar\mathbf{I}$, if $I = 0$, then $\mu = 0$.

488

To measure magnetic moments, we define a constant μ_n whose numerical value is fixed:

$$\mu_n \equiv \frac{e\hbar}{2Mc} \equiv \frac{\gamma\hbar}{g}$$

which is called *nuclear magneton*; it is the unit of (measuring of) nuclear magnetic moments: We express them in terms of it. We have then

$$\mu = \gamma P = \gamma \hbar I = \frac{eg}{2Mc}\hbar I = \mu_n g I \equiv \frac{\mu_n g}{\hbar} P .$$

The magnetic moment μ corresponds (or, is equivalent) to a tiny magnet: If put in a strong external field $\mathbf{H_0}$, it precesses (Fig. 17.2). The magnetic potential energy of a rod magnet in a static field $\mathbf{H_0}$ is

$$E = -\mu \cdot \mathbf{H_0}$$

where in our case we have

$$E = -\mu_n g \mathbf{I} \cdot \mathbf{H_0} = -\mu_n g m_I H_0 .$$

where m_I is the z-component (projection) of $\mathbf{I}$, and $\mathbf{H_0}$ is fully along the z-direction (Fig. 17.2). (In other words, since $\mathbf{P}$ is quantized as $P = \sqrt{I(I+1)}\hbar$, the energy is $E = -\gamma\hbar m_I H_0$.)

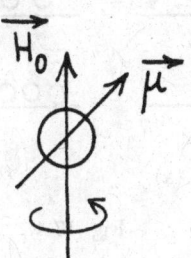

Fig. 17.2.

Since the selection rule for the spin magnetic quantum number m_I is $\Delta m_I = \pm 1$, the energy difference is

$$\Delta E = \pm \mu_n g H_0 .$$

The number m_I can take $2I + 1$ values, from $-I$ to $+I$. Therefore, if $I = 3/2$, the rod magnet can take four different (quantized) positions (states) (Fig. 17.3). Each of them has its own energy level, and so there are four

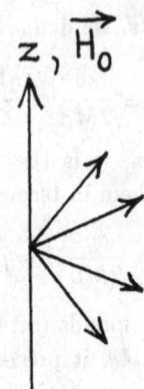

Fig. 17.3.

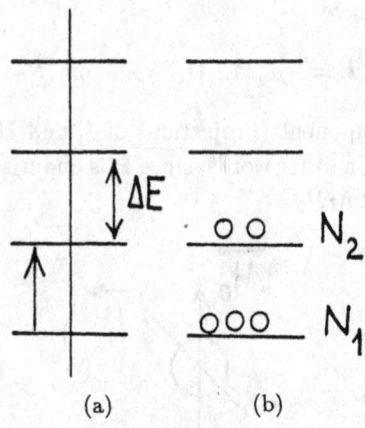

(a)   (b)

Fig. 17.4.

energy levels (Fig. 17.4(a)). The three ΔE energy differences are equal to each other. If we give an energy quantum $\Delta E = h\nu_0$ to the system, there will be a transition from a lower-energy level to a higher-energy level by absorption, *i.e.*, the particles jump from a level to another (Fig. 17.4(b)) because they absorb the supplied energy. According to the selection rule $\Delta m_I = \pm 1$, the transition occurs from a level to the next; a double jump is forbidden.

When we send electromagnetic waves to the system we can measure the energy that the system absorbs.

The nuclear magnetic moments take one of the $2I+1$ possible energy states in a strong magnetic field. The absorption and emission between these energy levels constitute the phenomenon of *nuclear magnetic resonance* (NMR).

Refer to Fig. 17.4(b): N_1 and N_2 are the populations of the corresponding levels (number of atoms per level). If we assume a Boltzmann distribution, we have $\Delta N = 10^{-6}$ for the transition, because ΔE is small. Thus the absorption (or emission) signal is too weak. To enhance it, we apply another field, H_1, which is weak and alternating, of high frequency ν. This rotating field $H_1(\nu)$ tries to pull μ always towards itself, and if we make $\nu \simeq \nu_0$ (where ν_0 is the frequency of precession of μ about $\mathbf{H}_0$), then $\mathbf{H}_1$ and μ stay together as the latter precesses about $\mathbf{H}_0$ (Fig. 17.5). That is, during the precession of μ, the force acting on it stays the same and at $\nu = \nu_0$ both $\mathbf{H}_1$ and μ rotate together (in a synchronized fashion) about $\mathbf{H}_0$ (where $\mathbf{H}_1$ stays in the xy-plane). There is a constant pull on μ by $\mathbf{H}_1$. Then μ gets from one energy state to another whereby we have nuclear absorption (NMR) at $\nu = \nu_0$. (And when it gets from a higher level to a lower one, it emits the quantum $h\nu_0$ (Fig. 17.6). Then we have nuclear emission.)

The field $\mathbf{H}_0$ is obtained from an electromagnet. The sample is in a test tube which is held between the poles of the electromagnet. $\mathbf{H}_1$ is obtained via a coil into which the tube is inserted (Fig. 17.7). Notice that $\mathbf{H}_0 \perp \mathbf{H}_1$.

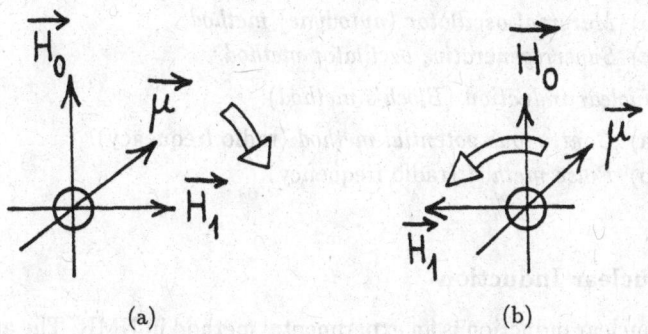

(a) (b)

Fig. 17.5.

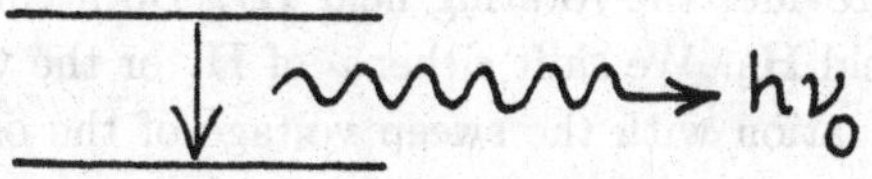

Fig. 17.6.

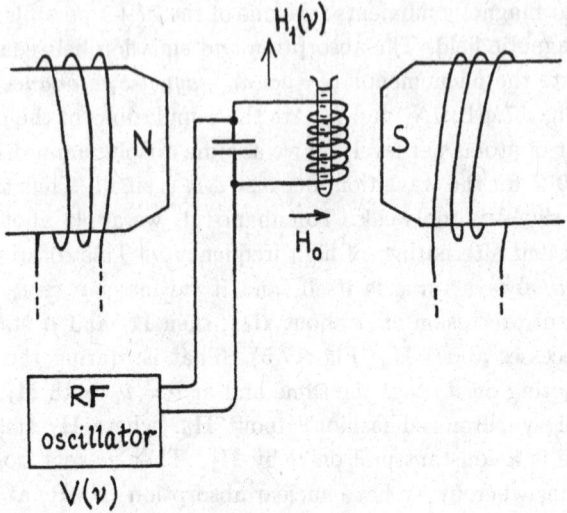

Fig. 17.7.

The phenomenon of resonance (at $\nu = \nu_0$) is observed in two ways:

(1) *Nuclear absorption (Purcell's method or Q-meter)*:

    (a) *Bridge method.*

    (b) *Marginal oscillator (autodyne) method.*

    (c) *Superregenerative oscillator method.*

(2) *Nuclear induction (Bloch's method)*:

    (a) *Continuous potential method* (radio frequency).

    (b) *Pulse method* (radio frequency).

17.2. Nuclear Induction

The nuclear induction is an experimental method in NMR. The apparatus is shown in Fig. 17.8. The sample is in a test tube which is inserted in a receiver coil. This coil is perpendicular to the coil of the radio-frequency supplier oscillator which provides the rotating field H_1. Both coils are perpendicular to the constant field H_0. We shift either ν of H_1 or the value of H_0,[1] so that there is synchronization with the sweep voltage of the oscillator. The radio-

---

[1]We choose it accordingly.

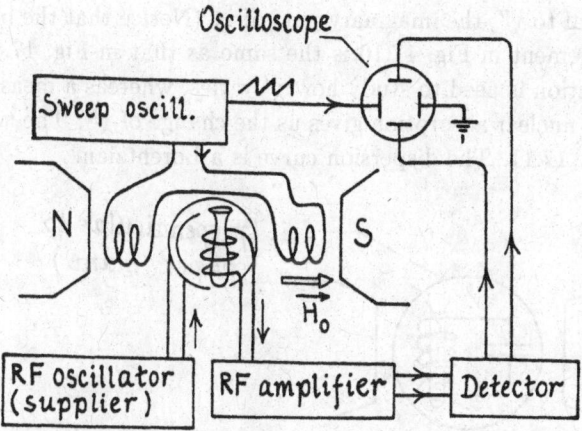

Fig. 17.8.

frequency amplifier amplifies the signal coming from the receiver coil and feeds it to the oscilloscope, on the screen of which we observe the resonance signal.

If the sweep voltage is continuous (and so is the potential obtained at the output of the detector), we have the *continuous potential method*. However, the alternating potential given by the radio-frequency oscillator may be in the form of discrete intermittent pulses (Fig. 17.9), in which case we have the *pulse method*.[2] The pulse method will be revisited below, in the discussion of the relaxation times.

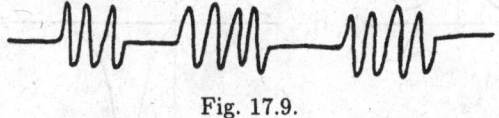

Fig. 17.9.

By the method of induction we measure the total magnetization

$$\vec{M} = \sum_{i=1}^{N} \mu_i = \chi \mathbf{H}_0$$

where N is the number of nuclei participating in the resonance, and χ is the magnetic susceptibility. It is a complex number: $\chi = \chi' - i\chi''$.

In the nuclear induction, the value read by the meter in Fig. 17.10 is proportional to χ', whereas in the nuclear absorption the resonance signal obtained

---

[2]The pulse method is used both in the induction and absorption.

is proportional to χ'', the imaginary part of χ. (Notice that the block diagram of the arrangement in Fig. 17.10 is the same as that in Fig. 17.8.) Thus the nuclear induction is used to study how χ' varies, whereas a measuring system that employs nuclear absorption gives us the change of χ''. The two curves are shown in Fig. 17.11. The dispersion curve is a Lorentzian.

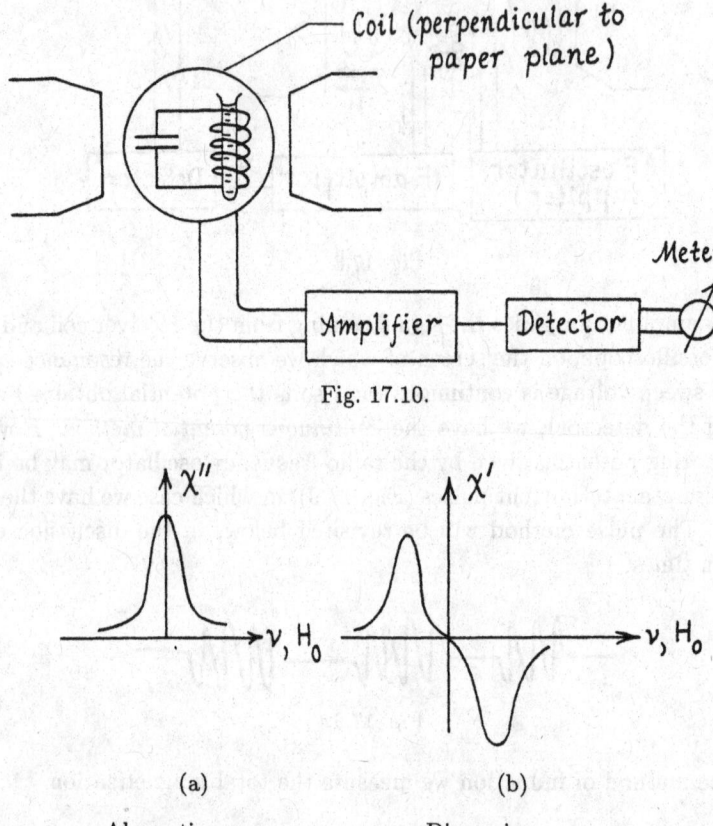

Fig. 17.10.

(a)   (b)

Absorption curve   Dispersion curve

Fig. 17.11.

As mentioned above, in the nuclear induction we change either ν or H_0. In the latter case we vary H_0 very slowly. At a certain H_0 value (which thereafter we hold fixed) the following condition is satisfied at ν_0:

$$h\nu_0 = \mu_n g H_0,$$

and the phenomenon of induction occurs. The same thing holds for the absorption, too: We change either ν or H_0.

17.3. The Bridge Method

Figure 17.12 shows the high-frequency alternating current bridge. The bridge is brought to resonance by the potential that comes from the radio-frequency oscillator. The equilibrium gets disturbed because of absorption. Thus we observe absorption.

For induction, the bridge is initially brought to equilibrium by the two variable capacitors shown in Fig. 17.12. Before the resonance frequency is reached, the bridge is shifted a bit off equilibrium. As the scan passes through the resonance frequency, we observe the dispersion curve shown in Fig. 17.11(b).

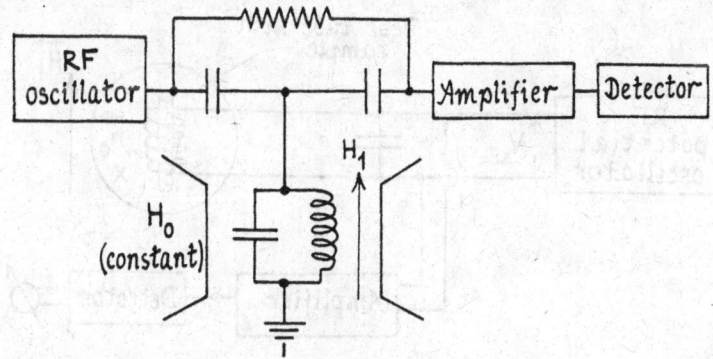

Fig. 17.12.

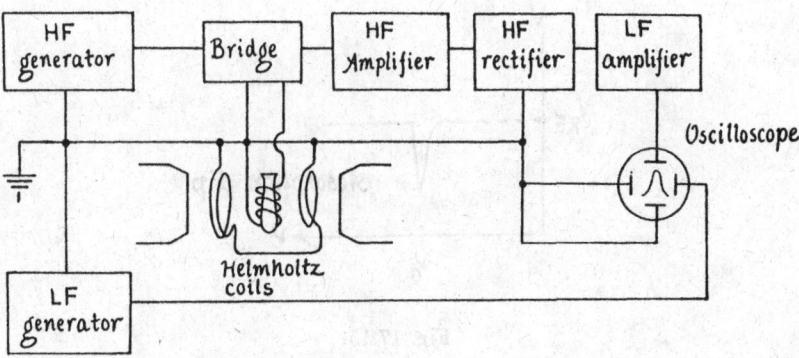

Fig. 17.13.

Purcell's method gives the spin and magnetic moment of stable and radioactive nuclei. The Purcell apparatus is shown in Fig. 17.13.

17.4. The Marginal Oscillator Method

This method is used for absorption only. Instead of a bridge, we use the apparatus shown in Fig. 17.14. An electromagnet provides $\mathbf{H}_0$. The radio-frequency potential[3] V_{RF} is sent to the coil that contains the tube with the sample. To make measurements, we measure the change in V_{RF}. When the apparatus is in equilibrium, the meter shows zero. The oscillator frequency ν is changed slowly. When we reach resonance, the nuclei make absorption and V_{RF} decreases (*absorption dip*) (Fig. 17.15). That is, the absorption occurs when ν reaches $\nu = \nu_0$.

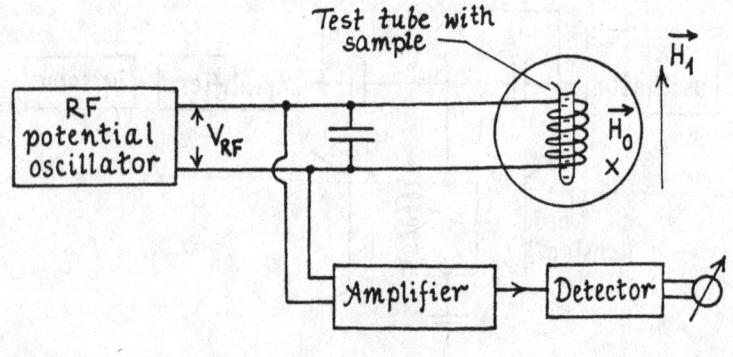

Fig. 17.14.

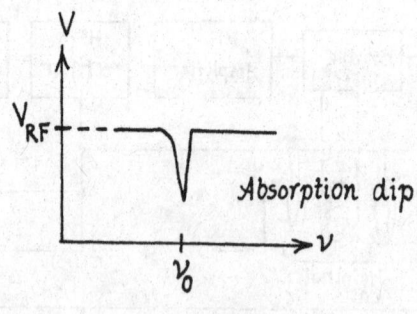

Fig. 17.15.

---

[3] A radio-frequency field is electronically generated.

Figure 17.16 shows the situation concerning the direction of the fields.

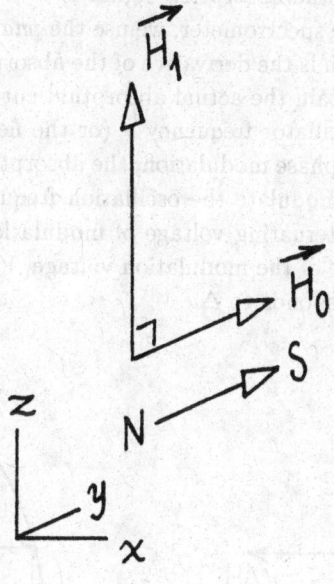

Fig. 17.16.

17.5. Phase Detection System

In general, it is difficult to change the oscillator frequency ν. Instead, we change H_0. How? We place two Helmholtz coils over the two poles of the electromagnet (Fig. 17.17) and thus we change H_0 by ΔH_0. For example, for $H_0 = 10$ kGauss we have $\Delta H_0 = 10$ Gauss.

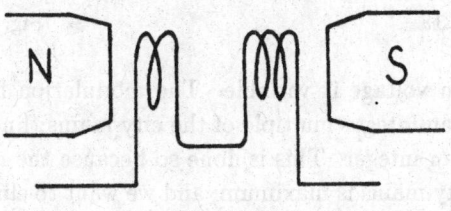

Fig. 17.17.

In the induction method we change H_0. But in the bridge and the marginal-oscillator method we change the frequency ν.

When we observe NMR, the voltage ΔV to be measured is of the order of microvolts. (Especially, the absorption signal is $\sim \mu$Volts.) Since there is a large noise voltage in the spectrometer, we use the *phase detection system*. We thus obtain a curve which is the derivative of the absorption curve (Fig. 17.18). By integrating it, we obtain the actual absorption curve.

We modulate the oscillator frequency ν (or the field H_0) with a very low frequency. If there is no phase modulation, the absorption curve looks like that in Fig. 17.19. We now modulate the oscillation frequency ν (or the field H_0 instead) with a small alternating voltage of modulation frequency ν_m, where $\nu_m \ll \nu_0$. The amplitude of the modulation voltage, V_m, is $V_m \ll V_0$ where V_0 is the voltage that corresponds to $\Delta\nu$.

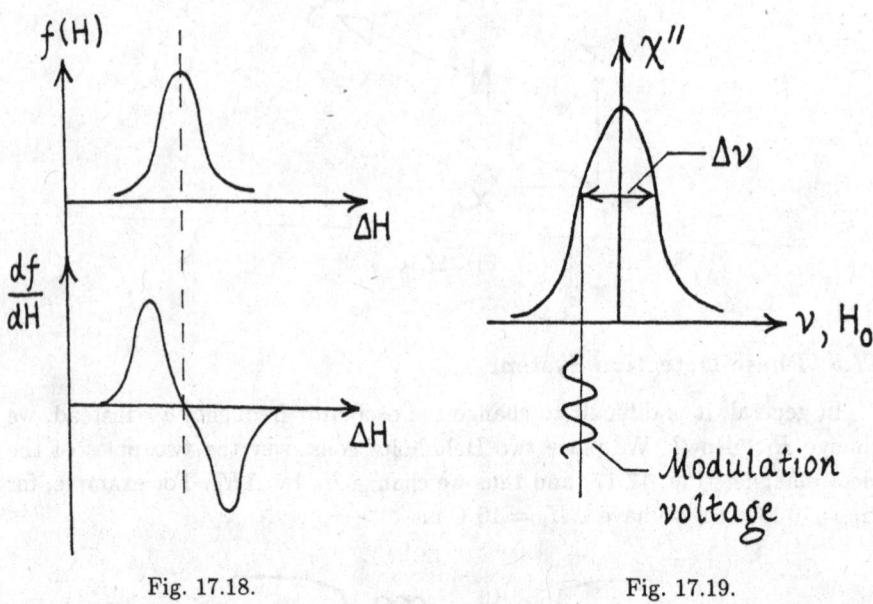

Fig. 17.18.   Fig. 17.19.

The modulation voltage is variable. The modulation frequency is chosen such that it is not an integer multiple of the city mains (line) frequency: $\nu_m \neq n\nu_{city}$, where n is an integer. This is done so because the noise voltage in the frequency of the city mains is maximum, and we want to eliminate it. So, since ν_{city} is 50 or 60 Hz, we take $\nu_m = 33$ or 77 Hz, or similar values.

Consider Fig. 17.20. Point A is the starting point. The frequency ν is shifted. The absorption potential is modulated by a frequency shown horizontally in Fig. 17.20. The whole situation is shown in Fig. 17.21.

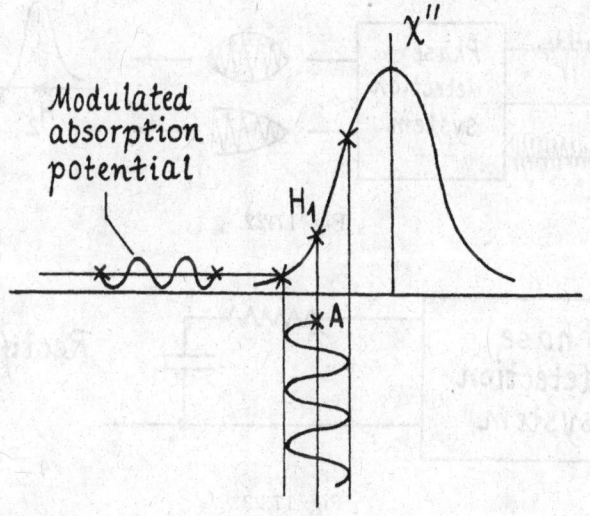

Fig. 17.20.

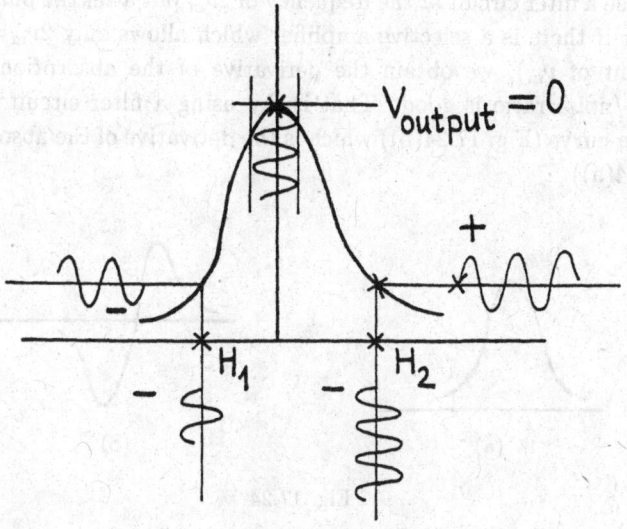

Fig. 17.21.

The signals before and after resonance are in opposite phase by 180°.

Figure 17.22 shows the curve obtained at the output. The frequency ν (or H_0) is shifted from the beginning. The signal on the left increases first, and then decreases. Figure 17.23 shows the rectifier.

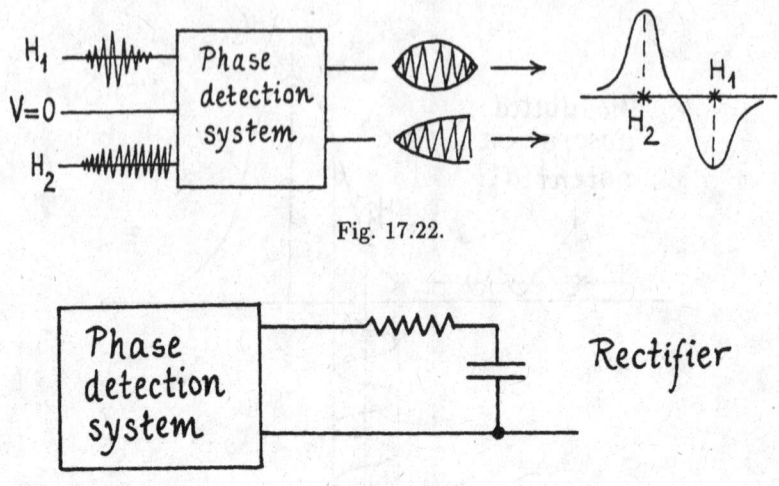

Fig. 17.22.

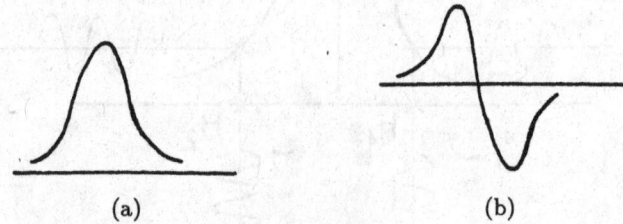

Fig. 17.23.

If we use a filter circuit at the frequency of $2\nu_m$ put after the phase detection system, or if there is a selective amplifier which allows only $2\nu_m$ (*i.e.*, double the amount of ν_m), we obtain the derivative of the absorption curve, and the signal/noise ratio is good. That is, by using a filter circuit we obtain a Lorentzian curve (Fig. 17.24(b)) which is the derivative of the absorption curve (Fig. 17.24(a)).

(a)   (b)

Fig. 17.24.

This way the signal/noise ratio is increased ~ 100 to 150 times.

If initially we have signal/noise = 3, the noise voltage is comparable to the signal voltage:

$$V_{\text{noise}} = \frac{1}{3} V_{\text{signal}},$$

and a good measurement is not possible. A curve shown in Fig. 17.25 appears on the oscilloscope screen: The signal is burdened with noise of comparable size. However, if we use the arrangement stated above, we can reduce noise. The noise over the signal becomes too feeble to disrupt the meaningful signal.

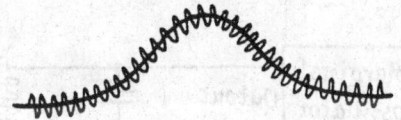

Fig. 17.25.

The block diagram is shown in Fig. 17.26. The output voltage produced by, and taken from the radio-frequency oscillator is modulated by ν_m. This voltage is supplied to a phase-shifting circuit. The signal coming from the other phase-shifting circuit is fed into a phase-detection system together with the detected signal which comes from the amplifier. The output of the phase-detection circuit is connected to a recorder.

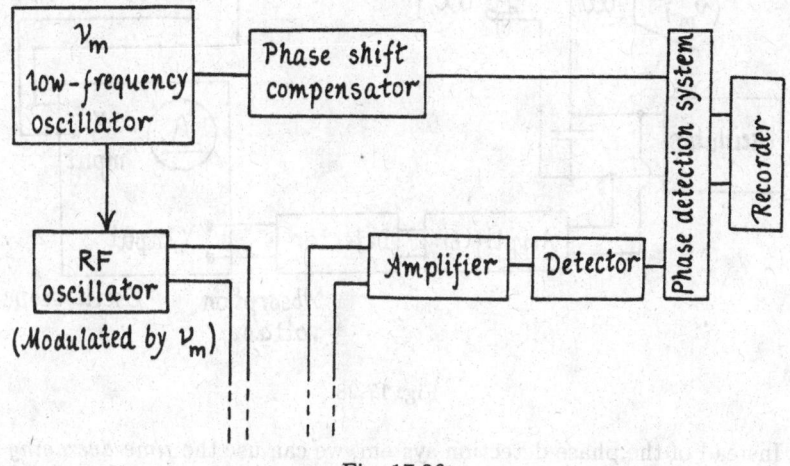

Fig. 17.26.

The phase shift is so properly chosen that the curve is put into a shape which is exactly the mathematical derivative of the actual absorption curve.

The inserted phase-shift circuit compensates for the phase shift that occurs in the other system. That is, the inserted phase-shift circuit is used to ensure that the signal given by the frequency oscillator has the same phase as the signal given by the phase-shifting device.

Figure 17.27 shows the principle of the technique. Figure 17.28 shows the full block diagram. By ν_m we achieve the change ΔH_0. The oscilloscope screen shows the absorption curve. Both in absorption and induction we change either H_0 or ν, and keep the other constant. The phase-detection system ensures increase in sensitivity.

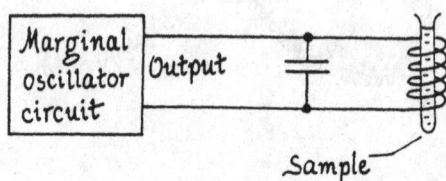

Fig. 17.27.

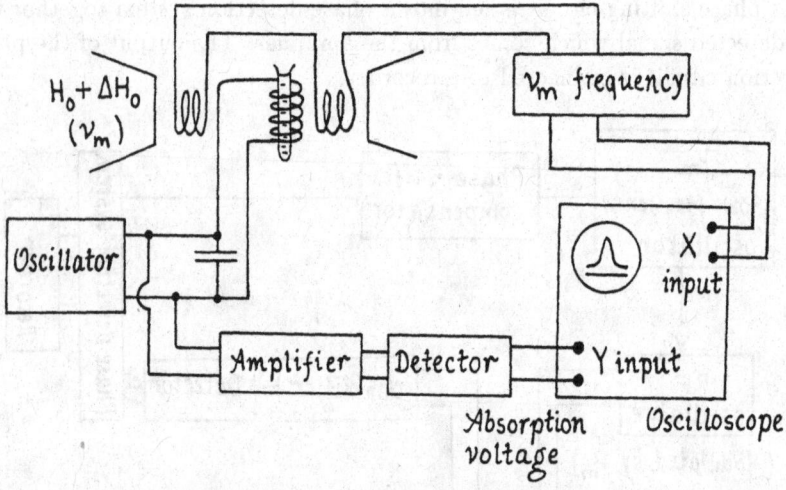

Fig. 17.28.

Instead of the phase detection system, we can use the *time-averaging computer system* (TAC). How does it work? We will discuss it in the next section.

17.6. TAC System

TAC can be applied to any measuring system.

Figure 17.29 shows the absorption curve with a sinusoidal noise voltage over it. In TAC the noise voltage changes: It does not have a constant phase and amplitude. We record the spectrum in the spectrometer for an interval of

1 sec. The signal coming from the spectrometer is fed into TAC (Fig. 17.30) where it is split into ten channels. The voltage is recorded at ten points, at one-second intervals (Fig. 17.31). An analog-digital system (ADS) is used to convert the voltage to the digital system. The ADS is connected to a memory which has ten sections (Fig. 17.32). The device makes addition — for example, it performs the operation 100,000 or 10^6 times. Upon addition, the signal-to-noise ratio increases proportionally to 10^6. That is, if $n = 10^6$, then upon addition we have

$$\frac{\text{Signal}}{\text{Noise}} \propto n.$$

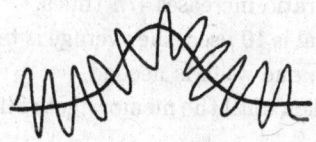

Fig. 17.29.

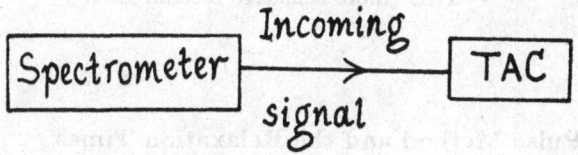

Fig. 17.30.

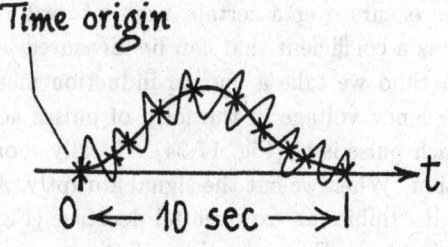

Fig. 17.31.

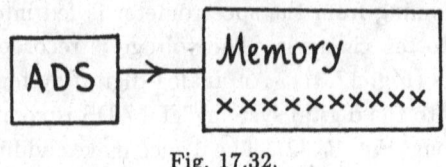

Fig. 17.32.

Since the noise voltage is random, we have

$$\text{Noise} \propto \frac{\text{Signal}}{\text{Noise}} \sqrt{\text{Noise}},$$

so that the signal/noise ratio increases $\sqrt{n}$ times.

If one counting interval is 10 μsec, the average is taken in two hours (100,000 averagings); if it is 1 sec, one week is needed.

Measurements are taken until the memory gets filled. The achieved increase becomes $300\sqrt{10^5}$.

So, the logic is as follows:

Oscillograph $\rightarrow$ Phase detection system (sensitive measurement)

$\rightarrow$ TAC (more sensitive measurement).

17.7. The Pulse Method and the Relaxation Times

Before the application of H_0, the mean energy of the magnetic moment is $\overline{E}(\mu) = 0$, and there is no splitting, nor any special orientation of μ in space. When μ is placed in H_0 (Fig. 17.33), the component $\mathcal{M}_z$ of the magnetization (the component in the direction of H_0) increases and then changes exponentially. This change occurs over a certain time interval τ_1, called *spin-lattice relaxation time*. It is a coefficient that can be measured by the pulse method.

In the pulse method we take a nuclear-induction measuring system. We send the radio-frequency voltage in the form of pulses acting on the system. The duration of each pulse is t_g (Fig. 17.34). Usually short pulses are used in the nuclear induction. When we cut the signal abruptly, $\mathcal{M}_z$ does not become zero immediately; it exhibits an exponential decrease (Fig. 17.35). The signal exponential is $\exp(-t/\tau_1)$. From the time of the exponential decay one can find the relaxation time τ_1.

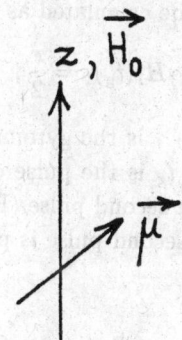

Fig. 17.33.

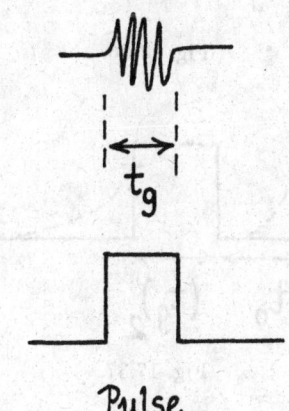

Pulse

Fig. 17.34.

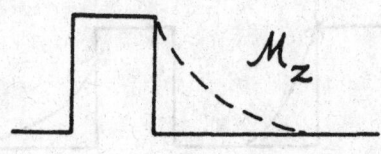

Fig. 17.35.

The relaxation time τ_2 can be calculated as follows: If we choose

$$\gamma H_1 (t_g)_1 = \frac{\pi}{2},$$

then we have a 90°-pulse. Here γ is the gyromagnetic ratio, H_1 is the radio-frequency field amplitude, and t_g is the pulse duration (Fig. 17.36). A time t_0 after the 90°-pulse we send a second pulse (Fig. 17.37), and obtain a pulse change according to τ_1. If the second pulse is proportional to $\pi/2$, we have a 90°–90° pulse.

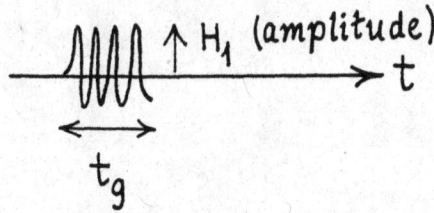

Fig. 17.36.

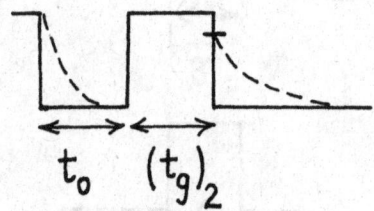

Fig. 17.37.

So, if there is the same condition for $(t_g)_2$, we can adjust the time t_0 so that the decay curve starts from different starting points (Fig. 17.38). Then the change of the amplitude becomes as in Fig. 17.39. This curve has the form of $\exp(t/\tau_2^*)$ where τ_2^* is the time constant of the exponential increase.

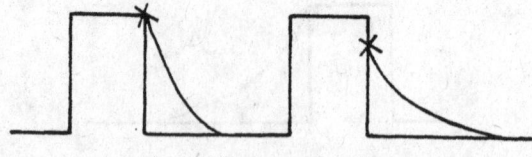

Fig. 17.38.

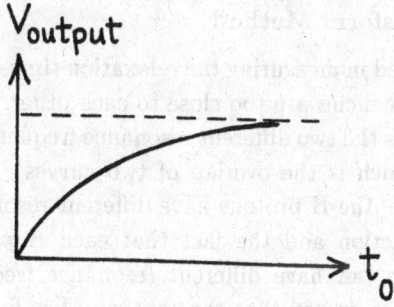

Fig. 17.39.

The relation is

$$\frac{1}{\tau_2^*} = \frac{1}{\tau_2} - \frac{1}{2\tau_1}$$

where τ_2^* and τ_1 can be measured and thus τ_2 can be found indirectly.

The relaxation time τ_2 is called *spin-spin relaxation time*. It cannot be measured directly. However, if we have a 90°–180° pulse (*i.e.*, if the duration of the second pulse is π), then τ_2 can be measured directly. Otherwise the value of τ_2^* is approximate, subject to the experimental conditions.

If we send two 90°–180° pulses, a *third* pulse shows up (Fig. 17.40) although it was not sent. This is the *spin echo*, first observed by Hahn. That is, the pulse hits the nuclear system and returns as an echo. The theoretical explanation of this phenomenon was made by Hahn.

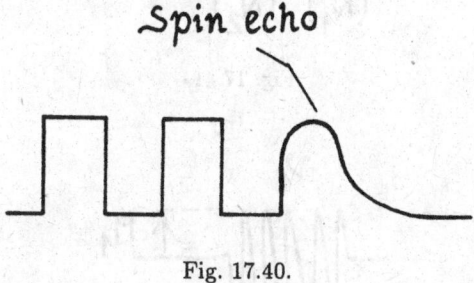

Fig. 17.40.

In the previous section we saw that when the oscillograph is connected to a phase-detection system, the sensitivity increases. Similarly, when we measure τ_1 by using the pulse method, we can make the sensitivity increase if we use the *Fourier transform method* (FTM).

17.8. Fourier Transform Method

This method is used in measuring the relaxation time of two nuclei together, whose resonance frequencies are too close to each other.

Figure 17.41 shows the two different resonance frequencies and the resulting absorption curve (which is the overlap of two curves). For example, in the CH_3CH_2OH molecule, the H protons have different resonance frequencies due to the mutual interaction and the fact that each H proton feels a slightly different H_0. So, we can have different resonance frequencies in the same molecule, in which case we say that the spectrum has fine structure (details).

In Fig. 17.41, N_1 and N_2 are the numbers of the nuclei having the corresponding resonance frequencies. We have two resonance frequencies very close to each other. If we send a pulse (Fig. 17.42), the peak on the left is at resonance exactly, whereas the peak on the right — being at a close frequency — makes forced oscillations (the system oscillates). As a result, we have an interference between the two, resulting in beats which in turn appear in the pulses of the output potential (Fig. 17.43). So, upon the application of the

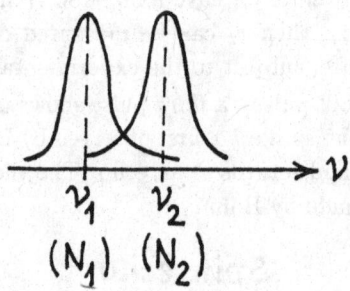

Fig. 17.41.

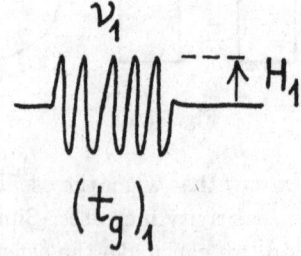

Fig. 17.42.

pulse shown in Fig. 17.42, we obtain an outcome depicted in Fig. 17.43. We can make the Fourier transformation (spectrum analysis) of this outcome pulse to obtain components shown in Fig. 17.44. By the use of electronic equipment, we can measure τ_1 and τ_2. The whole system is too complicated and intricate to discuss here in detail. We have outlined the method only, very laconically. For space reasons, we will not get into further analysis, even a telegraphically brief one, because a thorough treatment would take pages.

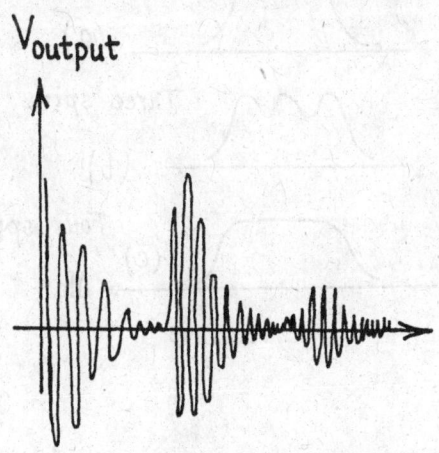

Fig. 17.43.

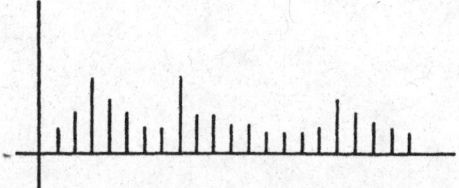

Fig. 17.44.

17.9. Multispin Systems

In general, many nuclear spins partake of the resonance in a system. If there are two such spins (two nuclei), the absorption curve has two peaks (Fig. 17.45(a)). If there are three, the curve is as in (b) of the same figure.

If there are four or more, we obtain a plateau (c) instead of resolvable peaks, due to the overlap.

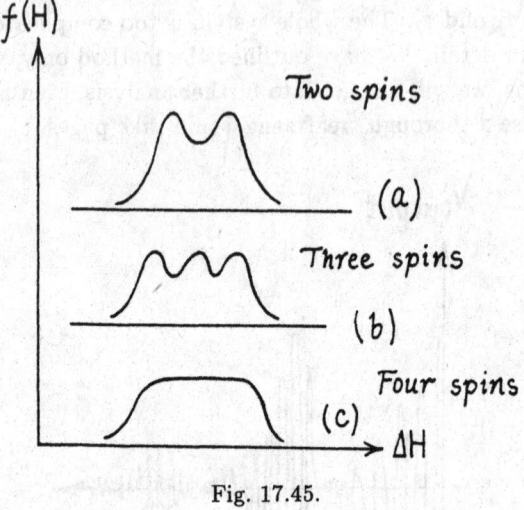

Fig. 17.45.

Chapter 18

APPLICATIONS OF NMR

18.1. General Discussion

NMR has many applications in physics, biochemistry, molecular biology, and pharmacy. The most noteworthy applications can be listed as follows:

(1) NMR frequency shift:
    (a) Chemical shift.
    (b) Knight shift (or shift due to conduction electrons).
(2) Measuring of relaxation times.
(3) Line shape of resonance curves.
(4) Magnetic moment measurements.
(5) Fine structure of spectral lines.

Atoms with small atomic number Z do not diffract X-rays. On the other hand, the neutron diffraction is a method with low sensitivity. That leaves NMR as the best method to investigate molecules by.

18.2. Chemical Shift

Consider the ethylalcohol molecule: CH_3-CH_2-OH. Due to its chemical molecular structure there are fine structures in its spectrum (because of the protons, $i.e.$, the H nuclei). At a strong and constant $H_0 = 10$ kGauss, the proton resonance frequency occurs at $\nu_0 = 42.5$ MHz. A rough observation gives the spectral line which is asymmetric, with a shoulder on its side (Fig. 18.1). Observation with a better spectrometer reveals the fact that the resonance line splits into three lines (Fig. 18.2) because of the three kinds of protons present.

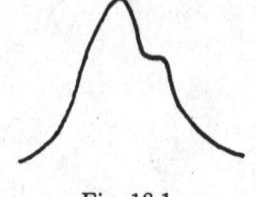

Fig. 18.1.

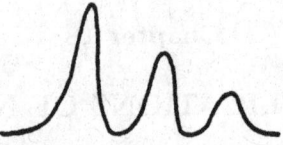

Fig. 18.2.

Each proton gives a resonance line as a detail of the overall resonance line. The areas under these three lines (intensities) are at the ratio of 3:2:1 (Fig. 18.3) because of the CH_3, CH_2, and OH groups in the molecule. In other words, an intensity three times as big means that there are three times as many protons participating in the absorption (resonance). So, the first line belongs to the protons of CH_3, the second line to CH_2, and the third line to OH.

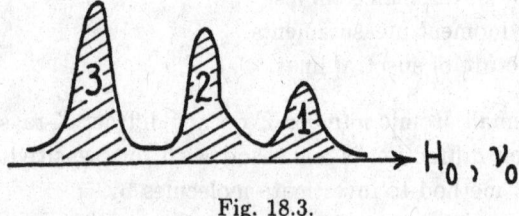

Fig. 18.3.

Figure 18.4 shows the frequency intervals and the corresponding fields. At 10 kGauss the shift is of the order of 0.01.

This triple line is called *fine structure*.

If we look at this fine structure with a spectrometer of a higher resolving power, we see that actually the first two lines split into three and four sublines respectively, due to spin-spin interaction (Fig.18.5). The interaction Hamiltonian is

$$\mathcal{H}_{spin-spin} = a\mathbf{I}_1 \cdot \mathbf{I}_2$$

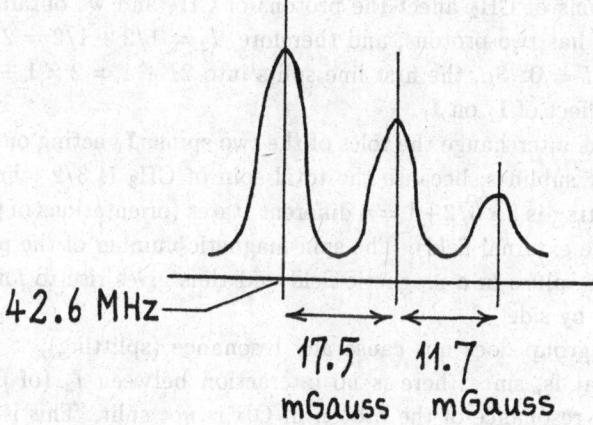

42.6 MHz

17.5
mGauss

11.7
mGauss

Fig. 18.4.

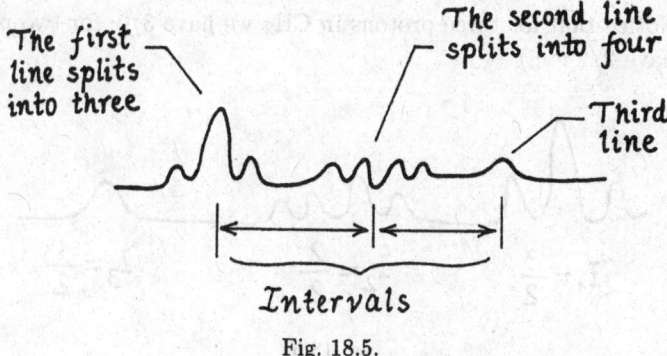

The first
line splits
into three

The second line
splits into four

Third
line

Intervals

Fig. 18.5.

where a is the spin-spin coupling coefficient (interaction strength). The Hamiltonian gives the spin-spin interaction energy of the fine structure which is due to the interaction between the spins:

$$\Delta E = am_1 m_2$$

which is obtained from the perturbation theory. Here I_1 is the nuclear spin of the resonating nucleus, I_2 is the total spin of the radical that causes the resonance, and $m_1(m_2)$ is the magnetic quantum number of the first (second) nucleus. ΔE is a small amount.

When a system has spin I, its magnetic quantum number can take $2I + 1$ values. Let us apply this to the number of lines:

The protons of CH_2 affect the protons of CH_3 and we obtain three lines. Indeed, CH_2 has two protons, and therefore, $I_2 = 1/2 + 1/2 = 2/2 = 1$. The spin of C is $I = 0$. So, the first line splits into $2I + 1 = 2 \times 1 + 1 = 3$ lines. This is the effect of $\mathbf{I}_2$ on $\mathbf{I}_1$.

Now let us interchange the roles of the two spins: $\mathbf{I}_1$ acting on $\mathbf{I}_2$ splits the line into four sublines, because the total spin of CH_3 is $3/2$ (three protons), and the splitting is $2 \times 3/2 + 1 = 4$ different states (orientations of protons with respect to the external field). The spin magnetic number of the protons takes four different values in a magnetic field and thus gives rise to four resonance sublines side by side.

The OH group does not cause any resonance (splitting); it stays single, unsplit. That is, since there is no interaction between I_2 (of CH_2) and I_3 (of OH), the resonance of the proton in OH is not split. This is called *rapid exchange* in quantum mechanics; that is, the exchange of energy is too rapid, and hence its average is measured as a single value.

Figure 18.6 shows the same spectrum with the total number spins of each group. (Notice that for three protons in CH_3 we have $3/2$; for two protons in CH_2 we have $2/2$ etc.)

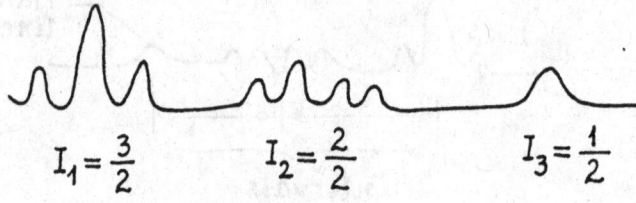

$$I_1 = \frac{3}{2} \qquad I_2 = \frac{2}{2} \qquad I_3 = \frac{1}{2}$$

Fig. 18.6.

If we further increase the resolving power of the spectrometer, we can observe a *hyperfine structure* over the fine structure, due to spin-orbit and quadrupole interactions. The situation is shown in Fig. 18.7. Notice the hyperfine splitting for CH_2. It can be explained by the interaction between quadrupole moments. The dwarfish little lines are due to the spin-orbit interaction.

By examining the chemical shift we can obtain information about the molecule examined. And as we determine the molecular structure, we can discover similar formulas and isomers. Information about isomers is obtainable from the chemical shift; information about crystals that exhibit variations in their crystalline structure is obtainable from the quadrupole interaction. That

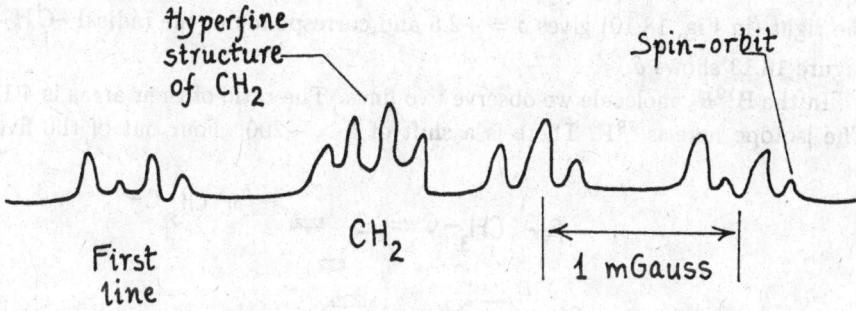

Fig. 18.7.

is, the quadrupole interaction gives states which are not the same physically, whereas the chemical shift gives differences in the molecular structure (*i.e.*, which type of structure is fit for which molecule).

The chemical shift can be measured at the order of $1/10^6$. The amount of shift, ΔH, satisfies the relation

$$\frac{\Delta H}{H_0} \times 10^6 = \delta$$

where for a shift of 10 mGauss we have $\delta = 10$.

The protons of the water molecule exhibit no shift: $\delta = 0$.

The chemical shift is a valuable informant about molecular bonds to be examined. One examines molecules with hydrogen protons belonging to different radicals. Then one determines the resonance frequencies and the shifts. For example, we can determine δ for the protons of the radical CH_3-C-. We do it for many kinds of molecules containing the radical and we take the average. We notice that δ lies in the intervals shown at the upper right corner of Fig. 18.8.

Figure 18.9 shows the dicetene molecule: It is probable that it has two isomers, as shown. The spectrum of this molecule was measured from low temperatures up to $+170°C$. The constant result was two lines (Fig. 18.10). These two lines so obtained were of the same area under the peak curves, which means that they both belong to the same type of molecular bond, *i.e.*, the two different formulas involve the same type of bond. This in turn means that the isomer in Fig. 18.9(a) does not exist. Had both isomers existed, we would have observed three lines as in Fig. 18.11, with areas at the ratio of 3:4:1. If only (a) had existed, the spectrum would have been as in Fig. 18.12, *i.e.*, 3:1. But we get that in Fig. 18.10 where the chemical shift is seen. The peak on the right (in Fig. 18.10) gives $\delta = +2.5$ and corresponds to the radical $-\overset{|}{C}H_2$. Figure 18.13 shows δ.

In the $B^{19}F_5$ molecule we observe two lines. The ratio of their areas is 4:1. The isotope here is ^{19}F. There is a shift of $\delta > -200$. Four out of the five

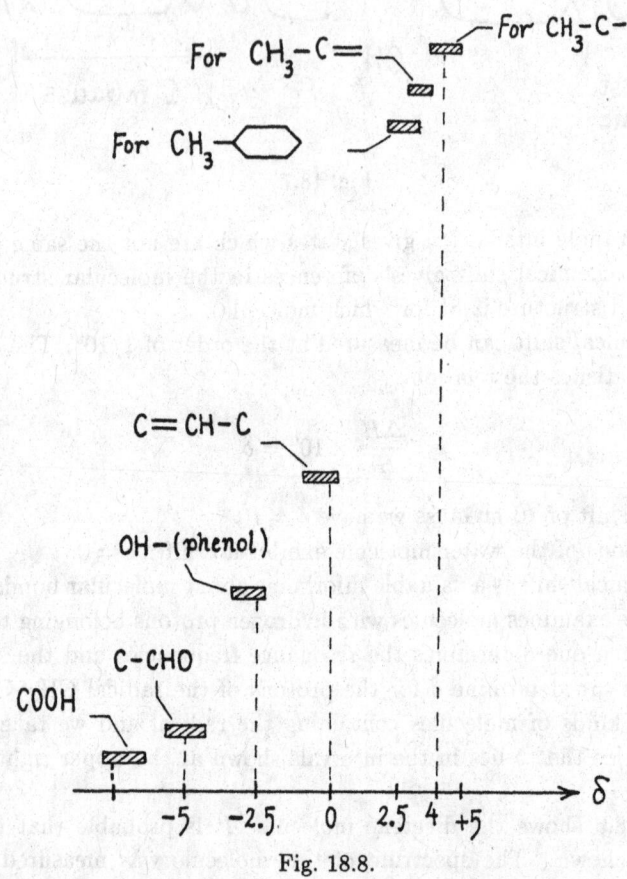

Fig. 18.8.

Fig. 18.9.

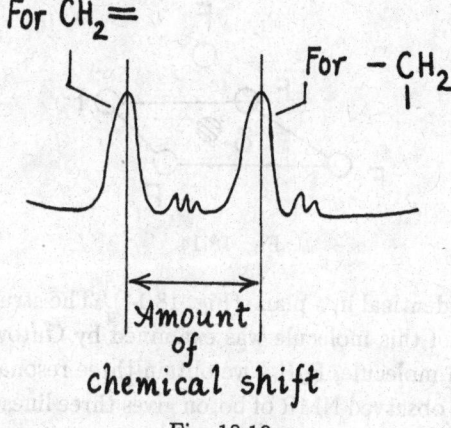

Fig. 18.10.

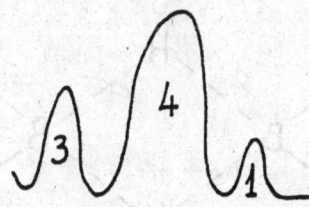

Fig. 18.11.

Fig. 18.12.

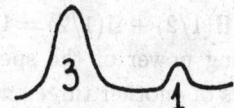

Fig. 18.13.

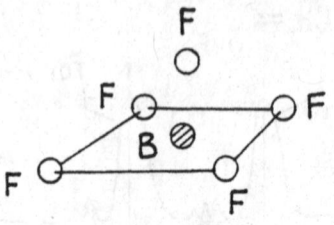

Fig. 18.14.

fluorine atoms are identical in a plane (Fig. 18.14). The structure is tetragonal. The fine structure of this molecule was examined by Gutowsky.[1]

For the diboron molecule, B_2H_6, we obtain three resonance lines. We have $\delta = \pm 10$. Since the observed NMR of boron gives three lines, we have $2I+1 = 3$ whence we must have $I = 1$. Each boron atom forms a bond with two H atoms (Fig. 18.15). The dashed lines represent weak bonds.

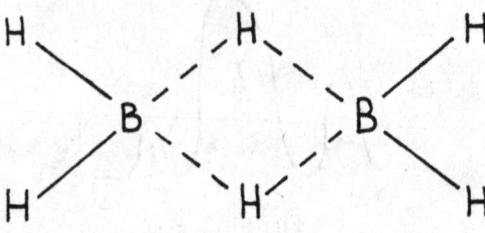

Fig. 18.15.

For the spin we have $I = H(1/2) + H(1/2) = 1$.

If we increase the resolving power of the spectrometer, we observe that each of the three lines consists of another three ones (Fig. 18.16), that is, each line splits into three. This splitting explains the weakness of the bond: The coupling constant is small. The splittings occur because of the interacting hydrogen atoms which are weakly bound to boron.

To summarize, we see that the NMR frequency depends on various effects. The chemical shift is utilized in the study of molecules and radicals, and the determination of their structure. The fine structure of the spectrum and pertinent phenomena can thus be studied.

The chemical structure of a molecule results in a small shift in the NMR frequency. This is the *chemical shift*, utilized in analyzing molecular structure. Molecules like H_2O, CH_2, ethylalcohol, and other radicals, including H, cannot

---

[1]See *Journal Chem. Phys.*, **22**, 26 (1954).

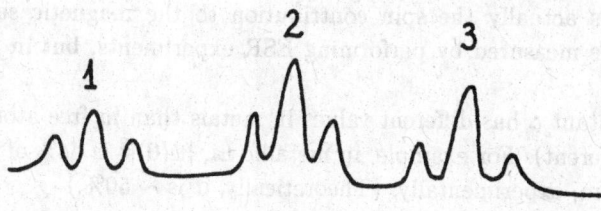

Fig. 18.16.

be studied by X-ray diffraction. So, we use the chemical shift to study molecular structure and to find out isomers, compound molecules, bonds, and chemical exchange.

18.3. Knight Shift

This is the effect where (at a certain frequency) the nuclear spin resonance occurs at a different (shifted) magnetic field in metals than in diamagnetic materials. The phenomenon can be used in studying conduction in metals.

A nucleus with spin $\mathbf{I}$ and gyromagnetic ratio γ_I interacts with an energy

$$E = I_z(-\gamma_I \hbar H_0 + a\overline{S}_z)$$

$\equiv$ (Interaction with the externally applied field)

$\quad$ +(Average hyperfine interaction between the nucleus and

$\qquad$ the conduction electrons).

$\overline{S}_z$ is the average spin of the conduction electrons where $\mathcal{M}_z = gN\mu_B\overline{S}_z = \chi H_0$ where χ is the Pauli susceptibility. Thus

$$E = \left(-\gamma_I\hbar + \frac{a\chi}{gN\mu_B}\right) H_0 I_z$$

$$= -\gamma_I\hbar H_0 I_z \left(1 + \frac{\Delta H}{H_0}\right)$$

where a is the interaction strength, *alias* hyperfine-coupling coefficient. The Knight shift is

$$\text{Shift} = -\frac{\Delta H}{H_0} \equiv \frac{a\chi}{gN\mu_B\gamma_I\hbar}.$$

Since the hyperfine energy is $a\mathbf{I} \cdot \mathbf{S}$, the shift is $\sim \chi|\psi(0)|^2/N$ where the spin susceptibility is multiplied by the ratio of the concentration of the conduction electrons at the origin (position of nucleus) to the average concentration.

Here χ is actually the spin contribution to the magnetic susceptibility, which can be measured by performing ESR experiments, but in general it is difficult.

The constant a has different values in metals than in free atoms (because $\psi(0)$ are different). For example, in metallic Li, $|\psi(0)|^2$ is 44% of the value in a free Li atom, experimentally. (Theoretically, it is $\sim 50\%$.)

The following table gives NMR Knight shifts in certain metals at room T.

Metal	Shift (%)	Metal	Shift (%)
^{7}Li	0.026	^{27}Al	0.162
^{23}Na	0.112	^{51}V	0.58
^{87}Rb	0.653	^{53}Cr	0.69
^{39}K	0.265	^{105}Pd	-3
^{63}Cu	0.237	^{207}Pb	1.47
^{195}Pt	-3.533	^{197}Au	1.4

The *Knight shift* (or *shift due to the conduction electrons*) was first observed by Knight in 1949.[2] This shift is very large compared to the chemical shift: It is of the order of 0.1% to 1%. It was first observed in CuCl (Fig. 18.17). When the measurement was made at $\nu_0 = 30$ MHz, the measured shift was 10.1 kHz.

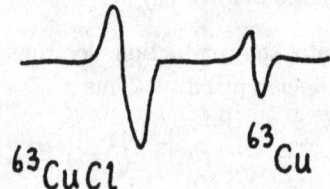

Fig. 18.17.

Whereas the chemical shift is a function of the magnetic field, the Knight shift has different properties. The Knight shift

(1) Occurs at constant fields H_0 and towards high frequencies (except for Pt where the opposite is true).

(2) Is very large.

---

[2]See *Phys. Rev.*, **76**, 1259 (1949).

(3) Is proportional to H_0.

(4) Depends too little on temperature.

(5) Is larger in nuclei with large mass number A.

(6) Is very large in elements whose conduction electrons are s-electrons.

Research with this method started in 1950 by Knight, Townes, and Herming. Two important studies are the following:

(a) The Al-Cu alloy was studied by Friedel.

(b) Cornell and Holmes added Na into liquid nitrogen and studied the shift due to the conduction electrons of Na. The population distribution (Fig. 18.18) was studied. Figure 18.19 shows the distribution. According to the Fermi distribution, the conduction band is partially filled (shaded area under the curve). Cornell was able to measure the distribution.

To summarize, we have to say the following about the Knight shift: As one measures the frequency of NMR, one observes a shift due to the conduction electrons in conductors: Electrons close to the Fermi surface result in paramagnetism. This shift is not observed in all metals. It is marked in heavy elements and is related to the change in the relaxation time τ_1. It can be utilized in studying metals, alloys, and ferromagnetic (and antiferromagnetic) materials, and in measuring the electronic charge density of the valence band.

ZnF_2 is paramagnetic. We can add ferromagnetic salts (MnF_2, FeF_2, CoF_2) to it at a certain concentration (solid solution). That is, we melt ZnF_2, add the salts as powders, and cool the system to form crystals. The ferromagnetic

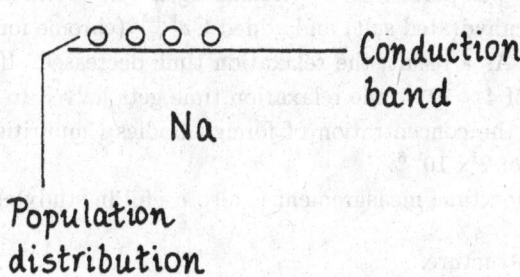

Fig. 18.18.

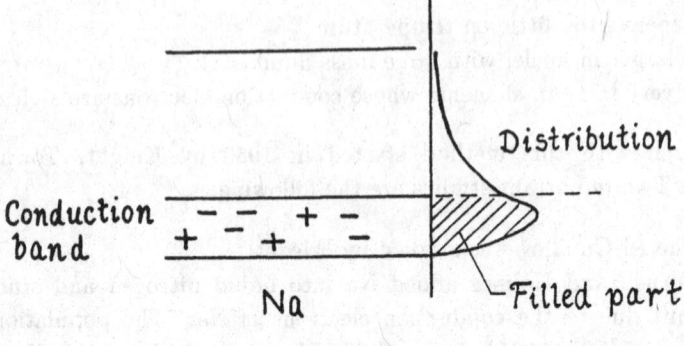

Fig. 18.19.

additives are now impurities in the main body, *i.e.*, we have a solid solution. The purpose here is to obtain materials of a better ferromagnetic quality, in the form of a solid solution (for high frequencies).

For more information about the Knight shift, see Kittel's book (listed in the References), pp. 514–516.

18.4. Measuring Relaxation Times

There are two relaxation times of interest: τ_1 (spin-lattice) and τ_2 (spin-spin). They give us information about energy exchange and the structure of the body. Measuring these relaxation times is important because one can deduce the relation between temperature and energy and study certain phenomena like the energy exchange between spin and lattice systems. It also supplies information about molecular vibrations and internal rotation.

The importance of τ_1 and τ_2 in NMR was first seen by Bloembergen in 1949. The first relaxation theory of NMR was formulated by Heitler and Teller. The first experiment was performed by Bloembergen who took $AlK(SO_4)_2.10H_2O$ (*i.e.*, alum, an enhydrated salt) and added Cr^{+++} (chrome ions) to it in a too small quantity. As a result, the relaxation time decreased. If we add Cr^{+++} ions at a ratio of 4×10^{-5}, the relaxation time gets halved to $\tau_1/2$. To obtain correct results, the concentration of foreign bodies (impurities) must be less than the order of 2×10^{-6}.

The relaxation time measurement is also useful in studying the following:

(1) Liquid structure.
(2) Internal rotation that can happen at a certain temperature in the molecules of solids.

(3) The proton-proton direction and distance in molecules containing H nuclei.

(4) The $CaSO_4.6H_2O$ molecule.

The interaction energy of two interacting dipoles which are close to each other is

$$\Delta E = \pm \frac{\mu^2}{r^3}(3\cos^3\theta - 1) \qquad (18.1)$$

where θ is the angle between $\mathbf{H}_0$ and the straight line that connects the two dipoles, and μ is the magnetic dipole moment (dipole strength). The plus (minus) sign holds for parallel (antiparallel) dipoles.

$\Delta E = h\Delta\nu$ can be measured (say, for the dipoles in the ethylalcohol molecule). Since μ is known, r can be found. The proton-proton distance r was so found in $CaSO_4.6H_2O$ by Pake. Thus the validity of Eq. (18.1) was satisfied. Figure 18.20 shows the proton-proton distance in the crystal lattice.

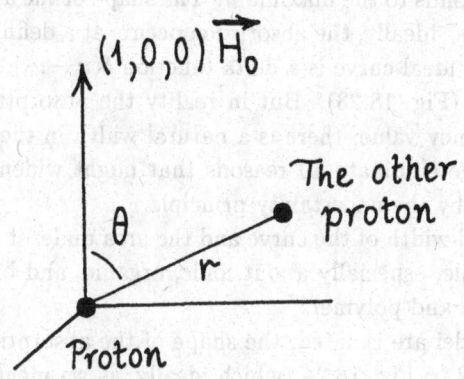

Fig. 18.20.

The dipole-dipole interaction gives two maxima (Fig. 18.21). We can find such an angle θ that ΔE is a minimum. That direction gives us the right and desired θ. So, θ is a variable angle and the proper value of it can be found.

In the PH_4I molecule, the internuclear distance P–H (*i.e.*, the separation between the P atom and the H atom) was measured by the same way. The result found was $r = 1.42$ Å.

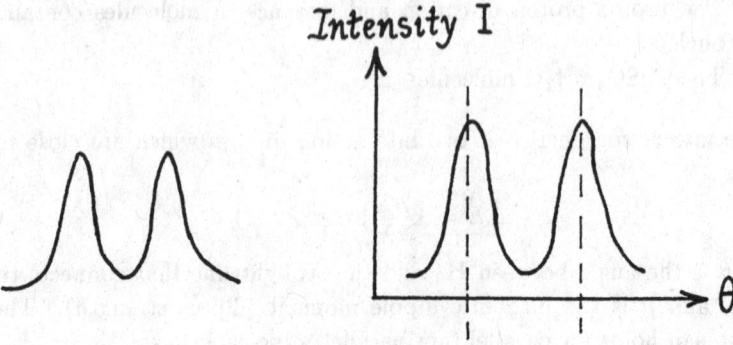

Fig. 18.21.

18.5. Resonance Curves and Line Shape

The absorption curve is a Gaussian (Fig. 18.22) where the resonance frequency ν_0 corresponds to the maximum. The shape of the area under the curve is called *line shape*. Ideally, the absorption occurs at a definite single frequency ν_0, and hence the ideal curve is a delta function $\delta(\nu - \nu_0)$, sharply peaking at a single point ν_0 (Fig. 18.23). But in reality the absorption is not observed at a single frequency value; there is a natural width in the line (curve) which prevails even if we eliminate all reasons that might widen the curve. It is a limit determined by the uncertainty principle.

The shape and width of the curve and the area under it give us information about the molecule, especially about ionic, organic, and biological molecules, as well as plastics and polymers.

When the nuclei are isolated, the shape of the absorption resonance curve is as in Fig. 18.22 or Fig. 18.24 (which ideally, as we mentioned, ought to be too sharp). But in reality other shapes are obtained because unpaired nuclei

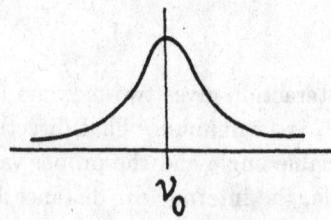

Fig. 18.22.

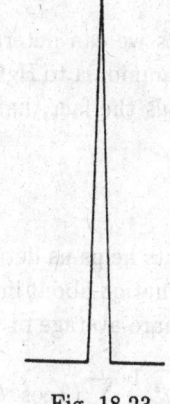

Fig. 18.23.

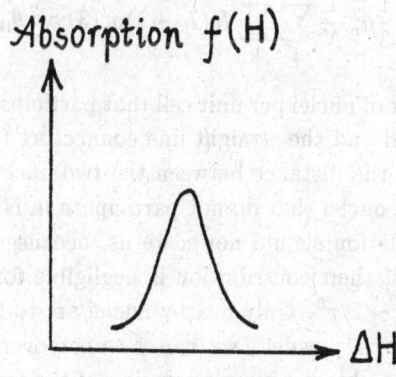

Fig. 18.24.

change the spectrum of each other. If we have two identical nuclei under H_0, separated by a distance r, and of spin $I = 1/2$, each of them feels an effective field given by

$$H_{\text{eff}} = H_0 \pm \Delta H = H_0 \pm \frac{3\mu\mu_{\text{n}}}{2r^3}(3\cos^2\theta - 1)$$

where μ is the magnetic moment.

The absorption line of a solid powder is the superposition of many curves averaged out. There is a widening due to the effect of the far nuclei, given by the exponential factor

$$\exp[-(\Delta H)^2/2b^2]$$

where b is a constant.

By studying absorption curves we can determine the molecular structure of solids. For example, if we add ammonia to $HgCl_2$ and examine the resulting sediment, an NMR research reveals the fact that the formula of the sediment is NH_2HgCl.

18.6. Second Moments

The method of second moments helps us determine molecular parameters. Here we use NMR to obtain information about internuclear distances. Namely, we use the *second moment* or square average of the absorption line

$$\overline{(\Delta H)^2} = \frac{3}{2}I(I+1)g^2\mu_n^2\frac{1}{N}\sum_{i>j}(3\cos^2\theta_{ij}-1)^2\frac{1}{r_{ij}^6}$$

$$+\frac{1}{3}\mu_n^2\frac{1}{N}\sum_i\sum_k I_k(I_k+1)g_k^2(3\cos^2\theta_{ik}-1)^2\frac{1}{r_{ik}^6}$$

where N is the number of nuclei per unit cell that participate in NMR; θ_{ij} is the angle between the field and the straight line connecting the i-th nucleus with the j-th nucleus; r_{ij} is the distance between the two nuclei; and the subscript k runs over all foreign nuclei that do not participate in NMR.

This monstrous relation should not scare us, because we need not extend the sums to far nuclei; their contribution is negligible for large r, becuase it drops dramatically as $\sim 1/r^6$. Only nearby nuclei are to be included.

If our specimen is a powder, we can average over the angle and set $\overline{(3\cos^2\theta-1)^2} = 4/5$ to obtain a simpler version of the formula:

$$\overline{(\Delta H)^2} = \frac{6}{5}I(I+1)g^2\mu_n^2\sum_{i,j}\frac{1}{r_{ij}^6} + \frac{4}{15}\mu_n^2\frac{1}{N}\sum_i\sum_k I_k(I_k+1)g_k^2\frac{1}{r_{ik}^6}. \quad (18.2)$$

This procedure was theoretical. Experimentally we have

$$\overline{(\Delta H)^2} = \frac{\displaystyle\int_0^\infty f(H)(\Delta H)^2 dH}{\displaystyle\int_0^\infty f(H)dH}$$

where $f(H)$ is the function representing the absorption curve, and ΔH is the difference away from the center of the symmetric absorption curve. Notice that the second moment is equal to the moment of inertia of the curve (numerator) divided by the area under the curve (denominator). The moment of inertia is taken about the symmetry axis of the curve.[3]

This method is sensitive because the second moment is $\sim 1/r^6$, and hence a 6% error made on the second moment results in only a 1% error on r, as we try to find internuclear distances. Researchers, however, should not be in a hurry to cheer up, because this method (and in general, NMR) cannot be used for nuclei with zero I. Of course it is true that one can find stable isotopes with nonzero spin for every element, but their abundance may not be sufficient.

A word of caution here: Eq. (18.2) holds only for nuclei with $I = 1/2$. If $I > 2$, then nuclear quadrupole moments come into play, causing splittings in the spectral lines of NMR. This affects the second moment.

Another limitation for Eq. (18.2) is that it holds for solid and rigid crystals (at the temperature of the laboratory where the experiment is being conducted). Also, the equation holds if the concentration of foreign paramagnetic atoms is 10^{-6} times less than the concentration of the nuclei under examination via NMR (because vibrations and rotations of the foreign atoms widen the absorption curve). There are more limitations and disadvantages (*e.g.*, knowing r_{ij} beforehand helps for better results),[4] and that is why this method is reserved only for molecules with H atoms where the X-ray diffraction fails in finding molecular parameters.

The value of the constant b in the widening factor is chosen arbitrarily but conveniently, subject to the conditions that reign in the unit cell.

18.7. Other Applications

Other applications of NMR are the following:

(1) Nuclear magnetic moments can be measured very sensitively.
(2) Examination of the fine structure of the spectral lines in NMR provides us with valuable information about molecular structure and nuclear physics. (This is partly discussed in Section 18.2 above.)
(3) Soft superconductors and unusual electron clouds in rare and complex compounds (molecules) can be studied by employing the knight shift.

Thus the merit of NMR as a molecular diagnostic probe is of great esteem, both in scientific research and professional practice.

---

[3] A device called integrator or a computer program for the calculation of the area is of use here.

[4] We should also know the position of nuclei other than the one whose absorption is being examined, in the unit cell. Especially important are the ones that have a mangetic moment.

18.8. Some Remarks on the Knight Shift

At a definite frequency, the resonance of a nuclear spin (NMR) is observed at a slightly shifted field in a metal than in diamagnetic solids. (See L.E. Drain, "*NMR in Metals*" in *Metallurgical Reviews*, **119** (1967).)

A nucleus of spin I and gyromagnetic ratio γ_I interacts with an energy

$$V \equiv E = I_z[-\gamma_I \hbar H_0 + a\overline{S}_z]$$

where a is the interaction strength, a constant. The first term expresses the interaction of I with H_0; the second term is just the mean hyperfine interaction between the conduction electrons and the nucleus (the nuclear spin actually). $\overline{S}_z$ is the mean spin of the conduction electrons which leads us to the Pauli spin susceptibility χ via the relation $M_z = \chi H_0 = gN\mu_B \overline{S}_z$. Hence

$$E = H_0 I_z \left[-\gamma_I \hbar + \frac{a\chi}{gN\mu_B} \right] = -\gamma_I \hbar H_0 I_z \left[1 + \frac{\Delta H}{H_0} \right],$$

and the shift is

$$\delta \equiv -\frac{\Delta H}{H_0} = \frac{a\chi}{gN\mu_B \gamma_I \hbar}.$$

The hyperfine interaction energy is $E = a\mathbf{I} \cdot \mathbf{S}$ where a is the hyperfine constant. We have

$$\delta \simeq \chi \frac{|\psi(0)|^2}{N},$$

i.e., χ augmented (scaled up) by the ratio of $|\psi(0)|^2$, the concentration of the conduction electrons *at the nucleus* ($r = 0$), to the *mean* concentration of the conduction electrons. The coupling constant a differs in metals than in free atoms (because $\psi(0)$, the wave functions at the position of the nucleus, are not the same).

Thus the effect of Knight shift is a means of studying conduction electrons. The shift in metallic Li gives $|\psi(0)|^2$ which is 44% of the value in free atomic Li. Theoretical calculations give $|\psi(0)|^2$(metallic Li) $\simeq (1/2)|\psi(0)|^2$(free Li). The spin contribution to the Pauli magnetic susceptibility χ can be found experimentally — by meticulous ESR techniques regarding conduction electrons and their spins — but in a few situations only, whereas the Knight-shift method is better, because it can probe metals, alloys, peculiar electronic systems (*e.g.*, $Na_x WO_3$), and soft and intermetallic superconductors.

Experimental results for δ in NMR in metallic elements (for some isotope nuclei) are tabulated on p. 520.

The hyperfine interaction is magnetic, between the nuclear spin (magnetic moment) in an atom and the spin of an electron. If we sit on the nucleus, we observe the electron moving (revolving) around us, creating a magnetic field at our position. This motional field (evaluated at $r = 0$, the nuclear position) interacts with $\mathbf{I}$. This electronic current about the nucleus exists if the electron's orbital angular momentum state exists. However, even if $\mathbf{L} = 0$ for the electron about the nucleus, there is still an electronic spin contribution which is nonzero. The electron still communicates with the nucleus.

Dirac's theory says that electron's Bohr magneton (magnetic moment) $\mu_B = e\hbar/2mc$ comes from the fact that the electron circulates in a current loop (moving charge) with speed c. The radius of the loop is $\tilde{\lambda} = \hbar/mc \sim 10^{-11}$ cm, the electron's Compton wavelength, and the current is $i = e/t = ec/\tilde{\lambda}$ where $1/t = $ revolutions/sec. This i creates $H = i/\tilde{\lambda}c = e/\tilde{\lambda}^2$. The nucleus can be found *inside* the electron (actually, inside a sphere of influence whose volume is $\tilde{\lambda}^3$ around the electron) with a probability $|\psi(0)|^2\tilde{\lambda}^3$ where the first multiplicand is the probability of finding the electron (whose wavefunction is ψ) at (the position of) the nucleus — ψ evaluated at $r = 0$. (We assume the nucleus at the origin.)

Thus the observer sitting on the nucleus experiences an *average* field (at the nucleus)

$$\overline{H}(0) = e\tilde{\lambda}|\psi(0)|^2 = \mu_B|\psi(0)|^2$$

with $\mu_B = (1/2)e\tilde{\lambda}$. The hyperfine interaction energy (spin part) is

$$E = -\boldsymbol{\mu}_I \cdot \overline{\mathbf{H}} = (-\boldsymbol{\mu}_I \cdot \boldsymbol{\mu}_B)|\psi(0)|^2 = \gamma\hbar\mu_B|\psi(0)|^2\mathbf{I} \cdot \mathbf{S} \qquad (18.3)$$

where I is the nuclear spin (in units of $\hbar$). The table below gives a for the ground state of some *free* atoms:

Nucleus	^{1}H	^{7}Li	^{23}Na	^{39}K	^{41}K
I	1/2	3/2	3/2	3/2	3/2
a (Gauss)	507	144	310	83	85
a (MHz)	1420	402	886	231	127

a in Gauss (as experienced by the spin of an electron) is $a/2\mu_B$, by definition.

Figure 18.25 shows **H** created by a charge circulating in a (circular) loop. The spin part of the magnetic hyperfine interaction with the nuclear spin is localized within the region of the space inside (or close to) the loop of current. **H** is averaged over a spherical shell that contains the loop, and the result is zero. (Why?) So, if $L = 0$ (s-electrons), the interaction is due to spin contribution only.

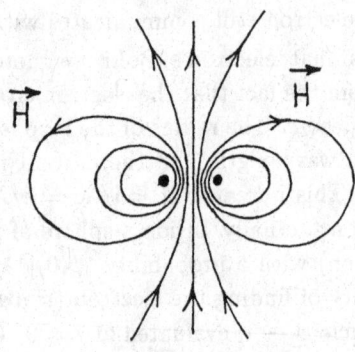

Fig. 18.25.

If **H** is strong, the energy-level spectrum of a free atom (or ion) is split (Zeeman splitting effect) as the atom is inside **H**. But the hyperfine interaction splits the electron levels further, providing an extra, additional splitting given by $\Delta E = a m_S m_I$, for strong **H**. The two possible transitions of the electron are given by the selection rules $\Delta m_I = 0$, $\Delta m_S = \pm 1$. The transition frequencies are $\omega_{1,2} = \gamma H_0 \pm (a/2\hbar)$.

The hyperfine interaction in an atom, say, hydrogen, may split the ground level. As the table above shows, the energy-level splitting in H is 1420 MHz, the famous 21-cm line of H, *i.e.*, the radio-frequency line of interstellar atomic H.

Figure 18.26 shows the energy-level diagram and the electron transitions. These are the energy states (levels) of a system with $S = I = 1/2$, lying in a field **H**. It is the case of the strong-field limit which is an approximation ($\mu_B H \gg a$ where $a > 0$). (Therefore the equality signs in Eq. (18.3) should be replaced by $\simeq$, the approximate equality.) The four levels are marked (labeled) by m_S and m_I. The selection rules given above belong to the strong electronic transitions. The figure does not show the nuclear transitions. These have $\Delta m_S = 0$ and ω(nuclear) $= a/2\hbar$, *i.e.*, $\omega(1 \to 2) = \omega(3 \to 4)$ for the nuclear transitions $1 \to 2$ and $3 \to 4$.

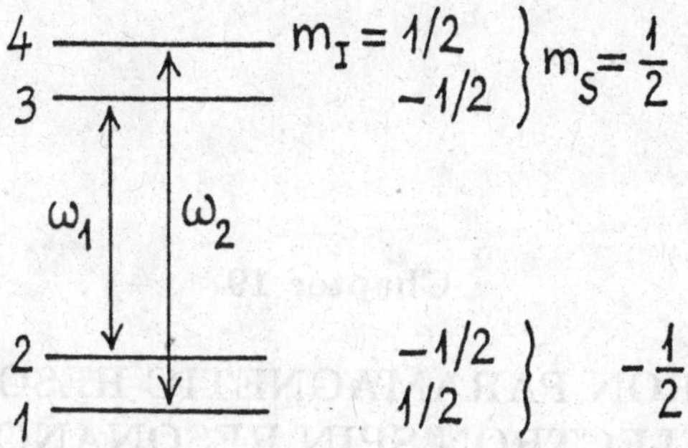

Fig. 18.26.

In the foregoing discussion notice that $\psi(0)$ is the electronic wavefunction at the nucleus. It has a meaning because we deal with quantum mechanics (where presence of electronic clouds is possible and probable anywhere).

In these discussions we kept the analysis simple, for instruction purposes, as it behooves a textbook. We avoided points and details that might be advanced enough to become subjects of doctorate dissertations or research papers and articles.

Chapter 19

ELECTRON PARAMAGNETIC RESONANCE
(ELECTRON SPIN RESONANCE)

19.1. The Method of ESR and Its Theory

We know that a nucleus with spin $\mathbf{I}$ has an angular momentum $\mathbf{P} = \hbar\mathbf{I}$ and a nuclear magnetic moment $\mu = \gamma\mathbf{P}$ where γ is the *gyromagnetic ratio* defined as $\gamma = e/2m_\mathrm{p}c$ where m_p is the mass of the proton. Further, the magnetic moment is $\mu = gI\mu_\mathrm{n}$ where g is the *Lande factor* and μ_n is the nuclear magneton given by $\mu_\mathrm{n} = eh/4\pi m_\mathrm{p}c = \hbar\gamma$.

Similarly, an orbiting atomic electron has a magnetic moment (ignoring spin) $\mu = \mu_\mathrm{B}\mathbf{L}$ where $\mathbf{L}$ is the orbital angular momentum and μ_B is the Bohr magneton given by $\mu_\mathrm{B} = eh/4\pi m_\mathrm{e}c$, with m_e being the mass of the electron.

Since the electrons have μ and spin, they exhibit a resonance similar to NMR. That is, since an electronic μ exists, the considerations about the nucleus (NMR) can be carried out for electrons, too. However, since $\mu_\mathrm{B} \gg \mu_\mathrm{n}$, the *electron spin resonance* (ESR) or *electron paramagnetic resonance* (EPR) falls into the region of microwaves (for 3–10 kGauss).

When a body is put in an external magnetic field $\mathbf{H}_0$, it exhibits a magnetization $\vec{\mathcal{M}}$ proportional to $\mathbf{H}_0$. The proportionality constant χ is the magnetic susceptibility, so that $\vec{\mathcal{M}} = \chi\mathbf{H}_0$.

The *paramagnetic magnetization* $\vec{\mathcal{M}}_\mathrm{par}$, which is the magnetization of the electrons, is

$$\vec{\mathcal{M}}_\mathrm{par} = \chi_\mathrm{par}\mathbf{H}_0$$

where χ_par is the *paramagnetic susceptibility*.

Langevin's theory explains paramagnetic properties by attributing a magnetic moment μ to every atom. If there are N atoms per cm^3, each having a magnetic moment μ, and if we assume a Boltzmann distribution, the Langevin equation for the magnetization reads

$$\vec{M} = N\mu \left[\coth\left(x - \frac{1}{x} \right) \right]$$

where $x \equiv \mu \cdot \mathbf{H}_0 / kT$. The numerator of x is the magnetic potential energy of a magnet, and the denominator is twice the amount of thermal energy per degree of freedom. The distribution is $\exp(-\Delta E/kT)$.

Here we did not consider the electronic quantum structure, but now we will. An electron has two angular momenta: An orbital angular momentum **l** that corresponds to its orbiting about the nucleus, and a spin **s** that is due to its revolution about its own axis just like the Earth (though this latter resemblance should not be taken too seriously, because **s** is a quantum-mechanical quantity and property associated with the particle and has no classical analogue). The two angular momenta combine (add up vectorially) to give $\mathbf{j} = \mathbf{l} + \mathbf{s}$, the total electronic[1] angular momentum. (And if there are many electrons, we have **J** instead of **j**.) The quantum theory of the Zeeman effect says that when an electron[2] is placed in an external field $\mathbf{H}_0$, it interacts with it, and the interaction energy is

$$E = \mu \cdot \mathbf{H}_0$$

where μ is the magnetic moment of the electron (or of the atom, if we disregard nuclear spin). (For a many-electron atom, μ is the total magnetic momentum of all its electrons — a vector sum of the individual moments of the electrons.)

The z-component of the total (atomic) magnetic moment is

$$\mu_z = g_L \mu_B m_J$$

where m_J is the magnetic quantum number (which takes $2J + 1$ values for a given J) and g_L is the Lande factor given by

$$g_L = \frac{3}{2} + \frac{S(S+1) - L(L+1)}{2J(J+1)}$$

where S is the spin quantum number of the vectorially total spin of the electrons, L is the angular momentum quantum number of the vectorially total orbital angular momentum of the electrons, and $\mathbf{J} = \mathbf{L} + \mathbf{S}$.

---

[1] Or atomic, if one disregards nuclear spin.
[2] Or an atom, if we disregard nuclear spin.

The magnetization is

$$\mathcal{M} = NJg_{\rm L}\mu_{\rm B}\left[\coth\left(x - \frac{1}{x}\right)\right]$$

where the product NJ can take the value $NJ = \pm1$. The energy is

$$E = g_{\rm L}\mu_{\rm B}H_0,$$

and the magnetic moment is $\mu = {\sf P}\mu_{\rm B}$ where ${\sf P}$ is the effective number of the Bohr magnetons, given by

$${\sf P} = g_{\rm L}\sqrt{J(J+1)}.$$

Obviously, $\mu = g_{\rm L}\mu_{\rm B}{\bf J}$. The Zeeman effect (due to the presence of ${\bf H}_0$) splits the energy into $2J + 1$ states.

19.2. Utility and Observation of ESR

The method of ESR is used to study some paramagnetic materials and to do research on the interaction between electrons and nuclear magnetic moments. It can also be used to study doped materials and measure impurities or free radicals in them.

The irradiation of plastics and polymers by X-rays or γ-rays, or their treatment with chemicals produces free radicals in them. These radicals (along with other foreign bodies) constitute the presence of impurities in the material, and they can cause some effects, *e.g.*, they reduce the relaxation time τ_1 by half if they are present at the ratio of 0.01%, as we saw in the previous chapter.

The concentration of these foreign traces is usually less than 10^{-6}, that is, $\Delta N/N < 10^{-6}$. By employing ESR we can measure the presence of the foreign material of the order of 10^{-3} μgr (*i.e.*, 10^{-9} gr) and we can even observe its structure! Certainly, we cannot do that by a qualitative analysis, and here is where the merit of ESR comes into play.

ESR is also a good method to calculate electron densities. Consider, for example, the HCl molecule. There is an electron exchange between the atoms (Fig. 19.1). We can measure the percentage of the excess or lack of electrons. As the two atoms share electrons, this percentage changes during the sharing and can be measured.

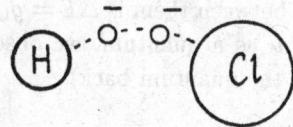

Fig. 19.1.

We do not observe ESR with every material. ESR is observed when the material has free electrons, and in the following cases:

(1) In cases where the (total) orbital quantum number is zero: $L = 0$.
(2) In metals (because they have free (conduction) electrons exhibiting ESR).
(3) In paramagnetic materials (where the total electronic magnetic moment is nonzero).
(4) In ferromagnetic and antiferromagnetic materials.
(5) In metals and semiconductors having electron bands that are not filled.
(6) In atoms with unfilled internal shells (*e.g.*, Fe, Co, Ni, where the orbits internal to the valence electrons are not full and they can still accept electrons). Also in ions with partially full electron shells (transition group, rare earths, actinides).
(7) In impure insulators containing impurities (foreign atoms) or atom vacancies. Also in materials having color centers and materials with defects after radiation or chemical reactions. (Here ESR is used to study color centers, biochemical effects, and polymerization.)
(8) In materials containing free chemical radicals and double radicals.
(9) In semiconductors with donor and acceptor levels (considered as impurities).
(10) In single molecules (O_2, NO, NO_2, ClO_2).

19.3. Theory vs. Experiment

ESR can be observed when $L = 0$. Then $\mathbf{J} = \mathbf{L} + \mathbf{S} = \mathbf{S}$, and theoretically, we find $g_L = 2$. The value found experimentally is $g_L = 2.00229$, fairly close to the theoretical value.

Figure 19.2 shows the two energy states (splitting) in $\mathbf{H}_0$, *i.e.*, as m_J takes the values $+1/2$ and $-1/2$ when $H_0 \neq 0$ is on. On the left side we see that there is one energy level for $H_0 = 0$ (when the field is off). At any $H_0 = H_0$ value, the split energy levels are equally separated from the initial single energy level, and the separation (difference) between them is $\Delta E = g_L \mu_B H_0$, *i.e.*, $\Delta E = f(H_0)$. If we supply this $\Delta E = h\nu$ as a quantum, we observe absorption and then emission (the system gives the quantum back).

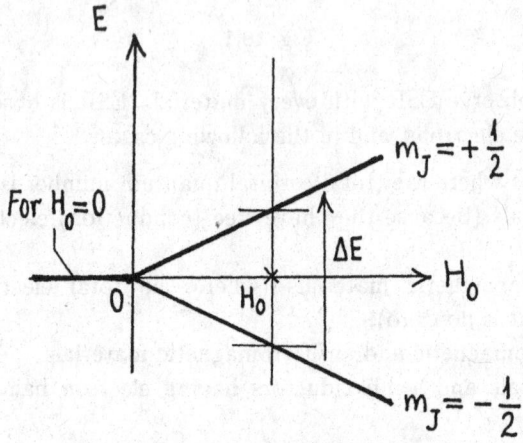

Fig. 19.2.

We need a strong H_0 field and a weak $H_1(\nu)$ field where $\mathbf{H}_0 \perp \mathbf{H}_1(\nu)$, that is, $\mathbf{H}_1$ is a weak field of high frequency (at frequency ν), perpendicular to $\mathbf{H}_0$. If $\nu \simeq \nu_L$ where ν_L is the Larmor frequency, then the nuclear magnetic moment μ precesses about $\mathbf{H}_0$ (Fig. 19.3).

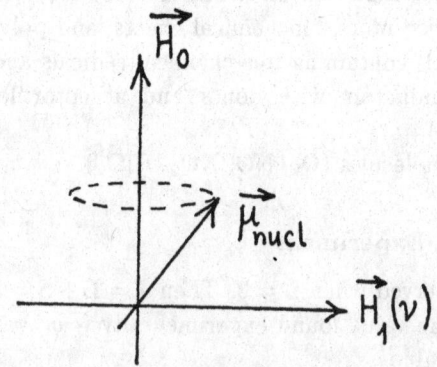

Fig. 19.3.

The unit of the magnetic moment for an electron is μ_B, the Bohr magneton, where $\mu_B = e\hbar/2m_e c$ where e is the charge of the electron. Since

$m_p \simeq 1836 \, m_e$, we have

$$\frac{\mu_B}{\mu_n} \simeq 1836$$

where μ_n is the nuclear magneton. Since the Bohr magneton is 1836 times larger than μ_n, the energy will be 1836 times as large, too, for the same value of g_L.

If we take $H_0 = 3600$ Gauss and $g_L = 2$, and put in the value of μ_B, we find for the resonance frequency the wave $\nu = 10,000$ MHz which is a high frequency, difficult to observe. It corresponds to $\lambda = 3$ cm of the electromagnetic spectrum, *i.e.*, to microwaves.

Let us now compare some theoretical and experimental results. When one works with rare earths (a special series in the periodic table with rare elements), for Ce, Gd, Dy, and Yb one finds a theoretical value of P close to its experimental counterpart. Rare earths are elements with incomplete $4f$-electrons and valence $5s$ and $5p$-electrons. The P values for some rare-earth ions are as follows:

Ion	P (experimental)	P (theoretical)
Ce^{+3}	2.4	2.54
Gd^{+3}	8	7.94

Here P is the effective number of Bohr magnetons, a coefficient given by $P = g_L \sqrt{J(J+1)}$, as we saw above. For the Fe group there is a difference: $P = 2\sqrt{S(S+1)}$. The reason is as follows: In the Fe group, vectors **L** and **S** are ferromagnetic. The internal local magnetic field is too high, and so it is added to H_0 as $H_0 + H_{int}$. Then $\mathbf{J} \neq \mathbf{L} + \mathbf{S}$, *i.e.*, the coupling of **L** with **S** breaks down. In **S** we observe the resonance corresponding to the spin angular momentum only. For some ions we have the following table:

Ion	P (theoretical)	P (experimental)	$2\sqrt{S(S+1)}$
Fe^{+3}, Mn^{+2}	5.92	5.90	5.92
V^{+3}	1.63	2.8	2.83
Fe^{+2}	6.70	5.4	4.9

19.4. Experimental Apparatus

Since the ESR frequency falls into the microwave region, we can observe it by using a microwave spectrometer.

For $H_0 = 3600$ Gauss we obtain $\nu = 10,000$ MHz. But if we make $H_0 = 3.6$ Gauss, we can still observe the phenomenon at $\nu = 10$ MHz, though with low sensitivity. Since the absorption is equal to $(1/2)\chi'' H_0^2$, *i.e.*, proportional to H_0^2, in going from 3600 to 3.6 Gauss (and consequently, from 10,000 to 10 MHz), the sensitivity becomes 10^{-6} times as great.

By using large electromagnets and high-technology superconductor electromagnets, we can achieve very strong[3] magnetic fields, as strong as 20,000 Gauss and up, in the laboratory. This way the frequency increases, but the disadvantage is that it is difficult to build electronic circuits operating at such high frequencies. The most convenient frequency to choose is $\nu = 10,000$ MHz.

The experiment was first performed by the Russian scientist Zavoisky in 1945.

In such a spectrometer we use a clystron oscillator as our source of electromagnetic waves. Since the produced electromagnetic waves are microwaves and thus exhibit properties of light, in order to prevent reflection and absorption, we use waveguide pipes. The wave resonator is designed so as not to disturb the magnetic field.

The set-up is shown in Fig. 19.4. It is the so-called "hybrid ring" arrangement. Electromagnetic waves travel from the clystron down pipe 1, and after getting branched around the ring, they arrive with the same phase to the cavity (wave-resonator box) where the material is placed, the absorption of which is to be measured. Pipes 1 and 2 are aligned on a straight line. The cavity is between the poles of the electromagnet. The ring has a radial extension on the left, at the end of which there is a balancing calibrator. Another extension, on the right, leads to the detector. The "matched load" principle is used here, where the load is balanced and the microwave system is set in equilibrium (just as it happens in the Wheatstone bridge). The waves arrive cophasic.

Figure 19.5 shows the detection mechanism. The electromagnetic waves strike the detector diode, and a current passes which is fed to the amplifier. The signal potential V_s leaves the amplifer and enters the phase detector, in

---

[3]In physics, when we say a "strong magnetic field", we compare it to that of the Earth. There are no absolutes in characterizing sizes; we always relate the characterization to something else, taken as a standard criterion. A "large chair" is larger than a chair of ordinary size in everyday experience. The characterization "tall man" can take different meanings, depending on whether we are talking about a basketball team or a group of Pygmies.

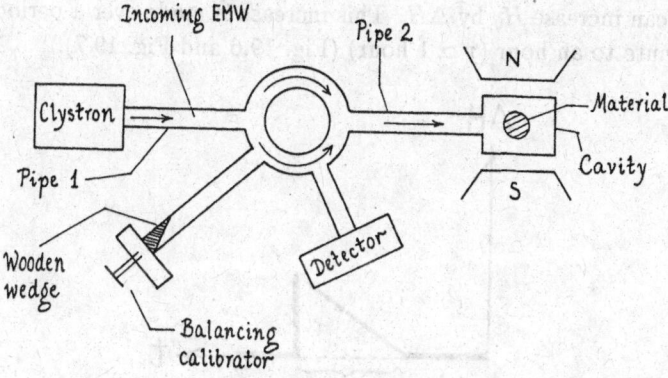

Fig. 19.4.

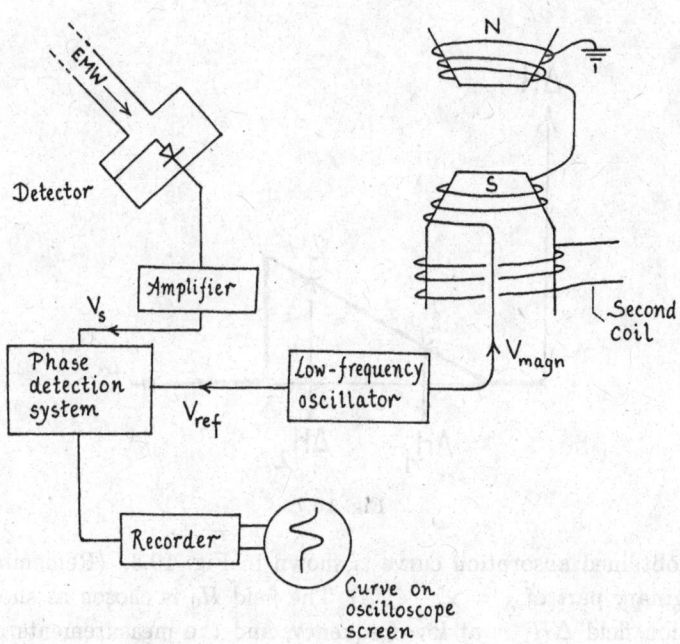

Fig. 19.5.

order to be compared with the incoming reference potential V_{ref}. The recorder gives a curve similar to the absorption curve. V_{magn} is the modulation potential given to the magnet. The south pole has a second coil wound around it, by which we can increase H_0 by ΔH. This increase is made over a period of time of one minute to an hour ($\tau \simeq 1$ hour) (Fig. 19.6 and Fig. 19.7).

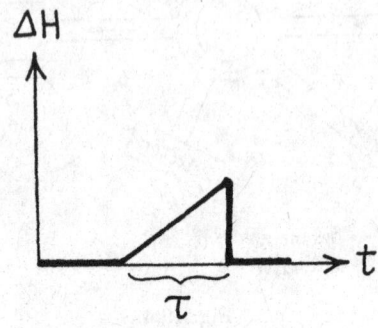

Fig. 19.6.

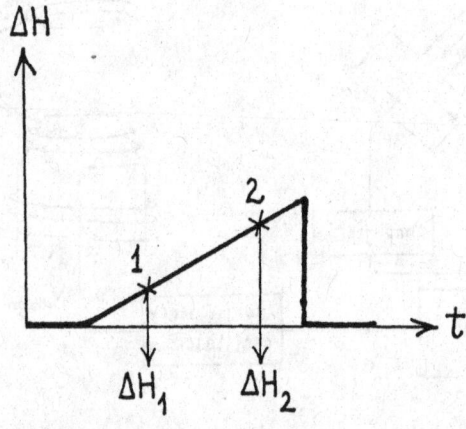

Fig. 19.7.

The obtained absorption curve is shown in Fig. 19.8. (Remember, χ'' is the imaginary part of $\chi = \chi' - i\chi''$.) The field H_0 is chosen as shown. The modulation field ΔH_1 is at low frequency, and the measurements are taken at point A. At point B (curve maximum) the absorption signal is zero. If we take the change of the absorption curve with respect to H, *i.e.*, the derivative $d\chi''/dH$, we obtain the curve in Fig. 19.9, which is the shape obtained on the screen of the recorder.

If a very wide magnetic-field sweep is done, miscellaneous fine structures are observed (due to interactions). However, it is very difficult to analyze

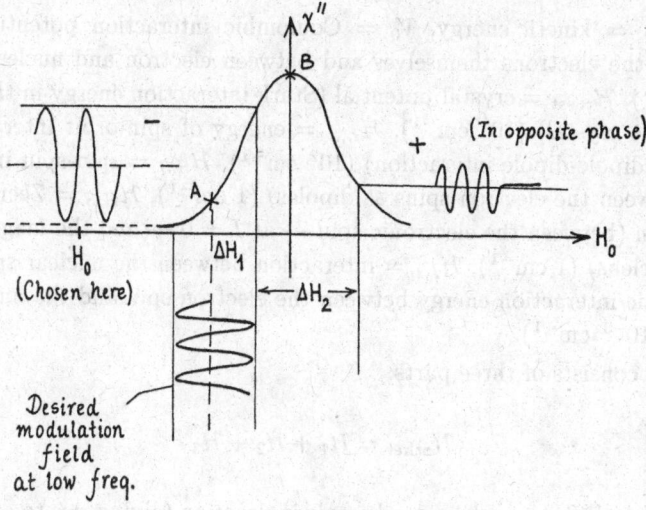

Fig. 19.8.

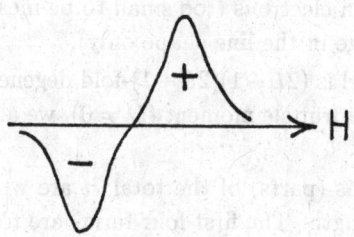

Fig. 19.9.

and understand the spectrum obtained via ESR. It appears as intricate and challenging as an elaborate solo movement of classical ballet would to a novice ballerina. For this reason, one proceeds theoretically, analyzing the spectrum in steps. The treatment is discussed below, in the following section.

19.5. Theoretical Analysis

To obtain information from the ESR spectrum, one inspects the fine structure of the spectral line and one studies this structure quantum mechanically.

The general expression for the Hamiltonian that gives the total energy of the electron, including all the interactions, is

$$\mathcal{H} = \mathcal{H}_0 + V_0 + \mathcal{H}_{\text{cryst}} + \mathcal{H}_{L,S} + \mathcal{H}_{S,S} + \mathcal{H}_{H,S} + \mathcal{H}_{I,I} + \mathcal{H}_{I,S} + \mathcal{H}_{\text{other}}$$

where $\mathcal{H}_0$ = kinetic energy, V_0 = Coulombic interaction potential energy (between the electrons themselves and between electron and nuclear charge) (10^5 cm^{-1}), $\mathcal{H}_{\text{cryst}}$ = crystal potential (Stark interaction energy in the electric field of the crystal) (10^4 cm^{-1}), $\mathcal{H}_{L,S}$ = energy of spin-orbit interaction (or magnetic dipole-dipole interaction) (10^2 cm^{-1}), $\mathcal{H}_{S,S}$ = spin-spin interaction part (between the electron spins as dipoles) (1 cm^{-1}), $\mathcal{H}_{H,S}$ = Zeeman-effect interaction (between the electronic spin — at $L = 0$ — and the magnetic field at the nucleus) (1 cm^{-1}), $\mathcal{H}_{I,I}$ = interaction between the nuclear spins, $\mathcal{H}_{I,S}$ = magnetic interaction energy between the electron spin and the nuclear spin (10^{-3} to 10^{-1} cm^{-1}).

$\mathcal{H}_{\text{other}}$ consists of three parts:

$$\mathcal{H}_{\text{other}} = \mathcal{H}_1 + \mathcal{H}_2 + \mathcal{H}_3$$

where $\mathcal{H}_1$ = electron-nucleus quadrupole interaction (giving rise to quadrupole fine structure; its energy varies), $\mathcal{H}_2$ = interaction between the nuclear magnetic moment and an external field (NMR) (10^{-4} cm^{-1}), and $\mathcal{H}_3$ = mutual exchange energy between electrons (too small to be measured; we can observe the corresponding change in the line shape only).

The ESR energy level is $(2L+1)(2S+1)$-fold degenerate. Further, if $I \geq 1$ and there is nonzero quadrupole moment ($Q \neq 0$), we also observe quadrupole fine structure.

Notice that the terms (parts) of the total $\mathcal{H}$ are written hierarchically, in decreasing order of strength. The first four terms are too large to be observed with ESR.

If one considers the energies after V_0, the largest energy is $\mathcal{H}_{\text{cryst}}$, having a coupling coefficient of the order of 10,000 cm^{-1}. $\mathcal{H}_{S,S}$ and $\mathcal{H}_{H,S}$ are both of the order of 1 cm^{-1} each.

The potential energy corresponding to the crystal Hamiltonian is $E_{\text{cryst}} = -eV_{\text{cryst}} = -eV(x, y, z)$ where the crystal potential is a function of the coordinates of the ions in the crystal, with which the electron interacts (Fig. 19.10). In terms of spherical harmonics, we have

$$V(x, y, z) = \sum_{n,m} \mathcal{A}_n^m r^n Y_n^m(\theta, \varphi)$$

where Y_n^m are the spherical harmonics, and r is the distance of the polar coordinate (magnitude of radius vector).

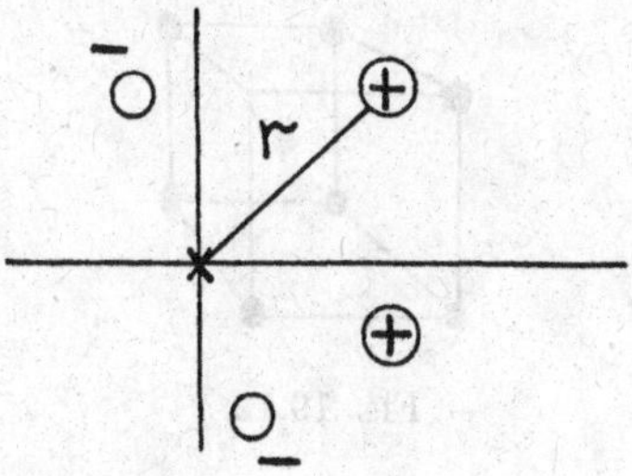

Fig. 19.10.

If there is no spherical symmetry, we write V in Cartesian coordinates[4]:

$$V = V_0 + \mathcal{A}x + \mathcal{B}y^2 + \mathcal{C}z^2 + \mathcal{D}(\text{Fourth order}).$$

If there is rotational symmetry about an axis (Fig. 19.11), *i.e.*, axial symmetry, then we have

$$V = \mathcal{A}(x^2 + y^2 - 2z^2) + \mathcal{D}(\text{Fourth order})$$

which means that the condition for axial symmetry is $\mathcal{A} = \mathcal{B} = -(1/2)\mathcal{C}$.

Cubic symmetry means $\mathcal{A} = \mathcal{B} = 0$, *i.e.*,

$$V = \mathcal{D}(x^4 + y^4 + z^4)$$

which is the potential for ions located at the corners and at the center of a cube (Fig. 19.12).

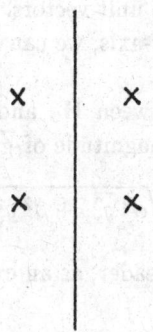

Fig. 19.11.

---

[4]Although attributed to the French philosopher Descartes, planar coordinates were in fact invented by a fifteen-century bishop, Nicholas of Oresme.

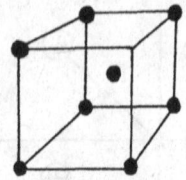

Fig. 19.12.

If there is a full spherical symmetry, we have $V = V_0 + 0 = V_0$, as though the potential were all concentrated at the center (origin).

Next we have $\mathcal{H}_{L,S} = \lambda \mathbf{L} \cdot \mathbf{S}$ where λ is the strength of the interaction (coupling constant). Here $\mathbf{L}$ couples (interacts) with $\mathbf{S}$, and the correspondig quantum-mechanical operators are used.

Next comes $\mathcal{H}_{S,S} = -\rho[(\mathbf{L} \cdot \mathbf{S})^2 + \frac{1}{2}\mathbf{L} \cdot \mathbf{S} - \frac{1}{3}L(L+1)S(S+1)]$ where ρ is the coupling coefficient for this interaction. This is an operational equation where $\mathbf{L}$ and $\mathbf{S}$ are *operators*, while L and S are *scalars*.

Next comes $\mathcal{H}_{\text{Zeeman}} = \mu_B(\mathbf{L} + 2\mathbf{S}) \cdot \mathbf{H}_0$. For $L = 0$ and $S = 1/2$ (and $\mathbf{H}_0$ along the z-direction), the ESR transition energy corresponding to $\mathcal{H}_{\text{Zeeman}}$ is

$$E = \mu_B H_0 g_{zz}$$

where g_{zz} is the matrix element of the quantum-mechanical tensor operator $\overleftrightarrow{\mathbf{g}}$ given by the sum of three dyadics:

$$\overleftrightarrow{\mathbf{g}} = \hat{\mathbf{x}}\, g_{xx}\, \hat{\mathbf{x}} + \hat{\mathbf{y}}\, g_{yy}\, \hat{\mathbf{y}} + \hat{\mathbf{z}}\, g_{zz}\, \hat{\mathbf{z}}$$

where $\hat{\mathbf{x}}$, $\hat{\mathbf{y}}$, $\hat{\mathbf{z}}$ are the Cartesian unit vectors.

If we choose $\mathbf{H}_0$ along the z-axis, we can take $g = g_{zz}$ (because the others are zero).

If the direction cosines between $\mathbf{H}_0$ and the Cartesian axes x, y, z are α, β, γ, then we have for the magnitude of $\overleftrightarrow{\mathbf{g}}$:

$$|\overleftrightarrow{\mathbf{g}}| = \sqrt{\alpha^2\, g_{xx}^2 + \beta^2\, g_{yy}^2 + \gamma^2\, g_{zz}^2}$$

which is to be verified by the reader, as an exercise in mathematical physics.

19.6. Applications of ESR

As mentioned above, ESR can be observed in elements with a single valence electron, in materials with unpaired electrons, in rare earths, in crystal lattices, in semiconductors (with donor and acceptor levels), in radicals, and in atoms having orbits with missing electrons. In crystals of simple structure the phenomenon can be explained pretty easily.

ESR can be used to study amorphous materials as well, but the phenomenon in such materials has not been explained in full yet.

The presence of foreign atoms in a material, even if it is of the order of 10^{-8}, can change the properties of the material, giving rise to absorption. For example, Si is an insulator if it is pure. But if it contains foreign atoms in its bulk, of the order of 10^{-8}, it becomes a semiconductor. In other words, foreign atoms (impurities) cause observable absorption which is a new property for the material. For alkali halides, these absorptions are called *optical-absorption bands* (Fig. 19.13). These bands are wide.

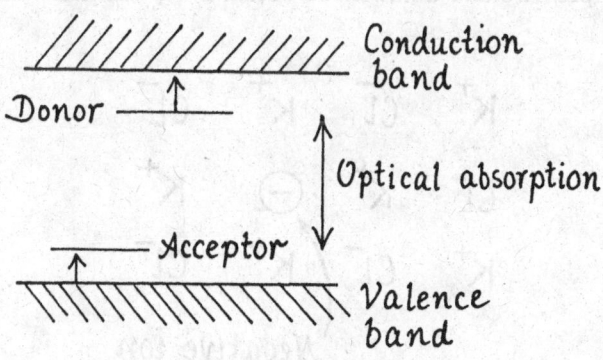

Fig. 19.13.

The points where a material contains impurities are called *color centers* or *chromocenters*.[5]

In insulators, besides the ionic-bond crystal structures that carry impurity atoms, there may also be pointlike crystal defects which similarly cause the appearance of optical-absorption energy bands. Thus the two most basic kinds of color centers are the following:

(1) Doping (foreign body) atoms.
(2) Pointlike crystal defects.

The examination of the color centers provides information about the chemical, physical, and electronic properties of the material, and about its quantum mechanics (perturbation terms).

---

[5] From the Greek χρωμόκεντρον = color center.

Out of the color centers the most important ones are the F-type[6] color centers. The second kind is the V-type[7] color centers. The third kind is the U-type[8] color centers. There are also M-type and R-type color centers, as well as other ones whose structure has not been much understood.

Let us give some examples about the structure of the color centers: Consider the KCl crystal which is ionic (Fig. 19.14). Color centers can be easily seen in ionic crystals. If a Cl atom abandons its place for some reason, a negative ion (anion) vacancy is created there. Now if an electron is captured by this ion vacancy, an F-type center is formed there: $F = \ominus + e^-$. The new situation is shown in Fig. 19.15.

We can have KCl + K (vapor) → F-type. That is, we send K vapor to KCl, and through pressure and diffusion we obtain F-type centers. How? As the

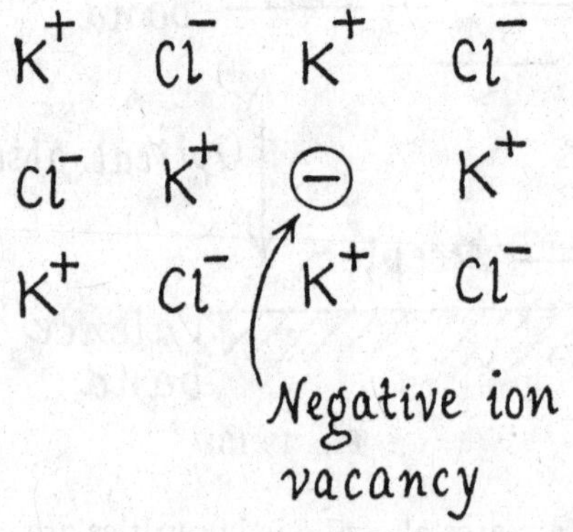

$$K^+ \quad Cl^- \quad K^+ \quad Cl^-$$
$$Cl^- \quad K^+ \quad \ominus \quad K^+$$
$$K^+ \quad Cl^- \quad K^+ \quad Cl^-$$

Negative ion
vacancy

Fig. 19.14.

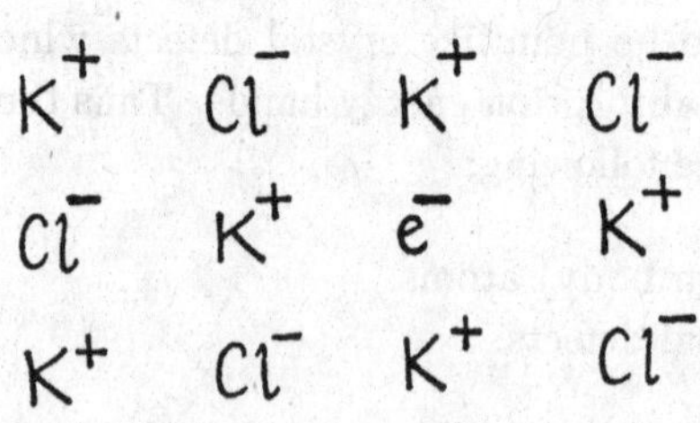

$$K^+ \quad Cl^- \quad K^+ \quad Cl^-$$
$$Cl^- \quad K^+ \quad e^- \quad K^+$$
$$K^+ \quad Cl^- \quad K^+ \quad Cl^-$$

Fig. 19.15.

---

[6] F stands for *Farbe* (= color, in German).
[7] V stands for *vacancy* (hole).
[8] U stands for *united*.

atoms (ions) in the crystal vibrate, a Cl^- ion gets displaced and approaches a K atom which is initially neutral. The Cl^- ion unites with the K atom, and the valence electron of the K atom is released free (for the K atom to become K^+ and combine, it should give off an electron). Now if this released electron takes the place of the Cl^- ion (*i.e.*, gets caught by the vacancy), an F-center is readily formed.

The atomic vibrations help in the capturing of the electron. The vacancies can capture also free electrons that wander through the crystal after having gotten free via absorption of thermal energy (thermal-noise electrons). Radiation (high-energy γ-rays) can also extract electrons and set them free inside the crystal. If these electrons are caught by vacancies, color centers are formed, whose moral author is the radiation this time.

If a K^+ ion abandons its place in the crystal lattice, it creates a positive vacancy shown in Fig. 19.16. If such ions position themselves interstitially throughout the KCl lattice in a mixed manner, *interstitial defects* are formed in the crystal. Such a defect is shown in Fig. 19.16.

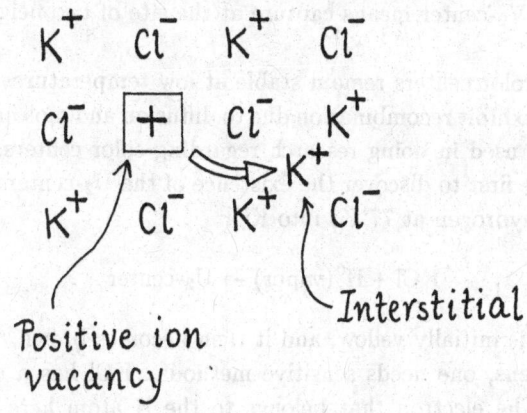

Fig. 19.16.

If an ionic crystal has equal numbers of positive ion vacancies and interstitials, we say that it has Frenkel type defects. (Silver halides have such defects, with interstitials in the vicinity of the ions.)

When an electron is caught by a positive ion vacancy, a V_1-type color center is formed: $V_1 = \boxplus + e^-$. But the situation is not stable. This is shown in Fig. 19.17(a) and (b). If next to the V_1-center two negative ions (*e.g.*, chlorines) lose their electron (say, after exposure to radiation), the captured electron can serve as a binding between the two chlorines, to form a U_2-type color center (Fig. 19.17(c) and (d)). This electron is used and shared by the two chlorines in common, and thus it smears to form a cloud of negative charge between them, jumping from one chlorine to the other.

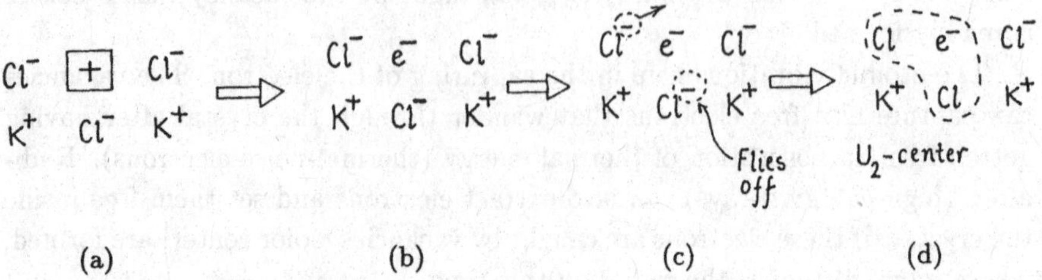

(a)   (b)   (c)   (d)

Fig. 19.17.

The vacancy color centers are V_1, V_2, V_3-type. The united color centers are U_1, U_2, U_3-type. The subscripts indicate the number of united ion sites. For example, a V_2-center means capture at the site of two neighboring atoms (ions).

In general, color centers remain stable at low temperatures. At high temperatures they exhibit recombination due to diffusion and molecular vibrations.

ESR can be used in doing research regarding color centers. For example, Spaeth[9] was the first to discover the existence of the U_2-centers, by letting H^0 vapor (atomic hydrogen at 77°K) into KCl:

$$KCl + H^0(\text{vapor}) \rightarrow U_2\text{-center}.$$

The crystal is initially yellow, and it then becomes brown. To understand how that happens, one needs sensitive methods. KCl has a cubic structure (Fig. 19.18). The electron that belongs to the H atom here does not form bonds with the atoms of the cube; it just stays bound to H. The H^0 atom is caught right at the center of the cube, thereby producing a U_2-center. ESR can be observed for this situation. The result is depicted in Fig. 19.19. Two absorption spectra are observed: The maximum on the left is the $h\nu_2$ line and occurs at 3.06 kGauss; the maximum on the right is the $h\nu_1$ line and occurs at 3.56 kGauss. This spectrum is explained as follows: When an unpaired electron combines with a proton, two fine structures are observed. The splitting into two is due to the interaction between the electron and the proton. In other words,

[9]J.M. Spaeth, *Journal de Physique*, 4, 147 (1967).

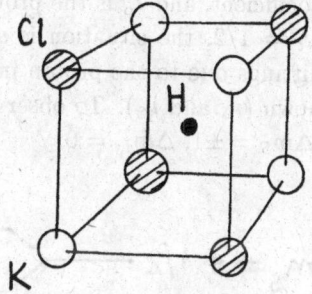

Fig. 19.18.

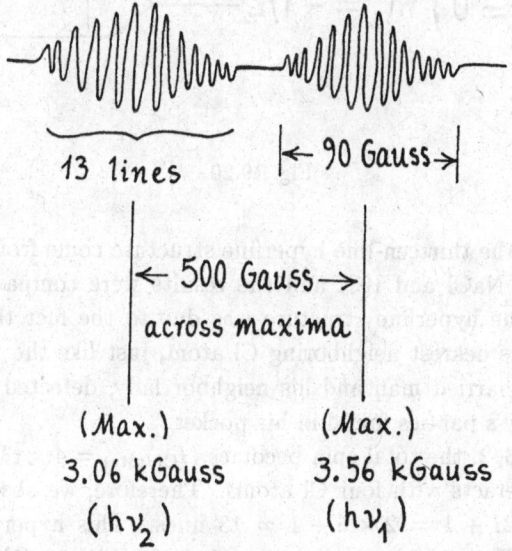

13 lines |← 90 Gauss →|

|← 500 Gauss →|
across maxima

(Max.)   (Max.)
3.06 kGauss   3.56 kGauss
($h\nu_2$)   ($h\nu_1$)

Fig. 19.19.

an unpaired electron interacts with a proton and gives a fine structure which is split into two due to the electron-proton interaction.

The magnetic-resonance interaction between the electron spin and the proton spin is

$$\mathcal{H}_{S,I_p} = A\mathbf{I}_p \cdot \mathbf{S}$$

where A is the interaction strength. The energy is

$$E_{S,I_p} = am_{I_p}m_S$$

where a is the coupling coefficient, and I_p is the proton spin.

Since $I_p = 1/2$ and $S = 1/2$, the situation is shown in Fig. 19.20. On the right we have the splittings due to the proton interaction. There are two allowed transitions, as shown (ν_1 and ν_2). To observe the fine structure with ESR, the conditions are $\Delta m_S = \pm 1$, $\Delta m_{I_p} = 0$.

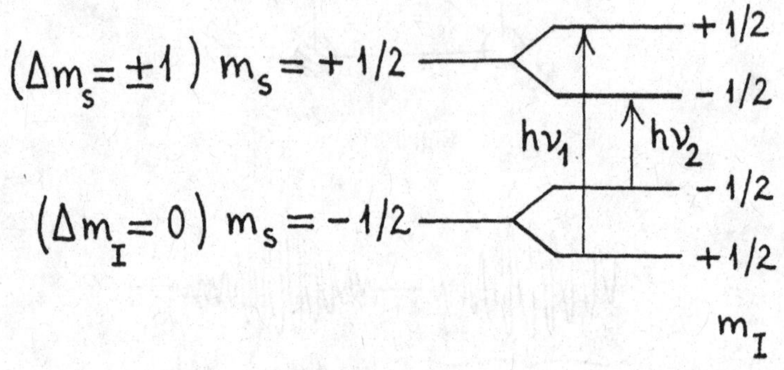

$$\left(\Delta m_s = \pm 1\right) \; m_s = +1/2 \begin{array}{l} +1/2 \\ -1/2 \end{array}$$

$$h\nu_1 \quad h\nu_2$$

$$\left(\Delta m_I = 0\right) \; m_s = -1/2 \begin{array}{l} -1/2 \\ +1/2 \end{array}$$

$$m_I$$

Fig. 19.20.

Where does the thirteen-line hyperfine structure come from? Experiments were done with NaCl and KCl and the results were compared. It was understood that the hyperfine structure was due to the fact that the electron interacts with its nearest neighboring Cl atom, just like the "secret" love affair between a married man and his neighbor lady, detected and manifested through the lady's panties found in his pocket.

Since $I_{Cl} = 3/2$, the total spin becomes $(I_{Cl})_{total} = 4 \times (3/2) = 6$ because the electron interacts with four Cl atoms. Therefore, we observe a hyperfine structure with $2I + 1 = 2 \times 6 + 1 = 13$ lines. This hyperfine structure is observed when $\mathbf{H}_0$ is along the direction $(1, 0, 0)$ (Fig. 19.21).

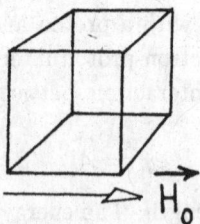

$$\vec{H}_0$$

Fig. 19.21.

For the U_2-center, the total Hamiltonian[10] is

$$\mathcal{H} = g_e \mu_B \mathbf{H}_0 \cdot \mathbf{S} - g_p \mu_n \mathbf{I}_p \cdot \mathbf{H}_0 + \mathcal{A}_p \mathbf{I}_p \cdot \mathbf{S}$$

$$+ \sum_i \left(\mathbf{I}_i \overset{\leftrightarrow}{\mathbf{A}}_i \mathbf{S} + \mathbf{I}_i \overset{\leftrightarrow}{\mathbf{Q}}_i \mathbf{I}_i - g_i \mu_n \mathbf{H}_0 \cdot \mathbf{I}_i \right)$$

where the first term is just the Zeeman interaction, the second term is the Zeeman energy of the nuclear spin, and the last term of the summand is the magnetic resonance of the Cl nuclei. The first two terms of the summand are tensor dyadics. All the vectorial and tensorial quantities here are operators. By using operators, we apply the perturbation theory and find the energy. There are four terms for the four Cl atoms that interact: Index i runs from 1 to 4, to include the nuclear spin of these four atoms. $\overset{\leftrightarrow}{\mathbf{A}}_i$ is the interaction tensor, and $\overset{\leftrightarrow}{\mathbf{Q}}_i$ is the quadrupole-moment tensor (which has one single component, in the direction of the spin).

This way one can determine the U_2-centers and their electron densities. However, ESR is not a sensitive method, nor can it be used to observe quadrupole resonances. A more sensitive method is ENDOR (see next chapter).

ESR was used to discover the structure of a V_K-type color center.[11] To produce and observe V_K-centers, we bombard KCl with X-rays, and explain the situation as follows (Fig. 19.22): A chlorine ion loses its electron (which goes away after exposure to radiation). A neighboring chlorine ion shares its electron with the chlorine that has lost its own electron. The two chlorines form a half-charge bond between them, with a single electron which shuttles between them to produce a cloud of an s-electron (smeared charge).

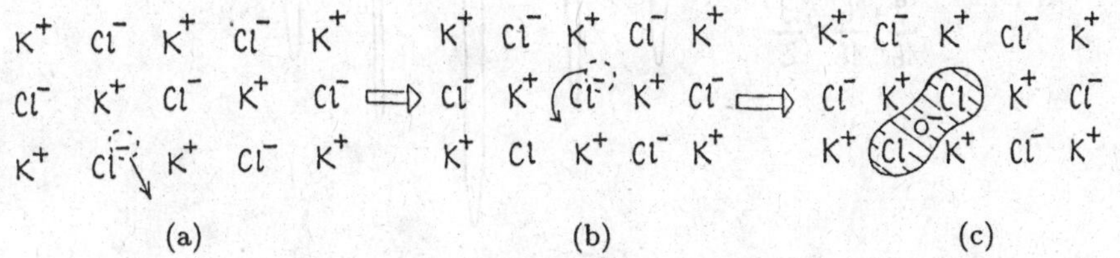

(a)　　　　　　　(b)　　　　　　　(c)

Fig. 19.22.

---

[10]In writing down the ESR perturbation, we should include the proton magnetic resonance and the proton-spin interaction.

[11]T.G. Castner and W. Känzig, *Journal Phys. Chem. of Solids*, **3**, 178 (1957).

That is,

$$KCl + X\text{-rays} \rightarrow Cl_2^-,$$

or

$$Cl^- + Cl^- \rightarrow Cl^- + Cl \rightarrow Cl_2^- + Free\ e^-.$$

After the departure of the free electron, we are left with Cl_2^- whose spectrum we obtain. (It is a singly-ionized chlorine molecule.)

The experiment must be done at low temperatures (*e.g.*, liquid-nitrogen temperature) because at high temperatures the bonds break down.

The obtained spectrum is shown in Fig. 19.23. The chlorine atoms are the isotopes ^{35}Cl and ^{37}Cl. This is the situation when H_0 is along the direction $(1, 1, 0)$. The V_K-center, as we interpret the spectrum, is formed between two neighboring chlorine atoms in the direction of a plane, not in the direction of the diagonal.

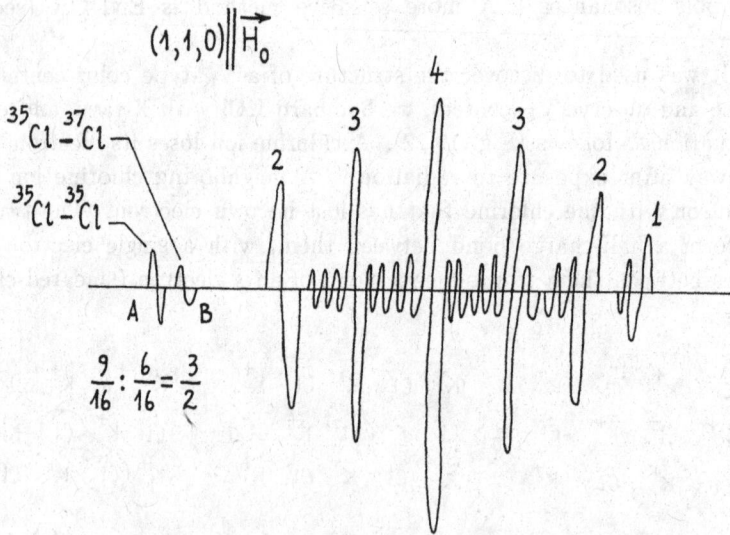

ESR spectrum of Cl_2^- ion within KCl
irradiated by X-rays

Fig. 19.23.

The total Hamiltonian is

$$\mathcal{H} = (\text{Zeeman effect for the electron spin}) + (\text{Spin-spin interaction})$$

$$= \mathcal{H}_{\text{Zeeman}} + \mathcal{H}_{I,I},$$

giving the energy

$$\hbar\omega = g_{zz}\mu_n H_0 + A_z(m_{I_1} + m_{I_2}).$$

We have $I_{\text{Cl}} = 3/2$ and $m_{I_{1,2}} = +3/2, +1/2, -1/2, -3/2$.

Let us tabulate the perturbation energy amplitude for every value that m_1 and m_2 can take:

(m_1, m_2)	$A(m_1 + m_2)$	Statistical weight
$\left(\dfrac{3}{2}, \dfrac{3}{2}\right)$	$\left(\dfrac{3}{2} + \dfrac{3}{2}\right)A = 3A$	1
$\left(\dfrac{3}{2}, \dfrac{1}{2}\right), \left(\dfrac{1}{2}, \dfrac{3}{2}\right)$	$2A$	2
$\left(\dfrac{3}{2}, -\dfrac{1}{2}\right), \left(\dfrac{1}{2}, \dfrac{1}{2}\right), \left(-\dfrac{1}{2}, \dfrac{3}{2}\right)$	A	3
$\left(\dfrac{3}{2}, -\dfrac{3}{2}\right), \left(\dfrac{1}{2}, -\dfrac{1}{2}\right), \left(-\dfrac{3}{2}, \dfrac{3}{2}\right), \left(-\dfrac{1}{2}, \dfrac{1}{2}\right)$	0	4
$\left(-\dfrac{3}{2}, -\dfrac{1}{2}\right), \left(-\dfrac{1}{2}, -\dfrac{1}{2}\right), \left(\dfrac{1}{2}, -\dfrac{3}{2}\right)$	$-A$	3
$\left(-\dfrac{3}{2}, -\dfrac{1}{2}\right), \left(-\dfrac{1}{2}, -\dfrac{3}{2}\right)$	$-2A$	2
$\left(-\dfrac{3}{2}, -\dfrac{3}{2}\right)$	$-3A$	1

Notice that for $2A$, we can have $(3/2, 1/2)$ or $(1/2, 3/2)$. If the number of the atoms that make absorption in state $(3/2, 3/2)$ is equal to the number of the atoms in states $(3/2, +1/2)$ and $(1/2, 3/2)$, then the number of the atoms in the last two states together is twice the number of the atoms in the first state, *i.e.*, the statistical weight is 2.

The small lines in the spectrum are due to the property of the Cl atom to have isotopes: ^{35}Cl and ^{37}Cl. We have

$$\frac{a(^{35}Cl)}{a(^{37}Cl)} = \frac{1}{0.83} \simeq \frac{1}{3/4}$$

where a is the abundance of the isotope. Either isotope of the atom can take part in the formation of a V_K-center. So, we have four possible combinations: (35, 35), (37, 37), (37, 35), and (35, 37). Since the two isotopes are found in nature at the ratio of 3/4 (*i.e.*, in N mol gram chlorine there are $(3/4)N$ atoms of one isotope and $(1/4)N$ atoms of the other isotope), the probability — or ratio — of finding a ^{35}Cl–^{35}Cl pair is

$$\frac{3}{4} \times \frac{3}{4} = \frac{9}{16} .$$

For ^{35}Cl–^{37}Cl:

$$\frac{3}{4} \times \frac{1}{4} = \frac{3}{16}$$

For ^{37}Cl–^{35}Cl: $\left.\right\} \frac{6}{16}$

$$\frac{1}{4} \times \frac{3}{4} = \frac{3}{16}$$

For ^{37}Cl–^{37}Cl:

$$\frac{1}{4} \times \frac{1}{4} = \frac{1}{16} .$$

Notice that in the spectrum (Fig. 19.23) the ratio of the size (intensity) of line A to that of line B is

$$\frac{9}{16} : \frac{6}{16} = \frac{3}{2}$$

because line A comes from the ^{35}Cl–^{35}Cl pair, and line B corresponds to the ^{35}Cl–^{37}Cl pair.

The line that corresponds to the ^{37}Cl–^{37}Cl pair $(1/4 \times 1/4 = 1/16)$ cannot be observed and is absent from the spectrum, because the signal/noise ratio is too small. However, do not be disappointed: We can observe it by employing ENDOR (see next chapter).

Since the coupling coefficients of ^{35}Cl and ^{37}Cl are different, different isotopes in combination give different resonances. Two separate lines mean two different coupling coefficients. Therefore, the energy must be written as follows:

$$\hbar\omega = g_{zz}\mu_n H_0 + \mathcal{A}_z^{35}m_1 + \mathcal{A}_z^{37}m_2$$
$$\mathcal{A}_z^{35} \cdot \frac{3}{2} + \mathcal{A}_z^{37} \cdot \frac{1}{2}$$
$$\mathcal{A}_z^{37} \cdot \frac{3}{2} + \mathcal{A}_z^{35} \cdot \frac{1}{2}.$$

To summarize in one sentence, ESR is a method that constitutes a good tool to probe into crystal imperfections. Its research field is wide.

19.7. EPR Effects

Consider a paramagnetic material at high enough temperature (disordered spins). Let the exchange interaction between two neighbor electron spins be of strength $\mathcal{J}$. The width of the spin resonance line is narrower — in general — than that for the interaction between dipoles. This is called *exchange narrowing* and resembles motional narrowing. The exchange (or flipping) frequency is

$$\frac{1}{\tau} = \omega_{exch} \simeq \frac{\mathcal{J}}{\hbar}.$$

The width of the exchange-narrowed line is

$$\Delta\omega \simeq \frac{(\Delta\omega)_0^2}{\omega_{exch}}$$

where $(\Delta\omega)_0^2 = \gamma^2 \overline{H_i^2}$ (with H_i being the perturbation field). $(\Delta\omega)_0$ is the static dipolar width when there is no exchange. A sharp exchange narrowing occurs in diphenyl-picryl-hydrazyl (a paramagnetic organic crystal). The half width of its resonance line is only 1.35 Gauss ($\sim$2% of the solely dipole width). This sharpness of the line makes this free radical a useful calibrator of magnetic fields.[12]

Some paramagnetic ions have crystal field splittings of their magnetic ground level of the order of $\sim 10^{10}$ Hz, a convenient frequency for microwave probing methods. The phenomenon is called *zero-field splitting*, studied extensively by Bleaney at Oxford University. The Mn^{++} ion (as an added impurity in crystals) has been studied where the ground-state splitting is roughly $\sim 10^8$ Hz.

19.8. Minimum Field

Refer to Fig. 19.24: How large must $\mathbf{H}_0$ be for the splitting to set on? Obviously, the splitting occurs after the application of $\mathbf{H}_0$, but does not start

---

[12]A.N. Holden, *Phys. Rev.*, **77**, 147 (1950).

abruptly; it takes place after some minimum value H_{min} (Fig. 19.24) which is determined by the natural width Γ of the split states. The lines that represent the states in Fig. 19.24 have some thickness Γ (due to the uncertainty principle) which determines H_{min}.

In Fig. 19.24 the thickness of the lines is exaggerated for clarity.

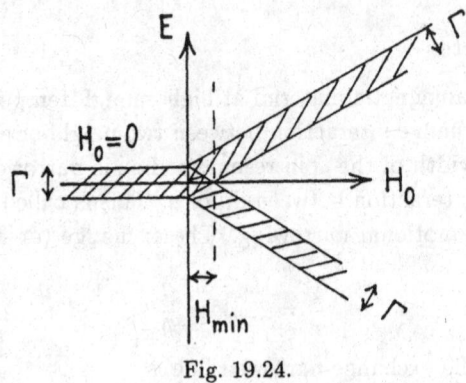

Fig. 19.24.

19.9. Ferromagnetic Resonance

It is a spin resonance at microwave frequencies, observed in ferromagnetic materials.[13] The total magnetic moment of the sample precesses about the static $\mathbf{H_0}$. The sample absorbs energy from the perpendicular radio-frequency field when its frequency is at resonance with the precession frequency. The total spin of the material is a vector (macroscopically) which is quantized in $\mathbf{H_0}$ (Zeeman effect), and the energy levels are split. Transitions only between adjacent levels are possible, because the selection rule is $\Delta m_S = \pm 1$.

We observe the following facts:

(1) The real and imaginary parts χ' and χ'' of the transverse susceptibility are large, because for a given $\mathbf{H_0}$, a ferromagnet is magnetized more than a paramagnet.

(2) Since $\vec{\mathcal{M}}$ is strong, a large demagnetizing field is needed.

(3) There is a pronounced coupling between the ferromagnetic electrons that masks the dipole contribution to the line width. Therefore, the ferromagnetic resonance line is narrow and sharp (less than 1 Gauss), provided that the other factors help.

---

[13] J.H.E. Griffiths, *Nature*, **158**, 670 (1946).

(4) The shape of the sample counts.

(5) Saturation occurs at low radio-frequency power.

(6) In a nuclear spin system we can reverse $\mathcal{M}_z$ or make it zero; but we cannot do that in a ferromagnetic spin system. The resonance excitation here destroys itself into spin waves (in different modes) even before $\mathcal{M}$ has a chance to be rotated by our action from its initial direction.

Figure 19.25 gives the line shape of FMR where the specimen is a polished sphere of yttrium iron garnet. We plot $4\pi\chi_x''$ versus H_0 (Gauss), at 3.33 GHz and 300°K, with the applied field H_0 being parallel to (1,1,1). The peak occurs at 1129.2 Gauss, and the line width at half power is 0.2 Gauss.

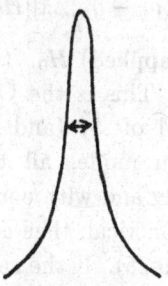

Fig. 19.25.

How does the shape of the sample affect the resonance frequency ω_0? Consider a three-dimensional sample. Let the *demagnetization coefficients* along three directions be h_x, h_y, and h_z, where $H_x' = -h_x\mathcal{M}_x$, $H_y' = -h_y\mathcal{M}_y$, $H_z' = -h_z\mathcal{M}_z$, with H' being the *demagnetizing field* (or magnetization removing field) due to the magnetization. H' tries to oppose to the applied external field H_0. The factors h are just constant numbers whose values depend on the geometry (shape) of the sample (sphere, slab, cylinder, cube, ellipsoid etc.). The local field H inside the sample is

$$H_x = (H_0)_x - h_x\mathcal{M}_x, \quad H_y = (H_0)_y - h_y\mathcal{M}_y, \quad H_z = (H_0)_z - h_z\mathcal{M}_z.$$

The torque has no contributions from the exchange field $\lambda\mathcal{M}$, nor from the Lorentz field $(4\pi/3)\mathcal{M}$, because if they are crossed to $\mathcal{M}$ we have $\mathcal{M} \times \mathcal{M} = 0$. If the static field H_0 is along z, the spin equation $d\mathcal{M}/dt = \gamma(\mathcal{M} \times \mathbf{H})$ consists of

$$\frac{d\mathcal{M}_x}{dt} = \gamma(\mathcal{M}_y H_z - \mathcal{M}_z H_y) = \gamma[H_0 + (h_y - h_z)\mathcal{M}]\mathcal{M}_y$$

$$\frac{d\mathcal{M}_y}{dt} = \gamma[\mathcal{M}(-h_x\mathcal{M}_x) - \mathcal{M}_x(H_0 - h_z\mathcal{M})] = -\gamma[H_0 + (h_x - h_z)\mathcal{M}]\mathcal{M}_x \ .$$

To a first-order approximation, $d\mathcal{M}_z/dt = d\mathcal{M}/dt = 0$. To solve the two equations above, we asume a time-dependent solution of the form $e^{-i\omega t}$. To have a solution, we should let the following determinant vanish:

$$\begin{vmatrix} i\omega & \gamma[H_0 + (h_y - h_z)\mathcal{M}] \\ -\gamma[H_0 + (h_x - h_z)\mathcal{M}] & i\omega \end{vmatrix} = 0$$

whence

$$\omega_0 = \gamma\sqrt{[H_0 + (h_y - h_z)\mathcal{M}][H_0 + (h_x - h_z)\mathcal{M}]}$$

is the *FMR frequency* in (the applied) H_0. (Only the positive result is considered for physical meaning.) This is the CGS result; in the MKS system write $\mu_0\vec{\mathcal{M}}$ everywhere instead of $\vec{\mathcal{M}}$ (and the h factors must be defined accordingly).[14] In the uniform mode, all the spins (magnetic moments) precess with this ω_0 cophasically and with equal amplitudes.

If the sample geometry is spherical, then $\omega_0 = \gamma H_0$ because $h_x = h_y = h_z$. The FMR line is sharp (Fig. 19.25). If the sample is a slab (and $\mathbf{H}_0$ normal to it), then $\omega_0 = \gamma(H_0 - 4\pi\mathcal{M})$ (in CGS) because $h_z = 4\pi$ and $h_x = h_y = 0$. (In MKS write μ_0 instead of 4π in ω_0.) But if $\mathbf{H}_0$ is in the xz-plane (along the slab), then $\omega_0 = \gamma\sqrt{H_0^2 + 4\pi H_0\mathcal{M}}$ (in CGS) because $h_y = 4\pi$ and $h_x = h_z = 0$. (Again, in MKS write μ_0 instead of 4π.)

Here γ can be found experimentally and thence g, the spectroscopic splitting factor, because $g = -\gamma\hbar/\mu_B$. Its values for some metals (at 300°K) are as follows:

Metal	g
Fe	2.10
Co	2.18
Ni	2.21

It should be noted that γ is also called *magnetogyric ratio*.

---

[14]In CGS, their sum is equal to 4π; in MKS, to 1.

19.10. Spin Wave Resonance[15]

An uniform radio-frequency magnetic field causes spin waves (of long wavelength) in thin ferromagnetic layers and films, by excitation. This can happen when the spins of the electrons on the sample surface experience different (anisotropic) fields, not the same as those seen by spins in the interior.

SWR is shown in Fig. 19.26 where H_0 is normal to the film. H_1 is the radio-frequency field, and $H = H_0 - 4\pi\mathcal{M}$ is the internal field. On the two surfaces the spins are kept nailed as shown (due to surface anisotropy and the forces thereof). H_1 is uniform and excites modes of spin waves whose number is an odd multiple of $\lambda/2$, as shown. (Modes with an even multiple do not interact with H_1 — their net interaction energy is none). The film thickness is d.

The *SWR frequency* is obtained if we take $\omega_0 = \gamma(H_0 - 4\pi\mathcal{M})$ from the previous section and include (by simple addition) the exchange contribution, which is Dk^2 where D is the proportionality constant (*spin wave exchange*

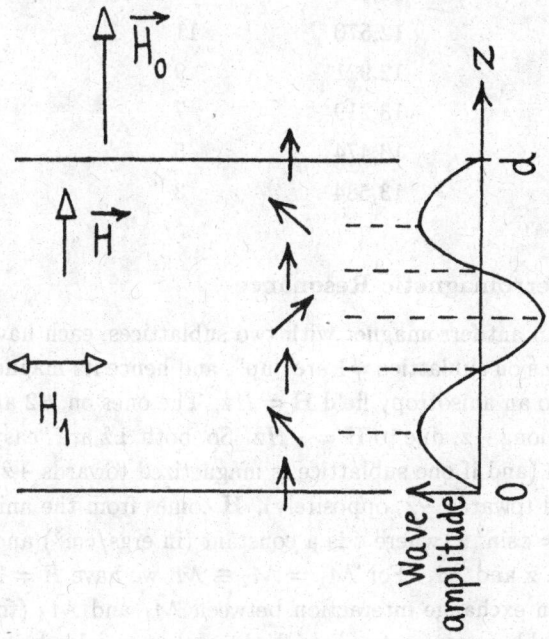

Fig. 19.26.

[15]C. Kittel, *Phys. Rev.*, **110**, 1295 (1958). M.H. Seavey, Jr. and P.E. Tannenwald, *Phys. Rev. Letters*, **1**, 108 (1958). P. Pincus, *Phys. Rev.*, **118**, 658 (1960). M. Okochi, *J. Phys. Soc. Japan*, **28**, 879 (1970) (angular dependence).

coefficient), and $k = n\pi/d$ is the wave vector (of a mode of n half-wavelengths) in a film of thickness d. The result for SWR in an applied H_0 is

$$\omega_0 = \gamma(H_0 - 4\pi\mathcal{M}) + Dk^2$$

where experimentally, we can take $kz \ll 1$.

Kittel (p. 521) has a graph of absorption intensity versus H_0 (in Gauss) and versus spin-wave order number, for the 80% Ni–20% Fe alloy at 9 GHz. (The order number is the number of half-wavelengths in d.) The absorption peaks in the SWR spectrum occur at the following values:

H_0 (Gauss)	Order number
11,226	17
11,732	15
12,191	13
12,570	11
12,921	9
13,219	7
13,474	5
13,584	3

19.11. Antiferromagnetic Resonance

Consider an antiferromagnet with two sublattices, each having spins fixed on it. The spins on sublattice #1 are "up", and hence its magnetization $\vec{\mathcal{M}}_1$ is along $\hat{z}$, due to an anisotropy field $\mathbf{H} = H\hat{z}$. The ones on #2 are "down", and hence $\vec{\mathcal{M}}_2$ is along $-\hat{z}$, due to $\mathbf{H} = -H\hat{z}$. So, both $\pm\hat{z}$ are "easy directions" of magnetization (and if one sublattice is magnetized towards $+\hat{z}$, the other will be magnetized towards $-\hat{z}$, oppositely). $\mathbf{H}$ comes from the anisotropy energy density $u(\theta) = c\sin^2\theta$, where c is a constant (in ergs/cm^3) and θ is the polar angle between $\hat{z}$ and $\vec{\mathcal{M}}_1$. For $\mathcal{M}_1 = \mathcal{M}_2 \equiv \mathcal{M}$, we have $H = 2c/\mathcal{M}$.

There is an exchange interaction between $\vec{\mathcal{M}}_1$ and $\vec{\mathcal{M}}_2$ (to be considered in the mean field approximation), with the exchange fields being $\mathbf{H}_1 = -\lambda\vec{\mathcal{M}}_2$ and $\mathbf{H}_2 = -\lambda\vec{\mathcal{M}}_2$, where $\lambda > 0$. Notice that $\mathbf{H}_1$ acts on (the spins of) sublattice #1, and $\mathbf{H}_2$ on #2.

If there is no external $\mathbf{H}_0$, the total field on #1 (and hence on $\vec{\mathcal{M}}_1$) is $\mathbf{H}_1 = -\lambda\vec{\mathcal{M}}_2 + H\hat{z}$, and the total field on #2 is $\mathbf{H}_2 = -\lambda\vec{\mathcal{M}}_1 - H\hat{z}$. If we let

$(\mathcal{M}_1)_z \equiv \mathcal{M}$ and $(\mathcal{M}_2)_z \equiv -\mathcal{M}$, we obtain a set of linear equations of motion:

$$(\dot{\mathcal{M}}_1)_x = \gamma[(\mathcal{M}_1)_y(\lambda\mathcal{M} + H) - \mathcal{M}(-\lambda(\mathcal{M}_2)_x)]$$

$$(\dot{\mathcal{M}}_1)_y = \gamma[\mathcal{M}(-\lambda(\mathcal{M}_2)_x) - (\mathcal{M}_1)_x(\lambda\mathcal{M} + H)]$$

$$(\dot{\mathcal{M}}_2)_x = \gamma[(\mathcal{M}_2)_y(-\lambda\mathcal{M} - H) - (-\mathcal{M})(-\lambda(\mathcal{M}_1)_y)]$$

$$(\dot{\mathcal{M}}_2)_y = \gamma[(-\mathcal{M})(-\lambda(\mathcal{M}_1)_x) - (\mathcal{M}_2)_x(-\lambda\mathcal{M} - H)].$$

For the sake of mathematical simplication, let $M_1 \equiv (\mathcal{M}_1)_x + i(\mathcal{M}_1)_y$ and $M_2 \equiv (\mathcal{M}_2)_x + i(\mathcal{M}_2)_y$. We guess an oscillatory solution of the form $e^{i\omega t}$ (for its time dependence) for the equations of motion and plug it in. After the operations, we get the set

$$-i\omega M_1 = -i\gamma[M_1(H + \lambda\mathcal{M}) + M_2(\lambda\mathcal{M})]$$

$$-i\omega M_2 = i\gamma[M_2(H + \lambda\mathcal{M}) + M_2(\lambda\mathcal{M})].$$

For this set to have a nontrivial solution, the following determinant should vanish:

$$\begin{vmatrix} \gamma(H + H_{\mathrm{ex}}) - \omega & \gamma H_{\mathrm{ex}} \\ \gamma H_{\mathrm{ex}} & \gamma(H + H_{\mathrm{ex}}) + \omega \end{vmatrix} = 0$$

where we have set $\lambda\mathcal{M} \equiv H_{\mathrm{ex}}$. The result is

$$\omega_0 = \sqrt{\gamma^2 H(H + 2H_{\mathrm{ex}})}$$

which is the *antiferromagnetic resonance frequency*.[16]

Figure 19.27 shows the effective fields in this resonance where $\vec{\mathcal{M}}_1$ experiences $\mathbf{H}_1$, and sublattice #2 feels $\mathbf{H}_2$. The crystal can be easily magnetized along both ends of its axis.

Figure 19.28 shows the stereochemical and magnetic structure of MnF_2, where the magnetic moments of the Mn atoms are marked (by arrows). The marked lattice points thus are the position of Mn^{++}, and the rest are of F^-. This substance is a well-known antiferromagnet whose H and H_{ex} were theoretically estimated by Keffer, as $H = 8800$ Gauss and $H_{\mathrm{ex}} = 540,000$ Gauss, so that $\sqrt{2HH_{\mathrm{ex}}} = 100,000$ Gauss, while the experimentally observed result is 93,000 Gauss.

---

[16]T. Nagamiya, *Prog. Theor. Phys.*, **6**, 342 (1951). C. Kittel, *Phys. Rev.*, **82**, 565 (1951). C. Kittel and F. Keffer, *Phys. Rev.*, **85**, 329 (1952).

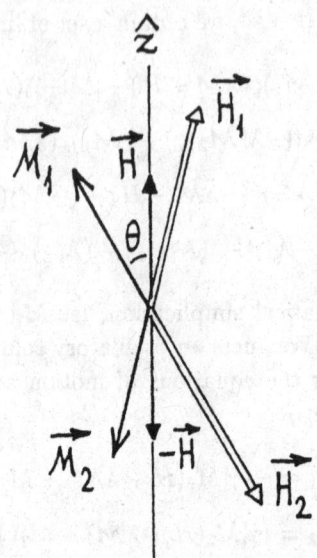

Fig. 19.27.

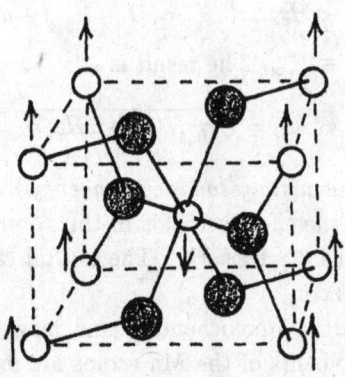

Fig. 19.28.

Figure 19.29 shows ω_0 for MnF_2 against T as observed and drawn by Johnson and Nethercot. Antiferromagnetic resonance frequency values for serveral molecular materials (crystals) (as extrapolated to 0°K) are tabulated by Richards, as follows:

Material	FeF$_2$	MnF$_2$	NiF$_2$	CoF$_2$	NiO	MnO
$\omega_0 \times 10^{10}$ Hz	158	26	93.3	85.5	109	82.8

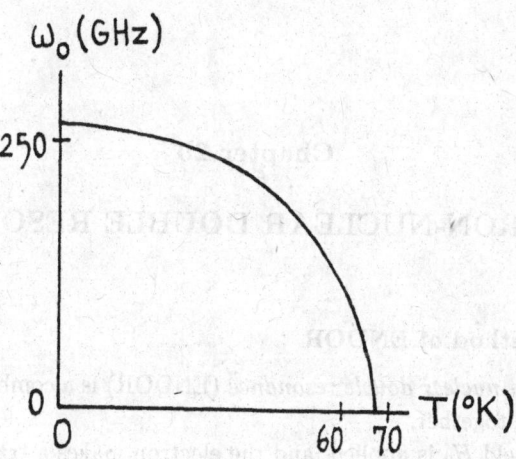

Fig. 19.29.

19.12. EPR Research

Today, EPR is a very broad research area with special topics and a highly specialized literature. For space reasons we will not elaborate on the subject any further.

19.13. Energy Units

It is worth noting that the quantity of energy in modern physics is measured and expressed in (units of) eV and also — by abuse — in Hz (units of frequency ν via $E = h\nu$) or cm (units of λ via $\nu = c/\lambda$), or even cm^{-1} (units of the wave number $\bar{\nu} = 1/\lambda$ or d/λ, which is the number of wavelengths per unit length or per length d).

It is easier and more practical to engineers to use Joules, to atomic physicists to use Hz, to particle physicists to use eV, and to spectroscopists (in dealing with spectral lines) to use cm^{-1}. This usage is conventional and an established habit.

Chapter 20

ELECTRON-NUCLEAR DOUBLE RESONANCE

20.1. The Method of ENDOR

The *electron-nuclear double resonance* (ENDOR) is a combination of NMR and ESR taken together.

In ESR, a field H_0 is applied, and the electron makes a transition between two energy levels, as shown in Fig. 20.1 where m_s is the spin magnetic quantum number. This transition gives ESR.

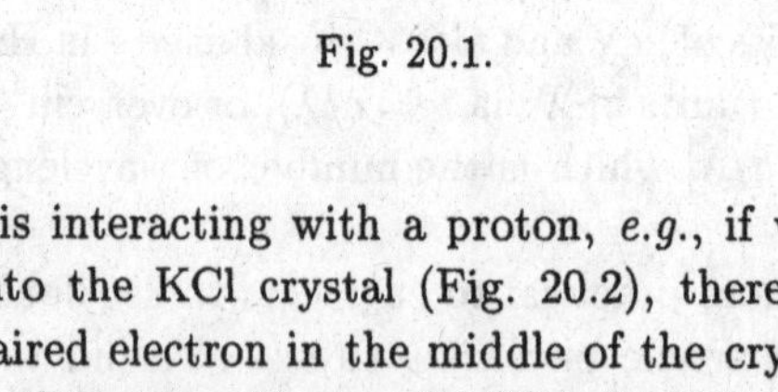

$$\text{—————} \bullet\ m_s = +1/2$$

Transition (ESR)

$$\text{—————} \bullet\ m_s = -1/2$$

Fig. 20.1.

If the electron is interacting with a proton, *e.g.*, if we introduce neutral hydrogen atoms into the KCl crystal (Fig. 20.2), there will be a hydrogen atom, *i.e.*, an unpaired electron in the middle of the crystal, giving rise to a color center. Then the energy levels split (Fig. 20.3(a)), forming four energy states. Since the selection rules are $\Delta m_s = \pm 1$ and $\Delta m_I = 0$, the transitions shown in Fig. 20.3(b) are possible.

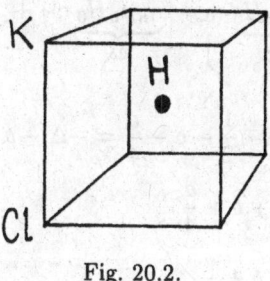

Fig. 20.2.

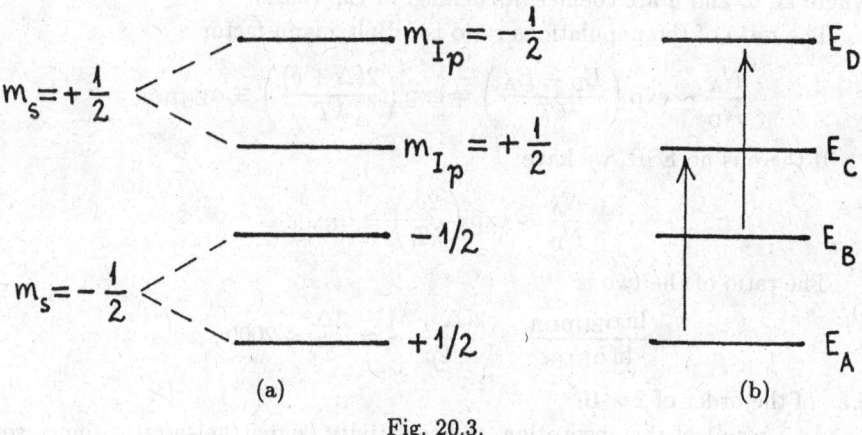

(a)   (b)

Fig. 20.3.

If we give the system many $h\nu_s$ quanta corresponding to ESR, many transitions occur, and we have equal populations: $N_B \simeq N_D$ and $N_A \simeq N_C$. And if we observe an absorption at the NMR frequency (*i.e.*, $h\nu_I$), the sensitivity increases, that is, the signal/noise ratio increases. This is called *Overhauser effect*. In general it is observed, that is, the signal/noise ratio increases, especially when the hyperfine interaction constant $\mathcal{A}_{I,S}$ is a scalar. But sometimes the opposite happens: If $\mathcal{A}_{I,S}$ is not isotropic (*i.e.*, if it is a tensor exhibiting different properties in different directions), then the signal/noise ratio decreases, and the Overhauser effect cannot be observed.

The interaction Hamiltonian of the system gives the following expression for the energy:

$$E = \underbrace{g_e \mu_B H_0}_{2\Delta} \, m_s - \underbrace{\gamma_n \mu_n H_0}_{2\delta} \, m_I + \underbrace{\mathcal{A}}_{a/4} \, m_s m_I, \qquad (20.1)$$

$$E_A = -2\Delta \cdot \frac{1}{2} - \delta - \frac{a}{4} = -\Delta - \delta - \frac{a}{4}$$

$$E_B = -\Delta + \delta + \frac{a}{4}$$

$$E_C = \Delta - \delta + \frac{a}{4}$$

$$E_D = \Delta + \delta - \frac{a}{4}$$

where Δ, δ, and a are coefficients defined in Eq. (20.1).

The ratio of the populations ratio is a Boltzmann factor:

$$\frac{N_A}{N_D} \sim \exp\left(\frac{E_D - E_A}{kT}\right) = \exp\left(\frac{2(\Delta + \delta)}{kT}\right) \equiv \alpha_{\text{ENDOR}}.$$

If there is no ESR, we have

$$\frac{N_A}{N_D} \sim \exp\left(\frac{2\delta}{kT}\right) \equiv \alpha_{\text{NMR}}.$$

The ratio of the two is

$$\frac{\ln \alpha_{\text{ENDOR}}}{\ln \alpha_{\text{NMR}}} = \frac{2(\Delta + \delta)}{2\delta} \simeq \frac{2\Delta}{\delta} \sim 2000,$$

i.e., of the order of 2×10^3.

As a result of the absorption, the sensitivity (signal/noise ratio) increases 2000 times. Due to the polarization (orientation) of spins, we have less thermal vibrations (their number decrease), and since the electrons interact with the nuclei (spin-spin interaction, *i.e.*, $\mathbf{I} \cdot \mathbf{S}$), the sensitivity increases.

With ENDOR, the signal/noise ratio increases 10^3 to 10^6 times. So, the resolving power of the spectrometer increases a lot, enabling us to sensitively measure hyperfine structures. This way we can determine electron densities and hyperfine structure constants (coupling coefficients). We can also measure the actual values of the wavefunctions.

Experimentally, if one considers the crystal defect shown in Fig. 20.2 and one applies the ENDOR method, one obtains two lines, as shown in Fig. 20.3(b).

20.2. Observation of ENDOR

ENDOR can be observed in two ways:

(1) Do ESR and observe the change in ESR.

(2) Do ESR and observe the NMR change.

The second method was used by Spaeth who did the experiment in 1967. The obtained spectrum for KCl (Fig. 20.2) was symmetric (Fig. 20.4). The central three-peak line belongs to ^{35}Cl. The direction of the field H_0 was chosen as shown in Fig. 20.5, *i.e.*, H_0 was parallel to the plane (1, 1, 0). The lower Cl atom showed a resonance different than that of the upper one. Notice that we have two isotopes: ^{35}Cl and ^{37}Cl. All the resonance lines were thrice peaked, as shown. The minus (plus) sign in the parenthesis indicates Cl atoms below (above) the middle line. The lower Cl atom gives a different resonance than the upper one.

One can determine the hyperfine-structure constants from here, for each nucleus.

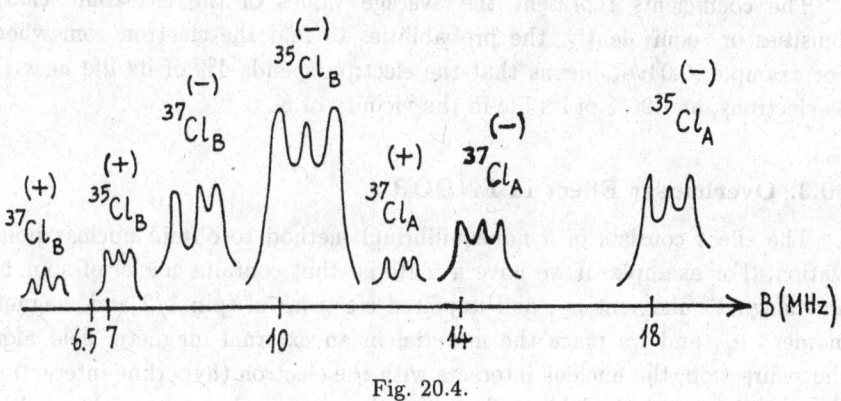

Fig. 20.4.

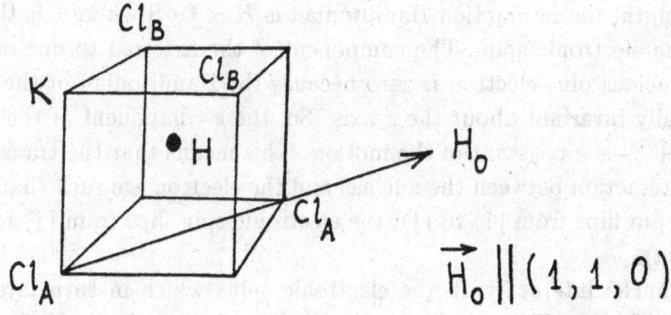

Fig. 20.5.

The orthogonalized wavefunction of the U_2-type color center is

$$\psi_{U_2} = 1.09 \times \left[\psi_H - \sum_{A=1}^{4} \left(0.07\psi_{3s}^{ClA} - 0.17\psi_{3p}^{ClA} \right) - \sum_{B=1}^{4} \left(0.04\psi_{3s}^{KB} - 0.06\psi_{3p}^{KB} \right) \right] .$$

For example, the last term tells us that we take the wavefunction of the $3p$-electron and sum for all K nuclei. Here ψ_H is the hydrogen wavefunction. We further have

$$\psi_{U_2} = N \left(\psi_H - \sum_{i,\alpha} \langle \psi_H | \psi_i^\alpha \rangle \psi_i^\alpha \right)$$

where the Dirac bracket means average (or expectation) value, and N is a coefficient. One sets up eight equations, finds the coefficients, and finally one finds the wavefunction. After that, one can find the energy.

The coefficients represent the average values of the electronic charge densities or, equivalently, the probabilities to find the electron somewhere. For example, $0.04\psi_{3s}$ means that the electron spends 4% of its life near the $3s$-electrons, and 96% of its life in the vicinity of K.

20.3. Overhauser Effect in ENDOR

The effect consists of a nonequilibrium method to obtain nuclear polarization. For example, if we have a material that contains nuclei of spin 1/2 and magnetic moment μ_n, and unpaired electrons of spin 1/2 and magnetic moment μ_e, and we place the material in an external magnetic field along the z-direction, the nucleus interacts with the electron (hyperfine interaction) through the magnetic field produced by the moving electron at (the position of) the nucleus. Apart from a proportionality constant that fixes the interaction strength, the interaction Hamiltonian is $\mathcal{H} \propto \mathbf{I} \cdot \mathbf{S}$ where $\mathbf{I}$ is the nuclear and $\mathbf{S}$ the electronic spin. The component of the external torque on the system of nucleus plus electron is zero because the Hamiltonian of the system is rotationally invariant about the z-axis. So, the z-component of the combined spins $I_z + S_z$ is a constant of the motion. This means that the transitions due to the interaction between the nucleus and the electron are such that when the nuclear spin flips from $|\uparrow\rangle$ to $|\downarrow\rangle$, the electronic spin flips from $|\downarrow\rangle$ to $|\uparrow\rangle$ (and vice versa).

The nuclei interact with the electronic spins which in turn interact with the crystal lattice. This sequence of interactions brings about the final thermal equilibrium of nuclear spins, so that

$$\frac{N_\uparrow}{N_\downarrow} \sim e^{2\mu_n H/kT}$$

where T is the lattice temperature, k is the Boltzmann constant, H is the external field, and $N_\uparrow (N_\downarrow)$ is the number of nuclei with spin up (down).

Likewise, for electrons (of energy $\pm \mu_e H$) we have

$$\frac{n_\uparrow}{n_\downarrow} \sim \frac{e^{\mu_e H/kT}}{e^{-\mu_e H/kT}} = e^{2\mu_e H/kT}.$$

The polarization is

$$\xi_n = \frac{N_\uparrow - N_\downarrow}{N_\uparrow + N_\downarrow} \quad \text{and} \quad \xi_e = \frac{n_\uparrow - n_\downarrow}{n_\uparrow + n_\downarrow}$$

where either ξ satisfies the relation $-1 \le \xi \le 1$. But $\xi_n \ll \xi_e$ because $\mu_e \gg \mu_n$. This means that the electrons are polarized while the nuclei are not as much, even in a strong H and at a low T. (With this situation we cannot perform nuclear-physics experiments on polarized nuclei.)

The system of nuclei plus electrons can be considered as a combination in contact with the lattice heat reservoir (Fig. 20.6). The transition probabilities are related by

$$\frac{P(|\uparrow\rangle_n |\downarrow\rangle_e \rightarrow |\downarrow\rangle_n |\uparrow\rangle_e)}{P(|\downarrow\rangle_n |\uparrow\rangle_e \rightarrow |\uparrow\rangle_n |\downarrow\rangle_e)} = e^{-(\mu_n H - \mu_e H + \mu_n H - \mu_e H)/kT}$$

$$= e^{-2(\mu_n - \mu_e)H/kT}$$

where $(E_\pm)_n = \mp \mu_n H$ and $(E_\pm)_e = \mp \mu_e H$. In equilibrium,

$$n_\uparrow N_\downarrow P(|\uparrow\rangle_n |\downarrow\rangle_e \rightarrow |\downarrow\rangle_n |\uparrow\rangle_e) = n_\downarrow N_\uparrow P(|\downarrow\rangle_n |\uparrow\rangle_e \rightarrow |\uparrow\rangle_n |\downarrow\rangle_e)$$

whence

$$\frac{n_\uparrow N_\downarrow}{n_\downarrow N_\uparrow} = e^{2(\mu_n - \mu_e)H/kT}.$$

But if we apply a radio-frequency field (at the ESR frequency), strong enough to saturate the electronic spin system, then $N_\uparrow = N_\downarrow$, and hence

$$\frac{n_\uparrow}{n_\downarrow} = e^{2(\mu_n - \mu_e)H/kT} \simeq e^{-\mu_e H/kT}$$

because $\mu_e \gg \mu_n$. This means that the nuclei are now polarized as if they had a magnetic moment as large as the electronic magnetic moment. This is the Overhauser effect by which we can obtain a large nuclear polarization in a nonequilibrium (but nevertheless steady-state) situation.

The interactions in the whole system are shown in Fig. 20.6.

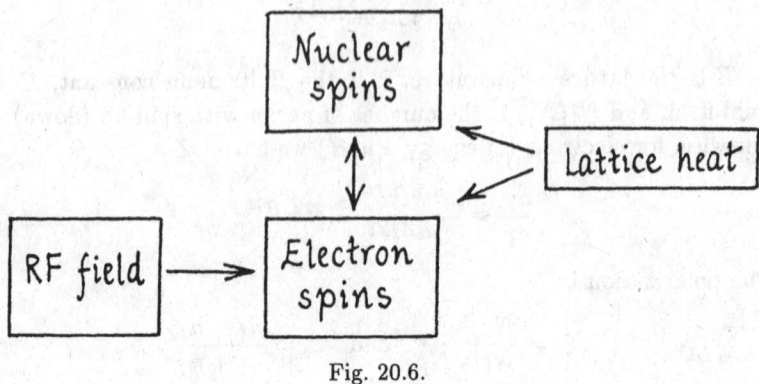

Fig. 20.6.

The polarization percentage is $\xi \times 100$.

Note that when we compare magnetic moments here, we take absolute values (because μ_e is negative).

20.4. Donors in Si and ENDOR

Phosphorus atoms become donors in Si. A P atom has five electrons; four of these form a diamagnetic covalent bond with the crystal. The remaining bound electron does not enter into any bond; it is paramagnetic and has $S = 1/2$. If the concentration of P in Si is above 10^{18} cm^{-3}, we obtain a single spectral line instead of two split ones, because of motional narrowing, as the donor electrons jump fast between donor atoms, *i.e.*, the fast jumping outruns the hyperfine splitting. The rate of jumping gets larger and larger as the concentration increases, because the wavefunctions of donor electrons overlap more.

Figure 20.7 shows the experimental hyperfine splitting (for strong **H**), *i.e.*, the ESR lines of donor P atoms in Si. When the concentration of P rises, a donor electron can jump around very fast, to dominate over the fine structure and suppress it. (See R.C. Fletcher, W.A. Yager, G.L. Pearson, and F.R. Merritt, *Phys. Rev.*, **95**, 844 (1954).)

The wavefunction ψ of the donor electron can cover not only the donor atom, but also Si atoms around, in fact many of them. Then the nuclear spins of ^{29}Si interact and provide extra hyperfine splittings which can be observed only via ENDOR, if the technical equipment is powerful enough. Such an observation helps us understand the wavefunctions of conduction electrons. (See G. Feher, *Phys. Rev.*, **114**, 1219 (1959), and W. Kohn, *Solid State Physics*, **5**, 257 (1957).)

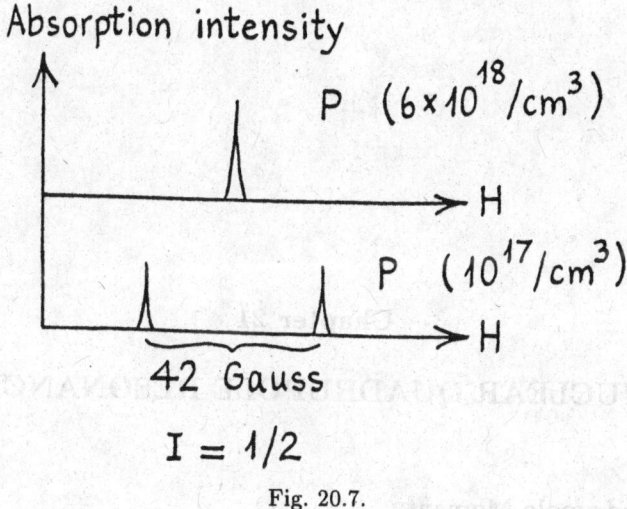

Fig. 20.7.

This is an example to a topic which is a subject of the ENDOR technique. In the sample considered the donor electron behaves paramagnetically, and its wavefunction spreads to cover many atoms as the electron moves around, hopping from a crystal site atom to another — and very fast so. And as the concentration of donors increases, this cloudlike electronic spread washes out the hyperfine splitting — a sort of averaging — because of the extensive overlap of the electronic wavefunctions. (This is connected to the electric conductivity of the material. Pertinent measurements are possible.)

Chapter 21

NUCLEAR QUADRUPOLE RESONANCE

21.1. Quadrupole Moment

A *quadrupole* is a charge configuration of four alternate positive and negative point charges located at the corners of a rigid square or parallelogram (Fig. 21.1).

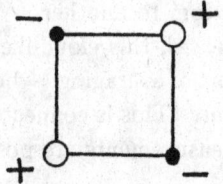

Fig. 21.1.

A quadrupole is equivalent to two equal and opposite dipoles $\vec{p}$ and $-\vec{p}$ located side by side and separated by some distance (Fig. 21.2).[1] A quadrupole forms an area A in space, which is spatially defined by the four fixed charges and their separations (Fig. 21.3).

We can define a quantity called *quadrupole moment* Q whose dimensions are those of an area. (Sometimes the quadrupole moment is defined as eQ where e is the electron charge.)

---

[1]They must be opposite. If they are in the same direction, the result is again a dipole, not a quadrupole.

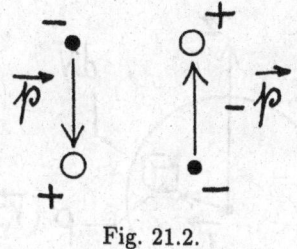

Fig. 21.2.

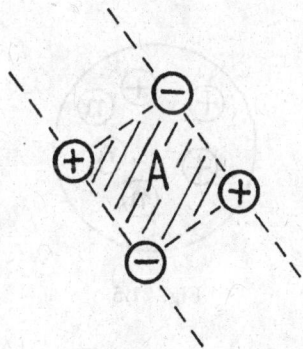

Fig. 21.3.

To have a nonzero Q, the area A must be nonzero.

The *nuclear quadrupole moment* is defined as

$$Q = \frac{1}{e} \iiint_{\substack{\text{Over} \\ \text{nucleus}}} \rho_{\text{nucl}}(\mathbf{r}_{\text{nucl}})(3z^2 - r_{\text{nucl}}^2) dV_{\text{nucl}}$$

where the integration is over the nuclear volume, r_{nucl} is the distance from the nucleus center (origin) to a volume element inside the nucleus, ρ_{nucl} is the nuclear charge density, and z is the coordinate (Fig. 21.4).

Since there is no negative charge in the nucleus, how can we talk about a nuclear quadrupole moment?

A nucleus consists of protons (positive charges) and neutrons (uncharged particles) (Fig. 21.5). But it *can* have a quadrupole moment, and this can be explained by considering moments. If the positive charge distribution in the nucleus is spherically symmetric (Fig. 21.6), then $Q = 0$. In order for a nucleus

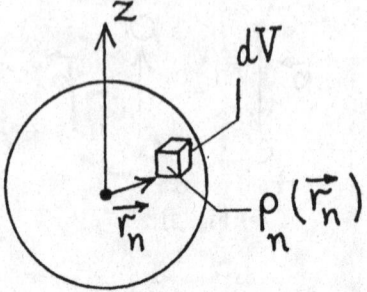

Fig. 21.4.

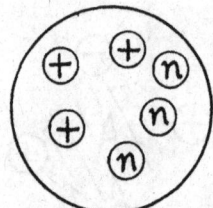

Fig. 21.5.

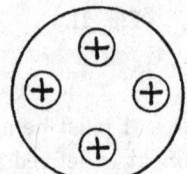

Fig. 21.6.

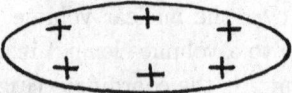

Fig. 21.7.

to have a nonzero Q, its shape must be that of an ellipsoid (Fig. 21.7). Such a charge distribution — though all positive — does create a nonzero Q.

The ellipsoidal nucleus can be thought as being equivalent to a sphere plus something additive that provides the equivalence (Fig. 21.8). This "something" has to be that shown in Fig. 21.9. (Notice the charges.) The situation is fully shown in Fig. 21.10. The radius r of the sphere considered satisfies the relation $a < r < b$ where a and b are the semiminor and semimajor axes of the ellipsoid respectively.

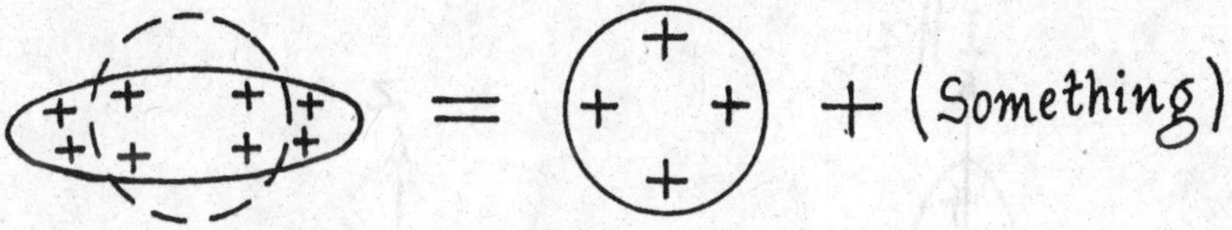

Fig. 21.8.

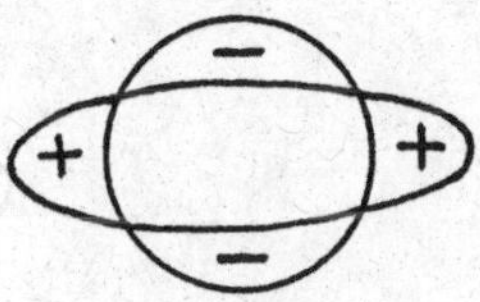

Fig. 21.9.

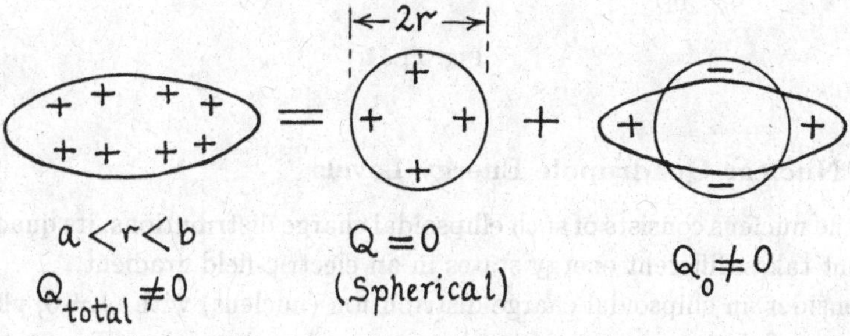

Fig. 21.10.

Therefore an ellipsoidal charge distribution is equivalent to a spherical charge distribution (where $Q_{sphere} = 0$) plus a convenient combination of an ellipsoid of revolution and a sphere (where $Q_0 \neq 0$). Thus the overall (total) quadrupole moment of the (first) ellipsoid becomes nonzero:

$$Q_{total} = Q_{sphere} + Q_{combination} = Q_{sphere} + Q_0 = 0 + Q_0 = Q_0 \neq 0.$$

If the nuclear spin I is zero, the quadrupole moment of the nucleus is zero. If the ellipsoidal nucleus is prolately elongated along the z-direction (which is taken along the direction of the spin **I**), then $Q > 0$. If the ellipsoid of revolution is oblate, as shown in the figure, then $Q < 0$ (Fig. 21.11(a) and (b) respectively).

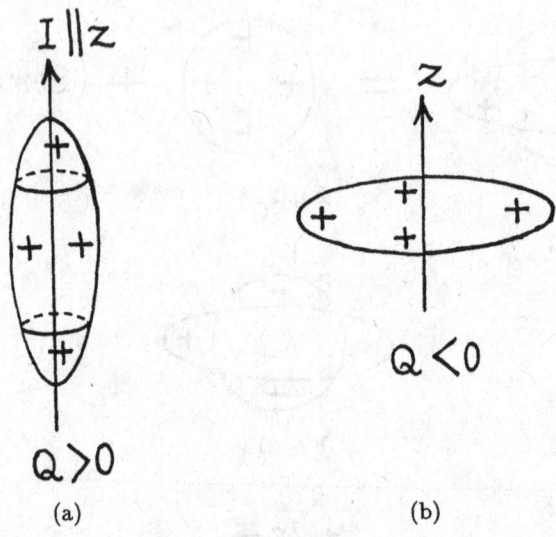

(a)   (b)

Fig. 21.11.

21.2. Nuclear Quadrupole Energy Levels

If the nucleus consists of such ellipsoidal charge distributions, its quadrupole moment takes different energy states in an electric-field gradient.

Consider an ellipsoidal charge distribution (nucleus) with $Q \neq 0$, placed in an electric field, say, in the region between active electrodes (Fig. 21.12). We will use a model to show that this distribution can take different energy states in the field.

The electric field gradient is $\vec{\nabla}\vec{\mathcal{E}}$ whose z-component is $|\vec{\nabla}\vec{\mathcal{E}}|_z = -\partial^2 V/\partial z^2$ $\equiv q_{zz}$.

For the electrode arrangement shown in Fig. 21.12, if a prolate ellipsoid is placed in between, there is an attractive force along the long axis of the ellipsoid, and if an oblate ellipsoid is placed, then the force is repulsive, as shown in Fig. 21.13. This resembles the case of putting a rod magnet opposite to, or along the magnetic field. Here we have the case of the maximum and minimum energy levels of two dipoles at right angles to each other (whereas in the case

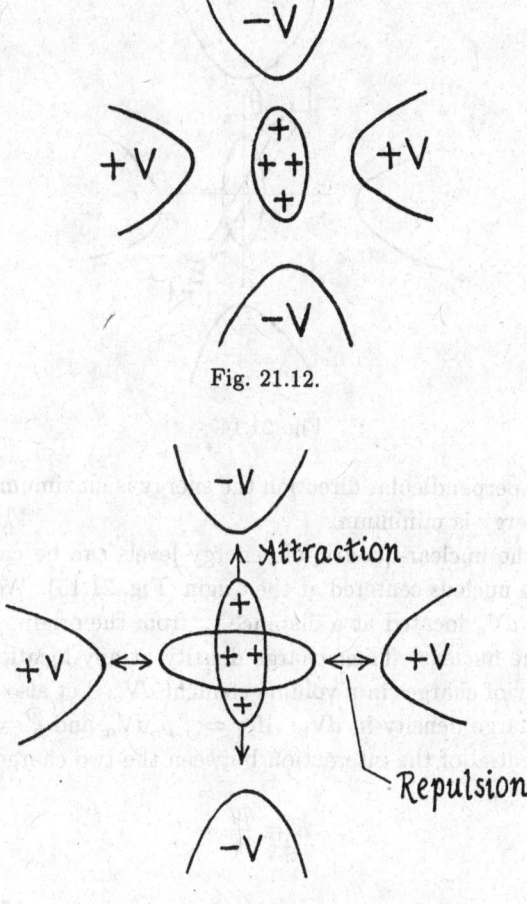

Fig. 21.12.

Fig. 21.13.

of the rod magnet we deal with the maximum and minimum energy states of a single dipole in the direction of the magnetic field and opposite to it).

Figure 21.14 shows the direction of the nuclear spin at an angle θ with the z-direction, and the corresponding energy E_1. It also shows the spin in the opposite direction (dashed line), with the corresponding potential energy E_{-1}. These two energies are the same, whereas in the case of a magnet this is not so. (A magnet in a magnetic field turns and orients itself so as to minimize the energy.)

The nuclear quadrupole moment takes maximum and minimum energy values, depending on whether it is perpendicular or parallel to the electric-field

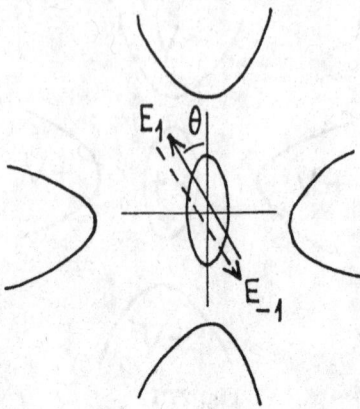

Fig. 21.14.

gradient. In the perpendicular direction the energy is maximum; in the parallel direction the energy is minimum.

In general, the nuclear-quadrupole energy levels can be calculated as follows: Consider a nucleus centered at the origin (Fig. 21.15). We take a nuclear volume element dV_n located at a distance r_n from the origin. There are electrons outside the nucleus. Their charge density at any location $\mathbf{r}_e$ is ρ_e. Let ρ_e be the density of charges in a volume element dV_e. Let also ρ_n be the (positive) nuclear charge density in dV_n. If $q = \int \rho_n dV_n$ and $q' = \int \rho_e dV_e$, then the electric potential of the interaction between the two charges q and q' is

$$E = \frac{qq'}{r}$$

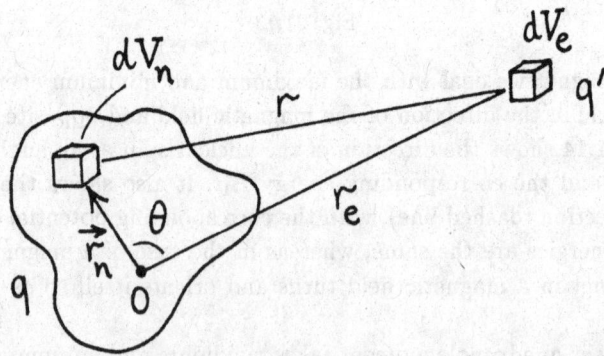

Fig. 21.15.

which is just the electrostatic (Coulombic) interaction energy, otherwise expressible as

$$E_{el} = \int_{Electronic} \int_{Nuclear} \frac{\rho_n dV_n \rho_e dV_e}{r} = \iint \frac{\rho_n \rho_e dV_n dV_e}{\sqrt{r_n^2 + r_e^2 - 2r_n r_e \cos \theta}}.$$

Since there are no electrons within the nucleus and in its immediate vicinity, we have $r_e > r_n$. The interaction energy can be written as

$$E_{el} = \iint \frac{\rho_n \rho_e dV_n dV_e}{r_e \sqrt{1 + \left(\frac{r_n}{r_e}\right)^2 - 2\frac{r_n}{r_e} \cos \theta}}.$$

For $r_n \ll r_e$ (far away from the nucleus) we expand $1/r$ into a series:

$$\frac{1}{r} = \frac{1}{r_e} P_0 + \frac{r_n}{r_e^2} P_1 + \left(\frac{r_n^2}{r_e^3}\right) P_2 + \ldots$$

$$= \frac{1}{r_e} \left[P_0 + \frac{r_n}{r_e} P_1 + \left(\frac{r_n}{r_e}\right)^2 P_2 + \left(\frac{r_n}{r_e}\right)^3 P_3 + \ldots \right].$$

We have then

$$E_{el} = \iint \frac{\rho_n \rho_e dV_e dV_n}{r_e} P_0 + \iint \ldots$$

where the first term is the electrostatic (Coulombic) interaction energy which considers all nuclear charges as being gathered at the origin (*i.e.*, the nucleus center) and interacting with all the electrons outside, that is, the mutual interaction between the nuclear charge and the cloud of the electrons around it. The second term is the electric-dipole term, *i.e.*, the potential energy of the electric interaction of a dipole moment. But since there is no negative charge in the nucleus, there is no electric dipole either (its electric dipole moment is zero). So, this second term (which contains P_1) is zero. The term with P_2 is the quadrupole interaction term. (And obviously, the first term which contains P_0 is the interaction energy of point charges.)

Here P_0, P_1, P_2 etc. are functions of θ, namely the Legendre polynomials:

$$P_0 = 1$$

$$P_1 = \cos \theta$$

$$P_2 = \frac{1}{2}(3\cos^2 \theta - 1)$$

etc. So, we have

$$E_0 = \iint \frac{\rho_n \rho_e dV_n dV_e}{r_e}$$

$$E_1 = \iint \frac{r_n \rho_n \rho_e \cos\theta dV_n dV_e}{r_e^2} = 0$$

$$E_2 \equiv E_Q = \iint \frac{r_n^2}{r_e^3} \rho_n \rho_e \frac{1}{2}(3\cos^2\theta - 1)dV_n dV_e$$

where E_Q is the quadrupole interaction energy. It can be written as a product of two tensors, that is, the quadrupole moment tensor times the electric field gradient tensor:

$$E_Q = \overset{\leftrightarrow}{\mathbf{Q}} \cdot \vec{\nabla}\vec{\mathcal{E}},$$

that is, in the form of a dyadic.

The angle θ is taken between the spin axis and the z-axis (Fig. 21.16), so that

$$\cos\theta = \frac{I_z}{I}.$$

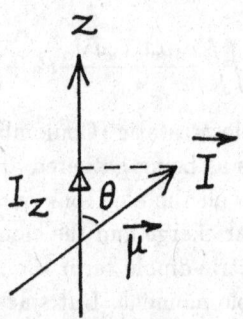

Fig. 21.16.

For a proton, $\mathbf{I} \| \boldsymbol{\mu}$.

Starting from the definition of the quadrupole moment

$$eQ \equiv \int_{\text{Nucleus}} \rho_n(r_n)(3z_n^2 - r_n^2)dV_n$$

and using this relation, one can find the nuclear quadrupole moment if one knows the z-coordinate of all the charges in the nucleus.

In calculating E_Q, one confronts two cases:

(a) The electric-field gradient $\vec{\nabla}\vec{\mathcal{E}}$ is symmetric with respect to an axis (axial symmetry). Then $q_{zz} \neq 0$ and $q_{xx} = q_{yy}$.

(b) The electric-field gradient is asymmetric. Then $q_{zz} \neq 0$ and $q_{xx} \neq q_{yy}$. In that case one defines the *asymmetry parameter* η as

$$\eta = \frac{q_{xx} - q_{yy}}{q_{zz}}$$

where $0 < \eta < 1$. (In general η is too small compared to unity.)

E_Q for case (a) is

$$E_Q = \frac{e^2 q_{zz} Q}{4I(2I-1)} [3m_I^2 - I(I+1)] \qquad (21.1)$$

where I is the nuclear spin and m_I is the magnetic quantum number which takes $2I+1$ values, from $-I$ to $+I$. Since m_I is squared in the formula, there is only one energy state for both the positive and negative values of m_I.

For example, consider ^{37}Cl with $I = 3/2$. The magnetic quantum number can take the values $m_I = -3/2, -1/2, 1/2, 3/2$. There is one energy for both $\pm 1/2$, and one for $\pm 3/2$ (Fig. 21.17).

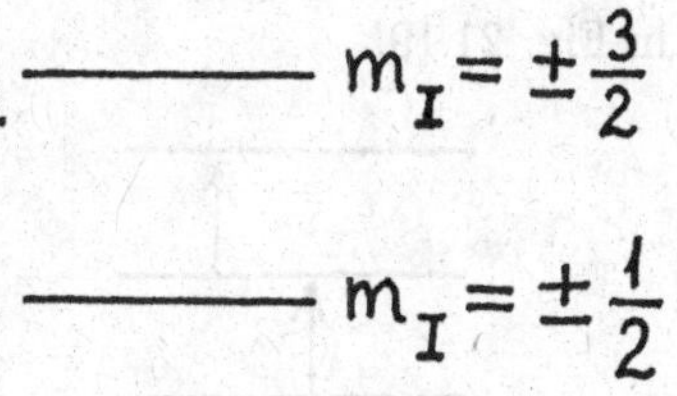

$$E_2 \text{———} m_I = \pm\frac{3}{2}$$

$$E_1 \text{———} m_I = \pm\frac{1}{2}$$

Fig. 21.17.

For a given nuclear spin I, there are $2I - 1$ energy levels for the nuclear quadrupole moment.

If $I = 1/2$, then $2I - 1 = 0$ and E_Q becomes infinite (see Eq. (21.1)) which is meaningless. In order to have a finite and nonzero E_Q, we must have $I \geq 1$.

21.3. Nuclear Quadrupole Resonance

If we give a certain amount of energy $h\nu$ to a nucleus with $I = 3/2$, we observe *nuclear quadrupole resonance* (NQR), that is, the system makes an absorption (Fig. 21.18). In other words, the effect of nuclear quadrupole resonance is the observed absorption when we give $\Delta E_Q(I) = h\nu$ to the nucleus.

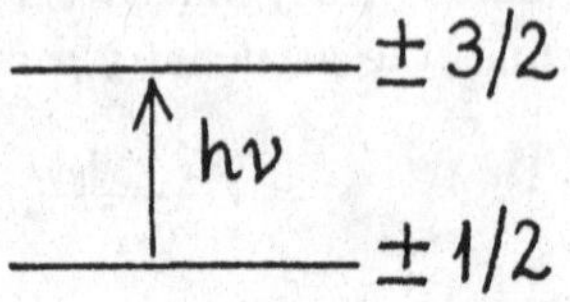

$$\underline{} \pm 3/2$$

$$\uparrow h\nu$$

$$\underline{} \pm 1/2$$

Fig. 21.18.

Any magnetic resonance spectrometer can be used here for measurements. But the energy we give must be at the proper quadrupole resonance frequency.

Example problem: By letting $e^2 q_{zz} Q \equiv \mathcal{A}$, calculate the NQR frequency ν_Q for $I = 3/2, 5/2$ and 2, in terms of $\mathcal{A}$. (The condition for allowed transitions, or selection rule, is $\Delta m_I = \pm 1$.)

Solution: We use the formula of Eq. (21.1) to calculate the energy levels for each given I. If we divide the difference(s) by h, we find the NQR frequency or frequencies. The actual calculation is left to the reader. One frequency will be found for $I = 3/2$, and two frequencies for $I = 2$. For $I = 5/2$ there are $2I - 1 = 4$ energy levels and hence three resonance frequencies. These three transitions are shown in Fig. 21.19.

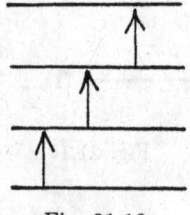

Fig. 21.19.

How can we obtain information about a molecule via NQR? Notice that

$$\vec{\nabla}\vec{\mathcal{E}} = \vec{\nabla}\vec{\mathcal{E}}_{\text{atom}} + \vec{\nabla}\vec{\mathcal{E}}_{\text{molecule}} .$$

If there is no axial symmetry, the field-gradient tensor (matrix)

$$\vec{\nabla}\vec{\mathcal{E}} = \begin{pmatrix} q_{xx} & q_{xy} & q_{xz} \\ q_{yx} & q_{yy} & q_{yz} \\ q_{zx} & q_{zy} & q_{zz} \end{pmatrix}$$

has nine components (matrix elements), out of which three are zero, leaving six independent components. These six components depend on the electronic distribution in the atom and on the ions around the nucleus whose nuclear quadrupole resonance is being examined.

For example, KCl has cubic structure (Fig. 21.20). Here $\vec{\nabla}\vec{\mathcal{E}} = 0$ because there is spherical symmetry.[2] Thus no quadrupole resonance can be observed. But for various symmetries various quadrupole resonance frequencies can be observed, the examination of which gives an idea about molecular structures.

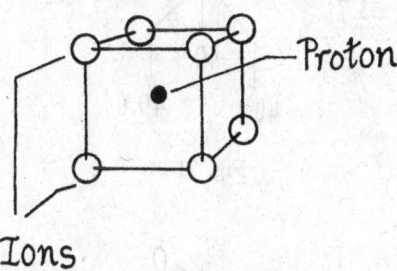

Fig. 21.20.

21.4. Examples

(1) Consider the trichloroacetylchloride molecule (Fig. 21.21). Here we have ^{37}Cl atoms. The observed frequencies are shown in the same figure. The value on the left corresponds to the Cl atom at position (I) while the bunch of values on the right correspond to the Cl atoms at positions (II), (III) and (IV).

(2) Figure 21.22 shows the situation for another molecule where two resonance lines are observed.

(3) Figure 21.23 shows the quadrupole resonance frequency for still another molecule. Here we obtain one single line because we have symmetry, that is, the two Cl atoms are similar and have the same frequency. Hence one line is observed.

(4) Figure 21.24 shows a peculiar situation. Since there are three Cl atoms, one would expect three lines, one for each Cl atom. However, we observe that the three lines are split into two each. (The bunch around 38 and 39 MHz corresponds to the two Cl atoms on the left, which

---

[2]Cubic-structured crystals exhibit spherical symmetry, and hence the field gradient vanishes for them.

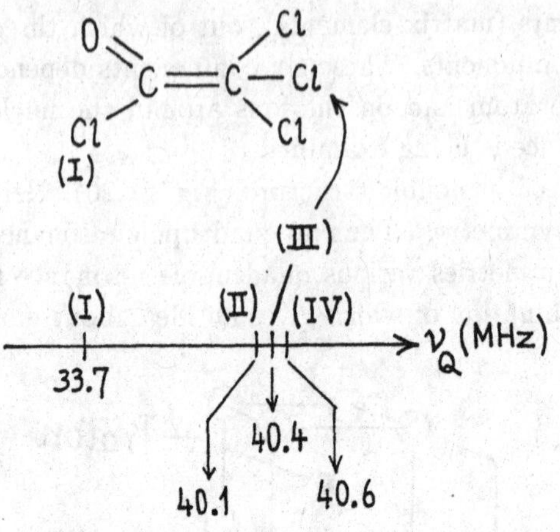

Fig. 21.21.

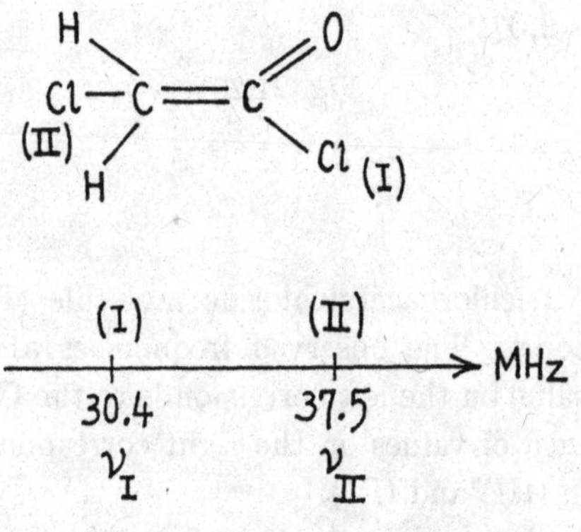

Fig. 21.22.

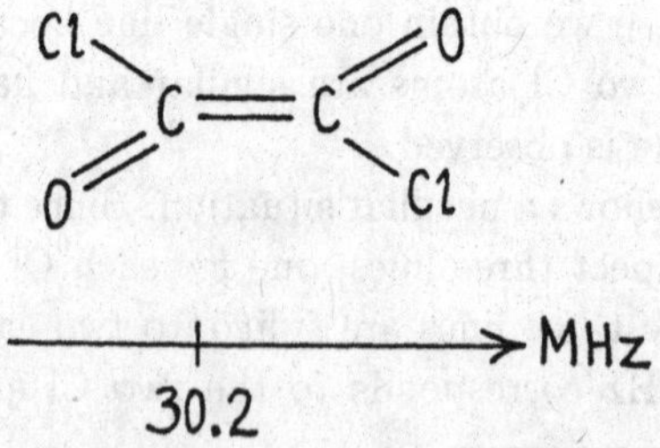

Fig. 21.23.

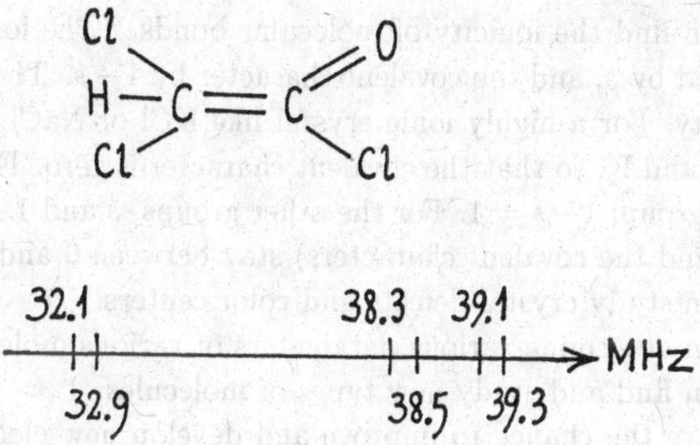

Fig. 21.24.

are close to each other.) This splitting tells us that this molecule has two different states inside the crystal. The corresponding electric-field gradients are different. In other words, the same crystal contains other molecules of the same kind, but of different $\vec{\nabla}\vec{\mathcal{E}}$. This means that this molecule has two different states (with respect to the electric structure) in the crystal structure. All the parameters are the same except $\vec{\nabla}\vec{\mathcal{E}}$, which is different. No other method besides NQR can disclose this situation and bring it to light. Therefore, by examining many resonance spectra, one can acquire knowledge about different molecules.

21.5. Utility of NQR

What is NQR good for? And what is it bad for? NQR can be used to do the following:

(1) We can determine the symmetries of the electric-field gradients in the crystal structure, and their value in various direction. The calculated electric-field gradients can be compared with those found experimentally.

(2) By doing measurements on similar molecules of the same nucleus, we can find out the differences in the molecular structure (*e.g.*, example (4) above).

(3) We can study molecular structure (crystals of $NaClO_3$, $GaSO_4.6H_2O$ etc.).

(4) We can examine the structure and kind of molecular bonds, and also the perturbed wavefunctions of the binding electrons.

(5) We can find the ionicity of molecular bonds. (The ionic character is denoted by s, and the covalent character by $1 - s$. Their sum is equal to unity. For a highly ionic crystal like KCl or NaCl, we have $s = 1$ for Cl and K, so that the covalent character is zero. For atoms of the alkali group, $1 - s = 1$. For the other groups, s and $1 - s$ (that is, the ionic and the covalent characters) stay between 0 and 1.)

(6) We can study crystal defects and color centers.

(7) We can determine various parameters of various molecules.

(8) We can find and study new types of molecules.

(9) We have the chance to improve and develop new electronic spectrometer circuits and methods.

(10) The NQR frequency depends on the temperature T as a result of the molecular vibrations. By studying the resonance frequency as a function of T (*i.e.*, its change with T), we can determine (a) the changes in the molecular structure, and (b) the rotations of the molecule inside a solid body.

Bayer's theory gives the change of ν_Q as a function of T (Fig. 21.25 for CH_2Cl–CH_2Cl). At some temperatures we observe sudden jumps which are due to defects in the molecular structure, or due to the rotation of the molecule inside the solid body. For example, for the antisymmetric molecule CH_2Cl–CH_2Cl (Fig. 21.26), the jump occurs at 170°K. This means that at 170°K the molecule rotates (by 9°) inside the crystal and hence $\vec{\nabla}\vec{\mathcal{E}}$ changes. This rotation is observed by the corresponding change in the graph of $\nu_Q = f(T)$, that is, by the jump in the frequency, as shown in Fig. 21.25.

The disadvantage of NQR is that certain nuclei have zero Q, and hence NQR cannot be observed. Also, it is not observed when a nucleus has $I < 1$. If

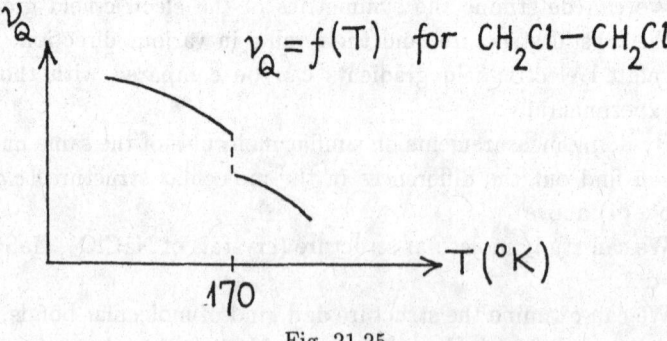

Fig. 21.25.

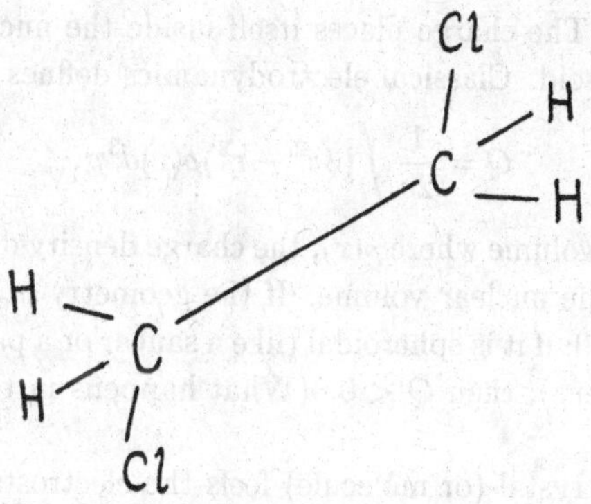

Fig. 21.26.

the number of protons in the nucleus is even (multiples of 4), then $Q = 0$, giving rise to no NQR. Apart from these limiting cases, NQR can be observed most of the times, specifically when the quadrupole moment of the nucleus is nonzero ($Q \neq 0$). So, NQR is observed for some nuclei, and not observed for other nuclei. There are periodic tables showing NMR and NQR observation.

The advantage of NQR is that unlike other molecular methods, it does not require strong magnets (which are large equipment), nor large power sources. It requires only an electronic apparatus which is not complicated, consisting of a few instruments. That makes NQR a preferred method in the molecular physics research. With a simple electronic spectrometer (even an old-type one with tubes) and an oscilloscope, one can observe NQR.

There are two measuring methods in NQR:

(1) The superregenerative-oscillator method.
(2) NMR spectrometers with $H_0 = 0$.

The first method is sensitive and simple (requiring a simple oscillator circuit). A detailed analysis of the method is outside the scope of this book.

As far as Q itself is concerned, it can be measured sensitively by employing the method of molecular beams and the experiments thereof. The fine structure of the molecular spectra can give Q, too.

21.6. Further Discussion on NQR

All nuclei that have spin $I \geq 1$ have an electric quadrupole moment Q (which measures the nuclear distortion off sphericity, in other words, how ellipsoidal the nuclear charge distribution is). Nonzero Q means ellipticity in

the nuclear shape. The charge places itself inside the nuclear volume such as
to produce an ellipsoid. Classical electrodynamics defines Q as

$$Q = \frac{1}{2e} \int (3z^2 - r^2)\rho(\mathbf{r})d^3r ,$$

integrated over the volume where $\rho(\mathbf{r})$, the charge density distribution function,
is nonzero: Here, the nuclear volume. If the geometry is egg-shaped (prolate
nucleus), then $Q > 0$; if it is spheroidal (like a saucer or a pancake, with pressed
poles) (oblate nucleus), then $Q < 0$. (What happens in the case of pulsating
nuclei?)

A nucleus in a crystal (or molecule) feels the electrostatic field that exists
around, in its environment, due to the charges present. If the field symmetry is
not cubic, but less, then we have split energy levels because of the interaction
between the ambient (local) field $\vec{\mathcal{E}}$ and Q. If the spin is I, there are $2I + 1$
split states. These can be directly observed, as a radio-frequency field (chosen
at the proper frequency) is able to cause transitions between them. NQR
means observation of such splittings (nuclear quadrupole states) *when there is
no static field H_0*.

When the molecular bond is covalent (I_2, Cl_2, Br_2 etc.), the quadrupole
splittings are large ($\sim 10^7$ to 10^8 Hz).

Figure 21.27 shows the orientation of a nuclear Q in the local electric field
of four ions (whose electrons are not drawn). We have the orientation for the
lowest energy (a) and the highest energy (b). Figure 21.28 shows the split
energy levels for $I = 1$.

Some nuclei have nonzero I. We can use them (their I) to get information
about the electric field inside a molecule with such nuclei. (For example, H has
$I = 1/2$; D has $I = 1$; Cl has $I = 3/2$ etc.) The electronic angular momentum
$\mathbf{P}$ (which has quantum number J) and the nuclear spin vector $\mathbf{I}$ are quantized
in space, so that the total atomic vector $|\mathbf{F}| = |\mathbf{P} + \mathbf{I}|$ can take values from
$J + I$ to $J - I$. If $J > I$, then $2I + 1$ different energy levels (spectral lines) are
possible, giving the quadrupole fine structure (splitting) which is due to the
spinning of the nucleus about itself *and* the electric field at (the position of)
the nucleus interacting with the spin. The field is due to the electrons around.

How can we calculate the quadrupole-interaction energy? Two charges
q_1 and q_2 separated by a distance r interact by the Coulomb potential energy
$V = q_1 q_2 / r$ (which is actually $E = qV$, with V being the electrostatic *potential*
of either charge on the other: $V = q/r$). In our case, we have a nucleus (nuclear
charge — discrete, or a continuous smear) and an electronic charge distribution
about it (again an assembly of discrete point charges — in which case the

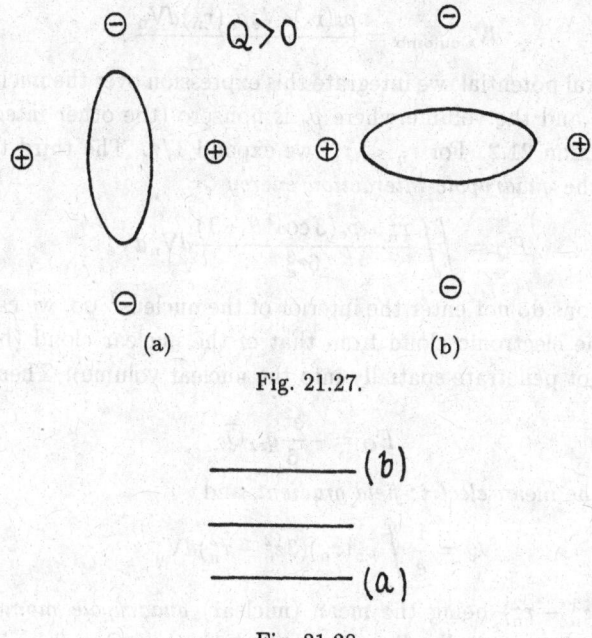

Fig. 21.27.

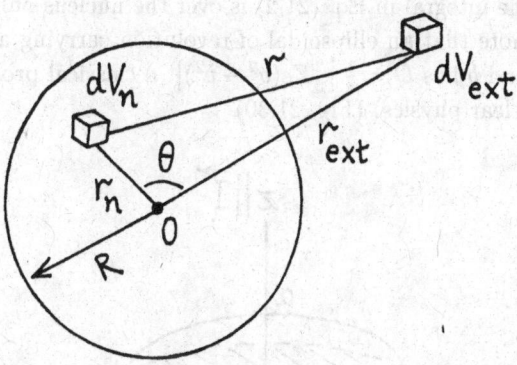

Fig. 21.28.

Fig. 21.29.

distribution can be written as an integral of $q\delta(\mathbf{r})$ instead of a sum of q_i — or a smooth, continuous distribution). Consider an exterior volume element dV_{ext} outside of the nucleus, at r_{ext}, where the electronic charge density is $\rho_e(\mathbf{r}_{ext})$, and an interior (nuclear) volume element dV_n where the nuclear charge density is $\rho_n(\mathbf{r}_n)$. The origin is at the nucleus center (Fig. 21.29). Then

$$dV_{\text{Coulomb}} = \frac{\rho_e(\mathbf{r}_e)dV_e\rho_n(\mathbf{r}_n)dV_n}{r}.$$

To find the total potential, we integrate this expression over the nuclear volume (one integral) and the volume where ρ_e is nonzero (the other integral). This is done in Section 21.2. For $r_n \ll r_e$, we expand $1/r$. The third term of the expansion is the *quadrupole-interaction energy*

$$E_Q = \iint \frac{r_n^2\rho_n\rho_e(3\cos^2\theta - 1)}{6r_e^3}dV_n dV_e.$$

But the electrons do not enter the interior of the nucleus[3]; so, we can separate the sum of the electronic cloud from that of the nuclear cloud (because the former does not penetrate spatially into the nuclear volume). Then

$$E_Q = -\frac{e^2}{6}\bar{q}_{zz}\overline{Q}$$

where $\bar{q}_{zz}$ is the mean *electric-field gradient*, and

$$Q = \frac{1}{e}\int \rho_n(\mathbf{r}_n)(3z_n^2 - r_n^2)dV_n, \tag{21.2}$$

with $\overline{Q} = \langle 3z_n^2 - r_n^2 \rangle$ being the mean (nuclear) *quadrupole moment*. If the nuclear charge is spherically distributed ($I < 1$), then $Q = 0$ by integration. For $Q > 0$ the nucleus is an ellipsoidal elongated along z, by virtue of its charge distribution. The integral in Eq. (21.2) is over the nucleus only.

We should note that an ellipsoidal of revolution carrying a charge Ze and having axes a and b has $Q = \frac{1}{e}\left[\frac{2}{5}Ze(a^2 - b^2)\right]$, a classical problem in electro-statics (and nuclear physics) (Fig. 21.30).

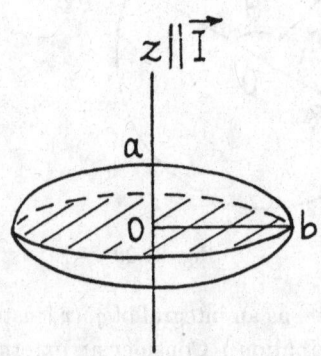

Fig. 21.30.

---

[3]This is usually so; but if it has enough energy, an electron may pass through the nucleus (which is not solid) without having a chance to interact with the nucleons.

The field gradient is a tensor (matrix) with nine components (elements) where

$$q_{zz} = (\vec{\nabla}\vec{\mathcal{E}})_{zz} = \left(\frac{\partial^2 V}{\partial z_n^2}\right)$$

is the gradient produced by the electrons, along the nuclear axis. Usually, the field gradient (produced by electrons) with respect to an axis parallel to the angular momentum has cylindrical symmetry. Hence the tensor $\vec{\nabla}\vec{\mathcal{E}}$ is diagonal, with only three nonzero elements:

$$\frac{\partial^2 V}{\partial x^2} = \frac{\partial^2 V}{\partial y^2} = \frac{1}{2}\frac{\partial^2 V}{\partial z^2} \neq 0$$

or $q_{xx} = q_{yy} = \frac{1}{2}q_{zz} \neq 0$ in

$$\begin{pmatrix} q_{xx} & 0 & 0 \\ 0 & q_{yy} & 0 \\ 0 & 0 & q_{zz} \end{pmatrix}$$

where z here is the principal axis along the nuclear spin angular momentum. Now we will write the (expression of the) field gradient along the principal axes, in terms of the electronic coordinates. If θ is the angle between the nuclear spin and z (or $\mathbf{J}$), and we choose $x \| x_{\text{nuclear}}$, then

$$\frac{\partial^2 V}{\partial z_n^2} = \frac{\partial^2 V}{\partial y^2}\sin^2\theta + \frac{\partial^2 V}{\partial z^2}\cos^2\theta = \left(\frac{3\cos^2\theta - 1}{2}\right)\frac{\partial^2 V}{\partial z^2}.$$

We call $Q' = \overline{\frac{\partial^2 V}{\partial z^2}}$ and have

$$E_Q = \frac{eQQ'}{4}\left(\frac{3\cos^2\theta - 1}{2}\right).$$

If $\vec{\mathcal{E}}$ has no cylindrical symmetry about $\mathbf{J}$ or z, then $\partial^2 V/\partial x^2 \neq \partial^2 V/\partial y^2$, and an *asymmetry parameter* can be defined as

$$\eta = \frac{\dfrac{\partial^2 V}{\partial x^2} - \dfrac{\partial^2 V}{\partial y^2}}{\dfrac{\partial^2 V}{\partial y^2}}.$$

For $I > 1$, we have $Q \neq 0$. If $\mathbf{J} \| \mathbf{I}$, the energy change (in the microwave spectrum) for symmetric molecules is

$$\Delta E = \frac{eQ}{c}\left[\frac{3K^2}{J(J+1)} - 1\right]\left[\frac{\frac{3}{4}C(C+1) - I(I+1)J(J+1)}{2I(2I-1)(2J+1)(2J+3)}\right]\frac{\partial^2 V}{\partial z_0^2}$$

where $C \equiv (J + I + 1)(J + I + 2) - J(J + 1) - I(I + 1)$, V is the potential due to electrons around the examined nucleus of interest, and z_0 is the axis of the symmetric-top molecule.

The experimental results regarding this topic are interesting. The central idea of the pertinent experimental activity is the following: Study the fine structure of the microwave spectrum and hence determine Q of the molecule. And do that for many molecules, or compare molecules having the same nucleus, $e.g.$, ^{35}Cl. If $QQ' = $ constant, any change in QQ' will be due (and proportional) to changes in $\vec{\nabla}\vec{\mathcal{E}}$. For example, $QQ' \simeq 0$ for NaCl (where $Q' \simeq 0$ because each Cl atom is symmetrically surrounded by eight atoms). Also $Q' \to 0$ for atomic (free) ^{35}Cl, as it can be verified by measurements taken by using the atomic-beam technique. (In ^{35}Cl the outermost orbit has seven electrons.) In FCl we have $Q' \neq 0$ because F is too electronegative. When the Cl bond is covalent (like in CH_2Cl, CH_3Cl, ClCN etc.), its Q' is close to the atomic Q' value. In TlCl the Cl bond is rather ionic, and hence Q' has its ionic value here. In SiH_3Cl the bonds are 50% ionic; thus Q' lies in the middle between its atomic and ionic values.

21.7. NQR Splittings

The fact that some nuclei have a Q is fixed by spectroscopic experimentation. Q interacts with the electric-field gradient at its (nuclear) position; as a result, there is an energy splitting. The phenomenon of the transitions between these different energy levels is NQR.[4] It cannot be observed for (nor can it occur in) nuclei with $I \leq 1$ and $Q = 0$. So, NQR depends on (a) the nuclear Q, (b) the nuclear $\mathbf{I}$, and (c) $\vec{\nabla}\vec{\mathcal{E}}$ due to electrons.

(a) Conceptually, $eQ = $ (Charge) $\times$ (Area). The nuclear charge inside the nucleus exhibits cylindrical symmetry with respect to the spin axis. The surface of an ellipsoid (of revolution about an axis) carries a surface charge Ze distributed (spread) on it (Fig. 21.31). Notice that if $a = b$, the ellipsoid degenerates into a sphere, and $Q = 0$ for spherical symmetry. For $a > b$ the ellipsoid is prolate along z, and $Q > 0$; for $a < b$ it is oblate, and $Q < 0$. Q is a volume integral that depends on all nuclear charges distributed in the nuclear volume, $i.e.$, it depends on the nuclear charge distribution ρ_n. Typically, $Q \sim 10^{-24}$ cm$^2 \equiv 1$ barn (where the term *barn* here is a unit so defined, not to be confused with farm barns).

---

[4]First observed by Krüger and Dehmelt. Q was first defined by Casimir.

(b) **I** depends on the number of nucleons (protons and neutrons) *and* their orientation. The experimental conclusions are as follows:

(i) Nuclei with nucleon (mass) number A *and* proton number Z both even, have $\mu = 0$ and total nuclear spin zero.

(ii) If a nucleus has A *and* Z odd, or A odd and Z even, then its spin is $I = (\text{Integer}) + \frac{1}{2}$, and $\mu \neq 0$.

(iii) If a nucleus has A even and Z odd, then $I = \text{integer}$, and $\mu \neq 0$.

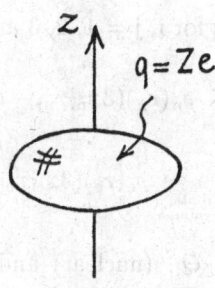

Fig. 21.31.

Since the nuclear forces (that govern nuclear phenomena and act inside the nucleus) are not well-understood, the theoretical reasons of these conclusions are unknown. Only models can be made and put to test to see if they work (agree with what the experiments have to say). The short-range strong forces and hence the internal nuclear structure are still not 100% clear yet. Schmidt's theory assumes that protons and neutrons can be lined up side by side in terms of their individual spin. The particle (nucleon) that is finally left unpaired gives the nuclear spin. (It is assumed that this nucleon has $s = 1/2$ and $l = 0, 1, 2, \ldots$) This theory is useful because the so calculated g_I value agrees with the corresponding experimental value. Of course, better results can be obtained if we can study experimentally the internal nuclear structure and hence build a correct model of nuclear structure (just like the atomic model where all forces are understood).

(c) $\vec{\nabla}\vec{\mathcal{E}}|_{\text{at nuclear position}}$ is due to the atomic electrons *and* the electrons of neighboring atoms in the same molecule, *and* other ions in the crystal, if there is a crystal to speak of. Q interacts with this gradient, as we saw above. The interaction Hamiltonian is

$$\mathcal{H}_Q = \int r_n^2 \rho_n(\mathbf{r}_n) \frac{1}{2}(3\cos^2\theta_{n,e} - 1)\frac{\rho_e(\mathbf{r}_e)}{|\mathbf{r}_e|^3} d\mathrm{V}_e \, d\mathrm{V}_n \, ,$$

with

$$eQ \equiv \int \rho_n(\mathbf{r_n}) \left[\left(\frac{3z_n^2}{|\mathbf{r_n}|^2} - 1 \right) |\mathbf{r_n}|^2 \right] dV_n .$$

If Ze is the nuclear charge, $\rho_n \propto Z$. That is, nuclei alike in shape and structure have the same $Q \propto Z$.

The field gradient can be found as follows: We write

$$eq = \int \rho_e(\mathbf{r_e})(3\cos^2\theta_{e,z} - 1)dV_e .$$

The Cartesian components for i, j = 1, 2, 3 are.

$$eQ_{ij} = \int_{\text{Nucleus}} \rho_n(\mathbf{r_n})(3x_{ni}x_{nj} - \delta_{ij}r_n^2)dV_n$$

$$eq_{ij} = \int_{\substack{\text{Electronic} \\ \text{region volume}}} \rho_e(\mathbf{r_e})(3x_{ei}x_{ej} - \delta_{ij}r_e^2)dV_e$$

in the tensorial form, where Q_{ij} (nuclear) and q_{ij} (electronic) are matrix elements (tensor components). The interaction is

$$\mathcal{H}_Q = -\frac{1}{6} \overset{\leftrightarrow}{\mathbf{Q}} : \vec{\nabla}\vec{\mathcal{E}}$$

in a dyadic form, where the 3×3 gradient tensor $\vec{\nabla}\vec{\mathcal{E}}$ has nine components. If we let

$$V_{ij} \equiv \frac{\partial^2 V}{\partial x_i \partial x_j} \equiv Q'_{ij} ,$$

the Laplace equation requires that $Q'_{xx} + Q'_{yy} + Q'_{zz} = 0$, if the gradient is due to electrons only external to the nucleus. The field tensor has then five nonzero components:

$$(\vec{\nabla}\vec{\mathcal{E}})_0 = \frac{Q'_{zz}}{2}$$

$$(\vec{\nabla}\vec{\mathcal{E}})_{\mp 1} = -\frac{1}{\sqrt{6}}(Q'_{xx} - iQ'_{yz})$$

$$(\vec{\nabla}\vec{\mathcal{E}})_{\pm 2} = \frac{1}{2\sqrt{6}}(Q'_{xx} - Q'_{yy} \pm 2iQ'_{xy}) .$$

The gradient is due to electrostatic charges. Now if we consider the principal axes X, Y, Z, these components become

$$(\vec{\nabla}\vec{\mathcal{E}})_0 = \frac{Q'_{ZZ}}{2} = \frac{1}{2}eq_{ZZ}$$

$$(\vec{\nabla}\vec{\mathcal{E}})_{\mp 1} = 0$$

$$(\vec{\nabla}\vec{\mathcal{E}})_{\pm 2} = \frac{1}{2\sqrt{6}}(Q'_{XX} - Q'_{YY}) = \frac{1}{2\sqrt{6}}e\eta q_{ZZ}$$

where $\eta = (Q'_{XX} - Q'_{YY})/Q'_{ZZ}$.

We take $|Q'_{ZZ}| > |Q'_{XX}| > |Q'_{YY}|$. Hence η varies as $0 \le \eta \le 1$. In general, the measured values are $\eta < 1$. If $\eta = 0$, there is axial symmetry about the Z-axis. Then the energy levels can be calculated from the matrix element

$$\langle m|\mathcal{H}_Q|m'\rangle = \frac{e^2qQ}{4I(2I-1)}[3m^2 - I(I+1)]\delta_{mm'}$$

where

$$\mathcal{H}_Q = \frac{e^2Q}{2I(2I-1)}[I_x^2q_{xx} + I_y^2q_{yy} + I_z^2q_{zz}],$$

and whence

$$E(m_I) = \frac{e^2qQ}{4I(2I-1)}[3m_I^2 - I(I+1)]$$

which is Eq. (21.1). Here $m_I = -I$ to $+I$, *i.e.*, it takes $2I + 1$ distinct values. But we have a degeneracy: $E(-I) = E(+I)$. So, if there is axial symmetry and $I = (\text{Integer}) + \frac{1}{2}$, there are $I + \frac{1}{2}$ states altogether. For example, if $I = 3/2$, we have $m_I = -3/2, -1/2, 1/2, 3/2$, four values. But since $+3/2$ and $-3/2$ are paired to one energy value, and likewise $+1/2$ with $-1/2$, we have only $3/2 + 1/2 = 2$ states.

For $I = 1/2$, $Q = 0$ and hence $E(m_I, I = 1/2) = 0$. The selection rule for transitions is $\Delta m_I = \pm 1$.

If there is no axial symmetry, the energy levels depend on η. The relations between E and η are the following *secular equations* (given for various I values):

I	Equation
3/2	$E^2 - 3\eta^2 - 9 = 0$
5/2	$E^3 - 7(3 + \eta^2)E^2 - 20(1 - \eta^2) = 0$
7/2	$E^4 - 42\left(1 + \dfrac{\eta^2}{3}\right)E^2 - 64(1 - \eta^2)E + 105\left(1 + \dfrac{\eta^2}{3}\right)^2 = 0$
$\vdots$	$\vdots$

The solution for $I = 3/2$ is exact:

$$E(\pm 3/2) = 3\mathcal{A}\sqrt{1 + \frac{\eta^2}{3}} \quad \text{and} \quad E(\pm 1/2) = -3\mathcal{A}\sqrt{1 + \frac{\eta^2}{3}}$$

where $\mathcal{A} \equiv e^2 qQ/12$.

The secular equations can be calculated by perturbation theory, for $\eta^2 \leq 1/4$. Bersohn did it for $\eta \leq 0.1$ and found

$$E(\pm 5/2) = \mathcal{A}\left(10 + \frac{5}{9}\eta^2 - \ldots \text{h.o.t.}\right)$$

where we can neglect h.o.t. in η. Das and Hahn found

$$E(\pm 3/2) = \mathcal{A}(-2 + 3\eta^2 + \ldots)$$

$$E(\pm 1/2) = \mathcal{A}\left(-8 - \frac{32}{9}\eta^2 - \ldots\right).$$

The solutions of the equations for $I = 5/2$, $7/2$, and $9/2$ were found by M.H. Cohen. For $I = 3/2$ we have

$$\Delta E_Q = 6\mathcal{A}\sqrt{1 + \frac{\eta^2}{3}},$$

with $\mathcal{A} = e^2 qQ/12$.

For $I = 5/2$ and $\eta < 0.1$ we have

$$\Delta E_Q(5/2 \rightarrow 3/2) = \mathcal{A}\left(12 - \frac{22}{9}\eta^2\right)$$

$$\Delta E_Q(3/2 \rightarrow 1/2) = \mathcal{A}\left(6 + \frac{5}{9}\eta^2\right).$$

For $I = 3/2$, η cannot be found. But it can be calculated for $5/2$.

NMR spectrometers with $H_0 = 0$ can be used in PQR (pure quadrupole resonance). By PQR we can sensitively measure Q_1/Q_2 for two nuclei. Figure 21.32 shows the block diagram of an NQR spectrometer. By studying the fine structure of a molecular spectrum, we can find Q, $q_{\text{molecular}}$, and q_{atomic}. Figure 21.33 shows the circuit (in principle) of the super-regenerative method used in NQR measurements.

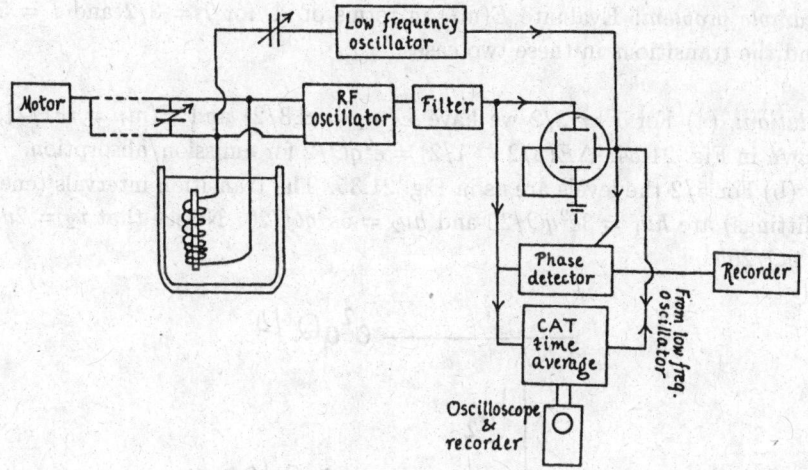

Fig. 21.32.

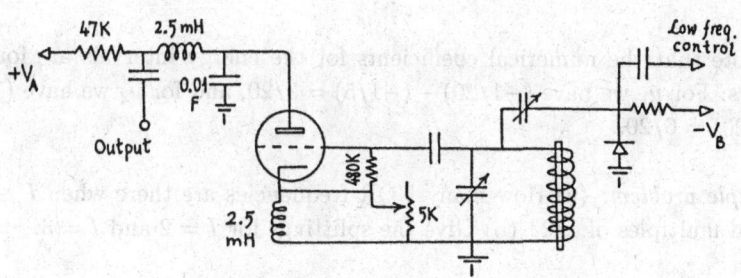

Fig. 21.33.

$$\underline{}\;\; E(\pm 3/2) = e^2 q Q /4$$

$$\updownarrow \;\; \Delta E$$

$$\underline{}\;\; E(\pm 1/2) = -e^2 q Q /4$$

Fig. 21.34.

Example problem: Evaluate $E(m_I)$ in terms of $\mathcal{A}$, for $I = 3/2$ and $I = 5/2$. Find the transitions in these two cases.

Solution: (a) For $I = 3/2$ we have $E(m_I = \pm 3/2)$ and $E(m_I = \pm 1/2)$, as shown in Fig. 21.34. $\Delta E(3/2 \rightleftarrows 1/2) = e^2 qQ/2$ for emission/absorption.

(b) For 5/2 the levels are as in Fig. 21.35. The transition intervals (energy splittings) are $h\nu_1 = 3e^2qQ/20$ and $h\nu_2 = 6e^2qQ/20$. Notice that $\nu_2 = 2\nu_1 = 3e^2qQ/20h$.

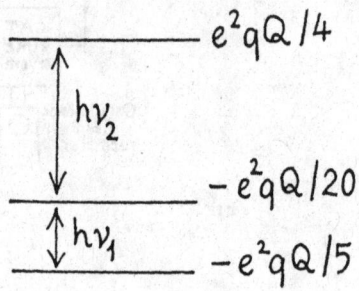

Fig. 21.35.

Note that the numerical coefficients for the energy intervals are found as follows: For ν_1 we have $(-1/20) - (-1/5) = 3/20$, and for ν_2 we have $(1/4) - (-1/20) = 6/20$.

Example problem: (a) How many NQR frequencies are there when I is equal to odd multiples of 1/2? (b) Give the splittings for $I = 2$ and $I = 3$.

Solution: (a) There are $I - \frac{1}{2}$ resonance frequencies where

$$\nu(m) = \frac{3e^2qQ}{4I(2I-1)h}(2|m|+1).$$

If $I =$ integer, then there are $I - 1$ frequencies.

(b) Call $e^2qQ \equiv a$. For $I = 2$ we have the states m $= -2, -1, 1, +2$. But there are actually two, m $= \pm 2$ and m $= \pm 1$ (because m^2 in the E_Q formula gives the same energy, so that the states are paired to common values). The upper state is

$$E(\pm 2) = \frac{a}{(4 \times 2)(2 \times 2 - 1)}[3 \times (2)^2 - 2(2+1)] = \frac{a}{4}.$$

The lower level is

$$E(\pm 1) = \frac{a}{(4 \times 2)(2 \times 2 - 1)}[3 \times (1)^2 - (2+1)] = -\frac{a}{8}.$$

See Fig. 21.36.

For $I = 3$ we have three states that are similarly calculated for m $= \pm 3$, ± 2, ± 1. See Fig. 21.37.

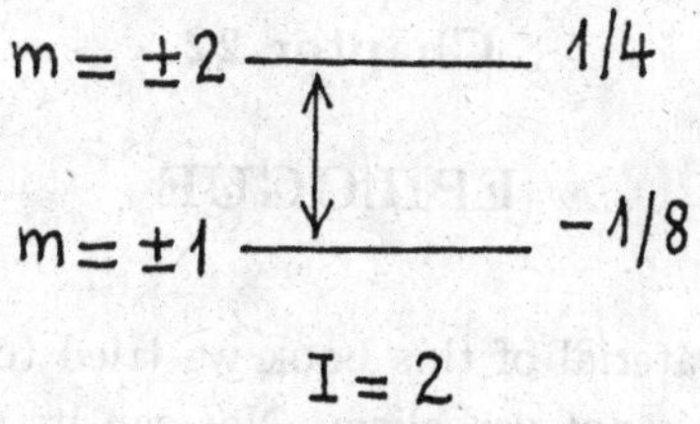

$$m = \pm 2 \underline{\hspace{2cm}} \quad 1/4$$

$$m = \pm 1 \underline{\hspace{2cm}} \quad -1/8$$

$$I = 2$$

Fig. 21.36.

$$m = \pm 3 \underline{\hspace{2cm}} \quad 15/60$$

$$m = \pm 2 \underline{\hspace{2cm}} \quad 0$$

$$m = \pm 1 \underline{\hspace{2cm}} \quad -9/60$$

$$I = 3$$

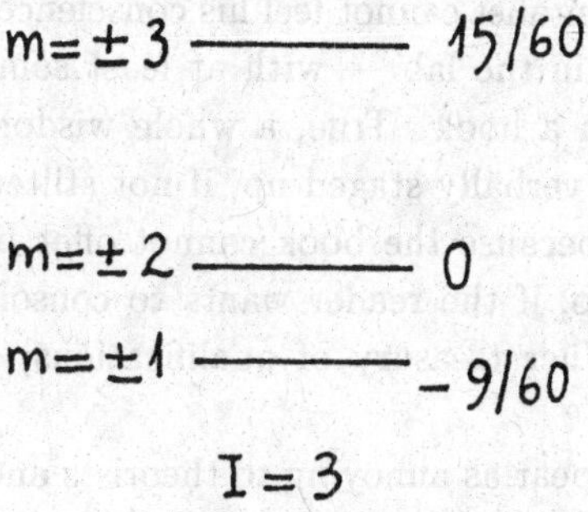

Fig. 21.37.

Chapter 22

EPILOGUE .

In the foregoing material of this book we tried to cook the meal by just chatting. But physics is not verbalism. Nor can its flavor be gotten by just reading books. One cannot get a feeling about physics unless one *enters the laboratory* as a scientist *to actually do physics*, or observe how nature works. Especially an experimentalist cannot feel his conscience at rest without getting himself acquainted — in the lab — with at least some experimental aspects of what he has read in a book. True, a whole wisdom can be learnt from a book, but only all too verbally staged up, if not stilted. There is going to be a missing dimension, because the book cannot offer first-hand experience; it only narrates facts. So, if the reader wants to consolidate the material well and commit it to his/her treasury of qualifications, he/she should attempt something more.

Our stance may appear as annoying to theorists and encouraging to experimentalists, but be it otherwise as it may, physics is definitely not a pencil and a piece of paper, or a computer. It is far more than that. Theory without the thrill of experiments is a dry plethora of words without empirical experience; and conversely, experimental physics alone without theory is an insipid, tasteless activity. But as the two complete each other, their combination leads to perfection and to the splendor of science and philosophy.

Theorists deal with complicated calculations, while experimentalists do physics and take measurements by using tools, instruments, and equipment. Both sides put in a vast amount of patience and invest time in their painstaking toils and efforts. The time and energy devoted require hours, days, months, and even years of hard work, sometimes at a neckbreaking rate. Diligence is generously dedicated to collecting and analyzing data, evaluating results, doing calculations, processing information, making amendments, and getting meaningful outcomes that await a responsible and consistent scientific interpretation. The uninitiated student should not think that this is an easy task. It is no less exhausting and difficult than the training of an operatic singer for a successful stage career, or of a professional soccer player, or of a military-base specialist. But the mysteries of being a physicist nowadays do not end here.[1] Besides physics, a modern physicist has to deal with other accompanying disciplines as well, and solve problems outside his main objective and subject. For example, a theorist may have to draw up efficient computer programs for his data processing, in order to facilitate and speed up his time-consuming complex calculations. He may have to learn the hardware, too. An experimentalist, on the other hand, faces the huge task of building, mastering, maintaining, servicing, and improving or extending his apparatus: Before he proceeds to his actual experimental project, *i.e.*, before he essentially lays hand on the mainly physical part of his research, he has to set up the machinery and keep it operable. This implies and entails parallel knowledge of current experimental technology, engineering, and techniques of practical machine sophistication, together with experience with the know-how of these systems. In other words, he simultaneously becomes a technician, repairman, electrician, electronics designer, plumber, material expert etc. He may have to fix a breakdown, to troubleshoot, to make decisions before he even attempts to start the experiment that involves the actual physics of his research.

Physics may be difficult, but the beauty of this minor art is that a serious theoretical project or a laboratory research is a matter of group work, a task for an enthusiastic and well-coordinated and directed team (if possible, by the employment of the *critical-path method* that minimizes the required time and optimizes the work load). It resembles, for instance, the gigantic task of the production of Verdi's *Rigoletto*, or the construction of a five-star deluxe hotel. In the former case the director has to design and control the difficult logistics of the production, lay out the staging, teach the actors and run the rehearsals,

---

[1] Up to about a hundred years ago, a physicist was also a philosopher, if he was to be considered as a cultured and well-rounded scientist in the literal sense of the word. Unfortunately, the fragmentation of scientific knowledge and the high degree of specialization in our modern times rendered that practically impossible, though there are some exceptions — rare ones — like, for example, Freeman Dyson, J.A. Wheeler, Richard P. Feynman, Isaac Asimov, Carl Sagan, Eugene Merzbacher, Engelbert Schücking, Léon Van Hove, Sydney Brenner, R. Buckminster Fuller, Philip Morrison, Wassily Leontief and others, renowned scientists with a well-pronounced philosophical charisma and radiance.

program the performances, and cooperate with a whole team of people — the conductor, the orchestra, the singers, the choir, the stage technicians, the lighting specialist, the set decorator, the technical assistants, the costume designer and tailors, the house administration etc. He has to bring all these professionals together in a favorable convergence, ensure a climate of harmonious collaboration, and create and keep up a cophasic, synchronized, and well-timed team work in the opera house, in order to finally produce a successful and impeccable stage performance, subject to a merciless criticism the day after. Similarly, in the case of the hotel construction, the contractor architect and the superintendent engineer have to work with consultants, civil engineers, foremen, technicians, workers, specialists, miscellaneous trades, and subcontractors. They have to mobilize the worksite, give management to the entire project, ensure smooth relations and constructive cooperation between all involved parties and with the contracting side, procure the materials, inspect and direct work, provide quality control and testing, overcome red-tape difficulties, solve incidental and accidental problems, schedule partial or parallel works, keep the accounts, make plans and set dates on a bar-chart basis, and finally hand over the hotel in a complete, perfect, safe, and operable condition.

The situation is functionally not different for a college lab team conducting research under the supervision of a professor who is the research project manager. Besides his teaching load, the professor has to design research programs, implement them with his staff, and publish the results in scientific journals.[2] This is a nontrivial task that takes a physics team led by the professor and consisting of senior scientists, researchers, research assistants, fellow graduate students, electronicians, computer programmers and operators, machine-shop

---

[2]This merciless "publish or die" mentality has been so cleverly devised to settle, that it serves two purposes simultaneously: It encourages, and paves the way for, fast and admirable progress through research, *and* keeps the academics busy enough to stay away from getting involved in politics, social and human issues, or exerting pressure on the government. The former purpose is advantageous and beneficial; the latter one may drag the scientists to a cold indifference and insensitivity, to mechanistically do science for the sake of science, instead of science for humanity. The adverse results reached the extent of having today scientists who dare to boldly assert that "the human element must be eliminated from science"! If they mean *human subjectivity* by that, then these words make sense; because science must indeed be divinely objective. But if they mean it literally, then their conceited and inhuman attitude should be faced with caution, because it is a dangerous temptation that may turn a scientist into a criminal. Further, many scientists, communicating with each other in a sealed language, and hurried away by the acquired and overwhelming speed of the progressing science, seem to be forgetting to turn their head back from time to time, to see where the rest of humanity is, to check if humanity is able to follow suit and keep abreast of them.

technicians, and helpers, in a cooperation under the professor's direction. It requires the availability of high-technology equipment, sophisticated and fully equipped lab installations, and auxiliary workshops with all the necessary support systems. We are talking about a costly professional scientific environment: A typical university physics laboratory network with all its might in equipment and accessories costs millions of U.S. dollars. The professor designs and programs the experiments, initiates new creative research programs, moves with the times and keeps up with the current developments in his field (by following a huge literature that keeps him informed of the state-of-the-art matters), finds sponsors interested in his research subjects and goals (so that he gets grants and money to keep the shop running), gets along with the department, chooses talented graduate students and fellows as assistants and degree candidates, teaches courses and deals with student problems as an advisor, maintains efficient contact with the industry, society, and mass media, offers his services to the nonacademic world as an expert consultant, publishes articles, gives talks, attends meetings, conferences, symposia, and social gatherings, makes himself available to visitors, cultivates friendship and keeps correspondence with colleagues, builds archives or computerized data retrieval systems for his activities, responds to auditing, writes up reports and research proposals to be submitted to sponsors who adopt his research project and want to support it (enterprises, scientific institutions, grant foundations, organizations, or the government), and tries to convince them that the ongoing research is useful, fruitful, and important. He may have to design, devise, and build from scratch the whole apparatus, or assign the task to graduate students as a partial requirement for their advanced degree. He should also be a sort of amateur businessman, as he is responsible for the finances of the project, the correct and wise allocation of the grants he receives, and the proper overall management of the economics of his group. All these necessitate a dynamic personality, a scientist of stature with a solid experience, commitment to science, and an established record of research interests, accomplishments, and publications. He must be an able, hard-working man of outstanding intelligence, adroitness, and sparkling intellectual acuity. Obviously, he cannot stand alone against this crushing volume of commitments — commitments for him, crushing to his assistants, that is. He is flanked by a research team consisting of sharp and wary people that feel at home with such an environment, just like an alert heart surgeon who is surrounded by the agile hands of physicians and nurses. It goes without saying that this strenuous and demanding work load requires a team governed or inspired by a strong spirit of group cooperation, and characterized by a disciplined programming, augmented professionalism, rich creativity, capacity to learn, and physical insight (if not aptitude and skill).[3]

Another point to stress here is the specter of *routine*. Routine means a work program or action method that one necessarily gets used to by the conditions. It is a state where the experimental activity loses all of its exciting character and originality, and wears out to become a drudgery of daily necessity, due to the lack of enthusiasm or the overfamiliarity with everyday actions. The activity fades to a dull labor without interest, and appears as if it were a repeated pattern of a monotonous, colorless, and burdensome regularity, without any extraordinary appeal and sensation. Its importance wanes and disappears. Experimental physics (and the theater, to speak of fine arts) is threatened the most by that, and the results would naturally be bad.

Experimental research is a hard task requiring too much patience and time, and a vast input of work. In fact, a full and thorough experiment takes *thirty years*. We are talking about a serious, detailed project in its full glory. It can even extend over the next generation which takes over and picks up the work from where the previous generation of physicists has left off. This is a rigorous research in its strict meaning. Of course, for sensible and practical reasons, no such long-range projects are designed or worked up in colleges for Ph.D. purposes, because a doctorate candidate cannot be condemned to such a long lock-in before he obtains his degree, nor can an M.S. candidate be tied up in an experimental activity planned to last for decades. These projects are meant for major research institutes (like, for example, CERN in Switzerland).

Theoretical work, too, does not fall short of severe difficulties like the ones described above. It is equally painful. But getting back to our principal point, we want to stress out time and again that even a theorist reader of a

---

[3]Finding a sponsor may sometimes be difficult, but in general, one can always find out people interested in nature's work just for fun or out of a real professional interest. There are parties who believe that a certain research activity is important and worth paying for, and they hence support it. These parties are government agencies and private-sector institutions, *e.g.*, educational foundations, industrial trusts, professional associations, private businesses and holdings, special enterprises, leagues, societies and clubs, embassies, memorial and charitable funds, and major establishments that offer research grants and fellowships in sciences, arts, and humanities. As far as hard science is concerned, one can name the U.S. National Science Foundation, the U.S. Department of Energy, the U.S. Navy, the Smithsonian Institution, the Riverside Research Institute, the IBM corporation, ITT, the Fulbright Foundation, the Rockefeller Foundation, UNESCO, NASA, and many others. In the particular case of molecular physics, a drug company, for example, may be interested in sponsoring a particular research, whether experimental or theoretical, and even sending employees for studies or training. Major public libraries and university libraries have catalogs, references, and directories listing such institutions.

book should have a slight feeling about the experimental aspects of molecular physics, because they are *sine qua non* parts of the education. However, this statement of ours should not be taken as dogmatic; it is only an opinion. What we really mean is: Do not be confined within a book only.

The reader of this book may have noticed that — as it happens in other similar textbooks — we have not departed from the well-known textbook philosophy: *Idealization of real life*. That is, a textbook — for instruction or illustration purposes — presents "textbook problems" only: Simple, idealized situations whose solution is either known, or can be found by reducing the problem to a manageable form, *i.e.*, bringing it down to a known configuration and solvable appearance by a series of simplifying assumptions, transformations, mappings, and mathematical tricks. This is further facilitated if the problem in question has a special symmetry (in general, it may not) that renders it relatively easy. So, textbook problems and theories are selected cases or made-up models and idealizations, and simplified versions of real situations encountered in the real physical world. (A real problem usually exhibits no inviting and encouraging elegance, and has too complex, utterly complex and messy a solution, or it may not have a solution at all.) This book has such idealizations (*e.g.*, the dumbbell model of a molecule, or easy, fabricated problems). The material is given mostly qualitatively and descriptively. We have barely touched the surface, giving the principles only, in an oversimplified way.

Despite the versatility of its material, a book cannot cover everything. Certain topics in this book are presented as outrageously brief sections, others are abridged to a few sentences only, and still others had to be done away with. Whole large topics, each with an enormous literature, had to be condensed to chapters in a representative manner, lest the treatment becomes scholastically lengthy and inappropriate for a textbook. All the same, it is hoped that the omissions will not degrade the quality of this work. Anyway, in our days one can learn more things from a scientific article in a journal or a preprint rather than from a book. Given the rapid progress of science today, by the time a textbook finds its way to the market, it may become obsolete! Indeed, the production of a good text takes time, and in the meantime it is overtaken by new developments that cannot be included all one by one. The pace is so fast, that the author hardly manages to enrich the book with the latest information and fresh data. And having frequent updated re-editions of a textbook is not an easy, nor a practical or feasible process. So, the interested reader should keep an eye on the research literature and the minutes of conferences and symposia in the field. It is with this hope that this book is delivered to the readers' hands and to the merciless irradiation of meticulous criticism, including biased carping.

As we march towards the dawning horizon of the 21st century, positive sciences parade along proudly, as though with a solemn chorale of an ancient drama, as the scientists indefatigably push boulders of work and contribution up the hill of their demanding duties, with their soul sometimes clouded by bitter agony and disappointment, and sometimes refreshed by a cherubic joy of success and achievement. Science is thus being transfused from generation to generation, and the yet unwritten history of modern science dematerializes in our minds like a wide, non-handmade and ethereal mural, before which a kneeling humanity offers thanks to God for the wonders of creation and pays due homage to those figures who contributed to science and human civilization.

It is with these feelings that I would like to join the procession of people who stop to marvel at the combination of divine gifts and human productivity.

Finally, I thank God who by his Divine Providence willed to vouchsafe unto me the writing of this humble book. Actually, it is not a book; it is an *attempt* to approach the subject.

ASSIGNMENT PROBLEMS

"But be ye doers of the word, and not hearers only, deceiving your own selves." (James 1:22)

ASSIGNMENT PROBLEMS

1. (a) What is the smallest particle (unit of matter) that still maintains the chemical properties and the chemical identity of the (bulk) material? Is it the atom or the molecule? (b) What is the least particle (part of matter) that still retains the physical properties of the bulk? Is it the atom, the molecule, or the crystal? Why? (c) Roughly estimated, how many air molecules are there in your room?

2. We have chemistry because we have electrons orbiting around the nucleus. We have different chemical elements because we have spin. On the atomic level there is no clear separation line to distinguish between physics and chemistry; an atom is a subject of both physics and chemistry, and the two sciences overlap in terms of jurisdiction microscopically.

Are these statements true or false? Discuss them with comments.

3. You have an unknown chemical substance in your hand or in a container. Is there any way to tell whether it is a chemical element or a molecular compound?

4. By using Avogadro's law, how can you prove that an oxygen molecule consists of two oxygen atoms?

5. (a) How can we draw the conclusion that the chlorine gas consists of Cl_2 molecules and not of Cl atoms? (b) How can we draw the conclusion that the chlorine molecule is Cl_2 and not, say, Cl_3?

6. What differences are there between a diatomic molecule of type A_2 and one of type AB?

7. What is the motional state of the molecules of (a) a gas, (b) a liquid, and (c) a solid?

8. List the physical and chemical differences between the hydrogen atom H and the hydrogen molecule H_2. What is hydrogen "in the making", [H]?

9. (a) On the atomic level, can we draw a sharp border line between physics and chemistry? Why? (b) What do we mean by saying that we have chemistry as a result of the electronic structure in atoms?

10. (a) Prove that the point group $\frac{\bar{4}}{m}m$ is equivalent to $\frac{4}{mmm}$. (b) Prove that $\frac{3}{m} = \bar{6}$. (c) The symmetry $\bar{1}mm$ is equivalent to $\frac{4}{m}$. True or false? (d) Two mirrors are placed at 45° to each other. Show that the point group of this system is $4mm$. What if the angle between the mirrors were 90°? 60°? (e) Prove that a four-legged square table (Fig. AP-1) exhibits a $4mm$ (tetragonal) symmetry.

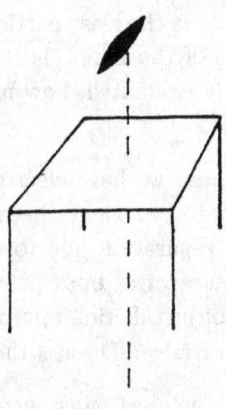

Fig. AP-1.

11. Draw diagrams to illustrate the symmetry groups $\frac{2}{m}$, 222, and $\frac{2}{m}\frac{2}{m}\frac{2}{m} = mmm$.

12. Prove that the SF_6 molecule (Fig. 1.6) has a symmetry point group $m3m(O_h)$.

13. Verify that in the two systems of notation, the alternating axes and the corresponding inversion axes are as follows:

Schönflies	International
S_1	$\bar{2} = m(C_s)$
S_2	$\bar{1}$
S_3	$\bar{6}$
S_4	$\bar{4}$
S_6	$\bar{3}$

14. Why is the knowledge of molecular symmetry so important?

15. An AB type molecule consists of atoms A and B whose atomic wavefunctions are ψ_s and ψ_p respectively. Calculate the molecular wavefunctions of the electrons that form the s–p bond in the molecule. Give a physical interpretation of the result.

16. The bond length between two atoms of the same kind (A_2) is related to the number of the electrons (multiplicity) that give rise to the binding. How? Explain briefly, by giving an example (*e.g.*, carbon atoms).

17. What determines the critical (limit or closest) distance of approach of two atoms, after which the formation of a bond is possible? Why is there no molecular bond whose length can be as large as your room?

18. To separate a CO molecule into carbon and oxygen atoms, an energy of 11 eV is needed. What is the binding energy per molecule? Should you treat the case classically or relativistically?

19. What is an heteropolar (ionic) bond? Give an example and explain.

20. The dissociation energy of H_2 is $D_e = 4.72$ eV, while that of H_2^+ is $D_e = 2.64$ eV. Why less?

21. What is the maximum separation of the ions for the NaCl molecule, if they can be considered as point charges? (Note: The potential energy of interaction must be at least -1.3 eV.)

22. The potential energy of a diatomic molecule is given by

$$V(r) = V_0 \left[\left(\frac{r_0}{r} \right)^{12} - 2 \left(\frac{r_0}{r} \right)^6 \right]$$

which is a function of the distance r between the atoms. This is called *Lennard-Jones potential*. (a) Show that r_0 is the interatomic separation when the potential energy is a minimum. (b) Show that the minimum potential energy is $V(r_0) = -V_0$. (c) Prove that the interatomic separation is $r_0/\sqrt[6]{2}$ when $V(r) = 0$. (d) Plot the graph of $V(r)$.

23. An approximation to the energy of H_2^+ can be obtained if one uses the *geometric mean variation function*

$$\psi = c_1(\psi_A \pm \psi_B) + c_2\sqrt{\psi_A\psi_B}$$

suggested by Radel.[1] Calculate the equilibrium value of r and the binding energy of the ion. (A computer program is needed.)

24. Discuss what happens if r is adiabatically reduced from very large values in H_2^+. (Hint: See Park, pp. 496–512.)

25. Compute the long-range interaction potential between an H atom (in the ground state) and a proton, by considering the fact that p^+ polarizes the atom.

26. In a diatomic molecule, let x be the distance over which the nuclei move in one period, and d be the size of the molecule. Show that the ratio x/d is

$$\frac{x}{d} \sim \sqrt[4]{(m/M)^3}$$

as the molecule vibrates, and

$$\frac{x}{d} \sim \frac{m}{M}$$

as the molecule rotates. (Read Park, pp. 243–247.)

27. (a) The muonic hydrogen molecule ion is an exotic molecule ion consisting of two protons bound by a muon μ^-. If $m_{\text{muon}} = 207m_{\text{electron}}$, find the separation of the two protons and the binding energy of the molecule ion. (b) Molecule ions consisting of a deuteron and a proton bound by a muon (like the ordinary ion HD^+) can be formed by stopping muons in liquid hydrogen containing traces of deuterium. What is the internuclear separation in the ion? (c) Refer to part (b). The nuclei may react and form ^3He. What becomes then of the muon?

28. Two identical oscillators are coupled together by a spring of constant κ. The force is proportional to the difference of the displacements $(x_1 - x_2)$, and the Hamiltonian is

---

[1] S.R. Radel *et al.*, *J. Chem. Phys.*, **50**, 3642 (1969).

$$\mathcal{H} = \frac{1}{2m}\left(p_1^2 + p_2^2\right) + \frac{1}{2}\kappa(x_1^2 + x_2^2) + \frac{1}{2}\alpha(x_1 - x_2)^2$$

$$= \frac{1}{2m}\left(p_1^2 + p_2^2\right) + \frac{1}{2}(\kappa + \alpha)(x_1^2 + x_2^2) - \alpha x_1 x_2$$

where α is a constant that gives the coupling strength. (a) Show that if the oscillators are initially in states n' and n, the coupling causes a first-order shift

$$E^{(1)} = \alpha\left[(n' + n + 1)\frac{\hbar}{2m\omega} \pm (\langle n'|x|n\rangle)^2\right]$$

where $\omega = \sqrt{\kappa/m}$. The sign depends on the symmetry of the unperturbed state. (b) Calculate the matrix element in part (a) by referring to a standard text on quantum mechanics. (c) Part (a) tells us that if one of the two identical oscillators is excited, we may have a first-order interaction between the two. Compute the long-range interaction potential between the two oscillators, if one is in state $n' = 1$ and the other in $n = 0$. (d) Prove that the expression for $\mathcal{H}$ written above is indeed the Hamiltonian of the coupled oscillators.

29. The Morse potential is

$$V(r) = V_0\left[e^{-2a(r-R)} - 2e^{-a(r-R)}\right],$$

a relation that gives the molecular energy (of a diatomic molecule) as a function of the internuclear separation r, and in terms of the equilibrium separation R, the binding (or dissociation) energy V_0, and some constant a (that can be chosen arbitrarily). This relation was first introduced by Morse in 1929. (a) Qualitatively, interpret physically the Morse potential, by referring to its explicit form written above. (b) Prove that $V(R) = -V_0$ and $V(r \to \infty) = 0$. (c) Prove that for small vibrations, the frequency is $(a/2\pi)\sqrt{2V_0/\tilde{\mu}}$ where $\tilde{\mu}$ is the reduced mass. (Read Schiff, pp. 445–455.)

30. A diatomic molecule consists of two masses, m_1 and m_2, separated by a distance r. The molecule rotates about an axis through its center of mass and perpendicular to r (*i.e.*, to the line joining the masses — actually the atoms). Prove that the system is equivalent to a single mass $\tilde{\mu}$ rotating at a distance r about the same axis, where

$$\tilde{\mu} = \frac{m_1 m_2}{m_1 + m_2}.$$

(In other words, two masses at a separation distance r rotating about their common center of mass can be reduced to, or thought as, a single equivalent mass $\tilde{\mu}$.)

31. Two point masses m_1 and m_2 are held by a linear spring of constant κ, and they vibrate at a frequency $f = (1/2\pi)\sqrt{\kappa/\tilde{\mu}}$ where $\tilde{\mu}$ is the reduced mass. (a) Show that the frequency of the system is indeed as given above. (b) If the two masses rotate about an axis through their center of mass and normal to the line between them, show that the moment of inertia of the system is $\tilde{\mu}r^2$ where r is the separation distance between the masses.

32. Show that the moment of inertia of a diatomic molecule about the center of mass of the two atoms of masses m_1 and m_2 separated by a distance d is $\tilde{\mu}d^2$ where $\tilde{\mu}$ is the reduced mass.

33. The rotation spectrum of HCl contains the following wavelengths (in μm): 60.4, 69.0, 80.4, 96.4, 120.4. Calculate the moment of inertia of the molecule.

34. Ignore the vibrational levels and show that the frequencies in a purely rotational spectrum are integer multiples of $h/2\pi\mathcal{I}$.

35. In the diatomic O_2 molecule the interatomic distance is 2×10^{-10} m. Find the moment of inertia about an axis through the center of mass and perpendicular to the line joining the two atoms.

36. The distance between atoms in H_2 is 0.074 nm. (a) What is the moment of inertia of H_2 about an axis passing through the center and perpendicular to the line joining the H atoms? (b) What are the energies of the rotational levels $J = 0$, $J = 1$, and $J = 2$? (c) What is the frequency of the photon emitted in the transition $J = 2 \rightarrow J = 0$?

37. A mass m is stuck to one end of a rigid massless rod which can rotate about an axis passing through its other end which is kept fixed. Solve the Schrödinger equation to prove that the eigenenergies of this system (rigid rotor) are

$$E = \frac{\hbar^2 J(J+1)}{2\mathcal{I}}$$

where $J = 0, 1, 2, \ldots$ and $\mathcal{I} = mr^2$, with $\mathcal{I}$ being the moment of inertia about the rotation axis, and r being the length of the rod.

38. Consider a rigid rotor which has zero rotational kinetic energy about a specific axis, and equal moments of inertia about the two perpendicular axes. Write down and solve the Schrödinger equation for this case.

39. The angular momentum of a dumbbell rotating about an axis normal to the massless rod connecting the two masses is $P = h\sqrt{J(J+1)}$ where the

rotational quantum number J can take integral values: $J = 0, 1, 2, \ldots$. The rotational kinetic energy of such an object is $E = (1/2)\mathcal{I}\omega^2 = (\mathcal{I}\omega^2)/2\mathcal{I} = P^2/2\mathcal{I}$ where ω is the angular velocity and $\mathcal{I}$ is the moment of inertia about the rotation axis. Show that the permitted energies of the rotator are restricted to the quantized values $E_J = J(J+1)\hbar^2/2\mathcal{I}$. (Note: Do not just plug in P; obtain the result by actually solving the Schrödinger equation.)

40. The interatomic distance in diatomic molecules is typically of the order of a few Å. Use the result of the previous problem to show that the spectrum of lines emitted (or absorbed) by diatomic molecules undergoing quantum transitions in their rotational states (pure rotation spectrum) lies in the far infrared and microwave regions. (The permitted transitions obey the selection rule that J changes by one integer at a time.)

41. (a) Derive the selection rules for transitions between rotational levels in a diatomic molecule (b) What difference, if any, is there in the selection rules for vibrational transitions between the case of an A_2 type and the case of an AB type diatomic molecule?

42. The vibrational frequency of H_2 is 1.29×10^{14} Hz. (a) Find the spacing of adjacent vibrational energy levels (in eV). In what region of the spectrum is it? (b) What is the wavelength of the emitted radiation in the transition from $v = 2$ to $v' = 1$ vibrational state? (c) From what initial values of v do transitions to the ground vibrational state give visible radiation?

43. The allowed energies of a simple harmonic oscillator whose classical oscillation frequency is f are $E_v = \left(v + \frac{1}{2}\right)hf$ where the vibrational quantum number is $v = 0, 1, 2, \ldots$ (integral values). (a) Prove that when transitions occur between adjacent energy levels, the photons emitted have the same frequency as the equivalent classical oscillator. (b) The minimum energy of the oscillator is $E_0 = (1/2)hf$ which is the zero-point energy. Is this valid classically?

44. Consider the $K^{35}Cl$ molecule. (a) Calculate the moment of inertia and the energies of the first four rotational levels. What do you observe? Why? (b) Calculate the wavelengths of rotational lines that are due to transitions between these levels.

45. Consider the rotational spectrum line which is due to the transition $J + 1 \rightarrow J$ in a diatomic molecule. Prove that the frequency of the line lies between the classical rotational frequencies of the molecule in the states $J + 1$ (upper) and J (lower). The rotational kinetic energy (of state J) is Eq. (10.1).

46. The dissociation energy of the H_2 molecule is 4.72 eV. This is the energy needed to completely separate the two atoms. Find the temperature at which the mean kinetic energy of H_2 equals this energy.

47. The equilibrium separation in RbCl is 2.89 Å and the dissociation energy is 3.96 eV. Assume that the Van der Waals attraction is almost equal to the zero-point vibrational energy. Give an estimate for the repulsive potential energy.

48. In HCl, for spring constant κ for the H–Cl bond is 470 Newton/m. What is the probability that a molecule is in the first excited state at 300°K?

49. In H_2, the spring constant is $\kappa = 573$ Newton/m. Neglect terms of anharmonicity in the potential energy expansion and find the vibrational level (*i.e.*, v) corresponding to the dissociation energy.

50. In ^{7}LiH, the equilibrium separation is 1.59 Å and the vibrational wavenumber is $\bar{\nu} = 1406$ cm^{-1}. (a) What is the energy of the first excited rotational level? (b) What is the ratio of the energy in part (a) to the energy of the first vibrational state? Comment on that. (c) What is the spring constant for the vibration of the molecule (at the frequency that corresponds to $\bar{\nu}$ given above)? (d) Since the rotational levels are very closely spaced together (compared to the vibrational levels), is it correct to say that a (pure) rotational interlevel transition is far easier and more favored than a vibrational one?

51. We have a collection of HCl molecules in their ground vibrational state v = 0. Find the ratio of the number of molecules in the ground rotational state ($J = 0$) to the number of those in the first excited state ($J = 1$) at 290°K. For the HCl molecule, $\mathcal{I} = 2.7 \times 10^{-47}$ kg m^2. (Note: Do not forget that the statistical weight of each J state is $2J + 1$.) (Answer: 0.4.)

52. Prove that (a) half of the molecules are vibrationally excited if $kT = h\nu$ for the vibrational states, and (b) almost half of the molecules rotate if kT becomes equal to the lowest rotational energy E_J. (Note: Each rotational state has a statistical weight of $2J + 1$.)

53. Consider the H_2 molecule. (a) At what temperature is kT equal to the energy of the first excited rotational level? (b) At what temperature is kT equal to the energy of the first vibrational state? ($\kappa = 573$ Newton/m.) (c) Which rotational level's energy lies closest to the energy for v = 1? (d) In the state $J = 1$, v = 1, how many times does the molecule rotate per one vibration?

54. Calculate the constant a of the Morse potential for Na_2 where $R = 3.07$ Å, $V_0 = E_{bind} = (1/4)$ eV, and $\Delta\bar{\nu}$ (for the fundamental vibration transition) $= 157.8$ cm^{-1}. (Answer: 0.84/Å.) Plot the Morse function for Na_2.

55. The Raman spectrum of the hydrogen gas gives $\bar{\nu} = 4159.2$ cm^{-1} for the fundamental vibration transition $(v = 0 \rightarrow v = 1)$. (a) The effective spring constant is

$$\kappa^* = \left.\frac{d^2V}{dr^2}\right|_{r=R}.$$

Evaluate it for ^{2}H. (Answer: 520 Newton/m.) (b) Find $\bar{\nu}$ for the same transition in D_2 and compare it to the experimental value 2990.3 cm^{-1}. (Note: D_2 is an hydrogen molecule consisting of two deuterium atoms.)

56. Consider the HCl molecule and refer to the table in Section 10.1. where J is tabulated against $\bar{\nu}$ and $\Delta\bar{\nu}$. Deduce the equilibrium separation between the H$^+$ and Cl$^-$ ions. Take the former as protons and the latter as ions of the isotope ^{35}Cl. Compare the result with the theoretical value (for the ground state) 1.27 Å which is a bit lower. Why?

57. Refer to the previous problem. The table shows that $\Delta\bar{\nu}$ decreases as J goes up. Explain the reason.

58. Consider Eqs. (11.4), (11.5) and (11.6). Show that the P and R-branches in the Fortrat diagram are both represented by the same equation in the case when J (or K) can take negative values.

59. Consider absorption spectroscopy with the HCl molecule. The apparatus is at room temperature where measurements are being made. The thermal energy per degree of freedom is $kT \sim (1/40)$ eV. How many degrees of freedom does the rotating HCl molecule have, if we exclude vibration and spinning about the longitudinal axis?

60. In Section 11.6 the three normal frequencies of CO_2 are given. These are observed in the infrared region interval of 1–15 μm (where more than forty different vibration bands are observed, and the three stated values are attributed to fundamental vibrations). And yet in the discussion we said that ν_1 is inactive in the infrared region. How do you resolve the contradiction?

61. How many vibrational modes (ways) are there in the linear CO_2 molecule? Which of them are observed in the infrared and which ones in the Raman spectrum? Explain and give reasons.

62. The rotational energy states of long, symmetric top type molecules are

$$E_{J,K} = Bh\,J(J+1) + h(A-C)K^2\,.$$

(a) What are the names of the constants A, B, and C? Which physical quantities of the molecule are they related to, and how? (b) What are J and K called? Which molecular physical quantities are they related to? How? (c) What are the selection rules pertinent to the energy expression given in part (a)? Explain.

63. What do we call selection rule? Give a brief explanation.

64. What is the utility of the absorption and emission spectroscopy? What things about molecules do we learn from spectroscopy?

65. What information about molecules can we obtain by using diffraction methods? Explain briefly how a diffraction method works.

66. Derive Bragg's law.

67. (a) Explain the method of X-ray diffraction. (b) By using this method, how can we deduce that a crystal has a face-centered cubic unit cell? Explain qualitatively. (c) In such a crystal, the side of the unit cell is $d = 5.2$ Å. The wavelength of the used X-rays is $\lambda = 1.56$ Å. (Notice that λ is comparable to d for good results.) What is the angle between the incident rays and the first-order rays diffracted by the plane 3, 0, 0?

68. In a NaCl crystal, a monochromatic beam of X-rays is diffracted according to Bragg's law. The wavelength of the beam is $\lambda = 0.97$ Å, and the first-order maximum is obtained at an angle of $10°$. (a) What is the lattice constant (spacing between principal planes) of NaCl? (b) At what angle will the second-order diffraction occur? (c) Is there a maximum order of diffraction? (Answer: (a) 2.82 Å. (b) $19°$.)

69. Monochromatic X-rays with an energy of 5 MeV hit a single KCl crystal whose lattice constant is 3.14 Å. At what angle with respect to the incident beam can we observe first-order Bragg reflection?

70. In the diffraction of X-rays, the incoming rays may not be monochromatic. (a) Explain how can you analyze them by a Bragg type X-ray spectrometer, and how can you determine their low-wavelength limit. (b) What parameters(s) must you vary so that this wavelength comes up to the surface of the crystal?

71. Electrons emitted by a hot filament are accelerated under a voltage of 30 kVolts and then they impinge (at normal incidence) upon a thin layer of calcite ($CaCO_3$, chalk) whose lattice constant is 3.03×10^{-8} cm. Assume that you can use Bragg's law. (a) Find the first-order diffraction angle of the electron beam. (b) In order to study the structure of the calcite crystal, would you rather employ electron diffraction or X-rays? Why? (c) What if you had an aluminum foil instead of calcite?

72. A narrow beam of thermal neutrons is incident upon a single NaCl crystal whose lattice constant is 2.82 Å. (a) At what angle with respect to the crystal surface must the Bragg planes be oriented in order to give a first-order diffraction for neutrons having an energy of 0.05 eV? (b) What is the angle between the incident and diffracted beams?

73. Explain briefly the principle of NMR. What is the practical utility of this method in molecular biology, biochemistry, pharmacy, medicine, chemistry, and physics?

74. In NMR, how do absorption and emission take place?

75. In a molecular-beam magnetic resonance experiment with NaF beams, one obtains a resonance minimum for the ^{19}F nucleus at 5.634×10^6 Hz inside a homogeneous magnetic field of 1408 Gauss. Find the nuclear magnetic moment of the ^{19}F nucleus.

76. Given that

$$e = (16021.0 \pm 0.6) \times 10^{-24} \text{ emu}$$

$$h = (6625.59 \pm 0.48) \times 10^{-30} \text{ erg sec}$$

$$m_{electron} = (9109.1 \pm 0.4) \times 10^{-31} \text{ gr}$$

$$c = 299792.5 \pm 0.3 \text{ km/sec}$$

$$\frac{m_{proton}}{m_{electron}} = 1836.10 \pm 0.03,$$

calculate the nuclear magneton μ_n and the root mean square error $\Delta\mu_n$.

77. Given that $\mu_n = (5050.5 \pm 0.4) \times 10^{-27}$ erg/Gauss, calculate $\mu_\perp$ for the following nuclei:

Nucleus	g	I
^{1}H	5.585	1/2
^{9}Be	0.785	3/2
^{10}B	0.6	3
^{13}C	1.405	1/2
^{14}N	0.404	1

78. Refer to the previous problem. Which of the listed nuclei must be in a strong field so that the NMR frequency be $\nu_0 = 30$ MHz?

79. What is EPR in principle? Explain briefly.

80. What is a color center? Explain.

81. Prove that the quadrupole moment Q has the dimensions of an area.

82. Prove that the quadrupole moment of an ellipsoid carrying a charge Ze is

$$Q = \frac{2}{5}Z(a^2 - b^2)$$

where a and b are the semimajor and semiminor axes of the ellipsoid of revolution.

83. Calculate the quadrupole moment of the linear electric quadrupole shown in Fig. AP-2 (which is actually two opposite dipoles). (Hint: Since the charge configuration is equivalent to two equal and opposite dipoles (Fig. AP-3), you have to prove that $Q = 0$ because no area is formed by this array of charges. Prove it explicitly.)

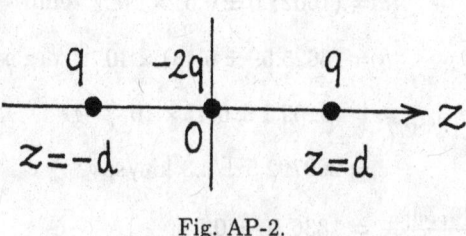

Fig. AP-2.

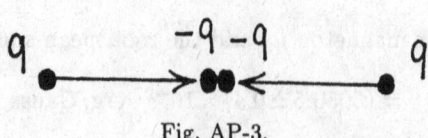

Fig. AP-3.

84. Find the quadrupole moment of the charge arrangement shown in Fig. AP-4.

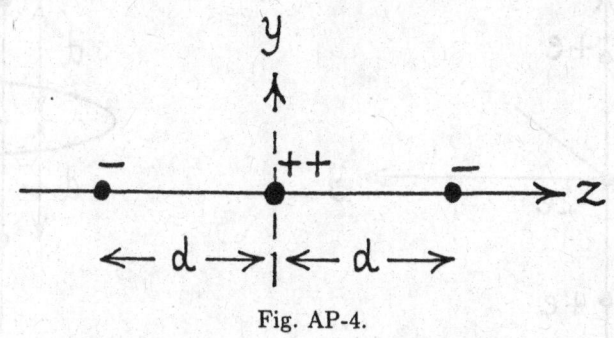

Fig. AP-4.

85. Calculate the quadrupole moment of the following charge distributions: (a) Three point charges on a line, as shown in Fig. AP-5(a). (b) Two equal and opposite point charges separated by a distance $2d$ and a uniformly charged ring of total charge $-e$, placed as shown in Fig. AP-5(b). (c) A sphere whose symmetric upper and lower caps are uniformly surface-charged with a total charge $+e$ each, while the middle zone is surface-charged with $-Ze$, as shown in Fig. AP-5(c). (All charges are on the outer surface.) The general expression for Q is

$$eQ = \int \rho(\mathbf{r})(3z^2 - r^2)d\mathrm{V} \,.$$

86. Consider the Spaeth case explained in Section 19.6. (a) What frequency was the ESR done at? (b) What is the electron–proton interaction coefficient $\mathcal{A} = ?$ (c) Calculate the hyperfine-structure coefficient for this case.

87. A coil of inductance L_0 connected in series to a resistor R_0 is completely filled with a sample that has a spin system with susceptibilities $\chi'(\omega)$ and $\chi''(\omega)$. (Here $\chi = \chi' + i\chi''$ for a linearly polarized radio-frequency field.) Prove that the inductance (at ω) takes the value $L = L_0[1 + 4\pi\chi'(\omega)]$ connected in series to an equivalent resistance $R = R_0 + 4\pi\omega\chi''(\omega)L_0$.

88. Consider a rotating reference frame of Cartesian coordinates where the angular velocity is ω and $d\hat{\mathbf{x}}/dt = \omega_y\hat{\mathbf{z}} - \omega_z\hat{\mathbf{y}}$ and so on. If $\mathbf{H}_0 = H_0\hat{\mathbf{z}}$ is the external static field (in which the nuclei are placed), show that the equation

$$\frac{d\vec{\mathcal{M}}}{dt} = \gamma\vec{\mathcal{M}} \times \mathbf{H}_0$$

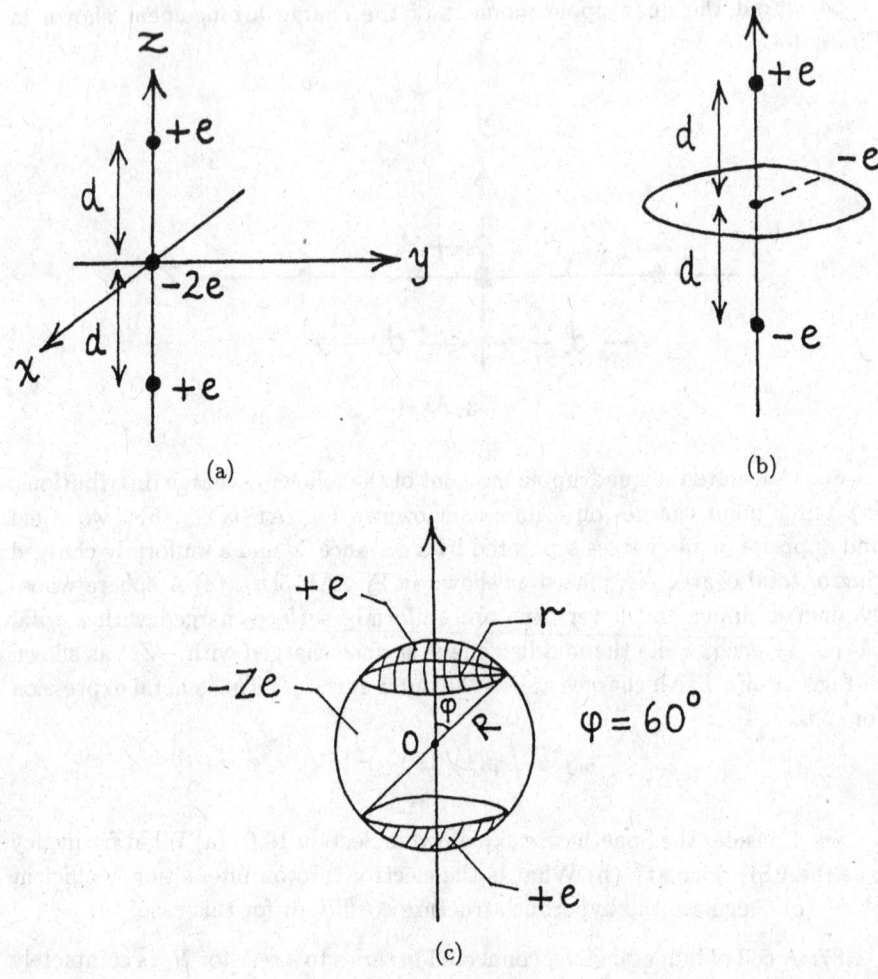

Fig. AP-5.

is equivalent to

$$\frac{d\vec{\mathcal{M}}}{dt}\bigg|_{\text{rot}} = \gamma\vec{\mathcal{M}} \times \left(\mathbf{H}_0 + \frac{\vec{\omega}}{\gamma}\right)$$

where "rot" means as viewed in the rotating frame. (Hint: Use the equation of vector addition in classical mechanics

$$\frac{d}{dt}\bigg|_{\text{fixed}} = \frac{d}{dt}\bigg|_{\text{rot}} + \vec{\omega}\times.$$

that operationally applies on any vector.)

89. Refer to the previous problem where the equation of motion for $\vec{\mathcal{M}}$ in a rotating frame is given. If $\vec{\omega} = -\gamma H_0 \hat{z}$, the static magnetic field vanishes in the rotating frame. (a) Suppose that one turns on a DC pulse $H\hat{x}$ that lasts for τ seconds. The magnetization was towards $\hat{z}$ in the beginning. Find τ if the magnetization flips to $-\hat{z}$ by the time the pulse is switched off. (Although incorrect, you may ignore relaxation, for the sake of simplicity.) (b) Express the pulse with respect to the fixed (laboratory) frame.

90. Consider a conduction electron in a metal. Its spin experiences an overall magnetic field due to the (hyperfine) interaction between the electron spin and the nuclear spin. This is a hyperfine effect on EPR in a metal. Suppose that

$$H_z = \frac{a}{N} \sum_{i=1}^{N} (I_z)_i$$

where H_z is the z-component that the j-th electron experiences, N is the number of nuclei that contribute to the effect, a is the interaction strength in $\mathcal{H} = a\mathbf{I} \cdot \mathbf{S}$, and $(I_z)_i$ is the z-component of the spin of the i-th nucleus, which can be either $+1/2$ or $-1/2$ equiprobably. (Notice that the sum runs over all nuclei that participate in the effect, not necessarily all the nuclei of the sample substance.) If N is too large, prove that the average of the field moments are

$$\overline{H_z^2} = a^2/4N \quad \text{and} \quad \overline{H_z^4} = 3a^4/16N^2 .$$

91. A two-level spin system, with populations N_1, N_2, and rates of transition $\dot{\rho}(1 \to 2)$, $\dot{\rho}(2 \to 1)$, is in a field $\mathbf{H}_0$ along $\hat{z}$. The system is in equilibrium at temperature T. A radio-frequency pulse is turned on, giving a transition rate $\dot{\rho}$. Define $N = N_1 + N_2$, $N' = N_1 - N_2$, and $1/\tau = \dot{\rho}(1 \to 2) + \dot{\rho}(2 \to 1)$. (a) What is $d\mathcal{M}_z/dt$ equal to? (b) Prove that when steady state sets on, $\mathcal{M}_z = \mathcal{M}_0/(1 + 2\dot{\rho}\tau)$. (Notice that if the quantity $2\dot{\rho}\tau$ is too small, the radio-frequency field has negligible effect, *i.e.*, it does not get the population distribution too much away from what it was in thermal equilibrium.) (c) At what rate is energy sucked from the field? (d) Discuss the case when $\dot{\rho} \to 1/2\tau$, *i.e.*, the *radio-field saturation* (by which we can measure τ).

92. Two particles of magnetic moments μ_1, μ_2, and spins $s_1 = s_2 \equiv s$, have zero orbital angular momentum and a total number spin $\mathbf{I}$. Find the total magnetic moment of this system when $I = I$ and when $I = 0$.

93. Two particles of masses m_1, m_2, and charges q_1, q_2 respectively, move circularly about their common center of mass. (a) What is the gyromagnetic ratio γ of this system? (b) What is g_l, the orbital factor? (c) Find γ and g_l for the following systems: (I) The n^0 and p^+ in the deuterium nucleus. (II) A negative muon ($m_\mu = 207\ m_e$) bound to a proton (an exotic muonic atom where we have μ^- instead of e^-). (III) The HCl molecule. (IV) The H_2 molecule. (c) Since cases (III) and (IV) of the previous part actually concern a three-body problem, is it legitimate to extend the substance of parts (a) and (b) to them?

94. The nuclear electric quadrupole moment of a proton is $Q = \overline{3z^2 - r^2}$ when $m_I = I$, *i.e.*, when $\mathbf{I}$ is fully aligned along the z-axis (full orientation). The wavefunction of the proton is $\psi = $ (Spatial part) $\times$ (Spin part) $= R(r)\ Y(\theta, \varphi)\chi$ where χ is the spin function, and the spatial (configurational) part is resolved to the radial and angular (spherical harmonic) part. (This is the full function when spin is included and not ignored.) (a) Prove that

$$Q = \left(-\frac{2l}{2l+3}\right)\overline{r^2} = \left(-\frac{2I-1}{2I+2}\right)\overline{r^2}$$

for $I = l \pm \frac{1}{2}$. (b) Prove that if $m_I \neq I$,

$$Q(m_I) = \left[\frac{3m_I - I(I+1)}{I(2I-1)}\right] Q(m_I = I).$$

(c) Is Q here classical or quantum mechanical? If the latter is true (if Q is an operator), is it not more accurate to write $Q = \langle 3z^2 - r^2 \rangle$?

95. Consider the following *hypothetical* molecular model (which is actually too rough and idealized to be correct): A molecule is taken as a sphere whose diameter is equal to the molecular size. Surface effects are considered, *i.e.*, an atom near the surface of the sphere is expected to be less bound than one further inside, because it has other atoms only on one side instead of all around. (This is only an assumption. It is not always true, because the bond strength may not always depend on the atomic position with respect to the center of the molecule.) In this model calculate the surface energy of the molecule. Assume that the binding energy of an atom is equal to the surface energy that corresponds to a slight bulge in the molecular surface when the atom is about to leave the molecule and fly off.

96. Does a spherical charge distribution (*e.g.*, a spherical nucleus) have a quadrupole moment? What if the nucleus is deformed (*e.g.*, elongated or ellipsoidal)?

97. Consider an ideal gas under normal conditions. Let each molecule be located at the center of a fictitious cube. (a) Prove that the side of this cube is $\sim 3 \times 10^{-7}$ cm and that it is ten times as big as the molecular diameter. (a) Re-answer part (a) by considering a mol of (liquid) H_2O. It occupies 18 cm^3. (Answer: $\sim 3 \times 10^{-10}$ m, almost the same as the molecular size.)

98. A gas consists of a number of molecules as big as the population of the United States (200 million). At standard conditions, what would the side of the gas container be, if the gas were placed in a cubic container? (Answer: $\sim 2 \times 10^{-4}$ cm.)

99. High vacuum of the order of 10^{-10} Torr (10^{-13} atm) can be achieved today in the laboratory. (Currently, this is the lowest attainable pressure.) At this pressure, how many molecules are there per cm^3? Take $T = 300°$K. (Answer: 2×10^6.)

100. Consider an O_2 molecule at 300°K. (a) Find its mean kinetic energy due to translation. (b) Find $\overline{v^2}$. (c) Find v_{rms}. (d) Find its momentum as it travels at v_{rms}. (e) Imagine one such molecule confined in a cubic box of 10 cm to a side. As the molecule travels back and forth, hitting opposite walls, calculate the mean force exerted on the walls. (f) Calculate the pressure in part (e). (g) If the monitored pressure is 1 atm, how many such molecules are there? How does this number compare with the true number of O_2 molecules occupying 1000 cm^3 at 300°K and 1 atm? (Answer: (a) 6×10^{-21} Joule. (b) 2.3×10^5 (m/sec)^2. (c) 483 m/sec. (d) 2.5×10^{-23} kg m/sec, single molecule. (e) 10^{-19} Nt. (f) 10^{-17} Nt/m^2. (g) 8×10^{21}. Three times as many.)

101. (a) Find the temperature at which $v_{rms}(O_2) = v_{rms}(H_2)$ at $T = 273°$K. (b) Find the temperature at which $v_{rms}(H_2)$ will equal the gravitational escape velocity from the Earth. (Answer: (a) 4368°K. (b) $10^{5°}$K.)

102. At a given temperature, different molecules have different v_{rms} (and hence different diffusion rates in the molecular vapor state). Why?

103. (a) Calculate the heat capacity c_V of H_2 and compare it with that of H_2O. (b) In calculating the molar c_V of a diatomic gas, one should take into account that there is an additional mean kinetic energy (degree of freedom) kT (per molecule) that accounts for the rotational motion. Keeping this in mind, calculate the molar c_V of a diatomic gas and compare the result with those tabulated below (which are experimental):

Gas		c_P	c_V	c_P/c_V
H_2	Diatomic	6.87	4.88	1.41
O_2		7.03	5.04	1.40
N_2		6.95	4.96	1.40
CO		6.97	4.98	1.40
CO_2	Polyatomic	8.83	6.80	1.30
SO_2		9.65	7.50	1.29
H_2S		8.27	6.20	1.34

$$\text{cal/mol}^\circ C$$

(Answer: (a) 10.4 Joules/gr°C. About twice as great. (b) 20.8 Joules/mol°K = 4.97 cal/mol°K.)

104. Refer to Problem 35. If the rotational kinetic energy of the molecule is $(1/2)kT$, find ω_{rms}, the root mean square angular velocity of the rotation, at 300°K.

105. In the kinetic theory of an ideal gas, when one discusses the kinetic-molecular level of the gas, one neglects gravity, *i.e.*, the gravitational effect on the molecular motion. Why?

106. Consider crystalline NaCl. The structure is cubic, and its density is 2.16 gr/cc. The atomic weights are 23 gr/atomgr for Na, and 35.5 gr/atomgr for Cl. Calculate the lattice constant of the crystal, that is, the spacing between adjacent (closest neighbor) atoms. (Answer: 2.82 Å.)

107. The kinetic theory of gases says that under normal conditions, the average speed of O_2 molecules is 460 m/sec. How much is the total kinetic energy of the molecules included in one mol of oxygen?

108. (a) Carbonated soft drinks (mostly water) come in aluminum cans of 12 fluid oz. each. To let CO_2 dissolve easily in water, the soda cans are filled and canned at low temperatures, say, 5°C. How much empty room must be left in the can to allow for the beverage expansion, if it is to be stored for consumption at room temperature (20°C)? (b) A beer bottle left under a blazing sun may blow up. Why?

109. Use arguments from the kinetic theory of gases to answer the following questions (if you like riddles): (a) The macroscopic behavior of a gas results from the overall bustling of its molecules. True or false? (b) Is the random

movement of dust particles in a room a Brownian motion? (c) Can the kinetic theory explain the phenomena of pressure, boiling, evaporation, osmosis, diffusion, and that the Moon has no atmosphere? How? Explain. (d) In general, $\overline{v^2} \neq (\bar{v})^2$. When are these two equal? In general, $\bar{v} \neq v_{\text{most probable}} \equiv v_{\text{mode}}$. Under what condition (on the speed distribution) are these two necessarily equal? (e) Could the motion of gas molecules be anisotropic? When? (f) How can we tell that gas molecules have a distribution of speeds instead of having all the same speed? (g) If we know the average temperature of a planet's atmosphere, can we have an idea about its composition? (h) Explain how a liquid with an open surface evaporates without any need of boiling. (i) Explain why chemical reactions usually take place easier at a high temperature. Is there any exception where the opposite happens?

110. Why does a molecule have no orbital angular momentum **L**? Can it have (a) an instantaneous (b) a permanent dipole moment? Why?

111. Consider two isotopes of the same atom, with masses m_1 and m_2. Their vibration frequencies are $\nu_1 \neq \nu_2$. Explain why the lighter isotope has a larger zero-point vibration energy.

112. Consider the charge distributions shown in Fig. AP-6. (a) For each of them, calculate the quadrupole moment Q. (b) Multipole-expand the potentials $\Phi_{(\mathrm{I})}$ and $\Phi_{(\mathrm{II})}$ and keep the leading term. Thus prove that they are

$$\Phi_{(\mathrm{I})} \simeq \frac{3qd^2(\cos^2\theta - \sin^2\theta\sin^2\varphi)}{r^3} = \frac{3qd^2\sin^2\varphi\cos\alpha\sin\alpha}{r^3}$$

$$\Phi_{(\mathrm{II})} \simeq \frac{qd^2(3z^2 - r^2)}{r^5} = 2qd^2\frac{P_2(\cos\theta)}{r^3}$$

where $P_2(\cos\theta)$ is the Legendre polynomial. Notice that to this order, the potential is dimensionally equal to charge over length (distance). (c) Compare $\Phi_{(\mathrm{II})}(r)$ with what can be calculated from Coulomb's law for three point charges. (d) Check the asymptotic form of $\Phi_{(\mathrm{II})}(r)$. This gives the situation at large $r \gg d$. (e) Calculate all the nonvanishing multipole moments q_{lm} (for all l) for the three charge distributions.

113. A nucleus with quadrupole moment $Q \neq 0$ is in a cylindrically symmetric $\vec{\mathcal{E}}$ where $(\partial\mathcal{E}_z/\partial z)_0 \neq 0$ along z at the origin (position of the center of the nucleus). (a) Prove that the energy of the quadrupole interaction is

$$E = -\frac{eQ}{4}\left(\frac{\partial\mathcal{E}_z}{\partial z}\right)_0$$

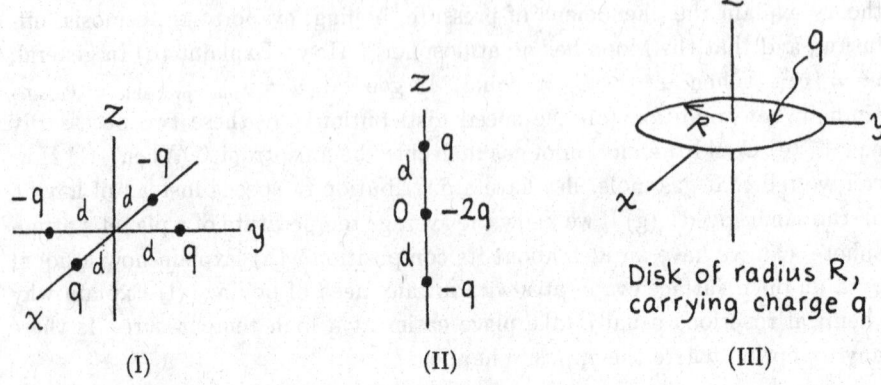

Fig. AP-6.

where the field gradient is evaluated at $z = 0$. (b) Given that $Q = 2\times10^{-24}$ cm^2 and $\nu = E/h = 10$ MHz, find the field gradient in terms of e/a_0^3 where $a_0 = \hbar^2/me^2 = 0.52$ Å is the Bohr radius.

114. (a) The nuclear charge density is approximately constant inside an ellipsoidal (spheroidal) nuclear volume of semimajor and semiminor axes a and b respectively, where the total charge is Ze. What is the quadrupole moment Q of the nucleus? (b) The nucleus of ^{153}Eu has $Z = 63$, $Q = 2.5\times10^{-24}$ cm^2, and a mean radius $R = (a + b)/2 = 7 \times 10^{-13}$ cm. Find the eccentricity $(a - b)/R$ of the nuclear shape.

115. (a) Given a charge density $\rho(\mathbf{r}) = \rho(r, \theta, \varphi) = \frac{1}{64\pi}r^2e^{-r}\sin^2\theta$ which has azimuthal symmetry (no φ-dependence), multipole-expand $\Phi(\mathbf{r})$ and express it in terms of the Legendre polynomials at large r (far away from the source distribution ρ). Note that this ρ belongs to the $2p$ state of H for $m_l = \pm1$. (b) Find $\Phi(\mathbf{r})$ at *any* $\mathbf{r}$. (c) Show that as $r \to 0$ (approaches the origin), we obtain

$$\Phi(\mathbf{r}) \simeq \frac{1}{4} - \frac{r^2}{120}P_2(\cos\theta),$$

if we ignore higher-order terms in r. (d) At the origin there is a nucleus with $Q = 10^{-24}$ cm^2. What is the quadrupole interaction energy and frequency in terms of e and a_0? (Show that it is typically of molecular order of magnitude.) (e) Is the treatment here classical or quantum mechanical?

116. A pointlike dipole (moment) $\vec{p}$ is at $\mathbf{r}_0$. Prove that it can be represented by a continuous charge density (function) $\rho(\mathbf{r}) = -\vec{p} \cdot \vec{\nabla}\delta(\mathbf{r} - \mathbf{r}_0)$ when we want to calculate $\Phi(\mathbf{r})$ or the interaction energy between $\vec{p}$ and an external field. (Hint: Use Dirac-function properties.)

117. A charge density $\rho(\mathbf{r})$ is in a constant external field $\vec{\mathcal{E}}_0(\mathbf{r})$ whose potential $\Phi_0(\mathbf{r})$ varies slowly and smoothly over the spatial region where $\rho \neq 0$. (a) State in words, what is the force $\mathbf{F}$ acting on ρ, if we write $\mathbf{F} =$ (Multipole expansion) $\times$ (Derivatives of $\vec{\mathcal{E}}_0$)? (Stop at the quadrupole term.) (b) Prove that

$$\mathbf{F} = q\vec{\mathcal{E}}_0(0) + \left\{ \vec{\nabla}\left[\vec{p} \cdot \vec{\mathcal{E}}_0(\mathbf{r})\right]\right\}_{r=0} + \left\{ \vec{\nabla}\left[\frac{1}{6}\sum_{i,j} Q_{ij} \frac{\partial (\mathcal{E}_0)_i(\mathbf{r})}{\partial x_j}\right]\right\}_{r=0} + \cdots$$

where again truncate the expansion at the quadrupole moment. (c) The energy (expansion)

$$E = q\Phi(0) - \vec{p} \cdot \vec{\mathcal{E}}(0) - \frac{1}{6}\sum_i \sum_j Q_{ij} \frac{\partial \mathcal{E}_i(0)}{\partial x_j} + \cdots$$

is a number, *not* a function of $\mathbf{r}$ (because the expression is evaluated at $r = 0$). Relate E to $\mathbf{F}$. (d) Calculate the torque $\vec{T}$. It is enough to prove that (for one Cartesian component)

$$T_x = [\vec{p} \times \vec{\mathcal{E}}_0(0)]_x + \frac{1}{3}\left[\frac{\partial}{\partial z}\left(\sum_i Q_{yi}(\mathcal{E}_0)_i\right) - \frac{\partial}{\partial y}\left(\sum_i Q_{zi}(\mathcal{E}_0)_i\right)\right]_{r=0} + \cdots$$

(e) The torque is nonzero if the given charge distribution $\rho(\mathbf{r})$ gives nonvanishing dipole and/or quadrupole moments. True or false?

118. The multipole moments of some $\rho(x, y, z)$ are q, $\vec{p}$, Q_{ij} etc. with respect to the (x, y, z) system, and q', $\vec{p'}$, Q'_{ij} etc. with respect to the (x', y', z') system, where $x_i \| x'_i$ with i $= 1, 2, 3$, and the origin O does not coincide with O'. (a) Relate the primed quantities to the unprimed quantities (moments). (b) If $q \neq 0$, is there a vector $\vec{OO'}$ that makes $\vec{p'} = 0$? (c) If $q \neq 0$ and $\vec{p} \neq 0$ (or only $\vec{p} \neq 0$), is there an $\vec{OO'}$ that makes $Q' = 0$?

119. Suppose that gaseous H_2O and H_2 are in equilibrium. If the velocity of the H_2O molecules (on the average) is 600 m/sec, what is the velocity of the H_2 molecules (on the average)?

120. A very good vacuum can be achieved by means of a combination of pumping and cooling the system to the temperature at which He becomes liquid (4°K). Assume that such a system has a residual pressure of 10^{-8} Nt/m^2 of molecular N_2. Assume also that the density is too low for all the N_2 to condense out, in spite of the extremely low temperature. Determine the number of molecules of N_2 per cm^3, the average kinetic energy of translation, and the rms speed.

121. (a) In the wave-mechanics picture of a molecule, what is oscillating? What determines the wavelength? (b) How does the de Broglie wavelength of a moving mass of 1 kg moving at 10 m/sec compare to the size of a molecule? (c) Why are the noble gases called inert? Does the shell model of electron configuration explain that? How? (d) Consider a quantum energy level of high angular momentum l (or J). What is rotating? (Explain for an atom and for a molecule.) Compared to a level with $l = 0$, is the electron closer to or farther from the nucleus? (e) Consider the probability distributions of s and p-electrons. Do they overlap? Is it ever possible to find an s-electron at a greater distance from the nucleus than a p-electron? (f) How does the Pauli principle determine binding and molecular structure, and consequently, chemical properties? (g) Explain what we mean by electron screening and its effect on the shell ordering and the energy of the outer electrons. (h) Why a diatomic molecule has many more energy levels than a free atom, a property which leads to a more crowded spectrum of emission lines? (i) Explain how we can use the rotational energy formula $E(J) = (\hbar^2/2\mathcal{I})J(J+1)$ and the bond length in an AB type molecule (between A and B atoms), to find the wavelength of a rotational transition of the molecule from J to J'. (j) If $T = 300°K$, can the speed of a molecule be considered as moderate?

122. (a) Practically all solids have a crystal structure which in many of them can be recognized by the use of X-rays only. True or false? (b) In a diatomic molecule the binding electrons are found mostly and most likely in the region between the two atoms, just like a tennis ball which spends most of its time between the two players during a match. True or false?

123. A diamagnetic material can also be called non-magnetic. True or false?

124. Prove that the molar diamagnetic susceptibility of atomic hydrogen is -2.36×10^{-6} cm^3/mol for its ground state ($1s$) which has $\psi = \left(\dfrac{1}{\sqrt{\pi a_0^3}}\right) e^{-r/a_0}$ (where $a_0 = \hbar^2/me^2 = 0.529 \times 10^{-8}$ cm) and charge density $\rho(\mathbf{r}) = \rho(r, \theta, \varphi) =$

$-e|\psi|^2$ (according to the quantum statistical interpretation of the wavefunction). (Prove first that $\overline{r^2} = 3a_0^2$ in $1s$, *i.e.*, for this state. See Ashcroft and Mermin, p. 649.)

125. Certain organic molecules have a triplet $(S = 1)$ excited state of energy kT_0 above a singlet $(S = 0)$ ground state, where T_0 is a temperature that characterizes the interaction energy $E_{\text{int}} = kT_0$ which tends to orient the spins preferentially. (a) Relate the mean magnetic moment $\overline{\mu}$ to the field H. (b) Prove that χ for $T \gg T_0$ does not depend on T_0 (approximately). (c) Plot the energy levels versus H, and the entropy versus H. How can we cool the system by adiabatic magnetization (not demagnetization)? (Read first Kittel, pp. 435–440 and 446–453, to refresh your knowledge about Langevin diamagnetism, magnetic susceptibility, paramagnetism, cooling by adiabatic demagnetization, and nuclear demagnetization.)

126. In KCl, assume that the ion-core repulsion potential is of the empirical form aR^{-n} where a and n are constants. (a) Evaluate a and n. (b) The experimental frequency of vibration (for small displacements) is 8.34×10^{12} Hz. Compare this value to what you would find if you used the rough results of part (a). (Hint: Read Solved Problem 82 first.)

127. Thermal neutrons are used for neutron scattering, to determine molecular structure. Low-energy thermal neutrons (at $T = 300°K$) have an average kinetic energy $E \sim 1/40$ eV. (a) Calculate their de Broglie wavelength. (Answer: $\lambda \sim 1.82$ Å.) (b) If they are in equilibrium with the mass of a medium (say, graphite or paraffin) at room temperature (300°K), what could their speed distribution be? (c) Calculate their most probable speed (given by $E \sim kT$).

128. Verify whether the following molecules have the indicated symmetries:

CH_3CHO: 1	NH_3: $3m$ and m
Trans CHCl–CHCl: $2/m$	IF_5: $4mm$
C_2H_4: mmm	SF_6: $m3m$
BCl_3: $\bar{6}2m$	CH_4: $\bar{4}3m$
C_6H_6: $6/mmm$	H_2O: mm
CH_2–C–CH_2: $\bar{4}2m$	

129. (a) Is the combined system A + B mentioned in Section 6.14 realistic and existent? Is the He atom an example to it? (b) Why most of the diatomic molecules (except O_2) are diamagnetic?

130. Consider the discussion for a one-electron atom on p. 285, concerning the ground state and the excited states of an atom. For a many-electron atom still in its ground state, can its valence electron be, say, an f-electron?

131. Prove that $\Delta E_{\mathrm{rot}} \propto 2BJ$. (Start from $E = BhJ(J+1)$.) Find ΔE_{rot} in getting, say, from $J = 7$ to $J = 8$. Is the answer proportional to $2BJ$ with $J = 7$?

132. An ion has a partially full shell of quantum number J, and Z more electrons in the inner filled shells. Find the ratio of the paramagnetic susceptibility (Curie's law) to the Larmor diamagnetic susceptibility, and prove that it is ~500 at room temperature. Note that the paramagnetic susceptibility depends on temperature, whereas the diamagnetic susceptibility does not. (Hint: Read Ashcroft and Mermin, Ch.31.)

133. Let a system have $S = 1$. Consider paramagnetism in this system with the given spin. (a) If the system's moment is μ and the concentration is N/V, set up and plot $\mathcal{M} = \mathcal{M}(T)$. (b) For $\mu H \ll kT$, show that the answer goes over to $\mathcal{M} \simeq (2\mu^2 NH)/(3kIV) \sim 1/T$.

134. (a) By using simple arguments from the theory of molecular forces, explain why you do not fall through the floor. (b) Are all the molecular forces eventually of electrostatic origin? (c) We can generate a uniform magnetic field by the following ways: (i) In the region between the large poles of an electromagnet, (ii) inside of a very long solenoid, somewhere in the middle, and (iii) at the center of Helmholtz coils. Which of these arrangements is the fittest for NMR experiments? Why? (d) When does an atom have an elongated shape, looking almost like a pencil? Does L, the orbital quantum number determine its shape? (e) Alkali halides have (almost all) the NaCl crystal structure, with internuclear separation (in Å) as follows:

	r_0		r_0
LiCl	2.57	LiF	2.01
NaCl	2.81	NaF	2.31
KCl	3.14	KF	2.67

If the F^- ion is taken as a hard sphere of (ionic) radius equal to 1.33 Å, estimate the radius of the Cl^- ion. Surprisingly enough, the hard-sphere idealization does not give wrong results, although the ions are not hard spheres at all (because these crystalline solids can be compressed). Explain the reason. (f) What is the effect of thermal agitation on infrared bands? (g) Why can we find a wide distribution of molecules among various rotational levels even at room temperature? (h) Why should the binding theory — as presented in this book — be taken in an idealized sense? Do advanced studies take it otherwise? (Investigate before answering.) (i) The lowest energy levels of an ion are $(2L+1)(2S+1)$-fold degenerate. The crystal field is given as $\mathcal{E}_{cryst} \sim |L|^2$. Let $\mathcal{E}_{cryst}$ perturbatively be stronger and dominate compared to the spin-orbit interaction (coupling). Prove that for $L = 1$ we obtain ground states which are only $2S + 1$ times degenerate, where the matrix of the operator $\mathbf{L}$ has all its elements zero: $(\mathbf{L}) \equiv 0$. (Why?) Interpret the situation physically, in words. (j) Is it possible to draw a sharp boundary line between molecular physics and solid state physics? How and where do they overlap? How do you define the territory of each? For example, what topics may be common to both and what other ones belong to the jurisdiction of molecular physics only? Do your instructor and fellow students or colleagues agree with your answer? (k) Compare a polyatomic molecule with the two political economic models of USA and EU. The former is a single-nation confederation of states almost fused together; the latter is a collection of nation-states (each with its own identity and features intact) linked together, but still with substantial freedom of action. In these modelings, are there common points and analogies? (l) Ponder the analogies and dissimilarities between molecular physics and political science in the use of the following terms: spectrum, interaction, potential, entropy, stability, pool of energy, polarization, technique, energy, level, bond, anharmonicity, resonance, exchange. (m) Since $m_{prot} \gg m_{el}$ in the H atom, can we take $\bar{\mu} \simeq m_{prot}$ and neglect m_{el}? (n) When the vibration amplitude is small, we can cut out the expansion series because the converging terms get smaller and smaller after each step (term). True or false?

135. A linear molecule of the CO_2 type has a charge distribution given by

$$\rho = \rho(x, y, z) = 2q\delta(x)\delta(y)\delta(z)[\delta(z - r_0) + \delta(z + r_0) - 2\delta(z)]$$

with $r_0 \ll 1$ (Fig. AP-7). Prove that this distribution gives rise to an electrostatic potential

$$\Phi = \Phi(r, \theta, \varphi) = \frac{2q}{r^3}r_0^2(3\cos^2\theta - 1).$$

136. The electric potential (at any point in space) produced by a single charge q located at $z = r_0$, for $r < r_0$, is

$$\Phi(\mathbf{r}) = \frac{q}{4\pi\varepsilon_0 r_0}\sum_{n=0}^{\infty}\left(\frac{r}{r_0}\right)^n P_n(\cos\theta),$$

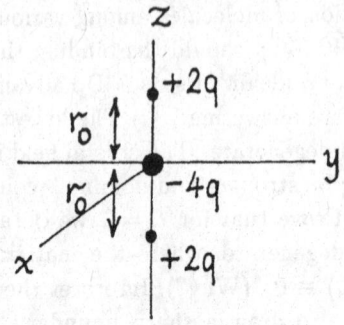

Fig. AP-7.

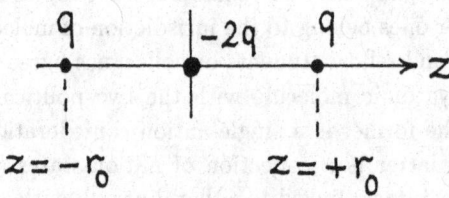

Fig. AP-8.

i.e., a multipole series expansion (with correction terms of higher-order approximations, to the main leading term that goes as $\sim r^{-1}$). Here $P_n(\cos\theta)$ are the Legendre polynomials. (a) Develop this potential for the array of charges shown in Fig. AP-8. The array is a lineal electric quadrupole, equivalent to two lineal and opposite dipoles as in Fig. AP-9. (b) Prove that the quadrupole moment of this array is zero ($Q = 0$). (Hint: No area is formed by the lineal array. Further, since it is equivalent to two lineal and opposite dipoles, the net quadrupole effect is obviously zero.) (c) Prove that although two lineal quadrupoles may be placed so that the quadrupole term is cancelled out, the octupole term survives. (d) Consider the possible configurations of charges shown in Fig. AP-10. Show that alternate positive and negative charges at the corners of a square or parallelogram give a (nonzero) quadrupole term (field), whereas alternate charges on the vertices of a cube give an electric octupole field. (e) Show that in the configuration of electric charges given in Fig. AP-10(a), the net electric quadrupole is nonzero, whereas there is no net monopole or dipole. (Hint: The total charge (monopole) is zero, by inspection. Actually, we have two opposite dipoles side by side, forming a square of nonzero area between them.)

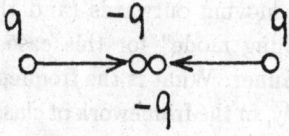

Fig. AP-9.

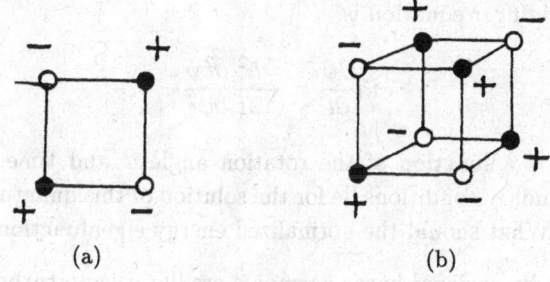

(a)   (b)

Fig. AP-10.

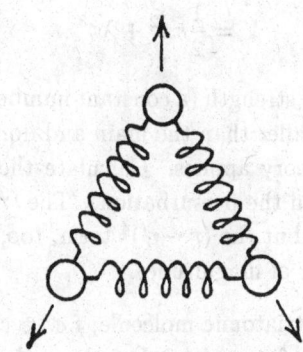

Fig. AP-11.

137. A planar triatomic molecule is made up of three identical atoms of equal mass m, that form an equilateral triangle (Fig. AP-11). It vibrates as if there were identical springs of constant κ between the atoms. Consider planar motion only. At rest (equilibrium), the atoms are separated by a distance d. (a) How may normal modes are there? Could you guess before you actually solve the problem? Which of them have zero frequency? Why? (b) A possible

normal mode is the symmetrical stretch-out of the molecule (see Fig. AP-11), with all three atoms moving outwards (and then inwards) at the same time. This is the "breathing mode" for this case, *i.e.*, the atoms move in synchronous phase and manner. What is the frequency of this mode? (c) Now solve the problem rigorously, in the framework of classical mechanics. Write the equations of motion and find the normal coordinates and frequencies thereof. Exhibit the solution in detail.

138. A molecule of moment of inertia $\mathcal{I}$ rotates about an axis like a rigid body. Its Schrödinger equation is

$$i\hbar \frac{\partial \psi}{\partial t} = -\frac{\hbar^2}{2\mathcal{I}} \frac{\partial^2 \psi}{\partial \varphi^2}$$

where $\psi(\varphi, t)$ is a function of the rotation angle φ and time t. (a) What should the boundary conditions be for the solution of this quantum mechanical equation? (b) What should the normalized energy eigenfunctions be?

139. A one-dimensional linear harmonic oscillator is perturbed by an extra potential term λx^3, so that the potential is

$$V = \frac{1}{2}\kappa x^2 + \lambda x^3$$

where λ is the perturbation strength (a constant number as a coefficient). Since the extra term is much smaller than the main and dominant (quadratic) term ($\lambda \ll \kappa$), perturbation theory applies. Calculate the change in each energy level to the second order (in the perturbation). The $(r - r')^3$ term contributes to the second-order effect, but the $(r - r')^4$ term, too, produces a comparable effect (of comparable order of magnitude).

140. Consider a linear triatomic molecule, *i.e.*, a central atom of mass m_1 and two end atoms, each of mass m_2. Let the molecule be moving (vibrating) in one dimension only, that is, only along its length (the line joining the atoms). There is no interaction (coupling) between the two end atoms. Find the normal coordinates and the modes of vibration for each atom. Assume a flexible, springlike bond of constant κ between the atoms. Why do the normal coordinates decouple the motion?

141. A linear triatomic molecule consists of three masses m_1, m_2, and m_3 in a row (where $m_1 = m_3 < m_2$). The two end masses are noninteracting and are just coupled to the middle mass by two identical springs of constant κ. This

dynamic system vibrates in the lateral (transverse) direction only. Discuss the motion and write the equations of motion. Use classical dynamics.

142. How does one approach the many-body problem? Is there an exact solution? How can we get around it? Explain in detail. Discuss briefly the three-body problem.

143. Consider an hydrogen-discharge tube at low pressure (*i.e.*, the probability of atomic collisions is low — the collisions are rare and biatomic only). Such tubes are used to study emission spectroscopy. The externally applied potential ΔV to the electrons is unimportant compared to the internal $V(r)$ in the atom (experienced by the atomic electrons). (a) Estimate and compare ΔV with $V(r)$. (b) So, what is the Hamiltonian of the atomic electrons? (c) What wavefunctions are fit for the experimental results? (d) All the energy levels of the hydrogen atom take definite values. So, is that why the spectrum of the hydrogen atom gives sharp spectral lines?

144. Refer to the theory in Section 4.1. (a) When the two atoms are still free, they are described by the (two separate) *atomic* wavefunctions ψ_1 and ψ_2, centered at atom #1 and 2 (or A and B) respectively. Does each of these functions describe the whole atom or just the outermost bound electron (of an hydrogenlike atom)? (b) If one of the atoms is, say, uranium, do we have to solve 92 Schrödinger equations (one for each electron in the atom) to reach ψ_1? (c) When the two atoms get bound into a formed molecule, the appropriate (*molecular*) wavefunction is ψ_{mol}. Is this pertinent to the molecule as a *whole*, or to the electron that mediates the bond? Or are the two schemes essentially the same thing?

145. (a) Qualitatively, how would you go about if you wanted to investigate stability in a diatomic molecule? (b) In studying molecular rotations, we see that the rotational energy is quantized, and yet it includes the moment of inertia, which is a classical mechanical quantity. How do you reconcile the quantum picture with the classical one? Is it accurate to say that one picture is actually the extension of the other? Are they indeed in contradiction to each other? Comment on the fact that the moment of inertia is included in a factor (rotational constant) whereas the quantization is in $J(J+1)$. Is this approach semiclassical?

146. Refer to Solved Problem 19. Why are the two electrons equivalent (identical and indistinguishable) upon interchange?

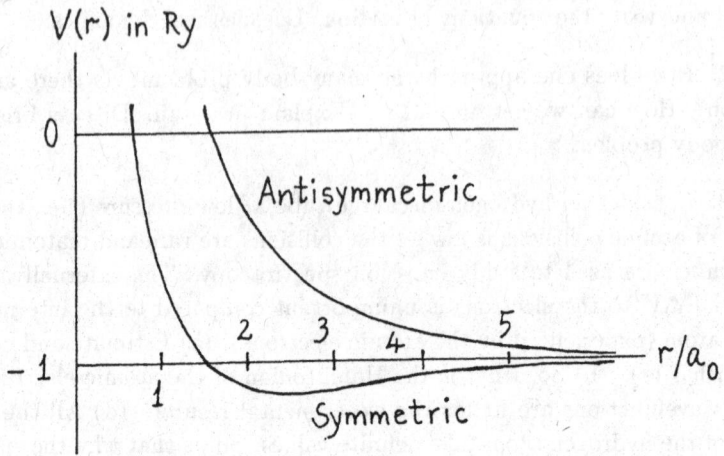

Fig. AP-12.

147. Consider the molecular H_2^+ ion. It consists of two protons and includes an electron. The total potential energy $V(r)$ of this system, as a function of the separation r of the two nuclei, decreases with increasing r, to a minimum at $r \sim 2.5a_0$. At this separation the system is stable (for $V(r)$ that corresponds to the symmetric wavefunction; the antisymmetric one decreases without a minimum) (Fig. AP-12). The total binding energy is 1.13 Rydberg unit. The ion dissociates gradually as r increases further, and eventually pulls itself apart into a proton p^+ and an H atom of energy -1 Rydberg unit. Thus the binding energy of the ion is 0.13 Rydberg. (a) Express it in eV. (Answer: 1.8 eV.) (b) Now consider the fact that the ion vibrates even in its ground state, about equilibrium position, at a (lowest) frequency ν_0 and with a zero-point (vibration) energy $E_0 = (1/2)h\nu_0$. Perform a rough calculation to estimate this E_0. Does your result agree with the value 0.1 eV (which gives the theoretical binding energy as 1.7 eV)? (c) The experimental value of the binding energy is ~ 2.65 eV. How can you use the variational principle to improve your result in part (b)? (d) Can we find E_0 experimentally?

148. Refer to Problem 28. (a) Perform the transformations

$$\xi_1 = \frac{1}{\sqrt{2}}(x_1 + x_2)$$

$$\xi_2 = \frac{1}{\sqrt{2}}(x_1 - x_2)$$

and thus calculate the energy levels of the system exactly, as contrasted to the approximate (perturbative) solution in quantum mechanics. (b) Note that the change of variables decouples the oscillators. How does this compare with the classical case? (c) Is there any other way to solve the classical version of the problem?

149. Consider the atomic hydrogen gas in a star. Find the long-range potential between a ground-state H atom and an excited atom in the $2s$ state, as they interact. Comment on how this particular molecular physics topic about the interaction (potential) serves astrophysics.

150. Consider two spherical conglomerates of atoms (they could be colloidal particles or micelles) of size (radius) d, and each of n atoms/cm^3. The interaction potential between atoms goes as $\sim r^{-6}$. (a) How does the interaction potential between two such conglomerates vary with R, the (variable) separation distance between the conglomerates? (b) Consequently, what is the force law $F \sim R^p$, *i.e.*, what is the power exponent $p =$? (c) What are the two limits (to which F goes over) for very large and very small R, that is, for $R \to \infty$ and 0?

151. Refer to Problem 135. (a) Since the electric potential is an additive scalar, we can add up the contributions of individual point charges, to find the total potential. But since in this case the total charge is zero, the total potential is zero to the first order. True or false? (b) How do we find the given result by collecting the contributions of the charges (summing the discrete elements or integrating over the continuous charge distribution)? To what order is the given result nonzero?

152. Consider molecular hydrogen and helium, that is, the isotopic molecules $^1H_2$, $^2H_2$, $^3He_2$, and $^4He_2$ in thermal equliibrium. Let the temperature T be high enough, so that the thermal (Boltzmann) factor can be taken as $e^{-E_J/kT} \simeq 1$ in the intensities of successive lines of the molecules' electronic band spectra. (If kT is large, the line intensity depends only on the degeneracy; its variation due to the exponential is negligible.) (a) Evaluate the ratio of intensities of adjacent spectral lines. For the atomic nuclei, you are given that the nuclear spins are $I(\text{proton}) = 1/2$ (for ^1H), $I(\text{deuteron}) = 1$ (for ^2H), $I(^3He) = 1/2$, and $I(^4He$ or alpha particle$) = 0$. (b) By considering wavefunction symmetrization and allowed J values, determine the lines that do not appear in the spectrum. (Hint: All these molecules are A_2 type. Upon interchange of the two atoms, ψ_{total} must stay symmetric if $I = $ integer, and

antisymmetric if $I =$ half integer. We can write

$$\psi_{\text{total}} = (\text{Vibrational part}) \times (\text{Rotational part}) \times (\text{Nuclear-spin part})$$

where the first part is symmetric anyway (because it depends on the absolute value of the variable interatomic separation only). So, the product of the latter two parts must stay symmetry or antisymmetric, depending on I.)

153. By resorting to experimental data and sources on optical methods in physics and chemistry, design a spectrograph to detect molecular spectra in the far infrared region. Give the parameters of the apparatus and describe it briefly in words. State its parts clearly, discuss the particulars of the source arrangement, and go over the sample and the detection technique to be used.

154. (a) By the NMR method we measure nuclear magnetic moments. In such an experiment we should know the magnetic field between the magnet poles (where the sample is to be put) very accurately. If the field is ~ 1 Tesla, how can we measure it in the gap between the two poles, and how accurately? Discuss accuracy for a pole diameter ~ 15 cm and a gap width ~ 3 cm. What limitations are there? (b) How large should the poles be for a typical homogeneous field in the gap? Should they be circular or rectangular? (Before answering, you might wish to consult NMR specialists and experimentalists.) (c) If the sample is hydrogen, how do you tune the other field to the sample frequency? (d) How can a theorist be of help in designing and programming an NMR experiment intelligently and feasibly, towards what is expected?

155. Write up a concise technical report to describe the set-up for an NMR research activity at a typical college laboratory. Discuss the features of the set-up, give typical sizes, and include design considerations. Make your report look professional, by including costs and using the proper essay-type scientific language and terminology — avoid slipshod work with an "oral" style and a utilitarian presentation. Type the report neatly, as you would if it were to be submitted to a board. To enrich it, you might also include suggestions about a computerized environment and block diagrams of sophisticated electronics used in industrial NMR research. Include error sources. To refine your answer, feel free to confer with professional physicists and discuss pertinent laboratory matters and practices with them.

156. Whereas atoms (monatomic substances or chemical elements) give simple spectral lines, molecules (molecular materials or chemical compounds) give complicated spectra with *bands*. Explain why.

157. In the upper atmosphere, molecular oxygen O_2 splits into two oxygen atoms by the protons that come from the Sun. The maximum wavelength of a proton causing this process is 1750 Å. What could the binding energy between the two oxygen atoms be in an O_2 molecule? (This phenomenon may be a good energy source for high-altitude flights; the solar radiation separates molecular oxygen into atomic oxygen, by rending apart O_2 molecules, and the atoms recombine into molecules, thereby releasing energy.) (Answer: 7.1 eV.)

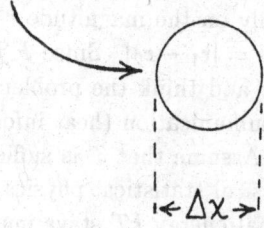

Fig. AP-13.

158. (a) A molecule of a certain typical size moves in space. Discuss the uncertainty Δx in its location (Fig. AP-13). (b) Beyond what sizes (that is, above what molecular diameters) are the statistical considerations about the molecule unimportant? (c) If an O_2 molecule is hit by H_2O molecules from around, should we care about any statistical considerations in this process? (d) Is it meaningful to speak of a de Broglie wavelength of a macromolecule (large biological molecule)?[2] (e) Consider a dilute solution of macromolecules at temperature T, placed in a centrifugal arrangement[3] rotating with a constant angular velocity ω. There is a centripetal acceleration $\omega^2 r$ acting on any molecule of mass m and keeping it in a circular orbit. However, this nonequilibrium situation can be thought as an equilibrium situation in the rotating reference system, by introducing an opposite *fictitious centrifugal* force $-m\omega^2 r$ acting radially outward (on m), according to D'Alembert's principle. Set up

---

[2]A *micelle* is such a large molecule encountered in biochemistry. It is a euphemism from the Greek word μικύλλιον (micyllium) which means tiny, too small, as a diminutive of the word μικρόν (micron = small, little). But in fact it is a *large* molecule (perhaps subjectively called tiny by the first researchers), and its correct Greek nomenclature is μεγαμόριον (megamorion = large molecule).

[3]Biologists and microbiologists employ ultracentrifuges to separate molecular substances of different densities in a mixture. Heavier substances accumulate further away from the axis of rotation, in the bottom of the test tube, and lighter ones stay as upper strata, closer to the rotation axis. This is how they separate high and low-density lipoproteins (HDL and LDL) in blood samples.

$\rho(r)$, the density of the molecules as a function of the radial distance r from the rotation axis. (f) Refer to the previous part. If the ratio $\rho_1(r_1)/\rho_2(r_2)$ can be measured optically, can we find the molecular weight of the macromolecules in the solution? How?

159. Consider the Lennard-Jones potential (between two atoms of mass m) given in Problem 22. It represents the mutual potential energy from which a force $F = -\partial V/\partial r$ is derivable, by which the atoms interact. Here r is the separation distance between the two atoms. (a) Derive $F(r)$. (Notice that it is radial, *i.e.*, it depends only on the magnitude r of $\mathbf{r} = \mathbf{r}_1 - \mathbf{r}_2$ and not on the direction, where $r = |\mathbf{r}| = |\mathbf{r}_1 - \mathbf{r}_2|$. Since F is independent of θ and φ, we can write x instead of r and think the problem as one-dimensional.) (b) Let the atoms exchange communication (heat information) with a large heat deposit at temperature T. Assume that T is sufficiently high, so as to allow the use of principles of classical statistical physics, but at the same time not too high, so that the thermal energy kT stays much lower than V_0 (where k is the Boltzmann constant and V_0 is the lowest value — minimum — of $V(r)$ where stability is possible for a molecule once formed by the combination of these two interacting atoms). Notice that the atoms do not interact by any potential other than $V(r)$ here. Also notice that $kT \ll V_0$ means that thermal fluctuations around (in the environment) do not disturb V_0, the lowest potential value — in other words, the stability is not affected by the ambient thermal energy so as to break down. Refer to a standard text in statistical mechanics and obtain enough information to calculate $\bar{r}(T)$, the average separation of the atoms (as a function of T). Obviously, $\bar{r}$ varies with T: As T rises, kT rises and threatens the stability, as the molecule starts vibrating (and has an average $\bar{r}$). (c) From the result of the previous part find

$$\alpha \equiv \frac{1}{\bar{r}} \frac{\partial \bar{r}}{\partial T},$$

which is the *thermal expansion coefficient* (for linear expansion) of a solid (bulk material) made of such molecules. Notice that by calculating the quantity α defined as above, we achieve a legitimate way of finding the expansion coefficient. Notice also that the procedure is basic and simple. (d) $V(r)$ given in Problem 22 is for two atoms at any separation r between them (and of course within a molecule, *i.e.*, for a diatomic molecule). Can we use the Lennard-Jones potential to describe the interaction between two *molecules* at a separation r (by treating the molecules as solid entities without internal structure)? What happens then at $r = r_0$ and $V = -V_0$? (Hint: To solve parts (b) and (c), expand $V(r)$ into a power series in r about its minimum. Take the potential function as given in Problem 22. You need not go to higher-order terms and struggle with a tedious algebra: Since T is low enough (small), keep expansion terms of lower order only, until you obtain a nonzero α — the first few lowest-order ones will suffice. Cut off the expansion after these sufficient terms. Also, you will encounter some integrals which you should evaluate by approximations, pulling in your mathematical abilities or getting help from a mathematician. Finally, it is advisbale to read Solved Problem 77 first.)

160. Electrons in molecules (and atoms) are not stationary. As an electron moves in a molecule, there is an instantaneous separation between a positive (nuclear) and a negative (electron) charge (within the molecule). As a result, the molecule has an electric dipole moment (at an instant of time), which, thus created by the motion of the electron, varies as $\vec{p}(t)$. This in turn creates a variable electric field at some distance r outside and away from the molecule. (a) Find this field $\mathcal{E}(r)$ as a function of r when $r \gg$ molecular size, *i.e.*, far away from the molecule (so that the latter can be taken as a small — instantaneous of course — dipole, and the dipole approximation for a dipole field far away at large distances holds). This is an electrostatics problem. (b) From $\mathcal{E}$ obtain V for large r. (c) Consider another molecule of polarizability α coming to within a distance r of the first molecule. Then a dipole moment is induced (provoked) in the second molecule (by the first). Find its dependence on r. (d) Consider the interaction between these two dipoles (electrostatic dipole-dipole interaction). Find its dependence on distance r, and prove that it turns out to be the Van der Waals potential (or force) which is identical with the long-range behavior (large-r part) of the Lennard-Jones potential given in Problem 22. (Note: Read first about the Van Der Waals interaction potential. Use any text in solid state physics.) (e) Is your result in part (d) the same as that in part (b)? Should it be?

161. (a) Talk to an experienced spectroscopist and find out how he/she deciphers a difficult and complicated molecular spectrum obtained experimentally in the laboratory. Ask him/her how such a band spectrum is interpreted. Then write a brief essay regardig your impressions. (b) Ponder, and comment on, the question of how molecular physics can help professionally other sciences. For example, think about how experimental molecular spectroscoy might be of service to archaeology, ecology, criminology, medicine, pharmacy, forensic medicine, astrophysics, industry and diagnostics in general.

162. Consider the long-range interaction -between two He atoms in their ground state. By using the variational method, calculate the interaction potential. Prove that it is $V(r) \sim -1/R^6$, though this result is debatable and only approximately right. To solve this problem, follow the steps below: (a) First read about the He atom. (There are good books — we recommend Park, Ch. 15.) (b) Take

$$\psi^{(0)} = \left(\frac{z'^3}{\pi}\right) e^{-z'(r_1+r_2)}$$

as the unperturbed wavefunction, where $Z'e$ is the *effective* nuclear charge due to screening (i.e., each e^- partially screens the nucleus from the other, and so, each e^- feels an effective nuclear charge $Z'e$). (c) Find Z' and then V. Note again that your result is not a confirmed answer that enjoys general agreement, but it makes sense.

163. Imagine a linear ionic crystal of $2N$ ions of charge q each, alternating in sign. The repulsive potential (between closest neighbors) is $V \sim r^{-n}$. Prove that at the equilibrium separation R we have

$$V(R) = \left(-\frac{2N}{R}\right)\left(\frac{n-1}{n}\right)\ln 2.$$

(Hint: Start from Eq. (3.29) in Kittel.)

164. (a) Why do the spectrum lines come close upon each other at the band head? (b) How satisfactory is the classical (mechanical) model and treatment of vibrations in the molecular model? (c) An *element molecule* is an atom itself. True or false? (d) What do we call *vibrational displacement*? (e) What are the spectroscopic findings for Li_2? (f) How do you explain the polar character of H_2O? How large is its valence angle? Why does the O atom in H_2O stay at an important position? How large is the distance between the two protons? (g) Consider the discussion on p. 227. Why is ψ taken as the combination of four atomic orbitals? (h) How do we draw the valence (bond) lines between the atoms in H_2O? (i) Consider the discussion on p. 223. Can a P_y and a p_z-electron form a bond? (j) In a metal, how far away from its master atom can an itinerant electron drift? Is the electric current flow effected by charge jumps from a location to the next, or by a free run of groups of electrons from one side of the conductor to the other? (k) Why can a metal bulk be thought as a large molecule? (l) What do we call *ionic mobility*? (m) What is a *band head*? Explain. (Not a marching band of course!) (n) What is a *resonant exchange*?

165. (a) Is the notion of rotation the same in the context of symmetry and in the context of energy? (b) A *crystal* consists of regular array of atoms or atom groups positioned orderly in space. True or false? (c) What is *nuclear quadrupole mass spectroscopy*? (d) What are the *liaison forces*? Are they the same as the cohesive forces (cohesion)? (e) There are atoms that are wanting in electrons. Does the NaCl molecule form for want of an electron by Na? Is that how atomic Na enters into a bond with other atoms? (f) What is a magnetic resonance? How is the detection made, and what is the importance of the results of the detection?

166. A particle of mass m moves in the one-dimensional well potential

$$V(x) = -V_0\delta(x - a) - V_0\delta(x + a).$$

If $V_0 > 0$, obtain transcendental equations of the two energy eigenvalues of the system. Estimate the splitting between the energy levels in the limit of large a. If there is an additional repulsive interaction $V = \lambda V_0/(2a)$ between the wells (atoms), show that, for a sufficiently small λ value, the system (molecule) is stable, if the particle (electron) is in the even-parity state. Sketch the total potential energy of the system, as a function of a.

167. (a) From a molecular physics point of view, describe the benzene molecule. (see *The Feynman Lectures on Physics*.) (b) When the ties that hold countries together are not strong enough to consolidate a union (political, economic, social or cultural), the corresponding conglomerate falls apart or does not form at all. Similarly, when the ties between atoms are weak, the molecule dissolves. True or false? (c) Spectroscopically, Λ in molecules is (and stands for) what L is for atoms. True or false? (d) When the spin-orbit effect is weaker, the ΛS-coupling is appropriate, and when the interaction is stronger, the Ω-coupling is better. True or false? Is the ΛS-coupling in molecules analogous to the LS-coupling in atoms? (e) The external field itself is a pool of energy for a molecule to take energy from. True or false?

168. Comment on whether the following statements are true or false: Nature's privacy dictates that there is always a lowest (least) vibrational energy in a molecule, as we cannot remove all the energy of the system (molecule). This is the zero-point energy $E = (1/2)h\nu_0$ that disallows us to reach $T = 0°$K exactly, because absolute zero means exactly zero internal energy in the system. There is always a leftover nonzero atomic vibrational energy.

169. (a) Consider the mathematical model of vibrations. How well does it serve molecular reality? (b) Consider two identical coupled oscillators (springs,

pendulums, diatomic molecules, tuning forks with beats — beating against each other). How real are they to be applicable to molecules? In mechanics and acoustics, there are two normal frequencies for such coupled oscillators, one higher and one lower than the natural frequency f_0 of either oscillator. True or false?

170. A current law is always replaced only by a new law that suits the situation and serves the needs better. Similarly, a theory can be overturned only by another theory that explains things better. A theory has always a proof; otherwise it is just an hypothesis or a speculation. The molecular theory is a standing theory that explains molecular phenomena sufficiently. On what basis? Does the route *observation* → *experiment* (repetition of the phenomenon) → *law of nature* → *theory* prevail in molecular physics as a science, or does the opposite way (*theory formulation* → *experimental test* of it) get the upper hand?

SOLVED PROBLEMS

"A picture is better than a thousand words."

Chinese proverb

SOLVED PROBLEMS

1. There are 6×10^{23} hydrogen atoms in one gram of hydrogen. Assume each hydrogen atom as a sphere of radius 5.3×10^{-11} m. (a) If all the atoms in one gram of hydrogen were lined up one next to the other like a rosary, what a length would they span? (b) If the same atoms were laid on a flat surface like marbles touching each other, what an area would they cover? (c) If they were all put together in space — so that the closest neighbors touch each other — what a volume would they fill?

Solution: (a) Just multiply the diameter of a single atom by the number of atoms. The length will be

$$l = 10.6 \times 10^{-11} \times 6 \times 10^{23} = 63.6 \times 10^{12} \text{ m} .$$

To get an idea about this length, let us compare it with the distance between the Earth and the Sun, which is 150 million kilometers:

$$\frac{63.6 \times 10^{12} \text{ m}}{150 \times 10^9 \text{ m}} = 424 .$$

So the atoms in 1 gr of hydrogen all lined up span a distance more than 400 times larger than the distance between the Earth and the Sun!

(b) The area of the square that hosts an atom is equal to the square of the diameter. The total area A of 6×10^{23} such squares is

$$A = (10.6 \times 10^{-11})^2 \times 6 \times 10^{23} = 6720 \text{ m}^2$$

649

which can be converted to acres.

(c) The volume of a cube that hosts an atom is equal to the cube of the diameter. The total volume V of 6×10^{23} such cubes is

$$V = (10.6 \times 10^{-11})^3 \times 6 \times 10^{23} = 0.7 \text{ cm}^3 .$$

2. The dimensions of the molecules of certain liquids (lighter than water) can be found as follows: A known volume of material is left floating on the water surface, until it spreads on the surface and forms a very thin film whose thickness is equal to the size of one molecule. Then one can measure the area over which the film extends, and since the volume is known, the film thickness can be found. In myristic acid the atoms in a molecule are bound together like a linear chain. When the acid forms a continuous thin film, the bound atoms line up vertically, up to a height equal to the size of a single molecule. In a special experiment, an amount of 0.03 mgr of myristic acid (whose density is 0.862 gr/cm^3) was left off on water. The continuous film that formed covered an area of 222 cm^2. (a) Find the size of a single molecule (*i.e.*, the length of the atomic chain in the molecule). (b) If the mass of a single molecule is 4.20×10^{-22} gr, find the total number of the molecules present, and hence find the cross-sectional area of each molecule as seen from above.

Solution: (a) If the thickness (height) of the film (= size of a molecule) is h, its volume will be $V = 222h$ (because $A = 222$ cm^2). Since ρ and m are known, we have $V\rho = m$ or

$$222h \times 0.862 = 3 \times 10^{-5}$$

whence $h = 1.57 \times 10^{-7}$ cm.

(b) Since the mass of a single molecule is known, there will be N molecules in 3×10^{-5} gr where

$$N = \frac{3 \times 10^{-5}}{4.2 \times 10^{-22}} = \frac{5}{7} \times 10^{17} .$$

N molecules cover an area of 222 cm^2. So the cross-sectional area of a single molecule is

$$A' = 222 : \left(\frac{5}{7} \times 10^{17} \right) = 3.11 \times 10^{-15} \text{ cm}^2 .$$

3. Structural methods, optical and radio-frequency spectroscopy, and NMR are good for finding interatomic separations r, whereas NQR and ESR are good for obtaining information about molecular polarity. True or false?

Solution: True.

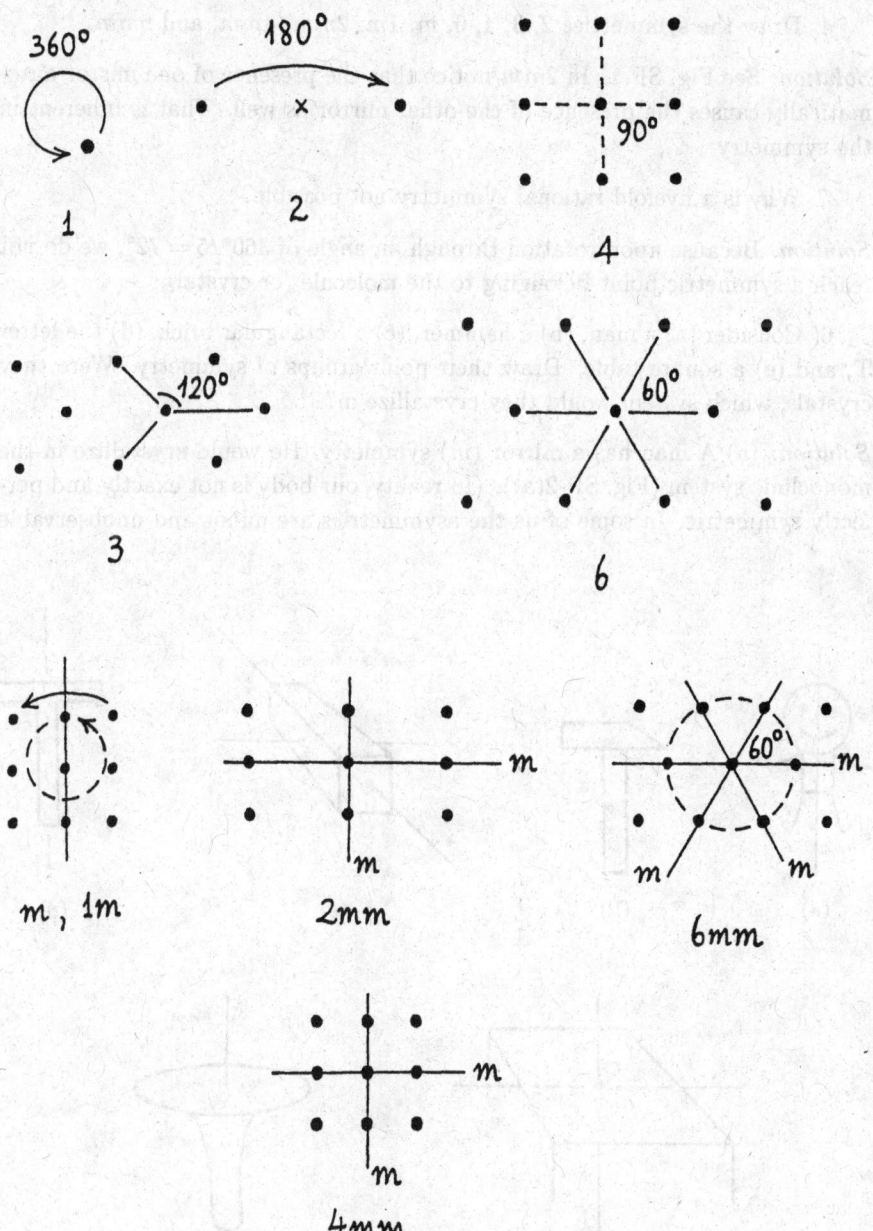

Fig. SP-1.

4. Draw the symmetries 2, 3, 4, 6, m, $1m$, $2mm$, $4mm$, and $6mm$.

Solution: See Fig. SP-1. In $2mm$ notice that the presence of one mirror automatically causes the presence of the other mirror as well. That is inherent in the symmetry.

5. Why is a fivefold rational symmetry not possible?

Solution: Because upon rotation through an angle of $360°/5 = 72°$, we do not reach a symmetric point belonging to the molecule (or crystal).

6. Consider (a) a man, (b) a hammer, (c) a rectangular brick, (d) the letter T, and (e) a square table. Draw their point groups of symmetry. Were they crystals, which system would they crystallize in?

Solution: (a) A man has a mirror (m) symmetry. He would crystallize in the monoclinic system (Fig. SP-2(a)). (In reality, our body is not exactly and perfectly symmetric. In some of us the asymmetries are minor and unobservable

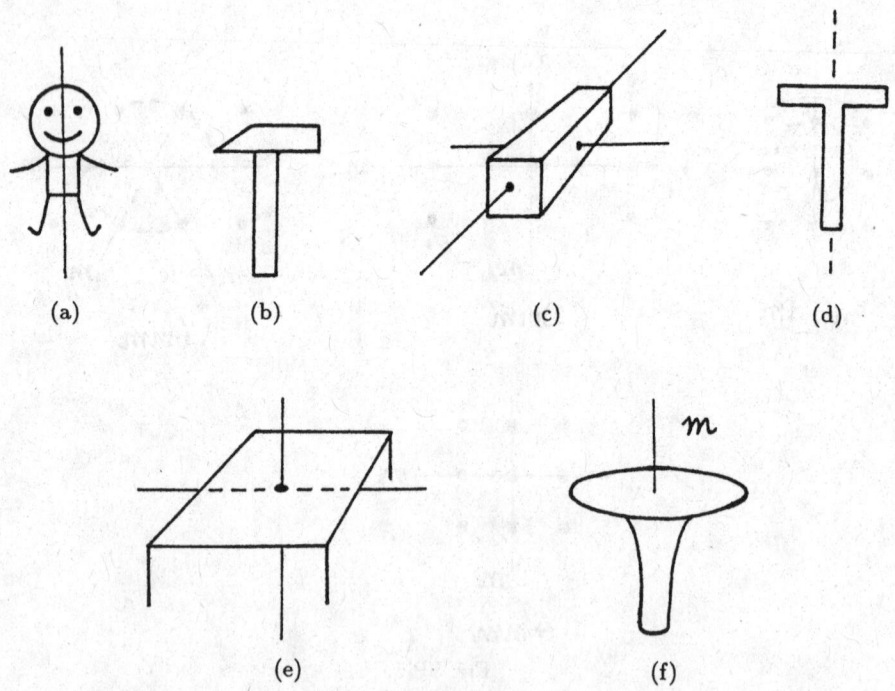

(a)   (b)   (c)   (d)

(e)   (f)

Fig. SP-2.

— like the impeccable physical build of a Bo Derek or a Raquel Welch body — while in others are pronounced, *e.g.*, one brow up and one down etc. In our case we consider an Apollo or a stylized picture of man.)

(b) Onefold rotation. Triclinic system (Fig. SP-2(b)).

(c) $\frac{2}{m}\frac{2}{m}\frac{2}{m} = mmm$. (Both twofold rotation and m.) Orthorhombic system (Fig. SP-2(c)).

(d) Mirror symmetry. Monoclinic (Fig. SP-2(d)).

(e) $4mm$ (fourfold rotation, twofold rotation, and m). Tetragonal (Fig. SP-2(e)). For a circular table (Fig. SP-2(f)) we have onefold rotation and infinite m planes.

7. A structure has a symmetry point group $\frac{3}{m}$. Which system does it crystallize in?

Solution: Threefold rotation and a mirror plane perpendicular to it. So, $\frac{3}{m} = \bar{6}$. Hexagonal system.

8. Draw the symmetries of (a) the water molecule, and (b) the ammonia molecule.

Solution: See Fig. SP-3.

9. Draw the symmetries of the benzene molecule.

Solution: See Fig. SP-4.

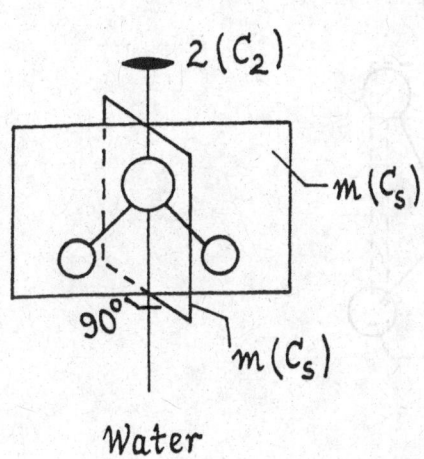

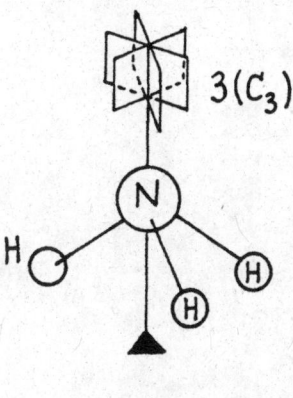

Fig. SP-3.

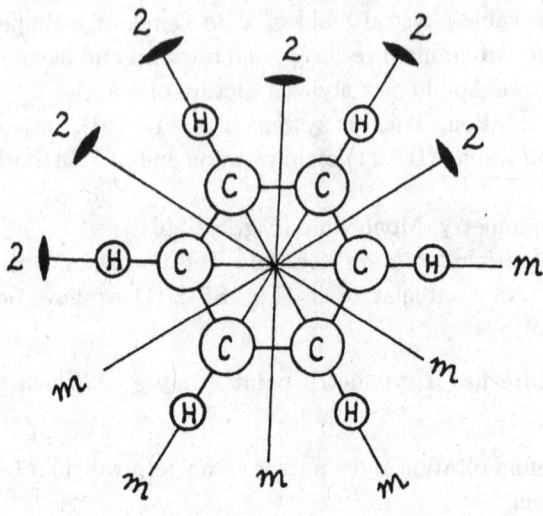

Fig. SP-4.

10. Draw a molecule having a fourfold alternating or inversion axis.

Solution: See Fig. SP-5.

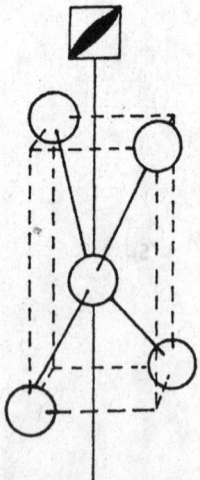

Fig. SP-5.

11. Draw all the possible rotational screw axes n_p up to 6_5, *i.e.*, from $n = 2$ to $n = 6$.

Solution: The axes 2, 2_1, 3, 3_1, and 3_2 are drawn in Fig. 2.30. The rest are in Fig. SP-6.

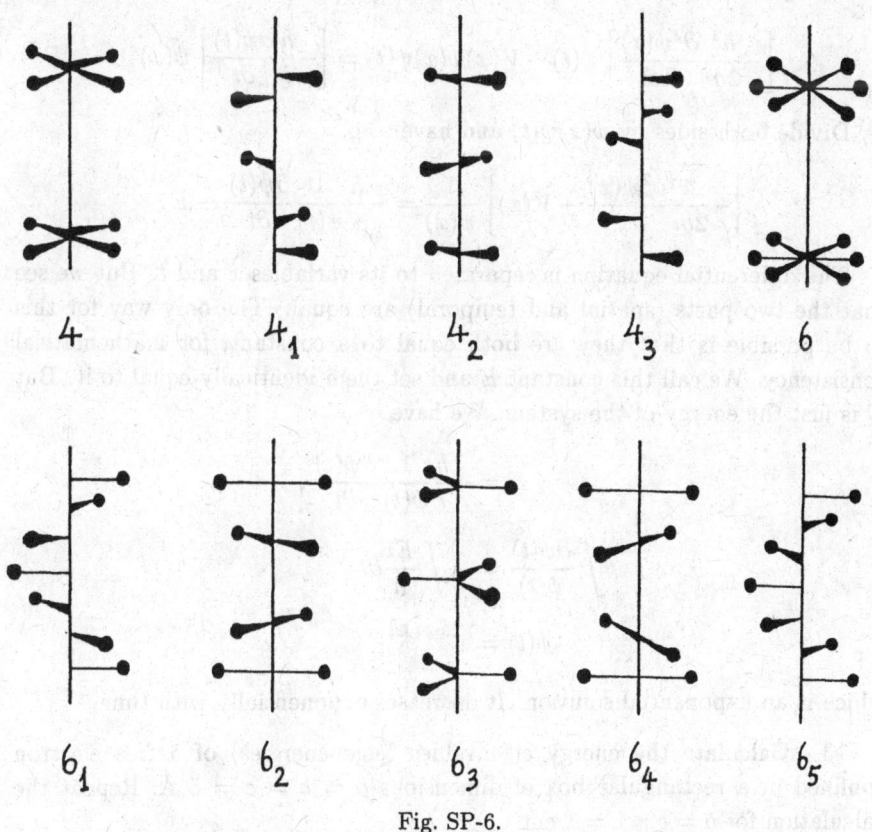

Fig. SP-6.

12. Find the temporal solution of the time-dependent, one-dimensional Schrödinger equation.

Solution: Write the wavefunction $\psi(x, t)$ as a product of a spatial (coordinate-dependent) and a temporal (time-dependent) part:

$$\psi(x, t) = \psi(x)\psi(t)$$

where ψ is separated now to its variables and has the form of a product.

The Schrödinger equation is $\mathcal{H}\psi = E\psi$ where

$$\mathcal{H} = -\frac{\hbar^2}{2m}\nabla_x^2 + V(x) = -\frac{\hbar}{i}\frac{\partial}{\partial t}$$

where ∇_x^2, V, and $\partial/\partial t$ are operators (which operate on ψ). We have

$$\left[-\frac{\hbar^2}{2m}\frac{\partial^2\psi(x)}{\partial x^2}\right]\psi(t) + V(x)\psi(x)\psi(t) = \left[-\frac{\hbar}{i}\frac{\partial\psi(t)}{\partial t}\right]\psi(x).$$

Divide both sides by $\psi(x)\psi(t)$ and have

$$\left[-\frac{\hbar}{2m}\frac{\partial^2\psi(x)}{\partial x^2} + V(x)\right]\frac{1}{\psi(x)} = -\frac{\hbar}{i}\frac{1}{\psi(t)}\frac{\partial\psi(t)}{\partial t} \equiv E.$$

The differential equation is separated to its variables x and t. But we see that the two parts (spatial and temporal) are equal. The only way for this to be possible is that they are both equal to a constant, for mathematical consistency. We call this constant E and set them identically equal to it. But E is just the energy of the system. We have

$$E = -\frac{\hbar}{i}\frac{1}{\psi(t)}\frac{\partial\psi(t)}{\partial t},$$

$$\int \frac{\partial\psi(t)}{\psi(t)} = -\int \frac{Ei}{\hbar}\partial t,$$

$$\psi(t) = e^{-2\pi i\frac{E}{\hbar}t}$$

which is an exponential solution. It decreases exponentially with time.

13.   Calculate the energy eigenvalues (eigenenergies) of a free electron confined in a rectangular box of dimensions $a = b = c = 5$ Å. Repeat the calculation for $a = b = c = 1$ cm.

Solution: Refer to Fig. SP-7. Since the electron is free, $V = 0$. We have

$$\frac{\hbar^2}{2m}\frac{\partial^2\psi}{\partial x^2} + E\psi = 0$$

where m is the electronic mass, ψ is the wavefunction of the free electron, and E is its energy (which is quantized to certain values). Call

$$k = \sqrt{2mE}/\hbar$$

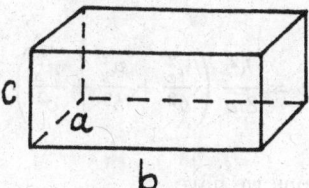

Fig. SP-7.

and have

$$\frac{\partial^2 \psi}{\partial x^2} + k^2 \psi = 0$$

whose solution is $\psi(x) = \mathcal{A}\sin(kx) + \mathcal{B}\cos(kx)$, with $\mathcal{A}$ and $\mathcal{B}$ being constants (to be determined from the boundary conditions).

Since $\psi(x = 0) \to 0$, we should have $\mathcal{B} = 0$. And since $\psi(a) = \mathcal{A}\sin(ka) = 0$, we have $ka = n\pi$ or

$$\frac{8\pi^2 mE}{h^2} = \frac{n^2 \pi^2}{a^2}$$

whence

$$E = \frac{h^2 n^2}{8ma^2}$$

where n is an integer (that gives the quantization of E): $n = 1, 2, 3, \dots$. Notice that the case is one-dimensional (only along x). The other dimensions are treated similarly; so, we can call the integer n_x for the x-direction, and n_y and n_z for the other two directions.

To find $\mathcal{A}$, we use the normalization condition[1]

$$\int \psi \psi^* d\mathrm{V} = \int \mathcal{A}^2 \sin^2\left(\frac{n\pi}{a}x\right) dx = 1$$

whence $\mathcal{A} = \sqrt{2/a}$, if we integrate over all space ($-\infty$ to $+\infty$). So,

$$\psi(x) = \sqrt{\frac{2}{a}}\sin\left(\frac{n\pi}{a}x\right).$$

The full function is

$$\psi(x, y, z) = \sqrt{\frac{8}{abc}}\sin\left(\frac{n_x \pi}{a}x\right)\sin\left(\frac{n_y \pi}{b}y\right)\sin\left(\frac{n_z \pi}{c}z\right)$$

where

$$k^2 = k_x^2 + k_y^2 + k_z^2$$

---

[1]We can always fix a constant by normalization.

and

$$E = \frac{h^2}{8m} \left(\frac{n_x^2}{a^2} + \frac{n_y^2}{b^2} + \frac{n_z^2}{c^2} \right).$$

For the given dimensions we have

$$E = \frac{(6.62 \times 10^{-27})^2}{8 \times 9.1 \times 10^{-28}} \times \frac{1}{(5 \times 10^{-8})^2} \times (n_x^2 + n_y^2 + n_z^2)$$

$$= 2.4 \times 10^{-12} \text{ erg} = 1.45 \text{ eV}.$$

And for the second box, $E = 9 \times 10^{-15}$ eV.

The situation concerns the free conduction electrons in a metal (energy bands).

14. (a) Consider the electron bound to the hydrogen atom. The wavefunctions of its lowest energy states are

$$\psi_{100} = \pi^{-1/2} \rho^{-3/2} e^{-r/\rho} \qquad (1s)$$

$$\psi_{200} = \pi^{-1/2} 2^{-5/2} \rho^{-3/2} \left(2 - \frac{r}{\rho} \right) e^{-r/2\rho} \qquad (2s).$$

Prove that the probability of finding an electron in the 1s state peaks (is maximum) for a value of r equal to the Bohr radius. (b) The probability of finding somewhere an electron of the 2s state exhibits two maxima. Find their values in terms of the Bohr radius. (c) Show that the transition $E_2 \rightarrow E_1$ is forbidden.

Solution: Since ψ^2 is the probability density, the probability is $\psi^2 dV$. After spherical normalization, the probability of finding an electron between r and $r + dr$ (*i.e.*, within a spherical shell of thickness dr) (Fig. SP-8) is

$$f(r)dr = 4\pi r^2 \psi^2 dr$$

where $r^2 dr$ is the radial element, and 4π comes from the integration over all φ angles (since we integrate over variables we are not interested in; we do not care about which particular φ we should be working at).

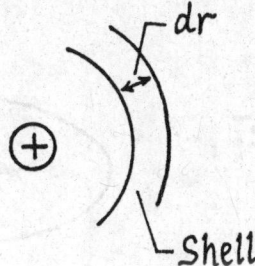

Fig. SP-8.

To find the maximum, we plug in ψ, take the derivative of f, and set it equal to zero.

$$f\,dr = 4\pi r^2 \left(\pi^{-1/2}\rho^{-3/2}e^{-r/\rho}\right)^2 dr\,,$$

$$\frac{\partial f(r)}{\partial r} = 0 = 8r\rho^{-3}e^{-2r/\rho} - 4\rho^{-3}r^2 e^{-2r/\rho}\left(\frac{2}{\rho}\right),$$

$$1 - \frac{r}{\rho} = 0$$

whence $\rho \equiv a_0/Z = r$, and $r = \rho = a_0$, since $Z = 1$ for hydrogen (Fig. SP-9). Hence $a_0 = $ Bohr radius.

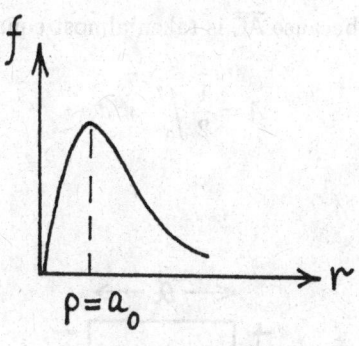

Fig. SP-9.

The rest of the problem can be solved similarly. It is left to the reader.

15. An electron describes a closed, elliptic trajectory. Calculate its magnetic moment in terms of the angular momentum and the Bohr magneton.

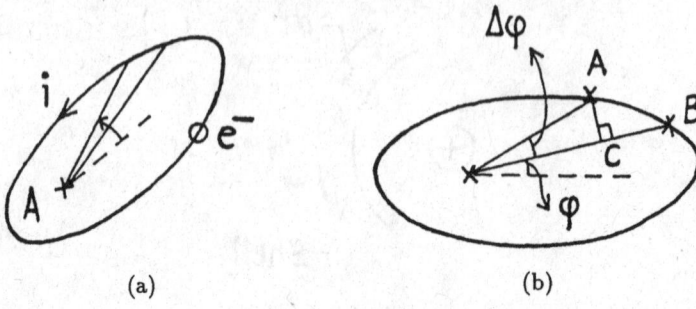

(a)   (b)

Fig. SP-10.

Solution: Refer to Fig. SP-10. The revolving electron forms a current loop. The current is

$$i = \frac{\Delta q}{\Delta t} = \frac{e}{c\tau}$$

where τ is the period of one revolution, and c is the speed of light that fixes the electromagnetic units. To the loop there corresponds a magnetic moment $\mu = iA$ where A is the area of the loop.[2] The situation is equivalent to a tiny magnet of length d and magnetic pole strength m (Fig. SP-11) whose magnetic dipole moment is $\mu = md$. We have

$$\mu = \frac{e}{c}\frac{A}{\tau}$$

where $\Delta A = \frac{1}{2}rr\Delta\varphi$ (because $\overline{AC}$ is taken almost equal to the arc $\overset{\frown}{AB}$ because it is small) and

$$A = \frac{1}{2}\int_0^{2\pi} r^2 d\varphi.$$

$$\xleftarrow{\quad} d \xrightarrow{\quad}$$
$$+ \boxed{} -$$
$$m \qquad\qquad m$$

Fig. SP-11.

[2]Actually, $\boldsymbol{\mu} = i\mathbf{A} = iA\hat{n}$ where $\hat{n}$ is a unit vector normal to the area (that defines the area vectorially).

The angular momentum is $L = m\omega r^2$ where m is the mass of the electron, and $\omega = \Delta\varphi/\Delta t$. So, $r^2 = L/m(\Delta\varphi/\Delta t)$, and

$$A = \int_0^{2\pi} \frac{L}{m\frac{d\varphi}{dt}} d\varphi = \frac{1}{2} \int_0^{\tau} \frac{L}{m} dt,$$

$$\mu = \frac{e}{c\tau} \frac{L\tau}{2m} \quad \text{and} \quad \mu = \frac{e}{2mc} \mathbf{L}.$$

The magnetic moment is related to the angular momentum where $L = l\hbar$, and hence

$$\mu = \frac{e\hbar}{2mc} l \equiv \mu_B l$$

where $\mu_B = e\hbar/2mc$ is the Bohr magneton.

We have $e = 4.8 \times 10^{-10}$ statCb, $\hbar = 1.0559 \times 10^{-27}$ erg sec, $c = 2.997 \times 10^{10}$ cm/sec, $m_e = 9.109 \times 10^{-28}$ gr. We find $\mu_B = 9.2732 \pm 0.0006 \times 10^{-21}$ erg/Gauss.

The (potential) energy of the dipole in an external field $\mathbf{H}_0$ is $V = -\mu \cdot \mathbf{H}_0$.

16. (a) Calculate the gravitational potential of an A_2 type molecule whose two atoms are of atomic weight about 250 each, and separated by a distance $r_0 \simeq 3$ Å. (b) Calculate the potential of the magnetic interaction (attractive or repulsive force) between two magnetic dipoles whose magnetic moments are equal to the Bohr magneton. (c) Calculate the interaction between two dipoles, each equal to the nuclear magneton, and compare the result to that of part (b). (d) Compare these potentials (for the same distance) to the Coulombic potential of an electron bound to the hydrogen atom.

Solution: (a) See Fig. SP-12. The gravitational force is

$$F = G\frac{m_1 m_2}{r_0^2} = -\frac{\partial V_{\text{grav}}}{\partial r}.$$

$$V_{\text{grav}} = -Gm_1 m_2 \int r^{-2} dr = Gm_1 m_2 r^{-1},$$

$$V_{\text{grav}} = \frac{6.7 \times 10^{-8} \times (250 \times 1.66 \times 10^{-24})^2}{3 \times 10^{-8}}$$

$$= 3.9 \times 10^{-44} \text{ erg} = 2.4 \times 10^{-32} \text{ eV}.$$

(b) $V = \mu \cdot \mathbf{H}$, and $H = -\mu/r^3$ for the side position. So,

$$V_{\text{magn}} = -\frac{2\mu^2}{r^3}$$

where the factor 2 comes from the fact that there are two dipoles. If they are antiparallel, they attract each other; if they are parallel, they repel each other (Fig. SP-13).

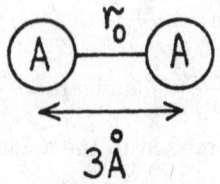

Fig. SP-12.

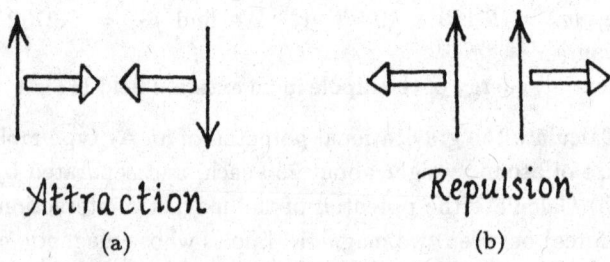

Attraction
(a)

Repulsion
(b)

Fig. SP-13.

By using $\mu_B = 0.928 \times 10^{-20}$ erg/Gauss, we have for the interacting dipoles

$$V_{\text{magn}} = \frac{(-2) \times (0.928 \times 10^{-20})^2}{(3 \times 10^{-8})^3} = -6.36 \times 10^{-16} \text{ erg} = -4 \times 10^{-4} \text{ eV}.$$

(c) The nuclear magneton is $\mu_n = \mu_B/1840$.

$$V_{\text{nucl}} = \frac{2\mu_n^2}{r^3} = \frac{2\mu_B^2}{(1840)^2 r^3} = \frac{1}{(1840)^2} V_{\text{magn}} = \frac{4 \times 10^{-4}}{(1840)^2} = 1.2 \times 10^{-10} \text{ eV},$$

that is, energy of nuclear dipoles $\ll$ interaction energy of the electron.

(d) $V_{\text{Coul}} = -\frac{e^2}{r} = -\frac{(4.8 \times 10^{-10})^2}{3 \times 10^{-8}} = -7.7 \times 10^{-12} \text{ erg} = -5 \text{ eV}.$

17.   The potentials of the attractive and repulsive forces in a diatomic molecule are given as $V_{\text{attr}} = -ar^{-m}$ and $V_{\text{rep}} = +br^{-n}$ where $n > m$ and a, $b > 0$ (constants). (a) Calculate the attractive and repulsive force. (b) Find the equilibrium separation r_0. (c) Find the binding energy V_0.

Solution: (a) $F_{\text{attr}} = -\frac{\partial V_{\text{attr}}}{\partial r} = amr^{-m-1}$ and $F_{\text{rep}} = bnr^{-n-1}$.

(b) To find r_0, take the total potential V_t and set its derivative equal to zero.

$$\frac{\partial V_t}{\partial r} = amr^{-m-1} - bnr^{-n-1} = 0,$$

$$r \equiv r_0 = \sqrt[n-m]{\frac{bn}{am}}.$$

(c) $V_0 = (-a)\left(\frac{bn}{am}\right)^{\frac{-m}{n-m}} + b\left(\frac{bn}{am}\right)^{\frac{-n}{n-m}} = (-a)(r_0)^{-m}\left[1 - \left(\frac{bn}{am}\right)^{\frac{-n}{n-m}}(\cdots)^{\frac{m}{n-m}}\right]$
$= (-a)r_0^{-m}\left(1 - \frac{m}{n}\right)$.

18. In the ethylene molecule (C_2H_4) the angle between neighboring C=H bonds is $\theta = 120°$. Calculate the s and p characters.

Solution: The wavefunction can be expressed as $\psi = s\psi_s + p\psi_p$. Refer to Fig. SP-14. We have $\psi_1 = s_1\psi_s + p_1\psi_p$ and $\psi_2 = s_2\psi_s + p_2\psi_p$.

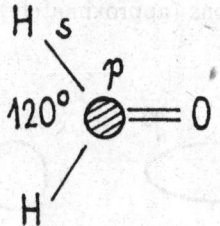

Fig. SP-14.

From the normalization condition

$$\int \psi_1\psi_1^* dV = 1$$

we find $s_1^2 + p_1^2 = 1$ and $s_2^2 + p_2^2 = 1$.

We assume that the two wavefunctions of the molecule are orthogonal. The orthogonality condition is

$$\int \psi_1\psi_2 dV = 0.$$

We have

$$\psi_1 = (1 - s_1^2)^{1/2}\psi_p + s_1^2\psi_s,$$
$$(1 - s_1^2)^{1/2}(1 - s_2^2)^{1/2}\cos\theta + s_1 s_2 = 0$$

with $s_1 = s_2$. The plane that passes through the two H atoms is perpendicular to the plane that passes through the other two H atoms. The H atoms are symmetric with respect to these planes. Hence their wavefunctions are of the same character. This gives the angle

$$\cos\theta = \frac{-s^2}{1-s^2}.$$

Now if we take $\theta = 120°$, we find $s = 0.33$ and $(1 - s^2)^{1/2} = 0.94$.

The wavefunctions pertinent to the molecule (which forms C–H bonds) are

$$\psi_{1,2} = 0.94\psi_p + 0.33\psi_s$$

which is an approximate result, because we took the wavefunction of H. One expects that the wavefunction of C is distorted. One should resort to quantum mechanics for a sufficient precision.

19. The valence electrons of two atoms are s and p electrons. Calculate the unperturbed wavefunctions (approximately) of the s–p bond between the two atoms (Fig. SP-15).

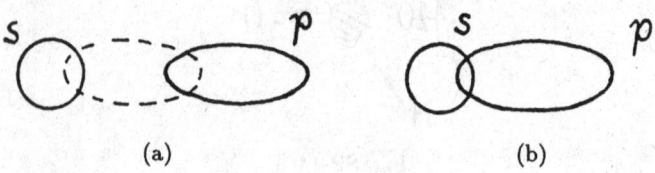

(a)   (b)

Fig. SP-15.

Solution: The wavefunction of s electrons is spherical, whereas that of p electrons is elliptic. We have to find the approximate expression of the wavefunctions that characterize s and p. There are two wavefunctions because there is a splitting in the energy states.

Let us indiciate the first electron by I and the second one by II. We write

$$\psi_1 = a_1\psi_s(\text{I}) + b_1\psi_p(\text{II})$$

$$\psi_2 = a_2\psi_s(\text{II}) + b_2\psi_p(\text{I})$$

$$\psi_{nlm} = (\text{Radial part}) \times (\text{Polar part}) \times (\text{Azimuthal part}) = \psi(r)\psi(\theta)\psi(\varphi)$$

where we are interested only in the θ-dependent angular part, $\psi(\theta)$.

The angular part of ψ_s is constant:

$$\psi_{10}(\theta) = \psi_{20}(\theta) = \frac{1}{\sqrt{2}}.$$

$$\psi_{210}(\theta) = \sqrt{\frac{3}{2}} \cos \theta.$$

The equivalence of electrons requires $a_1^2 = a_2^2 \equiv a^2$. That is, we cannot distinguish the electrons and tell which one belongs to whom. If they are interchanged, the same wavefunctions are obtained (the wavefunctions do not change). Therefore, from $a_1^2 + a_2^2 = 1$ we obtain $a^2 = 1/2$ and $a_1 = a_2 = 1/\sqrt{2}$.

We cannot determine the position of the two electrons; one can take the place of the other.

From the normalization condition we have

$$\int \psi_1^2 dV = 1 = a_1^2 + b_1^2$$

whence $a_1 = 1/\sqrt{2}$ and $b_1 = \pm 1/\sqrt{2}$.

From the orthogonality condition we have

$$\int \psi_1 \psi_2 dV = 0 = a^2 + b_1 b_2$$

whence $b_2 = -1/\sqrt{2}$. So,

$$\psi_{1,2} = \frac{1}{\sqrt{2}} \psi_s \pm \frac{1}{\sqrt{2}} \psi_p \quad \text{and} \quad \psi_{1,2}(\theta) = \frac{1}{\sqrt{2}} \pm \sqrt{\frac{3}{2}} \cos \theta.$$

Therefore, the ratio of the coefficients is 1 to $\sqrt{3}$. If one is taken $1/\sqrt{2}$ (for s), the other one will be $\sqrt{3/2}$ (for p).

20. Redo the previous problem for the s–p^2 bond.

Solution: There are three electrons, and hence three wavefunctions. We take

$$\psi_1 = a_1 \psi_s + b_1 \psi_{p_1} + c_1 \psi_{p_2} = a_1 \psi_s + b_1 \psi_{p_x} + c_1 \psi_{p_z}$$

where we see that there are two functions for p. We write the other ones, too:

$$\psi_2 = a_s \psi_s + b_2 \psi_{p_y} + c_2 \psi_{p_z}$$

$$\psi_3 = a_3 \psi_s + b_3 \psi_{p_y} + c_3 \psi_{p_z}.$$

There are three energy splittings.

Refer to Fig. SP-16. If ψ_1 is along the p_x-direction, the third term vanishes ($c_1 = 0$) because it has no effect.

The rest can be carried out by the reader.

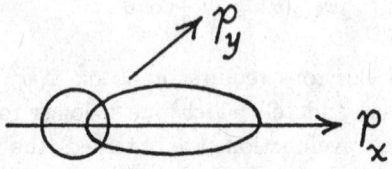

Fig. SP-16.

21. (a) The melting heat of ice is 80 kcal/kg. This heat supplied to ice breaks the intermolecular crystallic bonds and thus liquid water is obtained. If there are six bonds per molecule (to its nearest neighbors), calculate the bond energy. (b) In hydrocarbon fuels, the C–H binding energy is 4.28 eV per bond, and that of C–C is 3.6 eV per bond. Find the heat released by 1 kg of ethane (Fig. SP-17) when burned.

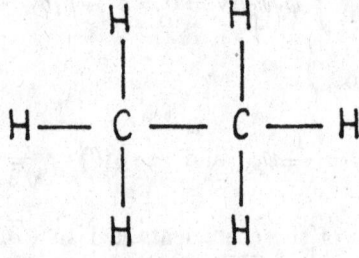

Fig. SP-17.

Solution: (a) This is a molecular-bond problem. We have 80 kcal/kg = 80 × 4180 Joules/kg = 3.76×10^{22} eV/mol (because 1 mol = 18 gr). One mol of ice has $6 \times 6.022 \times 10^{23}$ bonds. Hence we have $(3.76 \times 10^{22})/(3.61 \times 10^{24}) = 0.01$ eV per bond.

(b) A burning ethane molecule has six C–H and one C–C bonds broken. The released energy (per molecule) is

$$6 \times (4.28) + 3.6 = 29.28 \text{ eV}.$$

Since the molecular weight is 30, one kg of ethane contains $6.022 \times 10^{23} \times (1000/30)$ molecules. The overall produced heat is

$$29.28 \times 1.6 \times 10^{-19} \times 6.022 \times 10^{23} \times (1000/30) = 2.2 \times 10^4 \text{ kcal/kg}.$$

22. (a) Redo part (b) of the previous problem for 1 kg of methane. (b) One mol of CO gives off 84 kcal if burned. Find the binding energy.

Solution: (a) Methane is CH_4. Its molecular weight is 16. In 1 kg of it there are

$$\frac{1000 \text{ gr}}{16 \text{ gr}} \times 6.022 \times 10^{23} = 3.76 \times 10^{25} \text{ molecules}.$$

Each molecule has four C–H bonds, and each bond stores 4.28 eV. So, the answer is

$$3.76 \times 10^{25} \times 4 \times 4.28 \times 1.6 \times 10^{-19} = 10^8 \text{ Joules} = 2.5 \times 10^4 \text{ kcal}.$$

(b) 84 kcal/mol = 3.64 eV/atom. Since we have one bond per atom, the answer is 3.64 eV.

23. Two spherical clusters of mass M each consists of bound molecules of mass m each. The binding energy per molecule is E_0. The clusters move with a speed v and collide directly. What is the limit speed above which the collision smashes and blows up the clusters by breaking all the bonds?

Solution: There are M/m molecules per cluster. We have

$$\text{Kinetic energy} \geq \text{Total binding energy}$$

or

$$\frac{1}{2}Mv^2 = \frac{M}{m}E_0$$

for the smashing, where equality holds for the limit speed. Hence $v = \sqrt{2E_0/m}$ which is independent of the size of the cluster or the number of the molecules.

To get an idea about this speed, one can consider several molecules and calculate it, by plugging in m and E_0 of the molecule in question.

24. The lines of the far-infrared rotation spectrum of the hydrogen iodide molecule (HI) are observed at an interval $\Delta\bar{\nu} = 12.8$ cm^{-1}. Calculate the interatomic separation (bond length) H–I in this molecule.

Solution: There is a problem concerning molecular spectra. We have

$$E_{\text{rot}} = BhJ(J+1)$$

where $B = h/8\pi^2 \mathcal{I}$ = rotation constant, J = rotational quantum number, and $\mathcal{I} = \tilde{\mu}r_0^2$ = moment of inertia of the molecule, with $\tilde{\mu}$ being the reduced mass:

$$\frac{1}{\tilde{\mu}} = \frac{1}{m} + \frac{1}{m}.$$

The energy change upon transition from a rotational J state to another is

$$\Delta E = Bh[(J+1)(J+2) - J(J+1)] = 2Bh(J+1)$$

where $(\Delta E)_{min} = 2Bh$ for $J = 0$.

From $\bar{\nu} = 1/\lambda$, we have $\Delta\nu = c\Delta\bar{\nu}$ and $\Delta E = h\Delta\nu = 2Bh$ whence $2B = c\Delta\bar{\nu}$ and

$$B = \frac{h}{8\pi^2\mathcal{I}} = \frac{3 \times 10^{10} \times 12.8}{2}$$

whence $\mathcal{I} = \tilde{\mu}r_0^2 = 4.37 \times 10^{-40}$ gr cm^2.

Since the atomic weights of H and I are 1 and 127 respectively, we have

$$\tilde{\mu} = \frac{127 \times 1}{127 + 1} \times \frac{1}{6.02 \times 10^{23}} = 1.63 \times 10^{-24} \text{ gr},$$

and hence from $\mathcal{I} = \tilde{\mu}r_0^2$ we find $r_0 = 1.63$ Å.

25. (a) The following rotational transitions are observed in the carbonyl sulphur molecule ($^{16}O = {}^{12}C = {}^{32}S$):

J_1	J_2	ν (MHz)
1	2	24,325.92
2	3	36,488.82
3	4	48,651.64
4	5	60,814.08

The rotational energies are given by the expression

$$E_J = BhJ(J+1) - D_0 h[J(J+1)]^2$$

where D_0 is the centrifugal distortion coefficient (that gives the second-order correction term).[3] Calculate the rotation constant B and the coefficient D_0.

---

[3] During the rotation, a centrifugal force of different value acts on each atom (because the distance of each atom to the axis of rotation is different).

(b) The following rotational frequencies are observed in $^{16}O = {}^{12}C = {}^{34}S$:

J_1	J_2	ν (MHz)
1	2	23,732.33
3	4	47,462.40

Find B, D_0, and the moment of inertia of this molecule. (c) Calculate the spacings $r_1(O = C)$ and $r_2(C\text{--}S)$. Assume that the molecule is linear.

Solution: (a) For the transition $2 \to 1$ we have

$$\frac{\Delta E_J}{h} = \nu_{21} = B(2 \times 3 - 1 \times 2) - D_0(4 \times 9 - 1 \times 4) = 24,325.92 \times 10^6 \text{ Hz}.$$

$$\nu_{32} = B(12 - 6) - D_0(9 \times 16 - 4 \times 9) = 36,488.82 \times 10^6 \text{ Hz}.$$

$$\left. \begin{array}{l} 4B - 32D_0 = \nu_{21} \\ 6B - 108D_0 = \nu_{32} \end{array} \right\} \text{ find } D_0 \simeq 1 \text{ kHz and } B.$$

(b) The way is similar.

(c) See Fig. SP-18. The masses of ^{16}O, ^{12}C, ^{32}S, and ^{34}S are m_1, m_2, m_3, and m_4 respectively. Call $m^* = m_1 + m_2 + m_4$.

$$\mathcal{I}_1 = m_1 x_1^2 + m_2 x_2^2 + m_3 x_3^2$$

where the masses are known, and the distances are to be found. ($\mathcal{I}_1$ cannot be calculated from B.)

Apply Steiner's theorem of parallel axes now for the second molecule:

$$\mathcal{I}_2 + m^* x^2 = m_1 x_1^2 + m_2 x_2^2 + m_4 x_3^2.$$

$$\left. \begin{array}{l} m_1 x_1 + m_2 x_2 = m_3 x_3 \\ m_1(x_1 + x) + m_3(x_2 + x) = m_4(x_3 - x) \end{array} \right\}.$$

After the substitutions, one finds $x_2 = 0.52$ Å, $x_1 = 1.68$ Å, $r_1 = x_1 - x_2 = 1.16$ Å, and $r_2 = x_2 + x_3 = 1.55$ Å.

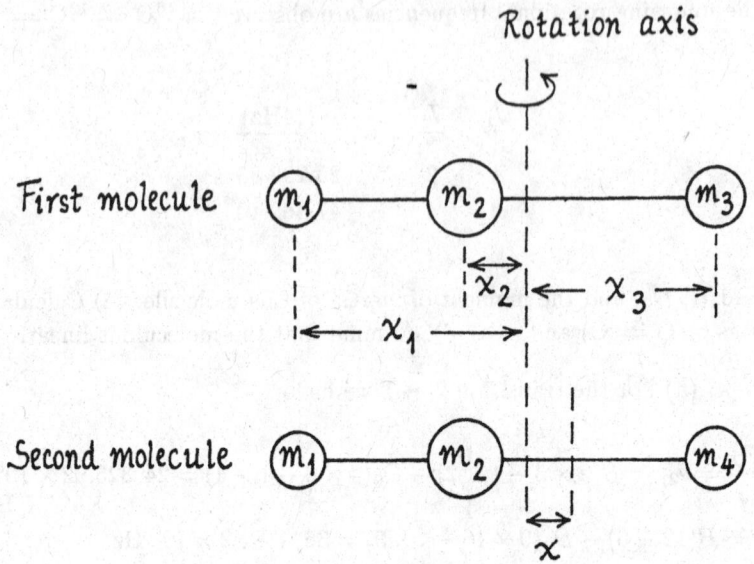

Fig. SP-18.

$$m^*x = m_4 x_3 - (m_1 x_1 + m_2 x_2),$$

$$x = \frac{m_4 - m_3}{m^*} x_3.$$

$$I_2 = m_1 x_1^2 + m_2 x_2^2 + \left[m_4 - \frac{m_4 - m_3}{m^*} \right] x_3^2,$$

$$I_2 - I_1 = M \left[m_4 - m_3 - \frac{m_4 - m_3}{m^*} \right] x_3^2.$$

We know that $m_1 = 16.000$, $m_2 = 12.00386$, $m_3 = 31.98089$, $m_4 = 33.97711$, and $M = 1/N_A = 1/(6.02486 \times 10^{23})$. Plug in these values and find $x_3 \simeq 1.03$ Å. Then find x_1 and x_2, and hence r_1 and r_2, and finally $x = 0.03$ Å.

26. The wave numbers of the spectrum lines of the HCl molecule in the infrared region are 20.5, 41, and 61.5 cm^{-1}. (a) Find the wavelengths λ. (b) Calculate the distance r_0 between H atom and the Cl atom.

Solution: (a) $\bar{\nu} = 1/\lambda = \lambda^{-1}$ whence $\lambda = (1/20.5)$ cm $= 0.04875$ cm $= 487.5 \ \mu$m.

(b) $E_J = BhJ(J+1)$ and $\Delta E_J = 2Bh(J_1 + 1)$ where J_1 is the quantum number of the lower energy level. We must have $\Delta J = 1$ for a transition. If $J_1 = 0$, we have

$$\Delta E_J = 2Bh = h\nu = h\frac{c}{\lambda} = \frac{2h^2}{8\pi^2 \mathcal{I}}$$

where $B = h/8\pi^2 \mathcal{I}$ and $\mathcal{I} = \tilde{\mu}r_0^2$ where $\tilde{\mu}$ is the reduced mass.

$$\mathcal{I} = \frac{h}{4\pi^2 c\bar{\nu}} = \frac{6.62 \times 10^{-27}}{39 \times 3 \times 10^{10} \times 20.5} = 2.73 \times 10^{-40} \text{ gr cm}^2 \ .$$

$$r_0 = \sqrt{\frac{\mathcal{I}}{\tilde{\mu}}} = \sqrt{\frac{\mathcal{I}}{\left(\frac{m_1 m_2}{m_1 + m_2}\right)\frac{1}{N_A}}}$$

where m_1 and m_2 are in fact the atomic masses of H and Cl, which are 1 and 35.5 respectively. N_A is the Avogadro number. If everything is plugged in, one finds $r_0 = 1.30$ Å.

27. A HgCl molecule makes a transition from the $J = 1$ rotational level to the $J = 0$ level, and emits a photon of $\lambda = 4.4$ cm. (a) Find the interatomic distance. (b) Find the number of rotations per second executed by the HCl molecule when the latter is in the state $J = 1$.

Solution: (a) This is a molecular rotation problem concerning rotational states and transitions. A molecular spectrum has three kinds of energy levels, and the total energy is the sum of them: $E = E_{\text{electronic}} + E_{\text{vibr}} + E_{\text{rot}}$. A diatomic molecule (consisting of two atoms of masses m_1 and m_2) rotates about an axis perpendicular to the (fictitious) line that joins the centers of the two atoms and passing through their common center of mass. Its moment of inertia about that axis is $\mathcal{I} = \tilde{\mu}d^2$ where

$$\tilde{\mu} = \frac{m_1 m_2}{m_1 + m_2} = \frac{3.33 \times 10^{-25} \times 0.588 \times 10^{-25}}{3.92 \times 10^{-25}} = 5 \times 10^{-26} \text{ kg}$$

because $m(\text{Hg}) = 3.33 \times 10^{-25}$ kg and $m(\text{Cl}) = 5.88 \times 10^{-26}$ kg.

A molecule rotating at an angular speed ω has an angular momentum $P = \mathcal{I}\omega$ where P is quantized as

$$P = \sqrt{J(J+1)}\hbar$$

with $J = 0, 1, 2, 3, \ldots$. A diatomic molecule has the following allowed angular speeds:

$$\omega = \sqrt{J(J+1)}\hbar/\mathcal{I}$$

whence the rotational kinetic energy becomes

$$E_{\text{rot}} = \frac{1}{2}\mathcal{I}\omega^2 = J(J+1)\hbar^2/\mathcal{I}.$$

The energy difference in the transition $J = 1 \rightarrow J = 0$ is

$$\Delta E_{\text{rot}} = h/4\pi^2\mathcal{I} = h\nu$$

whence

$$\frac{hc}{\lambda} = \frac{\hbar^2}{4\pi^2\tilde{\mu}d^2}$$

and

$$d^2 = \frac{h\lambda}{4\pi^2\tilde{\mu}c} = \frac{6.62 \times 10^{-34} \times 0.044}{4 \times 9.86 \times 5 \times 10^{-26} \times 3 \times 10^8} = 4.92 \times 10^{-20}$$

and $d = 2.22 \times 10^{-10}$ m.

(b) $\mathcal{I} = \tilde{\mu}d^2 = 2.46 \times 10^{-45}$ kg m^2,

$$\omega = \sqrt{J(J+1)}\frac{\hbar}{\mathcal{I}} = \frac{\sqrt{2} \times 6.62 \times 10^{-34}}{6.28 \times 2.46 \times 10^{-45}} = 6 \times 10^{10} \text{ rad/sec}$$

and the rotation frequency is $\nu_{\text{rot}} = \omega/2\pi = 0.95 \times 10^{10}$ rev/sec.

28. Consider the rotation band of the HBr molecule. The interatomic distance is 1.42×10^{-8} cm. Assume that the (heavy) Br atom stays still in space. Determine the interval between lines in the band (in cm^{-1}).

Solution: The rotation spectrum of a diatomic molecule depends mainly on its $\mathcal{I}$ (with respect to its rotation axis) where

$$\mathcal{I} = \tilde{\mu}r^2 = \frac{m_1 m_2}{m_1 + m_2}r^2 = \frac{1 \times 80}{1 + 80} \times 1.66 \times 10^{-24} \times (1.42 \times 10^{-8})^2$$

$$= 3.31 \times 10^{-40} \text{ gr cm}^2 \quad (\text{where } 1/N_A = 1.66 \times 10^{-24}).$$

The spectrum consists of equally spaced lines whose intervals are $h/4\pi^2\mathcal{I}c$. Thus the wave-number interval is

$$\Delta\bar{\nu} = \frac{h}{4\pi^2\mathcal{I}c} = \frac{6.62 \times 10^{-27}}{4 \times 9.86 \times 3.31 \times 10^{-40} \times 3 \times 10^{10}} = 16.9 \text{ cm}^{-1}.$$

If one observes the bands with an instrument of a high resolving power, one sees that the bands consist of lines which form a series.

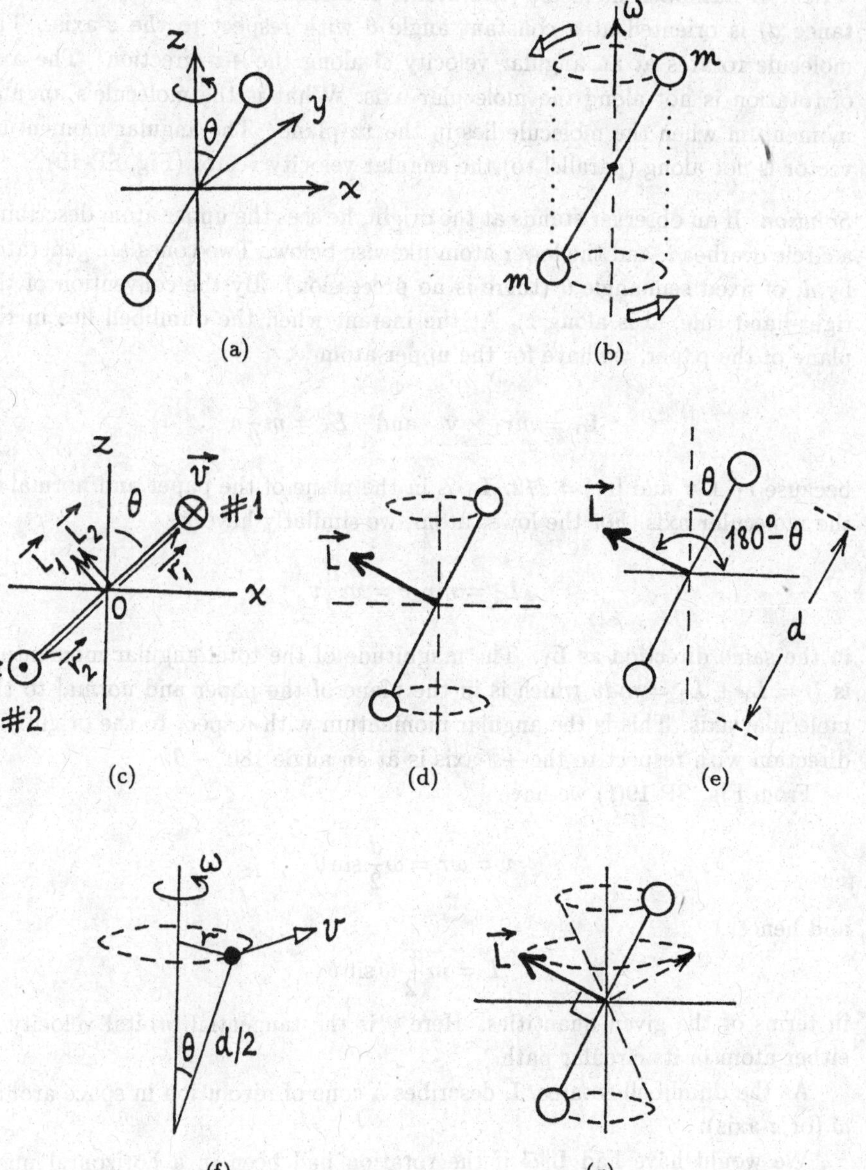

Fig. SP-19.

29. A dumbbell molecule (two atoms of mass m separated by a rigid distance d) is oriented at a constant angle θ with respect to the z-axis. The molecule rotates at an angular velocity $\vec{\omega}$ along the $+z$-direction. The axis of rotation is not along the molecular axis. What is the molecule's angular momentum when the molecule lies in the xz-plane? The angular momentum vector is not along (parallel to) the angular velocity vector (Fig. SP-19).

Solution: If an observer stands at the origin, he sees the upper atom describing a circle overhead, and the lower atom likewise below. Two cones are generated by d, of fixed semiangle θ (there is no precession). By the convention of the right-hand rule, $\vec{\omega}$ is along z. At the instant when the dumbbell lies in the plane of the paper, we have for the upper atom

$$\mathbf{L}_1 = m\mathbf{r}_1 \times \mathbf{v} \quad \text{and} \quad L_1 = m\frac{d}{2}v$$

because $\mathbf{r}_1 \perp \mathbf{v}$ and $|\mathbf{r}_1| = d/2$. $\mathbf{L}_1$ is in the plane of the paper and normal to the molecular axis. For the lower atom, we similarly have

$$L_2 = mr_2 v = m\frac{d}{2}v$$

in the same direction as $\mathbf{L}_1$. The magnitude of the total angular momentum is $L = L_1 + L_2 = mdv$ which is in the plane of the paper and normal to the molecular axis. This is the angular momentum with respect to the origin. Its direction with respect to the $+x$-axis is at an angle $180° - \theta$.

From Fig. SP-19(f) we have

$$v = \omega r = \omega\frac{d}{2}\sin\theta$$

and hence

$$L = m\frac{d^2}{2}\omega\sin\theta$$

in terms of the given quantities. Here v is the tangential orbital velocity of either atom in its circular path.

As the dumbbell rotates, $\mathbf{L}$ describes a cone of revolution in space around $\vec{\omega}$ (or z-axis).

We would have had $\mathbf{L}\|\vec{\omega}$ if the rotation had been in a horizontal plane (Fig. SP-20).

30. (a) Why a diatomic molecule has far more energy levels than an isolated atom (so that its emission line spectrum is richer than that of the atom)?

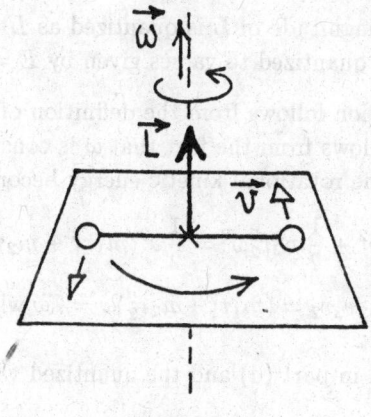

Fig. SP-20.

(b) Molecules give band spectra. True or false?

Solution: See text.

31. A diatomic molecule rotates about an axis normal to the interatomic separation line and passing through the center of mass, as shown in Fig. SP-21. (a) Prove that $m_1 r_1 = m_2 r_2$.

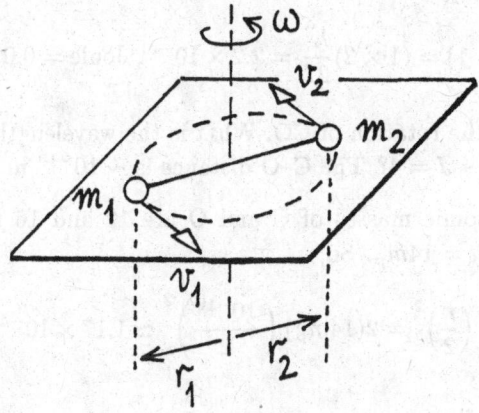

Fig. SP-21.

(b) Prove that $\omega = v_1/r_1 = v_2/r_2$. (c) Prove that the (total) kinetic energy of the molecule is $E = \frac{1}{2}m_1 v_1^2 + \frac{1}{2}m_2 v_2^2 = \frac{1}{2}\mathcal{I}\omega^2$ where $\mathcal{I} = m_1 r_1^2 + m_2 r_2^2$ is the moment of inertia of the two atoms about the axis of rotation. (d) Prove that the molecular angular momentum is $L = \mathcal{I}\omega = m_1 v_1 r_1 + m_2 v_2 r_2$, so that

$E = L^2/2\mathcal{I}$. (e) The magnitude of $\mathbf{L}$ is quantized as $L = \hbar\sqrt{J(J+1)}$. Prove that the energy is also quantized to values given by $E = J(J+1)\hbar^2/2\mathcal{I}$.

Solution: (a) The relation follows from the definition of the center of mass.

(b) The relation follows from the fact that ω is constant.

(c) From $v = \omega r$, the rotational kinetic energy becomes

$$E = \frac{1}{2}m_1 r_1^2 \omega^2 + \frac{1}{2}m_2 r_2^2 \omega^2 = \frac{1}{2}\omega^2(m_1 r_1^2 + m_2 r_2^2) = \frac{1}{2}\omega^2 \mathcal{I}.$$

(d) $L = m_1 v_1 r_1 + m_2 v_2 r_2 = (m_1 r_1^2 + m_2 r_2^2)\omega = \mathcal{I}\omega$ whence $E = (1/2)\mathcal{I}\omega^2 = (1/2)(\mathcal{I}\omega)^2/\mathcal{I} = L^2/2\mathcal{I}$.

(e) Use the relation in part (d) and the quantized value of L to obtain the expression for E.

32. In H_2, the protons are separated by 7.7×10^{-11} m. Calculate the energy of the first excited rotational state.

Solution: In the molecular-hydrogen rotation, the moment of inertia of the two protons about the rotational axis through the center of mass is

$$\mathcal{I} = 2m_p \left(\frac{r}{2}\right)^2 = 4.9 \times 10^{-48} \text{ kg m}^2.$$

For $J = 1$ we have

$$E(J = 1) = (1 \times 2)\frac{\hbar^2}{2\mathcal{I}} = 2.2 \times 10^{-21} \text{ Joule} = 0.014 \text{ eV}.$$

33. Consider the rotation of CO. What is the wavelength of the rotational transition $J = 5 \to J = 4$? The C–O distance is $\sim 10^{-10}$ m.

Solution: The atomic masses of C and O are 12 and 16 respectively. The *average* mass is $\overline{m} = 14m_p$. So,

$$\mathcal{I} = 2\overline{m} \left(\frac{r}{2}\right)^2 = 2(14m_p) \left(\frac{10^{-10}}{2}\right)^2 \simeq 1.17 \times 10^{-46} \text{ kg m}^2$$

or, more exactly,

$$\mathcal{I} = m_1 r_1^2 + m_2 r_2^2 = 1.15 \times 10^{-46} \text{ kg m}^2$$

where r_1 and r_2 can be found from the center of mass. Instead of doing that, we can use the *effective* (or reduced) mass

$$\tilde{\mu} = \frac{m_1 m_2}{m_1 + m_2} = 6.86 \, m_p.$$

$E = J(J+1)\hbar^2/2\mathcal{I}$ where we plug in everything and find E. The transition implies an energy difference

$$\Delta E = \frac{\hbar^2}{2\mathcal{I}}(5 \times 6 - 4 \times 5) = 4.79 \times 10^{-22} \text{ Joule}$$

which corresponds to $\lambda = hc/(\Delta E) = 4.22 \times 10^{-4}$ m (far infrared).

34. Consider a three-dimensional isotropic harmonic oscillator. Start from its Schrödinger equation and derive the relation that gives its quantized energies.

Solution: This is actually a quantum mechanics problem. A harmonic oscillator is a mass m on which a force $\mathbf{F} = -\kappa\mathbf{r}$ acts (where $\mathbf{r}$ is measured from the origin). This is the restoring force as the oscillator deviates from equilibrium. The force is derived from a potential V (by $F = -\partial V/\partial r$) where

$$V(r) = \frac{1}{2}\kappa r^2 = \frac{\kappa}{2}(x^2 + y^2 + z^2)$$

which is radial, but expressible in Cartesian coordinates. Given that $\psi = \psi(x, y, z)$, the Schrödinger equation is

$$\frac{\partial^2\psi}{\partial x^2} + \frac{\partial^2\psi}{\partial y^2} + \frac{\partial^2\psi}{\partial z^2} + \frac{8\pi^2 m}{h^2}\left[E - \frac{\kappa}{2}(x^2 + y^2 + z^2)\right]\psi = 0.$$

To solve this differential equation, we separate ψ into its variables:

$$\psi(x, y, z) = X(x)Y(y)Z(z).$$

Substitute that into the equation and divide both sides by XYZ to obtain

$$\frac{1}{X}\frac{d^2X}{dx^2} + \frac{1}{Y}\frac{d^2Y}{dy^2} + \frac{8\pi^2 m}{h^2}\left[E - \frac{\kappa}{2}(x^2 + y^2)\right] = -\frac{1}{Z}\frac{d^2Z}{dz^2} + \frac{8\pi^2 m}{h^2}\frac{\kappa}{2}z^2.$$

The left-hand side depends on x and y, and the right-hand side on z. But given that x, y, and z are independent variables, the two sides can be equal to each other if and only if they are both equal to a common constant which we call $(8\pi^2 m/h^2)C$ to have

$$\frac{d^2Z}{dz^2} + \frac{8\pi^2 m}{h^2}\left(C - \frac{\kappa}{2}z^2\right)Z = 0 \qquad (*)$$

and

$$\frac{1}{X}\frac{d^2X}{dx^2} + \frac{8\pi^2 m}{h^2}\left(E - C - \frac{\kappa}{2}x^2\right) = -\frac{1}{Y}\frac{d^2Y}{dy^2} - \frac{8\pi^2 m}{h^2}\frac{\kappa}{2}y^2 \qquad (**)$$

where the last equation is possible only if both sides are equal to another constant (just as above) which we call $(8\pi^2 m/h^2)B$. If we set $E = A + B + C$, we obtain from (**)

$$\frac{d^2Y}{dy^2} + \frac{8\pi^2 m}{h^2}\left(B - \frac{\kappa}{2}y^2\right)Y = 0$$

$$\frac{d^2X}{dx^2} + \frac{8\pi^2 m}{h^2}\left(A - \frac{\kappa}{2}x^2\right)X = 0.$$

The last two equations and (*) are nothing else but the Schrödinger equations of three *separate and independent* harmonic oscillators, each of which is one-dimensional (along x, y, and z). The energy eigenvalues will be equal to the possible values of the separation constants A, B, and C. The energy of a single, one-dimensional oscillator is

$$E_v = \left(v + \frac{1}{2}\right)h\nu$$

where ν is the frequency of the corresponding classical oscillator. Therefore, the eigenvalues of the three independent oscillator equations written above will be

$$A = \left(v_x + \frac{1}{2}\right)h\nu$$

$$B = \left(v_y + \frac{1}{2}\right)h\nu$$

$$C = \left(v_z + \frac{1}{2}\right)h\nu.$$

Since $v = v_x + v_y + v_z =$ integer, the energy levels of a three-dimensional oscillator are given by the formula

$$E_v = E(v_x, v_y, v_z) = A + B + C = \left(v_x + v_y + v_z + \frac{3}{2}\right)h\nu = \left(v + \frac{3}{2}\right)h\nu.$$

Obviously, in the case of the one-dimensional linear harmonic oscillator, the quantization factor lapses to $(v + \frac{1}{2})$.

35. If an harmonic oscillator obeys the selection rule $\Delta v = \pm 1$, prove that its corresponding classical oscillator performs a motion without harmonics.

Solution: The energy of the oscillator is

$$E_v = \left(v + \frac{3}{2}\right) h\nu_0$$

where ν_0 is the frequency of the corresponding classical oscillator. In a transition between levels E_v and $E_{v'}$, the frequency of the emitted or absorbed quantum is

$$\nu = \frac{E_v - E_{v'}}{h} = (v - v')\nu_0 .$$

Since the selection rule $\Delta v = \pm 1$ is valid, we have $\Delta v = (v - v') = \pm 1$ whence $\nu = \nu_0$. That is, under the condition of the selection rule, the frequency is exactly equal to the classical frequency ν_0, and not to its multiples. Therefore, the motion is escorted by no harmonics (overtones).

36. (a) Write the Schrödinger wave equation of the linear harmonic oscillator. (b) The solutions of this equation are the following wavefunctions:

$$\text{For } v = 0, \ \psi_0 = 2^{1/4}\pi^{-1/4}a^{1/2}e^{-a^2 x^2}$$

$$\text{For } v = 1, \ \psi_1 = 2^{5/2}\pi^{-1/4}a^{3/2}x \ e^{-a^2 x^2}$$

$$\text{For } v = 2, \ \psi_2 = 2^{-1/4}\pi^{-1/4}a^{1/2}(4a^2 x^2 - 1)e^{-a^2 x^2} .$$

Prove that these wavefunctions are normalized and orthogonal (to each other). (c) Show that for a linear oscillator, the transitions $E_0 \rightleftarrows E_1$ are possible, whereas the transitions $E_0 \rightleftarrows E_2$ are forbidden.

Solution: (a) $\nabla^2\psi + \frac{\hbar^2}{2m}(E - V)\psi = 0$ where $F = -\kappa x = -dV/dx$ (in one dimension), and $V = -\int F dx = \frac{1}{2}\kappa x^2$. Hence the Schrödinger equation of the harmonic oscillator is

$$\frac{d\psi^2}{dx^2} + \frac{2m}{\hbar^2}\left(E - \frac{1}{2}\kappa x^2\right)\psi = 0 .$$

Solving this equation is a long story.

(b) Normalization condition:

$$\int \psi^2 dx = 1$$

for every ψ.

$$\int_{-\infty}^{+\infty} 2^{1/2}\pi^{-1/2}a \ e^{-2a^2 x^2} dx = 2^{1/2}\pi^{-1/2}\sqrt{\frac{\pi}{2}}\frac{1}{a} = 1$$

where we set $2a^2 = \alpha$ and had the integral

$$\int_{-\infty}^{+\infty} e^{-\alpha x^2} dx = \sqrt{\pi/\alpha}.$$

Orthogonality condition:

$$\int \psi_0 \psi_1 dx = 0.$$

$$\int \psi_0 \psi_1 dx = (\text{Constant}) \times \int x\, e^{-2a^2 x^2} dx$$

which is zero by inspection (even without performing the integration one could tell that) because the integral is equal to two equal and opposite parts due to the nature of the integrand.

Orthogonality holds for any pair of different wavefunctions.

(c) The electric dipole moment is $\mathfrak{p} = ex$. The time-dependent wavefunction $\psi_{0,1}$ can be written as a linear combination of the form

$$\psi_{0,1} = a\psi_0 + b\psi_1$$

where the transition $0 \rightleftarrows 1$ mixes (couples) the states ψ_0 and ψ_1, as the system is in between during the transition. The fractions of admixture (for each contributing state) are a and b. (Both states participate; the oscillator is described by a bit of ψ_0 and a bit of ψ_1, and the situation changes in time as the transition is effected. Each state has some amount of the other state mixed in it.) The state $\psi_{0,1}$ gives a measure of the probability of finding the oscillator (say, the electron) in either of the energies E_0 and E_1.

The mean value of any (quantum mechanical) physical quantity u is

$$\bar{u} = \int \psi u \psi^* dV \equiv \langle u \rangle.$$

To find the mean value of the electric dipole moment, we should find $e\bar{x}$. That is,

$$\bar{\mathfrak{p}} = e\bar{x} = e\int \psi x \psi^* dx = e\int (a\psi_0 + b\psi_1)(a\psi_0^* + b\psi_1^*)x\, dx$$

$$= ab\int (\psi_0\psi_1^* + \psi_1\psi_0^*)x\, dx.$$

Since ψ_0 and ψ_1 are orthogonal, terms of different indices vanish. (In the most general case, a and b may be complex numbers. Then we will have aa^* and

bb^* above.) The time dependence of the wavefunction goes as $exp(-2\pi i E_0 t/h)$. Thus

$$\bar{x} = ab \int x\psi_0\psi_1^* \left[e^{-2\pi i \frac{E_0}{h}t} e^{2\pi i \frac{E_1}{h}t} + e^{2\pi i \frac{E_0}{h}t} e^{-2\pi i \frac{E_1}{h}t} \right] dx$$

$$= ab \int x\psi_0\psi_1^* \left[e^{2\pi i \left(\frac{E_1-E_0}{h}\right)t} + e^{-2\pi i \left(\frac{E_1-E_0}{h}\right)t} \right] dx$$

$$= ab\frac{1}{2} \cos\left[\frac{2\pi}{h}(E_1 - E_0)t \right] \int \psi_0\psi_1^* x \; dx$$

$$= \text{plug in the wavefunctions} = (\text{Constant}) \times \int x^2 e^{-2a^2 x^2} dx \neq 0,$$

nonzero because of x^2 in the integrand where $x^2 > 0$ always. As a result, we obtain a finite number: $\bar{x} \neq 0$. Therefore, the change has an amplitude

$$\frac{1}{2} \cos\left[\frac{2\pi}{h}(E_1 - E_0)t \right].$$

From the relation $(2\pi/h)(E_1 - E_0)t = \omega t$ we obtain the frequency ν of the periodic wave:

$$\nu = \frac{E_1 - E_0}{h}.$$

An amount of energy $\Delta E = E_1 - E_0$ is emitted or absorbed in the transition $0 \rightleftarrows 1$. Consider now the transition $0 \rightleftarrows 2$. We have

$$\bar{x}_{0,2} = \langle 0|x|2 \rangle \propto \int x(4a^2 x^2 - 1)e^{-2a^2 x^2} dx = 0,$$

vanishes because x (*i.e.*, an odd power of the variable) appears in the integrand, so that the integral is symmetric about the origin (it consists of two equal and opposite parts that cancel each other). Therefore, no absorption or emission takes place at a frequency $\nu = (E_2 - E_0)/h$. So, the transition $E_0 \rightleftarrows E_2$ is not possible (Fig. SP-22).

Whenever a $\Delta E = E_v - E_{v'}$ is of question, the transition is possible if $\Delta v = \pm 1$ (between adjacent values of v, or between adjacent vibrational levels); otherwise no other transition is possible, and this situation can be proven mathematically, as above.

37. A diatomic molecule can be thought as two masses m_1 and m_2 connected by a spring of constant κ (which is the chemical bond). The system

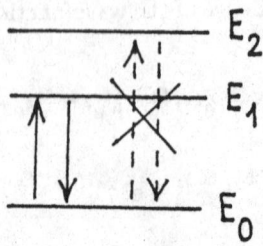

Fig. SP-22.

vibrates (molecular vibration) and can be imagined as carrying a single mass of effective (equivalent) value $\tilde{\mu}$ (that would produce the same effect as m_1 and m_2 did together). (a) Consider the CO molecule and find $m_1(C)$, $m_2(O)$, and $\tilde{\mu}$. (b) What is κ for CO, if the natural vibration frequency is 6.5×10^{13} cycles/sec? (c) Quantum mechanically, the vibration energies of CO are multiples of the fundamental value of 0.269 eV. Consider the fundamental vibration, and treating the system classically (in terms of the kinetic and potential energy for a spring), calculate the restoring force, the amplitude, and the velocity of the spring.

Solution: (a) Since the atomic masses of C and O are 12 and 16 (times the hydrogen mass m_H) respectively, $\tilde{\mu}$ of the molecule is

$$\tilde{\mu} = \frac{m_1 m_2}{m_1 + m_2} = \frac{12 \times 16}{12 + 16} m_H = \frac{12 \times 16}{12 + 16} \cdot \frac{1}{N_A} = 6.86 \, m_H.$$

So, $m_1(C) = 12 m_H$ and $m_2(O) = 16 m_H$ where $m_H = 1.67 \times 10^{-27}$ kg.

(b) $\kappa = \tilde{\mu}(\nu_0/2\pi)^2 = 1.23$ Nt/m.

(c) $V = \frac{1}{2}\kappa x^2 = 4.31 \times 10^{-20}$ Joule whence $x = 2.65 \times 10^{-10}$ m. $E = \frac{1}{2}\tilde{\mu}v^2 = 4.31 \times 10^{-20}$ Joule whence $v = 2740$ m/sec. $F = \kappa x = 3.3 \times 10^{-10}$ Nt at most.

38. The CO molecule has $\kappa = 1.9 \times 10^3$ Nt/m. (a) Find the natural vibration frequency of the molecule. (b) Find the splitting of the vibrational energy level. (c) Calculate the wavelength of the transition $v = 4 \rightarrow v = 0$.

Solution: (a) $\omega_0 = \sqrt{\kappa/\tilde{\mu}} = \sqrt{\frac{1.9 \times 10^3}{6.86 \times 1.67 \times 10^{-27}}} = 4.07 \times 10^{14}$ rad/sec.

(b) Vibration energy $E = \hbar\omega_0 = \frac{6.62 \times 10^{-34} \times 4.07 \times 10^{14}}{2\pi} = 4.29 \times 10^{-20}$ Joule = 0.268 eV.

(c) $\Delta E(v = 4 \rightarrow v = 0) = 4\hbar\omega_0 = h\nu = h\frac{c}{\lambda}$ whence

$$\lambda = \frac{hc}{E} = \frac{6.62 \times 10^{-34} \times 3 \times 10^8}{4 \times 4.29 \times 10^{-20}} = 1.16 \times 10^{-6} \text{ m} = 1.16 \ \mu\text{m (infrared)}.$$

39. The H_2 molecule is usually visualized as two protons (of mass m_p) connected by a spring of constant κ. Given that $\tilde{\mu} = \frac{1}{2}m_p$ and that the first excited vibrational state (v = 1) stands at 0.5 eV, what is κ? What is the natural vibration frequency?

Solution: The vibrational energy is quantized: $E_v = \left(v + \frac{1}{2}\right)\hbar\omega_0$. The energy quantum is

$$0.5 \text{ eV} = \hbar\omega_0 = \hbar\sqrt{\kappa/\tilde{\mu}}$$

where $\tilde{\mu}$ is known. Hence $\kappa = 486$ Nt/m and $\omega_0 = 7.63 \times 10^{14}$ rad/sec.

40. In H_2, the v = 1 level is $(1/2)$ eV up the ground level (v = 0). (a) What is the wavelength of the transition $0 \rightleftarrows 1$? (b) Collisions between molecules in the gas can excite molecules to v = 1 after a certain temperature. Find the gas temperature that triggers these excitations (threshold temperature).

Solution: (a) The given energy corresponds to the following λ:

$$\lambda = \frac{hc}{E} = 2.5 \ \mu\text{m (near infrared)}.$$

(b) The mean thermal energy kT of a flying molecule should be $\sim$ 30% of the excitation energy for high-energy molecules in the tail of the velocity distribution, so that collisions excite the upper levels. So, how hot should the molecular gas be?

Collisions start exciting the molecules from the moment when $kT = (30\%)$ $(1/2$ eV), that is, $T = 1740°$K.

41. Assume that $\kappa(O_2) = \kappa(H_2) = 486$ Nt/m. Since $m(O) = 16m(H)$, we have $\tilde{\mu}(O_2) = 16\tilde{\mu}(H_2)$. Given these data, find $E(v = 1)$ for O_2.

Solution: From $E = \left(v + \frac{1}{2}\right)\hbar\omega_0$ we have $E(v = 1) = \frac{3}{2}\hbar\omega_0$. Since $\omega_0 = \sqrt{\kappa/\tilde{\mu}}$ and $\tilde{\mu} = 16\tilde{\mu}'$, we have $\omega_0(O_2) = \frac{1}{4}\omega_0(H_2)$. Hence $\hbar\omega_0(O_2) = \frac{1}{4}\hbar\omega_0(H_2) = \frac{1}{4} \times (0.5 \text{ eV}) = 0.125$ eV, and $E(v = 1) = \frac{3}{2} \times 0.125 = 0.188$ eV.

42. The LiH molecule has a cubic crystal structure. Its density is 0.83. Assume that the interatomic distance is equal to the lattice constant (interatomic or intermolecular spacing in the crystal). Calculate the wavelength of the radiative transition $J = 1 \rightarrow J = 0$. (Answer: $\lambda \sim 10^3 \ \mu\text{m}$.)

Solution: Since $\rho = 0.83$ gr/cm^3, the volume per molecule is

$$\frac{8}{6.02 \times 10^{23}} \times \frac{1}{0.83} = 25 \times 10^{-24} \text{ cm}^3 = 25 \text{ (Å)}^3 .$$

Refer to Fig. SP-23. There are four molecules per unit cell (four Li and four H atoms). The volume of the unit cube is $d^3 = \frac{1}{4} \times 4 \times$ (Molecular volume) $= 25$ whence $d \simeq 2.92$ Å.

$$\mathcal{I} = \tilde{\mu} d^2 = 12.4 \times 10^{-40} \text{ gr cm}^2 \quad \text{and} \quad B = h/8\pi^2 \mathcal{I} = 6.76 \times 10^{10} .$$

$$\nu(1 \to 0) = \frac{\Delta E}{h} = 2BJ = (2B) \times 1 = 13.6 \times 10^{10} \text{ Hz} .$$

$$\lambda = 2200 \ \mu m .$$

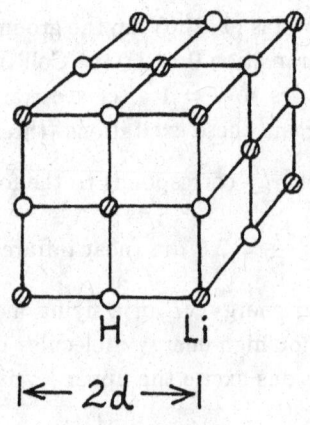

H   Li

$\vert\leftarrow 2d \rightarrow\vert$

Fig. SP-23.

Because of symmetry, we cannot choose the eighth of the cube. If we did, the symmetry would break.

43. In the HCl molecule, assume that Hooke's law is valid for the interatomic bond, and that $\kappa = 480$ Nt/m and $r_0 \simeq 1.29 \times 10^{-4}$ μm. (a) Find the wave numbers of the transition lines v = 0, $J = 0 \to$ v$' = 0$, $J' = 1$ and v = 0, $J = 0 \to$ v$' = 1$, $J = 0$. (The first transition is from the ground vibrational and rotational to the ground vibrational and first excited rotational level. The second transition is between the ground state and the first excited vibrational state — without any change in the rotational state.) (b) Find the energy differences (in ergs and eV) that correspond to these lines. (c) Find the frequencies

of these transitions. (d) By using the Boltzmann distribution, calculate the ratio of the number of molecules in the first excited state to the number of molecules in the ground state, at room temperature. (e) Use the same data to perform the calculations for the DCl (deuterium chloride) molecule.

Solution: $E_{\text{vibr}} + E_{\text{rot}} = h\nu \left(v + \frac{1}{2}\right) + BhJ(J+1)$.

$$\Delta E = h\nu + 2BhJ'$$

where J' refers to the upper level. The transitions are shown in Fig. SP-24:

$$v = 0,\, J = 0 \rightarrow v = 1,\, J = 1$$

$$v = 0,\, J = 0 \rightarrow v = 1,\, J = 0$$

$$v = 0,\, J = 0 \rightarrow v = 0,\, J = 1.$$

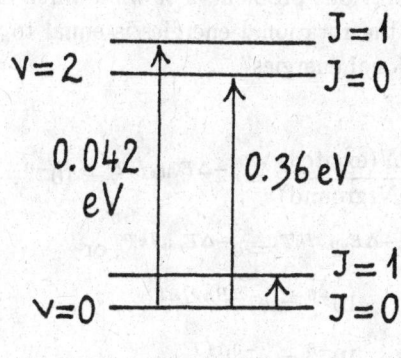

Fig. SP-24.

If we consider the situation as a simple harmonic motion, we have

$$\nu = \frac{1}{2\pi}\sqrt{\frac{\kappa}{\mu}}$$

$$= \frac{1}{6.28}\sqrt{\frac{480 \times 10^5 \times 10^{-2}}{\frac{35}{36} \times 1.62 \times 10^{-24}}} \simeq 90 \times 10^{12}\ \text{Hz}.$$

$$\lambda = \frac{c}{\nu} = \frac{3 \times 10^{10}}{90 \times 10^{12}} = 33 \times 10^2\ \mu m.$$

$$\Delta E(v = 0 \to v = 1) = h\nu = 6.62 \times 10^{-27} \times 90 \times 10^{12}$$

$$= 6 \times 10^{-14} \text{ erg} \simeq 0.040 \text{ eV}.$$

$$\Delta E(J = 0 \to J = 1) = 2BhJ' = 2 \times 3.3 \times 10^{11} \times 6.62 \times 10^{-27} \times 1$$

$$= 4.4 \times 10^{-15} \text{ erg} = 2.75 \times 10^{-3} \text{ eV}.$$

Consider $\Delta E_{\text{vibr}} = 0.36$ eV. At $T \simeq 300°$K we have

$$\frac{N(\text{excited})}{N(\text{ground})} = e^{-\Delta E/kT} = e^{-0.36/0.025} \simeq \frac{1}{e^{14}} \simeq 10^{-6}.$$

As we heat the gas, the molecules get excited to the upper vibrational level. Consider now the fraction as far as rotational states are concerned:

$$\frac{N(\text{excited})}{N(\text{ground})} = e^{-0.00275/0.025} \simeq e^{-0.11} \simeq 0.9.$$

44. Consider the previous problem. For what value of J does the population ratio concerning the rotational energies is equal to the population ratio concerning the vibrational energies?

Solution: We have

$$\frac{N(\text{excited})}{N(\text{ground})} = e^{-\Delta E_{\text{vibr}}/kT} = 10^{-6}.$$

We want to have $e^{-\Delta E_{\text{vibr}}/kT} = e^{-\Delta E_{\text{rot}}/kT}$, or

$$10^{-6} = e^{-2BhJ/kT},$$

$$10^{-6} = e^{-0.11J},$$

$$6 = 0.11J \log_{10} e,$$

$$J = \frac{6}{0.11 \times 0.43} = 127.$$

One can reshuffle the problem and phrase it as follows: For a given J, at what temperature do the two ratios become equal?

45. Consider the HCl molecule. (a) Calculate the ratio of the molecules in $J = 10$ to those with $J = 0$ at room T. (b) We have $\Delta E_{\text{vibr}} = h\nu = 0.35$ eV for $\lambda = 3.5$ μm. Calculate the $N(v')/N(v)$ ratio at room T ($kT \sim 1/40$ eV).

Solution: (a) $\frac{N(J'=10)}{N(J=0)} = \left(\frac{2\times 10+1}{2\times 0+1}\right) e^{-\Delta E/kT} = \frac{21}{1}e^{-143/25} = 21e^{-5.7} = 0.07.$

(b) $\frac{N(v')}{N(v)} = e^{-h\nu/kT} = e^{-0.35/(1/40)} < 10^{-6}$ which is a too small number. Therefore, practically almost all molecules are in the ground vibrational state v at room $T = 290°$K. Notice that the statistical weights are equal: $g(v') = g(v)$.

46. The rotation constants obtained from the rotation spectrum of the vinylene carbonate molecule are

$$A = 9346.40 \text{ MHz}$$

$$B = 4188.46 \text{ MHz}$$

$$C = 2891.54 \text{ MHz}.$$

(a) Calculate the moments of inertia of this molecule. (b) Discuss whether this molecule is planar or not.

Solution: (a) $\mathcal{I}_A = h/8\pi^2 A = 89,773 \times 10^{-40}$ gr cm^2, $\mathcal{I}_B = 200,334 \times 10^{-40}$ gr cm^2, and $\mathcal{I}_C = 290,188 \times 10^{-40}$ gr cm^2.

(b) Refer to Fig. SP-25. Is this molecule flat? Check to see if $\mathcal{I}_C = \mathcal{I}_A + \mathcal{I}_B$. Then it is flat. Indeed, we have $\mathcal{I}_A + \mathcal{I}_B = 290,180 \simeq 290,188$, within a relative error of

$$\frac{\Delta \mathcal{I}}{\mathcal{I}} = 2 \times 10^{-4}.$$

So, the molecule is (or can be taken as) planar within this error.

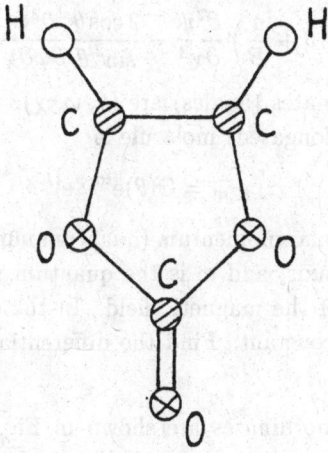

Fig. SP-25.

47. Two formulas are possible for the N_2O molecule: (a) N–O–N and (b) N $\equiv$ N–O. Since we have $\bar{\nu}_1 \simeq 1285$ cm^{-1}, $\bar{\nu}_2 \simeq 589$ cm^{-1}, and $\bar{\nu}_3 \simeq 2224$ cm^{-1} in the infrared spectrum, which formula is right? ($\lambda = 1/\bar{\nu}$.) (Fig. SP-26.)

(a)

(b)

Fig. SP-26.

Solution: In order for (a) to be right, we should have $\nu_1 = 0$. But we observe nonzero ν_1, ν_2, ν_3. Therefore, (b) is right. In (b) a change in the electric dipole takes place (and all three frequencies are observed).

(In case that ν_1 is not observed, then (a) is right.)

48. In the Euler coordinate system the Schrödinger equation for a prolate top type molecule can be written as follows:

$$\frac{1}{\sin\theta}\frac{\partial}{\partial\theta}\left(\sin\theta\frac{\partial\psi}{\partial\theta}\right) + \frac{1}{\sin^2\theta}\frac{\partial^2\psi}{\partial\varphi^2}$$

$$+ \left(\tan^2\theta + \frac{a}{B}\right)\frac{\partial^2\psi}{\partial\chi^2} - \frac{2\cos\theta}{\sin^2\theta}\frac{\partial^2\psi}{\partial\varphi\partial\chi} + \frac{E}{Bh}\psi = 0$$

where the Euler coordinates (angles) are (θ, φ, χ). In this system the wave-function of a prolate (elongated) molecule is

$$\psi_{J,K,m} = \Theta(\theta)e^{im\varphi}e^{iK\chi}$$

where J is the total angular momentum (quantum number), K is the projection of J on the molecular axis, and m is the quantum number of the projection of J on the direction of the magnetic field. In the equation, a is a constant and B is the rotation constant. Find the differential equation that gives the solutions for $\Theta(\theta)$.

Solution: The Euler coordinates are shown in Fig. SP-27. The solution is mathematically tedious, and we are too lazy to include it here. Why not make a nice pastime for the reader over the weekend?

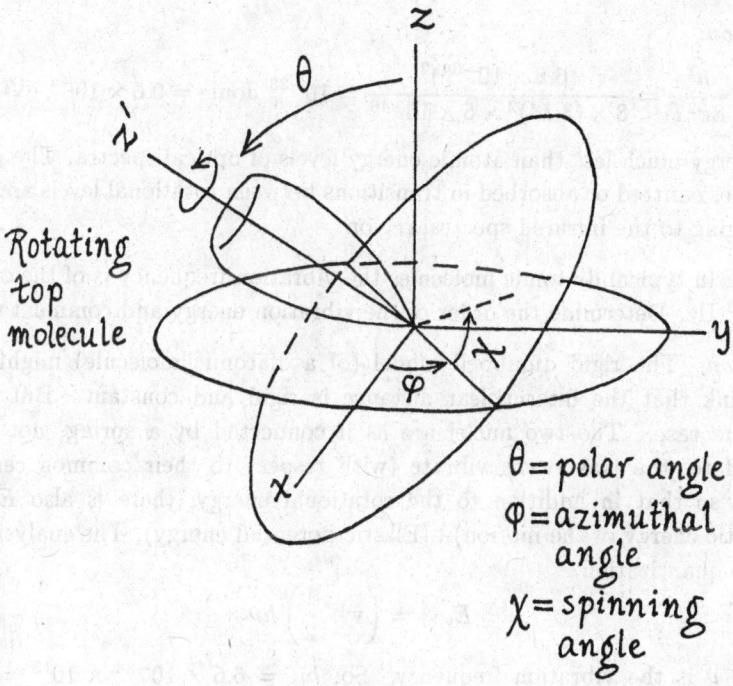

θ = polar angle
φ = azimuthal angle
χ = spinning angle

Fig. SP-27.

Since ψ is given, take its derivatives as dictated by the Schrödinger equation, plug them in, and divide the whole thing by ψ. You will reach a differential equation of the form

$$f\left(\frac{\partial\Theta(\theta)}{\partial\theta}, \frac{\partial^2\Theta(\theta)}{\partial\theta^2}\right) = 0$$

for the θ-part.

49. For the sake of simplicity, we usually analyze diatomic molecules viewed as dumbbells which rotate about an axis through the center of mass. The study of the rotation and the Schrödinger equation lead to the fact that the angular momentum and the motional energy are quantized:

$$E = \frac{h^2}{8\pi^2 \mathcal{I}} J(J+1).$$

For the O_2 molecule, $\mathcal{I} = 5 \times 10^{-46}$ kg m^2. Calculate the rotation constant and comment on it.

Solution:

$$\frac{h^2}{8\pi^2 \mathcal{I}} = \frac{(6.6 \times 10^{-34})^2}{8 \times (3.14)^2 \times 5 \times 10^{-46}} = 10^{-23} \text{ Joule} = 0.6 \times 10^{-4} \text{ eV},$$

an energy much less than atomic energy levels of optical spectra. The photon energies emitted or absorbed in transitions between rotational levels are small, belonging to the infrared spectral region.

50. In typical diatomic molecules the vibration frequency is of the order of $\sim 10^{13}$ Hz. Determine the order of the vibration energy and comment on it.

Solution: The rigid dumbbell model (of a diatomic molecule) might make us think that the internuclear distance is rigid and constant. But this is not the case. The two nuclei are as if connected by a spring, not a rod. Therefore, the two atoms vibrate (with respect to their common center of mass) so that in addition to the rotational energy, there is also $E_{\text{vibr}} =$ (Kinetic energy of the motion)+(Elastic potential energy). The analysis leads to the quantization

$$E_{\text{vibr}} = \left(v + \frac{1}{2}\right) h\nu$$

where ν is the vibration frequency. So, $h\nu = 6.6 \times 10^{-34} \times 10^{13} = 6.6 \times 10^{-21}$ Joule $= 0.04$ eV.

We notice that the typical vibrational energies — while still much lower than optical spectra energies — are larger than rotational energies.

51. A point charge q located at a distance r away from the coordinate origin produces an electric field at the origin. Show that the component along the direction of the z-axis of the field gradient is

$$(\vec{\nabla}\vec{\mathcal{E}})_{zz} = q\frac{3\cos^2\theta - 1}{r^3}.$$

Solution: The electric field is derivable from a potential V. The electric field gradient is

$$\text{Gradient} = \frac{\partial^2 V}{\partial z^2} = q_{zz}$$

where do not confuse the charge q with $q_{zz} \equiv (\nabla^2 V)_z = \partial^2 V/\partial z^2 =$ the z-component of the Laplacian of V. The gradient here is actually a matrix element, because $\vec{\nabla}\vec{\mathcal{E}}$ is a tensor (matrix). (This means that it may have different values along different directions — anisotropism — in the most general case.) By "gradient" above we mean the z-component of the full gradient.

For a point charge we have

$$V = \frac{q}{r} = \frac{9}{\sqrt{x^2 + y^2 + z^2}} = qa^{-1/2}$$

where $a \equiv x^2 + y^2 + z^2$. From Fig. SP-28, $\cos\theta = z/r$. Taking the second derivative of V, and plugging in $\cos\theta$, we have

$$\frac{\partial^2 V}{\partial z^2} = \frac{\partial}{\partial z}\left[\frac{\partial}{\partial z}(qa^{-1/2})\right] = q\frac{\partial}{\partial z}\left[-\frac{1}{2}a^{-3/2}\frac{\partial a}{\partial z}\right]$$

$$= q\frac{\partial}{\partial z}\left[\left(-\frac{1}{2}\right)(x^2 + y^2 + z^2)^{-3/2}2z\right] = q\frac{\partial}{\partial z}\left[-z(x^2 + y^2 + z^2)^{-3/2}\right]$$

$$= q\left[-(x^2 + y^2 + z^2)^{-3/2} + \left(\frac{3}{2}\right)2z^2(x^2 + y^2 + z^2)^{-5/2}\right]$$

$$= q[-r^{-3} + 3z^2 r^{-5}] = q\left[-r^{-3} + \frac{3z^2}{r^2}r^{-3}\right]$$

which is the given expression.

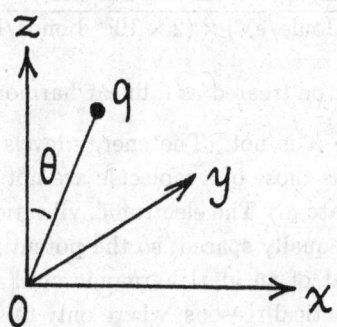

Fig. SP-28.

These considerations are important in relating the electric field gradient to the nuclear quadrupole moment eQ. The Hamiltonian of the nuclear quadrupole interaction energy is $\mathcal{H} = e\overset{\leftrightarrow}{Q}\cdot\vec{\nabla}\vec{\mathcal{E}}$, a dyadic. Here Q is the quadrupole moment of the atomic nucleus.

52. The heat of varporization of water is 539 cal/gr. Assume that every (liquid) water molecule is connected by six bonds to neighbor molecules. Calculate the bond energy in the water molecule.

Solution: The atomic weight of hydrogen is 1 and that of oxygen is 16. Since we have two hydrogen and one oxygen atom per molecule, the molecular weight is $2 \times 1 + 16 = 18$. (The atomic weight of an atom is the number of protons and neutrons in its nucleus. The electrons are ignored because their mass is 1836 times less than that of protons, and thus they do not contribute much to the atomic mass; the main contributors are the protons and neutrons, as heavier particles.) This means that 1 mol of H_2O has a mass of 18 gr. Since 1 mol consists of 6×10^{23} molecules, in 1 kg of water there are

$$\frac{1 \text{ kg}}{0.018 \text{ kg}} \times 6 \times 10^{23} \text{ molecules} \times \frac{6 \text{ bonds}}{\text{molecule}} = 2 \times 10^{26} \text{ bonds}.$$

To break all these bonds and make steam out of water, an energy equal to the heat of vaporization is needed (to be supplied to the water). This amount of energy divided by the number of bonds gives the energy per bond, *i.e.*, the bond energy (the energy of one bond, or the energy that corresponds to a single bond):

$$E = \frac{(539 \text{ kcal/kg}) \times (4180 \text{ Joules/kcal})}{(1.6 \times 10^{-19} \text{ Joule/eV}) \times (2 \times 10^{26} \text{ bonds/kg})} = 0.07 \text{ eV/bond}.$$

53. Can a molecule be treated as a linear harmonic oscillator?

Solution: No, because it is not. The energy levels of such an oscillator are equally spaced, whereas those of a molecule are not. (The same holds for the electronic levels of an atom.) The electronic, vibrational, and rotational levels of a molecule are not equally spaced; so the potential curve departs from the parabolic shape of that of an ideal harmonic oscillator. Even if it is a two-state molecule (for practical reasons, where only two states are of interest), it cannot be taken as a linear oscillator, because the latter has an infinite number of states, not only two. Only in the case where two atoms bind together to form a vibrating molecule whose vibrational levels are more closely spaced than the electronic levels, only then can the (partial) vibrational potential be taken as a linear-oscillator potential. That is, only the vibrational part of energy of a diatomic molecule can be approximated to that of a linear oscillator.

54. (a) How many gas molecules are there in 1 m^3 of a gas, considered ideal, under standard conditions? (b) Is this number large enough to constitute a

statistical system? (c) If $v_{rms} = 600$ m/sec at 300°K, find the molecular weight of the gas. (d) If a certain molecule moves with $v_x = 10$ m/sec and hits the walls of the container once in 10^{-3} sec, find the length of the side of the container along x. (Although incorrect, neglect collisions with other molecules en route.)

Solution: (a) The volume of 1 mol (at standard conditions) is 22.4 lt. Therefore, 1 m^3 contains 10^6 cm$^3/22.4 \times 10^3$ cm^3 = 44.6 mols, *i.e.*, (6.022 × 10^{23} molecules/mol) × 44.6 mols = 2.7×10^{25} molecules.

(b) Yes. Typically, it can be treated statistically.

(c) $v_{rms} = \sqrt{3kT/m}$ whence $m = 3.4 \times 10^{-26}$ kg which equals 21 units. (Here m is the molecular mass, *i.e.*, the mass of a single molecule.)

(d) $d = v_x t = 10 \times 10^{-3} = 10^{-2}$ m = 1 cm.

55. (a) The atmospheric temperature of Jupiter is 160°K, and the escape velocity for this planet is 60 km/sec. Can its atmosphere contain H_2? (b) At what temperature would the Earth lose its O_2 to the space?

Solution: (a) The molecular weight of H_2 is 2. At 160°K, we find $v_{rms} = 1.4$ km/sec. Since $v_{escape} \gg v_{rms}$, Jupiter is able to hold H_2.

(b) For the O_2 molecules to escape, we should have $v_{rms} = \frac{1}{6}v_{escape} = \frac{1}{6}\sqrt{\frac{2Gm}{r}} = 1860$ m/sec. From $\sqrt{3kT/m} = 1860$ we find $T = 4440°$K. (Here the molecular weight of O_2 is $M = 32$.)

56. (a) Find $v_{rms}/v_{most\ probable} = ?$ (b) A Maxwellian gas at $T = 350°$K has $N(v_{rms})/N(v = 100$ m/sec$) = 10$. Find the molecular mass of the gas. (c) Plot $v_{most\ probable}$ versus T. (d) Consider N_2 at room temperature and calculate $N(v_{sound})/N(v_{rms})$. (Note: $N(v)$ in general means the number of molecules having — or moving at — a speed v.)

Solution: (a)

$$\frac{v_{rms}}{v_{most\ probable}} = \frac{\sqrt{3kT/m}}{\sqrt{2kT/m}} = \sqrt{3/2}.$$

(b) The Maxwell distribution (of speeds) is (apart from a constant of proportionality)

$$N(v_{rms}) \propto v_{rms}^2 e^{-mv_{rms}^2/2kT} = \frac{3kT}{m}e^{-3/2}$$

$$N(v) \propto v^2 e^{-mv^2/2kT}.$$

For $v = 100$ m/sec we have

$$\frac{N(v_{rms})}{N(v)} = 10 = \frac{3kT}{mv^2}e^{-3/2+\frac{mv^2}{2kT}}$$

where $v = 100$ m/sec and $T = 350°$K. The exponential equation can be solved for m logarithmically.

(c) From $v_{most\ probable} = \sqrt{2kT/m}$, one can immediately see that $v_{most\ probable} \propto \sqrt{T}$.

(d) We can take N_2 as Maxwellian, with $m = 4.65 \times 10^{-26}$ kg (because the molecular weight is $M = 28$, *i.e.*, $m = 28$ units, where 1 unit $= 1.66 \times 10^{-27}$ kg). At $T = 300°$K (room temperature) we have $v_{rms} = \sqrt{3kT/m} = 517$ m/sec. The exponent is $mv_{rms}^2/2kT = 3/2$. On the other hand, for $v_{sound} = 340$ m/sec, we have $mv_{sound}^2/2kT \simeq 0.6$. From $N(v) \propto v^2\exp(-mv^2/2kT)$ we have the ratio of the number of molecules:

$$\frac{N(v_{sound})}{N(v_{rms})} = \sqrt{\frac{340}{517}}e^{-0.6+\frac{3}{2}} = 0.99 \quad \text{or} \quad 99\%.$$

57. In a room of dimensions $10 \times 10 \times 10$ m full of O_2, the temperature is $25°$C. (a) Find v_{rms} and $(v_x)_{rms}$. (b) If we roughly assume — though wrong — that a molecule does not collide with other ones, in how much time does it get across the room? (c) Find the number density of the O_2 molecules in the room. (d) An observer stands still in the room. His surface area is approximately 2 m^2. How many molecules hit him per unit time?

Solution: (a) Since $M(O_2) = 32$ units, we have $m = 5.3 \times 10^{-26}$ kg. At $T = 25° + 273° = 298°$K, we have $v_{rms} = \sqrt{3kT/m} = 482$ m/sec and $(v_x)_{rms} = \sqrt{(1/3)v_{rms}} = $ (from symmetry) $= \sqrt{kT/m} = 278$ m/sec.

(b) For a direct flight, $t = d/v_x = 10/278 = 0.035$ sec. (Notice that one need not specify the pressure in the room; it is immaterial and not needed in the calculation here. The only thing that we can deduce about pressure in this situation is that it should be too low, so that our assumption is almost valid, *i.e.*, it allows for direct flights without collisions en route.)

(c) From the gas law

$$n = \frac{N}{V} = \frac{P}{kT}$$

where $T = 298°$K, and we can adopt a typical low pressure here, say, $P \sim 10^{-10}$ atm $= 10^{-10} \times 1.013 \times 10^5$ Nt/m^2. Then $n \sim 2 \times 10^{15}$ molecules/m^3, a rather rare and lean density, since P is low.

(d) The molecular current (flow rate) is $J = nv_x$ molecules/m^2 sec. If the observer's area is A, he is hit by $nv_x A$ molecules/sec. With the given data we find $\sim 10^{18}$ molecules/sec. (How come the observer does not get crushed by this external pressure like a crumpled sheet of paper?)

58. At very high temperatures ($\sim 10^{12}$°K), v_{rms} of a gas molecule approaches the speed of light. Is then the formula $v_{rms} = c = \sqrt{3kT/m}$ valid in finding the corresponding T?

Solution: Although nothing stops us from using it, this formula does not exactly hold for relativistic speeds.

59. (a) Calculate c_P/c_V for a diatomic gas at room temperature, ignoring vibrations. (b) How many degrees of freedom does O_2 have at room temperature, not counting vibrations? (c) How many degrees of freedom does the methane molecule have, if it can rotate about three axes and vibrate (in several modes)?

Solution: (a) The degrees of freedom of a diatomic molecule are counted as follows: There are three translational and two rotational degrees, in total five degrees, neglecting vibrations. Hence $c_V = 5\mathcal{R}/2$. Also, $c_P = c_V + \mathcal{R} = 7\mathcal{R}/2$ and $c_P/c_V = 7/5$. (The vibrational degrees of freedom are neglected — and they can be easily treated so — because they are not excited at room temperature.)

(b) Since it is diatomic, it has 3 translational + 2 rotational = 5 degrees of freedom at room temperature.

(c) CH_4 has 3 translational + 3 rotational + 8 vibrational degrees = 14 degrees of freedom (CH_4 is not diatomic; so it has three possible axes of rotation, with moments of inertia of comparable order of magnitude. It has four C–H bonds that vibrate, each with two degrees: Potential and kinetic energy.)

60. Consider the N_2 gas. (a) How many degrees of freedom does it have at high temperatures? (b) Find c_P/c_V.

Solution: (a) We are to consider vibrational and rotational excitations. The molecule has 3 translational + 2 rotational + 2 vibrational degrees of freedom, seven in all. (At high temperatures the two vibrational degrees of freedom get excited — together with the translational and rotational ones. Two, because the vibration has kinetic energy $mv^2/2$ and potential energy $\kappa x^2/2$.)

(b) If i is the number of degrees of freedom, we have $c_V = iR/2$ and $c_P = (i+2)R/2$ whence $c_P/c_V = (i+2)/i$. With $i = 7$, we have in this case $c_P/c_V = 9/7$.

61. Consider the O_2 gas in equilibrium at room temperature. (a) Find the rotational energy of an O_2 molecule. (b) What is the energy per degree of freedom?

Solution: (a) The O_2 molecule has $i = 3$ translational $+ 2$ rotational $= 5$ degrees of freedom at $300°K$. So, the rotational energy is $2(kT/2) = kT = 4 \times 10^{-21}$ Joule.

(b) The internal energy of a diatomic gas is $U = 5NkT/2$ if there are N molecules. The internal energy per molecule is $5kT/2$. The energy per degree of freedom will be $5kT/2i = 5kT/(2 \times 5) = kT/2 = 2 \times 10^{-21}$ Joule.

62. Why does the solution of $CuSO_4$ in H_2O conduct electricity?

Solution: The solution of $CuSO_4$ in water is not molecular. It is ionic: $CuSO_4 \rightarrow Cu^{++} + SO_4^{-2}$. So, it conducts electricity.

63. A material has molecular weight M and density ρ. Find the following: (a) The mass of one molecule. (b) The number of molecules per gram. (c) The *molar volume* of the compound, *i.e.*, the volume of (the quantity of) one mol (of the material). (d) The *molecular volume*, *i.e.*, the volume occupied by a single molecule. (e) The number of molecules per cm^3 (number density). (f) The number of molecules per 1 cm^2 of the material.

Solution: (a) The absolute mass of a single molecule is $m = M/N_A$ where N_A is the Avogadro number. To check units, we have grams $=$ (gr/mol)/ (molecules/mol).

(b) In a compound, the number of molecules per unit mass is N_A/M.

(c) Since $1/\rho$ is the volume of (an amount of) 1 gr, the volume of one mol is M/ρ.

(d) $V_{molecular} = (M/\rho)/N_A$.

(e) The number of molecules per unit volume is $n = N_A/(M/\rho)$.

(f) There are $N_A\rho x/M = N_A m'/AM$ molecules/cm^2, where x is the thickness of the substance, m' is its total mass, and A is its area.

64. (a) When 1 kg of trinitrotoluene (TNT) explodes, a heat of 2×10^3 kcal is released. The molecular weight of this explosive material is 227. Find the energy per molecule upon explosion. (b) Air molecules (mean mass $28.8 \times 1.66 \times 10^{-24}$ gr) move at 5×10^4 cm/sec at $20°C$. How much energy corresponds to

a molecule? How much heat can 1 m^3 of air give? (1 mol of air weighs 28.8 gr and occupies 22.4 lt.)

Solution: (a) There are

$$\frac{6.023 \times 10^{23} \times 1000}{227} = 2.64 \times 10^{24}$$

molecules in 1000 gr of TNT. The energy per molecule is

$$\frac{2 \times 10^6 \text{ cal}}{2.64 \times 10^{24}} = 7.57 \times 10^{-19} \text{ cal} = 19.8 \text{ eV},$$

because 1 cal = 2.616×10^{19} eV.

(b) The energy of an air molecule is

$$\frac{1}{2} mv^2 = 14.4 \times 1.66 \times 10^{-24} \times 25 \times 10^8 \text{ erg} = 6 \times 10^{-14} \text{ erg} = 0.037 \text{ eV},$$

since 1 eV = 1.6×10^{-12} erg.

1 m^3 of air contains $10^3/22.4 = 44.6$ mols. The heat energy of the molecules in 1 m^3 of air is $44.6 \times 6.023 \times 10^{23} \times 0.037 = 9.94 \times 10^{23}$ eV = 38 kcal.

Another way of finding the same result is the following: The energy of one mol of gas is $(3/2)\mathcal{R}T$ where the gas constant is $\mathcal{R} = 1.98$ cal/°K mol, and $T = 20 + 273 = 293$°K (the absolute temperature). So, we have

$$\frac{3}{2} \times 1.98 \times 44.6 \times 293 = 38.8 \text{ kcal}.$$

65. Calculate the number of molecules per cm^3 of an ideal gas at 298°K, for (a) $P = 10$ bar, and (b) $P = 10^{-5}$ Torr.

Solution: (a) The number of molecules per mol is N_A. The volume of a mol is $V_{\text{mol}} = \mathcal{R}T/P$. The number of molecules per unit volume (number density) is $n = N_A/V_{\text{mol}} = N_A P/\mathcal{R}T$. Since $\mathcal{R} = 8.31 \times 10^7$ erg/°K mol and $P = 10$ bar = 10^7 dynes/cm^2, we have

$$n = \frac{6.023 \times 10^{23} \times 10^7}{8.31 \times 10^7 \times 298} = 2.43 \times 10^{20} \text{ molecules/cm}^3.$$

(b) 1 Torr = 1 mm Hg = 1333 dynes/cm^2.

$$n = \frac{6.023 \times 10^{23} \times 1333 \times 10^{-5}}{8.31 \times 10^7 \times 298} = 3.24 \times 10^{11} \text{ molecules/cm}^3.$$

66. Jean Perrin observed the Brownian motion of particles of a material whose density was 1.5 gr/cm^3. The diameter of the particles was 10^{-4} cm, and the liquid was water. (a) What was the molecular weight of the colloidal particles? (b) Find the rms speed of the particles at 27°C.

Solution: (a) The mass of a single particle is

$$m = \frac{4}{3}\pi r^3 \rho = 0.785 \times 10^{-12} \text{ gr}.$$

One mol of any material consists of N_A molecules (or atoms, if atomic). The molecular weight of the particles is then

$$M = mN_A = 0.785 \times 10^{-12} \times 6.023 \times 10^{23} = 4.7 \times 10^{11}.$$

(b) Perrin was able to determine the mass and mean speed of the particles, and hence calculate their mean kinetic energy. The value he found agreed with that given by the kinetic theory, $(3/2)\mathcal{R}T/N_A$. We can thus write

$$\frac{1}{2}m\overline{v^2} = \frac{3}{2}\frac{\mathcal{R}T}{N_A}$$

whence

$$\overline{v^2} = \frac{3\mathcal{R}T}{mN_A} = \frac{3 \times 8.31 \times 10^7 \times 300}{4.7 \times 10^{11}} = 0.16$$

and

$$\sqrt{\overline{v^2}} = 0.4 \text{ cm/sec}.$$

67. Define the *average polarizability* of a molecule.

Solution: When we diagonalize the polarizability matrix, we get

$$\begin{pmatrix} \alpha_1 & 0 & 0 \\ 0 & \alpha_2 & 0 \\ 0 & 0 & \alpha_3 \end{pmatrix}.$$

The molecule is free to choose any direction, because there are no rotational constraints. By symmetry we have $\alpha_1 = \alpha_2 \equiv \alpha_\perp$ (the same value for the two tensor elements which are perpendicular to the z-direction and in the same xy-plane). We call $\alpha_3 \equiv \alpha_\parallel$ (parallel to, or along the z-direction). Thus the answer is

$$\overline{\alpha} = \frac{\alpha_1 + \alpha_2 + \alpha_3}{3} = \frac{2\alpha_\perp + \alpha_\parallel}{3}.$$

68. (a) Write down the force and potential between two point charges, a charge and a dipole, and two dipoles. (b) Write down the mutual action of two coplanar dipoles. (c) Draw an octupole.

Solution: (a) See Fig. SP-29. The interaction is

$$
\begin{array}{lll}
\text{(I)} & F \propto q_1 q_2 / r^2 & V \propto q_1 q_2 / r \\
\text{(II)} & qp/r^3 & qp/r^2 \\
\text{(III)} & p_1 p_2 / r^4 & p_1 p_2 / r^3
\end{array}
$$

Fig. SP-29.

(b) Refer to Fig. SP-30. (I) is the general case.

Case	Force	Couple	Mutual potential energy
(II)	$F_r = \dfrac{3p_1 p_2}{4\pi\varepsilon_0 r^4}$ along r	0	$\dfrac{p_1 p_2}{4\pi\varepsilon_0 r^3}$ (maximum)
(III)	$F_r = -\dfrac{6p_1 p_2}{4\pi\varepsilon_0 r^4}$ along r	0	$-\dfrac{2p_1 p_2}{4\pi\varepsilon_0 r^3}$ (minimum or negative maximum)
(IV)	$F_\theta = \pm\dfrac{3p_1 p_2}{4\pi\varepsilon_0 r^4}$ normal to r	$\begin{cases} \dfrac{2p_1 p_2}{4\pi\varepsilon_0 r^3} \text{ on } p_1 \\ \text{clockwise} \\ \dfrac{p_1 p_2}{4\pi\varepsilon_0 r^3} \text{ on } p_2 \\ \text{clockwise} \end{cases}$	0

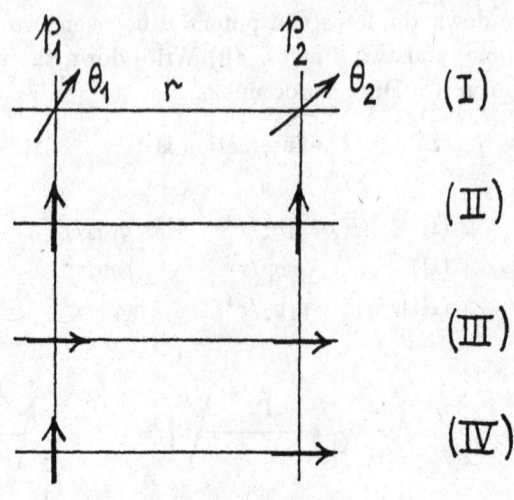

Fig. SP-30.

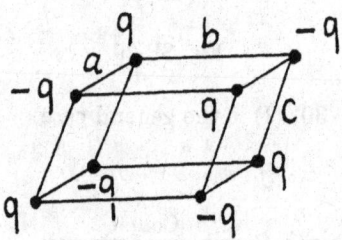

Fig. SP-31.

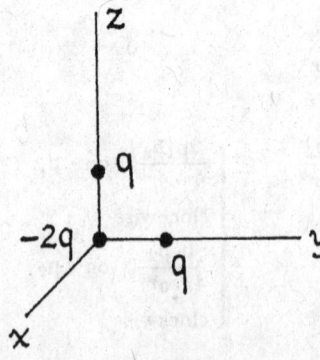

Fig. SP-32.

(c) See Fig. SP-31.

69. Three points charges are located as shown in Fig. SP-32, where q and q are at unit distance ($y = 1$ and $z = 1$) away from the origin. This model describes the water molecule. (a) What is the quadrupole moment of this arrangement? (b) What is the potential Φ at a point in space?

Solution: The charge distribution is discrete. One way of calculating the electrostatic potential is to calculate the contribution of each individual point charge and then add up the three contributions arithmetically (because Φ is a scalar). Another way is to use multipole expansion. From a set of discrete charges we can go over to a charge distribution (density) function $\rho(\mathbf{r}')$ by writing the appropriate delta functions that locate the charges, and then we can integrate continuously instead of summing. We have

$$\rho(\mathbf{r}') = \rho(x', y', z') = -2q\delta(x')\delta(y')\delta(z')$$
$$+q\delta(x')\delta(y'-1)\delta(z') + q\delta(x')\delta(y')\delta(z'-1).$$

(a) The quadrupole moment tensor Q has matrix elements

$$Q_{ij} = \iiint \rho(x', y', z')(3x'_i x'_j - r'^2 \delta_{ij})dx'\,dy'\,dz'$$

where we integrate between $\pm\infty$ because the medium is unbounded. If we put in ρ and calculate the matrix elements we find

$$Q_{xx} = Q_{11} = -2q$$

because $r'^2 = x'^2 + y'^2 + z'^2$ in the integrand, and $x'_i = x'_j$, so that $3x'_i x'_j = 3x'^2$, and because of the delta functions.

$$Q_{xy} = Q_{12} = \iiint 3x'y'\rho(x', y', z')dx'\,dy'\,dz'$$

turns out to be zero. All the off-diagonal elements of the Q matrix are zero.

$$Q_{yy} = \iiint (3y'^2 - x'^2 - y'^2 - z'^2)\rho(x', y', z')dx'\,dy'\,dz' = q$$

$$Q_{zz} = q.$$

Thus the quadrupole tensor for this (idealized) molecule is

$$Q = q \begin{pmatrix} -2 & 0 & 0 \\ 0 & 1 & 0 \\ 0 & 0 & 1 \end{pmatrix}.$$

(b) The monopole term of Φ vanishes, because of the equal amount of positive and negative charges that make the total charge zero. (The monopole is just the total charge.) The dipole term is

$$\vec{p} = \int \rho(\mathbf{r}')\mathbf{r}'d^3r' = \iiint dx'\,dy'\,dz'\,\rho(x',\,y',\,z')(x'\hat{x} + y'\hat{y} + z'\hat{z})$$

$$= q\hat{y} + q\hat{z} = q(\hat{y} + \hat{z}).$$

Thus the vector $\vec{p}$ has a magnitude q and is oriented at 45° with respect to the y-axis in the yz-plane (Fig. SP-33).

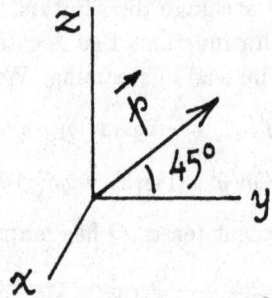

Fig. SP-33.

With $\mathbf{r}$ being the position vector of the observation point, we have

$$\Phi(\mathbf{r}) = \frac{q(\hat{y} + \hat{z})\cdot\mathbf{r}}{|\mathbf{r}|^3} + \frac{1}{2}\left(\frac{-2qx^2 + qy^2 + qz^2}{|\mathbf{r}|^5}\right)$$

$$= q\left[\frac{(\hat{y} + \hat{z})\cdot\mathbf{r}}{r^3} + \frac{1}{2}\left(\frac{-2x^2 + y^2 + z^2}{r^5}\right)\right]$$

$$= q\left[\frac{(\hat{y} + \hat{z})\cdot\mathbf{r}}{r^3} + \frac{1}{2}\left(\frac{-3x^2 + r^2}{r^5}\right)\right]$$

where we used $x_i x_j = x^2$ and $|\mathbf{r}| = r = \sqrt{x^2 + y^2 + z^2}$, and for the second term we applied

$$\frac{1}{2}\sum Q_{ij}\frac{x_i x_j}{r^5}.$$

Therefore, we have

$$\Phi = \frac{\vec{p}\cdot\mathbf{r}}{r^3} + \frac{q}{2}\left(\frac{r^2 - 3x^2}{r^5}\right)$$

where the dimensions of the second term are those of an inverse volume. Notice that x appears in the result because the water molecule is chosen to be in the yz-plane. The potential of the H_2O molecule looks like this to a first approximation. The oxygen atom at the center (origin) contributes nothing; only the off-center charges are responsble for the result.

Molecules are electrically *neutral*. They have no free charges. They can have dipole moments though, either permanent or induced.

70. Discuss the motion of a triatomic molecule $(m, 2m, m)$ that vibrates both longitudinally and laterally. Sketch the modes.

Solution: Let the molecule vibrate both in x and y-direction (Fig. SP-34). The Lagrangian is

$$\mathcal{L} = \frac{m}{2}(\dot{x}_1^2 + 2\dot{x}_2^2 + \dot{x}_3^2 + \dot{y}_1^2 + 2\dot{y}_2^2 + \dot{y}_3^2) - \frac{\kappa}{2}[(x_1 - x_2)^2 + (x_3 - x_2)^2]$$

$$-\frac{\kappa'}{2}[(-y_1 + y_2)^2 + (-y_1 + y_2 - y_3)^2 + (-y_3 + y_2)^2]$$

where the term with κ refers to the x-vibration and the term with κ' to the y-motion (vibration) (Fig. SP-35). We expand the y-vibration and get

$$-\frac{\kappa'}{2}[(y_1 - y_2) - (y_2 - y_3)]^2.$$

The equations of motion in the x-direction are

$$m\ddot{x}_1 + \kappa(x_1 - x_2) = 0$$

$$m\ddot{x}_3 + \kappa(x_3 - x_2) = 0$$

$$2m\ddot{x}_2 - \kappa(2x_2 - x_1 - x_3) = 0.$$

Since momentum must be conserved, we have $\ddot{x}_1 + 2\ddot{x}_2 + \ddot{x}_3 = 0$ whence only two modes of vibration in the x-direction exist. The normal modes are $x_1 - x_3$ with frequency $\sqrt{\kappa/m}$, and $x_1 - 2x_2 + x_3$ with frequency $\sqrt{2\kappa/m}$.

The motion in the y-direction is constrained by the conservation of linear *and* angular momentum. Hence only one mode can exist in the y-direction, with frequency $\sqrt{4\kappa'/m}$.

The possible modes are shown in Fig. SP-36. Notice that the vibration amplitudes (represented by the proportional lengths of the arrows) are such that momentum is conserved in the x-direction.

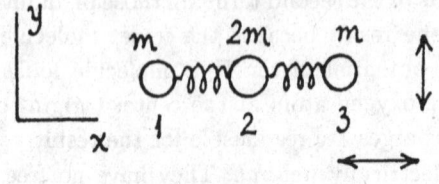

Fig. SP-34.

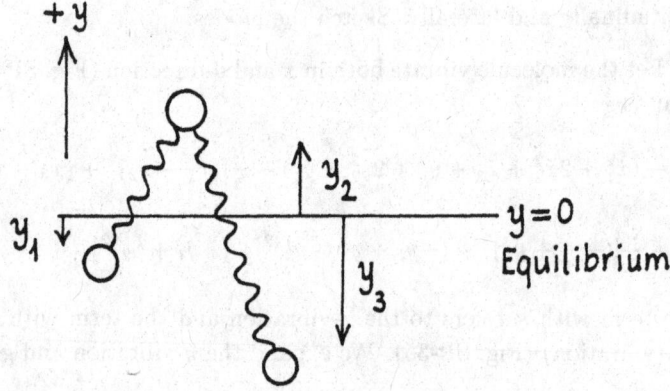

Fig. SP-35.

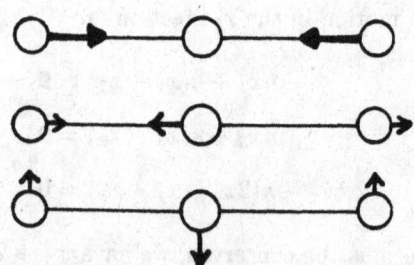

Fig. SP-36.

This was a classical-mechanics problem.

71. (a) The dissociation energy of the ^{1}H$_2$ molecule is 4.46 eV and that of the ^{2}H$_2$ molecule 4.54 eV. Calculate the zero-point energy of ^{1}H$_2$. (b) Can we say that an He atom is roughly a double H atom with two electrons revolving around a nucleus of $Z = 2$, without mutual interaction? (c) In the

far-infrared spectrum of HBr, the lines are separated by 17 cm^{-1}. Determine the interatomic separation in HBr.

Solution: (a) We have H_2 and D_2. The energy of a diatomic molecule is, approximately,

$$E = E' + \underbrace{\left(v + \frac{1}{2}\right) h\nu}_{\text{Vibration}} + \underbrace{BJ(J+1)}_{\text{Rotation}} + \text{h.o.t.} \qquad (*)$$

where the higher-order terms include contributions from ignored interactions. Here B is the rotation constant, and E' is a *negative* constant energy term that just shifts E by a constant amount and involves (absorbs) two dependences or contributions: The spatial motion of the two nuclei and their charges.

The dissociation energy is the difference (dip) between the lowest energy of the molecule (ground state, $v = 0$ and $J = 0$) and the energy of the two atoms when free and clear of any interaction. (We assume that the free atoms do not interact.) Thus

$$E_{\text{dissoc}} = -E' - \frac{1}{2}h\nu$$

where the second term on the right-hand side is the lowest vibrational energy (for $v = 0$). E' depends on the interatomic separation in the molecule and on the nuclear charges, but in the adiabatic approximation the nuclear motion can be dispensed with as negligible, and on the other hand, the nuclear masses can be ignored, too, as silent actors. In our problem here, the only difference between H_2 and D_2 is the nuclear mass; but since it is irrelevant, we have $E'(\text{proton}) = E'(\text{deuteron})$ whence

$$E_{\text{dissoc}}(D_2) - E_{\text{dissoc}}(H_2) = -\frac{1}{2}h[\nu(D_2) - \nu(H_2)] = 4.54 - 4.46 = 0.08 \text{ eV}$$

or

$$E_0(H)_2 - E_0(D_2) = 0.08 \text{ eV}$$

where $E_0 = (1/2)h\nu$ is the zero-point energy. We may now write

$$E_0(H_2)\left[1 - \frac{E_0(D_2)}{E_0(H_2)}\right] = 0.08 \text{ eV},$$

$$\frac{1}{2}h\nu(H_2)\left[1 - \frac{\nu(D_2)}{\nu(H_2)}\right] = 0.08 \text{ eV}$$

where $\nu = (1/2\pi)\omega = (1/2\pi)\sqrt{\kappa/\tilde{\mu}}$ for a vibrating molecule. Here κ is the spring constant (or force constant) and $\tilde{\mu}$ is the reduced mass of the diatomic system (molecule). Since the stiffness κ depends on the charge distribution only, we have $\nu = \nu(\tilde{\mu})$, so that

$$\frac{\nu(D_2)}{\nu(H_2)} = \frac{1}{\sqrt{2}}$$

and

$$E_0(H_2)\left[1 - \frac{1}{\sqrt{2}}\right] = \frac{1}{2}h\nu(H_2)\left[1 - \frac{1}{\sqrt{2}}\right] = 0.08 \text{ eV},$$

$$E_0(H_2) = \frac{1}{2}h\nu(H_2) = 0.27 \text{ eV},$$

which is the zero-point energy of H_2. By dividing by the Planck constant h, one can find the zero-point frequency ν.

Clearly, when the two atoms are far away from each other and non-interacting, $E = 0$ in (*), because $\psi_{mol} = 0$ (no molecule).

(b) Very crudely, yes. (Actually, this is an oversimplified model used to calculate the diamagnetic susceptibility of He and compare it with the experimental value.)

(c) Far-infrared outcome means rotational motion. The rotational energy levels are

$$E(J) = \frac{P^2}{2\tilde{\mu}r^2} = \frac{\hbar^2 J(J+1)}{2\tilde{\mu}r^2}$$

where P is the rotational angular-momentum operator, J is the molecular rotational quantum number associated with this P, and $\tilde{\mu}$ is the reduced (or effective) mass of the system (thought as an equivalent single-mass body of mass $\tilde{\mu}$):

$$\frac{1}{\tilde{\mu}} = \frac{1}{m_1} + \frac{1}{m_2} = \frac{1}{m(H)} + \frac{1}{m(Br)} \qquad (**)$$

and

$$\tilde{\mu} = \frac{m_1 m_2}{m_1 + m_2} \simeq m(H) \equiv m_p$$

because the mass of an H atom is too small compared to that of a Br atom: Since $m(H) \ll m(Br)$, we can neglect $1/m_2$ in (**) compared to $1/m_1$, because it is a small fraction. Hence $\tilde{\mu} \simeq m_1 \equiv$ (Mass of a proton)+(Mass of an electron).

But since m_e is too small compared to m_p (actually, $m_p = 1836m_e$), we can practically take $\tilde{\mu} = m_p$ only.

The selection rule for J is that it changes by $\Delta J = \pm 1$ where the upper (lower) sign corresponds to an absorption (emission). Thus in our emission spectrum, the energy interval between two adjacent rotational J states (in a transition from state J to $J - 1$) is

$$\Delta E = E(J_1) - E(J_2) = \frac{\hbar^2}{2\tilde{\mu}r^2}[J_1(J_1 + 1) - J_2(J_2 + 1)]$$

$$= \frac{\hbar^2}{2\tilde{\mu}r^2}[J(J + 1) - (J - 1)(J - 1 + 1)] = \frac{\hbar^2}{2\tilde{\mu}r^2}(2J) = \frac{\hbar^2 J}{\tilde{\mu}r^2},$$

and the interval between the lines is

$$2\pi\hbar c\Delta\bar{\nu} = \hbar^2/\tilde{\mu}r^2$$

(where $\bar{\nu} = 1/\lambda$) where we read off r and find it to be 1.4 Å.

72. Given the potential of a molecule, one can in principle put it into the Schrödinger equation, solve the equation, and obtain the molecular energy levels. The potential of a diatomic molecule of reduced mass $\tilde{\mu}$ is, approximately,

$$V(r) = -2V_0a\left(\frac{1}{r} - \frac{a}{2r^2}\right)$$

where V_0 is a constant and a is another constant that characterizes the binding scheme and has the dimensions of a length. (a) Plot the potential $V(r)$ and discuss what happens at $r = a$. What is V_0? (b) Expand the effective potential into a series and obtain the quantized energy levels. Consider small departures from equilibrium.

Solution: (b) To facilitate the mathematics, we make the transformation $\psi(r) = R(r)/r$, whereby with the new function $R(r)$ the Schrödinger equation becomes

$$-\frac{\hbar^2}{2\tilde{\mu}}\frac{d^2R}{dr^2} + {}`V`R = ER$$

where

$${}`V` \equiv -2V_0a\left(\frac{1}{r} - \frac{a}{2r^2}\right) + \frac{J(J + 1)\hbar^2}{2\tilde{\mu}r^2}$$

is the *effective potential*. A minimum in ‘V’ means stability and equilibrium in the two-body system (which in our case is the diatomic molecule). By setting the derivative equal to zero, $d'V'/dr = 0$, we find that the minimum occurs at $r = r_0$, a point about which we will expand the potential for small vibrations away from equilibrium.

Not by a stroke of genius, but just to further ease up the mathematics and make the expressions more elegant and manageable, we set $\rho \equiv r/a$ and $\rho_0 = r_0/a$, with ρ_0 being, from above,

$$\rho_0 = 1 + \frac{J(J+1)\hbar^2}{2\tilde{\mu}a^2 V_0} \equiv 1 + \xi .$$

The expansion is now

$$V(\rho) = -\frac{V_0}{1+\xi} + \frac{V_0}{(1+\xi)^3}(\rho - \rho_0)^2 + \text{h.o.t.}$$

which can be put into the Schrödinger equation:

$$\frac{\hbar^2}{2\tilde{\mu}}\frac{d^2 R}{dr^2} + \left[E + V_0 \left(\frac{1}{1+\xi} - \frac{(\rho-\rho_0)^2}{(1+\xi)^3} \right) \right] R = 0 .$$

But this is actually the differential equation of an harmonic oscillator. It gives

$$E + \frac{V_0}{1+\xi} = \hbar\sqrt{\frac{2V_0}{\tilde{\mu}a^2(1+\xi)^3}}\left(v + \frac{1}{2}\right) .$$

We expand for small ξ and set $\omega_0 \equiv \sqrt{2V_0/\tilde{\mu}a^2}$ to obtain the energy

$$E = -V_0 + \frac{J(J+1)\hbar^2}{2\tilde{\mu}a^2} + \hbar\omega_0\left(v + \frac{1}{2}\right) - \frac{3}{2}\frac{\hbar^3 J(J+1)\left(v + \frac{1}{2}\right)}{\tilde{\mu}^2 a^4 \omega_0}$$

with all the contributions: The first term on the right-hand side is a constant potential (shifting the energy by that amount); the second term is the rotational-energy part; the third term represents the vibrational energy; and the last term is the (coupled) rotational-vibrational part. Thus we obtained the rotational, vibrational, and rotational-vibrational energy levels for the molecule considered.

73. Consider an exotic muonic hydrogen atom (Fig. SP-37). It consists of a proton and a muon (μ-meson) revolving around it instead of an electron. (Such an atom is possible, with a proton capturing a muon μ^-. The muonic charge is the same as that of the electron, and $m(\mu^-) = 207m_e$.) This exotic

atom, together with a bare proton, may form a H_2^+ ion (just as a regular H atom would do) where e^- is replaced by μ^- (Fig. SP-38). Without getting into rigorous calculations, but by rather considering typical sizes and using simplistic estimation arguments, determine (a) a roughly estimated equilibrium separation distance r_0 for the two protons (H nuclei), (b) E_0, the zero-point energy of this molecule, and (c) the binding energy. To proceed with your estimations, you will need the following for the regular electronic H_2^+ ion: $r_0 = 1$ Å, $E_0 = 0.14$ eV, $E_{binding} = 2.7$ eV.

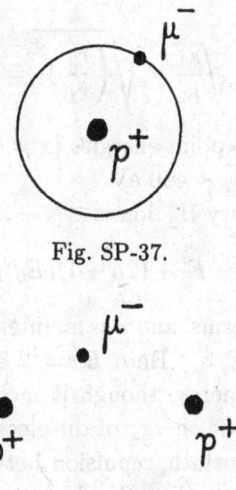

Fig. SP-37.

Fig. SP-38.

Solution: The muon is negatively charged, and equally so with the electron. This particle may replace the electron in a mu-mesic H atom. This atom in turn may form an H_2^+ ion with another proton.

In general, in a diatomic molecule, the (attractive) forces that bind the two nuclei are of electrostatic (Coulombic) origin, so that the (stable) internuclear separation r is $\sim$ atomic size (radius of orbit of outermost valence electron(s)). The two bound parties touch one another and cannot interpenetrate into each other, as any attempt for further approach would raise repulsion (between the nuclei of the same charge). So, one party can come close at most up to where the other one extends itself (and can recede at most up to the molecular break-up limit, beyond which the system dissociates).

In our case here, we simply consider a Bohr-type circular orbit for H. Since the orbit (Bohr) radius is $r \propto 1/m$, if we label the electronic atom as #1 and the muonic atom as #2, the ratio of their radii is $r_1/r_2 = m_2/m_1 = 207$ which immediately tells us that the size of the muonic atom is ~ 207 times smaller than that of the regular H atom, and so is the size of the muonic molecule ion compared to that of its electronic counterpart.

For a given principal quantum number n, the electronic energies compare as $E_1/E_2 = m_1/m_2 = 1/207$.

In the molecule, the restoring force upon vibration is $F = -\kappa r$ and depends on the (vibrating) charges, *i.e.*, $F \sim e^2/r^2$, electrostatically. So, $\kappa \sim e^2/r^3$. Since $E_{\text{vibr}} \propto \sqrt{\kappa}$, we have

$$\frac{(E_{\text{vibr}})_1}{(E_{\text{vibr}})_2} = \sqrt{\frac{\kappa_1}{\kappa_2}} = \sqrt{\left(\frac{r_2}{r_1}\right)^3} = \sqrt{\left(\frac{1}{207}\right)^3},$$

which is the ratio of the zero-point energies (v = 0). From the above we obtain $r_2 \sim 5 \times 10^{-11}$ cm and $(E_0)_2 \sim 400$ eV.

The energy of the ordinary H_2^+ ion is

$$E = -E' + (2n+1)(E_0)_1 + C$$

where C absorbs all minor terms, and n is an integer that ensures odd multiples of the zero-point energy $(E_0)_1$. Here $E' = 2.84$ eV and can be in general termed as satellite-charge energy, though it includes two contributions: The electronic energy (electrostatic energy of the electron in the electric field of the two protons) *and* the electrostatic repulsion between the two protons. (So, if the satellite charge is a muon, then the first contribution will be the muonic energy in the field of the two protons.) Thus $E_1'/E_2' = 1/200$ and for the muonic H_2^+ ion we have

$$E_2 \simeq -E_2' + (2n+1)(E_0)_2$$

where $E_2' = 200 \times E_1' = 200 \times 2.84 = 568$ eV, and hence $(E_{\text{binding}})_2 = E_2' - (E_0)_2 = 568 - 400 = 168$ eV.

Notice that $E_1' = (E_0)_1 + (E_{\text{binding}})_1 = 0.14 + 2.7 = 2.84$ eV.

Here we solved the problem with back-of-the-envelope calculations. If you want high accuracy, you will need a Fortran IV computer program. But since this is an ordinary homework or exam problem for practical-minded physicists, you need not go that far.

See also Assignment Problem 27.

74. Prove that in getting from a vibrational level to the next, the (vibrational) energy changes by $\Delta E = h\nu$ (or $\hbar\omega$).

Solution: We get from v to v + 1 (where v is the vibrational quantum number). We have

$$E = \hbar\omega\left(v + \frac{1}{2}\right)$$

and

$$\Delta E = E_2 - E_1 = E(v_2) - E(v_1) = \hbar\omega\left[\left(v_2 + \frac{1}{2}\right) - \left(v_1 + \frac{1}{2}\right)\right]$$

$$= \hbar\omega\left[(v+1) + \frac{1}{2} - \left(v + \frac{1}{2}\right)\right] = \hbar\omega\left[v + 1 + \frac{1}{2} - v - \frac{1}{2}\right] = \hbar\omega.$$

A simple supermarket calculation.

75. Prove that $\Delta E_{rot} \propto 2BJ$.

Solution: Let us say that we go from $J_1 = 7$ to $J_2 = 8$. We have

$$\Delta E = Bh[J_2(J_2 + 1) - J_1(J_1 + 1)]$$

$$= Bh[J(J+1) - (J-1)(J-1+1)] = 2BhJ$$

where in this case $J = 7$. That is, $J_1 = 7$ and $J_2 = J_1 + 1 = 8$.

76. Show that for $J = 0$ the outcome means no rotation of the molecule.

Solution: The rotational energy is

$$E_{rot} = \frac{P^2}{2\tilde{\mu}r_0^2} = \frac{J(J+1)\hbar^2}{2\tilde{\mu}r_0^2} = BhJ(J+1)$$

where the amount $J(J+1)$ is the *energy quantization*, with $J = 0, 1, 2, 3, \ldots$ (integer). Here P is the magnitude of the (*quantum mechanical*) angular momentum (operator) of the molecule. P is quantized, so that the energy is quantized, too. In quantum mechanics, P is an *operator*, operating on some appropriate wavefunction, to give the same function times its (the operator's) eigenvalues. The length of the quantized angular momentum (quantum mechanical vector operator) is $P = \hbar\sqrt{J(J+1)}$. Our system here is the molecule, and P is for the molecule here. Notice that E is quantized due to the factor $J(J+1)$ where J is the *quantum number* of (the operator) P.

Since $P^2 = \hbar^2 J(J+1)$, when $J = 0$, we have $E_{rot} = 0 \Rightarrow$ no rotational motion.

Note that the quantized angular momentum of the molecule (vector **P**) can have (take) several discrete spatial orientations labeled by the magnetic quantum number m_J (the quantized projection of the vector along an external magnetic field which is usually taken along the z-axis and with respect to which the discrete or quantized orientations of **P** are judged). The possible values run from $-J$ to J, that is, m_J can take the values $m_J = -J, -J+1, \ldots, J-1, J$. This means that for each J (for a given J value) there are $2J+1$ possible values of m_J, that is, $2J+1$ possible quantum states with the same energy. So, the multiplicity or *degeneracy* is $2J+1$. We say that E is $(2J+1)$-fold degenerate. The field is in the z-direction.

So, the molecular angular momentum is quantized (in J) in the sense that only certain states (and energies according to the J values) are possible and accessible upon rotation, and not continuous energy values in between. Nature forbids an energy continuum and allows a stepwise stature.

Also note that the moment of inertia of the molecule is $\mathcal{I} = (1/2)\tilde{\mu}r_0^2$ where $\tilde{\mu}$ is the (fictitious) effective mass (of a single equivalent rigid rotor like a rotating mace) and r_0 is the interatomic equilibrium separation that renders a molecular dumbbell model (diatomic molecule) stable and possible. Thus

$$E = \left(\frac{\hbar^2}{2\mathcal{I}}\right) J(J+1)$$

whence a small $\mathcal{I}$ means large spacings between the energy levels.

77.   Discuss the Lennard-Jones potential and the hard-sphere potential. Compare the two.

Solution: Consider a gas of N identical molecules (or atoms)[4] in a volume V at a temperature T. If T is high and the number density $n = N/\text{V}$ is low enough, we can apply kinetic theory and treat the case by using statistical mechanics.

The energy (Hamiltonian function) of the gas (system of N particles) is

$$\mathcal{H} = (\text{Kinetic energy}) + (\text{Potential energy}),$$

a function that gives the total energy of the system, if other interactions between molecules are ignored, so that no other terms are of importance (and hence are not included in $\mathcal{H}$). In this simplified discussion, no other complications need be considered besides the two terms (parts) in $\mathcal{H}$ above. Here the kinetic energy E is

$$E = \frac{1}{2m} \sum_{i=1}^{N} \mathbf{p}_i^2 = \sum_i \varepsilon_i$$

---

[4]The discussion works for atoms, too.

where the sum runs over all molecules (from 1 to N), ε is the kinetic energy of each molecule (summed for all molecules), $\mathbf{p}$ is the momentum of each molecule, and m is the molecular mass (assumed the same for all molecules, since they are identical and of the same kind). The potential energy V is due to the interaction between molecules, actually the sum of interactions V_{ij} (pair interactions) between the i-th and the j-th molecule. When we consider the interaction, we take two by two (in pairs), so that V_{ij} is between *two* (any two) molecules. Further, V_{ij} is assumed radial, depending only on the separation distance $r = |\mathbf{r}| = |\mathbf{r}_i - \mathbf{r}_j|$ between the two molecules: $V_{ij}(r)$. Then, since the potential energy is additive,

$$V = V_{12} + V_{13} + V_{14} + V_{15} + \cdots + V_{23} + V_{24} + \cdots + V_{N-1,N} = \frac{1}{2} \sum_{\substack{i,j=1 \\ (i \neq j)}}^{N} {}' V_{ij}$$

where the prime indicates a restricted sum (with $i \neq j$, because $V_{11} = V_{12} = \cdots = V_{NN} = 0$, as a molecule does not interact with itself), and the factor $1/2$ is put to avoid overcounting (double counting), because the molecules are identical and hence $V_{12} \equiv V_{21}$ etc. The sum runs over all the population (N molecules) of the gas, to cover all the contributions of the interaction.

Here $V(r)$ is given as in Fig. SP-39. It is strongly repulsive for small r values (when the molecules get close to each other) and weakly attractive as r grows large, as one would expect. If the molecules are simple (say, diatomic), $V(r)$ can be calculated quantum mechanically. Such a potential (as a function of r) is the *Lennard-Jones potential*, given also by the plot in Fig. SP-39 and by the function

$$V(r) = -V_0 r_0^6 \left(\frac{2}{r^6} - \frac{r_0^6}{r^{12}} \right)$$

where V_0 and r_0 are constants: $V_{\min} = -V_0$ occurs at $r = r_0$, *i.e.*, r_0 is the value of r at which (and for which) V reaches its deepest (minimum) value as a potential well. Notice that $V \to 0$ as $r \to \infty$ asymptotically, and the passage from $-V_0$ to $V = 0$ is in the form $V \sim r^{-6}$ as r grows large.

The Lennard-Jones potential is half empirical and half theoretical. Its $1/r^6$ dependence for large r can be theoretically checked and verified. This task is left to the reader (see Assignment Problem 160(d)).

Another pertinent theoretical potential is the *hard-sphere potential* whose graph is plotted in Fig. SP-40. The sketched curve gives V at any r. It is a very rough approximation and more distant from reality, but neverthe-

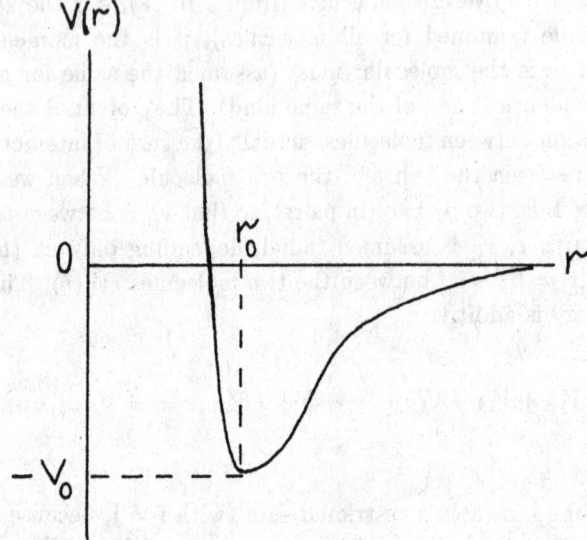

Fig. SP-39.

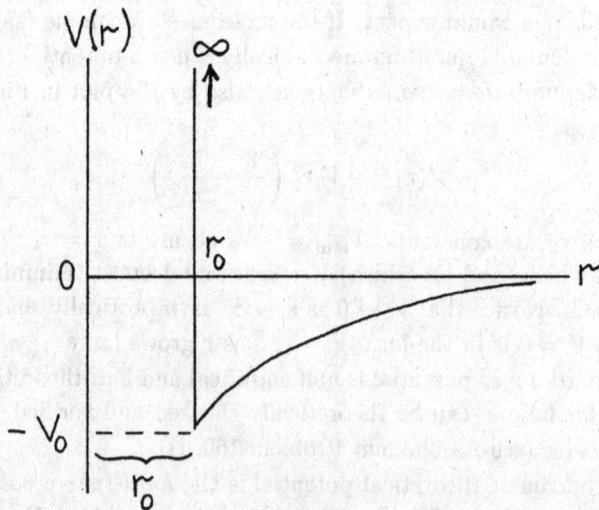

Fig. SP-40.

less manageable enough because of its relative mathematical simplicity and idealization:

$$V(r) = \begin{cases} -V_0 \left(\dfrac{r_0}{r}\right)^p & \text{for } r > r_0 \\ +\infty & \text{for } r \leq r_0 \end{cases}$$

where again, the constant parameter $-V_0$ is the lowest value that V ever takes. This piecewise potential is unrealistic, corresponding to no real case. It is just a mathematical fabrication, but not irrelevant theoretically. It is worthy of consideration as a *gedanken* model. For $r < r_0$ the potential barrier rises like a steep vertical wall, up to $r = r_0$. This means that each molecule behaves more or less like a hard, impenetrable sphere of diameter r_0. Beyond r_0, the molecules attract each other weakly (as the curve suggests for $r > r_0$). To make the model as realistic as possible, and for a worthwhile attempt, the proper empirical value for the power exponent is $p = 6$. Of course, other values can be taken, too, to shape the form of the curve (r^{-p} dependence) outside r_0 (for the attractive part of the potential), but choices with $p \neq 6$ seem to be unwise to take, and too far from the actual behavior of the molecules for $r > r_0$.

Comparing the two potentials, we see that for large r (outside the molecular size r_0) their qualitative behavior is not much different; one might say that they are almost the same (with their r^{-6} dependence and weak attraction common to both and dominating). The former potential is relatively more temperature compared to the latter for small r, and is much closer to reality. The repulsion is strong but slightly milder (curved), as if the wall were a bit softer, allowing some gentle penetration of the wavefunction of the other (approaching) molecule. Thus a partial interpenetration upon approach is possible.

The latter potential, clearly modified, represents an infinitely rigid wall, forbidding and severely repulsive, that defines the extent (radius) of the molecule sharply. (This feature is mathematical, not physical.) Outside this strong repulsion, V varies smoothly, and as r decreases, it comes to a sharp dip with an abrupt change into a resiliently repellent wall ($V \to \infty$) at r_0. We imagine here solid marbles without internal structure or voids, and with a sudden repulsive force at their surface. Molecules bounce back like marbles upon mutual encounter at r_0. The molecule "feels hard" at r_0 and "our touch" can go that far. This potential is an extreme case, but a convenient mathematical idealization. More or less the molecules are weakly attractive hard spheres here, whereas the former potential with the softer wall allows for a more flexible molecular boundary surface, not necessarily a definite sphere.

The potentials here are radial; they do not depend on the direction (on vector **r**).

78. By considering qualitative microscopic molecular behavior, discuss the Van der Waals pressure correction in a real gas of molecules.

Solution: The interaction between two gas molecules is weakly attractive for large intermolecular separation, and it reverses behavior, becoming strongly repulsive, when the separation r becomes comparable to the molecular size (diameter): $r \sim d$. When the temperature is low, the average kinetic energy of a molecule is small (because the molecule is slow). Then the weak intermolecular attraction (which holds for long r values) dominates significantly and tries to reduce the average separation $\bar{r}$ to values smaller than those it would have if there were no intermolecular interaction ($V = 0$). As this attraction prevails, the gas pressure decreases and becomes lower than what is proper for an ideal gas (where there is no intermolecular cohesion). Hence the pressure (of a real gas) receives a (negative) corrective term to account for this reduction.

If now the temperature is high, the average molecular kinetic energy is large, as the molecules move fast. Then the (already weak) intermolecular attraction is insignificant (with a potential energy that can be readily ignored if compared with the quantitatively more important term of the kinetic energy). The flying molecules are swift and more aggressive, and hence, as the attractive part wanes, the strong short-distance repulsive part dominates between the molecules as an appreciable interaction. As a result of these repulsive overtures, the gas pressure (of a real gas) is enhanced to exceed the value that we would expect for an ideal gas. Again, a correction is necessary, but this time the correction term for pressure in the equation of state is positive, to account for the increase. Be it negative or positive (depending on whether the temperature is low or high respectively), the correction term for the pressure exists in both cases for a real gas.

79. Refer to Problem 72. Verify that the dimensional analysis of the result gives an energy.

Solution: We will do it for the last term. The parentheses with the quantum numbers are dimensionless quantities. There remains

$$\frac{(\text{erg sec})^3}{\text{gr}^2\,\text{cm}^4\,\text{sec}^{-1}} = \frac{(\text{dyne cm sec})^3\,\text{sec}}{\text{gr}^2\text{cm}^4} = \frac{\text{gr}(\text{cm/sec}^2)\text{cm sec})^3\,\text{sec}}{\text{gr}^2\,\text{cm}^4} = \text{erg}.$$

BIBLIOGRAPHY

1. Main References

B. Bak, *Introduction to Molecular Spectra*, North Holland Publishing Co., Amsterdam (1954).

B. Chu, *Molecular Forces (Based on the Baker Lectures of Peter Debye)*, Interscience, New York (1967).

H.L. Dai and R.W. Field, eds., *Molecular Dynamics and Spectroscopy by Stimulated Emission Pumping*, World Scientific, Singapore (1993).

L. D'Ans, *Taschenbuch für Chemiker und Physiker, 3. Band: Eigenschaften von Atomen und Molekeln*, Springer-Verlag, Berlin (1970).

N. Davies, *Infrared Spectroscopy and Molecular Structure*, American Elsevier Co., New York (1965).

B.C. Eu, *Semiclassical Theories of Molecular Scattering*, Springer-Verlag, Berlin (1984).

G. Fieck, *Symmetry of Polycentric Systems — The Polycentric Tensor Algebra for Molecules*, Springer-Verlag, Berlin (1982).

H.D. Fösterling and H. Kuhn, *Moleküle und Molekülelanhäufungen*, Springer-Verlag, Berlin (1983).

K.H. Hellwege, *Einführung in die Physik der Molekeln*, Springer-Verlag, Berlin (1974).

G. Herzberg, *Atomic Spectra and Atomic Structure*, Dover Publications, New York (1944).

G. Herzberg, *Molecular Spectra and Molecular Structure*, 2nd ed., Van Nostrand, Princeton, N.J. (1950).

G. Herzberg, *Polyatomic Molecules*. Van Nostrand Reinhold Co., New York (1966).

G. Herzberg, *Spectra of Diatomic Molecules*, Van Nostrand, Princeton, N.J. (1950).

J.O. Hirschfelder, C. Curtiss, and R.B. Bird, *Molecular Theory of Gases and Liquids*, John Wiley, New York (1954).

T. Kakitani *et al.*, *A Unified Theory of Molecular Geometry, Electronic Structure, Molecular Vibrations, and Optical Absorption and Flourescence Spectral Curves of Conjugated Molecules* (Preprint paper, 1977).

H. Labhart, *Einführung in die Physikalische Chemie, 4. Teil: Molekülbau/ 5. Teil: Molekülspektroskopie*, Springer-Verlag, Berlin (1984–1987).

Landolt-Börnstein Numerical Data and Functional Relationships in Science and Technology, The 6th Edition, Volume I: Atomic and Molecular Physics, Springer-Verlag, Berlin (1980).

R.B. Leighton, *Principles of Modern Physics*, McGraw-Hill, New York (1959).

I. Lindgren and J. Morrison, *Atomic Many-Body Theory*, 2nd ed., Springer-Verlag, Berlin (1986).

H. Margenau and N. R. Kestner, *Theory of Intermolecular Forces*, Pergamon, Oxford (1969).

L. Marton, *Methods of Experimental Physics*, Academic Press, New York (1959).

C.A. Morrison, *Angular Momentum Theory Applied to Interactions in Solids*, Springer-Verlag, Berlin (1988).

L. Pauling, *The Nature of the Chemical Bond*, 3rd ed., Cornell Univ. Press, Ithaca, N.Y. (1960).

K.S. Pitzer, *Quantum Chemistry*, Prentice-Hall, Englewood Cliffs, N.Y. (1953).

P. Pyykkö, *Relativistic Theory of Atoms and Molecules*, Springer-Verlag, Berlin (1986).

A.A. Radzig and B.M. Smirnov, *Reference Data on Atoms, Molecules, and Ions*, Springer-Verlag, Berlin (1985).

N.F. Ramsey, *Molecular Beams*, Oxford Univ. Press, New York (1956).

A.E. Ruark and H.C. Urey, *Atoms, Molecules, and Quanta*, McGraw-Hill, New York (1930).

C. Saunders and R.E.D. Clark, *Atoms and Molecules Simply Explained*, Dover Publications, Inc., New York (1976).

R.A. Sawyer, *Experimental Spectroscopy*, Prentice-Hall, New York (1951).

J.C. Slater, *Introduction to Chemical Physics*, McGraw-Hill, New York (1939).

J.C. Slater, *Quantum Theory of Matter*, McGraw-Hill, New York (1951).

J.C. Slater, *Quantum Theory of Molecules and Crystals*, McGraw-Hill, New York (1965).

J.C. Slater, *Quantum Theory of Molecules and Solids*, vol. 1, McGraw-Hill, New York (1963).

C. P. Slichter, *Principles of Magnetic Resonances*, 3rd ed., Springer-Verlag, Berlin (1992).

S. Svanberg, *Atomic and Molecular Spectroscopy: Basic Aspects and Practical Applications*, 2nd ed., Springer-Verlag, Berlin (1992).

C.H. Townes and A.L. Schawlow, *Microwave Spectroscopy*, McGraw-Hill, N.Y. (1955).

P.J. Wheatley, *The Determination of Molecular Structure*, Clarendon Press, Oxford (1968).

D. Williams, *Molecular Physics*, Academic Press, New York (1962).

G.S. Zhdanov, *Crystal Physics*, Oliver and Boyd Publishing Co., Academic Press, N.Y. (1965).

2. Auxiliary Texts

N.W. Aschcroft and N.D. Mermin, *Solid State Physics*, Holt, Rinehart and Winston, New York (1976).

R.A. Becker, *Introduction to Theoretical Mechanics*, McGraw-Hill, New York (1954).

R. M. Eisberg, *Fundamentals of Modern Physics*, John Wiley, New York (1961).

E. Fermi, *Notes on Quantum Mechanics*, Univ. of Chicago Press (1961).

R.P. Feynman, R.B. Leighton, and M. Sands, *The Feynman Lectures on Physics*, Addison-Wesley, Reading, Mass. (1964).

H. Goldstein, *Classical Mechanics*, Addison-Wesley, Reading, Mass. (1950).

P.H. Groth, *Chemische Krystallographie*, 5 volumes, W. Englemann, Leipzig (1906).

International Tables for X-ray Crystallography, 3 volumes, Kynoch Press, Birmingham (1952–1962).

J.D. Jackson, *Classical Electrodynamics*, 2nd ed., John Wiley, New York (1975).

C. Kittel, *Introduction to Solid State Physics*, 5th ed., John Wiley, New York (1976).

N.F. Mott and I.N. Sneddon, *Wave Mechanics and Its Applications*, Clarendon Press, Oxford (1948).

D. Park, *Introduction to Quantum Theory*, 2nd ed., McGraw-Hill, New York (1974).

L. Pauling and E.B. Wilson, Jr., *Introduction to Quantum Mechanics*, McGraw-Hill, New York (1935).

F. Reif, *Fundamentals of Statistical and Thermal Physics*, McGraw-Hill, New York (1965).

F.K. Richtmyer, E.H. Kennard, and J.N. Cooper, *Introduction to Modern Physics*, 6th ed., McGraw-Hill, New York (1969) (pp. 489–525).

V. Rojansky, *Introduction to Quantum Mechanics*, Prentice-Hall, Englewood Cliffs, N.J. (1959).

L.I. Schiff, *Quantum Mechanics*, 3rd ed. (international student edition), McGraw-Hill-Kogakusha, Tokyo (1968) (pp. 445–455).

F.W. Sears, M.W. Zemansky, and H.D. Young, *University Physics*, 5th ed., Addison-Wesley, Reading, Mass. (1977) (pp. 779–782).

E. Segrè, *Nuclei and Particles*, W.A. Benjamin, New York (1965) (pp. 211–266 and 572–576).

H. Semat, *An Introduction to Atomic and Nuclear Physics*, 4th ed., Holt, Rinehart and Winston, New York (1962).

T. Triffet, *Mechanics: Point Objects and Particles*, John Wiley, New York (1968).

G.H. Wannier, *Elements of Solid State Theory*, Cambridge, London (1959).

3. Other References

A. Abragam, *The Principles of Nuclear Magnetism*, Oxford (1961).

A. Abragam and B. Bleaney, *EPR of Transition Ions*, Oxford (1970).

S.A. Altshuler and B.M. Kozyrev, *EPR in Compounds of Transition Elements*, 2nd ed., Halsted (1974).

G.E. Bacon and K. Lonsdale, *"Neutron Diffraction"* in *Reports on Progress in Physics*, The Physical Society of London (1953) (XVI, 1–61).

J.A. Barker and D. Henderson, *"The Fluid Phases of Matter"* in *Scientific American* (November 1981).

F. Beeching, *Electron Diffraction*, Methuen, London (1936).

I.B. Bersuker and V.Z. Polinger, *Vibronic Interactions in Molecules and Crystals*, Springer-Verlag, Berlin (1989).

B. Bleaney and K.W.H. Stevens, *"Paramagnetic Resonance"* in *Reports on Progress in Physics*, The Physical Society of London (1953) (XVI, 108).

M. Born and J.R. Oppenheimer, *Ann. Physik*, **84**, 457 (1927).

K.D. Bowers and J. Owen, *"Paramagnetic Resonance II"* in *Reports on Progress in Physics*, The Physical Society of London (1955) (IIXX, 304).

J.M. Bowman, ed., *Molecular Collison Dynamics*, Springer-Verlag, Berlin (1983).

W.H. Bragg and W.L. Bragg, *X-Rays and Crystal Structure*, George Bell and Sons, London (1925).

D.A. Bromley, *"Nuclear Molecules"* in *Scientific American* (December 1978).

R. Buckminster Fuller, *Synergetics 2*, Macmillan Publishing Co., New York (1979).

E.R. Caianiello, ed., *Lectures on the Many-Body Problem*, Academic Press, New York (1964).

S. Carrà and N. Rahman, eds., *From Molecular Dynamics to Combustion Chemistry* (Proceedings of a conference held in Trieste, Italy, 1991), World Scientific, Singapore (1992).

N. Chandrakumar and S. Subramanian, *Modern Techniques in High-Resolution FT-NMR*, Springer-Verlag, Berlin (1987).

B.D. Cullity, *Elements of X-Ray Diffraction*, Addison-Wesley Pub. Co., Reading, Mass. (1978).

V. Daniel, *Dielectric Relaxation*, Academic Press, New York (1967).

T.P. Das and E.L. Hahn, *"Nuclear Quadrupole Resonance Spectroscopy"* in *Solid State Physics*, supplement 1 (1958).

A.S. Davydov, *Theory of Molecular Excitons*, McGraw-Hill, New York (1962).

B.M. Deb, *Rev. Mod. Phys.*, **45**, 22 (1973).

P. Diehl *et al.*, eds., *NMR: Basic Principles and Progress*, vols. 27 and 28: *In Vivo Magnetic Resonance Spectroscopy*, Springer-Verlag, Berlin (1992).

L.E. Drain, *"Nuclear Magnetic Resonance in Metals"* in *Metallurgical Reviews*, Review 119 (1967).

H.J. Flechtner, *Die Welt in der Retorte*, Verlag Ullstein GmbH, Frankfurt/M-Berlin (1957).

H.F. Franzen, *Physical Chemistry of Inorganic Crystalline Solids*, Springer-Verlag, Berlin (1986).

A.P. French, *Vibrations and Waves*, Nelson, London (1971).

H. Frölich, *Theory of Dielectrics: Dielectric Constant and Dielectric Loss*, Oxford (1958).

G. Gamow, *Thirty Years That Shook Physics*, Doubleday, Garden City, N.Y. (1966).

C. Gavroglou, *Fritz London: A Scientific Biography*, Cambridge Univ. Press, ISBN 0 521 432731.

O.S. Heavens, *Optical Masers*, Methuen, London (1964).

E. Hirota, *High-Resolution Spectroscopy in Transient Molecules*, Springer-Verlag, Berlin (1985).

J.O. Hirschfelder, *Adv. Chem. Phys.*, **12**, 51 (1967).

A. Holden, *Nature of Solids*, Columbia Univ. Press (1965).

J.C. Kendrew, *"Three-Dimensional Structure of a Protein Molecule"* in *Scientific American*, No. 121 (1961).

H. Kopfermann, *Nuclear Moments*, Academic Press, New York (1958).

Landolt-Börnstein Collection, Springer-Verlag, Berlin (1987) (Vol. II/4: *Molecular Constants from Microwave Spectroscopy*. Vol. II/5: *Molecular Acoustics*. Vol. II/6: *Molecular Constants from Microwave, Molecular Beam, and ESR Spectroscopy*. Vol. II/7: *Structure Data of Free Polyatomic Molecules*. Vol. II/17: *Magnetic Properties of Free Radicals*.)
[The Landolt-Börnstein Collection contains among others the following: Results of magnetic investigations, data on spin Hamiltonian parameters (*g* factors and splitting parameters), biradicals, inorganic free and organic radicals, donor-acceptor complexes, susceptibility measurements (Curie constants, magnetic moments, transition temperatures), ESR measurements, figures on typical spectra, wave numbers of sharp single bands, influence of surrounding molecules, luminescence of organic substances, rotational constants, rotation-vibration interactions, *l*-type doubling constants, isotopic masses and mass ratios, dipole moments, nuclear quadrupole moments, hindered rotation constants, diagrams on the structural arrangement of molecules, molecular properties of matter (sound velocity, dispersion, absorption in gases, liquids and isotropic or quasi-isotropic solids with tables), molecular beam resonance, experimental methods, magnetic properties of coordination and organo-metallic transition metal compounds, organic anions and cations, polyradical, EPR figures, radical reaction rates in liquids (absolute and relative kinetic constants, equilibria and their temperature, solvent dependencies for more than 9000 reactions of 2100 species, radical production and detection, reaction mechanisms, and experimental procedures), carbon-centered radicals, nitroxyl-, oxyl-, peroxyl-, and related radicals, proton and electron transfer, diamagnetic susceptibility tables for 4000 substances, paramagnetic ions, non-conjugated carbon radicals, electronic structure of atoms and molecules (photoemission spectra and related data), research methodology, etc.]

I.R. Lapidus,"*One-Dimensional Model for the Diatomic Molecule*" in *Am. J. of Phys.*, **38**, 905 (1970).

W. Low, "*Paramagnetic Resonance in Solids*" in *Solid State Physics*, supplement 2 (1960).

D.S. McClure, "*Electronic Spectra of Molecules and Ions*" in *Solid State Physics*, 8, 1 (1959).

M. Mehring, *Principles of High-Resolution NMR in Solids*, 2nd ed., Springer-Verlag, Berlin (1983).

J.W. Orton, "*Paramagnetic Resonance Data*" in *Reports on Progress in Physics*, The Physical Society of London (1959) (XXII, 204).

G.E. Pake and T.L. Estle, *Physical Principles of Paramagnetic Resonance*, 2nd ed., W.A. Benjamin, New York (1973).

G.N. Patterson, *Molecular Flow of Gases*, John Wiley, New York (1956).

Z.G. Pinsker, *Electron Diffraction*, Butterworths Scientific Publications, London, (1953).

M.H. Pirenne, *The Diffraction of X-Rays and Electrons by Free Molecules*, Cambridge Univ. Press.

I.L. Pykett, "*NMR Imaging in Medicine*" in *Scientific American* (May 1982).

Y. Saito, *Inorganic Molecular Dissymetry*, Springer-Verlag, Berlin (1979).

M. Sargent III, M.O. Scully, and W.E. Lamb, Jr., *Laser Physics*, Addison-Wesley, Reading, Mass. (1974).

I.F. Silvera and J. Walrasen, "*The Stabilization of Atomic Hydrogen*" in *Scientific American* (January 1982).

C.P. Slichter, *Principles of Magnetic Resonance, With Examples from Solid State Physics*, Harper and Row, New York (1963).

C.P. Slichter, *Principles of Magnetic Resonance*, 2nd ed., Springer-Verlag, Berlin (1980).

T.P. Snow and J.M. Shull, *Physics*, West Publishing Co., St. Paul, MN (1986) (pp. 810–820).

K.J. Standley and R.A. Vaughan, *Electron Spin Relaxation in Solids*, Hilger, London (1969).

G. Turrell, *Infrared and Raman Spectra of Crystals*, Academic Press, New York (1972).

J.H. Van Vleck, *Theory of Electric and Magnetic Susceptibilities*, Oxford (1932).

A. Weber, ed., *Raman Spectropscopy of Gases and Liquids*, Springer-Verlag, Berlin (1979).

V.F. Weisskopf in W.E. Brittin, B.W. Downs, and J. Downs, eds., *Lectures in Theoretical Physics*, vol. 3, Interscience, New York (1961) (p. 80).

R.W.G. Wyckoff, *Crystal Structures*, 2nd ed., Interscience, New York (1963).

I.S. Zheludev, *Physics of Crystalline Dielectrics*, Plenum Press (1971).

4. Suggested References for Advanced and Specialized Reading

H. A. Bachor *et al.*, eds., *Atomic and Molecular Physics and Quantum Optics* (Proceedings of a summer school, Australian National University, Canberra, Australia, 1992), World Scientific, Singapore (1993).

Cheuk-Yiu Ng, ed., *Vacuum Ultraviolet Photoionization and Photodissociation of Molecules and Clusters*, World Scientific, Singapore (1991).

R.W. Damon, "*Ferromagnetic Resonance at High Power*" in *Magnetism* (I, 552).

S. Foner, "*Antiferromagnetic and Ferrimagnetic Resonance*" in *Magnetism* (I, 384).

C.W. Haas and H.B. Callen, "*Ferromagnetic Relaxation and Resonance Line Widths*" in *Magnetism* (I, 450).

V. Jaccarino, *"Nuclear Resonance in Antiferromagnets"* in *Magnetism* (IIA), edited by G.T. Rado and H. Suhl, Academic Press, New York.

W. Kolos, *Int. J. Quantum Chem.*, **1**, 169 (1967).

V. Kumar *et al.*, eds., *Clusters and Fullerenes (Proceedings of the Adriatico Research Conference*, ICTP, Trieste, Italy, 1992), World Scientific, Singapore (1993).

A.L. Lehninger, *Bioenergetics: The Molecular Basis of Biological Energy Transformations*, 2nd ed., W.A. Benjamin, Menlo Park (1971).

P.O. Löwdin and H. Yoshizumi, *Adv. Chem. Phys.*, **2** (1959).

D.C. Mattis, ed., *The Many-Body Problems*, World Scientific, Singapore (1993).

L. Pauling and J.Y. Beach, *Phys. Rev.*, **47**, 686 (1935).

A.M. Portis and R.H. Lindquist, *"Nuclear Resonance in Ferromagnetic Materials"* in *Magnetism* (IIA, 357).

M. Sparks, *Ferromagnetic Relaxation Theory*, McGraw-Hill, New York (1964).

P.W. Stephens, ed., *Physics and Chemistry of Fullerenes*, World Scientific, Singapore (1992).

C. Taliani *et al.*, eds., *Fullerenes: Status and Perspectives* (Proceedings of the First Italian Workshop held in Bologna, Italy, 1992), World Scientific, Singapore (1992).

G. J. F. Troup, *Masers and Lasers*, 2nd ed., Halsted (1973).

E.A. Turov and M.P. Petrov, *NMR in Ferro- and Antiferromagnetics*, Halsted (1972).

J.D. Watson, *Molecular Biology of the Gene*, 3rd ed., W.A. Benjamin, Menlo Park (1976).

All the following references are publications of Springer-Verlag, Berlin:

V.L. Broude, E.I. Rashba, and E.F. Sheka, *Spectroscopy of Molecular Excitons*, (1985).

C.D. Cantrell, ed., *Multiple-Photon Excitation and Dissociation of Polyatomic Molecules*, (1986).

M. Capitelli, ed., *Nonequilibrium Vibrational Kinetics*, (1986).

H.G. Elias, *Mega Molecules*, (1987).

N.D. Epiotis, *Unified Valence Bond Theory of Electronic Structure*, (1982).

F.K. Fong, ed., *Radiationless Processes in Molecules and Condensed Phases*, (1988).

F.A. Gianturco, *The Transfer of Molecular Energies by Collision — Recent Quantum Treatments*, (1979).

D.C. Hanna, M.A. Yuratich, and D. Cotter, *Nonlinear Optics of Free Atoms and Molecules*, (1979).

N. Nöth and B. Wrackmeyer, *Nuclear Magnetic Resonance Spectroscopy of Boron Compounds*, (1978).

I. Pockrand, *Surface Enhanced Raman Vibrational Studies at Solid/Gas Interfaces*, (1984).

P. Reineker *et al.*, eds., *Organic Molecular Aggregates*, (1983).

H. Sitter and M.A. Herman, *Molecular Beam Epitaxy — Crystallization of Semiconductor Films from Atomic and Molecular Beams*, (1988).

5. Conference Proceedings (Non-Exhaustive List)

All the entries below are publications of Springer-Verlag, Berlin. The year of publication appears at the end of each entry:

J.L. Ballot and M. Fabre de la Ripelle, eds., *Few-Body Problems in Particle, Nuclear, Atomic, and Molecular Physics* (Proceedings of the 11th European Conference on Few-Body Physics, held in Fontevraud, 1987), 1987.

G. Benedek, T.P. Martin, and G. Pacchioni, eds., *Elemental and Molecular Clusters* (Proceedings of the 13th International Summer School in Erice, Italy, 1987), 1988.

J. Broeckhove, L. Lathouwers, and P. van Leuven, eds., *Dynamics of Wave Packets in Molecular and Nuclear Physics* (Proceedings of the International Meeting held in Priorij Corsendonck, Belgium, 1985), 1986.

T. Dorfmüller and R. Pecora, eds., *Rotational Dynamics of Small and Macromolecules* (Proceedings of a workshop held in Bielefeld, Germany, 1986), 1987.

A. Ehrenberg, R. Rigler, A. Gräslund, and L. Nilsson, eds., *Structure, Dynamics, and Function of Biomolecules* (Proceedings of the First EBSA Workshop), 1987.

A. Heideman, A. Magerl, M. Prager, D. Richter, and T. Springer, eds., *Quantum Aspects of Molecular Motions* (Proceedings of the Third International Conference on Recent Progress in Many-Body Theories, held in Odenthal-Altenberg, Germany, 1983), 1984.

K. Kompa and S.D. Smith, eds., *Laser-Induced Processes in Molecules* (Proceedings of the European Physical Society, Edinburgh, Scotland, 1978), 1979.

A. Laubereau and M. Stockburger, eds., *Time-Resolved Vibrational Spectroscopy* (Proceedings of the Second International Conference held in Bayreuth, Germany, 1985), 1985.

G. Marowsky and V.V. Smirkov, eds., *Coherent Raman Spectroscopy: Recent Advances* (Proceedings of an international symposium held in Samarkand, USSR, 1990), 1992.

A.R.W. McKellar, T. Oka, and B.P. Stoicheff, eds., *Laser Spectroscopy* (Proceedings of the Fifth International Conference held in Alberta, Canada, 1981), 1981.

L.C. Pitchford, B.V. McKoy, A. Chutjian, and S. Trajmar, eds. *Swarm Studies and Inelastic Electron-Molecule Collisions* (Proceedings of the Fourth International Swarm Seminar in Tahoe City, Calif., 1985), 1987.

N.K. Rahman, ed., *Photons and Continuum States of Atoms and Molecules* (Proceedings of a workshop held in Cortona, Italy, 1986), 1987.

F. Yonezawa, ed., *Molecular Dynamics Simulations* (Proceedings of a symposium held in Japan, 1990), 1992.

6. Journals

Advances in Atomic and Molecular Physics, Academic Press, New York.

Advances in Multiphoton Processes and Spectroscopy, edited by S. H. Lin.

Few-Body Systems: Acta Physica Austriaca New Series (ISSN 0177-7963), Springer-Verlag, Wien.

Monatshefte für Chemie, Springer-Verlag, Wien.

Zeitschrift die Makromolekulare Chemie, Verlag Wepf, Basel.

Zeitschrift für Physik, Section D: *Atoms, Molecules, and Clusters* (ISSN 0178-7683).
[This new part of *Z. Physik* covers the entire field of atomic, molecular, cluster, and chemical physics in one single journal, thus obviating the need to search through several different journals. Modern research employs to an increasing extent similar techniques in these fields: LASER spectroscopy, ultraviolet and synchrotron radiation experiments, multiphoton processes etc. The papers published reflect this overlap. The focus is on free atoms, molecules and clusters and their properties and interactions as individual entities in gaseous, liquid, and solid environments. All aspects of atomic, molecular, and cluster structure, spectroscopy, dynamics, production, ionization, and fragmentation are covered. Other topics included are: Heavy-ion atomic physics; muonic, pionic, and other exotic atoms; hyperfine interactions; electron and positron scattering; collisions in experiment and theory; structure and stability calculations; statistical and dynamic theories of inter- and intra-molecular processes etc. The journal publishes articles, original reports, review papers, reprints, and short notes.]

Suggested Research Topics

The bibliography presented above serves another purpose as well: It is a source of topics that the instructor may assign to students for essays, term papers, research work, or just further reading. Very roughly, the following aspects of molecular physics might make or yield specific topics for such activities*:

— Molecular spectra
— Experimental spectroscopy and detection methods
— Interactions of molecules with photons, including NMR and relaxation
— Energy transfer in polyatomic molecules
— Inelastic quantum collisions (vibrational and rotational)
— Molecular collision processes and interactions (quasi-classical trajectory studies, energy transfer, scattering etc.)
— Experimentally obtained information on molecules, including hyperfine-structure constants
— Molecular alloys and aggregates
— Molecular structure and intermolecular forces
— Vibrational relaxation processes
— Polyatomic molecules and molecular fragments
— Studies of special molecules, including macromolecules and polymer molecules (polymer spectroscopy and rheology etc.)
— Excitation dynamics in molecular solids
— Magnetic studies with molecules
— Raman scattering cross-sections
— Behavior of macromolecules
— Molecular clouds in stars
— Molecular ionization and dissociation under nonequilibrium conditions
— Analytical theory of anharmonic oscillators
— Multiquantum transitions in collisions
— Vibrational energy transfer in collisions involving free radicals
— Dynamics of reactions involving excited molecules
— Vibrational excitation and dissociative attachment
— Vibrational distribution and rate constants for energy transfer
— Chemical reactivity of clusters
— Electric properties and quantum theory of polymers
— Light scattering from molecules
— Molecular beams and pertinent techniques

---

*The list is only a sample, not an exhaustive repertory.

— LASER irradiation of molecules and LASER-induced decompositions (LASER isotope separation by adiabatic inversion)
— Lasing molecules
— Attenuation of molecular beams by Maxwellian gases
— Molecular electrostatics, electrets, and dielectrics
— Molecular magnetism studies and susceptibilities
— Molecular partition functions
— Molecular dipole moments
— Macroscopic molecular phenomena (viscosity, osmosis, surface tension, adhesion, cohesion, diffusion, kinetic theory of gases, etc.) in classical physics
— Liquid gases
— Color centers
— Quantum theory of molecular bonds
— Molecular data and data analysis for space physics, astronomy, and astrophysics (spectroscopy satellites and molecular parameters)
— Molecular theory in atmospheric physics and environmental studies
— The correlation problem in atoms and simple molecules
— Microwave pressure broadening in molecules and dipole moments of symmetric-top molecules along applied electric fields (see G. Birnbaum's work)
— X-ray small-angle scattering in the determination of the size and shape of macromolecules in solutions
— The influence of water on the molecular mobility of macromolecules
— Acylderivate cyclic bindings.
— Time-resolved resonance Raman spectroscopy (general theory and applications to molecular systems, with numerical results)
— Molecular crystals
— Intramolecular vibration energy relaxation in large molecules
— Spectroscopy of intramolecular dynamics of highly excited vibrational levels in compounds
— Intermolecular interaction: Theory and experimental techniques and related methods
— Electron-molecule interactions
— Populated levels of molecules
— Spectral pattern recognition in molecular physics
— Rotation-vibration coupling
— Energy storage in molecules and molecular microstructure of matter
— Dynamics of molecular evaporation

— Scattering and attenuation in a molecular gas
— Maxwell-Boltzmann statistics in a molecular gas/kinetic theory
— Molecular beams technique
— Lifetimes in excited molecules
— Molecular beam studies of the dynamics of chemical processes
— Molecular mechanics and combination — critical parameters
— Structure and spectroscopy of molecules
— Spin-dependent scattering of electrons from molecules
— Interatomic and intermolecular forces
— Magnetic properties of free radicals (investigation of spin Hamiltonian parameters, g-factors, splitting, radical ions, organic radicals, inorganic free radicals, biradicals, donor-acceptor complexes)
— Magnetic properties of coordination and organometallic transition-metal compounds (susceptibility measurements, Curie constants, magnetic moments, transition temperatures) and ESR measurements
— Doppler-shift studies in molecules
— Large complex molecules in molecular biology
— Luminescence (data and wavenumbers of absorption and emission bands in organic substances, decay times, quantum efficiencies, figures on typical spectra, tables of information, wavenumbers of sharp single bands, influence of surrounding molecules etc.)
— Molecular constants from microwave spectroscopy (rotational, centrifugal distortion, rotation-vibration interaction, and l-type doubling constants, isotopic masses and mass ratios, electric dipole moments, nuclear quadrupole constants, hindered rotation constants, diagrams on the structural arrangement of molecules)
— Molecular acoustics (acoustic data based on the molecular properties of matter, sound velocity, dispersion and absorption in gases, liquids, and isotropic or quasi-isotropic solids, transmission in molecular media, tables and figures)
— Molecular constants from microwave, molecular beam, and ESR spectroscopy (double resonance, ESR, molecular-beam resonance)
— Structure data of free polyatomic molecules (data tables regarding molecules in the ground state and in excited electronic states. Experimental methods in microwave, infrared, Raman, and ultraviolet spectroscopy and electron diffraction, with results for inorganic and organic molecules)
— Atoms, inorganic radicals, and radicals in metal complexes

— Organic radicals, organic anion and cation radicals, polyradicals, organic N-centered and C-centered radicals
— Magnetic susceptibilities
— Electron Paramagnetic Resonance
— Radical reaction rates in liquids (accumulated data on reaction-rate constants for radicals, radical ions, and biradicals in liquids. Tables of absolute and relative kinetic constants, data on equilibria and their temperature, qualitative knowledge on radical reaction mechanism and the relevant experimental procedures. See Springer-Verlag's Landolt-Börnstein New Series, 1987, for volumes concerning data about 2100 species of compounds)
— Nitroxyls and radicals centered on S, P, O, Si, Ge, Sn, Pb, As, Sb and other heteroatoms
— Oxyl-, peroxyl-, and related radicals, nonconjugated carbon radicals
— Proton and electron transfer
— Classical and modern spectroscopy (and pertinent literature published to-date)
— Diamagnetic susceptibility (see Landolt-Börnstein New Series, 1987, for a volume containing present revised data, extended to earlier tables for 4000 substances, including diamagnetic and paramagnetic ions as well as some liquid organic mixtures)
— Electronic structure of atoms and molecules — photoemission spectra and related data
— Cohesive energy calculations by using the Lennard-Jones potential (*e.g.*, see Kittel, p. 102)
— Ionic crystals that employ the Coulomb attraction for binding (stability of NaCl structure) and possibility of ionic A^+A^- type crystals
— Study of solid molecular hydrogen
— Bivalent ionic crystals (see Kittel, p. 103)
— Cohesion in ionic solids (see M.P. Tosi, *Solid State Physics*, **16**, 1 (1964); J.O. Hirschfelder, ed., *Intermolecular Forces*, Interscience (1967))
— Hydrogen bonding (see M.D. Joesten and L. Schaad, *Hydrogen Bonding*, Dekker (1974))
— Molecular electrostatics and classical/quantum mechanical calculation of index of refraction (Clausius-Mossotti equation) (see Jackson, p. Ch. 4)
— MASER and LASER
— Classical molecular mechanics (Lagrangian formalism) in the framework of mathematical physics

— Thermodynamics and moleculear physics
— Measurement of radiative lifetimes of molecules
— Interatomic forces: The mechanics of a molecule, forces between atoms, forces in molecules, the hydrogen molecule ion (see Park, pp. 496–512).
— Interaction of molecules that have permanent and induced dipoles
— Fluorescence spectroscopic techniques and photodissociation dynamics of molecules in the vacuum UV region (accurate energetic and spectroscopic information on ions and neutral molecules)
— Major experimental developments in the studies of single vacuum UV photon ionization and dissociation processes of gaseous molecules (a general or detailed review)
— Recent progress in vacuum ultraviolet (VUV) photoionization and photodissociation processes (photoionization mass spectrometric studies of free radicals; photoelectron spectroscopy of short-living molecules; photoelectron-photoion coincidence studies of ion dissociation dynamics; applications of coherent vacuum UV to photofragment and photoionization spectroscopy; photon transfer in ionic states of hydrogen-bonded dimers — a photoelectron spectroscopic approach; spectroscopy and reaction dynamics using ultrahigh resolution VUV lasers; molecular beam photoionization and photoelectron-photoion coincidence studies of high-temperature molecules, transient species, and clusters; dispersed fluorescence as a probe of molecular photoionization dynamics; absorption and fluorescence studies of molecules and cluster; vibrationally resolved photoelectron angular distribution and branching ratios)
— vibrational spectroscopy and dynamics by stimulated emission pumping
— Multiphoton processes in molecules
— R-matrix theory of molecular processes
— Atoms, nuclear structure, and electron shells in molecules and ions
— Molecules in crystals
— Mass spectrometry in molecular physics
— Homogeneous IVR lifetimes and the mechanism of energy relaxation
— From molecular mechanics to combustion phenomena (a systematic determination of critical parameters and variables)
— Modeling complex reaction systems
— Itinerant electrons in the molecular structure
— Multiphoton spectroscopy of atoms, ions, and molecules in physics, chemistry, biology, biochemistry, and materials science (authoritative reviews and contributions by scientists in the field)

— Mass spectrometry and ion processes in atomic and molecular physics and LASER spectroscopy

— Molecular dynamics and spectroscopy by stimulated emission pumping (stimulated Raman; SEP intensities; ion dip SEP and IVR of compounds; DF of NO_2 and CS_2; SEP of C_3, NCO etc.; DF of C_2H_2; relaxation of SEP-populated levels of NO; FT-DFS; stepwise IVR; assignability of spectra at high energies; quantum chaos)

— Mathematics of coupled and forced oscillators

— Physico-chemical reaction dynamics with chaotic behavior in studying chaos in molecular chemistry and biochemistry

— Interaction of molecules with quantized radiation fields (molecules and non-classical states of light)

— Quantum mechanical cluster calculations in molecular and solid state studies (Hartree-Fock method, molecular surfaces, bulk defects and local crystal anomalies, embedded quantum cluster simulation of point defects and electronic band structures of ionic crystals)

— Molecular physics and condensed matter/solid state physics in modern technology (mathematical simplicity and elegance in complex systems; magnetism, liquid crystals, polymers, interfaces, wetting and adhesion etc.; propagation in disordered media; effects of double exchange in magnetic crystals; NMR modes in magnetic materials; fibrous structures and molecular films; asymmetric syntheses; surface tension and dynamics of drying; stratified molecular media; partial wetting; oil/water interfaces; conformations and attachments to interfaces and stability; neutron scattering experiments and interpretation; dynamics of fluctuations in liquid crystals; polymer solutions; hydro-dynamic properties of fluid phases of lipid/water)

— Collector's volumes, compendiums, and essential literature on molecular physics for specialists, connoisseurs, amateurs, and laymen

— List of basic highly specialized post-graduate readings and papers in molecular spectroscopic methods (advanced and state-of-the-art information)

— Quasi-elastic scattering of neutrons by dilute and ideal solutions of complex molecular compounds (*e.g.*, polymers)

— Flexible large molecules

— Theoretical methods of molecular statistics

— Colloidal molecular suspensions

— Fullerenes (Buckminster Fuller molecules and pertinent research: Superconducting and magnetic fullerides; atoms trapped inside fullerene cage;

chemically bonded complexes; helical carbon tubes; discovery, stability, and spectroscopy of the C_{60} molecule; larger fullerene molecules; solid fullerite and fullerides-structure and spectroscopy; magnetism and superconductivity in fullerides; endohedral complexes; bonds in fullerenes and nano-fibers; applications; physics and chemistry of clusters)

— Intermediate states of matter between molecules and solids (clusters; fullerenes; A_3C_{60} compounds; new classes of molecular groups; non-adiabatic mechanisms; Coulomb pseudopotential and screening in C_{60}; supershells; delayed electron emission from fullerenes; tight-binding molecular dynamics study of carbon clusters; STM studies of anomalous carbon structures)

— Molecular spectroscopy of fullerenes

— Fullerene studies (physical and chemical properties; higher and doped fullerenes; synthesis, photoelectron spectroscoy, electric and magnetic properties; fullerenes in outer space and nonlinear optical properties; physics and astrophysics studies; solved and unsolved problems in fullerene systematics; characterization and growth mechanism; analysis of the vibrational structure of emission and absorption of C_{60}; vibrational frequencies; wave-dispersed nonlinear spectroscopy in C_{60} thin films and fullerene nanowires; nanometer-scale clusters; fullerene-LASER interaction; neutron scattering studies of fullerenes; low-energy electronic excitation in K-doped C_{60} from Raman scattering excited at 1.16 eV; spectroscopic evidence of phase transition in fullerenes; rotational modes; electronic-structure studies of undoped and doped fullerenes; importance of studies in molecular electronics and spectroscopy, as well as condensed matter physics)

— Molecular studies of current importance for physicists, chemists, engineers, and high-temperature superconductivity scientists

— NMR: Basic principles and progress (spectroscopy, NMR studies in medicine and pharmacy-potential and limitations, probeheads and RF pulses, spectrum analysis)

— In vivo MR applications (NMR studies on metabolism of cells; individual nuclei and proton spectroscopy — experimental aspects; ^{13}C spectroscopy in humans; fluorine – ^{49}F NMR spectroscopy and imaging in vivo; sodium movements in muscles)

— special NMR techniques (surface coil spectroscopy; rotating-frame spectroscopy and spectroscopic imaging; depth-resolved spectroscopy (dress); image-guided volume-selective spectroscopy; NMR spectroscopy of human brain; localized spectroscopy using static magnetic-field gradients)

— spectral editing and kinetic measurements in NMR (homo- and heteronuclear cases in proton spectroscopy; specific methods using double quantum-
  coherence transfer spectroscopy; two-dimensional $^{31}P-{}^1H$ correlation
  spectroscopy in vivo; magnetization transfer techniques in medical
  measurements)

— Applications and uses of liquid crystals (chemical structure; molecular engineering and mixture formulation; nonlinear optical response; physical
  properties)

— History of molecular physics

— Theory of molecular rate processes in the presence of intense radiation
  (cooperative, chemical, and optical pumping)

— Industrial irradiation: Breakdown of biological molecules (sterilization) or
  polymer strengthening by fast electron-beam bombardment in the manufacturing of industrial products

The list of topics above is a challenge that gives the flavor of possible
fields of research or specialization for those interested in a professional career
in molecular physics or beyond. The instructor should be aware of the possibility that special interest towards such topics may surface on the part of the
students. The student's initiative should be encouraged towards such areas of
interest.

The same list gives ideas about possible topics for brief seminar-like talks,
organized in the framework of the course. Such in-class, ten-minute presenta-
tions could be assigned in lieu of term papers.

More thermatological suggestions are possible, but this sample list is sufficient as an instruction aid for those who plan to elaborate on relevant topics.
It is also an indicative passage from the world of the textbook and the staged
instruction problems to the real world outside.[†]

Notice that some of the suggested topics are undergraduate or graduate-
level issues, while others are too advanced, at a postdoctoral research level.[‡] We
also included some very special topics that lie beyond molecular physics, rather
in the domain of other sciences (such as chemistry or medicine), just to give an

---

[†]The transition from the academic life and course work to the world of industry and technology is not always smooth and painless. Academic research involves idealized scholastic
studies with historical analysis, whereas the practical world outside deals with applications,
matters of practice, and future issues and uses.

[‡]The readership of this book is assumed to consist of undergraduates, graduates, postgrad-
uates, researchers, and instructors mainly in molecular physics and chemistry, but also in
other relevant scientific disciplines, e.g., medicine, molecular biology, solid state and cluster
physics, etc.

idea about (a) how far the scientific merit of molecular physics can go, so as to influence other sciences or make molecular physics partially overlap with them (with common sections); (b) the extent of the applications of all knowledge concerning molecules, in areas of vast size (a fully comprehensive coverage is not possible here); and (c) the breadth and versatility of highly specialized scientific branches of expertise (even off the field of molecular physics but nevertheless stemming from or affected by it). Although very large molecules escape the jurisdiction of molecular physics, this science is the basis and the starting point of further specializations and refinements. Molecular physics is the foundation of the pertinent education, as it influences other sciences that deal with molecules.

INDEX

Absorption, 134
Absorption
  dip, 392, 496
  spectra, 373, 390
  spectrometry, 391
Adhesion, 44
Alternating symmetry axis, 77
Ammonia
  inversion, 339, 356
  molecule, 338, 355
Angular dependence
  (of wavefunction), 123
Angular momentum, 128, 130
Angular momentum,
  orbital, 112, 130
  total, 129, 130
Angular momentum
  operator, 362
  quantum number, 112, 284
Anharmonicity, 332
Anharmonicity coefficient, 332
Anisotropic crystal, 48
Antibonding electrons, 212
Antiferromagnetic resonance
  (AFMR), 469, 560
Antiferromagnetic resonance
  frequency, 561

Anti-Stokes lines, 62, 377
Asymmetric top, 359, 363
Asymmetry parameter
  (in molecules), 363, 366
Asymmetry parameter
  (in NQR), 591
Atom, 5, 15
Atomic
  binding, 146ff.
  lattice, 46
  physics, 8
  structure, 25
  wavefunction, 147, 166
Atomic orbitals, method of, 154
Average polarizability, SP-67
Avogadro, 3
Avogadro's number, 7
Azimuthal quantum number, 111

Band
  head, 61, 329, 373
  spectrum, 56, 283, 292, 304, 331
  system, 331
Bayer's theory, 586
BF_3 molecule, 67, 75
Benzene molecule, 196, 224, SP-9